Lecture Notes in Computer Science 8270

Commenced Publication in 1973
Founding and Former Series Editors:
Gerhard Goos, Juris Hartmanis, and Jan van Leeuwen

Kazue Sako   Palash Sarkar (Eds.)

Advances in Cryptology – ASIACRYPT 2013

19th International Conference on the Theory
and Application of Cryptology and Information Security
Bengaluru, India, December 1-5, 2013
Proceedings, Part II

 Springer

Volume Editors

Kazue Sako
NEC Corporation
Kawasaki, Japan
E-mail: k-sako@ab.jp.nec.com

Palash Sarkar
Indian Statistical Institute Kolkata, India
E-mail: palash@isical.ac.in

ISSN 0302-9743   e-ISSN 1611-3349
ISBN 978-3-642-42044-3   e-ISBN 978-3-642-42045-0
DOI 10.1007/978-3-642-42045-0
Springer Heidelberg New York Dordrecht London

CR Subject Classification (1998): E.3, D.4.6, F.2, K.6.5, G.2, I.1, J.1

LNCS Sublibrary: SL 4 – Security and Cryptology

Typesetting: Camera-ready by author, data conversion by Scientific Publishing Services, Chennai, India

Printed on acid-free paper

Springer is part of Springer Science+Business Media (www.springer.com)

Preface

It is our great pleasure to present the proceedings of Asiacrypt 2013 in two volumes of *Lecture Notes in Computer Science* published by Springer. This was the 19th edition of the International Conference on Theory and Application of Cryptology and Information Security held annually in Asia by the International Association for Cryptologic Research (IACR). The conference was organized by IACR in cooperation with the Cryptology Research Society of India and was held in the city of Bengaluru in India during December 1–5, 2013.

About one year prior to the conference, an international Program Committee (PC) of 46 scientists assumed the responsibility of determining the scientific content of the conference. The conference evoked an enthusiastic response from researchers and scientists. A total of 269 papers were submitted for possible presentations approximately six months before the conference. Authors of the submitted papers are spread all over the world. PC members were allowed to submit papers, but each PC member could submit at most two co-authored papers or at most one single-authored paper. The PC co-chairs did not submit any paper. All the submissions were screened by the PC and 54 papers were finally selected for presentations at the conference. These proceedings contain the revised versions of the papers that were selected. The revisions were not checked and the responsibility of the papers rests with the authors and not the PC members.

Selection of papers for presentation was made through a double-blind review process. Each paper was assigned three reviewers and submissions by PC members were assigned six reviewers. Apart from the PC members, 291 external reviewers were involved. The total number of reviews for all the papers was more than 900. In addition to the reviews, the selection process involved an extensive discussion phase. This phase allowed PC members to express opinion on all the submissions. The final selection of 54 papers was the result of this extensive and rigorous selection procedure. One of the final papers resulted from the merging of two submissions.

The best paper award was conferred upon the paper "Shorter Quasi-Adaptive NIZK Proofs for Linear Subspaces" authored by Charanjit Jutla and Arnab Roy. The decision was based on a vote among the PC members. In addition to the best paper, the authors of two other papers, namely, "Families of Fast Elliptic Curves from Q-Curves" authored by Benjamin Smith and "Key Recovery Attacks on 3-Round Even-Mansour, 8-Step LED-128, and Full AES^2" authored by Itai Dinur, Orr Dunkelman, Nathan Keller and Adi Shamir, were recommended by the Editor-in-Chief of the *Journal of Cryptology* to submit expanded versions to the journal.

A highlight of the conference was the invited talks. An extensive multi-round discussion was carried out by the PC to decide on the invited speakers. This resulted in very interesting talks on two different aspects of the subject. Lars Ramkilde Knudsen spoke on "Block Ciphers — Past and Present" a topic of classical and continuing importance, while George Danezis spoke on "Engineering Privacy-Friendly Computations," which is an important and a more modern theme.

Apart from the regular presentations and the invited talks, a rump session was organized on one of the evenings. This consisted of very short presentations on upcoming research results, announcements of future events, and other topics of interest to the audience.

We would like to thank the authors of all papers for submitting their research works to the conference. Such interest over the years has ensured that the Asiacrypt conference series remains a cherished venue of publication by scientists. Thanks are due to the PC members for their enthusiastic and continued participation for over a year in different aspects of selecting the technical program. External reviewers contributed by providing timely reviews and thanks are due to them. A list of external reviewers is provided in these proceedings. We have tried to ensure that the list is complete. Any omission is inadvertent and if there is an omission, we apologize to the person concerned.

Special thanks are due to Satyanarayana V. Lokam, the general chair of the conference. His message to the PC was to select the best possible scientific program without any other considerations. Further, he ensured that the PC co-chairs were insulated from the organizational work. This work was done by the Organizing Committee and they deserve thanks from all the participants for the wonderful experience. We thank Daniel J. Bernstein and Tanja Lange for expertly organizing and conducting the rump session.

The reviews and discussions were entirely carried out online using a software developed by Shai Halevi. At several times, we had to ask Shai for his help with some feature or the other of the software. Every time, we received immediate and helpful responses. We thank him for his support and also for developing the software. We also thank Josh Benaloh, who was our IACR liaison, for guidance on several issues. Springer published the volumes and made these available before the conference. We thank Alfred Hofmann and Anna Kramer and their team for their professional and efficient handling of the production process.

Last, but, not the least, we thank Microsoft Research; Google; Indian Statistical Institute, Kolkata; and National Mathematics Initiative, Indian Institute of Science, Bengaluru; for being generous sponsors of the conference.

December 2013

Kazue Sako
Palash Sarkar

Asiacrypt 2013

The 19th Annual International Conference on Theory and Application of Cryptology and Information Security

Sponsored by the *International Association for Cryptologic Research (IACR)*

December 1–5, 2013, Bengaluru, India

General Chair

Satyanarayana V. Lokam   Microsoft Research, India

Program Co-chairs

Kazue Sako   NEC Corporation, Japan
Palash Sarkar   Indian Statistical Institute, India

Program Committee

Michel Abdalla	École Normale Supérieure, France
Colin Boyd	Queensland University of Technology, Australia
Anne Canteaut	Inria Paris-Rocquencourt, France
Sanjit Chatterjee	Indian Institute of Science, India
Jung Hee Cheon	Seoul National University, Korea
Sherman S.M. Chow	Chinese University of Hong Kong, SAR China
Orr Dunkelmann	University of Haifa, Israel
Pierrick Gaudry	CNRS Nancy, France
Rosario Gennaro	City College of New York, USA
Guang Gong	University of Waterloo, Canada
Vipul Goyal	Microsoft Research, India
Eike Kiltz	University of Bochum, Germany
Tetsu Iwata	Nagoya University, Japan
Tanja Lange	Technische Universiteit Eindhoven, The Netherlands
Dong Hoon Lee	Korea University, Korea
Allison Lewko	Columbia University, USA
Benoit Libert	Technicolor, France
Dongdai Lin	Chinese Academy of Sciences, China
Anna Lysyanskaya	Brown University, USA
Subhamoy Maitra	Indian Statistical Institute, India

Willi Meier	University of Applied Sciences, Switzerland
Phong Nguyen	Inria, France and Tsinghua University, China
Kaisa Nyberg	Aalto University, Finland
Satoshi Obana	Hosei University, Japan
Kenny Paterson	Royal Holloway, University of London, UK
Krzysztof Pietrzak	Institute of Science and Technology, Austria
David Pointcheval	École Normale Supérieure, France
Manoj Prabhakaran	University of Illinois at Urbana-Champaign, USA
Vincent Rijmen	KU Leuven, Belgium
Rei Safavi-Naini	University of Calgary, Canada
Yu Sasaki	NTT, Japan
Nicolas Sendrier	Inria Paris-Rocquencourt, France
Peter Schwabe	Radboud University Nijmegen, The Netherlands
Thomas Shrimpton	Portland State University, USA
Nigel Smart	University of Bristol, UK
Francois-Xavier Standaert	Université Catholique de Louvain, Belgium
Damien Stehlé	École Normale Supérieure de Lyon, France
Willy Susilo	University of Wollongong, Australia
Tsuyoshi Takagi	Kyushu University, Japan
Vinod Vaikuntanathan	University of Toronto, Canada
Frederik Vercauteren	KU Leuven, Belgium
Xiaoyun Wang	Tsinghua University, China
Hoeteck Wee	George Washington University, USA and École Normale Supérieure, France
Hongjun Wu	Nanyang Technological University, Singapore

External Reviewers

Carlos Aguilar-Melchor
Masayuki Abe
Gergely Acs
Shashank Agrawal
Ahmad Ahmadi
Hadi Ahmadi
Mohsen Alimomeni
Joel Alwen
Prabhanjan Ananth
Gilad Asharov
Tomer Ashur
Giuseppe Ateniese
Man Ho Au
Jean-Philippe Aumasson
Pablo Azar

Foteini Baldimtsi
Subhadeep Banik
Paulo Barreto
Rishiraj Batacharrya
Lejla Batina
Anja Becker
Mihir Bellare
Fabrice Benhamouda
Debajyoti Bera
Daniel J. Bernstein
Rishiraj Bhattacharyya
Gaetan Bisson
Olivier Blazy
Céline Blondeau
Andrey Bogdanov

Alexandra Boldyreva
Joppe W. Bos
Charles Bouillaguet
Christina Boura
Elette Boyle
Fabian van den Broek
Billy Bob Brumley
Christina Brzuska
Angelo De Caro
Dario Catalano
André Chailloux
Melissa Chase
Anupam Chattopadhyay
Chi Chen
Jie Chen
Jing Chen
Yu Chen
Céline Chevalier
Ashish Choudhary
HeeWon Chung
Kai-Min Chung
Deepak Kumar Dalai
M. Prem Laxman Das
Gareth Davies
Yi Deng
Maria Dubovitskaya
François Durvaux
Barış Ege
Nicolas Estibals
Xinxin Fan
Pooya Farshim
Sebastian Faust
Nelly Fazio
Serge Fehr
Dario Fiore
Marc Fischlin
Georg Fuchsbauer
Eichiro Fujisaki
Jun Furukawa
Philippe Gaborit
Tommaso Gagliardoni
Martin Gagne
Steven Galbraith
David Galindo
Nicolas Gama

Sanjam Garg
Lubos Gaspar
Peter Gazi
Ran Gelles
Essam Ghadafi
Choudary Gorantla
Sergey Gorbunov
Dov S. Gordon
Louis Goubin
Matthew Green
Vincent Grosso
Jens Groth
Tim Güneysu
Fuchun Guo
Jian Guo
Divya Gupta
Sourav Sen Gupta
Benoît Gérard
Dong-Guk Han
Jinguang Han
Carmit Hazay
Nadia Heninger
Jens Hermans
Florian Hess
Shoichi Hirose
Viet Tung Hoang
Jaap-Henk Hoepmann
Dennis Hofheinz
Hyunsook Hong
Jin Hong
Qiong Huang
Tao Huang
Yan Huang
Fei Huo
Michael Hutter
Jung Yeon Hwang
Takanori Isobe
Mitsugu Iwamoto
Abhishek Jain
Stanislaw Jarecki
Mahavir Jhawar
Shoaquan Jiang
Ari Juels
Marc Kaplan
Koray Karabina

Aniket Kate
Jonathan Katz
Liam Keliher
Stéphanie Kerckhof
Hyoseung Kim
Kitak Kim
Minkyu Kim
Sungwook Kim
Taechan Kim
Yuichi Komano
Takeshi Koshiba
Anna Krasnova
Fabien Laguillaumie
Russell W.F. Lai
Adeline Langlois
Jooyoung Lee
Kwangsu Lee
Moon Sung Lee
Younho Lee
Tancrède Lepoint
Gaëtan Leurent
Anthony Leverrier
Huijia Rachel Lin
Feng-Hao Liu
Zhenhua Liu
Zongbin Liu
Adriana López-Alt
Atul Luykx
Vadim Lyubashevsky
Arpita Maitra
Hemanta Maji
Cuauhtemoc Mancillas-López
Kalikinkar Mandal
Takahiro Matsuda
Alexander May
Sarah Meiklejohn
Florian Mendel
Alfred Menezes
Kazuhiko Minematsu
Marine Minier
Rafael Misoczki
Amir Moradi
Tal Moran
Kirill Morozov
Pratyay Mukherjee

Yusuke Naito
María Naya-Plasencia
Gregory Neven
Khoa Nguyen
Antonio Nicolosi
Ivica Nikolić
Ryo Nishimaki
Ryo Nojima
Adam O'Neill
Cristina Onete
Elisabeth Oswald
Ilya Ozerov
Omkant Pandey
Tapas Pandit
Jong Hwan Park
Seunghwan Park
Michal Parusinski
Valerio Pastro
Arpita Patra
Goutam Paul
Roel Peeters
Christopher Peikert
Milinda Perera
Ludovic Perret
Thomas Peters
Christophe Petit
Duong Hieu Phan
Bertram Poettering
Joop van de Pol
Gordon Proctor
Emmanuel Prouff
Elizabeth Quaglia
Somindu C Ramanna
Mariana Raykova
Christian Rechberger
Francesco Regazzoni
Oscar Reparaz
Reza Reyhanitabar
Thomas Ristenpart
Damien Robert
Thomas Roche
Mike Rosulek
Sujoy Sinha Roy
Sushmita Ruj
Carla Ràfols

Santanu Sarkar
Michael Schneider
Dominique Schröder
Jacob Schuldt
Jae Hong Seo
Minjae Seo
Yannick Seurin
Hakan Seyalioglu
Setareh Sharifian
Abhi Shelat
Dale Sibborn
Dimitris E. Simos
Dave Singelee
William E. Skeith III
Boris Skoric
Adam Smith
Ben Smith
Hadi Soleimany
Katherine Stange
Douglas Stebila
John Steinberger
Ron Steinfeld
Mario Strefler
Donald Sun
Koutarou Suzuki
Yin Tan
Ying-Kai Tang
Sidharth Telang
Isamu Teranishi
R. Seth Terashima
Stefano Tessaro
Susan Thomson
Emmanuel Thomé
Gilles Van Assche
Konstantinos Vamvourellis
Alex Vardy
K. Venkata
Damien Vergnaud
Nicolas Veyrat-Charvillon
Gilles Villard
Ivan Visconti

Huaxiong Wang
Lei Wang
Meiqin Wang
Peng Wang
Pengwei Wang
Wenhao Wang
Gaven Watson
Carolyn Whitnall
Daniel Wichs
Michael J. Wiener
Shuang Wu
Teng Wu
Keita Xagawa
Haixia Xu
Rui Xue
Bohan Yang
Guomin Yang
Kan Yasuda
Takanori Yasuda
Kazuki Yoneyama
Hongbo Yu
Tsz Hon Yuen
Dae Hyun Yum
Aaram Yun
Hui Zhang
Liang Feng Zhang
Liting Zhang
Mingwu Zhang
Rui Zhang
Tao Zhang
Wentao Zhang
Zongyang Zhang
Colin Jia Zheng
Xifan Zheng
Hong-Sheng Zhou
Yongbin Zhou
Bo Zhu
Youwen Zhu
Vassilis Zikas
Paul Zimmermann

Organizing Committee

Raghav Bhaskar	Microsoft Research India, Bengaluru
Vipul Goyal	Microsoft Research India, Bengaluru
Neeraj Kayal	Microsoft Research India, Bengaluru
Satyanarayana V. Lokam	Microsoft Research India, Bengaluru
C. Pandurangan	Indian Institute of Technology, Chennai
Govindan Rangarajan	Indian Institute of Science, Bengaluru

Sponsors

Microsoft Research
Google
Indian Statistical Institute, Kolkata
National Mathematics Initiative, Indian Institute of Science, Bengaluru

Invited Talks

Block Ciphers – Past and Present

Lars Ramkilde Knudsen

DTU Compute, Denmark
lrkn@dtu.dk

Abstract. In the 1980s researchers were trying to understand the design of the DES, and breaking it seemed impossible. Other block ciphers were proposed, and cryptanalysis of block ciphers got interesting. The area took off in the 1990s where it exploded with the appearance of differential and linear cryptanalysis and the many variants thereof which appeared in the time after. In the 2000s AES became a standard and it was constructed specifically to resist the general attacks and the area of (traditional) block cipher cryptanalysis seemed saturated.... Much of the progress in cryptanalysis of the AES since then has come from side-channel attacks and related-key attacks.

Still today, for most block cipher applications the AES is a good and popular choice. However, the AES is perhaps not particularly well suited for extremely constrained environments such as RFID tags. Therefore, one trend in block cipher design has been to come up with ultra-lightweight block ciphers with good security and hardware efficiency. I was involved in the design of the ciphers Present (from CHES 2007), PrintCipher (presented at CHES 2010) and PRINCE (from Asiacrypt 2012). Another trend in block cipher design has been try to increase the efficiency by making certain components part of the secret key, e.g., to be able to reduce the number of rounds of a cipher.

In this talk, I will review these results.

Engineering Privacy-Friendly Computations

George Danezis [1,2]

[1] University College London
[2] Microsoft Research, Cambridge

Abstract. In the past few years tremendous cryptographic progress has been made in relation to primitives for privacy friendly-computations. These include celebrated results around fully homomorphic encryption, faster somehow homomorphic encryption, and ways to leverage them to support more efficient secret-sharing based secure multi-party computations. Similar break-through in verifiable computation, and succinct arguments of knowledge, make it practical to verify complex computations, as part of privacy-preserving client side program execution. Besides computations themselves, notions like differential privacy attempt to capture the essence of what it means for computations to leak little personal information, and have been mapped to existing data query languages.

So, is the problem of computation on private data solved, or just about to be solved? In this talk, I argue that the models of generic computation supported by cryptographic primitives are complete, but rather removed from what a typical engineer or data analyst expects. Furthermore, the use of these cryptographic technologies impose constrains that require fundamental changes in the engineering of computing systems. While those challenges are not obviously cryptographic in nature, they are nevertheless hard to overcome, have serious performance implications, and errors open avenues for attack.

Throughout the talk I use examples from our own work relating to privacy-friendly computations within smart grid and smart metering deployments for private billing, privacy-friendly aggregation, statistics and fraud detection. These experiences have guided the design of ZQL, a cryptographic language and compiler for zero-knowledge proofs, as well as more recent tools that compile using secret-sharing based primitives.

Table of Contents – Part II

Cryptographic Primitives

Analysis, Cryptanalysis and Passwords

Leakage-Resilient Cryptography

Two-Party Computation

Hash Functions

Table of Contents – Part I

Protocols

Theoretical Cryptography-II

Symmetric Key Cryptanalysis

Symmetric Key Cryptology: Schemes and Analysis

Side-Channel Cryptanalysis

New Generic Attacks against Hash-Based MACs

Gaëtan Leurent[1], Thomas Peyrin[2], and Lei Wang[2]

[1] Université Catholique de Louvain, Belgium
gaetan.leurent@uclouvain.be
[2] Nanyang Technological University, Singapore
thomas.peyrin@gmail.com,   wang.lei@ntu.edu.sg

Abstract. In this paper we study the security of hash-based MAC algorithms (such as HMAC and NMAC) above the birthday bound. Up to the birthday bound, HMAC and NMAC are proven to be secure under reasonable assumptions on the hash function. On the other hand, if an n-bit MAC is built from a hash function with a l-bit state ($l \geq n$), there is a well-known existential forgery attack with complexity $2^{l/2}$. However, the remaining security after $2^{l/2}$ computations is not well understood. In particular it is widely assumed that if the underlying hash function is sound, then a generic universal forgery attack should require 2^n computations and some distinguishing (*e.g.* distinguishing-H but not distinguishing-R) and state-recovery attacks should also require 2^l computations (or 2^k if $k < l$).

In this work, we show that above the birthday bound, hash-based MACs offer significantly less security than previously believed. Our main result is a generic distinguishing-H and state-recovery attack against hash-based MACs with a complexity of only $\tilde{O}(2^{l/2})$. In addition, we show a key-recovery attack with complexity $\tilde{O}(2^{3l/4})$ against HMAC used with a hash functions with an internal checksum, such as GOST. This surprising result shows that the use of a checksum might actually weaken a hash function when used in a MAC. We stress that our attacks are generic, and they are in fact more efficient than some previous attacks proposed on MACs instanciated with concrete hash functions.

We use techniques similar to the cycle-detection technique proposed by Peyrin *et al.* at Asiacrypt 2012 to attack HMAC in the related-key model. However, our attacks works in the single-key model for both HMAC and NMAC, and without restriction on the key size.

Keywords: NMAC, HMAC, hash function, distinguishing-H, key recovery, GOST.

1   Introduction

Message Authentication Codes (MACs) are crucial components in many security systems. A MAC is a function that takes a k-bit secret key K and an arbitrarily long message M as inputs, and outputs a fixed-length tag of size n bits. The tag is used to authenticate the message, and will be verified by the receiving party using the same key K. Common MAC algorithms are built from block

K. Sako and P. Sarkar (Eds.) ASIACRYPT 2013, Part II, LNCS 8270, pp. 1–20, 2013.

ciphers (*e.g.* CBC-MAC), from hash functions (*e.g.* HMAC), or from universal hash functions (*e.g.* UMAC). In this paper we study MAC algorithms based on hash functions.

As a cryptographic primitive, a MAC algorithm should meet some security requirements. It should be impossible to recover the secret key except by exhaustive search, and it should be computationally impossible to forge a valid MAC without knowing the secret key, the message being chosen by the attacker (existential forgery) or given as a challenge (universal forgery). In addition, cryptanalysts have also studied security notions based on distinguishing games. Informally, the distinguishing-R game is to distinguish a MAC construction from a random function, while the distinguishing-H game is to distinguish a known MAC construction (*e.g.* HMAC) instantiated with a known component (*e.g.* SHA-1) under a random key from the same construction instantiated with a random component (*e.g.* HMAC with a fixed input length random function).

One of the best known MAC algorithm is HMAC [2], designed by Bellare *et al.* in 1996. HMAC is now widely standardized (by ANSI, IETF, ISO and NIST), and widely deployed, in particular for banking processes or Internet protocols (*e.g.* SSL, TLS, SSH, IPSec). It is a single-key version of the NMAC construction, the latter being built upon a keyed iterative hash function H_K, while HMAC uses an unkeyed iterative hash function H:

$$\text{NMAC}(K_{out}, K_{in}, M) = H_{K_{out}}(H_{K_{in}}(M))$$
$$\text{HMAC}(K, M) = H(K \oplus \text{opad} \parallel H(K \oplus \text{ipad} \parallel M))$$

where opad and ipad are predetermined constants, and where K_{in} denotes the inner key and K_{out} the outer one.

More generally, a MAC algorithm based on a hash function uses the key at the beginning and/or at the end of the computation, and updates an l-bit state with a compression function. The security of MAC algorithms is an important topic, and both positive and negative results are known. On the one hand, there is a generic attack with complexity $2^{l/2}$ based on internal collisions and length extension [17]. This gives an existential forgery attack, and a distinguishing-H attack. One the other hand, we have security proofs for several MAC algorithms such as HMAC and sandwich-MAC [2,1,26]. Roughly speaking, the proofs show that some MAC algorithms are secure up to the birthday bound ($2^{l/2}$) under various assumptions on the compression function and hash function.

Thanks to those results, one may consider that the security of hash-based MAC algorithms is well understood. However, there is still a strong interest in the security above the birthday bound. In particular, it is very common to expect security 2^k for key recovery attacks if the hash function is sound; the Encyclopedia of Cryptography and Security article on HMAC states explicitly [16] "A generic key recovery attack requires $2^{n/2}$ known text-MAC pairs and 2^{n+1} time" (assuming $n = l = k$). Indeed, key recovery attacks against HMAC with a concrete hash function with complexity between $2^{l/2}$ and 2^l have been considered as important results [9,24,27]. Similarly, the best known dinstinguishing-H and state-recovery attacks have a complexity of 2^l (or 2^k if $k < l$), and distinguishing-H attacks on

Table 1. Comparison of our generic attacks on HMAC, and some previous attacks on concrete hash function. We measure the complexity as the number of calls to the compression function (*i.e.* number of messages times message length).

Function	Attack	Complexity	M. len	Notes	Ref
HMAC-MD5	dist-H, state rec.	2^{97}	2		[25]
HMAC-SHA-0	dist-H	2^{100}	2		[12]
HMAC-HAVAL (3-pass)	dist-H	2^{228}	2		[12]
HMAC-SHA-1 (43 first steps)	dist-H	2^{154}	2		[12]
HMAC-SHA-1 (58 first steps)	dist-H	2^{158}	2		[18]
HMAC-SHA-1 (61 mid. steps)	dist-H	2^{100}	2		[18]
HMAC-SHA-1 (62 mid. steps)	dist-H	2^{157}	2		[18]
Generic attacks:					
hash-based MAC (*e.g.* HMAC)	dist-H	$O(2^{l/2})$	$2^{l/2}$		Sec. 4
	state rec.	$\tilde{O}(2^{l/2})$	$2^{l/2}$		Sec. 5
	dist-H, state rec.	$O(2^{l-s})$	2^s	$s \leq l/4$	full version
HMAC with a checksum	key rec.	$O(l \cdot 2^{3l/4})$	$2^{l/2}$		Sec. 7
		$O(l \cdot 2^{3l/4})$	$2^{l/4}$		full version
HMAC-MD5*	dist-H, state rec.	$2^{67}, 2^{78}$	2^{64}		Sec. 4, 5
		$O(2^{96})$	2^{32}		full version
HMAC-HAVAL† (any)	dist-H, state rec.	$O(2^{202})$	2^{54}		full version
HMAC-SHA-1†	dist-H, state rec.	$O(2^{120})$	2^{40}		full version
HMAC-GOST*	key rec.	2^{200}	2^{128}		Sec. 7
		2^{200}	2^{64}		full version

* The MD5 and GOST specifications allow arbitrary-length messages
† The SHA-1 and HAVAL specifications limits the message length to 2^{64} bits (and 2^{64} bits is 2^{54} blocks)

HMAC with a concrete hash function with complexity between $2^{l/2}$ and 2^l have been considered as important results [12,18,25].

Our Contributions. In this paper we revisit the security of hash-based MAC above the birthday bound. We describe a generic distinguishing-H attack in the single key model and with complexity of only $O(2^{l/2})$ computations, thus putting an end to the long time belief of the cryptography community that the best generic distinguishing-H on NMAC and HMAC requires $\Omega(2^l)$ operations. Instead, we show that a distinguishing-H attack is not harder than a distinguishing-R attack. Our results actually invalidate some of the recently published cryptanalysis works on HMAC when instantiated with real hash functions [12,18,25]

Our method is based on a cycle length detection, like the work of Peyrin *et. al* [14], but our utilization is quite different and much less restrictive: instead of iterating many times HMAC with small messages and a special related-key, we will use only a few iterations with very long messages composed of almost the same message blocks to observe the cycle and deduct information from it. Overall, unlike in [14], our technique works in the single-key model, for both HMAC and NMAC, and can be applied for any key size. In addition, leveraging our new ideas, we provide a single-key internal state recovery for any hash-based MAC with only $O(l \cdot \log(l) \cdot 2^{l/2})$ computations.

We also introduce a different approach to reduce the length of the queried messages for distinguishing-H and internal-state-recovery attacks. Due to the limited space, we only give a quick overview of these shorter-message attacks, and refer the reader to the full version of this paper for more details.

Finally, this internal state recovery can be transformed into a single-key key recovery attack on HMAC with complexity $O(l \cdot 2^{3l/4})$ when instantiated with a hash function using a checksum, such as the GOST hash function [6]. A surprising corollary to our results is that incorporating a checksum to a hash function seems to actually reduce the security of the overall HMAC design.

We give an overview of our results, and a comparison with some previous analysis in Table 1.

The description of HMAC/NMAC algorithms and their security are given in Section 2 and we recall in Section 3 the cycle-detection ideas from [14]. Then, we provide in Section 4 the generic distinguishing-H attack. We give in Sections 5 an internal state recovery method, and finally describe our results when the hash function incorporates a checksum in Section 7.

2   Hash-Based MAC Algorithms

In this paper, we study a category of MAC algorithms based on hash functions, where the key is used at the beginning and at the end of the computation, as described in Figure 1. More precisely, we consider algorithms where: the message processing is done by updating an internal state x using a compression function h; the state is initialized with a key dependent value I_k; and the tag is computed from the last state x_p and the key K by an output function g.

$$x_0 = I_K \qquad x_{i+1} = h(x_i, m_i) \qquad \mathrm{MAC}_K(M) = g(K, x_p, |M|)$$

In particular, this description covers NMAC/HMAC [2], envelope-MAC [21], and sandwich-MAC [26]. The results described in Sections 4 and 5 can be applied to any hash-based MAC, but we focus on HMAC for our explanations because it is the most widely used hash-based MAC, and its security has been widely analyzed already. On the other hand the result of Section 7 is specific to MAC algorithms that process the key as part of the message, such as HMAC.

2.1   Description of NMAC and HMAC

A Hash Function. H is a function that takes an arbitrary length input message M and outputs a fixed hash value of size n bits.

Virtually every hash function in use today follows an iterated structure like the classical Merkle-Damgård construction [13,4]. Iterative hash functions are

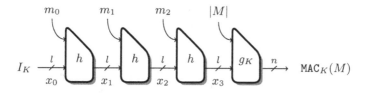

Fig. 1. Hash-based MAC. Only the initial value and the final transformation are keyed.

built upon successive applications of a so-called compression function h, that takes a b-bit message block and a l-bit chaining value as inputs and outputs a l-bit value (where $l \geq n$). An output function can be included to derive the n-bit hash output from the last chaining variable and from the message length. When $l > n$ (resp. when $l = n$) we say that the hash function is wide pipe (resp. narrow pipe).

The message M is first padded and then divided into blocks m_i of b bits each. Then, the message blocks are successively used to update the l-bit internal state x_i with the compression function h. Once all the message blocks have been processed, the output function g is applied to the last internal state value x_p.

$$x_0 = IV \qquad x_{i+1} = h(x_i, m_i) \qquad \mathsf{hash} = g(x_p, |M|)$$

The MAC algorithm NMAC [2] uses two l-bit keys K_{out} and K_{in}. NMAC replaces the public IV of a hash function H by a secret key K to produce a keyed hash function $H_K(M)$. NMAC is then defined as:

$$\mathrm{NMAC}(K_{out}, K_{in}, M) = H_{K_{out}}(H_{K_{in}}(M)).$$

The MAC algorithm HMAC [2] is a single-key version of NMAC, with $K_{out} = h(IV, K \oplus \mathsf{opad})$ and $K_{in} = h(IV, K \oplus \mathsf{ipad})$, where opad and ipad are b-bit constants. However, a very interesting property of HMAC for practical utilization is that it can use any key size and can be instantiated with an unkeyed hash function (like MD5, SHA-1, etc. which have a fixed IV):

$$\mathrm{HMAC}(K, M) = H(K \oplus \mathsf{opad} \parallel H(K \oplus \mathsf{ipad} \parallel M)).$$

where $\parallel$ denotes the concatenation operation. For simplicity of the description and without loss of generality concerning our attacks, in the rest of this article we assume that the key can fit in one compression function message block, i.e. $k \leq b$ (note that K is actually padded to b bits if $k < b$).

2.2 Security of NMAC and HMAC

In [1], Bellare proved that the NMAC construction is a pseudo-random function (PRF) under the sole assumption that the internal compression function h (keyed by the chaining variable input) is a PRF. The result can be transposed to HMAC as well under the extra assumption that h is also a PRF when keyed by the message input.

Concerning key recovery attacks, an adversary should not be able to recover the key in less than 2^k computations for both HMAC and NMAC. In the case of NMAC one can attack the two keys K_{out} and K_{in} independently by first finding an internal collision and then using this colliding pair information to brute force the first key K_{in}, and finally the second one K_{out}.

Universal or existential forgery attacks should cost 2^n computations for a perfectly secure n-bit MAC. However, the iterated nature of the hash functions used inside NMAC or HMAC allows a simple existential forgery attack requiring only $2^{l/2}$ computations [17]. Indeed, with $2^{l/2}$ queries, an adversary can first find an internal collision between two messages (M, M') of the same length during the first hash call. Then, any extra block m added to both of these two messages will lead again to an internal collision. Thus, the attacker can simply forge a valid MAC by only querying for $M \parallel m$ and deducing that $M' \parallel m$ will have the same MAC value.

Concerning distinguishers on HMAC or NMAC, two types have been discussed in the literature: Distinguishing-R and Distinguishing-H attacks, defined below:

Distinguishing-R. Let $\mathcal{F}_n$ be the set of n-bit output functions. We denote F_K the oracle on which the adversary $\mathcal{A}$ can make queries. The oracle is instantiated either with $F_K = \text{HMAC}_K$ (with K being a randomly chosen k-bit key) or with a randomly chosen function R_K from $\mathcal{F}_n$. The goal of the adversary is to distinguish between the two cases and its advantage is given by

$$Adv(\mathcal{A}) = |\Pr[\mathcal{A}(\text{HMAC}_K) = 1] - \Pr[\mathcal{A}(R_K) = 1]|.$$

Obviously the collision-based forgery attack detailed above gives directly a distinguishing-R attack on NMAC and HMAC. Thus, the expected security of HMAC and NMAC against distinguishing-R attacks is $2^{l/2}$ computations.

Distinguishing-H. The attacker is given access to an oracle HMAC_K and the compression function of the HMAC oracle is instantiated either with a known dedicated compression function h or with a random chosen function r from $\mathcal{F}_l^{b+l}$ (the set of $(b + l)$-bit to l-bit functions), which we denote HMAC_K^h and HMAC_K^r respectively. The goal of the adversary is to distinguish between the two cases and its advantage is given by

$$Adv(\mathcal{A}) = |\Pr[\mathcal{A}(\text{HMAC}_K^h) = 1] - \Pr[\mathcal{A}(\text{HMAC}_K^r) = 1]|.$$

The distinguishing-H notion was introduced by Kim *et al.* [12] for situations where the attacker wants to check which cryptographic hash function is embedded in HMAC. To the best of our knowledge, the best known generic distinguishing-H attack requires 2^l computations.

Related-Key Attacks. At Asiacrypt 2012 [14], a new type of generic distinguishing (distinguishing-R or distinguishing-H) and forgery attacks for HMAC was proposed. These attacks are in the related-key model and can apply even to wide-pipe proposals, but they only work for HMAC, and only when a special restrictive criterion is verified: the attacker must be able to force a specific difference between the inner and the outer keys (with the predefined values of opad and ipad in HMAC, this criterion is verified when $k = b$). The idea is to compare the cycle length when iterating the HMAC construction on small messages with a key K, and the cycle length when iterating with a key $K' = K \oplus \text{opad} \oplus \text{ipad}$.

Attacks on Instantiations with Concrete Hash Functions. Because of its widespread use in many security applications, HMAC has also been carefully scrutinized when instantiated with a concrete hash function, exploiting weaknesses of some existing hash function. In parallel to the recent impressive advances on hash function cryptanalysis, the community analyzed the possible impact on the security of HMAC when instantiated with standards such as MD4 [19], MD5 [20], SHA-1 [22] or HAVAL. In particular, key-recovery attacks have been found on HMAC-MD4 [9,24] and HMAC-HAVAL [27]. Concerning the distinguishing-H notion, one can cite for example the works from Kim et al. [12], Rechberger et al. [18] and Wang et al. [25].

However, to put these attacks in perspective, it is important to know the complexity of *generic* attacks, that work even with a good hash function.

3   Cycle Detection for HMAC

Our new attacks are based on some well-known properties of random functions.

3.1   Random Mapping Properties on a Finite Set

Let us consider a random function f mapping n bits to n bits and we denote $N = 2^n$. We would like to know the structure of the functional graph defined by the successive iteration of this function, for example the expected number of components, cycles, etc. First, it is easy to see that each component will contain a single cycle with several trees linked to it. This has already been studied for a long time, and in particular we recall two theorems from Flajolet and Odlyzko [7].

Theorem 1 ([7, Th. 2]). *The expectations of the number of components, number of cyclic points, number of terminal points, number of image points, and number of k-th iterate image points in a random mapping of size N have the asymptotic forms, as $N \to \infty$:*

(i) # Components: $\frac{1}{2} \log N$

(ii) # Cyclic nodes: $\sqrt{\pi N / 2}$

(iii) # Terminal nodes: $e^{-1} N$

(iv) # Image points: $(1 - e^{-1}) N$

(v) # k-th iterate images: $(1 - \tau_k) N$, with $\tau_0 = 0$, $\tau_{k+1} = e^{-1 + \tau_k}$.

In particular, a random mapping has only a logarithmic number of distinct components, and the number of cyclic points follows the square root of N.

By choosing a random starting point P and iterating the function f, one will follow a path in the functional graph starting from P, that will eventually connect to the cycle of the component in which P belongs, and we call **tail length** the number of points in this path. Similarly, we call **cycle length** the number of nodes in the cycle. Finally, the number of points in the non-repeating trajectory from P is called the **rho length**, and we call **α-node** of the path the node that connects the tail and the cycle.

Theorem 2 ([7, Th. 3]). *Seen from a random point in a random mapping of size N, the expectations of the tail length, cycle length, rho length, tree size, component size, and predecessors size have the following asymptotic forms:*

(i)   Tail length (λ): $\sqrt{\pi N/8}$ *(iv) Tree size: $N/3$*
(ii) Cycle length (μ): $\sqrt{\pi N/8}$ *(v)   Component size: $2N/3$*
(iii) Rho length ($\rho = \lambda + \mu$): $\sqrt{\pi N/2}$ *(vi) Predecessors size: $\sqrt{\pi N/8}$*

One can see that, surprisingly, in a random mapping most of the points tend to be grouped together in a single giant component, and there is a giant tree with a significant proportion of the points. The asymptotic expectation of the maximal features is given by Flajolet and Sedgewick [8].

Theorem 3 ([8, VII.14]). *In a random mapping of size N, the largest tree has an expected size of $\delta_1 N$ with $\delta_1 \approx 0.48$ and the largest component has an expected size of $\delta_2 N$ with $\delta_2 \approx 0.7582$.*

These statistical properties will be useful to understand the advantage of our attacks. We show the functional graph of a simple random-looking function in Figure 4 in the Appendix.

3.2   Using Cycle-Detection to Obtain Some Secret Information

In this article and as in [14], we will study the functional graph structure of a function to derive a distinguisher or obtain some secret information. More precisely, in [14] Peyrin *et al.* observed that the functional graph structure of HMAC was the same when instantiated with a key K or with a related key $K' = K \oplus \mathtt{ipad} \oplus \mathtt{opad}$ (note that in order to be able to query this related-key K', the key K has to be of size b or $b - 1$, which is quite restrictive). This is a property that should not exist for a randomly chosen function and they were able to detect this cycle structure by measuring the cycle length in both cases K and K', and therefore obtaining a distinguishing-R attack for HMAC in the related-key model. In practice, the attacker can build and observe the functional graph of HMAC by simply successively querying the previous n-bit output as new message input.

In this work, instead of studying the structure of the functional graph of HMAC directly, we will instead study the functional graph of the internal compression function h with a fixed message block: we denote $h_M(X) = h(X, M)$. We aim to obtain some information on h_M that we can propagate outside the HMAC structure. This is therefore perfectly suited for a distinguishing-H attack, which requires the attacker to exhibit a property of h when embedded inside the HMAC construction. We can traverse the functional graph of h_M by querying the HMAC oracle with a long message composed of many repetitions of the fixed message block M. The issue is now to detect some properties of the functional graph of h_M inside HMAC and without knowing the secret key. We explain how to do that in the next section.

4   Distinguishing-H Attack for Hash-Based MACs

In the rest of the article, we use the notation $[x]^k$ to represent the successive concatenation of k message blocks x, with $[x] = [x]^1$.

4.1   General Description

In order to derive a distinguishing-H attack, we need to do some offline computations with the target compression function h and use this information to compare online with the function embedded in the MAC oracle. We use the structure of the functional graph of $h_{[0]}$ to derive our attack (of course we can choose any fixed message block). We can travel in the graph by querying the oracle using consecutive [0] message blocks. However, since the key is unknown, we do not know where we start or where we stop in this graph. We have seen in the previous section that the functional graph of a random function is likely to have a giant component containing most of the nodes. We found that the cycle size of the giant component of $h_{[0]}$ is a property that can be efficiently tested.

More precisely, we first compute the cycle size of the giant component of $h_{[0]}$ offline; we denote it as L. Then, we measure the cycle size of the giant component of the unknown function by querying the MAC oracle with long messages composed of many consecutive [0] message blocks. If no length padding is used in the hash function, this is quite simple: we just compare $\text{MAC}([0]^{2^{l/2}})$ and $\text{MAC}([0]^{2^{l/2}+L})$. With a good probability, the sequence of $2^{l/2}$ zero block is sufficiently long to enter the cycle, and if the cycle has length L, the two MAC outputs will collide.

Unfortunately, this method does not work because the lengths of the messages are different, and thus the last message block with the length padding will be different and prevent the cycle collision to propagate to the MAC output. We will use a trick to overcome this issue, even though the basic method remains the same. The idea is to build a message M going twice inside the cycle of the giant component, so that we can add L [0] message blocks in the first cycle to obtain a message M_1 and L [0] message blocks in the second cycle to obtain M_2. This is depicted in Figure 2: M_1 will cycle L [0] message blocks in the first cycle (red dotted arrows), while M_2 will cycle L [0] message blocks in the second cycle (blue dashed arrows), and thus they will both have the same length overall.

To perform the distinguishing-H attack, the adversary simply randomly selects an initial message block m and query $M_1 = m \parallel [0]^{2^{l/2}} \parallel [1] \parallel [0]^{2^{l/2}+L}$ and $M_2 = m \parallel [0]^{2^{l/2}+L} \parallel [1] \parallel [0]^{2^{l/2}}$ to the MAC oracle, and deduce that the target function is used to instantiate the oracle if the two MAC values are colliding. The [1] message block is used to quit the cycle and randomize the entry point to return again in the giant component. We give below a detailed attack procedure and complexity analysis.

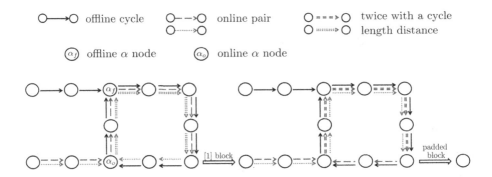

Fig. 2. Distinguishing-H attack

This attack is very interesting as the first generic distinguishing-H attack on HMAC and NMAC with a complexity lower than 2^l. However, we note that the very long message length might be a limitation. In theory this is of no importance and our attack is indeed valid, but in practice some hash functions forbid message inputs longer than a certain length. To address this issue we provide an alternative attack in the full version of this paper, using shorter messages, at the cost of a higher complexity.

4.2   Detailed Attack Process

1. (offline) Search for a cycle in the functional graph of $h_{[0]}$ and denote L its length.
2. (online) Choose a random message block m and query the HMAC value of the two messages $M_1 = m\|[0]^{2^{l/2}}\|[1]\|[0]^{2^{l/2}+L}$ and $M_2 = m\|[0]^{2^{l/2}+L}\|[1]\|[0]^{2^{l/2}}$.
3. If the HMAC values of M_1 and M_2 collide then output 1, otherwise output 0.

4.3   Complexity and Success Probability Analysis

We would like to evaluate the complexity of the attack. The first step will require about $2^{l/2}$ offline computations to find the cycle for the target compression function h. It is important to note that we can run this first step several times in order to ensure that we are using the cycle from the largest component of the functional graph of $h_{[0]}$. The second step makes two queries of about $2^{l/2} + 2^{l/2} + L \simeq 3 \cdot 2^{l/2}$ message blocks each. Therefore, the overall complexity of the attack is about $2^{l/2+3}$ compression function computations.

Next, we evaluate the advantage of the adversary in winning the distinguishing-H game defined in Section 2.2. We start with the case where the oracle is instantiated with the real compression function h. The adversary will output 1 if a collision happens between the HMAC computations of M_1 and M_2. Such a collision can happen when the following conditions are satisfied:

- The processing of the random block m sets us in the same component (of the functional graph of $h_{[0]}$) as in the offline computation. Since we ensured that the largest component was found during the offline computation, and since it has an average size of $0.7582 \cdot 2^l$ elements according to Theorem 2, this event will happen with probability 0.7582.
- The $2^{l/2}$ $[0]$ message blocks concatenated after m are enough to reach the cycle of the component, i.e. the tail length when reaching the first cycle is smaller than $2^{l/2}$. Since the average tail length is less than $2^{l/2}$ elements, this will happen with probability more than $1/2$.
- The processing of the block $[1]$ sets us in the same component (of the functional graph of $h_{[0]}$) as in the offline computation; again the probability is 0.7582.
- The $2^{l/2}$ $[0]$ message blocks concatenated after $[1]$ are enough to reach the cycle of the component; again this happens with probability $1/2$.

The collision probability is then $(0.7582)^2 \times 1/4 \simeq 0.14$ and thus we have that $\Pr[\mathcal{A}(\text{HMAC}_K^h) = 1] \geq 0.14$. In the case where the oracle is instantiated with a random function r, the adversary will output 1 if and only if a random HMAC collision happens between messages M_1 and M_2. Such a collision can be obtained if the processing of any of the last $[0]^{2^{l/2}}$ blocks of the two messages leads to an internal collision, therefore with negligible probability $2^{l/2} \cdot 2^{-l} = 2^{-l/2}$. Overall, the adversary advantage is equal to $Adv(\mathcal{A}) = \left|0.14 - 2^{-l/2}\right| \simeq 0.14$.

5 Internal State Recovery for NMAC and HMAC

This section extends the distinguishing-H attack in order to build an internal state recovery attack.

5.1 General Description

In order to extent the distinguishing-H attack to a state recovery attack, we observe that there is a high probability that the α-node reached in the online phase is the root of the giant tree of the functional graph of $h_{[0]}$. More precisely, we can locate the largest tree and the corresponding α-node in the offline phase, by repeating the cycle search a few times. We note that $\delta_1 + \delta_2 > 1$, therefore the largest tree is in the largest component with asymptotic probability one (see Theorem 3). Thus, assuming that the online phase succeeds, the α-node reached in the online phase is the root of the largest tree with asymptotic probability $\delta_1/\delta_2 \approx 0.63$. If we can locate the α-node of the online phase (i.e. we deduce its block index inside the queried message), we immediately get its corresponding internal state value from the offline computations.

Since the average rho length is $\sqrt{\pi 2^l/2}$ we can not use a brute-force search to locate the α-node, but we can use a binary search instead. We denote the length of the cycle of the giant component as L, and the follow the distinguishing-H attack to reach the cycle. We choose a random message block m and we query the two messages $M_1 = m \, \| \, [0]^{2^{l/2}} \, \| \, [1] \, \| \, [0]^{2^{l/2}+L}$ and $M_2 = m \, \| \, [0]^{2^{l/2}+L} \, \| \, [1] \, \| \, [0]^{2^{l/2}}$. If the MAC collide, we know that the state reached after processing $m \, \| \, [0]^{2^{l/2}}$ is located inside the main cycle. We use a binary search to find the smallest X so that the state reached after $m \, \| \, [0]^X$ is in the cycle.

The first step of the binary search should decide whether the node reached after $m \, \| \, [0]^{2^{l/2-1}}$ is also inside the first online cycle or not. More precisely, we check if the two messages $M_1' = m \, \| \, [0]^{2^{l/2-1}} \, \| \, [1] \, \| \, [0]^{2^{l/2}+L}$ and $M_2' = m \, \| \, [0]^{2^{l/2-1}+L} \, \| \, [1] \, \| \, [0]^{2^{l/2}}$ also give a colliding tag. If it is the case, then the node after processing $m \, \| \, 2^{l/2-1}$ is surely inside the cycle, which implies that the α-node is necessarily located in the first $2^{l/2-1}$ zero blocks. On the other hand, if the tags are not colliding, we cannot directly conclude that the α-node is located in the second half since there is a non-negligible probability that the α-node is in the first half but the $[1]$ block is directing the paths in M_1' and M_2' to distinct components in the functional graph of $h_{[0]}$ (in which case the tag are very likely to differ). Therefore, we have to test for collisions with several couples $M_1^u = m \, \| \, [0]^{2^{l/2-1}} \, \| \, [u] \, \| \, [0]^{2^{l/2}+L}$, $M_2^u = m \, \| \, [0]^{2^{l/2-1}+L} \, \| \, [u] \, \| \, [0]^{2^{l/2}}$, and if none of them collide we can safely deduce that the α-node is located in the second $2^{l/2-1}$ zero blocks. Overall, one such step reduces the number of the candidate nodes by a half and we simply continue this binary search in order to eventually obtain the position of the α-node with $\log_2(2^{l/2}) = l/2$ iterations.

5.2   Detailed Attack Process

1. (offline) Search for a cycle in the functional graph of $h_{[0]}$ and denote L its length.
2. (online) Find a message block m such that querying the two messages $M_1 = m \, \| \, [0]^{2^{l/2}} \, \| \, [1] \, \| \, [0]^{2^{l/2}+L}$ and $M_2 = m \, \| \, [0]^{2^{l/2}+L} \, \| \, [1] \, \| \, [0]^{2^{l/2}}$ leads to the same HMAC output. Let X_1 and X_2 be two integer variables, initialized to the values 0 and $2^{l/2}$ respectively.
3. (online) Let $X' = (X_1 + X_2)/2$. Select $\beta \log(l)$ distinct message blocks $[u]$, and for each of them query the HMAC output for messages $M_1^u = m \, \| \, [0]^{X'} \, \| \, [u] \, \| \, [0]^{2^{l/2}+L}$ and $M_2^u = m \, \| \, [0]^{X'+L} \, \| \, [u] \, \| \, [0]^{2^{l/2}}$. If at least one of the (M_1^u, M_2^u) pairs leads to a colliding HMAC output, then set $X_2 = X'$. Otherwise, set $X_1 = X'$. We use $\beta = 4.5$ as explained later.
4. (online) If $X_1 + 1 = X_2$ holds, output X_2 as the block index of the α-node. Otherwise, go back to the previous step.

5.3   Complexity and Success Probability Analysis

Complexity. We would like to evaluate the complexity of the attack. The first step will require about $2^{l/2}$ offline computations to find the cycle. Again, it is important to note that we can run this first step several times in order to ensure that we are using the cycle from the biggest component of the functional graph of $h_{[0]}$. The second step repeats the execution of the distinguishing-H attack from Section 4, which requires $6 \cdot 2^{l/2}$ computations for a success probability of 0.14, until it finds a succeeding message block m. Therefore, after trying a few m values, we have probability very close to 1 to find a valid one. The third and fourth steps will be executed about $l/2$ times (for the binary search), and each iteration of the third step performs $2 \cdot \beta \log(l)$ queries of about $2^{l/2} + 2^{l/2} + L \simeq 3 \cdot 2^{l/2}$ message blocks each. Therefore, the overall complexity of the attack is about $3\beta \cdot l \cdot \log(l) \cdot 2^{l/2}$ compression function computations.

Success Probability. Next we evaluate the success probability that the attacker recovers the internal state and this depends on the success probability of the binary search steps. We start with the case where the node after $m \parallel [0]^{X'}$ is inside the first online cycle. The third step will succeed as long as at least one of the (M_1^u, M_2^u) pairs collide on the output (we can omit the false positive collisions which happen with negligible probability). One pair (M_1^u, M_2^u) will indeed collide if:

- The random block $[u]$ sends both messages to the main component of the functional graph of $h_{[0]}$. Since it has an expected size of $\delta_2 \cdot 2^l$ (see Theorem 3), this is the case with probability δ_2^2.
- The $2^{l/2}$ $[0]$ message blocks concatenated after $[u]$ are enough to reach the cycle, i.e. the tail length when reaching the second cycle is smaller than $2^{l/2}$. Since the average tail length is smaller than $2^{l/2}$ elements, this will happen with probability $1/2$ for each message.

After trying $\beta \log(l)$ pairs, the probability that at least one pair collides is $1 - (1 - \delta_2^2/4)^{\beta \log(l)}$. If we use $\beta = -1/\log(1 - \delta_2^2/4) \approx 4.5$, this gives a probability of $1 - 1/l$. On the other hand, if the node after $m \parallel [0]^{X'}$ is not inside the cycle, the third step will succeed when no random collision occurs among the $\beta \log(l)$ tries, and such collisions happen with negligible probability. Overall, since there are $l/2$ steps in the binary search, the average success probability of the binary search is $(1 - 1/l)^{l/2} \geq e^{-1/2} \approx 0.6$.

Finally, the attack succeeds if the α-node is actually the root of the giant tree, as computed in the offline phase. This is the case with probability δ_1/δ_2, and the success probability of the full state recovery attack is $\delta_1/\delta_2 \cdot e^{-1/2} \approx 0.38$.

6   Internal State Recovery with Shorter Messages

In the full version of the paper, we give a an alternative internal-state recovery attack using shorter messages, that can also be used as a distinguishing-H attack with shorter messages. The attacks of Sections 4 and 5 have a complexity of $O(2^{l/2})$ using a small number of messages of length $2^{l/2}$; on the other hand the alternative attack has a complexity $O(2^{3l/4})$ using $2^{l/2}$ messages of length $2^{l/4}$. More generally, if the message size is limited to 2^s blocks ($s \leq l/4$), then the attack requires 2^{l-2s} messages. Due to space constraints, we only give a brief description of this attack here.

6.1   Entropy Loss in Collisions Finding

While the previous attacks are based on detecting cycles in the graph of a random function, this alternative attack is based on the fact that finding collisions by iterating a random function does not give a random collision: some particular collisions are much more likely to be found than others. This fact is well known in the context of collision search; for instance van Oorschot and Wiener [23] recommend to randomize the function regularly when looking for a large number of collisions. In this attack, we exploit this property to our advantage: first we use a collision finding algorithm to locate collisions in h_M with a fixed M; then we query the MAC oracle with messages with long repetitions of the block M and we detect collisions in the internal state; since the collisions found in this way are not randomly distributed, there is a good probability that we will reach one the collisions that was previously detected in the offline phase.

Actually, the attacks of Sections 4 and 5 can also be seen as following this stategy: we use a collision finding algorithm based on cycle detection (following Pollard's rho algorithm), and we know that with a good probability, the collision found will be the root of the giant tree. For the alternative attack, we use a collision finding algorithm similar to [23], but using using fixed length chains. In the full version of the paper, we study the entropy of the distribution of collisions found in this way, and we show that when using chains of length 2^s, we need about $2^{l/2-s}$ collisions in order to have a match between the online and offline steps. This translates to an attack complexity of 2^{l-s}, with $s \leq l/4$.

7   Key Recovery for HMAC Based on a Hash Function with an Internal Checksum

In this section we study HMAC used with a hash function with an internal checksum, such as GOST. We first show that the checksum does not prevent the distinguishing-H and state recovery attack, but more surprisingly the checksum actually allows to mount a full key-recovery attack significantly more efficient than exhaustive search.

A hash function with an internal checksum computes the sum of all message blocks, and uses this sum as an extra input to the finalization function. The sum can be computed for any group operation, but it will usually be an XOR sum or a modular addition. We use the XOR sum $\text{Sum}^{\oplus}$ to present our attack, but it is applicable with any group operation.

The checksum technique has been used to enhance the security of a hash function, assuming that controlling the checksum would be an additional challenge for an adversary. While previous work argued its effectiveness [10], our result reveals a surprising fact that incorporating a checksum into a hash function could even weaken its security in some applications such as HMAC.

A notable example of a hash function with a checksum is the GOST hash function, which has been standardized by the Russian Government [6] and by IETF [5]. HMAC-GOST has also been standardized by IETF [15] and is implemented in OpenSSL. GOST uses parameters $n = l = b$, and uses a separate call to process the message length, as follows:

$$x_0 = IV \qquad\qquad x_{i+1} = h(x_i, m_i) \qquad\qquad x_* = h(x_p, |M|)$$
$$\sigma_0 = 0 \qquad\qquad \sigma_{i+1} = \sigma_i \oplus m_i \qquad\qquad \mathsf{hash} = g(x_*, \sigma_p)$$

If this section we describe the attack on GOST-like functions following this structure; Figure 3 shows an HMAC computation with a GOST-like hash function. We give more general attacks when the output is computed as $g(x_p, |M|, \sigma_p)$ in the full version of this paper.

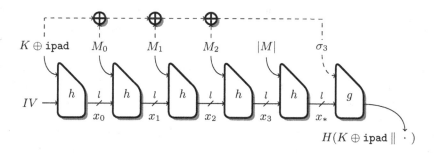

Fig. 3. HMAC based on a hash function with a checksum (dashed lines) and a length-padding block. We only detail the first hash function call.

7.1   General Description

In HMAC, $K \oplus \mathsf{ipad}$ is prepended to a message M, and $(K \oplus \mathsf{ipad}) \,\|\, M$ is hashed by the underlying hash function H. Therefore, the final checksum value is $\sigma_p = \mathsf{Sum}^\oplus((K \oplus \mathsf{ipad}) \,\|\, M) = K \oplus \mathsf{ipad} \oplus \mathsf{Sum}^\oplus(M)$. In this attack, we use the state recovery attack to recover the internal state x_* before the checksum is used and we take advantage of the fact that the value σ_p actually contains the key, but can still be controlled by changing the message. We use this to inject a known difference in the checksum, and to perform a kind of related key attack on the finalization function g, even though we have access to a single HMAC key.

More precisely, we use Joux's multicollision attack [11] to generate a large set of messages with the same value $\bar{x}$ for x_*, but with different values of the checksum. We detect MAC-collisions among those messages, and we assume that the collisions happens when processing the checksum of the internal hash function. For each such collision, we have $g(\bar{x}, K \oplus \texttt{ipad} \oplus \texttt{Sum}^{\oplus}(M)) = g(\bar{x}, K \oplus \texttt{ipad} \oplus \texttt{Sum}^{\oplus}(M'))$, and we compute the input difference $\Delta M = \texttt{Sum}^{\oplus}(M) \oplus \texttt{Sum}^{\oplus}(M')$.

Finally, we compute $g(\bar{x}, m)$ offline for a large set of random values m, and we collect collisions. Again, we compute the input difference $\Delta m = m \oplus m'$ for each collision, and we match Δm to the previously stored ΔM. When a match is found between the differences we look for the corresponding values and we have $K \oplus \texttt{ipad} \oplus \texttt{Sum}^{\oplus}(M) = m$ (or m') with high probability. This gives the value of the key K.

State Recovery with a Checksum. First, we note that the checksum σ does not prevent the state recovery attacks of Section 5; the complexity only depend on the size l of the state x. Indeed, the attack of Section 5 is based on detecting collisions between pairs of messages $M_1 = m \,\|\, [0]^{2^{l/2}} \,\|\, [k] \,\|\, [0]^{2^{l/2}+L}$ and $M_2 = m \,\|\, [0]^{2^{l/2}+L} \,\|\, [k] \,\|\, [0]^{2^{l/2}}$. Since the messages have the same checksum, a collision in the state will be preserved. More generally, the attacks can easily be adapted to use only message with a fixed sum. For instance, we can use random messages with two identical blocks in the attack of Section 5, and messages of the form $m \,\|\, m \,\|\, [0]^{2^{l/2}} \,\|\, [k] \,\|\, [k] \,\|\, [0]^{2^{l/2}+L}$ have a checksum of zero.

Recovering the State of a Short Message. Unfortunately, the state we recover will correspond to a rather long message (*e.g.* $2^{l/2}$ blocks), and all the queries based on this message will be expensive. In order to overcome this issue, we use the known state x_M after a long message M to recover the state after a short one. More precisely, we generate a set of $2^{l/4}$ long messages by appending a random message blocks m twice to M. Note that $\texttt{Sum}^{\oplus}(M \,\|\, m \,\|\, m) = \texttt{Sum}^{\oplus}(M)$. Meanwhile, we generate a set of $2^{3l/4}$ two-block messages $m_1 \| m_2$, with $m_1 \oplus m_2 = \texttt{Sum}^{\oplus}(M)$. We query these two sets to the HMAC oracle and collect collisions between a long and a short message. We expect that one collision correspond to a collision in the value x_* before the finalization function g. We can compute the value x_* for the long message from the known state x_M after processing M. This will correspond to the state after processing the message $m_1 \,\|\, m_2$ and its padding block, or equivalently, after processing the message $m_1 \,\|\, m_2 \,\|\, [2]$ (because the length block is processed with the same compression function). We can verify that the state is correctly recovered by generating a collision $m \| m, m' \| m'$ offline from the state x_*, and comparing $\texttt{HMAC}(m_1 \,\|\, m_2 \,\|\, [2] \,\|\, m \,\|\, m)$ and $\texttt{HMAC}(m_1 \,\|\, m_2 \,\|\, [2] \,\|\, m' \,\|\, m')$.

7.2   Detailed Attack Process

For simplicity of the description, we omit the padding block in the following description, and we refer to the previous paragraphs for the details of how to deal with the padding.

1. Recover an internal state value x_r after processing a message M_r through HMAC. Refer to Section 5 for the detailed process.
2. (online) Choose $2^{l/4}$ one-block random messages m, query $M_r \| m \| m$ to HMAC and store m and the corresponding tag.
3. (online) Choose $2^{3l/4}$ one-block random messages m, query $(\mathtt{Sum}^{\oplus}(M_r) \oplus m) \| m$ to HMAC and look for a match between the tag value and one of the stored tag values in Step 2.
   For a colliding pair $(\mathtt{Sum}^{\oplus}(M_r) \oplus m) \| m$ and $M_r \| m' \| m'$, denote $\mathtt{Sum}^{\oplus}(M_r) \oplus m \| m$ as M_1 and $h(h(x_r, m'), m')$ as x_1. Generate a collision $h(h(x_1, u), u) = h(h(x_1, u'), u')$. Query $M_1 \| u \| u$ and $M_1 \| u' \| u'$ and compare the tags. If they are equal, the internal state after processing M_1 (before the checksum block) is x_1.
4. (offline) Generate $2^{3l/4}$ messages that all collide on the internal state before the checksum block by Joux's multicollision. More precisely, choose $2^{l/2}$ random message m and compute $h(x_1, m)$ to find a collision $h(x_1, m_1) = h(x_1, m_1') = x_2$. Then iterate this procedure to find a collision $h(x_i, m_i) = h(x_i, m_i') = x_{i+1}$ for $i \le 3l/4$. Denote the value of $x_{3l/4+1}$ by $\bar{x}$.
5. (online) Query the set of messages in Step 4 to HMAC in order to collect tag collisions. For each collision M and M', compute the checksum difference $\Delta M = \mathtt{Sum}^{\oplus}(M) \oplus \mathtt{Sum}^{\oplus}(M')$, and store $(\mathtt{Sum}^{\oplus}(M), \Delta M)$.
6. (offline) Choose a set of $2^{3l/4}$ one-block random message m, compute $g(\bar{x}, m)$ and collect collisions. For each collision m and m', compute the difference $\Delta m = m \oplus m'$ and match Δm to the stored ΔM at Step 5. If a match is found, mark $\mathtt{Sum}^{\oplus}(M) \oplus \mathtt{ipad} \oplus m$ and $\mathtt{Sum}^{\oplus}(M) \oplus \mathtt{ipad} \oplus m \oplus \Delta m$ as potential key candidates.
7. (offline) filter the correct key from the potential candidates by verifying a valid message/tag pair.

7.3 Complexity and Success Probability Analysis

We need to evaluate the complexity of our key recovery attack.

Step 1: $O(l \cdot \log(l) \cdot 2^{l/2})$ **Step 2:** $2^{3l/4}$ **Step 3:** $2 \cdot 2^{3l/4}$

Step 4: $3l/4 \cdot 2^{l/2}$ **Step 5:** $3l/4 \cdot 2^{3l/4}$ **Step 6:** $2^{3l/4}$

Step 7: $O(1)$

Overall, the fifth step dominates the complexity, and the total complexity is about $3l/4 \cdot 2^{3l/4}$ compression function computations.

Next we evaluate the success probability of our method. The first step succeeds with a probability almost 1 after several trials. Steps 2 and 3 need to guarantee a collision between a long and a short message. Since there are 2^l pairs, one such collision occurs with a probability of $1 - (1 - 2^{-l})^{2^l} \approx 1 - 1/e \approx 0.63$. The success probability of producing no less than $2^{l/2}$ collisions at each of steps 5 and 6 is 0.5 since the expected number of collisions is $2^{l/2}$. Thus the overall success probability is no less than 0.16.

8   Conclusion

Our results show that the security of HMAC and hash-based MAC above the birthday bound is significantly weaker than previously expected. First, we show that distinguishing-H and state-recovery attacks are not much harder than a distinguishing-R attack, contrary to previous beliefs. Second, we show that the use of a checksum can allow a key-recovery attack against HMAC with complexity only $\tilde{O}(2^{3l/4})$. In particular, this attack is applicable to HMAC-GOST, a standard-ized construction.

We give a comparison of our attacks and previous attack against concrete in-stances of HMAC in Table 1, showing that some attacks against concrete instances are in fact less efficient than our generic attacks.

As future works, it would be interesting to find other applications of the internal state recovery for HMAC. Moreover, we expect further applications of the analysis of the functional graph, as it might be possible to use other distinguishing properties, such as the tail length, the distance of a node from the cycle, etc.

Acknowledgments. The authors would like to thank the anonymous referees for their helpful comments. Gaëtan Leurent is supported by the ERC project CRASH. Thomas Peyrin and Lei Wang are supported by the Singapore National Research Foundation Fellowship 2012 (NRF-NRFF2012-06).

References

1. Bellare, M.: New Proofs for NMAC and HMAC: Security Without Collision-Resistance. In: Dwork, C. (ed.) CRYPTO 2006. LNCS, vol. 4117, pp. 602–619. Springer, Heidelberg (2006)
2. Bellare, M., Canetti, R., Krawczyk, H.: Keying Hash Functions for Message Au-thentication. In: Koblitz, N. (ed.) CRYPTO 1996. LNCS, vol. 1109, pp. 1–15. Springer, Heidelberg (1996)
3. Brassard, G. (ed.): CRYPTO 1989. LNCS, vol. 435. Springer, Heidelberg (1990)
4. Damgård, I.: A Design Principle for Hash Functions. In: [3], pp. 416–427
5. Dolmatov, V.: GOST R 34.11-94: Hash Function Algorithm. RFC 5831 (Informa-tional) (March 2010)
6. FAPSI, VNIIstandart: GOST 34.11-94, Information Technology Cryptographic Data Security Hashing Function (1994) (in Russian)
7. Flajolet, P., Odlyzko, A.M.: Random Mapping Statistics. In: Quisquater, J.-J., Vandewalle, J. (eds.) EUROCRYPT 1989. LNCS, vol. 434, pp. 329–354. Springer, Heidelberg (1990)
8. Flajolet, P., Sedgewick, R.: Analytic Combinatorics. Cambridge University Press (2009)
9. Fouque, P.-A., Leurent, G., Nguyen, P.Q.: Full Key-Recovery Attacks on HMAC/NMAC-MD4 and NMAC-MD5. In: Menezes, A. (ed.) CRYPTO 2007. LNCS, vol. 4622, pp. 13–30. Springer, Heidelberg (2007)
10. Gauravaram, P., Kelsey, J.: Linear-XOR and Additive Checksums Don't Protect Damgård-Merkle Hashes from Generic Attacks. In: Malkin, T. (ed.) CT-RSA 2008. LNCS, vol. 4964, pp. 36–51. Springer, Heidelberg (2008)

11. Joux, A.: Multicollisions in Iterated Hash Functions. Application to Cascaded Constructions. In: Franklin, M. (ed.) CRYPTO 2004. LNCS, vol. 3152, pp. 306–316. Springer, Heidelberg (2004)
12. Kim, J., Biryukov, A., Preneel, B., Hong, S.: On the Security of HMAC and NMAC Based on HAVAL, MD4, MD5, SHA-0 and SHA-1 (Extended Abstract). In: De Prisco, R., Yung, M. (eds.) SCN 2006. LNCS, vol. 4116, pp. 242–256. Springer, Heidelberg (2006)
13. Merkle, R.C.: One Way Hash Functions and DES. In: [3], pp. 428–446
14. Peyrin, T., Sasaki, Y., Wang, L.: Generic Related-Key Attacks for HMAC. In: Wang, X., Sako, K. (eds.) ASIACRYPT 2012. LNCS, vol. 7658, pp. 580–597. Springer, Heidelberg (2012)
15. Popov, V., Kurepkin, I., Leontiev, S.: Additional Cryptographic Algorithms for Use with GOST 28147-89, GOST R 34.10-94, GOST R 34.10-2001, and GOST R 34.11-94 Algorithms. RFC 4357 (Informational) (January 2006)
16. Preneel, B.: HMAC. In: van Tilborg, H.C.A., Jajodia, S. (eds.) Encyclopedia of Cryptography and Security, 2nd edn., pp. 559–560. Springer (2011)
17. Preneel, B., van Oorschot, P.C.: MDx-MAC and Building Fast MACs from Hash Functions. In: Coppersmith, D. (ed.) CRYPTO 1995. LNCS, vol. 963, pp. 1–14. Springer, Heidelberg (1995)
18. Rechberger, C., Rijmen, V.: New Results on NMAC/HMAC when Instantiated with Popular Hash Functions. J. UCS 14(3), 347–376 (2008)
19. Rivest, R.L.: The MD4 Message Digest Algorithm. In: Menezes, A., Vanstone, S.A. (eds.) CRYPTO 1990. LNCS, vol. 537, pp. 303–311. Springer, Heidelberg (1991)
20. Rivest, R.L.: The MD5 message-digest algorithm. Request for Comments (RFC) 1320, Internet Activities Board, Internet Privacy Task Force (April 1992)
21. Tsudik, G.: Message Authentication with One-Way Hash Functions. In: INFOCOM, pp. 2055–2059 (1992)
22. U.S. Department of Commerce, National Institute of Standards and Technology: Secure Hash Standard (SHS) (Federal Information Processing Standards Publication 180-3) (2008), http://csrc.nist.gov/publications/fips/fips180-3/fips180-3_final.pdf
23. van Oorschot, P.C., Wiener, M.J.: Parallel Collision Search with Cryptanalytic Applications. J. Cryptology 12(1), 1–28 (1999)
24. Wang, L., Ohta, K., Kunihiro, N.: New Key-Recovery Attacks on HMAC/NMAC-MD4 and NMAC-MD5. In: Smart, N.P. (ed.) EUROCRYPT 2008. LNCS, vol. 4965, pp. 237–253. Springer, Heidelberg (2008)
25. Wang, X., Yu, H., Wang, W., Zhang, H., Zhan, T.: Cryptanalysis on HMAC/NMAC-MD5 and MD5-MAC. In: Joux, A. (ed.) EUROCRYPT 2009. LNCS, vol. 5479, pp. 121–133. Springer, Heidelberg (2009)
26. Yasuda, K.: "Sandwich" Is Indeed Secure: How to Authenticate a Message with Just One Hashing. In: Pieprzyk, J., Ghodosi, H., Dawson, E. (eds.) ACISP 2007. LNCS, vol. 4586, pp. 355–369. Springer, Heidelberg (2007)
27. Yu, H., Wang, X.: Full Key-Recovery Attack on the HMAC/NMAC Based on 3 and 4-Pass HAVAL. In: Bao, F., Li, H., Wang, G. (eds.) ISPEC 2009. LNCS, vol. 5451, pp. 285–297. Springer, Heidelberg (2009)

A   Additional Figures

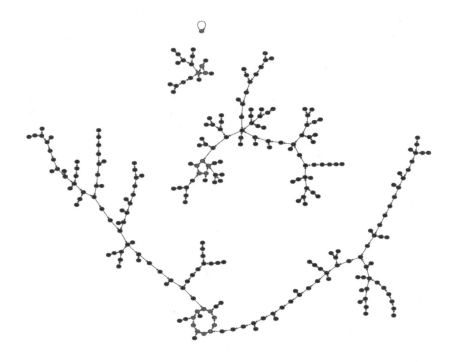

Fig. 4. Functional graph of Keccak (SHA-3) with 8-bit input and 8-bit output

Cryptanalysis of HMAC/NMAC-Whirlpool

Jian Guo[1], Yu Sasaki[2], Lei Wang[1], and Shuang Wu[1]

[1] Nanyang Technological University, Singapore
[2] NTT Secure Platform Laboratories, Japan
ntu.guo@gmail.com, sasaki.yu@lab.ntt.co.jp, {wang.lei,wushuang}@ntu.edu.sg

Abstract. In this paper, we present universal forgery and key recovery attacks on the most popular hash-based MAC constructions, *e.g.*, HMAC and NMAC, instantiated with an AES-like hash function Whirlpool. These attacks work with Whirlpool reduced to 6 out of 10 rounds in single-key setting. To the best of our knowledge, this is the first result on "original" key recovery for HMAC (previous works only succeeded in recovering the equivalent keys). Interestingly, the number of attacked rounds is comparable with that for collision and preimage attacks on Whirlpool hash function itself. Lastly, we present a distinguishing-H attack against the full HMAC- and NMAC-Whirlpool.

Keywords: HMAC, NMAC, Whirlpool, key recovery, universal forgery.

1 Introduction

AES (Advanced Encryption Standard) [6] is the probably most used block cipher nowadays, and it also inspires many designs for other fundamental primitives of modern cryptography, *e.g.*, hash function. As cryptographic algorithms for security applications, AES and AES-like primitives should receive continuous security analysis under various protocol settings. This paper discusses the security evaluation of these primitives in one notable setting; the MAC (Message Authentication Code) setting.

A MAC is a symmetric-key construction to provide integrity and authenticity for data. There are two popular approaches to build a MAC. The first approach is based on a block cipher or a permutation, *e.g.*, the well-known CBC (Cipher Block Chaining) MAC [1]. Such designs with an AES-like block cipher (or permutation) include CMAC-AES [28], PC-MAC-AES [19], ALPHA-MAC [7] and PELICAN-MAC [8]. A series of analysis results have been published on these AES-like block ciphers (or unkeyed permutations) under the CBC MAC setting. Refer to [12,13,32,4,9]. From a high-level view, cryptanalysts have managed to extend several analysis techniques devised on block cipher itself to also work in the CBC MAC setting, *e.g.*, [32,9] use the impossible differential attack. The second approach is based on a hash function. Such designs with an AES-like hash function include HMAC-Whirlpool and HMAC-Grøstl. Surprisingly, there is NO algorithmic analysis result yet on these AES-like hash functions in the MAC setting to our best knowledge, though a side-channel attack was published on HMAC-Whirlpool [33].

K. Sako and P. Sarkar (Eds.) ASIACRYPT 2013, Part II, LNCS 8270, pp. 21–40, 2013.

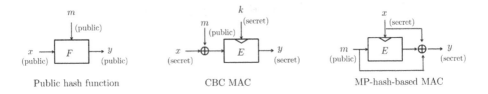

Fig. 1. Comparison of attack models

We briefly discuss the difficulty of applying the analysis techniques, which are devised to analyze public AES-like hash functions or to analyze AES-like block ciphers in the CBC MAC setting, to evaluate AES-like hash functions under the hash-based MAC setting. More precisely, we make a comparison of their model from an attacker's view by focusing on the underlying iterated small primitives; compression function of a hash function and block cipher of CBC MAC, which is also explained in Figure 1. A few new notations are introduced here: x is an internal state after processing previous message blocks, m is a current message block, y is an updated internal state, k is a secret key of block cipher, F is a compression function, and E is a block cipher.

For a hash function in public setting and in MAC setting, the main difference from an attacker's view is that x and y are public in the former setting, but are secret in the latter setting. Note that the effective analysis techniques rebound attack [18] and splice-and-cut preimage attack [25] on AES-like hash functions in public setting use a start-from-the-middle approach, which requires to know and to control the internal values of the compression function, and thus requires that x is public to the attacker. Therefore these techniques cannot be applied trivially in MAC setting.

For CBC MAC and hash-based MAC, the main difference is how a message block is injected to an internal state. CBC MAC uses a simple XOR sum $x \oplus m$, while hash-based MAC usually compresses x and m in a complicated process, e.g., the Miyaguchi-Preneel (MP) scheme $E_x(m) \oplus m \oplus x$. It affects the applicability of differential cryptanalysis. The attacker is able to derive the internal state difference Δx in the CBC MAC setting (i.e., randomize message block m to find a pair m and m' that leads to a collision on the input to E detectable from the colliding MAC outputs, and derive $\Delta x = m \oplus m'$). On the other hand, the internal state difference cannot be derived in the hash-based MAC setting except the collision case $\Delta x = 0$, which sets a constraint on the differentials of the underlying block cipher that can be exploited by an attacker.

This paper gives the first step on the algorithmic security evaluation of AES-like hash functions in the hash-based MAC setting. The main attack target is the Whirlpool hash function in the HMAC setting, which is motivated by the fact that both schemes are internationally standardized.

Whirlpool [24] was proposed by Barreto and Rijmen in 2000. Its compression function is built from an AES-like block cipher following Miyaguchi-Preneel mode.

Whirlpool has been standardized by ISO/IEC, and has been implemented in many cryptographic software libraries such as FreeOTFE and TrueCrypt. Its security has been evaluated and approved by NESSIE [20]. The first cryptanalysis result was published by Mendel et $al.$ in 2009 [18], which presented a collision attack on 4-round Whirlpool hash function (full version: 10 rounds). Later Lamberger et $al.$ extended the collision attack to 5 rounds [16]. After that, Sasaki published a (second) preimage attack on 5-round Whirlpool hash function in 2011 [25], and the complexity of his attack was improved by Wu et $al.$ in 2012 [31]. Later Sasaki et $al.$ extended the preimage attack to 6 rounds [27]. In addition to hash function attacks, several cryptanalysis results on the compression function of Whirlpool have also been published [16,27], and particularly a distinguisher on the full compression function was found [16].

HMAC [2] was proposed by Bellare et $al.$ in 1996. It has been standardized by ANSI, IETF, ISO and NIST, and widely deployed in SSL, TLS and IPsec. HMAC based on a hash function H takes a secret key K and a message M as input and is computed by

$$\text{HMAC}(K, M) = H(K \oplus \text{opad} \parallel H(K \oplus \text{ipad} \parallel M)),$$

where ipad and opad are two different public constants. HMAC is always viewed as a single-key variant of NMAC [2]. NMAC based on a hash function H takes two keys; the inner key K_{in} and the outer key K_{out}, and a message M as input, and is computed by

$$\text{NMAC}(K_{out}, K_{in}, M) = H_{K_{out}}(H_{K_{in}}(M)),$$

where the function $H_{K_{in}}(\cdot)$ stands for the hash funtion H with its initial value replaced by K_{in}, and similarly for $H_{K_{out}}(\cdot)$. The internal states $F(IV, K \oplus \text{opad})$ and $F(IV, K \oplus \text{ipad})$ of HMAC is equivalent to the K_{out} and the K_{in} of NMAC respectively, where F is the compression function and IV is the public initial value of H. This paper refers $F(IV, K \oplus \text{opad})$ and $F(IV, K \oplus \text{ipad})$ to as the equivalent outer key and the equivalent inner key respectively. Note that if these two equivalent keys are recovered, the attacker will be able to forge any message, resulting in a universal forgery attack on HMAC.

Our Contribution. We present universal forgery ($i.e.$, recover the two equivalent keys) and key recovery attacks on HMAC based on round-reduced Whirlpool, and a distinguishing-H attack on HMAC based on full Whirlpool. These attacks are also applicable to NMAC based on Whirlpool. All the results are summarized in Table 1. Interestingly, our attacks on the Whirlpool hash function in HMAC and NMAC setting reach attacked round numbers comparable to that in the public setting (even with respect to classical security notions; forgery and key recovery in MAC setting and collision and preimage attacks in public setting).

For HMAC and NMAC based on 5-round Whirlpool, we generate a structured collision on the first message block of the first call of hash function, which can be detected from the MAC output collisions and verified by the length extension property. For the structured collision, we know the differential path inside the block cipher $E_{K_{in}}$. Based on it, we apply a meet-in-the-middle attack to recover the value of K_{in}. After that, we apply two attacks. One is to recover the value of K_{out}, which results in a universal forgery attack on HMAC and a full-key recovery attack on NMAC. The attack of recovering K_{out} is similar with that of recovering K_{in}, except the procedure of finding target pairs. Instead of generating collisions as for recovering K_{in}, we will first recover the values of an intermediate chaining variable of the outer hash function, and then find a near collision on this intermediate chaining variable. The other attack is to recover the key of HMAC. Recall that $K_{in} = F(IV, K \oplus \texttt{ipad})$, recovering K from K_{in} is similar to inverting $F(IV, \cdot)$ to find a preimage of K_{in}. Thus we apply an attack similar with the splice-and-cut preimage attack to recover K from K_{in}. To our best knowledge, this is the *first* result of recovering the (original) key of HMAC, while previous results [11,22,23,29] only succeeded in recovering the equivalent keys.

We investigate the extension by one more round, namely 6-round Whirlpool, and find an interesting observation. More precisely, K_{out} can be recovered if a value of an intermediate chaining variable in the first call of hash function is recovered or leaked. Differently from the above attacks on 5 rounds, the procedure is based on generating a multi-near-collision on an intermediate chaining variable of the outer hash function. After K_{out} is recovered, we apply two attacks. One is to recover K_{in}, which results in a universal forgery attack on HMAC and a full-key recovery attack on NMAC. The other attack is to recover the key of HMAC. From a high-level overview, our observation reduces the problem of breaking the classical security notions (with significant impacts) universal forgery and key recovery to the problem of breaking a weak security notion (usually with rather limited impacts) internal-state recovery for HMAC and NMAC based on 6-round Whirlpool. We stress that such a reduction is not trivial. As an example, an internal-state recovery attack was published on HMAC/NMAC-MD5 in the single-key setting back to 2009 [30], but no universal forgery or key recovery attack is published on HMAC/NMAC-MD5 in the single-key setting yet to our best knowledge. Moreover, very recently Leurent et al. find a generic single-key internal-state recovery attack on HMAC and NMAC [17]. Combing their attack with our observation, we get universal forgery and key recovery attacks on HMAC and NMAC based on 6-round Whirlpool.

We would like to point out that the above universal forgery and key recovery attacks on round-reduced Whirlpool are also applicable in other hash-based MAC setting. More precisely, we can attack LPMAC and secret-suffix MAC with 6-round Whirlpool and Envelop MAC with 5-round Whirlpool, all in the single-key setting.

Lastly, we find a distinguishing-H attack on HMAC and NMAC with full Whirlpool, which in fact has wide applications besides Whirlpool. Recall HMAC and NMAC make two calls of hash function, and the outer hash function takes the inner hash outputs as input messages. Thus the outer hash function always processes n bits long messages, where n is the bit size of hash digests. Note that usually the length and the value of the padding bits are solely determined by

Table 1. Summarization of our results. These results are based on the minimization of max{data, time, memory}. More tradeoffs towards minimizing each parameter of data, time and memory are provided in the paper.

Our Result Summarization						
Attack Target	#Rounds	Attack mode	Complexity			Reference
			Time	Memory	Data	
HMAC-Whirlpool	5	universal forgery	2^{402}	2^{384}	2^{384}	Section 3
	5	key recovery	2^{448}	2^{377}	2^{321}	Section 3
	6	universal forgery	2^{451}	2^{448}	2^{384}	Section 4
	6	key recovery	2^{496}	2^{448}	2^{384}	Section 4
	10 (full)	distinguishing-H	2^{256}	2^{256}	2^{256}	Section 5
	10 (full)	distinguishing-H	2^{384}	2^{256}	2^{384}	[17]
NMAC-Whirlpool	5	key recovery	2^{402}	2^{384}	2^{384}	Section 3
	6	key recovery	2^{451}	2^{448}	2^{384}	Section 4
	10 (full)	distinguishing-H	2^{256}	2^{256}	2^{256}	Section 5
	10 (full)	distinguishing-H	2^{384}	2^{256}	2^{384}	[17]

Previous best results on Whirlpool hash function						
Whirlpool	5	collision attack	2^{120}	2^{64}	–	[16]
	6	preimage attack	2^{481}	2^{256}	–	[27]

the bit size of an input message. Therefore it is possible that the last block of the outer hash function of HMAC and NMAC contains fully padding bits and thus is with a constant value, and indeed this is the case for HMAC- and NMAC-Whirlpool. Our distinguishing-H attack can be applied with a complexity $2^{n/2}$ (n is 512 for Whirlpool). Our distinguisher has two advantages compared with Leurent *et al.*'s generic attack [17]. One is that our queried messages have practical length. The other one is that the complexity of our attack is significantly lower as long as the specification of the n-bit hash function restricts the input message with a block length shorter than $2^{n/2}$. Our distinguishing-H attack on HMAC- and NMAC- Whirlpool has a complexity of 2^{256}, while Leurent *et al.*'s attack has a complexity of at least 2^{384}.

Note that we focus on HMAC-Whirlpool using a 512-bit key and producing 512-bit MAC outputs in this paper. One may doubt the large size of the key and the tag. We would like to point out that besides pure theoretical research interests, evaluating such an instance of HMAC-Whirlpool also has practical impacts. This is due to the fact that ever since HMAC was designed and standardized, it has been widely implemented beyond the mere MAC applications. For example, the above instance of HMAC-Whirlpool will be used in HMAC-based Extract-and-Expand Key Derivation Function (HKDF) [15] if one instantiates this protocol with Whirlpool hash function, providing that Whirlpool is a long-stand secure hash function and has been implemented in many cryptographic software library. Based on these facts, HMAC-Whirlpool may have more applications in industry in the future, and thus deserves a careful security evaluation from the cryptography community in advance.

In the rest of the paper, Section 2 gives the specifications. Section 3 presents our attacks on HMAC and NMAC with 5-round Whirlpool. Section 4 describes our results on one more round. Section 5 provides a distinguishing-H attack on HMAC and NMAC with full Whirlpool. Finally we give conclusion and open discussions in Section 6.

2   Specifications

2.1   Whirlpool Hash Function [24]

The Whirlpool hash function follows the Merkle-Damgård structure and produces 512-bit digests. The input message M is padded by a '1', a least number of '0's, and 256-bit representation of the original message length, such that the padded message becomes a multiple of 512 bits.

The padded message is divided into 512-bit blocks and used in the iteration of compression functions. The compression function F is constructed based on a block-cipher E in Miyaguchi-Preneel mode (MP mode), *i.e.*, $F(C, M) = E_C(M) \oplus C \oplus M$. Starting from a constant initial value $C_0 = IV$, the chaining value is updated for each of the message block $C_{i+1} = F(C_i, M_i)$. After all message blocks are processed, the final chaining value is used as the hash value.

The underlying block cipher uses an AES-like structure with an 8×8 byte matrix. The round function of the key schedule consists of four operations, *i.e.*,

$$K_{i+1} = \text{AC} \circ \text{MR} \circ \text{SC} \circ \text{SB}(K_i), \text{ for } i \in \{0, 1, \ldots, 9\}.$$

- SubBytes(SB): apply an Sbox to each byte.
- ShiftColumns(SC): cyclically rotate the j-th column downwards by j bytes.
- MixRows(MR): multiply the state by an 8×8 MDS matrix.
- AddRoundConstant(AC): XOR a 512-bit round constant to the key state.

We denote the key state after SB, after SC and after MR in the $(i+1)$-th round of the key schedule by K_i^{SB}, K_i^{SC}, K_i^{MR} respectively.

The round function of the encryption is almost the same as the key schedule, except for the AddRoundKey(AK) operation, which XORs the key state to the data state, *i.e.*, the initial state is the XOR sum of the whitening key and the plaintext $S_0 = K_0 \oplus M$ and

$$S_{i+1} = \text{AK} \circ \text{MR} \circ \text{SC} \circ \text{SB}(S_i), \text{ for } i \in \{0, 1, \ldots, 9\}.$$

The final state S_{10} is used as the ciphertext. We denote the state after SB, after SC and after MR in the $(i+1)$-th round of the round encryption function by S_i^{SB}, S_i^{SC} and S_i^{MR} respectively.

2.2   HMAC and NMAC [2]

NMAC replaces the initial vector (IV) of a hash function H by a secret key K to produce a keyed hash function H_K. NMAC uses two secret key K_{in} and K_{out} and is defined by

$$\text{NMAC}_{K_{out}, K_{in}}(M) = H_{K_{out}}(H_{K_{in}}(M)).$$

K_{in} and K_{out} are usually referred to as the inner and the outer keys. Correspondingly $H_{K_{in}}$ and $H_{K_{out}}$ are referred to as the inner and the outer hash functions. HMAC is a single-key variant of NMAC. Denote the compression function by F.

$$\text{HMAC}_K(M) = \text{NMAC}_{F(IV,K\oplus\text{ipad}),F(IV,K\oplus\text{opad})}(M).$$

3   Attacks of HMAC and NMAC Based on 5-Round Whirlpool

In this section, we use one block long messages M to present our attack. Fig. 2 shows how HMAC/NMAC-Whirlpool processes M. Note that both M and T'' are one full block long, and thus an extra padding block P is appended in both the two calls of the hash function.

The attack starts with recovering value of (equivalent) K_{in}. We generate a structured collision on the internal state T'. Then for the collision, we get the differential path inside $E_{K_{in}}$, and recover some internal value of $E_{K_{in}}$ by a meet-in-the-middle (MitM) attack approach. Finally K_{in} is derived by a simple backward computations. Once K_{in} is recovered, we have two directions: 1) recover the value of (equivalent) K_{out} to amount a universal forgery for HMAC or to amount a full-key recovery for NMAC, and 2) recover the key of HMAC.

For the K_{out} recovery, note that T'' is public to the attacker now since K_{in} is recovered. We firstly derive the values of T''' with a technique similar to [26], and then obtain the values of $E_{K_{out}} \oplus K_{out}$: $T'' \oplus T'''$. Given that K_{out} has no difference, we search for a pair of messages that satisfies a pre-determined difference constraint on the outputs of $E_{K_{out}}$, and get the inside differential path. Finally we recover an internal value of $E_{K_{out}}$, and backwards compute the value of K_{out}.

For the key recovery of K, from $K_{in} = F(IV, K \oplus \text{ipad})$, we observe that $K \oplus \text{ipad}$ is a preimage of K_{in} regarding the Whirlpool compression function with a fixed chaining value $F(IV, \cdot)$. Note that the problem of inverting the compression function of Whirlpool has already been solved in [31] and [27] with splice-and-cut MitM approach. We use a similar approach to recover the value of $K \oplus \text{ipad}$ and then derive the value of K.

Moreover, we provide time-memory-data tradeoffs for recovering K_{in} and K_{out}.

3.1   How to Recover (Equivalent) K_{in}

In this section, we demonstrate the K_{in} recovery attack with optimizing its complexity for the key recovery of HMAC, and introduce the time-memory-data tradeoff in the next section.

Our attack is based on a 5-round differential path of the compression function, which is shown in Fig. 3. Each cell in this figure stands for a byte of the key or the state. Blank cells are non-active and cells with a dot inside are active. If the value of a byte is unknown, the cell is in white color. Red bytes are initially known from the message, tag or the recovered chaining value. Blue and green bytes

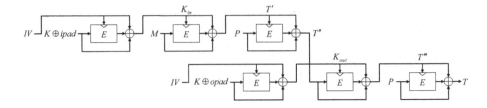

Fig. 2. HMAC/NMAC based on Whirlpool

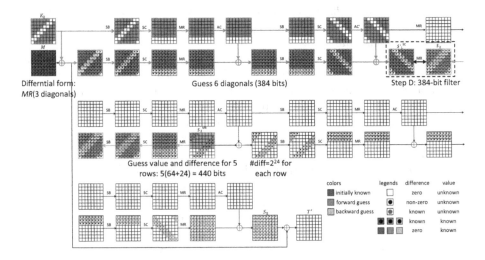

Fig. 3. Differential path for recovering K_{in} for HMAC and NMAC with 5-round Whirlpool

are the guessed bytes in the forward and backward directions of the MitM step. Moreover, some round functions are illustrated in equivalent expressions in this figure. The new operation AC' XORs the constant $MR^{-1}(RC_i)$ to the key state, where RC_i is used in the original AC operation, and it implies $AC \circ MR = MR \circ AC'$.

Produce a Structured Collision on T'. We use a structure of chosen messages in which any two messages satisfy a constraint of the differential form in Fig. 3. First we choose a set of 2^{192} values $\{M_1, M_2, \ldots, M_{2^{192}}\}$ such that the value of three specific rows of the messages take all possible 2^{192} values and all other bytes are chosen as constants. The positions of the three active rows are the top three rows in Fig. 3. Then, update the set by $M_i \leftarrow MR \circ SC(M_i)$ for $i = 1, 2, \ldots, 2^{192}$. This requires about 2^{192} computations. Note that for any two distinct indexes i_1 and i_2, $SC^{-1} \circ MR^{-1}(M_{i_1} \oplus M_{i_2})$ has three active rows in the pre-specified positions. Query the messages and obtain the corresponding tags $T_i = MAC(K, M_i)$, for $i = 1, 2, \ldots, 2^{192}$. Check if there is a collision of the tags. If a collision is found, we need to verify if it collides on T' by the length extension attack (*i.e.*, append a random message block M to each of the colliding messages M_{i_1} and M_{i_2}, and query $M_{i_1}\|M$ and $M_{i_2}\|M$ to see whether their tags collide). For a collision on T', it is ensured that the output difference of $E_{K_{in}}$ converted by $\text{SC}^{-1} \circ \text{MR}^{-1}$ has three active rows.

For a structure of 2^{192} messages generated by applying MR $\circ$ SC for each, we query them to MAC, store the corresponding tags and search for a collision. So it requires 2^{192} queries, 2^{192} computations, and 2^{192} memory. For one structure, we can make $\binom{2^{192}}{2} = 2^{383}$ pairs. After repeating the process for 2^{129} structures with different chosen constants, one collision is expected. The total number of queries is $2^{192+129} = 2^{321}$, the computational complexity is 2^{321} and the required memory is 2^{192}.

Recover K_{in}. Recall $T' = F(K_{in}, M) = E_{K_{in}}(M) \oplus M \oplus K_{in}$. For an inner collision on T', we know $\triangle T' = 0$. In the single-key attacks, the difference of K_{in} is also zero: $\triangle K_{in} = 0$. Thus the difference of the output of the block cipher can be computed as $\triangle E_{K_{in}}(M) = \triangle T' \oplus \triangle M \oplus \triangle K_{in} = \triangle M$. So we get $\triangle S_5 = \triangle M$, and thus $\text{SC}^{-1} \circ \text{MR}^{-1}(\triangle S_5)$ has three active rows. It ensures that the number of differences at each row of S_2^{MR} is at most 2^{24}. Now we describe the attack step by step.

Step A. Guessing in the forward direction

Guess the values of m diagonals of K_{in} (2^{64m} values) which are marked in blue, as in Fig. 3. Then we can determine the value of corresponding m diagonals in S'^{SC}_1. Now there are m known diagonals on the left side of the matching point - the MR operation in the second round. All the candidates are stored in a lookup table T_1.

Step B. Guessing in the backward direction

Guess the values and differences of n rows of S_2^{MR} ($2^{(64+24)n}$ candidates) which are marked in green, as in Fig. 3. Then we can determine the value of corresponding n (reverse) diagonals in S_2. Now there are n known diagonals on the right side of the matching point. All the candidates are stored in another lookup table T_2.

Step C. MitM matching across the MR operation

The technique of matching across an MDS transformation is already proposed and well-discussed in [25,31,27]. Here we directly give the result. For a 64-byte state, the bit size of the matching point is calculated as $64(m+n-8)$, where m and n are the number of known diagonals in both sides. Because we can match both of the value and difference on a 64-byte state, the bit size of the matching point is $128(m + n - 8)$. Therefore, the number of expected matches between T_1 and T_2 is $2^{64m+(64+24)n-128(m+n-8)} = 2^{-64m-40n+1024}$. Note that the matching candidate is a pair of (S'^{SC}_1, S_2) where all bytes are fully determined. Then, the corresponding K'^{AC}_1 is also fully determined. We use a pre-computation of complexity 2^{65} to build a table of size 2^{65}, which is used for (S'^{SC}_1, S_2) to determine the remaining two diagonals of the corresponding K_{in} by just a table lookup. More precisely, for all values of each unguessed diagonal of K_{in}, compute the corresponding diagonal values in S'^{SC}_1, and store them in a lookup table. The number of remaining candidates is also the number of suggested keys. The correctness of each suggested keys can be verified by the differential path from S_3 to S_5.

The total complexity of the attack is

$$2^{64m} + 2^{(64+24)n} + 2^{-64m-40n+1024}.$$

When $m = 6$ and $n = 5$, we get the complexity of about $2^{384} + 2^{440} + 2^{440} \approx 2^{441}$ computations. The sizes of T_1 and T_2 are 2^{384} and 2^{440} respectively. Since we only need to store one of them and leave the calculations of other direction "on the fly", the memory requirement is 2^{384}. Taking into account the phase to find the inner collision, the total time complexity for recovering K_{in} is 2^{441} time and 2^{384} memory, along with 2^{321} chosen queries. Recall that we chose the attack parameters by considering that the original key recovery attack on HMAC will require 2^{448} computations as we later show in Section 3.4. We balanced the time complexity and then reduced the memory and queries as much as possible.

3.2   Time-Memory-Data Tradeoff for K_{in} Recovery

For the differential path in Fig. 3, the number of active rows does not have to be three. Indeed, this derives a tradeoff between data (the number of queries) and time-memory. Intuitively, the more data we use, the more restricted differential path we can satisfy and thus time and memory can be smaller. On the other hand, data can be minimized by spending more time and memory. Let r be the number of active rows in Fig. 3. For a single structure, $\binom{2^{64r}}{2} = 2^{128r-1}$ pairs can be constructed with 2^{64r} queries. In the end, a collision can be found with $2^{513-64r}$ queries.

Then, the MitM phase is performed. The time complexity for the forward computation does not change, which is 2^{64m}, while the complexity for the backward computation is dependent on r, which is $2^{(64+8r)n}$. We can further introduce the tradeoff between time and memory, where their product takes a constant value. For simplicity, let us assume that $2^{64m} < 2^{(64+8r)n}$. The simple method computes the forward candidates with 2^{64m} computations and stores them. Then, the backward candidates are computed with $2^{(64+8r)n}$. Hence, the time is $2^{(64+8r)n}$ and the memory is 2^{64m}. Here, we divide the free bits for the forward computation into two parts; $64m - t$ and t. An attacker firstly guesses the value of t bits, and for each guess, computes the 2^{64m-t} forward candidates and stores them in a table with 2^{64m-t} entries. The backward computation does not change. Finally, the attack is iterated for 2^t guesses. In the end, the memory complexity becomes 2^{64m-t} while the time complexity becomes $2^{(64+8r)n+t}$ computations.

Let us demonstrate the impact of the time-memory-data tradeoff. In section 4.1, we aimed to achieve the time complexity of 2^{448}, and chose the parameter $(r, m, n, t) = (3, 6, 5, 0)$ which resulted in (data, time, memory) = $(2^{321}, 2^{441}, 2^{384})$. By choosing parameters $(r, m, n, t) = (3, 6, 5, 7)$, memory can be saved by 7 bits, i.e., (data, time, memory)= $(2^{321}, 2^{448}, 2^{377})$. Then let us consider the optimization from different aspects. First, we minimize the value max{data, time, memory}. We should choose $(r, m, n, t) = (2, 6, 5, 0)$, which results in (data, time, memory)$= (2^{384}, 2^{400}, 2^{384})$. Next, we try to minimize each of time, data, and memory complexities. If we minimize the time complexity, we should choose $(r, m, n, t) = (1, 6, 5, 0)$, which results in (data, time, memory)$= (2^{448}, 2^{384}, 2^{360})$. If we minimize the data complexity, we should choose $(r, m, n, t) = (4, 7, 5, t)$ which results in (data, time, memory)$= (2^{257}, 2^{480+t}, 2^{448-t})$. If we minimize the memory complexity, we should choose $(r, m, n, t) = (1, 6, 5, 144)$ which results in (data, time, memory)$= (2^{449}, 2^{504}, 2^{240})$.

3.3 How to Recover K_{out}

With the knowledge of K_{in}, we can calculate the value of T'' for any M at offline (refer to Fig. 2). Moreover, we can recover the value of T''' using a technique similar to [26]. Thus we are able to get the output value of $E_{K_{out}} \oplus K_{out}$: $T'' \oplus T'''$. For a pair of outputs of $E_{K_{out}} \oplus K_{out}$ that has a difference satisfying the constraint on the output difference of $E_{K_{in}}$ in Fig. 3, more precisely $\text{SC}^{-1} \circ \text{MR}^{-1}(\Delta(T'' \oplus T'''))$ has r active rows, the exactly same procedure of recovering K_{in} described in Section 3.1 can be applied to recover K_{out} in a straight-forward way. This section mainly describes the procedure of finding such a pair. Moreover, we provide a time-memory-data tradeoff for recovering K_{out}.

It is interesting to point out the difference for finding a target pair of recovering K_{in} and that of recovering K_{out}. For recovering K_{in}, we can freely choose the input M, but cannot derive the output differences of $E_{K_{in}}$ unless a collision occurs on the compression function. For recovering K_{out}, we cannot control the input T'', but can compute the output differences of $E_{K_{out}}$ easily since we know the values of both T'' and T'''.

Produce $(8-r)$-row Near Collision on $\text{SC}^{-1} \circ \text{MR}^{-1}(T'' \oplus T''')$. We need to recover the value of T''', which is as follows. Firstly, choose 2^x different random values X_i, calculate $Y_i = F(X_i, P)$ and store (X_i, Y_i) in a lookup table T_1 at offline. Secondly, choose 2^y different random values of M_i, query them to MAC, obtain $Z_i = \text{MAC}(K, M_i)$ and store (M_i, Z_i) in another lookup table T_2. Finally we match Y_i in T_1 and Z_j in T_2, and get $2^{x+y-511}$ matches on average. For each match, the internal state T''' of M_j is equal to X_i with a probability $1/2$. We stress that in fact we need to store only one of T_1 and T_2, and generate the other on the fly.

Next, we continue to search for a target pair. For each match of Y_i and Z_j, we compute the value of T'' of M_j, then compute $W = \text{SC}^{-1} \circ \text{MR}^{-1}(T'' \oplus X_i)$, and store them in a lookup Table T_3 to find $(8 - r)$-row near collisions on W. Recall the recovered value of T''' of a message is correct with a probability of $1/2$. Thus we need to generate 4 near collisions to ensure that one is a target pair. It implies $2^{2(x+y-511)-1} = 4 \times 2^{64(8-r)}$, and thus $2x + 2y + 64r = 1537$.

In total, the data complexity is 2^y queries, the time complexity is $2^x + 2^{x+y-511}$, and the memory requirement is $\min\{2^x, 2^y\}$.

Time-memory-data Tradeoff. The attack contains two tradeoffs. The first one is for finding a target pair, which is described above. The second one is for MitM phase, which is described in Section 3.2. Note that the MitM has to be applied to all the 4 near collisions, and so the time complexity of the tradeoff for MitM phase is increased by 4 times. Both tradeoffs depend on the parameter r, and thus we first determine the value of r, and analyze the two tradeoffs together. We provide the parameters that minimize the time complexity, the data complexity, or the memory complexity. Note that recovering K_{out} needs to recover K_{in} first, and so we should also take the tradeoff results on recovering K_{in} into account. Let us minimize the value max{data, time, memory}. Considering that the same MitM procedure of recovering K_{in} is used for recovering K_{out}, we just need to minimize that of recovering K_{in}, and obtain that (data, time, memory) $= (2^{384}, 2^{402}, 2^{384})$. If we minimize the time complexity, we should choose parameters $(r, m, n, t) = (1, 6, 5, 0)$ for recovering K_{in} and $(x, y, r, m, n, t) = (360, 397, 1, 6, 5, 0)$ for recovering K_{out}, which results in (data, time, memory) $= (2^{448}, 2^{386}, 2^{360})$. If we minimize the data complexity, we should choose parameters $(r, m, n, t) = (4, 7, 5, t)$ for recovering K_{in} and $(x, y, r, m, n, t) = (448, 225, 3, 6, 5, 0)$ for recovering K_{out}, which results in (data, time, memory) $= (2^{257}, 2^{480+t}, 2^{448-t})$. If we minimize the memory complexity, then the time and data are dominated by the K_{in} recovery, and thus (data, time, memory) $= (2^{449}, 2^{504}, 2^{240})$ by choosing the parameters given in Section 3.2.

3.4 Key Recovery for HMAC

As previously mentioned, we will recover the key of HMAC based on the splice-and-cut preimage attacks on the compression function with a fixed input chaining variable $F(IV, \cdot)$.

The attack model for the preimage attack on hash functions and the one for the key recovery attack on HMAC are slightly different. For a given hash value, there are two possibilities: 1) there exists one or more preimages; 2) no preimage exists. For the first case, the aim of the attacker is to find *any one* of the preimages, instead of all of them. The second case may occur when the size of input is restricted. In our sub-problem, *i.e.*, for a compression function with fixed chaining value, the sizes of the input message and the hash digest are the same. Thus for a random output, the probability that no preimage exists is not negligible: about e^{-1}.

For a key recovery attack, the solution (the secret key) always exists. However, the attacker has to go over *all* possible preimages to ensure that the correct key is covered. In the process of the MitM attack, sometimes there is some entropy loss in the initial structure, *i.e.*, the attacker only looks for the preimage in a subspace. When the size of the input space is bigger than the output space, a preimage attack is still possible with entropy loss. If such a preimage attack is used for key recovery, the real key could be missed.

In the preimage attacks of [31] and [27], no entropy is lost and all the possible values of the state can be covered. Thus the key recovery attack based on this preimage attack can always succeed. The complexity is 2^{448} time and 2^{64} memory.

Recall the discussion in Section 3.1, where K_{in} is recovered with (data, time, memory) $= (2^{321}, 2^{448}, 2^{377})$. Together with the preimage attack on the compression function, the original key for HMAC with 5-round Whirlpool is recovered with (data, time, memory) $= (2^{321}, 2^{448}, 2^{377})$.

3.5 Summary

In short, we have solved three sub-problems: (1) Recover K_{in} with 2^{384} chosen queries, 2^{400} time and 2^{384} memory. Then with the knowledge of K_{in}, we can solve another two sub-problems:(2) Recover K_{out} with 2^{303} known message-tag pairs, 2^{402} time and 2^{384} memory; and (3) Recover the key of HMAC from K_{in} with 2^{448} time and 2^{64} memory. The time-memory-data tradeoff exists in (1), and we can optimize its complexity depending on the final goal; (2) or (3).

Combining (1) and (2) with optimized trade-off, we have a key recovery attack on NMAC and a universal forgery attack on HMAC with 2^{384} chosen queries, 2^{402} time, and 2^{384} memory. Combining (1) and (3) with optimized trade-off, we have a key recovery attack on HMAC with 2^{321} chosen queries, 2^{448} time, and 2^{377} memory.

4   Analysis of HMAC and NMAC Based on 6-Round Whirlpool

This section presents how to extend an attack of recovering an intermediate chaining variable of the inner hash function to universal forgery and key recovery attacks for HMAC and NMAC with 6-round Whirlpool. Note that a generic single-key attack of recovering such an intermediate chaining variable for HMAC and NMAC has been published by Leurent et al. [17]. It takes around a complexity of 2^{384} blocks for all queries to recover an internal state value of a short message, e.g. one block long. Thus combining their results with our analysis, we get universal forgery and key recovery attacks on HMAC with 6-round Whirlpool.

In the rest of this section, we denote by M_1 the message whose intermediate chaining value is recovered by the attacker. We start with recovering K_{out}, which is depicted in Fig. 4.

Produce a 3-near-collision on $MR^{-1}(T'' \oplus T''')$. We append random messages M' to M_1, and query them to MAC. Note that we are able to compute their values of T'' at offline. We recover the values of T''' for those messages in the same way as we did for 5-round Whirlpool. After that, we compute $W = MR^{-1}(T'' \oplus T''')$, and search three messages that all collide on specific 56 bytes of W as shown in Fig. 4. We call such three messages 3-near-collision.

With 2^x online queries and 2^y offline computations, $2^{x+y-511}$ values of T''' are recovered, and the same number of W are collected. Note that around $\sqrt[3]{3!} \cdot 2^{\frac{2}{3}448} \approx 2^{300}$ values of W are necessary to find a target 3-near-collision [10]. Moreover, we need to generate 8 such 3-near-collisions to ensure one is indeed

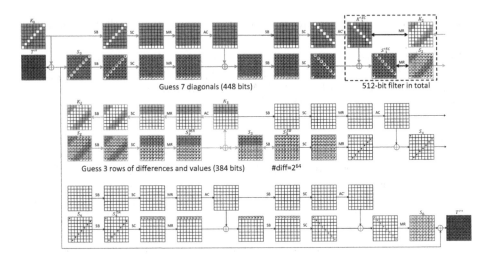

Fig. 4. How to recover K_{out} for HMAC and NMAC with 6-round Whirlpool

our target, since each value of recovered T''' is correct with a probability $1/2$. So we get $2^{x+y-511} = 2^{303}$, which implies that $x + y = 814$.

Recover K_{out}. A pair of messages from a 3-near-collision follows the differential path in Fig. 4 such that only one (reversed) diagonal of S_4 is active. Thus the number of possible differences in S_3^{SB} is 2^{64}. Denote three messages of a 3-near-collision as m_1, m_2 and m_3. Denote the values of the states S'^{SC}_1 and S_2 as L_i and R_i for m_i. We will apply the meet-in-the-middle attack two times, one for the pair (m_1, m_2) and the other for the pair (m_2, m_3).

Step A. Guessing in the forward direction

Guess the values of m diagonals of K_0 as shown in Fig. 4 (2^{64m} values, marked in blue) and determine the value of corresponding m diagonals in K'^{SC}_1 and S'^{SC}_1.

Step B. Guessing in the backward direction

Guess the values and differences of n diagonals of S_2 (2^{128n} values) which are marked in green. Then we can determine the value of corresponding n rows in S_2^{MR}. After the injection of K_3, we only know the difference in S_3. Since the number of possible differences of S_3^{SB} is only 2^{64}. According to rebound attack, we expect 2^{64} solutions for each guess of S_2. XOR 2^{64} values of S_3 and the guessed value of S_2^{MR}, and obtain 2^{64} values for the top n rows of K_3 and n diagonals of K_2. In total, the number of candidates on the right side of the MitM part is $2^{128n+64}$.

Step C. MitM matching across MR on both the key and the state

For the differential path between m_1 and m_2, the value and difference of S_2 are in fact R_1 and $R_1 \oplus R_2$. Once we have matched the value and difference of the state, i.e., $MR(L_1) = R_1$ and $MR(L_1 \oplus L_2) = R_1 \oplus R_2$, it is equivalent to match both the values $\text{MR}(L_1) = R_1$ and $\text{MR}(L_2) = R_2$. For the second differential path between m_2 and m_3, we only need to match another state $\text{MR}(L_3) = R_3$, since $\text{MR}(L_2) = R_2$ is already fulfilled. We come to an observation that the size of the matching point (filter size) is actually $(1 + 3) \times 64 \times (m + n - 8)$ bits, i.e., one from the key, three from the 3-near-collision. The expected number of matches (suggested keys) is $2^{64m+128n+64-256(m+n-8)} = 2^{2112-192m-128n}$.

The overall complexity to recover K_{out} is

$$2^{64m} + 2^{128n+64} + 2^{2112-192m-128n}.$$

When $m = 7$ and $n = 3$, the complexity is 2^{448} time and memory.

Note that the above procedure will be applied by 8 times. Thus the time complexity is 2^{451}. By setting $y = 451$, we get $x = 363$, and thus the data complexity is dominated by the recovery of an intermediate chaining variable of the inner hash function [17], namely 2^{384}.

Universal Forgery and Key Recovery. After K_{out} is recovered, almost the same procedure can be applied to recover K_{in}. So we get universal forgery on HMAC and full-key recovery on NMAC. Note that for recovering K_{in}, it is easy to verify whether a obtained 3-near-collision is our target. Thus the total complexity is dominated by recovering K_{out}, and we get (data, time, memory)=$(2^{384}, 2^{451}, 2^{448})$. Moreover, we apply the splice-and-cut preimage attack to recover K from K_{out} according to [27], which takes a time complexity of 2^{496} and a memory requirement of 2^{64}. Thus the total complexity of recovering K is (data, time, memory)=$(2^{384}, 2^{496}, 2^{448})$.

5 Distinguishing-H Attack on Full HMAC/NMAC-Whirlpool

In this section, we present a distinguishing-H attack on HMAC-Whirlpool, which is also applicable to NMAC-Whirlpool in a straightforward way. First, recall the definition of distinguishing-H [14]. A distinguisher D is to identify an oracle being either HMAC-Whirlpool or another primitive built by replacing the compression function F of HMAC-Whirlpool to a random function R with the same domain and range. For a hash function with n-bit digests, it is believed that a generic distinguishing-H attack requires 2^n complexity if the hash function is ideal.

We observe that during the computation of the outer Whirlpool in HMAC-Whirlpool, the last message block is always a constant denoted as P, more precisely $P = 1 0^{500} 1 0^{10}$ where 0^l means l consecutive 0s. This is because of the equal size of message block and hash digest and the padding rule of Whirlpool. The input messages to the outer Whirlpool consist of one block of $K \oplus \text{opad}$ and one block of the inner Whirlpool digest, and thus are always two full blocks long (namely 1024 bits), which are padded with one more block. Note that the padded block P, which is the last message block of the outer Whirlpool, is

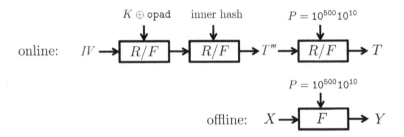

Fig. 5. Distinguishing-H attack on `HMAC-Whirlpool`

solely determined by the bit length of the input messages, and thus is always a constant. Based on the observation, we launch a distinguishing-H attack.

We first explain the overview of the attack. In the online phase, query random messages M to the oracle, and receive tag values T. In the offline phase, choose random values X (this simulates the value of T'''), and compute $Y = F(X, P)$. As depicted in Fig. 5, if the compression function of the oracle is F, two events lead to the occurrence of $Y = T$: one is $X = T'''$; and the other is $F(X, P) = F(T''', P)$ under $X \neq T'''$. If the compression function of the oracle is R, only one event leads to the occurrence of $Y = T$: $F(X, P) = R(T''', P)$. Therefore, the probability of the event $Y = T$ in the former case is (roughly) twice higher than that in the latter case. Thus, by counting the number of occurrence of $Y = T$, the compression function being either F or R can be distinguished. A detailed attack procedure is described below.

Online Phase. Send 2^{256} different messages M, which are one block long after padding, to the oracle. Receive the responses T and store them.

Offline Phase. Choose a random value as X, and compute $Y = F(X, P)$. Match Y to the set of Ts stored in the online phase. If a match is found, terminate the procedure, and output 1. Otherwise, choose another random value of X and repeat the procedure. After 2^{256} trials, if no match is found, terminate the procedure and output 0.

The complexity is 2^{256} online queries and 2^{256} offline computations. The memory cost is 2^{256} tag values. Next we evaluate the advantage of the distinguisher. Denote by D^F the case D interacts with `HMAC-Whirlpool`, and by D^R the other case. The advantage of the distinguisher $\mathrm{Adv}_{\mathsf{D}}^{\mathrm{Ind-H}}$ is defined as

$$\mathrm{Adv}_{\mathsf{D}}^{\mathrm{Ind-H}} := |\Pr[\mathsf{D}^F = 1] - \Pr[\mathsf{D}^R = 1]|.$$

In the case of D^F, the probability of $X = T'''$ is $1 - (1 - 2^{-512})^{2^{512}} \approx 1 - 1/e \approx 0.63$ since there are 2^{512} pairs of (X, T'''). The probability of $Y = T$ and $X \neq T'''$ is $(1 - 1/e) \times 1/e \approx 0.23$. Therefore, $\Pr[\mathsf{D}^F = 1]$ is 0.86 ($= 0.63 + 0.23$). In the other case, $\Pr[D^R = 1]$ is 0.63 by a similar evaluation. Overall, the advantage of the distinguisher is 0.23 ($= 0.86 - 0.63$).

Note that a trivial Data-Time tradeoff exists with the same advantage, Data$\times$ Time $= 2^{512}$.

Remarks on Applications. We emphasize that the above distinguishing-H attack has wide applications besides Whirlpool. For example, there are 11 out of 12 collision resistance PGV modes [21] including well-known Matyas-Meyer-Oseas mode and Miyaguchi-Preneel mode such that the chaining variable and the message block have equal bit size due to either the feed-forward or the feed-backward operations. If a hash function HF is built by iterating one of those PGV compression function schemes in the popular (strengthened) Merkle-Damgård domain extension scheme, the last message block of the input messages to the outer HF in HMAC or NMAC setting is always a constant, and thus the above distinguishing-H attack is applicable.

6   Conclusion and Open Discussions

In conclusion, we presented the first forgery and key recovery attacks against HMAC and NMAC based on the Whirlpool hash function reduced to 5 out of 10 rounds in single key setting, and 6 rounds in related-key setting. In addition to HMAC and NMAC, our attacks apply to other MACs based on reduced Whirlpool, such as LPMAC, secret-suffix MAC and Envelop MAC. We also gave a distinguishing-H attack against the full HMAC- and NMAC-Whirlpool.

As open discussion, it is interesting to see if the techniques presented in this paper are useful to analysis of other AES-like hash functions in hash-based MAC setting. First let us have a closer look at our analysis of the underlying AES-like block cipher in a hash function. One main and crucial strategy is restricting the differences to appear only in the encryption process and thus keep the key schedule process identical between the pair messages. For example, Whirlpool uses Miyaguchi-Preneel scheme $E_C(M) \oplus M \oplus C$ (notations follows Section 2), and the differences is introduced only by M. Recall through our analysis, C is kept the same during finding target message pairs. The main reason of this strategy is that a difference introduced from the keys propagates in both the key schedule and the encryption process, which usually makes it harder to analyze. For example, in our analysis on HMAC-Whirlpool, we need to derive the differential path in the encryption process, which becomes much harder when differences also propagate in the key schedule. Moreover, as briefly explained in Section 1, differently from that in CBC MAC setting, one cannot derive a difference on intermediate hash variable ΔC except $\Delta C = 0$. Thus the difference has to be introduced from M. After an investigation on proven secure PGV schemes [21], we find that our analysis approach is applicable to other three schemes besides Miyaguchi-Preneel scheme: $E_C(M) \oplus M$ (well known as Matyas-Meyer-Oseas scheme), $E_C(C \oplus M) \oplus M$ and $E_C(C \oplus M) \oplus C \oplus M$.

It is also interesting to see if the strategies proposed to analyze MD4-like hash functions (designed in a framework differently from AES) can be applied to AES-like hash functions from a high-spirit level, in hash-based MAC setting. There are two strategies to analyze MD4-family hash function in hash-based MAC setting to the best of our knowledge. The first one was proposed by Contini and Yin [5]. Their strategy heavily relies on one design character of MD4-like hash function: a message block is splitted into words, and these words are injected into the hash process sequentially. More precisely, an attacker can fix the beginning message words that have been ensured to satisfy the first steps of differential path, and randomize the other message words. *Unfortunately*, this strategy seems not promising to be applied to AES-like hash functions, because the latter injects the whole message block into the hash process at the same time, and moreover a byte difference propagates to the whole state very quickly due to the wide trail design of AES. The other strategy was proposed by Wang *et al.* [30]. Their strategy uses two message blocks and each block have differences. Firstly they generate a high-probability differential path on the second compression function $(\Delta C, \Delta M) \to \Delta C' = 0$, where C' is the output of the second compression function. Secondly they randomize the first message block to generate pairs of the compression function outputs that can satisfy ΔC, and each such pair can be obtained by a birthday bound complexity. Finally these pairs will be filtered out using the high-probability differential path on the second compression function, and exploited to amount further attacks. *Interestingly*, this strategy seems applicable to AES-like hash functions in MAC setting. One may build a high-probability related-key differential path on an AES-like compression function, *e.g.*, using the local collisions between the key schedule and the encryption process functions which has been found on AES [3] and on Whirlpool [27]. If it is achieved, then Wang *et al.*'s strategy seems to be applicable. Note that previous constraint $\Delta C = 0$ is now removed, and thus this strategy has a potentiality to be applied to more PGV schemes such as $E_M(C) \oplus C$ (well known as Davies-Meyer scheme).

As our result is the first step in this research topic, we expect that future works will provide deeper understanding of the security of AES-like hash functions in MAC setting.

Acknowledgements. We would like to thank Jiqiang Lu, and anonymous reviewers for their helpful comments. This research was initially started from a discussion at the second Asian Workshop on Symmetric Key Cryptography (ASK 2012). We would like to thank the organizers of ASK12. Jian Guo, Lei Wang and Shuang Wu are supported by the Singapore National Research Foundation Fellowship 2012 (NRF-NRFF2012-06).

References

1. ISO/IEC 9797-1. Information Technology-security techniques-data integrity mechanism using a cryptographic check function employing a block cipher algorithm. International Organizatoin for Standards

2. Bellare, M., Canetti, R., Krawczyk, H.: Keying Hash Functions for Message Authentication. In: Koblitz, N. (ed.) CRYPTO 1996. LNCS, vol. 1109, pp. 1–15. Springer, Heidelberg (1996)
3. Biryukov, A., Khovratovich, D., Nikolić, I.: Distinguisher and Related-Key Attack on the Full AES-256. In: Halevi, S. (ed.) CRYPTO 2009. LNCS, vol. 5677, pp. 231–249. Springer, Heidelberg (2009)
4. Bouillaguet, C., Derbez, P., Fouque, P.-A.: Automatic Search of Attacks on Round-Reduced AES and Applications. In: Rogaway, P. (ed.) CRYPTO 2011. LNCS, vol. 6841, pp. 169–187. Springer, Heidelberg (2011)
5. Contini, S., Yin, Y.L.: Forgery and Partial Key-Recovery Attacks on HMAC and NMAC Using Hash Collisions. In: Lai, X., Chen, K. (eds.) ASIACRYPT 2006. LNCS, vol. 4284, pp. 37–53. Springer, Heidelberg (2006)
6. Daemen, J., Rijmen, V.: The Design of Rijndael: AES - The Advanced Encryption Standard. Springer (2002)
7. Daemen, J., Rijmen, V.: A New MAC Construction ALRED and a Specific Instance ALPHA-MAC. In: Gilbert, H., Handschuh, H. (eds.) FSE 2005. LNCS, vol. 3557, pp. 1–17. Springer, Heidelberg (2005)
8. Daemen, J., Rijmen, V.: The Pelican MAC Function. IACR Cryptology ePrint Archive, 2005:88 (2005)
9. Dunkelman, O., Keller, N., Shamir, A.: ALRED Blues: New Attacks on AES-Based MAC's. IACR Cryptology ePrint Archive, 2011:95 (2011)
10. Feller, W.: An introduction to probability theory and its applications, 3rd edn., vol. 1. Wiley, New York (1967)
11. Fouque, P.-A., Leurent, G., Nguyen, P.Q.: Full Key-Recovery Attacks on HMAC/NMAC-MD4 and NMAC-MD5. In: Menezes, A. (ed.) CRYPTO 2007. LNCS, vol. 4622, pp. 13–30. Springer, Heidelberg (2007)
12. Huang, J., Seberry, J., Susilo, W.: On the Internal Structure of ALPHA-MAC. In: Nguyên, P.Q. (ed.) VIETCRYPT 2006. LNCS, vol. 4341, pp. 271–285. Springer, Heidelberg (2006)
13. Huang, J., Seberry, J., Susilo, W.: A five-round algebraic property of AES and its application to the ALPHA-MAC. IJACT 1(4), 264–289 (2009)
14. Kim, J., Biryukov, A., Preneel, B., Hong, S.: On the Security of HMAC and NMAC Based on HAVAL, MD4, MD5, SHA-0 and SHA-1 (Extended Abstract). In: De Prisco, R., Yung, M. (eds.) SCN 2006. LNCS, vol. 4116, pp. 242–256. Springer, Heidelberg (2006)
15. Krawczyk, H.: RFC: HMAC-based Extract-and-Expand Key Derivation Function (HKDF) (May 2010), https://tools.ietf.org/html/rfc5869.txt
16. Lamberger, M., Mendel, F., Rechberger, C., Rijmen, V., Schläffer, M.: Rebound Distinguishers: Results on the Full Whirlpool Compression Function. In: Matsui, M. (ed.) ASIACRYPT 2009. LNCS, vol. 5912, pp. 126–143. Springer, Heidelberg (2009)
17. Leurent, G., Peyrin, T., Wang, L.: New Generic Attacks Against Hash-based MACs. In: Sako, K., Sarkar, P. (eds.) ASIACRYPT 2013. LNCS, vol. 8270, pp. 1–20. Springer, Heidelberg (2013)
18. Mendel, F., Rechberger, C., Schläffer, M., Thomsen, S.S.: The Rebound Attack: Cryptanalysis of Reduced Whirlpool and Grøstl. In: Dunkelman, O. (ed.) FSE 2009. LNCS, vol. 5665, pp. 260–276. Springer, Heidelberg (2009)
19. Minematsu, K., Tsunoo, Y.: Provably Secure MACs from Differentially-Uniform Permutations and AES-Based Implementations. In: Robshaw, M. (ed.) FSE 2006. LNCS, vol. 4047, pp. 226–241. Springer, Heidelberg (2006)

20. NESSIE. New European Schemes for Signatures, Integrity, and Encryption. IST-1999-12324, http://cryptonessie.org/

21. Preneel, B., Govaerts, R., Vandewalle, J.: Hash Functions Based on Block Ciphers: A Synthetic Approach. In: Stinson, D.R. (ed.) CRYPTO 1993. LNCS, vol. 773, pp. 368–378. Springer, Heidelberg (1994)

22. Rechberger, C., Rijmen, V.: On Authentication with HMAC and Non-random Properties. In: Dietrich, S., Dhamija, R. (eds.) FC 2007 and USEC 2007. LNCS, vol. 4886, pp. 119–133. Springer, Heidelberg (2007)

23. Rechberger, C., Rijmen, V.: New Results on NMAC/HMAC when Instantiated with Popular Hash Functions. J. UCS 14(3), 347–376 (2008)

24. Rijmen, V., Barreto, P.S.L.M.: The WHIRLPOOL Hashing Function. Submitted to NISSIE (September 2000)

25. Sasaki, Y.: Meet-in-the-Middle Preimage Attacks on AES Hashing Modes and an Application to Whirlpool. In: Joux, A. (ed.) FSE 2011. LNCS, vol. 6733, pp. 378–396. Springer, Heidelberg (2011)

26. Sasaki, Y.: Cryptanalyses on a Merkle-Damgård Based MAC — Almost Universal Forgery and Distinguishing-H Attacks. In: Pointcheval, D., Johansson, T. (eds.) EUROCRYPT 2012. LNCS, vol. 7237, pp. 411–427. Springer, Heidelberg (2012)

27. Sasaki, Y., Wang, L., Wu, S., Wu, W.: Investigating Fundamental Security Requirements on Whirlpool: Improved Preimage and Collision Attacks. In: Wang, X., Sako, K. (eds.) ASIACRYPT 2012. LNCS, vol. 7658, pp. 562–579. Springer, Heidelberg (2012)

28. Song, J.H., Poovendran, R., Lee, J., Iwata, T.: The AES-CMAC Algorithm (June 2006)

29. Wang, L., Ohta, K., Kunihiro, N.: New Key-Recovery Attacks on HMAC/NMAC-MD4 and NMAC-MD5. In: Smart, N.P. (ed.) EUROCRYPT 2008. LNCS, vol. 4965, pp. 237–253. Springer, Heidelberg (2008)

30. Wang, X., Yu, H., Wang, W., Zhang, H., Zhan, T.: Cryptanalysis on HMAC/NMAC-MD5 and MD5-MAC. In: Joux, A. (ed.) EUROCRYPT 2009. LNCS, vol. 5479, pp. 121–133. Springer, Heidelberg (2009)

31. Wu, S., Feng, D., Wu, W., Guo, J., Dong, L., Zou, J. (Pseudo) Preimage Attack on Round-Reduced Grøstl Hash Function and Others. In: Canteaut, A. (ed.) FSE 2012. LNCS, vol. 7549, pp. 127–145. Springer, Heidelberg (2012)

32. Yuan, Z., Wang, W., Jia, K., Xu, G., Wang, X.: New Birthday Attacks on Some MACs Based on Block Ciphers. In: Halevi, S. (ed.) CRYPTO 2009. LNCS, vol. 5677, pp. 209–230. Springer, Heidelberg (2009)

33. Zhang, F., Shi, Z.J.: Differential and Correlation Power Analysis Attacks on HMAC-Whirlpool. In: ITNG, pp. 359–365. IEEE Computer Society (2011)

Lattice-Based Group Signatures
with Logarithmic Signature Size

Fabien Laguillaumie[1,3], Adeline Langlois[2,3],
Benoît Libert[4], and Damien Stehlé[2,3]

[1] Université Claude Bernard Lyon 1
[2] École Normale Supérieure de Lyon
[3] LIP (U. Lyon, CNRS, ENS Lyon, INRIA, UCBL),
46 Allée d'Italie, 69364 Lyon Cedex 07, France
[4] Technicolor, 975 Avenue des Champs Blancs, 35510 Cesson-Sévigné, France

Abstract. Group signatures are cryptographic primitives where users
can anonymously sign messages in the name of a population they belong
to. Gordon *et al.* (Asiacrypt 2010) suggested the first realization of group
signatures based on lattice assumptions in the random oracle model. A
significant drawback of their scheme is its linear signature size in the
cardinality N of the group. A recent extension proposed by Camenisch
et al. (SCN 2012) suffers from the same overhead. In this paper, we describe the first lattice-based group signature schemes where the signature
and public key sizes are essentially logarithmic in N (for any fixed security level). Our basic construction only satisfies a relaxed definition of
anonymity (just like the Gordon *et al.* system) but readily extends into a
fully anonymous group signature (*i.e.*, that resists adversaries equipped
with a signature opening oracle). We prove the security of our schemes
in the random oracle model under the SIS and LWE assumptions.

Keywords: Lattice-based cryptography, group signatures, anonymity.

1 Introduction

Group signatures are a core cryptographic primitive that paradoxically combines
the properties of authenticity and anonymity. They are useful in many real-
life applications including trusted computing platforms, auction protocols or
privacy-protecting mechanisms for users in public transportation.

Parties involved in such a system are a special entity, called the group manager, and group members. The manager holds a master secret key, generates a
system-wide public key, and administers the group members, by providing to
each of them an individual secret key that will allow them to anonymously sign
on behalf of the group. In case of dispute, the manager (or a separate authority) is able to determine the identity of a signer via an opening operation. This
fundamental primitive has been extensively studied, from both theoretical and
practical perspectives: It has been enriched with many useful properties, and

K. Sako and P. Sarkar (Eds.) ASIACRYPT 2013, Part II, LNCS 8270, pp. 41–61, 2013.

it has been implemented in the contexts of trusted computing (using privacy-preserving attestation [12]) and of traffic management (e.g., the Vehicle Safety Communications project of the U.S. Dept. of Transportation [29]).

Group signatures were originally proposed by Chaum and van Heyst [18] and made scalable by Ateniese *et al.* in [3]. Proper security models were introduced in [5] and [6,30] (for dynamic groups), whereas more intricate and redundant properties were considered hitherto. The model of Bellare *et al.* [5] requires two main security properties called *full anonymity* and *full traceability*. The former notion means that signatures do not leak the identities of their originators, whereas the latter implies that no collusion of malicious users can produce a valid signature that cannot be traced to one of them. Bellare *et al.* [5] proved that trapdoor permutations suffice to design group signatures, but their theoretical construction was mostly a proof of concept. Nevertheless, their methodology has been adapted in practical constructions: Essentially, a group member signs a message by verifiably encrypting a valid membership certificate delivered by the authority, while hiding its identity. While numerous schemes (e.g., [3,13,15,7]) rely on the random oracle model (ROM), others are proved secure in the standard model (e.g., [5,6,9,10,25]). Except theoretical constructions [5,6], all of these rely on the Groth-Sahai methodology to design non-interactive proof systems for specific languages involving elements in bilinear groups [27]. This powerful tool led to the design of elegant compact group signatures [10,25] whose security relies on pairing-related assumptions. The resulting signatures typically consist in a constant number of elements of a group admitting a secure and efficient bilinear map.

LATTICES AND GROUP SIGNATURES. Lattices are emerging as a promising alternative to traditional number-theoretic tools like bilinear maps. They lead to asymptotically faster solutions, thanks to the algorithmic simplicity of the involved operations and to the high cost of the best known attacks. Moreover, lattice-based schemes often enjoy strong security guarantees, thanks to worst-case/average-case connections between lattice problems, and to the conjectured resistance to quantum computers.

While numerous works have been (successfully) harnessing the power of lattices for constructing digital signatures (see [36,23,17,33,8,34] and references therein), only two works addressed the problem of efficiently realizing lattice-based group signatures. The main difficulty to overcome is arguably the scarcity of efficient and expressive non-interactive proof systems for statements involving lattices, in particular for statements on the witnesses of the hard average-case lattice problems. This state of affairs contrasts with the situation in bilinear groups, where powerful non-interactive proof systems are available [26,27].

In 2010, Gordon *et al.* [24] described the first group signature based on lattice assumptions using the Gentry *et al.* signature scheme [23] as membership certificate, an adaptation of Regev's encryption scheme [43] to encrypt it, and a zero-knowledge proof technique due to Micciancio and Vadhan [39]. While elegant in its design principle, their scheme suffers from signatures and public keys of sizes linear in the number of group members, making it utterly inefficient in comparison with constructions based on bilinear maps [7] or the strong RSA assumption [3]. Quite recently, Camenisch *et al.* [16] proposed anonymous attribute token systems, which can be seen as generalizations of group signatures. One of their schemes improves upon [24] in that the group public key has constant size[1] and the anonymity property is achieved in a stronger model where the adversary is granted access to a signature opening oracle. Unfortunately, all the constructions of [16] inherit the linear signature size of the Gordon *et al.* construction. Thus far, it remained an open problem to break the linear-size barrier. This is an important challenge considering that, as advocated by Bellare *et al.* [5], one should expect practical group signatures not to entail more than poly-logarithmic complexities in the group sizes.

OUR CONTRIBUTIONS. We describe the first lattice-based group signatures featuring sub-linear signature sizes. If t and N denote the security parameter and the maximal group size, the public keys and signatures are $\widetilde{\mathcal{O}}(t^2 \cdot \log N)$ bit long. Notice that no group signature scheme can provide signatures containing $o(\log N)$ bits (such signatures would be impossible to open), so that the main improvement potential lies in the $\widetilde{\mathcal{O}}(t^2)$ factor. These first asymptotically efficient (in t and $\log N$) lattice-based group signatures are a first step towards a practical alternative to the pairing-based counterparts. The security proofs hold in the ROM (as for [24,16]), under the Learning With Error (LWE) and Short Integer Solution (SIS) assumptions. While our basic system only provides anonymity in a relaxed model (like [24]) where the adversary has no signature opening oracle, we show how to upgrade it into a fully anonymous group signature, in the anonymity model of Bellare *et al.* [5]. This is achieved at a minimal cost in that the signature length is only increased by a constant factor. In contrast, Camenisch *et al.* [16, Se. 5.2] achieve full anonymity at the expense of inflating their basic signatures by a factor proportional to the security parameter.

CONSTRUCTION OVERVIEW. Our construction is inspired by the general paradigm from [5] consisting in *encrypting* a membership certificate under the authority's public key while providing a *non-interactive proof* that the ciphertext encrypts a valid *certificate* belonging to some group member. Nevertheless, our scheme differs from this paradigm in the sense that it is not the certificate itself which is encrypted. Instead, a temporary certificate, produced at each signature generation, is derived from the initial one and encrypted, with a proof of its validity.

We also depart from the approach of [24] at the very core of the design, *i.e.*, when it comes to provide evidence that the encrypted certificate corresponds to a legitimate group member. Specifically, Gordon *et al.* [24] hide their certificate, which is a GPV signature [23, Se. 6], within a set of $N - 1$ (encrypted) GPV pseudo-signatures that satisfy the same verification equation without being short vectors. Here, to avoid the $\mathcal{O}(N)$ factor in the signature size, we take a different approach which is reminiscent of the Boyen-Waters group signature [9]. Each group member is assigned a unique ℓ-bit identifier $\mathrm{id} = \mathrm{id}[1] \ldots \mathrm{id}[\ell] \in \{0, 1\}^\ell$,

---

[1] This can also be achieved with [24] by replacing the public key by a hash thereof, and appending the key to the signature.

where $\ell = \lceil \log_2 N \rceil$. Its certificate is an extension of a Boyen signature [8] consisting in a *full* short basis of a certain lattice (instead of a single vector), which allows the signer to generate *temporary certificates* composed of a pair $\mathbf{x}_1, \mathbf{x}_2 \in \mathbb{Z}^m$ of discrete Gaussian vectors such that

$$\mathbf{x}_1^T \cdot \mathbf{A} + \mathbf{x}_2^T \cdot (\mathbf{A}_0 + \sum_{1 \le i \le \ell} \mathrm{id}[i] \cdot \mathbf{A}_i) = \mathbf{0} \bmod q. \tag{1}$$

Here, q is a small bit length integer and $\mathbf{A}, \mathbf{A}_0, \ldots, \mathbf{A}_\ell \in \mathbb{Z}_q^{m \times n}$ are part of the group public key. Our choice of Boyen's signature [8] as membership certificate is justified by it being one of the most efficient known lattice-based signatures proven secure in the standard model, and enjoying a simple verification procedure corresponding to a relation for which we can design a proof of knowledge. A signature proven secure in the standard model allows us to obtain an easy-to-prove relation that does not involve a random oracle. As noted for example in [3,14,15], signature schemes outside the ROM make it easier to prove knowledge of a valid message-signature pair in the design of privacy-preserving protocols.

We encrypt $\mathbf{x}_2 \in \mathbb{Z}^m$ as in [24], using a variant of the dual-Regev encryption scheme [23, Se. 7]: the resulting ciphertext is $\mathbf{c}_0 = \mathbf{B}_0 \cdot \mathbf{s} + \mathbf{x}_2$, where $\mathbf{B}_0 \in \mathbb{Z}_q^{m \times n}$ is a public matrix and $\mathbf{s}$ is uniform in $\mathbb{Z}_q^n$. Then, for each $i \in [1, \ell]$, we also compute a proper dual-Regev encryption $\mathbf{c}_i$ of $\mathrm{id}[i] \cdot \mathbf{x}_2$ and generate a non-interactive OR proof that $\mathbf{c}_i$ encrypts either the same vector as $\mathbf{c}_0$ or the $\mathbf{0}$ vector.

It remains to prove that the encrypted vectors $\mathbf{x}_2$ are part of a signature satisfying (1) without giving away the $\mathrm{id}[i]$'s. To this end, we choose the signing matrices $\mathbf{A}_i$ orthogonally to the encrypting matrices $\mathbf{B}_i$, as suggested in [24]. Contrarily to the case of [24], the latter technique does not by itself suffice to guarantee the well-formedness of the $\mathbf{c}_i$'s. Indeed, we also need to prove properties about the noise vectors used in the dual-Regev ciphertexts $\{\mathbf{c}_i\}_{1 \le i \le \ell}$. This is achieved using a modification of Lyubashevsky's protocol [32,34] to prove knowledge of a solution to the Inhomogeneous Short Integer Solution problem (ISIS). This modification leads to a Σ-protocol which is zero-knowledge when the transcript is conditioned on the protocol not aborting. As the challenge space of this Σ-protocol is binary, we lowered the abort probability so that we can efficiently apply the Fiat-Shamir heuristic to a parallel repetition of the basic protocol. In the traceability proof, the existence of a witness extractor will guarantee that a successful forger will either yield a forgery for Boyen's signature or a short non-zero vector in the kernel of one of the matrices $\{\mathbf{A}_i\}_{1 \le i \le \ell}$. In either case, the forger allows the simulator to solve a SIS instance.

In the fully anonymous variant of our scheme, the difficulty is to find a way to open adversarially-chosen signatures. This is achieved by implicitly using a "chosen-ciphertext-secure" variant of the signature encryption technique of Gordon *et al.* [24]. While Camenisch *et al.* [16] proceed in a similar way using Peikert's technique [40], we use a much more economical method borrowed from the Agrawal *et al.* [1] identity-based cryptosystem. In our basic system, each $\mathbf{c}_i$ is of the form $\mathbf{B}_i \cdot \mathbf{s} + p \cdot \mathbf{e}_i + \mathrm{id}[i] \cdot \mathbf{x}_2$, where p is an upper bound on $\mathbf{x}_2$'s coordinates, and can be decrypted using a short basis $\mathbf{S}_i$ such that $\mathbf{S}_i \cdot \mathbf{B}_i = \mathbf{0} \bmod q$. Our fully anonymous system replaces each $\mathbf{B}_i$ by a matrix $\mathbf{B}_{i,\mathsf{VK}}$ that depends on the verification key VK of a one-time signature. In the proof of full anonymity, the reduction will be able to compute a trapdoor for all matrices $\mathbf{B}_{i,\text{VK}}$, except for one specific verification key $\text{VK}^\star$ that will be used in the challenge phase. This will provide the reduction with a backdoor allowing it to open all adversarially-generated signatures.

OPEN PROBLEMS. The schemes we proposed should be viewed as proofs of concept, since instantiating them with practical parameters would most likely lead to large keys and signature sizes. It is an interesting task to replace the SIS and LWE problems by their ring variants [35,41,37], to attempt to save linear factors in the security parameter t. The main hurdle in that direction seems to be the design of appropriate zero-knowledge proofs of knowledge for the LWE and ISIS relations (see Section 2.2).

As opposed to many pairing-based constructions, the security of our scheme is only proven in the random oracle model: We rely on the Fiat-Shamir heuristic to remove the interaction in the interactive proof systems. This is because very few lattice problems are known to belong to NIZK. The problems considered in the sole work on this topic [42] seem ill-fitted to devise group signatures. As a consequence, the security proofs of all known lattice-based group signatures are conducted in the random oracle model. Recently suggested multi-linear maps [22] seem like a possible direction towards solving this problem. However, currently known instantiations [22,19] rely on assumptions that seem stronger than LWE or SIS.

2   Background and Definitions

We first recall standard notations. All vectors will be denoted in bold lower-case letters, whereas bold upper-case letters will be used for matrices. If $\mathbf{b}$ and $\mathbf{c}$ are two vectors of compatible dimensions and base rings, then their inner product will be denoted by $\langle \mathbf{b}, \mathbf{c} \rangle$. Further, if $\mathbf{b} \in \mathbb{R}^n$, its euclidean norm will be denoted by $\|\mathbf{b}\|$. This notation is extended to any matrix $\mathbf{B} \in \mathbb{R}^{m \times n}$ with columns $(\mathbf{b}_i)_{i \leq n}$ by $\|\mathbf{B}\| = \max_{i \leq n} \|\mathbf{b}_i\|$. If $\mathbf{B}$ is full column-rank, we let $\widetilde{\mathbf{B}}$ denote the Gram-Schmidt orthogonalisation of $\mathbf{B}$.

If D_1 and D_2 are two distributions over the same countable support S, then their statistical distance is defined as $\Delta(D_1, D_2) = \frac{1}{2} \sum_{x \in S} |D_1(x) - D_2(x)|$. A function $f(n)$ is said negligible if $f(n) = n^{-\omega(1)}$. Finally, the acronym PPT stands for probabilistic polynomial-time.

2.1   Lattices

A (full-rank) lattice L is the set of all integer linear combinations of some linearly independent basis vectors $(\mathbf{b}_i)_{i \leq n}$ belonging to some $\mathbb{R}^n$. For a lattice L and a real $\sigma > 0$, we define the Gaussian distribution of support L and parameter σ by $D_{L,\sigma}[\mathbf{b}] \sim \exp(-\pi \|\mathbf{b}\|^2 / \sigma^2)$, for all $\mathbf{b}$ in L. We will extensively use the fact that samples from $D_{L,\sigma}$ are short with overwhelming probability.

Lemma 1 ([4, Le. 1.5]). *For any lattice $L \subseteq \mathbb{R}^n$ and $\sigma > 0$, we have* $\text{Pr}_{\mathbf{b} \hookleftarrow D_{L,\sigma}}[\|\mathbf{b}\| \leq \sqrt{n}\sigma] \geq 1 - 2^{-\Omega(n)}$.

As shown by Gentry *et al.* [23], Gaussian distributions with lattice support can be sampled from efficiently, given a sufficiently short basis of the lattice.

Lemma 2 ([11, Le. 2.3]). *There exists a* PPT *algorithm* GPVSample *that takes as inputs a basis* $\mathbf{B}$ *of a lattice* $L \subseteq \mathbb{Z}^n$ *and a rational* $\sigma \geq \|\widetilde{\mathbf{B}}\| \cdot \Omega(\sqrt{\log n})$, *and outputs vectors* $\mathbf{b} \in L$ *with distribution* $D_{L,\sigma}$.

Cash *et al.* [17] showed how to use GPVSample to randomize the basis of a given lattice. The following statement is obtained by using [11, Le. 2.3] in the proof of [17].

Lemma 3 (Adapted from [17, Le. 3.3]). *There exists a* PPT *algorithm* RandBasis *that takes as inputs a basis* $\mathbf{B}$ *of a lattice* $L \subseteq \mathbb{Z}^n$ *and a rational* $\sigma \geq \|\widetilde{\mathbf{B}}\| \cdot \Omega(\sqrt{\log n})$, *and outputs a basis* $\mathbf{C}$ *of* L *satisfying* $\|\widetilde{\mathbf{C}}\| \leq \sqrt{n}\sigma$ *with probability* $\geq 1 - 2^{-\Omega(n)}$. *Further, the distribution of* $\mathbf{C}$ *is independent of the input basis* $\mathbf{B}$.

Let $m \geq n \geq 1$ and $q \geq 2$. For a matrix $\mathbf{A} \in \mathbb{Z}_q^{m \times n}$, we define the lattice $\Lambda_q^{\perp}(\mathbf{A}) = \{\mathbf{x} \in \mathbb{Z}^m : \mathbf{x}^T \cdot \mathbf{A} = \mathbf{0} \bmod q\}$. We will use an algorithm that jointly samples a uniform $\mathbf{A}$ and a short basis of $\Lambda_q^{\perp}(\mathbf{A})$.

Lemma 4 ([2, Th. 3.2]). *There exists a* PPT *algorithm* TrapGen *that takes as inputs* 1^n, 1^m *and an integer* $q \geq 2$ *with* $m \geq \Omega(n \log q)$, *and outputs a matrix* $\mathbf{A} \in \mathbb{Z}_q^{m \times n}$ *and a basis* $\mathbf{T_A}$ *of* $\Lambda_q^{\perp}(\mathbf{A})$ *such that* $\mathbf{A}$ *is within statistical distance* $2^{-\Omega(n)}$ *to* $U(\mathbb{Z}_q^{m \times n})$, *and* $\|\widetilde{\mathbf{T_A}}\| \leq \mathcal{O}(\sqrt{n \log q})$.

Lemma 4 is often combined with the sampler from Lemma 2. Micciancio and Peikert [38] recently proposed a more efficient approach for this combined task, which should be preferred in practice but, for the sake of simplicity, we present our schemes using TrapGen. Lemma 4 was later extended by Gordon *et al.* [24] so that the columns of $\mathbf{A}$ lie within a prescribed linear vector subspace of $\mathbb{Z}_q^n$ (for q prime). For the security proof of our fully anonymous scheme, we will use an extension where the columns of the sampled $\mathbf{A}$ lie within a prescribed *affine* subspace of $\mathbb{Z}_q^n$. A proof is given in [31, Appendix C].

Lemma 5. *There exists a* PPT *algorithm* SuperSamp *that takes as inputs integers* $m \geq n \geq 1$ *and* $q \geq 2$ *prime such that* $m \geq \Omega(n \log q)$, *as well as matrices* $\mathbf{B} \in \mathbb{Z}_q^{m \times n}$ *and* $\mathbf{C} \in \mathbb{Z}_q^{n \times n}$ *such that the rows of* $\mathbf{B}$ *span* $\mathbb{Z}_q^n$. *It outputs* $\mathbf{A} \in \mathbb{Z}_q^{m \times n}$ *and a basis* $\mathbf{T_A}$ *of* $\Lambda_q^{\perp}(\mathbf{A})$ *such that* $\mathbf{A}$ *is within statistical distance* $2^{-\Omega(n)}$ *to* $U(\mathbb{Z}_q^{m \times n})$ *conditioned on* $\mathbf{B}^T \cdot \mathbf{A} = \mathbf{C}$, *and* $\|\widetilde{\mathbf{T_A}}\| \leq \mathcal{O}(\sqrt{mn \log q \log m})$.

Finally, we also make use of an algorithm that extends a trapdoor for $\mathbf{A} \in \mathbb{Z}_q^{m \times n}$ to a trapdoor of any $\mathbf{B} \in \mathbb{Z}_q^{m' \times n}$ whose top $m \times n$ submatrix is $\mathbf{A}$.

Lemma 6 ([17, Le. 3.2]). *There exists a* PPT *algorithm* ExtBasis *that takes as inputs a matrix* $\mathbf{B} \in \mathbb{Z}_q^{m' \times n}$ *whose first* m *rows span* $\mathbb{Z}_q^n$, *and a basis* $\mathbf{T_A}$ *of* $\Lambda_q^{\perp}(\mathbf{A})$ *where* $\mathbf{A}$ *is the top* $m \times n$ *submatrix of* $\mathbf{B}$, *and outputs a basis* $\mathbf{T_B}$ *of* $\Lambda_q^{\perp}(\mathbf{B})$ *with* $\|\widetilde{\mathbf{T_B}}\| \leq \|\widetilde{\mathbf{T_A}}\|$.

For the sake of simplicity, we will assume that when the parameter conditions are satisfied, the distributions of the outputs of TrapGen and SuperSamp are *exactly* those they are meant to approximate, and the probabilistic norm bounds of Lemmas 1 and 3 *always* hold.

2.2 Computational Problems

The security of our schemes provably relies (in the ROM) on the assumption that both algorithmic problems below are hard, *i.e.*, cannot be solved in polynomial time with non-negligible probability and non-negligible advantage, respectively.

Definition 1. *Let m, q, β be functions of a parameter n. The Short Integer Solution problem $\mathsf{SIS}_{m,q,\beta}$ is as follows: Given $\mathbf{A} \hookleftarrow U(\mathbb{Z}_q^{m \times n})$, find $\mathbf{x} \in \Lambda_q^\perp(\mathbf{A})$ with $0 < \|\mathbf{x}\| \leq \beta$.*

Definition 2. *Let q, α be functions of a parameter n. For $\mathbf{s} \in \mathbb{Z}_q^n$, the distribution $A_{q,\alpha,\mathbf{s}}$ over $\mathbb{Z}_q^n \times \mathbb{Z}_q$ is obtained by sampling $\mathbf{a} \hookleftarrow U(\mathbb{Z}_q^n)$ and (a noise) $e \hookleftarrow D_{\mathbb{Z},\alpha q}$, and returning $(\mathbf{a}, \langle \mathbf{a}, \mathbf{s} \rangle + e)$. The Learning With Errors problem $\mathsf{LWE}_{q,\alpha}$ is as follows: For $\mathbf{s} \hookleftarrow U(\mathbb{Z}_q^n)$, distinguish between arbitrarily many independent samples from $U(\mathbb{Z}_q^n \times \mathbb{Z}_q)$ and the same number of independent samples from $A_{q,\alpha,\mathbf{s}}$.*

If $q \geq \sqrt{n}\beta$ and $m, \beta \leq \mathsf{poly}(n)$, then standard worst-case lattice problems with approximation factors $\gamma = \widetilde{\mathcal{O}}(\beta\sqrt{n})$ reduce to $\mathsf{SIS}_{m,q,\beta}$ (see, e.g., [23, Se. 9]). Similarly, if $\alpha q = \Omega(\sqrt{n})$, then standard worst-case lattice problems with approximation factors $\gamma = \mathcal{O}(\alpha/n)$ quantumly reduce to $\mathsf{LWE}_{q,\alpha}$ (see [43], and also [40,11] for partial dequantizations). Note that we use the discrete noise variant of LWE from [24].

We will make use of a non-interactive zero-knowledge proof of knowledge (NIZPoK) protocol, which can be rather directly derived from [32,34], for the following relation corresponding to an inhomogenous variant of the SIS relation:

$$R_{\mathsf{ISIS}} = \left\{ (\mathbf{A}, \mathbf{y}, \beta; \mathbf{x}) \in \mathbb{Z}_q^{m \times n} \times \mathbb{Z}_q^n \times \mathbb{Q} \times \mathbb{Z}^m : \mathbf{x}^T \cdot \mathbf{A} = \mathbf{y}^T \wedge \|\mathbf{x}\| \leq \beta \right\}.$$

The protocol, detailed in Section 2.3, is derived from the parallel repetition of a Σ-protocol with binary challenges. We call $\mathsf{Prove}_{\mathsf{ISIS}}$ and $\mathsf{Verify}_{\mathsf{ISIS}}$ the PPT algorithms run by the Prover and the Verifier when the scheme is rendered non-interactive using the Fiat-Shamir heuristic (*i.e.*, the challenge is implemented using the random oracle $H(\cdot)$). Algorithm $\mathsf{Prove}_{\mathsf{ISIS}}$ takes $(\mathbf{A}, \mathbf{y}, \beta; \mathbf{x})$ as inputs, and generates a transcript $(\mathsf{Comm}, \mathsf{Chall}, \mathsf{Resp})$. Algorithm $\mathsf{Verify}_{\mathsf{ISIS}}$ takes $(\mathbf{A}, \mathbf{y}, \beta)$ and such a transcript as inputs, and returns 0 or 1. The scheme has completeness error $2^{-\Omega(n)}$: if $\mathsf{Prove}_{\mathsf{ISIS}}$ is given as input an element of R_{ISIS}, then given as input the output of $\mathsf{Prove}_{\mathsf{ISIS}}$, $\mathsf{Verify}_{\mathsf{ISIS}}$ replies 1 with probability $\geq 1 - 2^{-\Omega(m)}$ (over the randomness of Prove). Also, there exists a PPT algorithm $\mathsf{Simulate}_{\mathsf{ISIS}}$ that, by reprogramming the random oracle $H(\cdot)$, takes $(\mathbf{A}, \mathbf{y}, \beta)$ as input and generates a transcript $(\mathsf{Comm}, \mathsf{Chall}, \mathsf{Resp})$ whose distribution is within statistical distance $2^{-\Omega(m)}$ of the genuine transcript distribution. Finally, there also exists a PPT algorithm $\mathsf{Extract}_{\mathsf{ISIS}}$ that given access to a time T algorithm $\mathcal{A}$ that generates transcripts accepted by $\mathsf{Verify}_{\mathsf{ISIS}}$ with probability ε, produces, in time $\mathrm{Poly}(T, 1/\varepsilon)$ a vector $\mathbf{x}'$ such that $(\mathbf{A}, \mathbf{y}, \mathcal{O}(\beta \cdot m^2); \mathbf{x}') \in R_{\mathsf{ISIS}}$.

We will also need a NIZKPoK protocol for the following language:

$$R_{\mathsf{LWE}} = \left\{ (\mathbf{A}, \mathbf{b}, \alpha; \mathbf{s}) \in \mathbb{Z}_q^{m \times n} \times \mathbb{Z}_q^m \times \mathbb{Q} \times \mathbb{Z}_q^n : \|\mathbf{b} - \mathbf{A} \cdot \mathbf{s}\| \leq \alpha q \sqrt{m} \right\}.$$

As noted in [34], we may multiply $\mathbf{b}$ by a parity check matrix $\mathbf{G} \in \mathbb{Z}_q^{(m-n) \times m}$ of $\mathbf{A}$ and prove the existence of small $\mathbf{e} \in \mathbb{Z}^m$ such that $\mathbf{e}^T \cdot \mathbf{G}^T = \mathbf{b}^T \cdot \mathbf{G}^T$. This may be done with the above NIZKPoK protocol for R_{ISIS}. We call $\mathsf{Prove}_{\mathsf{LWE}}$, $\mathsf{Verify}_{\mathsf{LWE}}$, $\mathsf{Simulate}_{\mathsf{LWE}}$ and $\mathsf{Extract}_{\mathsf{LWE}}$ the obtained PPT algorithms.

2.3 Proof of Knowledge of an ISIS Solution

In [32], Lyubashevsky described an identification scheme whose security relies on the hardness of the SIS problem. Given a public vector $\mathbf{y} \in \mathbb{Z}_q^n$ and a matrix $\mathbf{A} \in \mathbb{Z}_q^{m \times n}$, the prover holds a short secret $\mathbf{x}$ and generates an interactive witness indistinguishable proof of knowledge of a short vector $\mathbf{x}'^T \in \mathbb{Z}^m$ such that $\mathbf{x}'^T \cdot \mathbf{A} = \mathbf{y}^T \bmod q$. A variant was later proposed in [34], which enjoys the property of being zero-knowledge (when the distribution of the transcript is conditioned on the prover not aborting). We present an adaptation of [34, Fig. 1] (still enjoying the same zero-knowledgedness property): the secret is a single vector, the challenges are binary (which we use for the extraction vector), and we increase the standard deviation of the commited vector to lower the rejection probability (we use a parallel repetition of the basic scheme, and want the probability that there is a reject among all the parallel iterations to be sufficiently away from 1).

Assume the prover P wishes to prove knowledge of an $\mathbf{x}$ such that $\mathbf{y}^T = \mathbf{x}^T \cdot \mathbf{A} \bmod q$ and $\|\mathbf{x}\| \leq \beta$, where $\mathbf{y}$ and $\mathbf{A}$ are public. The protocol takes place between the prover P and the verifier V and proceeds by the parallel repetition of a basic Σ-protocol with binary challenges. We set $\sigma = \Theta(\beta m^{3/2})$ and M_L as specified by [34, Th. 4.6]. Thanks to our larger value of σ, we obtain (by adapting [34, Le. 4.5]) that M_L is now $1 - \Omega(1/m)$.

1. The prover P generates a commitment $\mathsf{Comm} = (\mathbf{w}_i)_{i \leq t}$ where, for each $i \leq t$, $\mathbf{w}_i \in \mathbb{Z}_q^n$ is obtained by sampling $\mathbf{y}_i \hookleftarrow D_{\mathbb{Z}^m, \sigma}$ and computing $\mathbf{w}_i^T = \mathbf{y}_i^T \cdot \mathbf{A} \bmod q$. The message Comm is sent to V.
2. The verifier V sends a challenge $\mathsf{Chall} \hookleftarrow \{0, 1\}^t$ to P.
3. For $i \leq t$, the prover P does the following.

   a. Compute $\mathbf{z}_i = \mathbf{y}_i + \mathsf{Chall}[i] \cdot \mathbf{x}$, where $\mathsf{Chall}[i]$ denotes the ith bit of Chall.
   b. Set $\mathbf{z}_i$ to $\perp$ with probability $\min(1, \frac{\exp(-\pi \|\mathbf{z}\|^2/\sigma^2)}{M_L \cdot \exp(-\pi \|\mathsf{Chall}[i] \cdot \mathbf{x} - \mathbf{z}\|^2/\sigma^2)})$.

   Then P sends the response $\mathsf{Resp} = (\mathbf{z}_i)_{i \leq t}$ to V.

4. The verifier V checks the transcript $(\mathsf{Comm}, \mathsf{Chall}, \mathsf{Resp})$ as follows:

a. For $i \leq t$, set $d_i = 1$ if $\|\mathbf{z}_i\| \leq 2\sigma\sqrt{m}$ and $\mathbf{z}_i^T \cdot \mathbf{A} = \mathbf{w}_i^T + \mathsf{Chall}[i] \cdot \mathbf{y}^T$. Otherwise, set $d_i = 0$.
b. Return 1 (and accept the transcript) if and only if $\sum_{i \leq t} d_i \geq 0.65t$.

The protocol has completeness error $2^{-\Omega(t)}$. Further, by [34, Th. 4.6], the distribution of the transcript conditioned on $\mathbf{z}_i \neq \bot$ can be simulated efficiently. Note that if we implement the challenge phase with a random oracle, we can compute the $\mathbf{z}_i$'s for increasing value of i, and repeat the whole procedure if $\mathbf{z}_i = \bot$ for some i. Thanks to our choice of σ, for any $t \leq \mathcal{O}(m)$, the probability that $\mathbf{z}_i = \bot$ for some i is $\leq c$, for some constant $c < 1$. Thanks to this random oracle enabled rejection, the simulator produces a distribution that is within statistical distance $2^{-\Omega(m)}$ to the transcript distribution.

Finally, the modified protocol provides special soundness in that there is a simple extractor that takes as input two valid transcripts $(\mathsf{Comm}, \mathsf{Chall}, \mathsf{Resp})$, $(\mathsf{Comm}, \mathsf{Chall}', \mathsf{Resp}')$ with distinct challenges $\mathsf{Chall} \neq \mathsf{Chall}'$ and obtains a witness $\mathbf{x}'$ such that $\mathbf{x}'^T \cdot \mathbf{A} = \mathbf{y}^T \bmod q$ and $\|\mathbf{x}'\| \leq \mathcal{O}(\sigma\sqrt{m}) \leq \mathcal{O}(\beta m^2)$.

2.4 Group Signatures

This section recalls the model of Bellare, Micciancio and Warinschi [5], which assumes static groups. A group signature scheme $\mathcal{GS}$ consists of a tuple of four PPT algorithms $(\mathsf{Keygen}, \mathsf{Sign}, \mathsf{Verify}, \mathsf{Open})$ with the following specifications:

- Keygen takes 1^n and 1^N as inputs, where $n \in \mathbb{N}$ is the security parameter, and $N \in \mathbb{N}$ is the maximum number of group members. It returns a tuple $(\mathsf{gpk}, \mathsf{gmsk}, \mathbf{gsk})$ where gpk is the *group public key*, gmsk is the group manager secret key, and $\mathbf{gsk}$ is an N-vector of secret keys: $\mathsf{gsk}[j]$ is the signing key of the j-th user, for $j \in \{0, \dots, N-1\}$.
- Sign takes the group public key gpk, a signing key $\mathsf{gsk}[j]$ and a message $M \in \{0,1\}^*$ as inputs. Its output is a signature $\Sigma \in \{0,1\}^*$ on M.
- Verify is deterministic and takes the group public key gpk, a message M and a putative signature Σ of M as inputs. It outputs either 0 or 1.
- Open is deterministic and takes as inputs the group public key gpk, the group manager secret key gmsk, a message M and a valid group signature Σ w.r.t. gpk. It returns an index $j \in \{0, \dots, N-1\}$ or a special symbol $\bot$ in case of opening failure.

The group signature scheme must be *correct*, *i.e.*, for all integers n and N, all $(\mathsf{gpk}, \mathsf{gmsk}, \mathbf{gsk})$ obtained from Keygen with $(1^n, 1^N)$ as input, all indexes $j \in \{0, \dots, N-1\}$ and $M \in \{0,1\}^*$: $\mathsf{Verify}(\mathsf{gpk}, M, \mathsf{Sign}(\mathsf{gpk}, \mathsf{gsk}[j], M)) = 1$ and $\mathsf{Open}(\mathsf{gpk}, \mathsf{gmsk}, M, \mathsf{Sign}(\mathsf{gpk}, \mathsf{gsk}[j], M)) = j$, with probability negligibly close to 1 over the internal randomness of Keygen and Sign.

Bellare *et al.* [5] gave a unified security model for group signatures in static groups. The two main security requirements are *traceability* and *anonymity*. The former asks that no coalition of group members be able to create a signature that cannot be traced to one of them. The latter implies that, even if all group members' private keys are given to the adversary, signatures generated by two distinct members should be computationally indistinguishable.

$\mathbf{Exp}_{\mathcal{GS},\mathcal{A}}^{\text{trace}}(n, N)$

$(\text{gpk}, \text{gmsk}, \text{gsk}) \leftarrow \text{Keygen}(1^n, 1^N)$
$\text{st} \leftarrow (\text{gmsk}, \text{gpk})$
$\mathcal{C} \leftarrow \emptyset \; ; \; K \leftarrow \varepsilon \; ; \; Cont \leftarrow \textbf{true}$
while $(Cont = \textbf{true})$ do
$\quad (Cont, \text{st}, j) \leftarrow \mathcal{A}^{\mathcal{GS}.\text{Sign}(\text{gsk}[\cdot], \cdot)}(choose, \text{st}, K)$
$\quad$ if $Cont = \textbf{true}$ then $\mathcal{C} \leftarrow \mathcal{C} \cup \{j\}$;
$\quad K \leftarrow \text{gsk}[j]$
$\quad$ end if
end while;
$(M^\star, \Sigma^\star) \leftarrow \mathcal{A}^{\mathcal{GS}.\text{Sign}(\text{gsk}[\cdot], \cdot)}(guess, \text{st})$
if $\text{Verify}(\text{gpk}, M^\star, \Sigma^\star) = 0$ then Return 0
if $\text{Open}(\text{gmsk}, M^\star, \Sigma^\star) = \perp$ then Return 1
if $\exists j^\star \in \{0, \ldots, N-1\}$ such that
$\quad (\text{Open}(\text{gmsk}, M^\star, \Sigma^\star) = j^\star) \wedge (j^\star \notin \mathcal{C})$
$\quad \wedge ((j^\star, M^\star) \text{ not queried by } \mathcal{A})$
then Return 1 else Return 0

$\mathbf{Exp}_{\mathcal{GS},\mathcal{A}}^{\text{anon-b}}(n, N)$

$(\text{gpk}, \text{gmsk}, \text{gsk}) \leftarrow \text{Keygen}(1^n, 1^N)$
$(\text{st}, j_0, j_1, M) \leftarrow \mathcal{A}(choose, \text{gpk}, \text{gsk})$
$\Sigma^\star \leftarrow \text{Sign}(\text{gpk}, \text{gsk}[j_b], M)$
$b' \leftarrow \mathcal{A}(guess, \text{st}, \Sigma^\star)$
Return b'

Fig. 1. Random experiments for anonymity and full traceability

Anonymity. Anonymity requires that, without the group manager's secret key, an adversary cannot recognize the identity of a user given its signature. More formally, the attacker, modeled as a two-stage adversary (*choose* and *guess*), is engaged in the first random experiment depicted in Figure 1. The *advantage* of such an adversary $\mathcal{A}$ against a group signature $\mathcal{GS}$ with N members is defined as $\mathbf{Adv}_{\mathcal{GS},\mathcal{A}}^{\text{anon}}(n, N) = \left| \Pr[\mathbf{Exp}_{\mathcal{GS},\mathcal{A}}^{\text{anon-1}}(n, N) = 1] - \Pr[\mathbf{Exp}_{\mathcal{GS},\mathcal{A}}^{\text{anon-0}}(n, N) = 1] \right|$.

In our first scheme, we consider a *weak anonymity* scenario in which the adversary is not allowed to query an opening oracle. This relaxed model is precisely the one considered in [24], and was firstly introduced in [7]. Nonetheless, we provide in Section 5 a variant of our scheme enjoying chosen-ciphertext security. The adversary is then granted an access to an opening oracle that can be called on any string except the challenge signature $\Sigma^\star$.

Definition 3 (Weak and full anonymity, [5,7]). *A group signature scheme $\mathcal{GS}$ is said to be* weakly anonymous *(resp.* fully anonymous*) if for all polynomial $N(\cdot)$ and all* PPT *adversaries $\mathcal{A}$ (resp.* PPT *adversaries $\mathcal{A}$ with access to an opening oracle which cannot be queried for the challenge signature), $\mathbf{Adv}_{\mathcal{GS},\mathcal{A}}^{\text{anon}}(n, N)$ is a negligible function in the security parameter n.*

Full traceability. Full traceability ensures that all signatures, even those created by a coalition of users *and* the group manager, pooling their secret keys together, can be traced to a member of the forging coalition. Once again, the attacker is modeled as a two-stage adversary who is run within the second experiment described in Figure 1. Its success probability against $\mathcal{GS}$ is defined as $\mathbf{Succ}_{\mathcal{GS},\mathcal{A}}^{\text{trace}}(n, N) = \Pr[\mathbf{Exp}_{\mathcal{GS},\mathcal{A}}^{\text{trace}}(n, N) = 1]$.

Definition 4 (Full traceability, [5]). *A group signature scheme $\mathcal{GS}$ is said to be* fully traceable *if for all polynomial $N(\cdot)$ and all* PPT *adversaries $\mathcal{A}$, its success probability $\boldsymbol{Succ}^{trace}_{\mathcal{GS},\mathcal{A}}(n,N)$ is negligible in the security parameter n.*

3 An Asymptotically Shorter Lattice-Based Group Signature

At a high level, our key generation is based on the variant of Boyen's lattice signatures [8] described in [38, Se. 6.2]: Boyen's secret and verification keys respectively become our secret and public keys, whereas Boyen's message space is mapped to the users' identity space. There are however several additional twists in Keygen. First, each group member is given a *full* short basis of the public lattice associated to its identity, instead of a single short lattice vector. The reason is that, for anonymity and unlinkability purposes, the user has to generate each group signature using a *fresh* short lattice vector. Second, we sample our public key matrices $(\mathbf{A}_i)_{i\leq\ell}$ orthogonally to publicly known matrices $\mathbf{B}_i$, similarly to the group signature scheme from [24]. These $\mathbf{B}_i$'s will be used to publicly verify the validity of the signatures. They are sampled along with short trapdoor bases, using algorithm SuperSamp, which become part of the group signature secret key. These trapdoor bases will be used by the group authority to open signatures.

To anonymously sign M, the user samples a Boyen signature $(\mathbf{x}_1,\mathbf{x}_2)$ with its identity as message, which is a temporary certificate of its group membership. It does so using its full trapdoor matrix for the corresponding lattice. The user then encrypts $\mathbf{x}_2$, in a fashion that resembles [24], using Regev's dual encryption scheme from [23, Se. 7.1] with the $\mathbf{B}_i$'s as encryption public keys. Note that in all cases but one ($\mathbf{c}_0$ at Step 2), the signature is not embedded in the encryption noise as in [24], but as proper plaintext. The rest of the signing procedure consists in proving in zero-knowledge that these are valid ciphertexts and that the underlying plaintexts indeed encode a Boyen signature under the group public key. These ZKPoKs are all based on the interactive proof systems recalled in Sections 2.2 and 2.3. These were made non-interactive via the Fiat-Shamir heuristic with random oracle $H(\cdot)$ taking values in $\{0,1\}^t$. The message M is embedded in the application of the Fiat-Shamir transform at Step 6 of the signing algorithm.

The verification algorithm merely consists in verifying all proofs of knowledge concerning the Boyen signature embedded in the plaintexts of the ciphertexts.

Finally, the group manager can open any signature by decrypting the ciphertexts (using the group manager secret key) and then recovering the underlying Boyen signature within the plaintexts: this reveals which public key matrices $\mathbf{A}_i$ have been considered by the signer, and therefore its identity.

The scheme depends on several functions m, q, p, α and σ of the security parameter n and the group size $N(=2^\ell)$. They are set so that all algorithms can be implemented in polynomial time and are correct (Theorem 1), and so that the security properties (Theorems 2 and 3) hold, in the ROM, under the SIS and LWE hardness assumptions for parameters for which these problems enjoy reductions from standard worst-case lattice problems with polynomial approximation factors. More precisely, we require that:

- parameter m is $\Omega(n \log q)$,
- parameter σ is $\Omega(m^{3/2}\sqrt{\ell n \log q} \log m)$ and $\leq n^{\mathcal{O}(1)}$,
- parameter p is $\Omega((\alpha q + \sigma)m^{5/2})$,
- parameter α is set so that $\alpha^{-1} \geq \Omega(pm^3 \log m)$ and $\leq n^{\mathcal{O}(1)}$,
- parameter q is prime and $\Omega(\ell + \alpha^{-1}\sqrt{n\ell})$ and $\leq n^{\mathcal{O}(1)}$.

For example, we may set $m = \tilde{\mathcal{O}}(n)$, $\sigma = \tilde{\mathcal{O}}(n^2\sqrt{\ell})$, $p = \tilde{\mathcal{O}}(n^{9/2}\sqrt{\ell})$ as well as $\alpha^{-1} = \tilde{\mathcal{O}}(n^{15/2}\sqrt{\ell})$ and $q = \tilde{\mathcal{O}}(\ell + n^8\sqrt{\ell})$.

Keygen$(1^n, 1^N)$: Given a security parameter $n > 0$ and the desired number of group members $N = 2^\ell \in \mathsf{poly}(n)$, choose parameters q, m, p, α and σ as specified above and make them public. Choose a hash function $H : \{0,1\}^* \to \{0,1\}^t$ for some $t \in [\Omega(n), n^{\mathcal{O}(1)}]$, which will be modeled as a random oracle. Then, proceed as follows.

1. Run TrapGen$(1^n, 1^m, q)$ to get $\mathbf{A} \in \mathbb{Z}_q^{m\times n}$ and a short basis $\mathbf{T_A}$ of $\Lambda_q^\perp(\mathbf{A})$.
2. For $i = 0$ to ℓ, sample $\mathbf{A}_i \hookleftarrow U(\mathbb{Z}_q^{m\times n})$ and compute $(\mathbf{B}_i, \mathbf{S}_i') \leftarrow$ SuperSamp$(1^n, 1^m, q, \mathbf{A}_i, \mathbf{0})$. Then, randomize $\mathbf{S}_i'$ as $\mathbf{S}_i \leftarrow$ RandBasis$(\mathbf{S}_i', \Omega(\sqrt{mn \log q} \log m))$.[2]
3. For $j = 0$ to $N - 1$, let $\mathsf{id}_j = \mathsf{id}_j[1]\ldots\mathsf{id}_j[\ell] \in \{0,1\}^\ell$ be the binary representation of id_j and define: $\mathbf{A}_{\mathsf{id}_j} = \begin{bmatrix} \mathbf{A} \\ \mathbf{A}_0 + \sum_{i=1}^\ell \mathsf{id}_j[i]\mathbf{A}_i \end{bmatrix} \in \mathbb{Z}_q^{2m\times n}$.
   Then, run $\mathbf{T}_{\mathsf{id}_j}' \leftarrow$ ExtBasis$(\mathbf{A}_{\mathsf{id}_j}, \mathbf{T_A})$ to get a short delegated basis $\mathbf{T}_{\mathsf{id}_j}'$ of $\Lambda_q^\perp(\mathbf{A}_{\mathsf{id}_j})$. Finally, run $\mathbf{T}_{\mathsf{id}_j} \leftarrow$ RandBasis$(\mathbf{T}_{\mathsf{id}_j}', \Omega(m\sqrt{\ell n \log q} \log m))$.[2]
   The j-th member's private key is $\mathsf{gsk}[j] := \mathbf{T}_{\mathsf{id}_j}$.
4. The group manager's private key is $\mathsf{gmsk} := \{\mathbf{S}_i\}_{i=0}^\ell$ and the group public key is defined to be $\mathsf{gpk} := (\mathbf{A}, \{\mathbf{A}_i, \mathbf{B}_i\}_{i=0}^\ell)$. The algorithm outputs $(\mathsf{gpk}, \mathsf{gmsk}, \{\mathsf{gsk}[j]\}_{j=0}^{N-1})$.

Sign$(\mathsf{gpk}, \mathsf{gsk}[j], M)$: To sign a message $M \in \{0,1\}^*$ using the private key $\mathsf{gsk}[j] = \mathbf{T}_{\mathsf{id}_j}$, proceed as follows.

1. Run GPVSample$(\mathbf{T}_{\mathsf{id}_j}, \sigma)$ to get $(\mathbf{x}_1^T|\mathbf{x}_2^T)^T \in \Lambda_q^\perp(\mathbf{A}_{\mathsf{id}_j})$ of norm $\leq \sigma\sqrt{2m}$.
2. Sample $\mathbf{s}_0 \hookleftarrow U(\mathbb{Z}_q^n)$ and encrypt $\mathbf{x}_2 \in \mathbb{Z}_q^m$ as $\mathbf{c}_0 = \mathbf{B}_0 \cdot \mathbf{s}_0 + \mathbf{x}_2 \in \mathbb{Z}_q^m$.
3. Sample $\mathbf{s} \hookleftarrow U(\mathbb{Z}_q^n)$. For $i = 1$ to ℓ, sample $\mathbf{e}_i \hookleftarrow D_{\mathbb{Z}^m, \alpha q}$ and compute $\mathbf{c}_i = \mathbf{B}_i \cdot \mathbf{s} + p \cdot \mathbf{e}_i + \mathsf{id}_j[i] \cdot \mathbf{x}_2$, which encrypts $\mathbf{x}_2 \in \mathbb{Z}_q^m$ (resp. $\mathbf{0}$) if $\mathsf{id}_j[i] = 1$ (resp. $\mathsf{id}_j[i] = 0$).
4. Generate a NIZKPoK π_0 of $\mathbf{s}_0$ so that $(\mathbf{B}_0, \mathbf{c}_0, \sqrt{2}\sigma/q; \mathbf{s}_0) \in R_{\mathsf{LWE}}$ (see Section 2.2).
5. For $i = 1$ to ℓ, generate a NIZKPoK $\pi_{\mathsf{OR},i}$ of $\mathbf{s}$ and $\mathbf{s}_0$ so that either:
   (i) $((\mathbf{B}_i|\mathbf{B}_0), p^{-1}(\mathbf{c}_i - \mathbf{c}_0), \sqrt{2}\alpha; (\mathbf{s}^T| - \mathbf{s}_0^T)^T) \in R_{\mathsf{LWE}}$ (the vectors $\mathbf{c}_i$ and $\mathbf{c}_0$ encrypt the same $\mathbf{x}_2$, so that $p^{-1}(\mathbf{c}_i - \mathbf{c}_0)$ is close to the $\mathbb{Z}_q$-span of $(\mathbf{B}_i|\mathbf{B}_0)$);

---

[2] These randomisation steps are not needed for the correctness of the scheme but are important in the traceability proof.

(ii) or $(\mathbf{B}_i, p^{-1}\mathbf{c}_i, \alpha; \mathbf{s}) \in R_{\mathsf{LWE}}$ (the vector $\mathbf{c}_i$ encrypts $\mathbf{0}$, so that $p^{-1}\mathbf{c}_i$ is close to the $\mathbb{Z}_q$-span of $\mathbf{B}_i$).

This can be achieved by OR-ing two proofs for R_{LWE}, and making the resulting protocol non-interactive with the Fiat-Shamir heuristic.[3]

6. For $i = 1$ to ℓ, set $\mathbf{y}_i = \mathrm{id}_j[i]\mathbf{x}_2 \in \mathbb{Z}^m$ and generate a NIZKPoK π_K of $(\mathbf{e}_i)_{1 \leq i \leq \ell}, (\mathbf{y}_i)_{1 \leq i \leq \ell}, \mathbf{x}_1$ such that, for $i \in [1, \ell]$

$$\mathbf{x}_1^T \mathbf{A} + \sum_{i=0}^{\ell} \mathbf{c}_i^T \mathbf{A}_i = \sum_{i=1}^{\ell} \mathbf{e}_i^T (p\mathbf{A}_i) \quad \text{and} \quad \mathbf{e}_i^T (p\mathbf{A}_i) + \mathbf{y}_i^T \mathbf{A}_i = \mathbf{c}_i^T \mathbf{A}_i \quad (2)$$

with $\|\mathbf{e}_i\|, \|\mathbf{y}_i\|, \|\mathbf{x}_1\| \leq \max(\sigma, \alpha q)\sqrt{m}$ for all i. This is achieved using $\mathsf{Prove}_{\mathsf{ISIS}}$ in order to produce a triple $(\mathsf{Comm}_K, \mathsf{Chall}_K, \mathsf{Resp}_K)$, where $\mathsf{Chall}_K = H(M, \mathsf{Comm}_K, (\mathbf{c}_i)_{0 \leq i \leq \ell}, \pi_0, (\pi_{\mathrm{OR},i})_{1 \leq i \leq \ell})$.

The signature consists of

$$\Sigma = \big((\mathbf{c}_i)_{0 \leq i \leq \ell}, \pi_0, (\pi_{\mathrm{OR},i})_{1 \leq i \leq \ell}, \pi_K\big). \quad (3)$$

Verify$(\mathsf{gpk}, M, \Sigma)$: Parse Σ as in (3). Then, return 1 if $\pi_0, (\pi_{\mathrm{OR},i})_{1 \leq i \leq \ell}, \pi_K$ properly verify. Else, return 0.

Open$(\mathsf{gpk}, \mathsf{gmsk}, M, \Sigma)$: Parse gmsk as $\{\mathbf{S}_i\}_{i=0}^{\ell}$ and Σ as in (3). Compute $\mathbf{x}_2$ by decrypting $\mathbf{c}_0$ using $\mathbf{S}_0$. For $i = 1$ to ℓ, use $\mathbf{S}_i$ to determine which one of the vectors $p^{-1}\mathbf{c}_i$ and $p^{-1}(\mathbf{c}_i - \mathbf{x}_2)$ is close to the $\mathbb{Z}_q$-span of $\mathbf{B}_i$. Set $\mathrm{id}[i] = 0$ in the former case and $\mathrm{id}[i] = 1$ in the latter. Eventually, output $\mathrm{id} = \mathrm{id}[1]\ldots\mathrm{id}[\ell]$.

All steps of the scheme above can be implemented in polynomial-time as a function of the security parameter n, assuming that $q \geq 2$ is prime, $m \geq \Omega(n \log q)$, $\sigma \geq \Omega(m^{3/2}\sqrt{\ell n \log q} \log m)$ (using Lemmas 2 and 3), and $\alpha q \geq \Omega(1)$ (using Lemma 2). Under some mild conditions on the parameters, the scheme above is correct, *i.e.*, the verifier accepts honestly generated signatures, and the group manager successfully opens honestly generated signatures. In particular, multiplying the ciphertexts by the $\mathbf{S}_i$ modulo q should reveal $p \cdot \mathbf{e}_i + \mathrm{id}_j[i] \cdot \mathbf{x}_2$ over the integers, and $\|\mathrm{id}_j[i] \cdot \mathbf{x}_2\|_\infty$ should be smaller than p.

Theorem 1. *Let us assume that $q \geq 2$ is prime and that we have $m \geq \Omega(n \log q)$, $\sigma \geq \Omega(m^{3/2}\sqrt{\ell n \log q} \log m)$, $\alpha^{-1} \geq \Omega(pm^{5/2} \log m\sqrt{n \log q})$ as well as $q \geq \Omega(\alpha^{-1} + \sigma m^{5/2} \log m\sqrt{n \log q})$. Then, the group signature scheme above can be implemented in time polynomial in n, is correct, and the bit-size of the generated signatures in $\mathcal{O}(\ell t m \log q)$.*

4   Security

We now focus on the security of the scheme of Section 3.

---

[3] The disjunction of two relations that can be proved by Σ-protocols can also be proved by a Σ-protocol [20,21].

Anonymity. Like in [24,7], we use a relaxation of the anonymity definition, called weak anonymity and recalled in Definition 3. Analogously to the notion of IND-CPA security for public-key encryption, the adversary does not have access to a signature opening oracle. We show that the two versions (for $b = 0, 1$) of the anonymity security experiment recalled in Figure 1 are indistinguishable under the LWE assumption. We use several intermediate hybrid experiments called $G_b^{(i)}$, and show that each of these experiments is indistinguishable from the next one. At each step, we only change one element of the game (highlighted by an arrow in Figure 2), to finally reach the experiment $G^{(4)}$ where the signature scheme does not depend on the identity of the user anymore.

Theorem 2. *In the random oracle model, the scheme provides weak anonymity in the sense of Definition 3 under the $\mathsf{LWE}_{q,\alpha}$ assumption. Namely, for any PPT adversary $\mathcal{A}$ with advantage ε, there exists an algorithm $\mathcal{B}$ solving the $\mathsf{LWE}_{q,\alpha}$ problem with the same advantage.*

Proof. We define by G_0 the experiment of Definition 3 with $b = 0$ and by G_1 the same experiment with $b = 1$. To show the anonymity of the scheme, we prove that G_0 and G_1 are indistinguishable. We use several hybrid experiments named $G_b^{(1)}$, $G_b^{(2)}$, $G_b^{(3)}$ and $G^{(4)}$ (described in Figure 2), where b is either 0 or 1.

Lemma 7. *For each $b \in \{0,1\}$, G_b and $G_b^{(1)}$ are statistically indistinguishable.*

We only change the way we generate $(\mathbf{x}_1^T | \mathbf{x}_2^T)^T$, by using the fact that one way to generate it is to first sample $\mathbf{x}_2$ from $D_{\mathbb{Z}^m,\sigma}$ and then generate $\mathbf{x}_1$ from $D_{\mathbb{Z}^m,\sigma}$ such that $(\mathbf{x}_1^T | \mathbf{x}_2^T) \cdot \mathbf{A}_{id_{j_b}} = 0 \bmod q$ (by using the trapdoor $\mathbf{T_A}$). This change is purely conceptual and the vector $(\mathbf{x}_1^T | \mathbf{x}_2^T)^T$ has the same distribution anyway. The two experiments are thus identical from $\mathcal{A}$'s view and $\mathbf{x}_2$ is chosen independently of the signer's identity in the challenge phase.

Lemma 8. *For each $b \in \{0,1\}$, $G_b^{(1)}$ and $G_b^{(2)}$ are statistically indistinguishable.*

The differences are simply: Instead of generating the proofs $\{\pi_{\mathsf{OR},i}\}_{i=1}^\ell$ and π_K using the witnesses, we simulate them (see Section 2.2).

Lemma 9. *For each $b \in \{0,1\}$, if the $\mathsf{LWE}_{q,\alpha}$ problem is hard, then the experiments $G_b^{(2)}$ and $G_b^{(3)}$ are computationally indistinguishable.*

Proof. This proof uses the same principle as the proof of [24, Claim 1]: We use the adversary $\mathcal{A}$ to construct a PPT algorithm $\mathcal{B}$ for the $\mathsf{LWE}_{q,\alpha}$ problem. We consider an LWE instance $(\mathbf{B}', \mathbf{z}) \in \mathbb{Z}_q^{m\ell \times (n+1)}$ such that $\mathbf{B}' = (\mathbf{B}_1', \ldots, \mathbf{B}_\ell')$ and $\mathbf{z} = (\mathbf{z}_1, \ldots, \mathbf{z}_\ell)$ with $\mathbf{B}_i' \in \mathbb{Z}_q^{m \times n}$ and $\mathbf{z}_i \in \mathbb{Z}_q^m$. The component $\mathbf{z}$ is either uniform in $\mathbb{Z}_q^{m\ell}$, or of the form $\mathbf{z} = \mathbf{B}' \cdot \mathbf{s} + \mathbf{e}$ where $\mathbf{e}$ is sampled from $D_{\mathbb{Z}^{m\ell},\alpha q}$.

We construct a modified Keygen algorithm using this LWE instance: It generates the matrix $\mathbf{A}$ with a basis $\mathbf{T_A}$ of $\Lambda_q^\perp(\mathbf{A})$. Instead of generating the $\mathbf{B}_i$'s genuinely, we pick $\mathbf{B}_0$ uniformly in $\mathbb{Z}^{m \times n}$ and set $\mathbf{B}_i = \mathbf{B}_i'$ for $1 \le i \le \ell$. For $0 \le i \le \ell$, we compute $(\mathbf{A}_i, \mathbf{T}_i) \leftarrow \mathsf{SuperSamp}(1^n, 1^m, q, \mathbf{B}_i, \mathbf{0})$. Then, for each

Experiment G_b

- Run Keygen; give gpk $= (\mathbf{A}, \{\mathbf{A}_i, \mathbf{B}_i\}_i)$ and gsk $= \{\mathbf{T}_{id_j}\}_j$ to $\mathcal{A}$.
- $\mathcal{A}$ outputs j_0, j_1 and a message M.
- The signature of user j_b is computed as follows:
  1. $(\mathbf{x}_1^T | \mathbf{x}_2^T)^T \hookleftarrow \mathsf{GPVSample}(\mathbf{T}_{id_{j_b}}, \sigma)$;
     we have $(\mathbf{x}_1^T | \mathbf{x}_2^T) \cdot \mathbf{A}_{id_{j_b}} = \mathbf{0} \mod q$.
  2. Choose $\mathbf{s}_0 \hookleftarrow U(\mathbb{Z}_q^n)$, compute $\mathbf{c}_0 = \mathbf{B}_0 \cdot \mathbf{s}_0 + \mathbf{x}_2 \in \mathbb{Z}_q^m$.
  3. Choose $\mathbf{s} \hookleftarrow U(\mathbb{Z}_q^n)$, and for $i = 1$ to ℓ, choose $\mathbf{e}_i \hookleftarrow D_{\mathbb{Z}^m, \alpha q}$ and compute $\mathbf{c}_i = \mathbf{B}_i \cdot \mathbf{s} + p \cdot \mathbf{e}_i + id_{j_b}[i] \cdot \mathbf{x}_2$.
  4. Generate π_0.
  5. Generate $\{\pi_{OR,i}\}_i$.
  6. Generate π_K.

Experiment $G_b^{(1)}$

- Run Keygen; give gpk $= (\mathbf{A}, \{\mathbf{A}_i, \mathbf{B}_i\}_i)$ and gsk $= \{\mathbf{T}_{id_j}\}_j$ to $\mathcal{A}$.
- $\mathcal{A}$ outputs j_0, j_1 and a message M.
- The signature of user j_b is computed as follows:
  → 1. Sample $\mathbf{x}_2 \hookleftarrow D_{\mathbb{Z}^m, \sigma}$ and, using $\mathbf{T}_{\mathbf{A}}$, sample $\mathbf{x}_1 \hookleftarrow D_{\mathbb{Z}^m, \sigma}$ conditioned on $(\mathbf{x}_1^T | \mathbf{x}_2^T) \cdot \mathbf{A}_{id_{j_b}} = \mathbf{0} \mod q$.
  2. Choose $\mathbf{s}_0 \hookleftarrow U(\mathbb{Z}_q^n)$, compute $\mathbf{c}_0 = \mathbf{B}_0 \cdot \mathbf{s}_0 + \mathbf{x}_2 \in \mathbb{Z}_q^m$,
  3. Choose $\mathbf{s} \hookleftarrow U(\mathbb{Z}_q^n)$, and for $i = 1$ to ℓ, choose $\mathbf{e}_i \hookleftarrow D_{\mathbb{Z}^m, \alpha q}$ and compute $\mathbf{c}_i = \mathbf{B}_i \cdot \mathbf{s} + p \cdot \mathbf{e}_i + id_{j_b}[i] \cdot \mathbf{x}_2$.
  4. Generate π_0.
  5. Generate $\{\pi_{OR,i}\}_i$.
  6. Generate π_K.

Experiment $G_b^{(2)}$

- Run Keygen; give gpk $= (\mathbf{A}, \{\mathbf{A}_i, \mathbf{B}_i\}_i)$ and gsk $= \{\mathbf{T}_{id_j}\}_j$ to $\mathcal{A}$.
- $\mathcal{A}$ outputs j_0, j_1 and a message M.
- The signature of user j_b is computed as follows:
  1. Sample $\mathbf{x}_2 \hookleftarrow D_{\mathbb{Z}^m, \sigma}$; sample $\mathbf{x}_1 \hookleftarrow D_{\mathbb{Z}^m, \sigma}$, conditioned on $(\mathbf{x}_1^T | \mathbf{x}_2^T) \cdot \mathbf{A}_{id_{j_b}} = \mathbf{0} \mod q$.
  2. Choose $\mathbf{s}_0 \hookleftarrow U(\mathbb{Z}_q^n)$ and compute $\mathbf{c}_0 = \mathbf{B}_0 \cdot \mathbf{s}_0 + \mathbf{x}_2 \in \mathbb{Z}_q^m$,
  3. Choose $\mathbf{s} \hookleftarrow U(\mathbb{Z}_q^n)$, and for $i = 1$ to ℓ, choose $\mathbf{e}_i \hookleftarrow D_{\mathbb{Z}^m, \alpha q}$ and compute $\mathbf{c}_i = \mathbf{B}_i \cdot \mathbf{s} + p \cdot \mathbf{e}_i + id_{j_b}[i] \cdot \mathbf{x}_2$.
  4. Generate π_0.
  → 5. Simulate $\{\pi_{OR,i}\}_i$.
  → 6. Simulate π_K.

Experiment $G_b^{(3)}$

- Run Keygen; give gpk $= (\mathbf{A}, \{\mathbf{A}_i, \mathbf{B}_i\}_i)$ and gsk $= \{\mathbf{T}_{id_j}\}_j$ to $\mathcal{A}$.
- $\mathcal{A}$ outputs j_0, j_1 and a message M.
- The signature of user j_b is computed as follows:
  1. Sample $\mathbf{x}_2 \hookleftarrow D_{\mathbb{Z}^m, \sigma}$ Sample $\mathbf{x}_1 \hookleftarrow D_{\mathbb{Z}^m, \sigma}$ conditioned on $(\mathbf{x}_1^T | \mathbf{x}_2^T) \cdot \mathbf{A}_{id_{j_b}} = \mathbf{0} \mod q$.
  2. Choose $\mathbf{s}_0 \hookleftarrow U(\mathbb{Z}_q^n)$ and compute $\mathbf{c}_0 = \mathbf{B}_0 \cdot \mathbf{s}_0 + \mathbf{x}_2 \in \mathbb{Z}_q^m$,
  → 3. For $i = 1$ to ℓ, choose $\mathbf{z}_i \hookleftarrow U(\mathbb{Z}_q^m)$ and compute $\mathbf{c}_i = \mathbf{z}_i + id_{j_b}[i] \cdot \mathbf{x}_2$.
  4. Generate π_0.
  5. Simulate $\{\pi_{OR,i}\}_i$.
  6. Simulate π_K.

Experiment $G^{(4)}$

- Run Keygen; give gpk $= (\mathbf{A}, \{\mathbf{A}_i, \mathbf{B}_i\}_i)$ and gsk $= \{\mathbf{T}_{id_j}\}_j$ to $\mathcal{A}$.
- $\mathcal{A}$ outputs j_0, j_1 and a message M.
- The signature of user j_b is computed as follows:
  → 1. Sample $\mathbf{x}_2 \hookleftarrow D_{\mathbb{Z}^m, \sigma}$.

2. Choose $\mathbf{s}_0 \hookleftarrow U(\mathbb{Z}_q^n)$ and compute $\mathbf{c}_0 = \mathbf{B}_0 \cdot \mathbf{s}_0 + \mathbf{x}_2 \in \mathbb{Z}_q^m$,
→ 3. For $i = 1$ to ℓ, choose $\mathbf{z}_i \hookleftarrow U(\mathbb{Z}_q^m)$ and set $\mathbf{c}_i = \mathbf{z}_i$.
4. Generate π_0.
5. Simulate $\{\pi_{OR,i}\}_i$.
6. Simulate π_K.

Fig. 2. Experiments G_b, $G_b^{(1)}$, $G_b^{(2)}$, $G_b^{(3)}$ and $G^{(4)}$

$j \in [0, N-1]$, we define $\mathbf{A}_{id_j}$ as in the original Keygen algorithm, and compute a trapdoor $\mathbf{T}_{id_j}$ using $\mathbf{T_A}$. The adversary $\mathcal{A}$ is given gpk and $\{gsk_j\}_j$. In the challenge phase, it outputs j_0, j_1 and a message M. By [24], this Keygen algorithm and the one in all the experiments are statistically indistinguishable. Then, the signature is created on behalf of the group member j_b. Namely, $\mathcal{B}$ first chooses $\mathbf{x}_2 \hookleftarrow D_{\mathbb{Z}^m, \sigma}$ and finds $\mathbf{x}_1$ such that $(\mathbf{x}_1^T | \mathbf{x}_2^T)^T \cdot \mathbf{A}_{id_{j_b}} = \mathbf{0} \mod q$. Then it chooses $\mathbf{s}_0 \hookleftarrow U(\mathbb{Z}_q^n)$ and computes $\mathbf{c}_0 = \mathbf{B}_0 \cdot \mathbf{s}_0 + \mathbf{x}_2 \in \mathbb{Z}_q^m$. Third, it computes

$c_i = p \cdot \mathbf{z}_i + id_{j_b}[i] \cdot x_2$ (with the $\mathbf{z}_i$ of the LWE instance). Then it generates π_0 and simulates the $\pi_{\mathsf{OR},i}$'s and π_K proofs.

We let $\mathcal{D}_{\mathsf{LWE}}$ denote this experiment when $\mathbf{z} = \mathbf{s} \cdot \mathbf{B}'^T + \mathbf{e}$: This experiment is statistically close to $G_b^{(2)}$. Then, we let $\mathcal{D}_{rand}$ denote this experiment when $\mathbf{z}$ is uniform: It is statistically close to $G_b^{(3)}$. As a consequence, if the adversary $\mathcal{A}$ can distinguish between the experiments $G_b^{(2)}$ and $G_b^{(3)}$ with non-negligible advantage, then we can solve the $\mathsf{LWE}_{q,\alpha}$ problem with the same advantage.

Lemma 10. *For each* $b \in \{0,1\}$, $G_b^{(3)}$ *and* $G^{(4)}$ *are indistinguishable.*

Between these two experiments, we change the first and third steps. In the former, we no longer generate $\mathbf{x}_1$ and, in the latter, c_i is uniformly sampled in $\mathbb{Z}_q^m$. These changes are purely conceptual. Indeed, in experiment $G_b^{(3)}$, $\mathbf{x}_1$ is not used beyond Step 1. In the same experiment, we also have $c_i = \mathbf{z}_i + id_{j_b}[i]$. Since the $\mathbf{z}_i$'s are uniformly sampled in $\mathbb{Z}_q^m$, the c_i's are also uniformly distributed in $\mathbb{Z}_q^m$. As a consequence, the c_i's of $G_b^{(3)}$ and the c_i's of $G^{(4)}$ have the same distribution. In $G_b^{(4)}$, we conclude that $\mathcal{A}$'s view is exactly the same as in experiments $G_b^{(3)}$. Since the experiment $G^{(4)}$ no longer depends on the bit $b \in \{0,1\}$ that determines the signer's identity, the announced result follows. $\square$

Traceability. The proof of traceability relies on the technique of [1,8] and a refinement from [28,38], which is used in order to allow for a smaller modulus q.

A difference with the proof of [24] is that we need to rely on the knowledge extractor of a proof of knowledge π_K. Depending on whether the extracted witnesses $\{\mathbf{e}_i, \mathbf{y}_i\}_{i=1}^\ell$ of relation (2) satisfy $\mathbf{y}_i = id_j[i]\mathbf{x}_2$ for all i or not, we need to distinguish two cases. The strategy of the reduction and the way it uses its given $\mathsf{SIS}_{q,\beta}$ instance will depend on which case is expected to occur. The proof is provided is the long version of this article [31, Section 4.2].

Theorem 3. *Assume that* $q > \log N$, $p \geq \Omega((\alpha q + \sigma)m^{5/2})$ *and* $\beta \geq \Omega(\sigma m^{7/2}\sqrt{\log N} + p\alpha q m^{5/2})$. *Then for any* PPT *traceability adversary* $\mathcal{A}$ *with success probability* ε, *there exists a* PPT *algorithm* $\mathcal{B}$ *solving* $\mathsf{SIS}_{m,q,\beta}$ *with probability* $\varepsilon'' \geq \frac{\varepsilon'}{2N} \cdot \left(\frac{\varepsilon'}{q_H} - 2^{-t}\right) + \frac{\varepsilon'}{2\log N}$, *where* $\varepsilon' = \varepsilon - 2^{-t} - 2^{-\Omega(n)}$ *and* q_H *is the number of queries to the random oracle* $H : \{0,1\}^* \to \{0,1\}^t$.

5   A Variant with Full (CCA-)Anonymity

We modify our basic group signature scheme to reach the strongest anonymity level (Definition 3), in which the attacker is authorized to query an opening oracle. This implies the simulation of an oracle which opens adversarially-chosen signatures in the proof of anonymity. To this end, we replace each $\mathbf{B}_i$ from our previous scheme by a matrix $\mathbf{B}_{i,\mathsf{VK}}$ that depends on the verification key VK of a strongly unforgeable one-time signature. The reduction will be able to compute a trapdoor for all these matrices, except for one specific verification key $\mathsf{VK}^\star$

that will be used in the challenge phase. This will provide the reduction with a backdoor allowing it to open all adversarially-generated signatures.

It is assumed that the one-time verification keys VK belong to $\mathbb{Z}_q^n$ (note that this condition can always be enforced by hashing VK). Following Agrawal *et al.* [1], we rely on a full-rank difference function $H_{vk} : \mathbb{Z}_q^n \to \mathbb{Z}_q^{n \times n}$ such that, for any two distinct $\mathbf{u}, \mathbf{v} \in \mathbb{Z}_q^n$, the difference $H_{vk}(\mathbf{u}) - H_{vk}(\mathbf{v})$ is a full rank matrix.

Keygen$(1^n, 1^N)$: Given a security parameter $n > 0$ and the desired number of members $N = 2^\ell \in \mathsf{poly}(n)$, choose parameters as in Section 3 and make them public. Choose a hash function $H : \{0,1\}^* \to \{0,1\}^t$, that will be seen as a random oracle, and a one-time signature $\Pi^{\text{ots}} = (\mathcal{G}, \mathcal{S}, \mathcal{V})$.

1. Run $\mathsf{TrapGen}(1^n, 1^m, q)$ to get $\mathbf{A} \in \mathbb{Z}_q^{m \times n}$ and a short basis $\mathbf{T_A}$ of $\Lambda_q^\perp(\mathbf{A})$.
2. For $i = 0$ to ℓ, repeat the following steps.
   a. Choose uniformly random matrices $\mathbf{A}_{i,1}, \mathbf{B}_{i,0}, \mathbf{B}_{i,1} \in \mathbb{Z}_q^{m \times n}$.
   b. Compute $(\mathbf{A}_{i,2}, \mathbf{T}_{i,2}) \leftarrow \mathsf{SuperSamp}(1^n, 1^m, q, \mathbf{B}_{i,1}, 0^{n \times n})$ such that $\mathbf{B}_{i,1}^T \cdot \mathbf{A}_{i,2} = 0 \bmod q$ and discard $\mathbf{T}_{i,2}$, which will not be needed.
   Define $\mathbf{A}_i = \begin{bmatrix} \mathbf{A}_{i,1} \\ \mathbf{A}_{i,2} \end{bmatrix} \in \mathbb{Z}_q^{2m \times n}$.
   c. Run $(\mathbf{B}_{i,-1}, \mathbf{S}'_i) \leftarrow \mathsf{SuperSamp}(1^n, 1^m, q, \mathbf{A}_{i,1}, -\mathbf{A}_{i,2}^T \cdot \mathbf{B}_{i,0})$ to obtain $\mathbf{B}_{i,-1} \in \mathbb{Z}_q^{m \times n}$ such that $\mathbf{B}_{i,-1}^T \cdot \mathbf{A}_{i,1} + \mathbf{B}_{i,0}^T \cdot \mathbf{A}_{i,2} = 0 \bmod q$.
   d. Compute a re-randomized trapdoor $\mathbf{S}_i \leftarrow \mathsf{RandBasis}(\mathbf{S}'_i)$ for $\mathbf{B}_{i,-1}$.
   For any string VK, if the matrix $H_{vk}(\mathsf{VK})$ is used to define $\mathbf{B}_{i,\mathsf{VK}} = \begin{bmatrix} \mathbf{B}_{i,-1} \\ \mathbf{B}_{i,0} + \mathbf{B}_{i,1} H_{vk}(\mathsf{VK}) \end{bmatrix} \in \mathbb{Z}_q^{2m \times n}$, we have $\mathbf{B}_{i,\mathsf{VK}}^T \cdot \mathbf{A}_i = 0 \bmod q$ for all i.
3. For $j = 0$ to $N - 1$, let $\mathrm{id}_j = \mathrm{id}_j[1] \dots \mathrm{id}_j[\ell] \in \{0,1\}^\ell$ be the binary representation of id_j and define: $\mathbf{A}_{\mathrm{id}_j} = \begin{bmatrix} \mathbf{A} \\ \mathbf{A}_0 + \sum_{i=1}^{\ell} \mathrm{id}_j[i] \mathbf{A}_i \end{bmatrix} \in \mathbb{Z}_q^{3m \times n}$.
   Then run $\mathbf{T}_{\mathrm{id}_j} \leftarrow \mathsf{ExtBasis}(\mathbf{T_A}, \mathbf{A}_{\mathrm{id}_j})$ to get a short delegated basis $\mathbf{T}_{\mathrm{id}_j} \in \mathbb{Z}^{3m \times 3m}$ of $\Lambda_q^\perp(\mathbf{A}_{\mathrm{id}_j})$ and define $\mathsf{gsk}[j] := \mathbf{T}_{\mathrm{id}_j}$.
4. Finally, define $\mathsf{gpk} := (\mathbf{A}, \{\mathbf{A}_i, (\mathbf{B}_{i,-1}, \mathbf{B}_{i,0}, \mathbf{B}_{i,1})\}_{i=0}^\ell, H, \Pi^{\text{ots}})$ and $\mathsf{gmsk} := \{\mathbf{S}_i\}_{i=0}^\ell$. The algorithm outputs $(\mathsf{gpk}, \mathsf{gmsk}, \{\mathsf{gsk}[j]\}_{j=0}^{N-1})$.

Sign$(\mathsf{gpk}, \mathsf{gsk}[j], M)$: To sign a message $M \in \{0,1\}^*$ using the private key $\mathsf{gsk}[j] = \mathbf{T}_{\mathrm{id}_j}$, generate a one-time signature key pair $(\mathsf{VK}, \mathsf{SK}) \leftarrow \mathcal{G}(1^n)$ for Π^{ots} and proceed as follows.

1. Run $\mathsf{GPVSample}(\mathbf{T}_{\mathrm{id}_j}, \sigma)$ to get $(\mathbf{x}_1^T | \mathbf{x}_2^T)^T \in \Lambda_q^\perp(\mathbf{A}_{\mathrm{id}_j})$ of norm $\leq \sigma\sqrt{3m}$.
2. Sample $\mathbf{s}_0 \hookleftarrow U(\mathbb{Z}_q^n)$ and encrypt $\mathbf{x}_2 \in \mathbb{Z}_q^{2m}$ as $\mathbf{c}_0 = \mathbf{B}_{0,\mathsf{VK}} \cdot \mathbf{s}_0 + \mathbf{x}_2 \in \mathbb{Z}_q^{2m}$.
3. Sample $\mathbf{s} \hookleftarrow U(\mathbb{Z}_q^n)$. For $i = 1$ to ℓ, sample $\mathbf{e}_i \leftarrow D_{\mathbb{Z}^{2m}, \alpha q}$ and a random matrix $\mathbf{R}_i \in \mathbb{Z}^{m \times m}$ whose columns are sampled from $D_{\mathbb{Z}^m, \sigma}$. Then, compute $\mathbf{c}_i = \mathbf{B}_{i,\mathsf{VK}} \cdot \mathbf{s} + p \cdot [\mathbf{e}_i | \mathbf{e}_i \cdot \mathbf{R}_i] + \mathrm{id}_j[i] \cdot \mathbf{x}_2$, which encrypts $\mathbf{x}_2 \in \mathbb{Z}_q^{2m}$ (resp. 0^{2m}) if $\mathrm{id}_j[i] = 1$ (resp. $\mathrm{id}_j[i] = 0$).
4. Generate a NIZKPoK π_0 of $\mathbf{s}_0$ so that $(\mathbf{B}_0, \mathbf{c}_0, \sqrt{2}\sigma/q; \mathbf{s}_0) \in R_{\mathsf{LWE}}$.
5. For $i = 1$ to ℓ, generate a NIZKPoK $\pi_{\mathsf{OR},i}$ of $\mathbf{s}$ and $\mathbf{s}_0$ so that either:
   (i) $((\mathbf{B}_{i,\mathsf{VK}} | \mathbf{B}_{0,\mathsf{VK}}), p^{-1}(\mathbf{c}_i - \mathbf{c}_0), \sqrt{2}\alpha; (\mathbf{s}^T | -\mathbf{s}_0^T)^T) \in R_{\mathsf{LWE}}$ (the vectors $\mathbf{c}_i$ and $\mathbf{c}_0$ encrypt the same $\mathbf{x}_2$, so that the vector $p^{-1}(\mathbf{c}_i - \mathbf{c}_0)$ is close to the $\mathbb{Z}_q$-span of $(\mathbf{B}_{i,\mathsf{VK}} | \mathbf{B}_{0,\mathsf{VK}})$);

(ii) or $(\mathbf{B}_{i,\mathsf{VK}}^T, p^{-1}\mathbf{c}_i, \alpha; \mathbf{s}) \in R_{\mathsf{LWE}}$ ($p^{-1}\mathbf{c}_i$ is close to the $\mathbb{Z}_q$-span of $\mathbf{B}_{i,\mathsf{VK}}^T$).

6. For $i = 1$ to ℓ, set $\mathbf{y}_i = \mathrm{id}_j[i]\mathbf{x}_2 \in \mathbb{Z}^{2m}$ and generate a NIZKPoK π_K of $(\mathbf{e}_i)_{1 \le i \le \ell}, (\mathbf{y}_i)_{1 \le i \le \ell}$, $\mathbf{x}_1$ such that: $\mathbf{x}_1^T \mathbf{A} + \sum_{i=0}^{\ell} \mathbf{c}_i^T \mathbf{A}_i = \sum_{i=1}^{\ell} \mathbf{e}_i^T (p \cdot \mathbf{A}_i)$ and $\mathbf{e}_i^T (p \cdot \mathbf{A}_i) + \mathbf{y}_i^T \mathbf{A}_i = \mathbf{c}_i^T \mathbf{A}_i$ with $\|\mathbf{x}_1\| \le \sigma\sqrt{m}$ and $\|\mathbf{y}_i\| \le \sigma\sqrt{2m}$ for each $i \in \{1, \dots, \ell\}$.

This is achieved using $\mathsf{Prove}_{\mathsf{ISIS}}$, giving a triple $(\mathsf{Comm}_K, \mathsf{Chall}_K, \mathsf{Resp}_K)$, where $\mathsf{Chall}_K = H(M, \mathsf{Comm}_K, (\mathbf{c}_i)_{0 \le i \le \ell}, \pi_0, (\pi_{\mathsf{OR},i})_{1 \le i \le \ell})$.

7. Compute $sig = \mathcal{S}(\mathsf{SK}, (\mathbf{c}_i)_{0 \le i \le \ell}, \pi_0, (\pi_{\mathsf{OR},i})_{1 \le i \le \ell}, \pi_K))$.

The signature consists of

$$\Sigma = (\mathsf{VK}, \mathbf{c}_0, \mathbf{c}_1, \dots, \mathbf{c}_\ell, \pi_0, \pi_{\mathsf{OR},1}, \dots, \pi_{\mathsf{OR},\ell}, \pi_K, sig). \tag{4}$$

Verify(gpk, M, Σ): Parse the signature Σ as in (4). Then, return 0 in the event that $\mathcal{V}(\mathsf{VK}, sig, (\mathbf{c}_i)_{0 \le i \le \ell}, \pi_0, (\pi_{\mathsf{OR},i})_{1 \le i \le \ell}, \pi_K)) = 0$. Then, return 1 if all proofs $\pi_0, (\pi_{\mathsf{OR},i})_{1 \le i \le \ell}, \pi_K$ properly verify. Otherwise, return 0.

Open$(gpk, gmsk, M, \Sigma)$: Parse gmsk as $\{\mathbf{S}_i\}_{i=0}^{\ell}$ and Σ as in (4). For $i = 0$ to ℓ, compute a trapdoor $\mathbf{S}_{i,\mathsf{VK}} \leftarrow \mathsf{ExtBasis}(\mathbf{S}_i, \mathbf{B}_{i,\mathsf{VK}})$ for $\mathbf{B}_{i,\mathsf{VK}}$. Using the delegated basis $\mathbf{S}_{0,\mathsf{VK}} \in \mathbb{Z}^{2m \times 2m}$ (for which we have $\mathbf{S}_{0,\mathsf{VK}} \cdot \mathbf{B}_{0,\mathsf{VK}} = 0 \bmod q$), compute $\mathbf{x}_2$ by decrypting $\mathbf{c}_0$. Then, using $\mathbf{S}_{i,\mathsf{VK}} \in \mathbb{Z}^{2m \times 2m}$, determine which vector among $p^{-1}\mathbf{c}_i \bmod q$ and $p^{-1}(\mathbf{c}_i - \mathbf{x}_2) \bmod q$ is close to the $\mathbb{Z}_q$-span of $\mathbf{B}_{i,\mathsf{VK}}$. Set $\mathrm{id}[i] = 0$ in the former case and $\mathrm{id}[i] = 1$ in the latter case. Eventually, output $\mathrm{id} = \mathrm{id}[1] \dots \mathrm{id}[\ell]$.

In [31, Appendix D], we prove the following theorems.

Theorem 4. *The scheme provides full anonymity in the ROM if the* $\mathsf{LWE}_{q,\alpha}$ *assumption holds and if the one-time signature is strongly unforgeable.*

Theorem 5. *Assuming that* $q > \log N$, *the scheme is fully traceable in the ROM under the* $\mathsf{SIS}_{q,\beta}$ *assumption. More precisely, for any* PPT *traceability adversary* $\mathcal{A}$ *with success probability* ε, *there exists an algorithm* $\mathcal{B}$ *solving the* $\mathsf{SIS}_{q,\beta}$ *problem with probability at least* $\frac{1}{2N} \cdot \left(\varepsilon - \frac{1}{2^t}\right) \cdot \left(\frac{\varepsilon - 1/2^t}{q_H} - \frac{1}{2^t}\right)$, *where* q_H *is the number of queries to* $H : \{0,1\}^* \to \{0,1\}^t$.

Acknowledgements. We thank Daniele Micciancio and Khoa Nguyen for helpful discussions. Parts of the research described in this work were underwent while the third author was visiting ENS de Lyon, under the INRIA invited researcher scheme. The last author was partly supported by the Australian Research Council Discovery Grant DP110100628.

References

1. Agrawal, S., Boneh, D., Boyen, X.: Efficient lattice (H)IBE in the standard model. In: Gilbert, H. (ed.) EUROCRYPT 2010. LNCS, vol. 6110, pp. 553–572. Springer, Heidelberg (2010)

2. Alwen, J., Peikert, C.: Generating shorter bases for hard random lattices. Theor. Comput. Science 48(3), 535–553 (2011)

3. Ateniese, G., Camenisch, J.L., Joye, M., Tsudik, G.: A practical and provably secure coalition-resistant group signature scheme. In: Bellare, M. (ed.) CRYPTO 2000. LNCS, vol. 1880, pp. 255–270. Springer, Heidelberg (2000)

4. Banaszczyk, W.: New bounds in some transference theorems in the geometry of number. Math. Ann. 296, 625–635 (1993)

5. Bellare, M., Micciancio, D., Warinschi, B.: Foundations of group signatures: Formal definitions, simplified requirements, and a construction based on general assumptions. In: Biham, E. (ed.) EUROCRYPT 2003. LNCS, vol. 2656, pp. 614–629. Springer, Heidelberg (2003)

6. Bellare, M., Shi, H., Zhang, C.: Foundations of group signatures: The case of dynamic groups. In: Menezes, A. (ed.) CT-RSA 2005. LNCS, vol. 3376, pp. 136–153. Springer, Heidelberg (2005)

7. Boneh, D., Boyen, X., Shacham, H.: Short group signatures. In: Franklin, M. (ed.) CRYPTO 2004. LNCS, vol. 3152, pp. 41–55. Springer, Heidelberg (2004)

8. Boyen, X.: Lattice mixing and vanishing trapdoors: A framework for fully secure short signatures and more. In: Nguyen, P.Q., Pointcheval, D. (eds.) PKC 2010. LNCS, vol. 6056, pp. 499–517. Springer, Heidelberg (2010)

9. Boyen, X., Waters, B.: Compact group signatures without random oracles. In: Vaudenay, S. (ed.) EUROCRYPT 2006. LNCS, vol. 4004, pp. 427–444. Springer, Heidelberg (2006)

10. Boyen, X., Waters, B.: Full-domain subgroup hiding and constant-size group signatures. In: Okamoto, T., Wang, X. (eds.) PKC 2007. LNCS, vol. 4450, pp. 1–15. Springer, Heidelberg (2007)

11. Brakerski, Z., Langlois, A., Peikert, C., Regev, O., Stehlé, D.: On the classical hardness of learning with errors. In: Proc. of STOC, pp. 575–584. ACM (2013)

12. Brickell, E.: An efficient protocol for anonymously providing assurance of the container of a private key. Submitted to the Trusted Computing Group (2003)

13. Camenisch, J., Lysyanskaya, A.: Dynamic accumulators and application to efficient revocation of anonymous credentials. In: Yung, M. (ed.) CRYPTO 2002. LNCS, vol. 2442, pp. 61–76. Springer, Heidelberg (2002)

14. Camenisch, J., Lysyanskaya, A.: A signature scheme with efficient protocols. In: Cimato, S., Galdi, C., Persiano, G. (eds.) SCN 2002. LNCS, vol. 2576, pp. 268–289. Springer, Heidelberg (2003)

15. Camenisch, J., Lysyanskaya, A.: Signature schemes and anonymous credentials from bilinear maps. In: Franklin, M. (ed.) CRYPTO 2004. LNCS, vol. 3152, pp. 56–72. Springer, Heidelberg (2004)

16. Camenisch, J., Neven, G., Rückert, M.: Fully anonymous attribute tokens from lattices. In: Visconti, I., De Prisco, R. (eds.) SCN 2012. LNCS, vol. 7485, pp. 57–75. Springer, Heidelberg (2012)

17. Cash, D., Hofheinz, D., Kiltz, E., Peikert, C.: Bonsai trees, or how to delegate a lattice basis. In: Gilbert, H. (ed.) EUROCRYPT 2010. LNCS, vol. 6110, pp. 523–552. Springer, Heidelberg (2010)

18. Chaum, D., van Heyst, E.: Group signatures. In: Davies, D.W. (ed.) EUROCRYPT 1991. LNCS, vol. 547, pp. 257–265. Springer, Heidelberg (1991)

19. Coron, J.-S., Lepoint, T., Tibouchi, M.: Practical multilinear maps over the integers. In: Canetti, R., Garay, J.A. (eds.) CRYPTO 2013, Part I. LNCS, vol. 8042, pp. 476–493. Springer, Heidelberg (2013)

20. Cramer, R., Damgård, I., Schoenmakers, B.: Proof of partial knowledge and simplified design of witness hiding protocols. In: Desmedt, Y.G. (ed.) CRYPTO 1994. LNCS, vol. 839, pp. 174–187. Springer, Heidelberg (1994)

21. Damgård, I.: On Σ-protocols (2010), http://www.daimi.au.dk/~ivan/Sigma.pdf (manuscript)

22. Garg, S., Gentry, C., Halevi, S.: Candidate multilinear maps from ideal lattices. In: Johansson, T., Nguyen, P.Q. (eds.) EUROCRYPT 2013. LNCS, vol. 7881, pp. 1–17. Springer, Heidelberg (2013)

23. Gentry, C., Peikert, C., Vaikuntanathan, V.: Trapdoors for hard lattices and new cryptographic constructions. In: Proc. of STOC, pp. 197–206. ACM (2008)

24. Gordon, S.D., Katz, J., Vaikuntanathan, V.: A group signature scheme from lattice assumptions. In: Abe, M. (ed.) ASIACRYPT 2010. LNCS, vol. 6477, pp. 395–412. Springer, Heidelberg (2010)

25. Groth, J.: Fully anonymous group signatures without random oracles. In: Kurosawa, K. (ed.) ASIACRYPT 2007. LNCS, vol. 4833, pp. 164–180. Springer, Heidelberg (2007)

26. Groth, J., Ostrovsky, R., Sahai, A.: Perfect non-interactive zero knowledge for NP. In: Vaudenay, S. (ed.) EUROCRYPT 2006. LNCS, vol. 4004, pp. 339–358. Springer, Heidelberg (2006)

27. Groth, J., Sahai, A.: Efficient non-interactive proof systems for bilinear groups. In: Smart, N.P. (ed.) EUROCRYPT 2008. LNCS, vol. 4965, pp. 415–432. Springer, Heidelberg (2008)

28. Hohenberger, S., Waters, B.: Short and stateless signatures from the RSA assumption. In: Halevi, S. (ed.) CRYPTO 2009. LNCS, vol. 5677, pp. 654–670. Springer, Heidelberg (2009)

29. VSC Project IEEE P1556 Working Group. Dedicated short range communications (dsrc) (2003)

30. Kiayias, A., Yung, M.: Secure scalable group signature with dynamic joins and separable authorities (IJSN) 1(1/2), 24–45 (2006)

31. Laguillaumie, F., Langlois, A., Libert, B., Stehlé, D.: Lattice-based group signatures with logarithmic signature size. Cryptology ePrint Archive, Report 2013/308 (2013), http://eprint.iacr.org/2013/308

32. Lyubashevsky, V.: Lattice-based identification schemes secure under active attacks. In: Cramer, R. (ed.) PKC 2008. LNCS, vol. 4939, pp. 162–179. Springer, Heidelberg (2008)

33. Lyubashevsky, V.: Fiat-Shamir with aborts: Applications to lattice and factoring-based signatures. In: Matsui, M. (ed.) ASIACRYPT 2009. LNCS, vol. 5912, pp. 598–616. Springer, Heidelberg (2009)

34. Lyubashevsky, V.: Lattice signatures without trapdoors. In: Pointcheval, D., Johansson, T. (eds.) EUROCRYPT 2012. LNCS, vol. 7237, pp. 738–755. Springer, Heidelberg (2012)

35. Lyubashevsky, V., Micciancio, D.: Generalized compact knapsacks are collision resistant. In: Bugliesi, M., Preneel, B., Sassone, V., Wegener, I. (eds.) ICALP 2006. LNCS, vol. 4052, pp. 144–155. Springer, Heidelberg (2006)

36. Lyubashevsky, V., Micciancio, D.: Asymptotically efficient lattice-based digital signatures. In: Canetti, R. (ed.) TCC 2008. LNCS, vol. 4948, pp. 37–54. Springer, Heidelberg (2008)

37. Lyubashevsky, V., Peikert, C., Regev, O.: On ideal lattices and learning with errors over rings. In: Gilbert, H. (ed.) EUROCRYPT 2010. LNCS, vol. 6110, pp. 1–23. Springer, Heidelberg (2010)

38. Micciancio, D., Peikert, C.: Trapdoors for lattices: Simpler, tighter, faster, smaller. In: Pointcheval, D., Johansson, T. (eds.) EUROCRYPT 2012. LNCS, vol. 7237, pp. 700–718. Springer, Heidelberg (2012)

39. Micciancio, D., Vadhan, S.P.: Statistical zero-knowledge proofs with efficient provers: Lattice problems and more. In: Boneh, D. (ed.) CRYPTO 2003. LNCS, vol. 2729, pp. 282–298. Springer, Heidelberg (2003)

40. Peikert, C.: Public-key cryptosystems from the worst-case shortest vector problem. In: Proc. of STOC, pp. 333–342. ACM (2009)

41. Peikert, C., Rosen, A.: Efficient collision-resistant hashing from worst-case assumptions on cyclic lattices. In: Halevi, S., Rabin, T. (eds.) TCC 2006. LNCS, vol. 3876, pp. 145–166. Springer, Heidelberg (2006)

42. Peikert, C., Vaikuntanathan, V.: Noninteractive statistical zero-knowledge proofs for lattice problems. In: Wagner, D. (ed.) CRYPTO 2008. LNCS, vol. 5157, pp. 536–553. Springer, Heidelberg (2008)

43. Regev, O.: On lattices, learning with errors, random linear codes, and cryptography. J. ACM 56(6) (2009)

The Fiat–Shamir Transformation
in a Quantum World

Özgür Dagdelen, Marc Fischlin, and Tommaso Gagliardoni

Technische Universität Darmstadt, Germany
oezguer.dagdelen@cased.de, marc.fischlin@gmail.com,
tommaso@gagliardoni.net
www.cryptoplexity.de

Abstract. The Fiat-Shamir transformation is a famous technique to turn identification schemes into signature schemes. The derived scheme is provably secure in the random-oracle model against classical adversaries. Still, the technique has also been suggested to be used in connection with quantum-immune identification schemes, in order to get quantum-immune signature schemes. However, a recent paper by Boneh et al. (Asiacrypt 2011) has raised the issue that results in the random-oracle model may not be immediately applicable to quantum adversaries, because such adversaries should be allowed to query the random oracle in superposition. It has been unclear if the Fiat-Shamir technique is still secure in this quantum oracle model (QROM).

Here, we discuss that giving proofs for the Fiat-Shamir transformation in the QROM is presumably hard. We show that there cannot be black-box extractors, as long as the underlying quantum-immune identification scheme is secure against active adversaries and the first message of the prover is independent of its witness. Most schemes are of this type. We then discuss that for some schemes one may be able to resurrect the Fiat-Shamir result in the QROM by modifying the underlying protocol first. We discuss in particular a version of the Lyubashevsky scheme which is provably secure in the QROM.

1 Introduction

The Fiat-Shamir transformation [19] is a well-known method to remove interaction in three-move identification schemes between a prover and verifier, by letting the verifier's challenge ch be determined via a hash function H applied to the prover's first message com. Currently, the only generic, provably secure instantiation is by modeling the hash function H as a random oracle [5,33]. In general, finding secure instantiations based on *standard* hash functions is hard for some schemes, as shown in [22,7]. However, these negative results usually rely on peculiar identification schemes, such that for specific schemes, especially more practical ones, such instantiations may still be possible.

THE QUANTUM RANDOM-ORACLE MODEL. Recently, the Fiat-Shamir transformation has also been applied to schemes which are advertised as being based

K. Sako and P. Sarkar (Eds.) ASIACRYPT 2013, Part II, LNCS 8270, pp. 62–81, 2013.

on quantum-immune primitives, e.g., [28,3,23,12,13,35,30,34,25,1,11,2,17]. Interestingly, the proofs for such schemes still investigate classical adversaries only. It seems unclear if (and how) one can transfer the proofs to the quantum case. Besides the problem that the classical Fiat-Shamir proof [33] relies on rewinding the adversary, which is often considered to be critical for quantum adversaries (albeit not impossible [39,38]), a bigger discomfort seems to lie in the usage of the random-oracle model in presence of quantum adversaries.

As pointed out by Boneh et al. [8] the minimal requirement for random oracles in the quantum world should be *quantum access*. Since the random oracle is eventually replaced by a standard hash function, a quantum adversary could evaluate this hash function in superposition, while still ignoring any advanced attacks exploiting the structure of the actual hash function. To reflect this in the random-oracle model, [8] argue that the quantum adversary should be also allowed to query the random oracle in superposition. That is, the adversary should be able to query the oracle on a state $|\varphi\rangle = \sum_x \alpha_x |x\rangle |0\rangle$ and in return would get $\sum_x \alpha_x |x\rangle |H(x)\rangle$. This model is called the quantum random-oracle model (QROM).

Boneh et al. [8] discuss some classical constructions for encryption and signatures which remain secure in the QROM. They do not cover Fiat-Shamir signatures, though. Subsequently, Boneh and Zhandry [41,40,9] investigate further primitives with quantum access, such as pseudorandom functions and MACs. Still, the question about the security of the Fiat-Shamir transform in the QROM raised in [8] remained open.

FIAT-SHAMIR TRANSFORM IN THE QROM. Here, we give evidence that conducting security proofs for Fiat-Shamir transformed schemes and black-box adversaries is hard, thus yielding a negative result about the provable security of such schemes. More specifically, we use the meta-reduction technique to rule out the existence of quantum extractors with black-box access to a quantum adversary against the converted (classical) scheme. If such extractors would exist then the meta-reduction, together with the extractor, yields a quantum algorithm which breaks the active security of the identification scheme. Our result covers *any* identification scheme, as long as the prover's initial commitment in the scheme is independent of the witness, and if the scheme itself is secure against active quantum attacks where a malicious verifier may first interact with the genuine prover before trying to impersonate or, as we only demand here, to compute a witness afterwards. Albeit not quantum-immune, the classical schemes of Schnorr [36], Guillou and Quisquater [24], and Feige, Fiat and Shamir [18] are conceivably of this type (see also [4]). Quantum-immune candidates are, for instance, [31,27,26,30,35,2].

Our negative result does not primarily rely on the rewinding problem for quantum adversaries; our extractor may rewind the adversary (in a black-box way). Instead, our result is rather based on the adversary's possibility to hide actual queries to the quantum random oracle in a "superposition cloud", such that the extractor or simulator cannot elicit or implant necessary information for such queries. In fact, our result reveals a technical subtlety in the QROM which previous works [8,40,41,9] have not addressed at all, or at most implicitly.

It refers to the question how a simulator or extractor can answer superposition queries $\sum_x \alpha_x |x\rangle |0\rangle$.

A possible option is to allow the simulator to reply with an arbitrary quantum state $|\psi\rangle = \sum_x \beta_x |x\rangle |y_x\rangle$, e.g., by swapping the state from its local registers to the ancilla bits for the answer in order to make this step unitary. This seems to somehow generalize the classical situation where the simulator on input x returns an arbitrary string y for $H(x)$. Yet, the main difference is that returning an arbitrary state $|\psi\rangle$ could also be used to eliminate some of the input values x, i.e., by setting $\beta_x = 0$. This is more than what the simulator is able to do in the classical setting, where the adversary can uniquely identify the preimage x to the answer. In the extreme the simulator in the quantum case, upon receiving a (quantum version of) a classical state $|x\rangle |0\rangle$, could simply reply with an (arbitrary) quantum state $|\psi\rangle$. Since quantum states are in general indistinguishable, in contrast to the classical case the adversary here would potentially continue its execution for inputs which it has not queried for.

In previous works [8,41,40,9] the simulator specifies a classical (possibly probabilistic) function h which maps the adversary query $\sum_x \alpha_x |x\rangle |0\rangle$ to the reply $\sum_x \alpha_x |x\rangle |h(x)\rangle$. Note that the function h is not given explicitly to the adversary, and that it can thus implement keyed functions like a pseudorandom function (as in [8]). This basically allows the simulator to freely assign values $h(x)$ to each string x, without being able to change the input values. It also corresponds to the idea that, if the random oracle is eventually replaced by an actual hash function, the quantum adversary can check that the hash function is classical, even if the adversary does not aim to exploit any structural weaknesses (such that we still hide h from the adversary).

We thus adopt the approach of letting the simulator determine the quantum answer via a classical probabilistic function h. In fact, our impossibility hinges on this property but which we believe to be rather "natural" for the aforementioned reasons. From a mere technical point of view it at least clearly identifies possible venues to bypass our hardness result. In our case we allow the simulator to specify the (efficient) function h adaptively for each query, still covering techniques like programmability in the classical setting. Albeit this is sometimes considered to be a doubtful property [20] this strengthens our impossibility result in this regard.

POSITIVE RESULTS. We conclude with some positive result. It remains open if one can "rescue" plain Fiat-Shamir for schemes which are not actively secure, or to prove that alternative but still reasonably efficient approaches work. However, we can show that the Fiat-Shamir technique in general *does* provide a secure signature scheme in the QROM if the protocol allows for oblivious commitments. Roughly, this means that the honest verifier generates the prover's first message com obliviously by sampling a random string and sends com to the prover. In the random oracle transformed scheme the commitment is thus computed via the random oracle, together with the challenge. Such schemes are usually not actively secure against malicious verifiers. Nonetheless, we stress that in order to derive a secure signature scheme via the Fiat-Shamir transform, the underlying identification scheme merely needs to provide passive security and honest-verifier zero-knowledge.

To make the above transformation work, we need that the prover is able to compute the response for commitments chosen obliviously to the prover. For some schemes this is indeed possible if the prover holds some trapdoor information. Albeit not quantum-immune, it is instructive to look at the Guillou-Quisquater RSA-based proof of knowledge [24] where the prover shows knowledge of $w \in \mathbb{Z}_N^*$ with $w^e = y \bmod N$ for $x = (e, N, y)$. For an oblivious commitment the prover would need to compute an e-th root for a given commitment $R \in \mathbb{Z}_N^*$. If the witness would contain the prime factorization of N, instead of the e-th root of y, this would indeed be possible. As a concrete allegedly quantum-immune example we discuss that we can still devise a provably secure signature version of Lyubashevsky's identification scheme [29] via our method. Before, Lyubashevsky only showed security in the classical random-oracle model, despite using an allegedly quantum-immune primitive.

RELATED WORK. Since the introduction of the quantum-accessible random-oracle model [8], several works propose cryptographic primitives or revisit their security against quantum algorithms in this stronger model [40,41,9]. In [15], Damgård et al. look at the security of cryptographic protocols where the underlying primitives or even parties can be queried by an adversary in a superposition. We here investigate the scenario in which the quantum adversary can only interact classically with the classical honest parties, except for the locally evaluable random oracle.

In a concurrent and independent work, Boneh and Zhandry [10] analyze the security of signature schemes under quantum chosen-message attacks, i.e., the adversary in the unforgeability notion of the signature scheme may query the signing oracle in superposition and, eventually, in the quantum random oracle model. Our negative result carries over to the quantum chosen-message attack model as well, since our impossibility holds even allowing only classical queries to the signing oracle. Moreover, while the authors of [10] show how to obtain signature schemes secure in the quantum-accessible signing oracle model, starting with schemes secure in the classical sense, we focus on signature schemes and proofs of knowledge derived from identification schemes via the Fiat-Shamir paradigm.

2   Preliminaries

We first describe (to the level we require it) quantum computations and then recall the quantum random-oracle model of Boneh et al. [8]. We also introduce the notion of Σ-protocols to which the Fiat-Shamir transformation applies. In the full version of this paper [14], we recall the definition of signature schemes and its security.

2.1   Quantum Computations in the QROM

We first briefly recall facts about quantum computations and set some notation; for more details, we refer to [32]. Our description follows [8] closely.

QUANTUM SYSTEMS. A quantum system A is associated to a complex Hilbert space $\mathcal{H}_A$ of finite dimension and with an inner product $\langle \cdot | \cdot \rangle$. The state of the system is given by a (class of) normalized vector $|\varphi\rangle \in \mathcal{H}_A$ with Euclidean norm $\| |\varphi\rangle \| = \sqrt{\langle \varphi | \varphi \rangle} = 1$. The joint or composite quantum state of two quantum systems A and B over spaces $\mathcal{H}_A$ and $\mathcal{H}_B$, respectively, is given through the tensor product $\mathcal{H}_A \otimes \mathcal{H}_B$. The product state of $|\varphi_A\rangle \in \mathcal{H}_A$ and $|\varphi_B\rangle \in \mathcal{H}_B$ is denoted by $|\varphi_A\rangle \otimes |\varphi_B\rangle$. We sometimes simply write $|\varphi_A\rangle |\varphi_B\rangle$ or $|\varphi_A, \varphi_B\rangle$. An n-qubit system is associated in the joint quantum system of n two-dimensional Hilbert spaces. The standard orthonormal computational basis $|x\rangle$ for such a system is given by $|x\rangle = |x_1\rangle \otimes \cdots \otimes |x_n\rangle$ for $x = x_1 \ldots x_n \in \{0,1\}^n$. We often assume that any (classical) bit string x is encoded into a quantum state as $|x\rangle$, and vice versa we sometimes view such a state simply as a classical state. Any pure n-qubit state $|\varphi\rangle$ can be expressed as a superposition in the computational basis as $|\varphi\rangle = \sum_{x \in \{0,1\}^n} \alpha_x |x\rangle$ where α_x are complex amplitudes obeying $\sum_{x \in \{0,1\}^n} |\alpha_x|^2 = 1$.

QUANTUM COMPUTATIONS. Evolutions of quantum systems are described by unitary transformations with $\mathbb{I}_A$ being the identity transformation on register A. For a composite quantum system over $\mathcal{H}_A \otimes \mathcal{H}_B$ and a transformation U_A acting only on $\mathcal{H}_A$, it is understood that $U_A |\varphi_A\rangle |\varphi_B\rangle$ is a simplification of $(U_A \otimes \mathbb{I}_B) |\varphi_A\rangle |\varphi_B\rangle$. Note that any unitary operation and, thus, any quantum operation, is invertible.

Information can be extracted from a quantum state $|\varphi\rangle$ by performing a positive-operator valued measurement (POVM) $M = \{M_i\}_i$ with positive semi-definite measurement operators M_i that sum to the identity $\sum_i M_i = \mathbb{I}$. Outcome i is obtained with probability $p_i = \langle \varphi | M_i | \varphi \rangle$. A special case are projective measurements such as the measurement in the computational basis of the state $|\varphi\rangle = \sum_x \alpha_x |x\rangle$ which yields outcome x with probability $|\alpha_x|^2$. Measurements can refer to a subset of quantum registers and are in general not invertible.

We model a quantum algorithm $\mathcal{A}_Q$ with access to oracles $O_1, O_2, \ldots$ by a sequence of unitary transformations $U_1, O_1, U_2, \ldots, O_{T-1}, U_T$ over $m = \text{poly}(n)$ qubits. Here, oracle function $O_i : \{0,1\}^a \to \{0,1\}^b$ maps the final $a + b$ qubits from basis state $|x\rangle |y\rangle$ to $|x\rangle |y \oplus O_i(x)\rangle$ for $x \in \{0,1\}^a$ and $y \in \{0,1\}^b$. This mapping is inverse to itself. We can let the oracles share (secret) state by reserving some qubits for the O_i's only, on which the U_j's cannot operate. Note that the algorithm $\mathcal{A}_Q$ may also receive some (quantum) input $|\psi\rangle$. The adversary may also perform measurements. We sometimes write $\mathcal{A}_Q^{|O_1(\cdot)\rangle, |O_2(\cdot)\rangle, \ldots}(|\psi\rangle)$ for the output.

To introduce asymptotics we assume that $\mathcal{A}_Q$ is actually a sequence of such transformation sequences, indexed by parameter n, and that each transformation sequence is composed out of quantum systems for input, output, oracle calls, and work space (of sufficiently many qubits). To measure polynomial running time, we assume that each U_i is approximated (to sufficient precision) by members of a set of universal gates (say, Hadamard, phase, CNOT and $\pi/8$; for sake of concreteness [32]), where at most polynomially many gates are used. Furthermore, $T = T(n)$ is assumed to be polynomial, too.

QUANTUM RANDOM ORACLES. We can now define the quantum random-oracle model by picking a random function H for a given domain and range, and letting (a subset of) the oracles O_i evaluate H on the input in superposition, namely those O_i's which correspond to hash oracle queries. In this case the quantum adversary can evaluate the hash function in parallel for many inputs by querying the oracle about $\sum_x \alpha_x |x\rangle$ and obtaining $\sum_x \alpha_x |H(x)\rangle$, appropriately encoded as described above. Note that the output distribution $\mathcal{A}_Q^{|O_1(\cdot)\rangle, |O_2(\cdot)\rangle, \dots}(|\psi\rangle)$ now refers to the $\mathcal{A}_Q$'s measurements and the choice of H (and the random choices for the other oracles, if existing).

2.2 Classical Interactive Proofs of Knowledge

Here, we review the basic definition of Σ-protocols and show the classical Fiat-Shamir transformation which converts the interactive Σ-protocols into non-interactive proof of knowledge (PoK) protocols (in the random-oracle model). Let $\mathcal{L} \in \mathcal{NP}$ be a language with a (polynomially computable) relation $\mathcal{R}$, i.e., $x \in \mathcal{L}$ if and only if there exists some $w \in \{0,1\}^*$ such that $\mathcal{R}(x,w) = 1$ and $|w| = poly(|x|)$ for any x. As usual, w is called a witness for $x \in \mathcal{L}$ (and x is sometimes called a "theorem" or statement). We sometimes use the notation $\mathcal{R}_\lambda$ to denote the set of pairs (x,w) in $\mathcal{R}$ of some complexity related to the security parameter, e.g., if $|x| = \lambda$.

Σ-PROTOCOLS. The well-known class of Σ-protocols between a prover $\mathcal{P}$ and a verifier $\mathcal{V}$ allows $\mathcal{P}$ to convince $\mathcal{V}$ that it knows a witness w for a public theorem $x \in \mathcal{L}$, without giving $\mathcal{V}$ non-trivially computable information beyond this fact. Informally, a Σ-protocol consists of three messages $(\mathsf{com}, \mathsf{ch}, \mathsf{rsp})$ where the first message com is sent by $\mathcal{P}$ and the challenge ch is sampled uniformly from a challenge space by the verifier. We write $(\mathsf{com}, \mathsf{ch}, \mathsf{rsp}) \leftarrow \langle \mathcal{P}(x,w), \mathcal{V}(x) \rangle$ for the randomized output of an interaction between $\mathcal{P}$ and $\mathcal{V}$. We denote individual messages of the (stateful) prover in such an execution by $\mathsf{com} \leftarrow \mathcal{P}(x,w)$ and $\mathsf{rsp} \leftarrow \mathcal{P}(x,w,\mathsf{com},\mathsf{ch})$, respectively. Analogously, we denote the verifier's steps by $\mathsf{ch} \leftarrow \mathcal{V}(x,\mathsf{com})$ and $d \leftarrow \mathcal{V}(x,\mathsf{com},\mathsf{ch},\mathsf{rsp})$ for the challenge step and the final decision.

Definition 1 (Σ-Protocol). *A Σ-protocol $(\mathcal{P}, \mathcal{V})$ for an $\mathcal{NP}$-relation $\mathcal{R}$ satisfies the following properties:*

COMPLETENESS. *For any security parameter λ, any $(x,w) \in \mathcal{R}_\lambda$, any $(\mathsf{com}, \mathsf{ch}, \mathsf{rsp}) \leftarrow \langle \mathcal{P}(x,w), \mathcal{V}(x) \rangle$ it holds $\mathcal{V}(x,\mathsf{com},\mathsf{ch},\mathsf{rsp}) = 1$.*

PUBLIC-COIN. *For any security parameter λ, any $(x, w) \in \mathcal{R}_\lambda$, and any com $\leftarrow$ $\mathcal{P}(x, w)$, the challenge ch $\leftarrow \mathcal{V}(x, \mathsf{com})$ is uniform on $\{0, 1\}^{\ell(\lambda)}$ where ℓ is some polynomial function.*

SPECIAL SOUNDNESS. *Given $(\mathsf{com}, \mathsf{ch}, \mathsf{rsp})$ and $(\mathsf{com}, \mathsf{ch}', \mathsf{rsp}')$ for $x \in \mathcal{L}$ (with ch $\neq$ ch') where $\mathcal{V}(x, \mathsf{com}, \mathsf{ch}, \mathsf{rsp}) = \mathcal{V}(x, \mathsf{com}, \mathsf{ch}', \mathsf{rsp}') = 1$, there exists a PPT algorithm Ext (the extractor) which for any such input outputs a witness $w \leftarrow \mathsf{Ext}(x, \mathsf{com}, \mathsf{ch}, \mathsf{rsp}, \mathsf{ch}', \mathsf{rsp}')$ for x satisfying $\mathcal{R}(x, w) = 1$.*

HONEST-VERIFIER ZERO-KNOWLEDGE (HVZK). *There exists a PPT algorithm Sim (the zero-knowledge simulator) which, on input $x \in \mathcal{L}$, outputs a transcript $(\mathsf{com}, \mathsf{ch}, \mathsf{rsp})$ that is computationally indistinguishable from a valid transcript derived in a $\mathcal{P}$-$\mathcal{V}$ interaction. That is, for any polynomial-time quantum algorithm $\mathcal{D} = (\mathcal{D}_0, \mathcal{D}_1)$ the following distributions are indistinguishable:*

- *Let $(x, w, \mathsf{state}) \leftarrow \mathcal{D}_0(1^\lambda)$. If $\mathcal{R}(x, w) = 1$, then $(\mathsf{com}, \mathsf{ch}, \mathsf{rsp}) \leftarrow \langle \mathcal{P}(x, w), \mathcal{V}(x) \rangle$; else, $(\mathsf{com}, \mathsf{ch}, \mathsf{rsp}) \leftarrow \bot$.
   Output $\mathcal{D}_1(\mathsf{com}, \mathsf{ch}, \mathsf{rsp}, \mathsf{state})$.*
- *Let $(x, w, \mathsf{state}) \leftarrow \mathcal{D}_0(1^\lambda)$. If $\mathcal{R}(x, w) = 1$, then $(\mathsf{com}, \mathsf{ch}, \mathsf{rsp}) \leftarrow \mathsf{Sim}(x)$; else, $(\mathsf{com}, \mathsf{ch}, \mathsf{rsp}) \leftarrow \bot$.
   Output $\mathcal{D}_1(\mathsf{com}, \mathsf{ch}, \mathsf{rsp}, \mathsf{state})$.*

Here, state can be a quantum state.

FIAT-SHAMIR (FS) TRANSFORMATION. The Fiat-Shamir transformation of a Σ-protocol $(\mathcal{P}, \mathcal{V})$ is the same protocol but where the computation of ch is done as ch $\leftarrow H(x, \mathsf{com})$ instead of $\leftarrow \mathcal{V}(x, \mathsf{com})$. Here, H is a public hash function which is usually modeled as a random oracle, in which case we speak of the Fiat-Shamir transformation of $(\mathcal{P}, \mathcal{V})$ in the random-oracle model. Note that we include x in the hash computation, but all of our results remain valid if x is omitted from the input. If applying the FS transformation to a (passively-secure) identification protocol one obtains a signature scheme, if the hash computation also includes the message m to be signed.

2.3   Quantum Extractors and the FS Transform

QUANTUM EXTRACTORS IN THE QROM. Next, we describe a black-box quantum extractor. Roughly, this extractor should be able to output a witness w for a statement x given black-box access to the adversarial prover. There are different possibilities to define this notion, e.g., see the discussion in [38]. Here, we take a simple approach which is geared towards the application of the FS transform to build secure signature schemes. Namely, we assume that, if a quantum adversary $\mathcal{A}_Q$ on input x and with access to a quantum-accessible random oracle has a non-negligible probability of outputting a valid proof $(\mathsf{com}, \mathsf{ch}, \mathsf{rsp})$, then there is an extractor $\mathcal{K}_Q$ which on input x and with black-box access to $\mathcal{A}_Q$ outputs a valid witness with non-negligible probability, too.

We need to specify how the extractor simulates the quantum-accessible random oracle. This time we view the extractor $\mathcal{K}_Q$ as a sequence of unitary transformations $U_1, U_2, U_3, \ldots$, interleaved with interactions with the adversary $\mathcal{A}_Q$, now represented as the sequence of (stateful) oracles $O_1, O_2, \ldots$ to which $\mathcal{K}_Q$ has access to. Here each O_i corresponds to the local computations of the adversary until the "next interaction with the outside world". In our case this will be basically the hash queries $|\varphi\rangle$ to the quantum-accessible random oracle. We stipulate $\mathcal{K}_Q$ to write the (circuit description of a) classical function h with the expected input/output length, and which we assume for the moment to be deterministic, in some register before making the next call to an oracle. Before this call is then actually made, the hash function h is first applied to the quantum state $|\varphi\rangle = \sum_x \alpha_x |x\rangle |0\rangle$ of the previous oracle in the sense that the next oracle is called with $\sum_x \alpha_x |x\rangle |h(x)\rangle$. Note that we can enforce this behavior formally by restricting $\mathcal{K}_Q$'s steps $U_1, U_2, \ldots$ to be of this described form above.

At some point the adversary will return some classical proof (com, ch, rsp) for x. To allow the extractor to rewind the adversary we assume that the extractor can invoke another run with the adversary (for the same randomness, or possibly fresh randomness, appropriately encoded in the behavior of oracles). If the reduction asks to keep the same randomness then since the adversary only receives classical input x, this corresponds to a reset to the initial state. Since we do not consider adversaries with auxiliary quantum input, but only with classical input, such resets are admissible.

For our negative result we assume that the adversary does not perform any measurements before eventually creating the final output, whereas our positive result also works if the adversary measures in between. This is not a restriction, since in the meta-reduction technique we are allowed to choose a specific adversary, without having to consider more general cases. Note that the intrinsic "quantum randomness" of the adversary is fresh for each rewound run but, for our negative result, since measurements of the adversary are postponed till the end, the extractor can re-create the same quantum state as before at every interaction point. Also note that the extractor can measure any quantum query of the adversary to the random oracle but then cannot continue the simulation of this instance (unless the adversary chose a classical query in the first place). The latter reflects the fact that the extractor cannot change the quantum input state for answering the adversary's queries to the random oracle.

In summary, the black-box extractor can: (a) run several instances of the adversary from the start for the same or fresh classical randomness, possibly reaching the same quantum state as in previous executions when the adversary interacts with external oracles, (b) for each query to the QRO either measure and abort this execution, or provide a hash function h, and (c) observe the adversary's final output. The black-box extractor cannot, for instance, interfere with the adversary's program and postpone or perform additional measurements, nor rewind the adversary between interactions with the outside world, nor tamper with the internal state of the adversary. As a consequence, the extractor cannot observe the adversary's queries, but we still allow the extractor to access queries if these are classical. In particular, the extractor may choose h adaptively but not based on quantum queries (only on classical queries). We motivate this model with the observation that, in meaningful scenarios, the extractor should only be able to give a classical description of h, which is then "quantum-implemented" by the adversary $\mathcal{A}_Q$ through a "quantum programmable oracle gate"; the gate itself will be part of the adversary's circuit, and hence will be outside the extractor's influence. Purification of the adversary is also not allowed, since this would discard those adversaries which perform measurements, and would hence hinder the notion of black-box access.

For an interesting security notion computing a witness from x only should be infeasible, even for a quantum adversary. To this end we assume that there is an efficient instance generator Inst which on input 1^λ outputs a pair $(x, w) \in \mathcal{R}$ such that any polynomial-time quantum algorithm on (classical) input x returns some classical string w' with $(x, w') \in \mathcal{R}$, is negligible (over the random choices of Inst and the quantum algorithm). We say Inst is a *hard instance generator for relation* $\mathcal{R}$.

Definition 2 (Black-Box Extractor for Σ-Protocol in the QROM). *Let $(\mathcal{P}, \mathcal{V})$ be a Σ-protocol for an $\mathcal{NP}$-relation $\mathcal{R}$ with hard instance generator Inst. Then a black-box extractor $\mathcal{K}_Q$ is a polynomial-time quantum algorithm (as above) such that for any quantum adversary $\mathcal{A}_Q$ with quantum access to oracle H, it holds that, if*

$$\mathrm{Prob}\left[\mathcal{V}^H(x, \mathsf{com}, \mathsf{ch}, \mathsf{rsp}) = 1 \text{ for } (x, w) \leftarrow \mathsf{Inst}(1^\lambda); (\mathsf{com}, \mathsf{ch}, \mathsf{rsp}) \leftarrow \mathcal{A}_Q^{|H\rangle}(x)\right] \not\approx 0$$

is not negligible, then

$$\mathrm{Prob}\left[(x, w') \in \mathcal{R} \text{ for } (x, w) \leftarrow \mathsf{Inst}(1^\lambda); w' \leftarrow \mathcal{K}_Q^{\mathcal{A}_Q}(x)\right] \not\approx 0$$

is also not negligible.

For our negative (and our positive) results we look at special cases of black-box extractors, denoted *input-respecting* extractors. This means that the extractor only runs the adversary on the given input x. All known extractors are of this kind, and in general it is unclear how to take advantage of executions for different x'.

ON PROBABILISTIC HASH FUNCTIONS. We note that we could also allow the extractor to output a description of a *probabilistic* hash function h to answer each random oracle call. This means that, when evaluated for some string x, the reply is $y = h(x; r)$ for some randomness r (which is outside of the extractor's control). In this sense a query $|\varphi\rangle = \sum_x \alpha_x |x\rangle |0\rangle$ in superposition returns $|\varphi\rangle = \sum_x \alpha_x |x\rangle |h(x; r_x)\rangle$ for independently chosen r_x for each x.

We can reduce the case of probabilistic functions h to deterministic ones, if we assume quantum-accessible pseudorandom functions [8]. These functions are indistinguishable from random functions for quantum adversaries, even if queried in superposition. In our setting, in the deterministic case the extractor incorporates the description of the pseudorandom function for a randomly chosen key κ into the description of the deterministic hash function, $h'(x) = h(x; \mathsf{PRF}_\kappa(x))$. Since the hash function description is not presented to the adversary, using such derandomized hash functions cannot decrease the extractor's success probability significantly. This argument can be carried out formally by a reduction to the quantum-accessible pseudorandom function, i.e., by forwarding each query $|\varphi\rangle$ of the QROM adversary to the random or pseudorandom function oracle, and evaluating h as before on x and the oracle's reply. Using a general technique in [41] we can even replace the assumption about the pseudorandom function and use a q-wise independent function instead.

3 Impossibility Result for Quantum-Fiat-Shamir

We use meta-reductions techniques to show that, if the Fiat-Shamir transformation applied to the identification protocol would support a knowledge extractor, then we would obtain a contradiction to the active security. That is, we first build an all-powerful quantum adversary $\mathcal{A}_Q$ successfully generating accepted proofs. Coming up with such an adversary is necessary to ensure that a black-box extractor $\mathcal{K}_Q$ exists in the first place; Definition 2 only requires $\mathcal{K}_Q$ to succeed *if* there is some successful adversary $\mathcal{A}_Q$. The adversary $\mathcal{A}_Q$ uses its unbounded power to find a witness w to its input x, and then uses the quantum access to the random oracle model to "hide" its actual query in a superposition. The former ensures that that our adversary is trivially able to construct a valid proof by emulating the prover for w, the latter prevents the extractor to apply the rewinding techniques of Pointcheval and Stern [33] in the classical setting. Once we have designed our adversary $\mathcal{A}_Q$ and ensured the existence of $\mathcal{K}_Q$, we wrap $\mathcal{K}_Q$ into a reduction $\mathcal{M}_Q$ which takes the role of $\mathcal{A}_Q$ and breaks active security. The (quantum) meta-reduction now plays against the honest prover of the identification scheme "on the outside", using the extractor "on the inside". In this inner interaction $\mathcal{M}_Q$ needs to emulate our all-powerful adversary $\mathcal{A}_Q$ towards the extractor, but this needs to be done efficiently in order to make sure that the meta-reduction (with its inner interactions) is efficient.

In the argument below we assume that the extractor is input-respecting (i.e., forwards x faithfully to the adversary). In this case we can easily derandomize the adversary (with respect to classical randomness) by "hardwiring" a key of a random function into it, which it initially applies to its input x to recover the same classical randomness for each run. Since the extractor has to work for all adversaries, it in particular needs to succeed for those where we pick the function randomly but fix it from thereon.

3.1 Assessment

Before we dive into the technical details of our result let us re-assess the strength and weaknesses of our impossibility result:

1. The extractor has to choose a classical hash function h for answering QRO queries. While this may be considered a restriction in general interactive quantum proofs, it seems to be inevitable in the QROM; it is rather a consequence of the approach where a quantum adversary mounts attacks in a classical setting. After all, both the honest parties as well as the adversary expect

a classical hash function. The adversary is able to check this property easily, even if it treats the hash function otherwise as a black box (and may thus not be able to spot that the hash function uses (pseudo)randomness). We remark again that this approach also complies with previous efforts [8,41,40,9] and our positive result here to answer such hash queries.

2. The extractor *can* rewind the quantum adversary to any point before the final measurement. Recall that for our impossibility result we assume, to the advantage of the extractor, that the adversary does not perform any measurement until the very end. Since the extractor can re-run the adversary from scratch for the same classical randomness, and the "no-cloning restriction" does not apply to our adversary with classical input, the extractor can therefore easily put the adversary in the same (quantum) state as in a previous execution, up to the final measurement. However, because we consider *black-box* extractors, the extractor can only influence the adversary's behavior via the answers it provides to $\mathcal{A}_Q$'s external communication. In this sense, the extractor may always rewind the adversary to such communication points. We also allow the extractor to measure and abort at such communication points.

3. The extraction strategy by Pointcheval and Stern [33] in the purely classical case *can* be cast in our black-box extractor framework. For this the extractor would run the adversary for the same classical randomness twice, providing a lazy-sampling based hash function description, with different replies in the i-th answers in the two runs. The extractor then extracts the witness from two valid signatures. This shows that a different approach than in the classical setting is necessary for extractors in the QROM.

3.2   Prerequisites

Witness-Independent Commitments. We first identify a special subclass of Σ-protocols which our result relies upon:

Definition 3 (Σ-protocols with witness-independent commitment). *A Σ-protocol has* witness-independent commitments *if the prover's commitment* com *does not depend on the witness w. That is, we assume that there is a PPT algorithm* COM *which, on input x and some randomness r, produces the same distribution as the prover's first message for input (x, w).*

Examples of such Σ-protocols are the well known graph-isomorphism proof [21], the Schnorr proof of knowledge [37], or the recent protocol for lattices used in an anonymous credential system [11]. A typical example of non-witness-independent commitment Σ-protocol is the graph 3-coloring ZKPoK scheme [21] where the prover commits to a random permutation of the coloring.

We note that perfectly hiding commitments do not suffice for our negative result. We need to be able to generate (the superposition of) all commitments without knowledge of the witness.

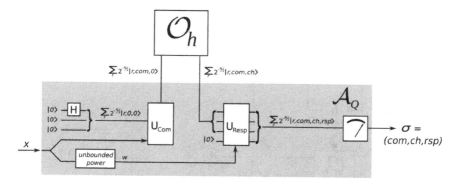

Fig. 1. The canonical adversary

WEAK SECURITY AGAINST ACTIVE QUANTUM ADVERSARIES. We next describe the underlying security of (non-transformed) Σ-protocols against a weak form of active attacks where the adversary may use quantum power but needs to eventually compute a witness. That is, we let $\mathcal{A}_Q^{\mathcal{P}(x,w)}(x)$ be a quantum adversary which can interact classically with several prover instances. The prover instances can be invoked in sequential order, each time the prover starts by computing a fresh commitment com $\leftarrow \mathcal{P}(x, w)$, and upon receiving a challenge ch $\in \{0, 1\}^\ell$ it computes the response rsp. Only if it has returned this response $\mathcal{P}$ can be invoked on a new session again. We say that the adversary *succeeds in an active attack* if it eventually returns some w' such that $(x, w') \in \mathcal{R}$.

For an interesting security notion computing a witness from x only should be infeasible, even for a quantum adversary. To this end we assume that there is an efficient instance generator Inst which on input 1^λ outputs a pair $(x, w) \in \mathcal{R}$ such that any polynomial-time quantum algorithm on (classical) input x returns some classical string w' with $(x, w') \in \mathcal{R}$, is negligible (over the random choices of Inst and the quantum algorithm). We say Inst is a *hard instance generator for relation* $\mathcal{R}$.

Definition 4 (Weakly Secure Σ-Protocol Against Active Quantum Adversaries). *A Σ-protocol $(\mathcal{P}, \mathcal{V})$ for an $\mathcal{NP}$-relation $\mathcal{R}$ with hard instance generator Inst is weakly secure against active quantum adversaries if for any polynomial-time quantum adversaries $\mathcal{A}_Q$ the probability that $\mathcal{A}_Q^{\mathcal{P}(x,w)}(x)$ succeeds in an active attack for $(x, w) \leftarrow$ Inst(1^λ) is negligible (as a function of λ).*

We call this property weak security because it demands the adversary to compute a witness w', instead of passing only an impersonation attempt. If the adversary finds such a witness, then completeness of the scheme implies that it can successfully impersonate. In this sense we put more restrictions on the adversary and, thus, weaken the security guarantees.

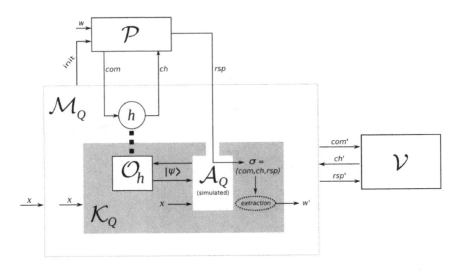

Fig. 2. An overview of our meta-reduction

3.3 The Adversary and the Meta-reduction

ADVERSARY. Our (unbounded) adversary works roughly as follows (see Figure 1). It receives as input a value x and first uses its unbounded computational power to compute a random witness w (according to uniform distributions of coin tosses ω subject to $\mathsf{Inst}(1^n; \omega) = (x, w)$, but where ω is a random function of x). Then it prepares all possible random strings $r \in \{0, 1\}^N$ (where $N = \mathrm{poly}(n)$) for the prover's algorithm in superposition. It then evaluates (a unitary version of) the classical function $\mathrm{COM}()$ for computing the prover's commitment on this superposition (and on x) to get a superposition of all $|r\rangle |\mathsf{com}_{x,r}\rangle$. It evaluates the random oracle H on the com-part, i.e., to be precise, the hash values are stored in ancilla bits such that the result is a superposition of states $|r\rangle |\mathsf{com}_{x,r}\rangle |H(x, \mathsf{com}_{x,r})\rangle$. The adversary computes, in superposition, responses for all values and finally measures in the computational basis, yielding a sample $(r, \mathsf{com}_{x,r}, \mathsf{ch}, \mathsf{rsp}_{x,w,r})$ for $\mathsf{ch} = H(x, \mathsf{com}_{x,r})$ where r is uniform over all random strings; it outputs the transcript $(\mathsf{com}, \mathsf{ch}, \mathsf{rsp})$.

THE META-REDUCTION. We illustrate the meta-reduction in Figure 2. Assume that there exists a (quantum) black-box extractor $\mathcal{K}_Q$ which on input x, sampled according to Inst, and which is also given to $\mathcal{A}_Q$, is able to extract a witness w to x by running several resetting executions of $\mathcal{A}_Q$, each time answering $\mathcal{A}_Q$'s (only) random oracle query $|\varphi\rangle$ by supplying a classical, possibly probabilistic function h. We then build a (quantum) meta-reduction $\mathcal{M}_Q$ which breaks the weak security of the identification scheme in an active attack when communicating with the classical prover.

    The quantum meta-reduction $\mathcal{M}_Q$ receives as input the public statement x. It forwards it to $\mathcal{K}_Q$ and waits until $\mathcal{K}_Q$ invokes $\mathcal{A}_Q(x)$, which is now simulated by $\mathcal{M}_Q$. For each (reset) execution the meta-reduction skips the step where the adversary would compute the witness, and instead immediately computes the same superposition query $|r\rangle |com_{x,r}\rangle$ as $\mathcal{A}_Q$ and outputs it to $\mathcal{K}_Q$. When $\mathcal{K}_Q$ creates (a description of) the possibly probabilistic function h we let $\mathcal{M}_Q$ initiate an interaction with the prover to receive a classical sample $com_{x,r}$, on which it evaluates h to get a challenge ch. Note that $\mathcal{M}_Q$ in principle does not need a description of h for this, but only a possibility to compute h once. The meta-reduction forwards the challenge to the prover to get a response rsp. It outputs (com, ch, rsp) to the reduction. If the reduction eventually outputs a potential witness w' then $\mathcal{M}_Q$ uses this value w' to break the weak security.

3.4 Analysis

For the analysis note that the extractor's perspective in each execution is identical in both cases, when interacting with the actual adversary $\mathcal{A}_Q$, or when interacting with the meta-reduction $\mathcal{M}_Q$. The reason is that the commitments are witness-independent such that the adversary (using its computational power to first compute a witness) and the meta-reduction computing the commitments without knowledge of a witness, create the same distribution on the query to the random oracle. Since up to this point the extractor's view is identical in both runs, its distribution on h is also the same in both cases. But then the quantum adversary internally computes, in superposition over all possible random strings r, the challenge ch $\leftarrow h(x, com_{x,r})$ and the response $rsp_{x,w,r}$ for $x, w,$ and ch. It then measures r in the computational basis, such that the state collapses to a classical tuple $(com_{x,r}, ch, rsp_{x,w,r})$ over uniformly distributed r. Analogously, the meta-reduction, upon receiving h (with the same distribution as in $\mathcal{A}_Q$'s attack), receives from the prover a commitment $com_{x,r}$ for a uniformly distributed r. It then computes ch $\leftarrow h(x, com_{x,r})$ and obtains $rsp_{x,w,r}$ from the prover, which is determined by x, w, r and ch. It returns $(com_{x,r}, ch, rsp_{x,w,r})$ for such a uniform r.

In other words, $\mathcal{M}_Q$ considers only a single classical execution (with r sampled at the outset), whereas $\mathcal{A}_Q$ basically first runs everything in superposition and only samples r at the very end. Since all the other computations in between are classical, the final results are identically distributed. Furthermore, since the extractor is input-respecting, the meta-reduction can indeed answer all runs for the very same x with the help of the external prover (which only works for x). Analogously, the fact that the adversary always chooses, and uses, the same witness w in all runs, implies that the meta-reduction can again rely on the external prover with the single witness w.

Since the all-powerful adversary succeeds with probability 1 in the original experiment, to output a valid proof given x and access to a quantum random oracle only, the extractor must also succeed with non-negligible probability in extracting a witness. Hence, $\mathcal{M}_Q$, too, succeeds with non-negligible probability in an active attack against weak security. Furthermore, since $\mathcal{K}_Q$ runs in polynomial time, $\mathcal{M}_Q$ invokes at most a polynomial number of interactions with the external prover. Altogether, we thus obtain the following theorem:

Theorem 1 (Impossibility Result). *For any Σ-protocol $(\mathcal{P}, \mathcal{V})$ with witness-independent commitments, and which is weakly secure against active quantum adversaries, there does not exist an input-preserving black-box quantum knowledge extractor for $(\mathcal{P}, \mathcal{V})$.*

We note that our impossibility result is cast in terms of proofs of knowledge, but can be easily adapted for the case of signatures. In fact, the adversary $\mathcal{A}_Q$ would be able to compute a valid proof (i.e., a signature) for any given message m which it receives as additional input to x.

OUR META-REDUCTION AND CLASSICAL QUERIES TO THE RANDOM ORACLE. One might ask why the meta-reduction does not apply to the Fiat-Shamir transform when adversaries have only classical access to the random oracle. The reason is the following: if the adversary made a classical query about a single commitment (and so would the meta-reduction), then one could apply the rewinding technique of Pointcheval and Stern [33] changing the random oracle answers, and extract the underlying witness via special soundness of the identification scheme. The quantum adversary here, however, queries the random oracle in a superposition. In this scenario, as we explained above, the extractor is not allowed to "read" the query of the adversary unless it makes the adversary stop. In other words, the extractor cannot measure the query and then keep running the adversary until a valid witness is output. This intrinsic property of black-box quantum extractors, hence, makes "quantum" rewinding impossible. Note that rewinding in the classical sense —as described by Pointcheval and Stern [33]— is still possible, as this essentially means to start the adversary with the same random coins. One may argue that it might be possible to measure the query state without disturbing $\mathcal{A}_Q$'s behavior significantly, but as we already pointed out, this would lead to a non-black-box approach —vastly more powerful than the classical read-only access.

ON THE NECESSITY OF ACTIVE SECURITY. If we drop the requirement on active security we can indeed devise a solution based on quantum-immune primitives. Namely, we use the (classical) non-interactive zero-knowledge proofs of knowledge of De Santis and Persiano [16] to build the following three-move scheme: The first message is irrelevant, e.g., we let the prover simply send the constant 0 (potentially padded with redundant randomness), making the commitment also witness-independent. In the second message the verifier sends a random string which the prover interprets as a public key pk of a dense encryption scheme and a common random string crs for the NIZK. The prover encrypts the witness under pk and gives a NIZK that the encrypted value forms a valid witness for the public value x. The verifier only checks the NIZK proof.

The protocol is clearly not secure against active (classical) adversaries because such an adversary can create a public key pk via the key generation algorithm, thus, knowing the secret key and allowing the adversary to recover the witness from a proof by the prover. It is, however, honest-verifier zero-knowledge, even against quantum distinguishers if the primitives are quantum-secure, because then the IND-CPA security and the simulatability of the NIZK hide the witness and allow for a simulation. We omit a more formal argument here, as it will be covered as a special case from our general result in the next section.

4   Positive Results for Quantum-Fiat-Shamir

In Section 3.4 we have sketched a generic construction of a Σ-protocol based on NIZKPoKs [16] which can be converted to a secure NIZK-PoK against quantum adversaries in the QROM via the Fiat-Shamir (FS) paradigm. While the construction is rather inefficient and relies on additional primitives and assumptions, it shows the path to a rather efficient solution: drop the requirement on active security and let the (honest) verifier choose the commitment obliviously, i.e., such that it does not know the pre-image, together with the challenge. If the prover is able to use a trapdoor to compute the commitment's pre-image then it can complete the protocol as before.

4.1 Σ-Protocols with Oblivious Commitments

The following definition captures the notion of Σ-protocols with oblivious commitments formally.

Definition 5 (Σ-protocols with Oblivious Commitments). *A Σ-protocol $(\mathcal{P}, \mathcal{V})$ has* oblivious commitments *if there are PPT algorithms* COM *and* SMPLRND *such that for any $(x, w) \in \mathcal{R}$ the following distributions are statistically close:*

- *Let* com $=$ COM$(x; \rho)$ *for* $\rho \leftarrow \{0, 1\}^\lambda$, ch $\leftarrow \mathcal{V}(x, \text{com})$, *and* rsp $\leftarrow \mathcal{P}(x, w, \text{com}, \text{ch})$. *Output* $(x, w, \rho, \text{com}, \text{ch}, \text{rsp})$.
- *Let* $(x, w, \rho, \text{com}, \text{ch}, \text{rsp})$ *be a transcript of a protocol run between $\mathcal{P}(x, w)$ and $\mathcal{V}(x)$, where $\rho \leftarrow$ SMPLRND(x, com).*

Note that the prover is able to compute a response from the given commitment com without knowing the randomness used to compute the commitment. This is usually achieved by placing some extra trapdoor into the witness w. For example, for the Guillou-Quisquater RSA based proof of knowledge [24] where the prover shows knowledge of $w \in \mathbb{Z}_N^*$ with $w^e = y \bmod N$ for $x = (e, N, y)$, the prover would need to compute an e-th root for a given commitment $R \in \mathbb{Z}_N^*$. If the witness would contain the prime factorization of N, instead of the e-th root of y, this would indeed be possible.

Σ-protocols with oblivious commitments allow to move the generation of the commitment from the prover to the honest verifier. For most schemes this infringes with active security, because a malicious verifier could generate the commitment "non-obliviously". However, the scheme remains honest-verifier zero-knowledge, and this suffices for deriving secure signature schemes. In particular, using random oracles one can hash into commitments by computing the random output of the hash function and running COM$(x; \rho)$ on this random string ρ to sample a commitment obliviously.

In the sequel we therefore often identify ρ with $\text{COM}(x; \rho)$ in the sense that we assume that the hash function maps to $\text{COM}(x; \rho)$ directly. The existence of SMPLRND guarantees that we could "bend" this value back to the actual pre-image ρ. In fact, for our positive result it would suffice that the distributions are computationally indistinguishable for random $(x, w) \leftarrow \text{Inst}(1^n)$ against quantum distinguishers.

4.2   FS Transformation for Σ-Protocols with Oblivious Commitments

We explain the FS transformation for schemes with oblivious commitments for signatures only; the case of (simulation-sound) NIZK-PoKs is similar, the difference is that for signatures the message is included in the hash computation for signature schemes. For sake of concreteness let us give the full description of the transformed signature scheme. We note that for the transformation we also include a random string r in the hash computation (chosen by the signer). Jumping ahead, we note that this source of entropy ensures simulatability of signatures; for classical Σ-protocols this is usually given by the entropy of the initial commitment but which has been moved to the verifier here. Recall from the previous section that we simply assume that we can hash into commitments directly, instead of going through the mapping via COM and SMPLRND.

Construction 2. *Let $(\mathcal{P}, \mathcal{V})$ be a Σ-protocol for relation $\mathcal{R}$ with oblivious commitments and instance generator Inst. Then construct the following signature scheme $\mathcal{S} = (SKGen, Sig, SVf)$ in the (quantum) random-oracle model:*

KEY GENERATION. *$SKGen(1^\lambda)$ runs $(x, w) \leftarrow \text{Inst}(1^\lambda)$ and returns $sk = (x, w)$ and $pk = x$.*

SIGNING. *For message $m \in \{0, 1\}^*$ the signing algorithm Sig^H on input sk, picks random $r \stackrel{\$}{\leftarrow} \text{RND}$ from some superpolynomial space, computes $(\textsf{com}, \textsf{ch}) = H(pk, m, r)$, and obtains $\textsf{rsp} \leftarrow \mathcal{P}(pk, sk, \textsf{com}, \textsf{ch})$. The output is the signature $\sigma = (r, \textsf{com}, \textsf{ch}, \textsf{rsp})$.*

VERIFICATION. *On input pk, m, and $\sigma = (r, \textsf{com}, \textsf{ch}, \textsf{rsp})$ the verification algorithm Vf^H outputs 1 iff $\mathcal{V}(pk, \textsf{com}, \textsf{ch}, \textsf{rsp}) = 1$ and $(\textsf{com}, \textsf{ch}) = H(pk, m, r)$; else, it returns 0.*

Note that one can shorten the signature size by simply outputting $\sigma = (r, \textsf{rsp})$. The remaining components $(\textsf{com}, \textsf{ch})$ are obtained by hashing the tuple (pk, m, r). Next, we give the main result of this section saying that the Fiat-Shamir transform on Σ-protocols with oblivious commitments yield a quantum-secure signature scheme.

Theorem 3. *If Inst is a hard instance generator for the relation $\mathcal{R}$ and the Σ-protocol $(\mathcal{P}, \mathcal{V})$ has oblivious commitments, then the signature scheme in Construction 2 is existentially unforgeable under chosen message attacks against quantum adversaries in the quantum-accessible random-oracle model.*

The idea is roughly as follows. Assume for the moment that we are only interested in key-only attacks and would like to extract the secret key from an adversary $\mathcal{A}_Q$ against the signature scheme. For given x we first run the honest-verifier zero-knowledge simulator of the Σ-protocol to create a transcript $(\mathsf{com}^\star, \mathsf{ch}^\star, \mathsf{rsp}^\star)$. We choose another random challenge $\mathsf{ch}' \leftarrow \{0,1\}^\ell$. Then, we run the adversary, injecting $(\mathsf{com}^\star, \mathsf{ch}')$ into the hash replies. This appropriate insertion will be based on techniques developed by Zhandry [41] to make sure that superposition queries to the random oracle are harmless. With sufficiently large probability the adversary will then output a proof $(\mathsf{com}^\star, \mathsf{ch}', \mathsf{rsp}')$ from which we can, together with $(\mathsf{com}^\star, \mathsf{ch}^\star, \mathsf{rsp}^\star)$ extract a witness due to the special-soundness property. Note that, if this extraction fails because the transcript $(\mathsf{com}^\star, \mathsf{ch}^\star, \mathsf{rsp}^\star)$ is only simulated, we could distinguish simulated signatures from genuine ones. We can extend this argument to chosen-message attacks by simulating signatures as in the classical case. This is the step where we take advantage of the extra random string r in order to make sure that the previous adversary's quantum hash queries have a negligible amplitude in this value (x, m, r). Using techniques from [6] we can show that changing the oracle in this case does not change the adversary's success probability significantly.

The full proof with preliminary results appears in the full version [14].

Moreover, we also discuss a concrete instantiation based on Lyubashevsky's lattice-based scheme [29] in the full version [14] to show that one can use our technique in principle, and how it could be used for other schemes.

Acknowledgments. We thank the anonymous reviewers for some valuable comments. We also thank Dominique Unruh for useful discussions on black-box extractors. Marc Fischlin is supported by the Heisenberg grant Fi 940/3-1 of the German Research Foundation (DFG). Özgür Dagdelen and Tommaso Gagliardoni are supported by the German Federal Ministry of Education and Research (BMBF) within EC-SPRIDE. This work was also supported by CASED (www.cased.de).

References

1. Abdalla, M., Fouque, P.-A., Lyubashevsky, V., Tibouchi, M.: Tightly-secure signatures from lossy identification schemes. In: Pointcheval, D., Johansson, T. (eds.) EUROCRYPT 2012. LNCS, vol. 7237, pp. 572–590. Springer, Heidelberg (2012)
2. Asharov, G., Jain, A., López-Alt, A., Tromer, E., Vaikuntanathan, V., Wichs, D.: Multiparty computation with low communication, computation and interaction via threshold FHE. In: Pointcheval, D., Johansson, T. (eds.) EUROCRYPT 2012. LNCS, vol. 7237, pp. 483–501. Springer, Heidelberg (2012)
3. Barreto, P.S.L.M., Misoczki, R.: A new one-time signature scheme from syndrome decoding. Cryptology ePrint Archive, Report 2010/017 (2010), http://eprint.iacr.org/
4. Bellare, M., Palacio, A.: GQ and Schnorr identification schemes: Proofs of security against impersonation under active and concurrent attacks. In: Yung, M. (ed.) CRYPTO 2002. LNCS, vol. 2442, pp. 162–177. Springer, Heidelberg (2002)

5. Bellare, M., Rogaway, P.: Random oracles are practical: A paradigm for designing efficient protocols. In: Ashby, V. (ed.) ACM CCS 1993: 1st Conference on Computer and Communications Security. pp. 62–73. ACM Press (November 1993)

6. Bennett, C.H., Bernstein, E., Brassard, G., Vazirani, U.V.: Strengths and weaknesses of quantum computing. SIAM J. Comput. 26(5), 1510–1523 (1997)

7. Bitansky, N., Dachman-Soled, D., Garg, S., Jain, A., Kalai, Y.T., López-Alt, A., Wichs, D.: Why "Fiat-shamir for proofs" lacks a proof. In: Sahai, A. (ed.) TCC 2013. LNCS, vol. 7785, pp. 182–201. Springer, Heidelberg (2013)

8. Boneh, D., Dagdelen, Ö., Fischlin, M., Lehmann, A., Schaffner, C., Zhandry, M.: Random oracles in a quantum world. In: Lee, D.H., Wang, X. (eds.) ASIACRYPT 2011. LNCS, vol. 7073, pp. 41–69. Springer, Heidelberg (2011)

9. Boneh, D., Zhandry, M.: Quantum-secure message authentication codes. Cryptology ePrint Archive, Report 2012/606 (2012), http://eprint.iacr.org/

10. Boneh, D., Zhandry, M.: Secure signatures and chosen ciphertext security in a quantum computing world. Cryptology ePrint Archive, Report 2013/088 (2013), http://eprint.iacr.org/

11. Camenisch, J., Neven, G., Rückert, M.: Fully anonymous attribute tokens from lattices. In: Visconti, I., De Prisco, R. (eds.) SCN 2012. LNCS, vol. 7485, pp. 57–75. Springer, Heidelberg (2012)

12. Cayrel, P.-L., Lindner, R., Rückert, M., Silva, R.: Improved zero-knowledge identification with lattices. In: Heng, S.-H., Kurosawa, K. (eds.) ProvSec 2010. LNCS, vol. 6402, pp. 1–17. Springer, Heidelberg (2010)

13. Cayrel, P.-L., Véron, P., El Yousfi Alaoui, S.M.: A zero-knowledge identification scheme based on the q-ary syndrome decoding problem. In: Biryukov, A., Gong, G., Stinson, D.R. (eds.) SAC 2010. LNCS, vol. 6544, pp. 171–186. Springer, Heidelberg (2011)

14. Dagdelen, Ö., Fischlin, M., Gagliardoni, T.: The fiat-shamir transformation in a quantum world. Cryptology ePrint Archive, Report 2013/245 (2013), http://eprint.iacr.org/

15. Damgård, I., Funder, J., Nielsen, J.B., Salvail, L.: Superposition attacks on cryptographic protocols. Cryptology ePrint Archive, Report 2011/421 (2011), http://eprint.iacr.org/

16. De Santis, A., Persiano, G.: Zero-knowledge proofs of knowledge without interaction (extended abstract). In: FOCS, pp. 427–436. IEEE Computer Society (1992)

17. Ducas, L., Durmus, A., Lepoint, T., Lyubashevsky, V.: Lattice signatures and bimodal gaussians. In: Canetti, R., Garay, J.A. (eds.) CRYPTO 2013, Part I. LNCS, vol. 8042, pp. 40–56. Springer, Heidelberg (2013)

18. Feige, U., Fiat, A., Shamir, A.: Zero-knowledge proofs of identity. Journal of Cryptology 1(2), 77–94 (1988)

19. Fiat, A., Shamir, A.: How to prove yourself: Practical solutions to identification and signature problems. In: Odlyzko, A.M. (ed.) CRYPTO 1986. LNCS, vol. 263, pp. 186–194. Springer, Heidelberg (1987)

20. Fischlin, M., Lehmann, A., Ristenpart, T., Shrimpton, T., Stam, M., Tessaro, S.: Random oracles with(out) programmability. In: Abe, M. (ed.) ASIACRYPT 2010. LNCS, vol. 6477, pp. 303–320. Springer, Heidelberg (2010)

21. Goldreich, O., Micali, S., Wigderson, A.: How to prove all NP-statements in zero-knowledge and a methodology of cryptographic protocol design. In: Odlyzko, A.M. (ed.) CRYPTO 1986. LNCS, vol. 263, pp. 171–185. Springer, Heidelberg (1987)

22. Goldwasser, S., Kalai, Y.T.: On the (in)security of the Fiat-Shamir paradigm. In: 44th FOCS Annual Symposium on Foundations of Computer Science, pp. 102–115. IEEE Computer Society Press (October 2003)

23. Gordon, S.D., Katz, J., Vaikuntanathan, V.: A group signature scheme from lattice assumptions. In: Abe, M. (ed.) ASIACRYPT 2010. LNCS, vol. 6477, pp. 395–412. Springer, Heidelberg (2010)
24. Guillou, L.C., Quisquater, J.-J.: A "Paradoxical" identity-based signature scheme resulting from zero-knowledge. In: Goldwasser, S. (ed.) CRYPTO 1988. LNCS, vol. 403, pp. 216–231. Springer, Heidelberg (1990)
25. Güneysu, T., Lyubashevsky, V., Pöppelmann, T.: Practical lattice-based cryptography: A signature scheme for embedded systems. In: Prouff, E., Schaumont, P. (eds.) CHES 2012. LNCS, vol. 7428, pp. 530–547. Springer, Heidelberg (2012)
26. Kawachi, A., Tanaka, K., Xagawa, K.: Concurrently secure identification schemes based on the worst-case hardness of lattice problems. In: Pieprzyk, J. (ed.) ASIACRYPT 2008. LNCS, vol. 5350, pp. 372–389. Springer, Heidelberg (2008)
27. Lyubashevsky, V.: Lattice-based identification schemes secure under active attacks. In: Cramer, R. (ed.) PKC 2008. LNCS, vol. 4939, pp. 162–179. Springer, Heidelberg (2008)
28. Lyubashevsky, V.: Fiat-Shamir with aborts: Applications to lattice and factoring-based signatures. In: Matsui, M. (ed.) ASIACRYPT 2009. LNCS, vol. 5912, pp. 598–616. Springer, Heidelberg (2009)
29. Lyubashevsky, V.: Lattice signatures without trapdoors. In: Pointcheval, D., Johansson, T. (eds.) EUROCRYPT 2012. LNCS, vol. 7237, pp. 738–755. Springer, Heidelberg (2012)
30. Melchor, C.A., Gaborit, P., Schrek, J.: A new zero-knowledge code based identification scheme with reduced communication. CoRR abs/1111.1644 (2011)
31. Micciancio, D., Vadhan, S.P.: Statistical zero-knowledge proofs with efficient provers: Lattice problems and more. In: Boneh, D. (ed.) CRYPTO 2003. LNCS, vol. 2729, pp. 282–298. Springer, Heidelberg (2003)
32. Nielsen, M.A., Chuang, I.L.: Quantum Computation and Quantum Information. Cambridge University Press (2000)
33. Pointcheval, D., Stern, J.: Security arguments for digital signatures and blind signatures. Journal of Cryptology 13(3), 361–396 (2000)
34. Sakumoto, K.: Public-key identification schemes based on multivariate cubic polynomials. In: Fischlin, M., Buchmann, J., Manulis, M. (eds.) PKC 2012. LNCS, vol. 7293, pp. 172–189. Springer, Heidelberg (2012)
35. Sakumoto, K., Shirai, T., Hiwatari, H.: Public-key identification schemes based on multivariate quadratic polynomials. In: Rogaway, P. (ed.) CRYPTO 2011. LNCS, vol. 6841, pp. 706–723. Springer, Heidelberg (2011)
36. Schnorr, C.P.: Efficient identification and signatures for smart cards. In: Brassard, G. (ed.) CRYPTO 1989. LNCS, vol. 435, pp. 239–252. Springer, Heidelberg (1990)
37. Schnorr, C.P.: Efficient signature generation by smart cards. Journal of Cryptology 4(3), 161–174 (1991)
38. Unruh, D.: Quantum proofs of knowledge. In: Pointcheval, D., Johansson, T. (eds.) EUROCRYPT 2012. LNCS, vol. 7237, pp. 135–152. Springer, Heidelberg (2012)
39. Watrous, J.: Zero-knowledge against quantum attacks. In: Kleinberg, J.M. (ed.) 38th ACM STOC Annual ACM Symposium on Theory of Computing, pp. 296–305. ACM Press (May 2006)
40. Zhandry, M.: How to construct quantum random functions. In: IEEE Annual Symposium on Foundations of Computer Science, pp. 679–687. IEEE Computer Society (2012)
41. Zhandry, M.: Secure identity-based encryption in the quantum random oracle model. In: Safavi-Naini, R., Canetti, R. (eds.) CRYPTO 2012. LNCS, vol. 7417, pp. 758–775. Springer, Heidelberg (2012)

On the Security of One-Witness Blind Signature Schemes

Foteini Baldimtsi and Anna Lysyanskaya[*]

Department of Computer Science, Brown University, Providence, RI, USA
{foteini,anna}@cs.brown.edu

Abstract. Blind signatures have proved an essential building block for
applications that protect privacy while ensuring unforgeability, i.e., electronic cash and electronic voting. One of the oldest, and most efficient
blind signature schemes is the one due to Schnorr that is based on his famous identification scheme. Although it was proposed over twenty years
ago, its unforgeability remains an open problem, even in the random-
oracle model. In this paper, we show that current techniques for proving
security in the random oracle model do not work for the Schnorr blind
signature by providing a meta-reduction which we call "personal nemesis adversary". Our meta-reduction is the first one that does not need
to reset the adversary and can also rule out reductions to interactive
assumptions. Our results generalize to other important blind signatures,
such as the one due to Brands. Brands' blind signature is at the heart of
Microsoft's newly implemented UProve system, which makes this work
relevant to cryptographic practice as well.

Keywords: Blind signatures, meta-reduction technique, unforgeability,
random oracle model.

1 Introduction

In a blind signature scheme, first introduced by Chaum in 1982 [16], a user can
have a document signed without revealing its contents to the signer, and in such
a way that the signer will not be able to recognize it later, when he sees the
signature. Blind signatures have proven to be a very useful building block in
applications requiring both anonymity and unforgeability, such as e-cash and
anonymous credentials [12–15, 27].

Transactions that ensure unforgeability without violating privacy are of growing interest to cryptographic practice. The European Union E-Privacy Directive [31] limits the scope of the data that organizations are allowed to collect;
so, to make sure that it is not in violation of this directive, an online bank or
vendor interacting with a user has an incentive to learn as little as possible
about this user. Therefore, industry leaders such as Microsoft and IBM [30, 36]
have been developing, implementing and promoting cryptographic software tools

---

[*] This work was supported by NSF grants 0964379 and 1012060.

K. Sako and P. Sarkar (Eds.) ASIACRYPT 2013, Part II, LNCS 8270, pp. 82–99, 2013.

that promise the best of both worlds: unforgeability for banks and vendors, and privacy for users.

As a result, research on blind signatures has flourished, and provably secure solutions have been proposed based on well-established theoretical complexity assumptions in the standard model [14, 3, 22, 24] while some of these have been adapted for practical use by IBM [14]. However, schemes in the standard model either require exponentiation in the RSA group or bilinear pairings, which are typically considerably slower than, say, elliptic curve operations.

Thus, more efficient solutions that are provably secure in the random-oracle (RO) model [8] remain of practical importance [2, 9, 6]. Some of the earliest proposed schemes [12, 35, 23] do not have proofs of security even in the RO model; in fact, the security properties of the Schnorr blind signature is an important open problem. Moreover, Microsoft's UProve proposal [29, 30] is based on one of the unproven blind signatures, namely the one due to Brands [12]. UProve is currently part of a pilot project by NSTIC (National Strategy for Trusted Identities in the Cyberspace) that will be used quite extensively in a situation that will potentially affect millions of people [1]. Therefore, the security properties of these unproven but important blind signatures is a natural topic to look at.

In a nutshell, a blind signature scheme is secure if it satisfies two key properties: one-more unforgeability, which means that an adversary cannot produce more signatures than have been issued; and blindness, which means that an adversary cannot link a particular signature to a particular signing instance [33, 34].

The Schnorr blind signature scheme is the most efficient of all the blind signature schemes proposed in the literature given that it can be implemented using elliptic curves without pairings. It is constructed from the corresponding identification protocol via the Fiat-Shamir heuristic and some blinding operations. However, the security of this important scheme is an open problem. If the Schnorr identification scheme is not secure (i.e., after some number of interactions with the prover, the adversary can impersonate him), then the blind Schnorr signature is not one-more unforgeable. It is known that the Schnorr identification scheme cannot be proven secure under the discrete-logarithm assumption using black-box reductions in the standard model [32], so at the very least, it seems that Schnorr blind signatures require that we assume the security of Schnorr identification (also studied by Bellare and Palacio [7]). Perhaps an even stronger assumption may be reasonable. Can we prove it secure under this or a stronger assumption?

To make this question more interesting, let us make it more general. Let us consider not just the Schnorr blind signature, but in general the blind variants of all Fiat-Shamir based signature schemes constructed along the lines described above: the signer acts as the prover in an identification protocol. And let us see if they can be proven secure under any reasonable assumption (by reasonable, we mean an assumption that is not obviously false), not just specific ones.

PS Reduction. Pointcheval and Stern showed that we can prove the security of blind signature schemes in the RO model when the underlying identification scheme is a witness-indistinguishable proof protocol for proving knowledge of a secret key, such that many secret keys are associated with the same public key [33, 34]. Their result does not apply to the original Schnorr blind signature, in which there is a single secret key corresponding to the public key. Other important blind signatures to which it does not apply are the ones due to Brands' (which are at the heart of Microsoft's UProve), and the ones based on the GQ signature [12, 23].

The idea of the Pointcheval-Stern reduction (also called "an oracle replay reduction") is to replay the attack polynomially many times with different random oracles in order to make the attacker successfully forge signatures. More precisely, we first run the attack with random keys, tapes and oracle f. Then, we randomly choose an index j and we replay with same keys and random tapes but with a new, different oracle f' such that the first $j - 1$ answers are the same as before. We expect that, with non-negligible probability we will obtain two different signatures, σ, σ' of the same message m and we will be able to use them to solve a hard algorithmic problem (usually the one underlying the blind signature scheme) in polynomial time. This proof technique works for standard (i.e. not blind) versions of the Schnorr, Brands and GQ signatures. They also showed that it works for a modification of Schnorr blind signature which is less efficient than the original Schnorr's. A very natural question is: can it work for the original Schnorr blind signature and its generalizations, such as the Brands or GQ blind signatures?

Our Results. Let us take a closer look at oracle replay reductions, as used by Pointcheval and Stern. Their reduction can be modeled as a Turing machine that has a special tape that is used specifically for answering random oracle queries; it always uses the next unused value when answering, afresh, the next random oracle query. We call this type of reductions: *Naive RO replay reductions* and as we will discuss in Section 3.1 it can be used to model every known reduction for proving the security of digital signature schemes. Our result is that, in fact, naive RO replay reductions cannot be used to prove security of generalized Schnorr blind signatures, no matter how strong an assumption we make. Our result also holds for interactive assumptions or even if we assume the security of the blind signature scheme itself! Put another way, any such reduction can be used in order to break the underlying assumption.

Meta-reductions. In our proof we make use of the "meta-reduction" method [10]: a separation technique commonly used to show impossibility results in cryptography. Let $\mathcal{A}$ be an adversary who breaks the unforgeability of generalized Schnorr blind signatures with non-negligible probability. We will use a meta-reduction (which we call "personal nemesis adversary") to show that there *cannot* exist a naive RO replay reduction, $\mathcal{B}$, which turns $\mathcal{A}$ into a successful adversary for *any* hard assumption that may be considered. We do that by transforming $\mathcal{B}$ through the meta-reduction into an algorithm that breaks the underlying assumption, without relying on the existence of a successful adversary.

What makes our technique particularly interesting is that for the *first time* we introduce a meta-reduction (our personal nemesis adversary) that does not need to reset the reduction $\mathcal{B}$, as it is usually done when using the meta-reduction paradigm [19]. For example, our personal nemesis adversary could reset the reduction $\mathcal{B}$, get an additional signature and return this signature back to $\mathcal{B}$ as his forgery. However, this resetting makes things more complicated since the two executions are correlated. Our technique, instead, is much simpler: the personal nemesis adversary, $p\mathcal{A}$, will simply interact with the reduction $\mathcal{B}$ the way an actual adversary would (but taking advantage of powers not available to an adversarial algorithm, such as remembering its prior state if and when the reduction resets it, and having access to the reduction's random oracle tape), without resetting it at any time. When $\mathcal{B}$ halts, if it succeeded in breaking the assumption (as it should with non-negligible probability, or it wouldn't be a valid security reduction), $p\mathcal{A}$ has succeeded too — but *without* assuming the existence of an actual adversary that breaks the security of the underlying signature scheme.

What are the implications of our results on the security of Schnorr blind signatures and generalizations? We must stress that our results do not in fact constitute an attack, and so for all we know, these schemes might very well be secure. However, we have essentially ruled out all *known* approaches to proving their security. So in order to give any security guarantee on these signature schemes, the cryptographic community would have to come up with radically new techniques.

Related Work. Schnorr and Jakobsson [18] proved security of the Schnorr blind signature in the combined random oracle and generic group model which is very restricted. Fischlin and Schröder [22] show that proving security of a broad class of blind signature schemes (which, in particular, includes what we refer to as generalized Schnorr blind signatures) via black-box reductions in the standard model is as hard as solving the underlying hard problem. Their technique uses the meta-reduction paradigm to show that black-box reductions for this type of blind signatures can be turned into solvers for hard non-interactive assumptions. However, their result does not rule out reductions in the RO model, and is technically very different from ours for that reason.

Rafael Pass studied the assumptions needed for proving security of various cryptographic schemes [32]. In particular, relevant to our work, he considers the Schnorr identification scheme and variants, and a category of blind signatures called "unique blind signatures." Pass considers whether so-called *r-bounded-round* assumptions are strong enough to prove, in a black-box fashion in the standard model, the security of certain schemes when repeated more than r times. His results apply to Schnorr blind signatures (and their generalizations) in the following way: he shows that no so-called bounded-round assumption can imply secure composition of the Schnorr identification scheme using black-box reductions (and therefore the Schnorr blind signature).

Here is how our work goes beyond what was shown by Pass [32] for "unique blind signatures." First of all, we do not limit our consideration to *r-bounded-round* assumptions but we show that our result applies for every possible intractability assumption. Thus, we rule out the existence of a very special type of reduction, the naive RO replay one, that models all the known reductions for proving security of digital signatures, *irrespective of assumption*. As an example, consider the One More Discrete Logarithm assumption (OMDL) [6] which has been used to prove security of the Schnorr identification scheme against active attacks [7]. Our result directly implies that Schnorr blind signature cannot be proven secure under the OMDL assumption in the RO model. Finally, our result applies even after *just one signature was issued* whereas Pass' result questions the security of schemes when repeated more than r times.

The meta-reduction technique has been used to analyze security of Schnorr signatures. Paillier and Vergnaud [28] showed that the security of Schnorr signatures cannot be based on the difficulty of the one more discrete logarithm problem in the standard model. Fischlin and Fleischhacker [20] extended their result by showing that the security of Schnorr signatures cannot be based to the discrete logarithm problem without programming the random oracle. Their work is also relevant to ours since the meta-reduction they define also doesn't need to reset the reduction[1]. However, their result holds only for reductions to the discrete logarithm problem and applies to non-programming reductions while our naive RO replay reductions fall somewhere in between the programmable and non-programmable setting (see Section 3.1 for a discussion about programmability). Finally, their result only holds for a very limited class of reductions: those that run a single copy of the adversary which makes our work much broader.

2   Generalized Blind Schnorr Signature

First we explicitly define the class of blind signatures that our result applies to. For a complete presentation of all the necessary building blocks please refer to our full version [5].

In the signature scheme described by Schnorr [35] the signer's secret key is an exponent x, while his public key is $h = g^x$. A signature on a message m is obtained, via the Fiat-Shamir heuristic, from the Schnorr identification protocol, i.e. the three-round proof of knowledge of x. Thus, a signature on a message m is of the form $\sigma = (a, r)$ such that $g^r = ah^{H(m,a)}$, where H is a hash function that is modeled as a random oracle in the security proof. A blind issuing protocol was proposed for this signature back in the 1980s [18], and, on a high level, it works by having the user "blind" the value a he receives from the signer into some unrelated a', then the user obtains $c' = H(m, a')$ and, again, "blinds" it into some unrelated c which he sends to the signer. The signer responds with r which the user, again, "blinds" into r' such that (a', r') are a valid signature on m.

$$
\begin{array}{lll}
\text{Signer}(q, g, h = g^x) & & \text{User}(q, g, h, m) \\
\hline
y \leftarrow \mathbb{Z}_q,\ a = g^y & \xrightarrow{\ a\ } & \\
& \xleftarrow{\ c\ } & \alpha, \beta \leftarrow \mathbb{Z}_q,\ a' = ag^\alpha h^\beta,\ c' = H(m, a'),\ c = c' + \beta \\
r = y + cx \bmod q & \xrightarrow{\ r\ } & g^r \overset{?}{=} ah^c,\ r' = r + \alpha,\ \text{output } r', c'
\end{array}
$$

---

[1] This is a result that Fischlin and Fleischhacker [20] obtained after the first version of our manuscript appeared on eprint [5]; our result is in fact the first in which a meta-reduction works without resetting the reduction $\mathcal{B}$.

The signature is: $\sigma(m) = (a', c', r')$ and the verification checks whether $g^{r'} = a'h^{c'}$.

Ever since this protocol was proposed, its security properties were an open problem. Okamoto proposed a modification [26]; Pointcheval and Stern proved security of this modification [33, 34]. Our work studies this blind signature and its generalizations, defined as follows:

Definition 1 (Generalized Blind Schnorr Signature). *A blind signature scheme (Gen, S, U, Verify) is called Generalized Blind Schnorr Signature if:*

1. *$(pk, sk) \in R_L$ is a unique witness relation for a language $L \in \mathcal{NP}$.*
2. *There exists a Σ-protocol (P, V) for R_L such that for every $(pk, sk) \in R_L$ the prover's algorithm, $P(pk, sk)$, is identical to the signer's blind signing algorithm $S(pk, sk)$.*
3. *Let $Sign(pk, sk, m)$ be the signing algorithm implicitly defined by (S, U). Then, there exists a Σ-protocol $P(pk, sk)$, $V(pk)$ such that, in the random oracle (RO) model, a signature $\sigma = (a, c, r)$, where $c = H(m, a)$ is distributed identically to a transcript of the Σ-protocol.*
4. *There exists an efficient algorithm that on input (pk, sk) a "valid tuple" (a, c, r) and a value c', computes r' s.t. (a, c', r') is a valid tuple. (By "valid tuple" we mean a signature for which the verification equation holds.) Note that no additional information about a is required, such as, e.g. its discrete logarithm.*

Let us now see why Schnorr's blind signature falls under the generalized blind Schnorr signature category. (1) The secret/public key pair is an instance of the DL problem which is a unique witness relation; (2) the signer's side is identical to the prover's side of the Schnorr identification scheme, which is known to be a Σ-protocol; (3) the signature $\sigma(m) = (a', c', r')$ is distributed identically to the transcript of the Schnorr identification protocol since a' comes uniformly at random from G; c' is truly random in the RO model, and r' is determined by α (4) finally, for a tuple (a, c, r) and a value c' one can compute $r' = r - cx + c'x$ so that (a, c', r') is still a valid tuple.

The definition also captures other well-known blind signature schemes, such as the blind GQ [23] and Brands [12] (for Brands also see Section 4).

3   Security of Generalized Blind Schnorr Signatures

We first define a general class of RO reductions and then prove that generalized blind Schnorr signature schemes cannot be proven unforgeable, and thus secure, using these reductions.

3.1   Naive RO Replay Reductions

We first explicitly describe the type of reductions that we rule out.

Definition 2 (Naive RO replay reduction). *Let $\mathcal{B}$ be a reduction in the random-oracle model that can run an adversary $\mathcal{A}$, and may also reset $\mathcal{A}$ to a* *previous state, causing $\mathcal{A}$ to forget $\mathcal{B}'s$ answers to its most recent RO queries. We assume, without loss of generality, that if $\mathcal{A}$ has already queried the RO on some input x, and hasn't been reset to a state that is prior to this query, then $\mathcal{A}$ does not make a repeat query for x.*

We say that $\mathcal{B}$ is a naive RO replay reduction if: $\mathcal{B}$ has a special random tape for answering the RO queries as follows: when $\mathcal{A}$ queries the RO, $\mathcal{B}$ retrieves the next value v from its RO tape, and replies with $c = f(b, v)$ where b is the input to the reduction, and f is some efficiently computable function.

Discussion. Let us now take a closer look at known reductions for proving security of signatures in the RO model and see whether they fall under the naive RO replay reduction category. We first look at the reduction given by Pointcheval and Stern [33] for proving security of blind signatures. Their reduction could be easily modeled as a naive RO replay reduction with f being the identity function on its second input. PS reductions are *perfect* since they always create a signature. The same holds for the reduction given by Abe and Okamoto [4]. To convince the reader that our way of modeling reductions in the RO model is a very natural one, let us also look at the reduction given by Coron [17] proving the security of full domain hash (FDH) RSA signature. Coron's reduction works as follows: the reduction, $\mathcal{B}$, gets as input (N, e, y) where (N, e) is the public key and y is a random element from $\mathbb{Z}_N^*$ and tries to find $x = y^d \bmod n$. $\mathcal{B}$ runs an adversary $\mathcal{A}$, who can break the signature, with input the public key. As usual, $\mathcal{A}$ makes RO and signing queries which $\mathcal{B}$ answers. Whenever $\mathcal{A}$ makes an RO query, $\mathcal{B}$ picks a random $r \in \mathbb{Z}_n^*$ and either returns $h = r^e \bmod N$ with probability p or returns $h = yr^e \bmod N$ with probability $1 - p$. So, it is pretty straightforward that we could model Coron's reduction as a naive RO replay reduction by interpreting the contents of an RO tape as r and the output of a p-biased coin flip (return either r^e or yr^e). Other well-known reductions used in the literature to prove security of digital signatures in the RO model can be modeled as naive RO replay reductions as well [9, 6, 8].

Programmability. Let us compare naive RO replay reductions with other previously defined types. Non-programmable random-oracle reductions [25] do not give the reduction the power to set the answers to the RO queries; instead these answers are determined by some truly random function. Naive RO replay reductions can be more powerful than that: they can, in fact, answer the adversary's queries in some way they find convenient, by applying the function f to the next value of their RO tape. However, they are not as powerful as the general programmable RO reductions: naive RO replay reductions are not allowed, for example, to compute an answer to an RO query as a function of the contents of the query itself. Fischlin et al. [21] also consider an intermediate notion of programmability, called "random re-programming reductions", which are incomparable to ours.

3.2 Theorem for Perfect Naive RO Replay Reduction

Our first result is on a simpler class of reductions called "perfect". We will extend it to non-perfect reductions in Section 3.3.

Definition 3 (Perfect-Naive RO replay reduction). *A naive RO replay reduction $\mathcal{B}$ is called* perfect naive RO replay reduction *if $\mathcal{B}$ always gives valid responses to $\mathcal{A}$, i.e. its behavior is identical to that of the honest signer.*

We show that *perfect* naive RO replay reductions cannot be used to prove security of generalized blind Schnorr signature schemes.

Theorem 1. *Let $(Gen, S, U, Verify)$ be a generalized blind Schnorr signature scheme. Assume that there exists a polynomial-time perfect naive RO replay reduction $\mathcal{B}$ such that $\mathcal{B}^{\mathcal{A}}$ breaks an interactive intractability assumption C for every $\mathcal{A}$ that breaks the unforgeability of the blind signature (S, U). Then, C can be broken in polynomial time.*

Proof of Theorem for Perfect Naive RO Replay Reduction. We start by introducing some terminology. Note that the reduction $\mathcal{B}$ is given black-box access to $\mathcal{A}$ and is allowed to run $\mathcal{A}$ as many times as it wishes, and instead of running $\mathcal{A}$ afresh every time, it may reset $\mathcal{A}$ to some previous state. At the same time, $\mathcal{B}$ is interacting with its own challenger C; we do not restrict C in any way.

Consider how $\mathcal{B}$ runs $\mathcal{A}$. $\mathcal{B}$ must give to $\mathcal{A}$ some public key pk for the signature scheme as input. Next, $\mathcal{B}$ runs the blind signing protocol with $\mathcal{A}$; recall that a generalized blind Schnorr signing protocol always begins with a message a from the signer to the user. When $\mathcal{B}$ runs $\mathcal{A}$ again, it can choose to give it the same (pk, a) or different ones. It is helpful for the description of the adversary we give, as well as for the analysis of the interaction, to somehow organize various calls that $\mathcal{B}$ makes to $\mathcal{A}$.

Every time that $\mathcal{B}$ runs $\mathcal{A}$, it either runs it "anew", providing a new public key pk and first message a, or it "resets" it to a previous state, in which some pk and a have already been given to $\mathcal{A}$. In the latter case, we say that $\mathcal{A}$ has been "reincarnated", and so, an *incarnation* of $\mathcal{A}$ is defined by (pk, a). Note that $\mathcal{B}$ may reincarnate $\mathcal{A}$ with the same (pk, a) several times. In this case, we say that this incarnation is *repeated*. Thus, if this is the ith time that $\mathcal{A}$ has been reset to a previous state for this specific (pk, a), then we say that this is the ith repeat of the (pk, a) incarnation. Without loss of generality, $\mathcal{B}$ never runs $\mathcal{A}$ anew with (pk, a) that it has used (i.e., if $\mathcal{B}$ has already created an incarnation for (pk, a), it does not create another one).

Let us consider what happens once $\mathcal{A}$ receives (pk, a). The signing protocol, in which $\mathcal{A}$ is acting as the user, expects $\mathcal{A}$ to send to $\mathcal{B}$ the challenge c. Additionally, $\mathcal{A}$ is free to make any random oracle queries it chooses. Once $\mathcal{B}$ receives c, the signing protocol expects it to send to $\mathcal{A}$ the response r. After that, the security game allows $\mathcal{A}$ to either request another signature, or to output a one-more signature forgery, i.e., a set of signatures (one more than it was issued);

also, again, $\mathcal{A}$ can make RO queries. The adversaries that we consider in the sequel will not request any additional signatures, but will, at this point, output two signatures (or will fail).

Note that, if $\mathcal{B}$ is a perfect naive RO replay reduction, then it will always provide to $\mathcal{A}$ a valid response r to the challenge c; while if it is not perfect, then it may, instead, provide an invalid response, or stop running $\mathcal{A}$ at this point altogether. Thus, a particular run can be:

- Uncompleted: no valid response, r, was given by $\mathcal{B}$ at the end of the protocol (cannot happen if $\mathcal{B}$ is perfect).
- Completed but unsuccessful: a valid r was given but $\mathcal{A}$ was not able to output a forgery.
- Completed and successful: a valid r was given and $\mathcal{A}$ did output a forgery.

The technique we follow to prove our theorem is the following. We first define a special adversary which we call the *super adversary*, $s\mathcal{A}$, who exists if it is easy to compute the signing key for this signature scheme from the corresponding verification key. We do not show how to construct such an adversary (because we do not know how to infer the signing key for generalized blind Schnorr, and in fact we generally assume that it is impossible to do so in polynomial time); instead, we construct another adversary, the *personal nemesis adversary*, $p\mathcal{A}$, whose behavior, as far as the reduction $\mathcal{B}$ can tell, will be identical to $s\mathcal{A}$.

Note that, generally, an adversary is modeled as a deterministic circuit, or a deterministic non-uniform Turing machine: this is because, inside a reduction, its randomness can be fixed. Thus, we need $s\mathcal{A}$ to be deterministic. Yet, we need to make certain randomized decisions. Fortunately, we can use a pseudorandom function for that. Thus, $s\mathcal{A}$ is parametrized by s, a seed to a pseudorandom function[2] $F_s : \{0,1\}^* \to \{0,1\}^k$. Additionally, it is parametrized by two messages m_1, m_2: signatures on these messages will be output in the end.

Consider $s\mathcal{A}_{s,m_1,m_2}$ that interacts with a signer as follows:

Definition 4 (Perfect super adversary $s\mathcal{A}_{s,m_1,m_2}$). *On input the system parameters:*

1. *Begin signature issue with the signer and receive (pk, a).*
2. *Find sk.*
3. *Use sk to compute the signatures: pick a_1, a_2 and make two RO queries (m_1, a_1) and (m_2, a_2). Produce two forged signatures for m_1, m_2, denote them as σ_1 and σ_2 (remember that $s\mathcal{A}$ is deterministic so if reincarnated he makes the same RO queries).*
4. *Resume the signature protocol with the signer: send to the signer the value $c = F_s(trans)$ where trans is the current transcript between $s\mathcal{A}_{s,m_1,m_2}$, the RO and the signer, and receive from the signer the value r in response (which will always be valid for the perfect naive RO reduction $\mathcal{B}$).*
5. *Output the two message-signature pairs, (m_1, σ_1) and (m_2, σ_2).*

---

[2] We know that if $\mathcal{B}$ exists then secure signatures exist which imply one way functions existence and PRFs existence, so this is not an extra assumption.

Note that when sA executes the signature issue protocol with the signer it computes c as a pseudorandom function of its current transcript with the RO and the signer. Thus, there is only a very small probability (of about 2^{-k}) for sA to send the same c in another run.

The next lemma follows directly from the definition of a reduction $\mathcal{B}$:

Lemma 1. *If a perfect naive RO replay reduction $\mathcal{B}$ exists, then $\mathcal{B}^{sA(\cdot)}$ (pk, system params) solves the assumption C.*

Lemma 1 works even if the assumption C is an interactive one. That is why, sA and pA are defined in such a way that they do not reset the reduction $\mathcal{B}$.

Next, we define the personal nemesis adversary, pA. Similarly to sA, it is parametrized by (s, m_1, m_2); and so we denote it pA_{s,m_1,m_2}. To the reduction $\mathcal{B}$, pA_{s,m_1,m_2} will look exactly the same as sA_{s,m_1,m_2}, even though pA_{s,m_1,m_2} cannot compute sk. Instead, pA_{s,m_1,m_2} looks inside the reduction $\mathcal{B}$ itself; this is why we call pA_{s,m_1,m_2} "$\mathcal{B}$'s personal nemesis":

Definition 5 (Perfect $\mathcal{B}$'s personal nemesis adversary pA_{s,m_1,m_2}). *On input the system parameters, pA_{s,m_1,m_2} performs a "one-more" forgery attack, using the following special powers: (1) pA_{s,m_1,m_2} has full access to $\mathcal{B}$'s random oracle tape; (2) in case pA_{s,m_1,m_2} is rewound, he remembers his previous state.*

pA_{s,m_1,m_2} performs the one-more forgery for $\ell = 1$. Thus, he runs one signature issuing session with the signer and then outputs two valid signatures. Specifically, in his ith incarnation, pA does the following:

1. *Begin signature issue with the signer, and receive (pk, a).*
2. *Do nothing (pA cannot find sk).*
3. *– If (pk, a) is the same as in some previous incarnation j then make the same RO queries as the last time this incarnation was run (sA remembers the previous RO queries; obviously it will receive different c_1, c_2 than before).*
   – If (pk, a) is a new tuple, then this is a new incarnation; do the following:
      * *If pA has already computed the sk for this pk, then use this power to forge two signatures on (m_1, m_2); call the resulting signatures σ_1 and σ_2,*
      * *else (if sk not already known), pA computes two signatures using its special access to $\mathcal{B}$ by looking in advance what the next c_1, c_2 are going to be, then picking random [3] r_1, r_2 and solving for a_1, a_2 using the third property of generalized blind Schnorr signatures and the simulator from the underlying Σ-protocol. pA makes two RO queries of the form $(m_1, a_1), (m_2, a_2)$ and gets c_1, c_2 in response. Call the resulting signatures σ_1 and σ_2.*
4. *Resume the signature issue protocol with the signer: send to the signer the value $c = F_s(trans)$ where trans is the current transcript between pA, the RO and the signer, and receive from the signer the value r in response (which will be valid for the perfect naive RO reduction $\mathcal{B}$).*

---

[3] Recall that pA uses a PRF that takes as input its current state in order to make each random choice.

5.   – *If this is the first time for this incarnation, then output the two message-signature pairs, (m_1, σ_1) and (m_2, σ_2) (completed and successful run).*

   – *If this is a repeat of some incarnation j, and the value $c = F_s(trans) \neq c_j$, where c_j is the corresponding value from incarnation j, then using r and r_j, property 3 of generalized blind Schnorr signatures and the extractability of the Σ-protocol, compute sk (if you don't already know it for this pk). Next, compute σ_1 and σ_2 consistent with the RO queries from incarnation j, using property 4 of generalized blind Schnorr signatures (completed and successful run).*

   – *If i is a repeat of j, and the value $c = F_s(trans) = c_j$, then fail (completed and unsuccessful run).*

Given the definition above it becomes clear why our naive RO reductions are not allowed to compute answers to the RO queries as a function of the query itself. It is important that the personal nemesis adversary has full access to the reduction's special RO tape and he should able to see what the next answer would be *before* forming his query. In particular, on the second case of Step 3 in Definition 5, $p\mathcal{A}$ first looks into $\mathcal{B}$'s RO tape to see what is the next c_1, c_2 and then formulates his RO query which depends on c_1, c_2. In this case, our analysis would break if the answer to the query was computed as a function of the content of the query itself.

Lemma 2. *If $\mathcal{B}$ is a perfect naive RO replay reduction, then $\mathcal{B}$'s view in interacting with $p\mathcal{A}_{s,m_1,m_2}$ is indistinguishable from its view when interacting with $s\mathcal{A}_{s,m_1,m_2}$.*

Proof. In order to prove this, we will analyze the behavior of $s\mathcal{A}$ and $p\mathcal{A}$ step by step, as they were defined, and we will show that $\mathcal{B}$ receives indistinguishable views when interacting with $s\mathcal{A}_s$ or $p\mathcal{A}_s$ with all but negligible probability (to simplify notation we will omit writing the messages m_1, m_2 to the parameters given to the adversaries). We begin by defining $s\mathcal{A}_{Rand}$ and $p\mathcal{A}_{Rand}$ who behave exactly as $s\mathcal{A}_s$ and $p\mathcal{A}_s$ do but using a truly random source instead of the pseudorandom function F_s. We will use the following hybrid argument: $s\mathcal{A}_s \approx s\mathcal{A}_{Rand} \approx p\mathcal{A}_{Rand} \approx p\mathcal{A}_s$.

Let us first argue that $s\mathcal{A}_s \approx s\mathcal{A}_{Rand}$. This follows by a straightforward reduction that contradicts the pseudorandomness of F_s. Similarly, it holds that $p\mathcal{A}_{Rand} \approx p\mathcal{A}_s$. We prove that $s\mathcal{A}_{Rand} \approx p\mathcal{A}_{Rand}$ by examining step by step the behavior of $s\mathcal{A}_{Rand}$ and $p\mathcal{A}_{Rand}$.

1. In the first step, both $s\mathcal{A}_{Rand}$ and $p\mathcal{A}_{Rand}$ begin the signature issuing with the Signer and wait for him to respond with (pk, a). For $\mathcal{B}$ there is no difference whether talking to $s\mathcal{A}_{Rand}$ or $p\mathcal{A}_{Rand}$.

2. In the second step there is no interaction with $\mathcal{B}$.

3. Here we have two different cases on $p\mathcal{A}_{Rand}$'s behavior depending on whether the current incarnation is *repeated* or not. In both cases the interaction between $p\mathcal{A}_{Rand}$ and $\mathcal{B}$ consists of $p\mathcal{A}_{Rand}$ making two RO queries where $p\mathcal{A}_{Rand}$ either makes two RO queries on fresh values that it computed on

the current step or makes the same RO queries as in the *repeated* incarnation (so, there is no difference for $\mathcal{B}$). Thus, in Step 3, no matter who $\mathcal{B}$ is talking to, $\mathcal{B}$ receives two RO queries distributed identically.

4. Step 4 is identical for both $s\mathcal{A}_{Rand}$ and $p\mathcal{A}_{Rand}$. Just send $c = R(trans)$, where R is a random function and receive the value r in response.

5. Since r will always be a valid response (recall that $\mathcal{B}$ is perfect), $s\mathcal{A}_{Rand}$ will always output two message-signature pairs, (m_1, σ_1) and (m_2, σ_2). $p\mathcal{A}_{Rand}$ will also output (m_1, σ_1) and (m_2, σ_2), which are distributed identically to the ones output by $s\mathcal{A}_{Rand}$ unless it is the case that the incarnation is a repeat of j and $c = R(trans) = c_j$. In that case $p\mathcal{A}_{Rand}$ fails. The probability that $c = R(trans) = c_j$ is only $2^{-\Theta(k)}$. Thus, with probability $1 - 2^{-\Theta(k)}$ $\mathcal{B}$'s view is identical no matter whether he is talking to $s\mathcal{A}_{Rand}$ or $p\mathcal{A}_{Rand}$.

So, by the hybrid argument we defined at the beginning of the proof, it holds that $s\mathcal{A}_s \approx p\mathcal{A}_s$. □

Remark: we don't explicitly exploit blindness and in fact our result would go through even if a signature could be linkable to an issuing instance. For example, including the first message of the signer into the RO query would produce a contrived scheme in which the resulting signatures are linkable to the issuing instance; yet it would not affect our negative result.

3.3 Theorem for Non-perfect Naive RO Replay Reductions

Let us apply our result to a broader class of reductions by removing the requirement that our reduction be perfect, i.e. always outputs valid responses. Instead, we will require an upper bound L on the number of times that the reduction can invoke the adversary which is independent of $\mathcal{A}$'s success probability. Note that, of course, $\mathcal{B}$'s success probability needs to depend on $\mathcal{A}$'s success probability. However, the number of times it invokes $\mathcal{A}$ need not; in fact known reductions (such as Coron or Pointcheval and Stern) as a rule only invoke the adversary a constant number of times.

Definition 6 (L-Naive RO replay reduction). *A naive RO replay reduction $\mathcal{B}$ is called L-naive RO replay reduction if there is a polynomial upper bound L on how many time $\mathcal{B}$ resets $\mathcal{A}$; this upper bound is a function of the number of RO queries that $\mathcal{A}$ makes, but otherwise is independent of $\mathcal{A}$, in particular, of $\mathcal{A}$'s success probability.*

Our previous analysis wouldn't work for the *L-naive RO replay reduction*. Think of the scenario where $p\mathcal{A}$ receives a message a from $\mathcal{B}$ for the first time but is not given a valid r at the end. Then in the repeat of this incarnation, $p\mathcal{A}$ will have to make the same two RO queries he did before and output forgeries if given a valid r at the end. But, given the definitions of $\mathcal{B}$ and $p\mathcal{A}$ we gave before, $p\mathcal{A}$ will now get different c_1 and c_2 for his RO queries and thus he will not be able to output the same forgeries he had prepared before.

What changes in our new analysis is that: (a) $p\mathcal{A}$ is also given write access to $\mathcal{B}$'s RO tape, and (b) both $p\mathcal{A}$ and $s\mathcal{A}$ will be successful in producing a forgery with probability only $1/(\binom{L}{2} + L)$.

Theorem 2. *Let $(Gen, S, U, Verify)$ be a generalized blind Schnorr signature scheme. Suppose that there exists a polynomial-time L-naive RO replay reduction $\mathcal{B}$ such that $\mathcal{B}^{\mathcal{A}}$ breaks an intractability assumption C for every $\mathcal{A}$ that breaks the unforgeability of the blind signature (S, U). Then, C can be broken in polynomial time.*

This theorem rules out a broader class of security reductions. If we look back to our running example of Schnorr blind signatures, this theorem shows that under any assumption (DL, security of Schnorr identification, etc.) we cannot find an L-naive RO replay reduction to prove its security.

Proof of Theorem for L-naive RO Replay Reduction. Similar to what we did before, we first define the *super adversary* $s\mathcal{A}_{s,m_1,m_2,L}$ who knows L and works as follows:

Definition 7. *[Super adversary $s\mathcal{A}_{s,m_1,m_2,L}$] On input the system parameters:*

1. *Begin signature issue with the signer and receive (pk, a). Decide whether this is going to be a successful incarnation: choose "successful" with probability $1/(\binom{L}{2} + L)$ and "unsuccessful" with probability $1 - 1/(\binom{L}{2} + L)$.*
2. *Find sk.*
3. *Use sk to compute the signatures: pick a_1, a_2 and make two RO queries (m_1, a_1) and (m_2, a_2). Produce two forged signatures for m_1, m_2, denote them as σ_1 and σ_2.*
4. *Resume the signature protocol with the signer: send to the signer the value $c = F_s((trans))$ where trans is the current transcript between $s\mathcal{A}$, the RO and the signer, and receive from the signer the value r in response.*
5. *– If r is not valid, then this was an uncompleted run, then fail.*
   – If r valid (completed run) and in Step 1 it was decided that this is a successful incarnation, output the two message-signature pairs, (m_1, σ_1) and (m_2, σ_2). Otherwise fail.

The next lemma (similar to Lemma 1) follows from the definition of $\mathcal{B}$:

Lemma 3. *If an L-naive RO replay reduction $\mathcal{B}$ exists, then $\mathcal{B}^{s\mathcal{A}(\cdot)}$ $(pk, system\ params)$ solves the assumption C.*

Now we are going to define the personal nemesis adversary, $p\mathcal{A}_{s,m_1,m_2,L}$.

Definition 8 ($\mathcal{B}$'s personal nemesis adversary $p\mathcal{A}_{s,m_1,m_2,L}$). *On input the system parameters, $p\mathcal{A}_{s,m_1,m_2,L}$ performs a "one-more" forgery attack, using the following special powers: (1) $p\mathcal{A}_{s,m_1,m_2,L}$ has full read and write access to $\mathcal{B}$'s random oracle tape; (2) in case $p\mathcal{A}_{s,m_1,m_2,L}$ is rewound, it does remember his previous state.*

$pA_{s,m_1,m_2,L}$ performs the one-more forgery for $\ell = 1$. Thus, it runs one signature issuing session with the signer and then outputs two valid signatures with probability $1/(\binom{L}{2} + L)$. Specifically, in his i^{th} incarnation, $pA_{s,m_1,m_2,L}$ does the following:

1. Begin signature issue with the signer, and receive (pk, a).
2. Do nothing.
3. – If (pk, a) is received for the first time, then this is a new incarnation; do the following:
    - If pA has already found sk for this pk, then use this power to forge two signatures on (m_1, m_2) (still required to make two RO queries); call these signatures σ_1 and σ_2,
    - else, pA guesses (i_1, i_2) where $i_1(\leq i_2)$ denotes the repeat where c_1 will be given in response to pA's next RO query; and i_2 is pA's guess for the first completed repeat of this incarnation. Then, pA randomly picks v_1, v_2, computes $c_1 = f(v_1), c_2 = f(v_2)$, picks r_1, r_2, solves for a_1, a_2 using the third property of generalized blind Schnorr signatures and the simulator from the underlying Σ-protocol and computes two signatures σ_1, σ_2.
   – pA makes two RO queries of the form $(m_1, a_1), (m_2, a_2)$ (the two RO queries are always the same for a specific incarnation).
   – If this is the repeat incarnation i_1, and $\mathcal{B}$ wants a fresh answer to the query (m_1, a_1) then write v_1 on $\mathcal{B}$'s RO tape; else (if this isn't repeat i_1) write a random v_1'.
   – If this is the repeat incarnation i_2 then write v_2 on $\mathcal{B}$'s RO tape; else (if this isn't repeat i_2) write a random v_2'.
4. Resume the signature issue protocol with the signer: send to the signer the value $c = F_s(trans)$ where F_s is a PRF and $trans$ is the current transcript between pA, the RO and the signer, and wait to receive the value r as a response from the signer.
5. – If r is valid (completed run):
    - If already know the secret key, sk, then output (m_1, σ_1) and (m_2, σ_2) with probability $1/(\binom{L}{2} + 2)$ or else fail.
    - If this is the first time for this incarnation, then output the two message-signature pairs, (m_1, σ_1) and (m_2, σ_2).
    - If this is the second successful repeat for this incarnation and the value $c = F_s(trans) \neq c_j$, where c_j is the corresponding value from the jth run of this incarnation, then using r and r_j solve for sk using property 4 of generalized Schnorr signatures. Next, compute σ_1 and σ_2 consistent with the RO queries from this incarnation.
    - If this is the second successful repeat for this incarnation but $c = F_s(trans) = c_j$, then fail (unsuccessful run).
    - If the guess (i_1, i_2) was correct (that is, this is repeat i_2 of this incarnation, it was successful, and $\mathcal{B}$'s answer to (m_1, a_1) was the same as in incarnation i_1; and in incarnation i_1, $\mathcal{B}$ wanted a fresh answer to the (m_1, a_1) RO query) then output the two message-signature pairs, (m_1, σ_1) and (m_2, σ_2).

> • *If the guess (i_1, i_2) was wrong then fail (unsuccessful run).*
> – *If r is not valid or r was not received then fail.*

Lemma 4. *If $\mathcal{B}$ is an L-naive RO replay reduction, then $\mathcal{B}$'s view in interacting with $p\mathcal{A}_{s,m_1,m_2}$ is indistinguishable from its view when interacting with $s\mathcal{A}_{s,m_1,m_2}$.*

The proof is similar to the one of Lemma 2 and can be found in the full version of the paper [5].

4   Brands' Blind Signature Scheme

Here we show that our results apply to the blind signature scheme given by Brands [11]. Let us first describe his construction. G is a group of order q, where q a k-bit prime, and g is a generator of the group. The signer holds a secret key $x \leftarrow \mathbb{Z}_q$ and the corresponding public key $h = g^x$, while the user knows signer's public key h as well as g, q. $\mathcal{H}$ is a collision resistant hash function. The signature issuing protocol works as follows:

Signer (g, h, x)		User(g, h)
		$m = g^\alpha$
	$\xleftarrow{\quad \alpha \quad}$	
$w \in_R \mathbb{Z}_q,\ z \leftarrow m^x,\ a \leftarrow g^w,\ b \leftarrow m^w$	$\xrightarrow{\ z, a, b\ }$	
		$s, t \in_R \mathbb{Z}_q,\ m' \leftarrow m^s g^t,\ z' \leftarrow z^s h^t$
		$u, v \in_R \mathbb{Z}_q,\ a' \leftarrow a^u g^v,\ b' \leftarrow a^{ut} b^{us} (m')^v$
	$\xleftarrow{\quad c \quad}$	$c' \leftarrow \mathcal{H}(m', z', a', b'),\ c \leftarrow c'/u \bmod q$
$r \leftarrow w + cx \bmod q$	$\xrightarrow{\quad r \quad}$	$h^c a \overset{?}{=} g^r,\ z^c b \overset{?}{=} m^r,\ r' \leftarrow ur + v \bmod q$

A signature on m' is $\sigma(m') = (z', a', b', c', r')$. Anyone can verify a signature by first computing $c' = \mathcal{H}(m', z', a', b')$ and then checking whether the following equations hold: $h^{c'} a' \overset{?}{=} g^{r'}$, $(z')^{c'} b' \overset{?}{=} (m')^{r'}$.

4.1   Security of Brands' Blind Signatures

Corollary 1. *If there exists a perfect or an L-naive RO replay reduction $\mathcal{B}$ that solves any intractability assumption C using an adversary $\mathcal{A}$ that breaks the unforgeability of Brands' signature, then assumption C can be solved in polynomial time with non-negligible probability.*

In order for this corollary to hold we need to show that Brands' blind signature is a generalized blind Schnorr signature. We can show this by inspecting one by one the needed requirements: (1) Brands public/secret key pair is $(h = g^x, x)$, which is a unique witness relation for $L = \{h : g^x = h\} \in \mathcal{NP}$, (2) the signer's side of Brands blind signature is the same as the prover's side in Schnorr's identification scheme, which is known to be a Σ-protocol, (3) Brands blind signature is of the form $\sigma(m') = ((z', a', b'), c', r')$ which has identical distribution to a transcript of a Σ-protocol, as we will explain below (4) given the secret key x and a valid transcript of Brands scheme: $(\hat{a}, c'_1, r'_1)$, where $\hat{a} = (z', a', b')$, then $\forall\ c'_2$ we can compute r_2' as: $r_2' = r_1' - c_1'x + c_2'x$ so that $(\hat{a}, c_2', r_2')$ is still a valid transcript. Brands blind signature is indeed a Σ-protocol: (a) it is a three-round protocol, (b) for any h and any pair of accepting conversations $(\hat{a}, c_1', r_1')$ and $(\hat{a}, c_2', r_2')$ where $c_1' \neq c_2'$ one can efficiently compute x such that $h = g^x$ and (c) there exists a simulator S who on input h and a random c' picks r', m and z, solves for a', b', so he can output an accepting conversation of the form $((z', a', b'), c', r')$.

Thus, by applying Theorems 1 and 2, we rule out perfect and L-naive RO replay reductions for Brands' blind signatures.

Pointcheval and Stern [33] suggest that for their proof approach to work, the public key of the scheme should have more than one secret key associated with it. One could modify Brands' scheme similarly to how the original Schnorr blind signature was modified to obtain the variant that Pointcheval and Stern proved secure. In the full version [5] we propose such a modification; the public key of the signer will be of the form $H = G_1^{w_1} G_2^{w_2}$ where (H, G_1, G_2) are public and (w_1, w_2) are the secret key. As a blind signature, the resulting signature scheme is inferior, in efficiency, to the provably secure variant of the Schnorr blind signature. As far as its use in an electronic cash protocol is concerned, it is still an open problem whether provable guarantees against double-spending can be given for our modification of Brands.

References

1. http://www.nist.gov/nstic/pilot-projects2012.html (2012)
2. Abe, M.: A secure three-move blind signature scheme for polynomially many signatures. In: Pfitzmann, B. (ed.) EUROCRYPT 2001. LNCS, vol. 2045, pp. 136–151. Springer, Heidelberg (2001)
3. Abe, M., Fuchsbauer, G., Groth, J., Haralambiev, K., Ohkubo, M.: Structure-preserving signatures and commitments to group elements. In: Rabin, T. (ed.) CRYPTO 2010. LNCS, vol. 6223, pp. 209–236. Springer, Heidelberg (2010)
4. Abe, M., Okamoto, T.: Provably Secure Partially Blind Signatures. In: Bellare, M. (ed.) CRYPTO 2000. LNCS, vol. 1880, pp. 271–286. Springer, Heidelberg (2000)
5. Baldimtsi, F., Lysyanskaya, A.: On the security of one-witness blind signature schemes. Cryptology ePrint Archive, Report 2012/197 (2012)
6. Bellare, M., Namprempre, C., Pointcheval, D., Semanko, M.: The one-more-rsa-inversion problems and the security of chaum's blind signature scheme. Journal of Cryptology 16, 185–215 (2003)
7. Bellare, M., Palacio, A.: GQ and schnorr identification schemes: Proofs of security against impersonation under active and concurrent attacks. In: Yung, M. (ed.) CRYPTO 2002. LNCS, vol. 2442, pp. 162–177. Springer, Heidelberg (2002)
8. Bellare, M., Rogaway, P.: Random oracles are practical: A paradigm for designing efficient protocols. In: ACM-CCS 1993, pp. 62–73. ACM (1993)
9. Boldyreva, A.: Threshold signatures, multisignatures and blind signatures based on the gap-diffie-hellman-group signature scheme. In: Desmedt, Y.G. (ed.) PKC 2003. LNCS, vol. 2567, pp. 31–46. Springer, Heidelberg (2002)
10. Boneh, D., Venkatesan, R.: Breaking RSA may not be equivalent to factoring. In: Nyberg, K. (ed.) EUROCRYPT 1998. LNCS, vol. 1403, pp. 59–71. Springer, Heidelberg (1998)

11. Brands, S.: An efficient off-line electronic cash system based on the representation problem. In CWI Technical Report CS-R9323

12. Brands, S.: Untraceable off-line cash in wallets with observers. In: Stinson, D.R. (ed.) CRYPTO 1993. LNCS, vol. 773, pp. 302–318. Springer, Heidelberg (1994)

13. Camenisch, J., Hohenberger, S., Lysyanskaya, A.: Compact E-Cash. In: Cramer, R. (ed.) EUROCRYPT 2005. LNCS, vol. 3494, pp. 302–321. Springer, Heidelberg (2005)

14. Camenisch, J., Lysyanskaya, A.: An efficient system for non-transferable anonymous credentials with optional anonymity revocation. In: Pfitzmann, B. (ed.) EUROCRYPT 2001. LNCS, vol. 2045, pp. 93–118. Springer, Heidelberg (2001)

15. Chaum, D., Fiat, A., Naor, M.: Untraceable electronic cash. In: Goldwasser, S. (ed.) CRYPTO 1988. LNCS, vol. 403, pp. 319–327. Springer, Heidelberg (1990)

16. Chaum, D.: Blind signatures for untraceable payment. In: CRYPTO 1982, pp. 199–203 (1982)

17. Coron, J.-S.: On the exact security of full domain hash. In: Bellare, M. (ed.) CRYPTO 2000. LNCS, vol. 1880, pp. 229–235. Springer, Heidelberg (2000)

18. Schnorr, C.P., Jakobsson, M.: Security of discrete log cryptosystems in the random oracle + generic model. In: The Mathematics of Public-Key Cryptography, The Fields Institute (1999)

19. Fischlin, M.: Black-box reductions and separations in cryptography. In: Mitrokotsa, A., Vaudenay, S. (eds.) AFRICACRYPT 2012. LNCS, vol. 7374, pp. 413–422. Springer, Heidelberg (2012)

20. Fischlin, M., Fleischhacker, N.: Limitations of the meta-reduction technique: The case of schnorr signatures. In: Johansson, T., Nguyen, P.Q. (eds.) EUROCRYPT 2013. LNCS, vol. 7881, pp. 444–460. Springer, Heidelberg (2013)

21. Fischlin, M., Lehmann, A., Ristenpart, T., Shrimpton, T., Stam, M., Tessaro, S.: Random oracles with(out) programmability. In: Abe, M. (ed.) ASIACRYPT 2010. LNCS, vol. 6477, pp. 303–320. Springer, Heidelberg (2010)

22. Fischlin, M., Schröder, D.: On the impossibility of three-move blind signature schemes. In: Gilbert, H. (ed.) EUROCRYPT 2010. LNCS, vol. 6110, pp. 197–215. Springer, Heidelberg (2010)

23. Guillou, L.C., Quisquater, J.-J.: A practical zero-knowledge protocol fitted to security microprocessor minimizing both transmission and memory. In: Günther, C.G. (ed.) EUROCRYPT 1988. LNCS, vol. 330, pp. 123–128. Springer, Heidelberg (1988)

24. Hazay, C., Katz, J., Koo, C.-Y., Lindell, Y.: Concurrently-secure blind signatures without random oracles or setup assumptions. In: Vadhan, S.P. (ed.) TCC 2007. LNCS, vol. 4392, pp. 323–341. Springer, Heidelberg (2007)

25. Nielsen, J.B.: Separating random oracle proofs from complexity theoretic proofs: The non-committing encryption case. In: Yung, M. (ed.) CRYPTO 2002. LNCS, vol. 2442, pp. 111–126. Springer, Heidelberg (2002)

26. Okamoto, T.: Provably secure and practical identification schemes and corresponding signature schemes. In: Brickell, E.F. (ed.) CRYPTO 1992. LNCS, vol. 740, pp. 31–53. Springer, Heidelberg (1993)

27. Okamoto, T., Ohta, K.: Universal electronic cash. In: Feigenbaum, J. (ed.) CRYPTO 1991. LNCS, vol. 576, pp. 324–337. Springer, Heidelberg (1992)

28. Paillier, P., Vergnaud, D.: Discrete-log-based signatures may not be equivalent to discrete log. In: Roy, B. (ed.) ASIACRYPT 2005. LNCS, vol. 3788, pp. 1–20. Springer, Heidelberg (2005)

29. Paquin, C.: U-prove cryptographic specification v1.1. In Microsoft Technical Report (February 2011), http://connect.microsoft.com/site1188

30. Paquin, C.: U-prove technology overview v1.1. In Microsoft Technical Report (February 2011), http://connect.microsoft.com/site1188
31. European Parliament and Council of the European Union. Directive 2009/136/ec. In Official Journal of the European Union (2009)
32. Pass, R.: Limits of provable security from standard assumptions. In: STOC, pp. 109–118 (2011)
33. Pointcheval, D., Stern Provably, J.: secure blind signature schemes. In: Kim, K.-C., Matsumoto, T. (eds.) ASIACRYPT 1996. LNCS, vol. 1163, pp. 252–265. Springer, Heidelberg (1996)
34. Pointcheval, D., Stern, J.: Security arguments for digital signatures and blind signatures. Journal of Cryptology 13, 361–396 (2000)
35. Schnorr, C.-P.: Efficient identification and signatures for smart cards. In: Brassard, G. (ed.) CRYPTO 1989. LNCS, vol. 435, pp. 239–252. Springer, Heidelberg (1990)
36. IBM Security Team. Specification of the identity mixer cryptographic library, version 2.3.0. In IBM Research Report (2010)

Unconditionally Secure and Universally Composable Commitments from Physical Assumptions

Ivan Damgård[1,*] and Alessandra Scafuro[2,**]

[1] Dept. of Computer Science, Aarhus University, Denmark
[2] Dept. of Computer Science, UCLA, USA

Abstract. We present a constant-round unconditional black-box compiler that transforms any ideal (i.e., statistically-hiding and statistically-binding) straight-line extractable commitment scheme, into an extractable and equivocal commitment scheme, therefore yielding to UC-security [9]. We exemplify the usefulness of our compiler by providing two (constant-round) instantiations of ideal straight-line extractable commitment based on (malicious) PUFs [36] and *stateless* tamper-proof hardware tokens [26], therefore achieving the first unconditionally UC-secure commitment with malicious PUFs and stateless tokens, respectively. Our constructions are secure for adversaries creating arbitrarily malicious stateful PUFs/tokens.

Previous results with malicious PUFs used either computational assumptions to achieve UC-secure commitments or were unconditionally secure but only in the indistinguishability sense [36]. Similarly, with stateless tokens, UC-secure commitments are known only under computational assumptions [13,24,15], while the (not UC) unconditional commitment scheme of [23] is secure only in a weaker model in which the adversary is not allowed to create stateful tokens.

Besides allowing us to prove feasibility of unconditional UC-security with (malicious) PUFs and stateless tokens, our compiler can be instantiated with any ideal straight-line extractable commitment scheme, thus allowing the use of various setup assumptions which may better fit the application or the technology available.

Keywords: UC-security, hardware assumptions, unconditional security, commitment scheme.

* The authors acknowledge support from the Danish National Research Foundation and The National Science Foundation of China (under the grant 61061130540) for the Sino-Danish Center for the Theory of Interactive Computation, within which part of this work was performed; and also from the CFEM research center (supported by the Danish Strategic Research Council) within which part of this work was performed.
** Work done while visiting Aarhus University. Research supported in part by NSF grants CCF-0916574; IIS-1065276; CCF-1016540; CNS-1118126; CNS-1136174; and Defense Advanced Research Projects Agency through the U.S. Office of Naval Research under Contract N00014-11-1-0392.

K. Sako and P. Sarkar (Eds.) ASIACRYPT 2013, Part II, LNCS 8270, pp. 100–119, 2013.

1   Introduction

Unconditional security guarantees that a protocol is secure even when the adversary is unbounded. While it is known how to achieve unconditional security for multi-party functionalities in the plain model assuming honest majority [4,14], obtaining unconditionally secure *two-party* computation is impossible in the plain model. In fact, for all non-trivial two-party functionalities, achieving unconditional security requires some sort of (physical) setup assumption.

Universally composable (UC) security [9] guarantees that a protocol is secure even when executed concurrently with many other instances of any arbitrary protocol. This strong notion captures the real world scenarios, where typically many applications are run concurrently over the internet, and is therefore very desirable to achieve. Unfortunately, achieving UC-security in the plain model is impossible [11].

Hence, constructing 2-party protocols which are unconditionally secure or universally composable requires the employment of some setup. One natural research direction is to explore which setup assumptions suffice to achieve (unconditional) UC-security, as well as to determine whether (or to what extent) we can *reduce* the amount of trust in a third party. Towards this goal, several setup assumptions have been explored by the community.

In [12] Canetti et. al show that, under computational assumptions, any functionality can be UC-realized assuming the existence of a trusted Common Reference String (CRS). Here, the security crucially relies on the CRS being honestly sampled. Hence, security in practice would typically rely on a third party sampling the CRS honestly and security breaks down if the third party is not honest. Similar arguments apply to various assumptions like "public-key registration" services [3,10].

Another line of research explores "physical" setup assumptions. Based on various types of noisy channels, unconditionally secure Bit Commitment (BC) and Oblivious Transfer (OT) can be achieved [16,17] for two parties, but UC security has not been shown for these protocols and in fact seems non-trivial to get for the case of [17].

In [26] Katz introduces the assumption of the existence of tamper-proof hardware tokens. The assumption is supported by the possibility of implementing tamper-proof hardware using current available technology (e.g., smart cards). A token is defined as a physical device (a wrapper), on which a player can upload the code of any functionality, and the assumption is that any adversary cannot tamper with the token. Namely, the adversary has only black-box access to the token, i.e., it cannot do more then observing the input/output characteristic of the token. The main motivation behind this new setup assumption is that it allows for a *reduction of trust*. Indeed in Katz's model tokens are not assumed to be trusted (i.e., produced by a trusted party) and the adversary is allowed to create a token that implements an arbitrary malicious function instead of the function dictated by the protocol. (However, it is assumed that once the token is sent away to the honest party, it cannot communicate with its creator. This assumption is necessary, as otherwise we are back to the plain model).

A consequence of this model is that the security of a player now depends only on its *own token* being good and holds even if tokens used by other players are not genuine! This new setup assumptions has gained a lot of interest and several works after [26] have shown that unconditional UC-security is possible [31,24], even using a single *stateful* token [21,22]. Note that a stateful token, in contrast with a *stateless* token, requires an updatable memory that can be subject to reset attacks. Thus, ensuring tamper-proofness for a stateful token seems to be more demanding than for a stateless token, and hence having protocols working with stateless tokens is preferable.

However, the only constructions known for stateless tokens require computational assumptions [13,29,24,15] and a non-constant number of rounds (if based on one-way functions only). In fact, intuitively it seems challenging to achieve unconditional security with stateless tokens: A stateless token runs always on the same state, thus an unbounded adversary might be able to extract the secret state after having observed only a polynomial number of the token's outputs. This intuition is confirmed by [23] where it is proved that unconditional OT is impossible using stateless tokens. On the positive side, [23] shows an unconditional commitment scheme (not UC) based on stateless tokens, but the security of the scheme holds only if the adversary is not allowed to create malicious stateful tokens. This is in contrast with the standard tamper-proof hardware model, where the adversary is allowed to construct any arbitrary malicious (hence possibly stateful) token. Indeed, it seems difficult in practice to detect whether an adversary sends a stateless or a stateful token. Therefore, the question of achieving unconditional commitments (UC-secure or not) in the standard stateless token model (where an adversary possibly plays with stateful tokens) is still open.

In this work we provide a positive answer showing the first UC-secure unconditional commitment scheme with stateless tokens.

Following the approach of [26], Brzuska et al. in [7] propose a new setup assumption for achieving UC security, which is the existence of Physically Uncloneable Functions (PUFs). PUFs have been introduced by Pappu in [38,37], and since then have gained a lot of interest for cryptographic applications [2,42,1,40]. A PUF is a physical noisy source of randomness. In other words a PUF is a device implementing a function whose behavior is unpredictable even to the manufacturer. The reason is that even knowing the exact manufacturing process there are parameters that cannot be controlled, therefore it is assumed infeasible to construct two PUFs with the same challenge-response behavior. A PUF is noisy in the sense that, when queried twice with the same challenge, it can output two different, although close, responses. Fuzzy extractors are applied to PUF's responses in order to reproduce a unique response for the same challenge. The "PUF assumption" consists in assuming that PUFs satisfy two properties: 1) unpredictability: the distribution implemented by a PUF is unpredictable. That is, even after a polynomial number of challenge/response pairs have been observed, the response on any new challenge (sufficiently far from the ones observed so far) is unpredictable; this property is *unconditional*; 2) uncloneability: as a PUF is the output of a physical uncontrollable manufacturing process, it is assumed that creating two identical PUFs is hard even for the manufacturer. This property is called hardware uncloneability. Software uncloneability corresponds to the hardness of modeling the function implemented by the PUF and is enforced by unpredictability (given that the challenge/response space of the PUF is adequately large). Determining whether (or to what extent) current PUF candidates actually satisfy the PUF assumption is an active area of research (e.g., [27,5]) but is out of the scope of this work. For a survey on PUF's candidates the reader can refer to [30], while a security analysis of silicon PUFs is provided in [27].

Designing PUF-based protocols is fundamentally different than for other hardware tokens. This is due to the fact that the functional behavior of a PUF is unpredictable even for its creator. Brzuska et al. modeled PUFs in the UC-setting by formalizing the ideal PUF functionality. They then provided constructions for Unconditional UC Oblivious Transfer and Bit Commitment. However, their UC-definition of PUFs assumes that all PUFs are *trusted*. Namely, they assume that even a malicious player creates PUFs honestly, following the prescribed generation procedure. This assumption seems too optimistic as it implies that an adversary must not be capable of constructing hardware that "looks like" a PUF but that instead computes some arbitrary function. The consequence of assuming that all PUFs are trusted is that the security of a player depends on the PUFs created by other players. (Indeed, in the OT protocol of [7], if the receiver replaces the PUF with hardware implementing some predictable function, the security of the sender is violated).

In [36] Ostrovsky et al. extend the ideal PUF functionality of [7] in order to model the adversarial behavior of creating and using "malicious PUFs". A malicious PUF is a physical device for which the security properties of a PUF are not guaranteed. As such, it can be a device implementing *any* function chosen by the adversary, so that the adversary might have full control on the answers computed by its own "PUF". Similarly to the hardware-token model, a malicious PUF cannot communicate with the creator once is sent away. A malicious PUF can, of course, be stateful. The major advantage of the malicious PUF model is that the security of a player depends only on the goodness of its own PUFs. Obviously, the price to pay is that protocols secure in presence of malicious PUFs are more complex than protocols designed to deal only with honest PUFs. Nevertheless, [36] shows that even with malicious PUFs it is possible to achieve UC-secure computations relying on computational assumptions. They also show an unconditional commitment scheme which is secure only in the indistinguishability sense. Achieving unconditional UC-secure commitments (and general secure computations) is left as an open problem in [36].

In this paper, we give a (partial) positive answer to this open problem by providing the first construction of unconditional UC-secure Bit Commitment in the malicious PUFs model. Whether unconditional OT (and thus general secure computation) is possible with malicious PUFs is still an interesting open question. Intuitively, since PUFs are stateless devices, one would think to apply the arguments used for the impossibility of unconditional OT with stateless tokens [23]. However, due to the unpredictability property of PUFs which holds unconditionally, such arguments do not carry through. Indeed, as long as there is at least one honest PUF in the system, there is enough entropy, and this seems to defeat the arguments used in [23]. On the other hand, since a PUF is in spirit just a "random function", it is not clear how to implement the OT functionality when only one of the players uses honest PUFs.

Van Dijk and Rührmair in [19] also consider adversaries who create malicious PUFs, that they call "bad PUFs" and they consider only the stand-alone setting. They show that unconditional OT is impossible in the bad PUF model but this impossibility proof works assuming that also honest parties play with bad PUFs. Thus, such impossibility proof has no connection to the question of achieving unconditional OT in the malicious PUF model (where honest parties play with honest PUFs).

Our Contribution. In this work we provide a tool for constructing UC-secure commitments given any straight-line extractable commitment. This tool allows us to show feasibility results for unconditional UC-secure protocols in the stateless token model and in the malicious PUF model. More precisely, we provide an unconditional black-box compiler that transforms any ideal (i.e., statistically hiding and binding) straight-line extractable commitment into a UC-secure commitment. The key advantage of such compiler is that one can implement the ideal extractable commitment with the setup assumption that is more suitable to the application and the technology available.

We then provide an implementation of the ideal extractable commitment scheme in the malicious PUFs model of [36]. Our construction builds upon the (stand-alone) unconditional BC scheme shown in [36] [1] which is not extractable. By plugging our extractable commitment scheme in our compiler we obtain the first unconditional UC-secure commitment with malicious PUFs.

We then construct ideal extractable commitments using stateless tokens. We use some of the ideas employed for the PUF construction, but implement them with different techniques. Indeed, while PUFs are intrinsically unpredictable and even having oracle access to a PUF an unbounded adversary cannot predict the output on a new query, with stateless tokens we do not have such guarantee. Our protocol is secure in the standard token model, where the adversary has no restriction and can send malicious stateful tokens. By plugging such protocol in our compiler, we achieve the first unconditional UC-secure commitment scheme with stateless tokens. Given that unconditional OT is impossible with stateless tokens, this result completes the picture concerning feasibility of unconditional UC-security in this model.

Related Work. Our compiler can be seen as a generalization of the black-box trapdoor commitment given by Pass and Wee [39] which is secure only in the

---

[1] For completeness, we would like to mention that [41] claims an "attack" on such construction. Such "attack" however arises only due to misunderstanding of conventions used to write protocol specifications and does not bear any security threat. The reader can refer to the discussion of [35] (full version of [36]) at Pag. 7, paragraph "On [RvD13]", line 20–40 for more details.

computational stand-alone setting. Looking ahead to our constructions of extractable commitment, the idea of querying the hardware token with the opening of the commitment was first used by Müller-Quade and Unruh in [32,33], and later by Chandran et al. in [13]. The construction of [13] builds UC-secure multiple commitments on top of extractable commitments. Their compiler requires computational assumptions, logarithmic number of rounds and crucially uses cryptographic primitives in a non-black box manner.

Remark 1. In the rest of the paper it is assumed that even an unbounded adversary can query the PUF/token only a polynomial number of times. We stress that this is not a restriction of our work but it is a necessary assumption in all previous works achieving unconditional security with PUFs and stateless tokens (see pag.15 of [8] for PUFs, and pag. 5 of [23] for stateless tokens). Indeed, if we allowed the adversary to query the PUF/token on all possible challenges, then she can derive the truth table implemented by the physical device.

2 Definitions

Notation. We denote the security parameter by n, and the property of a probabilistic algorithm whose number of steps is polynomial in its security parameter, by PPT. We denote by $(v_A, v_B) \leftarrow \langle A(a), B(b) \rangle(x)$ the local outputs of A and B of the random process obtained by having A and B activated with independent random tapes, interacting on common input x and on (private) auxiliary inputs a to A and b to B. When the common input x is the security parameter, we omit it. If A is a probabilistic algorithm we use $v \xleftarrow{\$} A(x)$ to denote the output of A on input x assigned to v. We denote by $\text{view}_A(A(a), B(b))(x)$ the view of A of the interaction with player B, i.e., its values is the transcript $(\gamma_1, \gamma_2, ..., \gamma_t; r)$, where the γ_i's are all the messages exchanged and r is A's coin tosses. We use notation $[n]$ to denote the set $\{1, \ldots, n\}$. Let P_1 and P_2 be two parties running protocol (A,B) as sub-routine. When we say that party "P_1 runs $\langle A(\cdot), B(\cdot) \rangle(\cdot)$ with P_2" we always mean that P_1 executes the procedure of party A and P_2 executes the procedure of party B. In the following definitions we assume that the adversary has auxiliary information, and assume that parties are stateful.

2.1 Ideal Extractable Commitment Scheme

We denote by $\mathcal{F}_{\text{aux}}$ an auxiliary set-up functionality accessed by the real world parties (and by the extractor).

Definition 1 (Ideal Commitment Scheme in the $\mathcal{F}_{\text{aux}}$-hybrid model). *A commitment scheme is a tuple of PPT algorithms* Com = (C, R) *having access to an ideal set-up functionality* $\mathcal{F}_{\text{aux}}$, *implementing the following two-phase functionality. Given to* C *an input* $b \in \{0, 1\}$, *in the first phase called commitment phase,* C *interacts with* R *to commit to the bit* b. *We denote this interaction by* $((\mathbf{c}, d), \mathbf{c}) \leftarrow \langle \mathsf{C}(\mathsf{com}, b), \mathsf{R}(\mathsf{recv}) \rangle$ *where* c *is the transcript of the (possibly interactive) commitment phase and d is the decommitment data. In the* *second phase, called* decommitment phase, C *sends* (b, d) *and* R *finally outputs* "accept" *or* "reject" *according to* $(\mathbf{c}, d, b)$. *In both phases parties could access to* $\mathcal{F}_{\text{aux}}$. Com $= (\mathsf{C}, \mathsf{R})$ *is an* ideal *commitment scheme if it satisfies the following properties.*

Completeness. *For any* $b \in \{0, 1\}$, *if* C *and* R *follow their prescribed strategy then* R *accepts the commitment* $\mathbf{c}$ *and the decommitment* (b, d) *with probability 1.*

Statistically Hiding. *For any malicious receiver* R^* *the ensembles* $\{\text{view}_{\mathsf{R}^*}$ $(\mathsf{C}(\text{com}, 0), \mathsf{R}^*) (1^n)\}_{n \in \mathbb{N}}$ *and* $\{\text{view}_{\mathsf{R}^*}(\mathsf{C}(\text{com}, 1), \mathsf{R}^*) (1^n)\}_{n \in \mathbb{N}}$ *are statistically indistinguishable, where* $\text{view}_{\mathsf{R}^*} (\mathsf{C}(\text{com}, b), \mathsf{R}^*)$ *denotes the view of* R^* *restricted to the commitment phase.*

Statistically Binding. *For any malicious committer* C^*, *there exists a negligible function* ϵ, *such that* C^* *succeeds in the following game with probability at most* $\epsilon(n)$: *On security parameter* 1^n, C^* *interacts with* R *in the commitment phase obtaining the transcript* $\mathbf{c}$. *Then* C^* *outputs pairs* $(0, d_0)$ *and* $(1, d_1)$, *and succeeds if in the decommitment phase,* $\mathsf{R}(\mathbf{c}, d_0, 0) = \mathsf{R}(\mathbf{c}, d_1, 1) = \text{accept}$.

In this paper the term *ideal* is used to refer to a commitment which is statistically-hiding and statistically-binding.

Definition 2 (Interface Access to an Ideal Functionality $\mathcal{F}_{\text{aux}}$). *Let* $\Pi = (P_1, P_2)$ *be a two-party protocol in the* $\mathcal{F}_{\text{aux}}$-*hybrid model. That is, parties* P_1 *and* P_2 *need to query the ideal functionality* $\mathcal{F}_{\text{aux}}$ *in order to carry out the protocol. An algorithm* M *has* interface access *to the ideal functionality* $\mathcal{F}_{\text{aux}}$ *w.r.t. protocol* Π *if all queries made by either party* P_1 *or* P_2 *to* $\mathcal{F}_{\text{aux}}$ *during the protocol execution are observed (but not answered) by* M, *and* M *has oracle access to* $\mathcal{F}_{\text{aux}}$. *Consequently,* $\mathcal{F}_{\text{aux}}$ *can be a non programmable and non PPT functionality.*

Definition 3 (Ideal Extractable Commitment Scheme in the $\mathcal{F}_{\text{aux}}$ model). IdealExtCom $= (\mathsf{C}_{\text{ext}}, \mathsf{R}_{\text{ext}})$ *is an* ideal extractable *commitment scheme in the* $\mathcal{F}_{\text{aux}}$ *model if* $(\mathsf{C}_{\text{ext}}, \mathsf{R}_{\text{ext}})$ *is an ideal commitment and there exists a straight-line strict polynomial-time extractor* E *having* interface access *to* $\mathcal{F}_{\text{aux}}$, *that runs the commitment phase only and outputs a value* $b^* \in \{0, 1, \bot\}$ *such that, for all malicious committers* C^*, *the following properties are satisfied.*

Simulation: *the view generated by the interaction between* E *and* C^* *is identical to the view generated when* C^* *interacts with the honest receiver* R_{ext}: $\text{view}_{\mathsf{C}^*}^{\mathcal{F}_{\text{aux}}}(\mathsf{C}^*(\text{com}, \cdot), \mathsf{R}_{\text{ext}}(\text{recv})) \equiv \text{view}_{\mathsf{C}^*}^{\mathcal{F}_{\text{aux}}}(\mathsf{C}^*(\text{com}, \cdot), E)$

Extraction: *let* $\mathbf{c}$ *be a valid transcript of the commitment phase run between* C^* *and* E. *If* E *outputs* $\bot$ *then probability that* C^* *will provide an accepting decommitment is negligible.*

Binding: *if* $b^* \neq \bot$, *then probability that* C^* *decommits to a bit* $b \neq b^*$ *is negligible.*

2.2   Physically Uncloneable Functions

Here we recall the definition of PUFs taken from [7]. A Physically Uncloneable Function (PUF) is a noisy physical source of randomness. A PUF is evaluated with a physical stimulus, called the *challenge*, and its physical output, called the *response*, is measured. Because the processes involved are physical, the function implemented by a PUF can not necessarily be modeled as a mathematical function, neither can be considered computable in PPT. Moreover, the output of a PUF is noisy, namely, querying a PUF twice with the same challenge, could yield to different outputs.

A PUF-family $\mathcal{P}$ is a pair of (not necessarily efficient) algorithms Sample and Eval. Algorithm Sample abstracts the PUF fabrication process and works as follows. Given the security parameter in input, it outputs a PUF-index id from the PUF-family satisfying the security property (that we define soon) according to the security parameter. Algorithm Eval abstracts the PUF-evaluation process. On input a challenge s, it evaluates the PUF on s and outputs the response σ. A PUF-family is parametrized by two parameters: the bound on the noisy of the answers d_{noise}, and the size of the range rg. A PUF is assumed to satisfy the bounded noise condition, that is, when running $\text{Eval}(1^n, \text{id}, s)$ twice, the Hamming distance of any two responses σ_1, σ_2 is smaller than $d_{\text{noise}}(n)$. We assume that the challenge space of a PUF is the set of strings of a certain length.

Security Properties. We assume that PUFs enjoy the properties of *uncloneability* and *unpredictability*. Unpredictability is modeled in [7] via an entropy condition on the PUF distribution. Namely, given that a PUF has been measured on a polynomial number of challenges, the response of the PUF evaluated on a new challenge has still a significant amount of entropy. The following definition automatically implies uncloneability (see [8] pag. 39 for details).

Definition 4 (Unpredictability). *A (rg, d_{noise})-PUF family $\mathcal{P} = $ (Sample, Eval) for security parameter n is $(d_{\text{min}}(n), m(n))$-unpredictable if for any $s \in \{0, 1\}^n$ and challenge list $\mathcal{Q} = (s_1, \ldots, s_{\text{poly}(n)})$, one has that, if for all $1 \leq k \leq \text{poly}(n)$ the Hamming distance satisfies $\text{dis}_{\text{ham}}(s, s_k) \geq d_{\text{min}}(n)$, then the average min-entropy satisfies $\tilde{H}_\infty(\text{PUF}(s)|\text{PUF}(\mathcal{Q})) \geq m(n)$, where $\text{PUF}(\mathcal{Q})$ denotes a sequence of random variables $\text{PUF}(s_1), \ldots, \text{PUF}(s_{\text{poly}(n)})$ each one corresponding to an evaluation of the PUF on challenge s_k. Such a PUF-family is called a $(rg, d_{\text{noise}}, d_{\text{min}}, m)$- PUF family.*

Fuzzy Extractors. The output of a PUF is noisy. That is, querying the PUF twice with the same challenge, one might obtain two distinct responses σ, σ', that are at most d_{noise} apart in hamming distance. Fuzzy extractors of Dodis et al. [20] are applied to the outputs of the PUF, to convert such noisy, high-entropy measurements into *reproducible* randomness. Very roughly, a fuzzy extractor is a pair of efficient randomized algorithms (FuzGen, FuzRep), and it is applied to PUFs 'responses as follows. FuzGen takes as input an ℓ-bit string, that is the PUF's response σ, and outputs a pair (p, st), where st is a uniformly distributed string, and p is a public helper data string. FuzRep takes as input the PUF's noisy response σ' and the helper data p and generates the very same string st obtained with the original measurement σ.

The security property of fuzzy extractors guarantees that, if the min-entropy of the PUF's responses are greater than a certain parameter m, knowledge of the public data p only, without the measurement σ, does not give any information on the secret value st. The correctness property, guarantees that all pairs of responses σ, σ' that are close enough, i.e., their hamming distance is less then a certain parameter t, will be recovered by FuzRep to the same value st generated by FuzGen. In order to apply fuzzy extractors to PUF's responses it is sufficient to pick an extractor whose parameters match with the parameter of the PUF being used. For formal definitions of fuzzy extractors and PUFs the reader is referred to the full version [18].

Ideal Functionalities for Malicious PUFs and Stateless Tokens. We follow the malicious PUF model introduced in [36]. In this model, the adversary is allowed to create arbitrarily malicious PUFs. Very informally, a malicious PUF is any physical device that "looks like" a PUF but it implements an arbitrary malicious, possibly stateful, function. Such adversarial behaviour has been modeled in [36] by extending the ideal functionality proposed in [7]. We denote by $\mathcal{F}_{\mathsf{PUF}}$ the ideal functionality for malicious PUF. A stateless token is a wrapper that can be programmed with any arbitrary stateless function. Tokens are modeled by [26,13] as the ideal functionality $\mathcal{F}_{\mathsf{wrap}}$. For lack of space, the formal definitions of the functionalities $\mathcal{F}_{\mathsf{PUF}}$ and $\mathcal{F}_{\mathsf{wrap}}$ together with the UC-definition are provided in the full version [18].

3   The Compiler

In this section we show how to transform any ideal extractable commitment scheme into a protocol that UC-realizes the $\mathcal{F}_{\mathsf{com}}$ functionality, unconditionally. Such transformation is based on the following building blocks.

Extractable Blobs. "Blob" was used in [6] to denote a commitment. In this paper a blob is a *pair* of bit commitments such that the actual bit committed in the blob is the xor of the pair. The representation of a bit as its exclusive-or allows to prove equality of the bits committed in two blobs using commitments as black boxes. Let IdealExtCom be any ideal extractable commitment scheme satisfying Def. 3. If the commitment phase of IdealExtCom is interactive then the blob is the pair of transcripts obtained from the interaction. Procedures to create a blob of a bit b, and to reveal the bit committed in the blob, are the following.

> Blob(b): Committer picks bits b^0, b^1 uniformly at random such that $b = b^0 \oplus b^1$. It commits to b^0, b^1 (in parallel) running IdealExtCom as sub-routine and obtains commitment transcripts $\mathbf{c}^0, \mathbf{c}^1$, and decommitments d^0, d^1. Let $\mathbf{B} = (\mathbf{c}^0, \mathbf{c}^1)$ be the **blob** of b.

OpenBlob($\mathbf{B}$): Committer sends $(b^0, d^0), (b^1, d^1)$ to Receiver. Receiver accepts iff d^0, d^1 are valid decommitments of b^0, b^1 w.r.t. transcripts $(\mathbf{c}^0, \mathbf{c}^1)$ and computes $b = b^0 \oplus b^1$.

A blob inherits the properties of the commitment scheme used as sub-protocol. In particular, since IdealExtCom is used as sub-routine, each blob is statistically hiding/binding and straight-line extractable.

Equality of Blobs. Given the representation of a bit commitment as a blob, a protocol due to Kilian [28] allows to prove that two committed bits (two blobs) are equal, without revealing their values. We build upon this protocol to construct a "simulatable" version, meaning that (given some trapdoor) a simulator can prove equality of two blobs that are *not* equal. Let $\mathbf{B}_i, \mathbf{B}_j$ be two blobs. Let $b_i = (b_i^0 \oplus b_i^1)$ be the bit committed in $\mathbf{B}_i$, and $b_j = (b_j^0 \oplus b_j^1)$ be the bit committed in $\mathbf{B}_j$. Let P denote the prover and V the verifier. In the following protocol P proves to V that $\mathbf{B}_i$ and $\mathbf{B}_j$ are the commitment of the same bit (i.e., $b_i = b_j$).

BobEquality($\mathbf{B}_i, \mathbf{B}_j$)
1. V uniformly chooses $e \in \{0, 1\}$ and commits to e using IdealExtCom.
2. P sends $y = b_i^0 \oplus b_j^0$ to V.
3. V reveals e to P.
4. P reveals b_i^e and b_j^e (i.e., P sends decommitments d_i^e, d_j^e to V). V accepts iff $y = b_i^e \oplus b_j^e$.

Protocol BobEquality satisfies the following properties. **Soundness:** if $b_i \neq b_j$, any malicious prover P^* convinces V with probability negligibly close to $1/2$, that is the probability of guessing the challenge e. Here we are using the statistically hiding property of the ideal commitment IdealExtCom used to commit e. **Privacy:** If $b_i = b_j$ then after executing the protocol, the view of any verifier V^*, is independent of the actual value of b_i, b_j (given that $\mathbf{B}_i, \mathbf{B}_j$ were secure at the beginning of the protocol). **Simulation:** there exists a straight-line strictly PPT simulator SimFalse such that, for any $(\mathbf{B}_i, \mathbf{B}_j)$ that are not equal (i.e., $b_i \neq b_j$), for any malicious verifier V^*, produces a view that is statistically close to the case in which $(\mathbf{B}_i, \mathbf{B}_j)$ are equal (i.e., $b_i = b_j$) and V^* interacts with the honest P. The above properties are formally proved in the full version [18]. Note that the protocol uses blobs in a black-box way. Note also, that a blob can be involved in a single proof only.

3.1 Unconditional UC-secure Commitments from Ideal Extractable Commitments

We construct unconditional UC-secure commitments using *extractable* blobs and protocol BobEquality as building blocks. We want to implement the following idea. The committer sends two blobs of the same bit and proves that they are equal running protocol BobEquality. In the decommitment phase, it opens only one blob (a similar technique is used in [25], where instead the commitment scheme is crucially used in a non black-box way). The simulator extracts the bit of the committer by exploiting the extractability property of blobs. It equivocates by committing to the pair 0, 1 and cheating in the protocol BobEquality. In the opening phase, it then opens the blob corresponding to the correct bit. Because soundness of BobEquality is only 1/2 we amplify it via parallel repetition.

Specifically, the committer will compute n pairs of (extractable) blobs. Then it proves equality of each pair of blobs by running protocol BobEquality with the receiver. The commitment phase is successful if all equality proofs are accepting. In the decommitment phase, the committer opens one blob for each pair. The receiver accepts if the committer opens one blob for each consecutive pair and all revealed blobs open to the same bit. The construction is formally described in Fig. 1.

Protocol UCComCompiler

Committer's input: $b \in \{0, 1\}$.
Commitment Phase
1. Committer: run Blob(b) $2n$ times. Let $\mathbf{B}_1, \ldots, \mathbf{B}_{2n}$ be the **blobs** obtained.
2. Committer $\Leftrightarrow$ Receiver: for $i = 1$; $i = i + 2$; $i \leq 2n - 1$; run BobEquality($\mathbf{B}_i, \mathbf{B}_{i+1}$).
3. Receiver: if all equality proofs are accepting, accept the commitment phase.

Decommitment Phase
1. Committer: for $i = 1$; $i = i + 2$; $i \leq 2n - 1$; randomly choose $\ell \in \{i, i+1\}$ and run OpenBlob($\mathbf{B}_\ell$) with the Receiver.
2. Receiver: 1) check if Committer opened one blob for each consecutive pair; 2) if all n blobs open to the same bit b, output b and accept. Else output reject.

Fig. 1. UCComCompiler: Unconditional UC Commitment from any Ideal Extractable Commitment

Theorem 1. *If* IdealExtCom *is an ideal extractable commitment scheme in the* $\mathcal{F}_{\mathsf{aux}}$*-hybrid model, then protocol in Fig. 1 is an unconditionally UC-secure bit commitment scheme in the* $\mathcal{F}_{\mathsf{aux}}$*-hybrid model.*

Proof Sketch. To prove UC-security we have to show a straight-line simulator Sim which correctly simulates the view of the real-world adversary $\mathcal{A}$ and extracts her input. Namely, when simulating the malicious committer in the ideal world, Sim internally runs the real-world adversarial committer $\mathcal{A}$ simulating the honest receiver to her, so to extract the bit committed to by $\mathcal{A}$, and play it in the ideal world. This property is called extractability. When simulating the malicious receiver in the ideal world, Sim internally runs the real-world adversarial receiver $\mathcal{A}$ simulating the honest committer to her, without knowing the secret bit to commit to, but in such a way that it can be opened as any bit. This property is called equivocality. In the following, we briefly explain why both properties are achieved.

Straight-line Extractability. It follows from the straight-line extractability and binding of IdealExtCom and from the soundness of protocol BobEquality. Roughly, Sim works as follows. It plays the commitment phase as an honest receiver (and running the straight-line extractor of IdealExtCom having access to $\mathcal{F}_{\text{aux}}$). If all proofs of BobEquality are *successful*, Sim extracts the bits of each consecutive pair of blobs and analyses them as follows. Let $b \in \{0,1\}$. If all extracted pairs of bits are either (b,b) or $(\bar{b},b)$, (i.e. there are no pairs like $(\bar{b},\bar{b})$), it follows that, the only bit that $\mathcal{A}$ can successfully decommit to, is b. In this case, Sim plays the bit b in the ideal world. If there is at least a pair (b,b) and a pair $(\bar{b},\bar{b})$, then $\mathcal{A}$ cannot provide any accepting decommitment (indeed, due to the binding of blobs, $\mathcal{A}$ can only open the bit b from one pair, and the bit $\bar{b}$ from another pair, thus leading the receiver to reject). In this case Sim sends a random bit to the ideal functionality. If all the pairs of blobs are not equal, i.e., all pairs are either $(\bar{b},b)$ or $(b,\bar{b})$, then $\mathcal{A}$ can later decommit to any bit. In this case the simulator fails in the extraction of the bit committed, and it aborts. Note that, this case happens only when *all* the pairs are not equal. Thus $\mathcal{A}$ was able to cheat in all executions of BobEquality. Due to the soundness of BobEquality, this event happens with probability negligible close to 2^{-n}.

Straight-line Equivocality. It follows from the simulation property of BobEquality. Sim prepares n pairs of not equal blobs. Then it cheats in all executions of BobEquality, by running the straight-line simulator associated to this protocol. In the decommitment phase, after having received the bit to decommit to, for each pair, Sim reveals the blob corresponding to the correct bit.

Note that, in both cases Sim crucially uses the extractor associated to IdealExtCom, that in turn uses the access to $\mathcal{F}_{\text{aux}}$. The formal proof of the above theorem can be found in the full version [18].

In Section 4 we show an implementation of IdealExtCom with malicious PUFs, while in Section 5, we show how to implement IdealExtCom using stateless token. By plugging such implementations in protocol UCComCompiler we obtain the first unconditional UC-secure commitment scheme with malicious PUFs (namely, in the $\mathcal{F}_{\text{PUF}}$-hybrid model), and stateless tokens (namely, in the $\mathcal{F}_{\text{wrap}}$-hybrid model).

4   Ideal Extractable Commitment from (Malicious) PUFs

In this section we show a construction of ideal extractable commitment in the $\mathcal{F}_{\text{PUF}}$ model. Our construction builds upon the ideal commitment scheme presented in [36]. For simplicity, in the informal description of the protocol we omit mentioning the use of fuzzy extractors.

Ideal Commitment Scheme in the $\mathcal{F}_{\mathsf{PUF}}$ Model (from [36]). The idea behind the protocol of [36], that we denote by $\mathsf{CPuf} = (\mathsf{C}_{\mathsf{CPuf}}, \mathsf{R}_{\mathsf{CPuf}})$, is to turn Naor's commitment scheme [34] which is statistically binding but only computationally hiding, into statistically hiding and binding, by replacing the PRG with a (possibly *malicious*) PUF. Roughly, protocol CPuf goes as follows. At the beginning of the protocol, the committer creates a PUF, that we denote by $\mathcal{P}_S$. It preliminary queries $\mathcal{P}_S$ with a random string s (of n bits) to obtain the response σ_S (of $rg(3n)$ bits, where rg is the PUF's range) and finally sends the PUF $\mathcal{P}_S$ to the receiver. After receiving the PUF, the receiver sends a random string r (i.e., the first round of Naor's commitment) to the committer. To commit to a bit b, the committer sends $\mathbf{c} = \sigma_S \oplus (r \wedge b^{|r|})$ to the receiver. In the decommitment phase, the committer sends (b, s) to the receiver, who checks the commitment by querying $\mathcal{P}_S$ with s. For the formal description of CPuf the reader can refer to [36] or to the full version of this work [18].

Our Ideal Extractable Commitment Scheme in the $\mathcal{F}_{\mathsf{PUF}}$ Model. We transform CPuf into a *straight-line extractable* commitment using the following technique. We introduce a new PUF $\mathcal{P}_R$, sent by the receiver to the committer at the beginning of the protocol. Then we force the committer to query the PUF $\mathcal{P}_R$ with the opening of the commitment computed running CPuf. An opening of protocol CPuf is the value σ_S[2]. This allows the extractor, who has access to the interface of $\mathcal{F}_{\mathsf{PUF}}$, to extract the opening. The idea is that, from the transcript of the commitment (i.e., the value $\mathbf{c} = \sigma_S \oplus (r \wedge b)$) and the queries made to $\mathcal{P}_R$, (the value σ_S) the bit committed if fully determined.

To force the committer to query $\mathcal{P}_R$ with the correct opening, we require that it commits to the answer σ_R obtained by $\mathcal{P}_R$. Thus, in the commitment phase, the committer runs two instances of CPuf. One instance, that we call ComBit, is run to commit to the secret bit b. The other instance, that we call $\mathsf{ComResp}$, is run to commit to the response of PUF $\mathcal{P}_R$, queried with the opening of ComBit. In the decommitment phase, the receiver gets $\mathcal{P}_R$ back, along with the openings of both the bit and the PUF-response. Then it queries $\mathcal{P}_R$ with the opening of ComBit, and checks if the response is consistent with the string committed in $\mathsf{ComResp}$. Due to the unpredictability of PUFs, the committer cannot guess the output of $\mathcal{P}_R$ on the string σ_S without querying it. Due to the statistically binding of CPuf, the committer cannot postpone querying the PUF in the decommitment phase. Thus, if the committer will provide a valid decommitment, the extractor would have observed the opening already in the commitment phase with all but negligible probability.

However, there is one caveat. The unpredictability of PUFs is guaranteed only for queries that are sufficiently apart from each other. Which means that, given a challenge/response pair (c, r), the response on any strings c' that is "close" in hamming distance to c ("close" means that $\mathsf{dis}_{\mathsf{ham}}(c, c') \leq d_{\mathsf{min}}$), could be predictable. Consequently, a malicious committer could query the PUF with a

---

[2] In the actual implementation we require the committer to query $\mathcal{P}_R$ with the output of the fuzzy extractor st_S, i.e., $(st_S, p_S) \leftarrow \mathsf{FuzGen}(\sigma_S)$.

string that is only "close" to the opening. Then, given the answer to such a query, she could predict the answer to the actual opening, *without* querying the PUF. Hence, the extraction fails.

We overcome this problem by using Error Correcting Codes, in short ECC. The property of an ECC with distance parameter dis, is that any pair of strings having hamming distance dis, decodes to a unique string. Therefore, we modify the previous approach asking the committer to query PUF $\mathcal{P}_R$ with the *encoding* of the opening, i.e., Encode(σ_S). In this way, all queries that are "too close" in hamming distance decode to the same opening, and the previous attack is defeated. Informally, hiding and biding follow from hiding and binding of CPuf. Extractability follows from the statistically biding of CPuf, the unpredictability of $\mathcal{P}_R$ and the properties of ECC. The protocol is formally described in Fig. 2. In the full version [18] we discuss the parameters of the PUF and how to prevent the replacement of a honest PUF by the adversary, and we provide the proof of the following theorem.

Theorem 2. *If* CPuf *is an Ideal Commitment in the* $\mathcal{F}_{\mathsf{PUF}}$-*model, then* ExtPuf *is an Ideal* Extractable *Commitment in the* $\mathcal{F}_{\mathsf{PUF}}$ *model.*

5   Ideal Extractable Commitments from Stateless Tokens

In this section we show how to construct ideal extractable commitments from stateless tokens. We first construct an ideal commitment scheme. Then, we use it as building block for constructing an ideal *extractable* commitment.

Ideal Commitment Scheme in the $\mathcal{F}_{\mathsf{wrap}}$ *Model.* Similarly to the construction with PUFs, we implement Naor's commitment scheme by replacing the PRG with a stateless token.

In the construction with PUFs, the PRG was replaced with a PUF that is inherently unpredictable. Now, we want to achieve statistically hiding using *stateless* token. The problem here is that we do not have unpredictability for free (as it happens with PUFs). Thus, we have to program the stateless token with a function that is, somehow, unconditionally unpredictable. Clearly, we cannot construct a token that implements a PRG. Indeed, after observing a few pairs of input/output, an unbounded receiver can extract the seed, compute all possible outputs, and break hiding. We use a point function following [23] . A point function f is a function that outputs always zero, except in a particular point x, in which it outputs a value y. Formally, $f : \{0,1\}^n \to \{0,1\}^m$ such that $f(x) = y$ and $f(x') = 0$, for all $x' \neq x$.

Thus, we adapt Naor's commitment scheme as follows. The committer picks a n-bit string x and a $3n$-bit string y and creates a stateless token that on input x outputs y, while it outputs 0 on any other input. The stateless token is sent to the receiver at the beginning of the protocol. After obtaining the token, receiver sends the Naor's first message, i.e., the random value r, to the committer. The committer commits to the bit b by sending $\mathbf{c} = y \oplus (r \wedge b^{|r|})$. In the decommitment

---

Protocol ExtPuf

ECC = (Encode, Decode) is a (N, L, d_{min}^1) error correcting code, where $L = \ell = 3n$. Parties use PUF family: $\mathcal{P}^1 = (rg^1, d_{noise}^1, d_{min}^1, m^1)$, with challenge size L. (FuzGen^1, FuzRep^1) is a $(m^1, \ell^1, t^1, \epsilon^1)$-fuzzy extractor of appropriate matching parameters. Protocol CPuf = (C_{CPuf}, R_{CPuf}) is run as sub-routine. Committer's Input: $b \in \{0, 1\}$.

Commitment Phase

1. Receiver R_{ExtPuf}: create PUF $\mathcal{P}_R$ and send it to C_{ExtPuf}.
2. **Commitment of the Secret Bit: ComBit.**
   $C_{ExtPuf} \Leftrightarrow R_{ExtPuf}$: run $\langle C_{CPuf}(com, b), R_{CPuf}(com) \rangle$ so that C_{ExtPuf} commits to bit b. Let $(st_S, p_S) \leftarrow$ FuzGen(σ_S) be the value obtained by C_{ExtPuf}, after applying the fuzzy extractor to the answer obtained from its own PUF $\mathcal{P}_S$ when running protocol ComBit.
3. Committer C_{ExtPuf}: Query $\mathcal{P}_R$ with Encode(st_S) and obtain response σ_R. If $\sigma_R = \bot$ (i.e., PUF $\mathcal{P}_R$ aborts), set $\sigma_R \leftarrow 0$. Compute $(st_R, p_R) \leftarrow$ FuzGen$^1(\sigma_R)$.
4. **Commitment of $\mathcal{P}_R$'s Response: ComResp.**
   $C_{ExtPuf} \Leftrightarrow R_{ExtPuf}$: run $\langle C_{CPuf}(com, st_R || p_R), R_{CPuf}(com) \rangle$ so that C_{ExtPuf} commits to the string $st_R || p_R$.

Decommitment Phase

1. $C_{ExtPuf} \Leftrightarrow R_{ExtPuf}$: run $\langle C_{CPuf}(open, b), R_{CPuf}(open) \rangle$ and $\langle C_{CPuf}(open, st_R || p_R), R_{CPuf}(open) \rangle$.
2. Committer C_{ExtPuf}: send PUF $\mathcal{P}_R$ back to R_{ExtPuf}.
3. Receiver R_{ExtPuf}: If both decommitments are successfully completed, then R_{ExtPuf} gets the bit b' along with the opening st_S' for ComBit and string $st_R' || p_R'$ for ComResp.
   Check validity of st_R': query $\mathcal{P}_R$ with Encode(st_S') and obtain σ_R'. Compute $st_R'' \leftarrow$ FuzRep$^1(\sigma_R', p_R')$. If $st_R'' = st_R'$, then accept and output b. Else, reject.

Fig. 2. ExtPuf: Ideal **Extractable** Commitment in the $\mathcal{F}_{PUF}$ model

phase, the committer sends x, y, b. The receiver queries the token with x and obtains a string y'. If $y = y'$ the receiver accepts iff $\mathbf{c} = y' \oplus (r \wedge b)$.

The statistically binding property follows from the same arguments of Naor's scheme. The token is sent away before committer can see r. Thus, since x is only n bits, information theoretically the committer cannot instruct a malicious token to output y' adaptively on x. Thus, for any malicious possibly *stateful* token, binding is preserved. The statistically hiding property holds due to the fact that x is secret. A malicious receiver can query the token with any polynomial number of values x'. But, whp she will miss x, and thus she will obtain always 0.

The above protocol is denoted by CTok and is formally described in Fig. 3. We stress that, this is the first construction of unconditional commitment scheme that is secure even against malicious *stateful* tokens.

From Bit Commitment to String Commitment. To commit to a ℓ-bit string using one stateless token only is sufficient to embed ℓ pairs $(x_1, y_1), \ldots, (x_\ell, y_\ell)$ in the token $\mathcal{T}_C$ and to require that for each i, $x_i \in \{0,1\}^n$ and $y_i \in \{0,1\}^{3\ell n}$. Namely, $\mathcal{T}_C$ grows linearly with the size of the string to be committed. Then, execute protocol CTok for each bit of the string in parallel. The receiver accepts the string iff all bit commitments are accepting.

Protocol CTok. Committer's Input: $b \in \{0,1\}$.

Commitmen Phase

1. Committer C_{CTok}: pick $x \xleftarrow{\$} \{0,1\}^n$, $y \xleftarrow{\$} \{0,1\}^{3n}$. Create token $\mathcal{T}_C$ implementing the point function $f(x) = y$; $f(x') = 0$ for all $x' \neq x$. Send $\mathcal{T}_C$ to R_{CTok}.
2. Receiver R_{CTok}: pick $r \xleftarrow{\$} \{0,1\}^{3n}$. Send r to C_{CTok}.
3. Committer C_{CTok}: Send $c = y \oplus (r \wedge b^{3n})$ to R_{CTok}.

Decommitment Phase

1. Committer C_{CTok}: send (b,x) to R_{CTok}.
2. Receiver R_{CTok}: query $\mathcal{T}_C$ with x and obtain y. If $b = 0$, check that $c = y$. Else, check that $y = c \oplus r$. If the check passes, **accept** and output b, else reject.

Fig. 3. CTok: Ideal Commitments in the $\mathcal{F}_{wrap}$ model

Ideal Extractable Commitment in the $\mathcal{F}_{wrap}$ model. Extractability is achieved as before. The receiver sends a token $\mathcal{T}_R$ to the committer. The committer is required to query $\mathcal{T}_R$ with the opening of the commitment (namely, the value y) and then commit to the token's response. In the decommitment phase, the committer opens both the commitment of the bit and of the token's response. The receiver then checks that the latter value corresponds to the response of $\mathcal{T}_R$ on input the opening of the commitment of the bit. Note that here the receiver can check the validity of the token's response without physically possessing the token.

We now need to specify the function computed by token $\mathcal{T}_R$. Such function must be resilient against an unbounded adversary that can query the stateless token an arbitrary polynomial number of times.

The function, parameterized by two independent MAC keys k_{rec}, k_{tok}, takes as input a commitment's transcript (r, c), a MAC-tag σ_{rec} and an opening y. The function checks that y is a valid opening of (r, c), and that σ_{rec} is a valid MAC-tag computed on (r, c) with secret key k_{rec} (i.e., $\sigma_{rec} = \mathsf{Mac}(k_{rec}, r||c)$). If both checks are successful, the function outputs the MAC-tag computed on the opening y (i.e., $\sigma_{tok} = \mathsf{Mac}(k_{tok}, y)$). Due to the unforgeability of the MAC, and the statistically binding property of the commitment scheme CTok, a malicious committer can successfully obtain the answer to exactly one query. Note that, a malicious committer can perform the following attack. Once it receives the string r from the receiver, it picks strings y_0 and y_1 such that $r = y_0 \oplus y_1$ and sends the commitment $c = y_0$ to the receiver, obtaining the MAC of c. With the

---

Protocol ExtTok

$(\mathsf{Gen}, \mathsf{Mac}, \mathsf{Vrfy})$ is a one-time unconditional MAC. Protocol $\mathsf{CTok} = (\mathsf{C_{CTok}}, \mathsf{R_{CTok}})$ is run as sub-routine. Committer's Input: $b \in \{0, 1\}$.

Commitment Phase

1. Receiver $\mathsf{R_{ExtTok}}$: pick MAC-keys: k_{rec}, k_{tok}. Create token $\mathcal{T_R}$ implementing the following functionality. On input a tuple $(r||\mathbf{c}, \sigma_{rec}, y)$: if $\mathsf{Vrfy}(k_{rec}, r||\mathbf{c}, \sigma_{rec}) = 1$ and $(\mathbf{c} = y$ OR $\mathbf{c} = y \oplus r)$ then output $\sigma_{tok} = \mathsf{Mac}(k_{tok}, y)$ else output $\bot$. Send $\mathcal{T_R}$ to the sender $\mathsf{C_{ExtTok}}$.

   Commitment of the Secret Bit: ComBit.

2. $\mathsf{C_{ExtTok}} \Leftrightarrow \mathsf{R_{ExtTok}}$: run $\langle \mathsf{C_{CTok}}(\mathsf{com}, b), \mathsf{R_{CTok}}(\mathsf{com}) \rangle$ so that $\mathsf{C_{CTok}}$ commits to bit b. Let $(r, \mathbf{c})$ be the transcript of such commitment phase. Let y be the opening of $\mathbf{c}$.

3. Receiver $\mathsf{R_{ExtTok}}$: compute $\sigma_{rec} \leftarrow \mathsf{Mac}(k_{rec}, r||\mathbf{c})$. Send σ_{rec} to Committer $\mathsf{C_{ExtTok}}$.

4. Committer $\mathsf{C_{ExtTok}}$: query $\mathcal{T_R}$ with $q = (r||\mathbf{c}, \sigma_{rec}, y)$ and obtain σ_{tok}. If token $\mathcal{T_R}$ aborts, set $\sigma_{tok} = 0^n$.

   Commitment of $\mathcal{T_R}$'s Response: ComResp.

   $\mathsf{C_{ExtTok}} \Leftrightarrow \mathsf{R_{ExtTok}}$: run $\langle \mathsf{C_{CTok}}(\mathsf{com}, \sigma_{tok}), \mathsf{R_{CTok}}(\mathsf{com}) \rangle$ so that $\mathsf{C_{ExtTok}}$ commits to the response σ_{tok} received from $\mathcal{T_R}$.

Decommitment Phase

1. $\mathsf{C_{ExtTok}} \Leftrightarrow \mathsf{R_{ExtTok}}$: opening of both commitments. Run $\langle \mathsf{C_{CTok}}(\mathsf{open}, b), \mathsf{R_{CTok}}(\mathsf{open}) \rangle$ and $\langle \mathsf{C_{CTok}}(\mathsf{open}, \sigma_{rec}), \mathsf{R_{CTok}}(\mathsf{open}) \rangle$.

2. Receiver $\mathsf{R_{ExtTok}}$: If both decommitment are successfully completed, then $\mathsf{R_{ExtTok}}$ gets the bit b' along with the opening y' for ComBit and string σ'_{tok} for ComResp. If $\mathsf{Vrfy}(k_{tok}, r||y', \sigma'_{tok}) = 1$ then $\mathsf{R_{ExtTok}}$ accept and output b'. Else, reject.

Fig. 4. ExtTok: Ideal **Extractable** Commitment in the $\mathcal{F}_{wrap}$ model

commitment so computed and the tag, it can query token $\mathcal{T_R}$ twice with each valid opening. In this case, the committer can extract the MAC key, and at the same time baffling the extractor that observes two valid openings. The observation here is that, due to the binding of CTok, for a commitment $\mathbf{c}$ computed in such a way, the malicious committer will not be able, in the decommitment phase, to provide a valid opening. (The reason is that whp she cannot instruct its token to output neither y_0 or y_1). Thus, although the extractor fails and outputs $\bot$, the decommitment will not be accepting. Thus extractability is not violated.

As final step, the committer commits to the token's response σ_{tok}, using the scheme CTok. (If the token of the receiver aborts, the committer sets σ_{tok} to the zero string). In the decommitment phase, the receiver first checks the validity of both commitments (commitment of the bit, commitment of the answer σ_{tok}). Then, given the opening of the bit, it checks that σ_{tok} is a valid MAC computed under key k_{tok} on such opening.

Binding follows from the binding of CTok and the unforgeability of MAC. Hiding still follows from the hiding of CTok. Indeed, the answer of $\mathcal{T_R}$ sent by the malicious receiver, is not forwarded to the receiver, but is committed using the ideal commitment CTok. Furthermore, if $\mathcal{T}_R$ selectivly aborts, the committer does not halt but it continues committing to the zero-string. The receiver will get its token's answer in clear only in the decommitment phase when the bit has been already revealed. The formal description of the above protocol, that we denote by ExtTok, is shown in Fig. 4.

Theorem 3. *Protocol* ExtTok *is an ideal* extractable *commitment in the* $\mathcal{F}_{wrap}$ *model.*

The proof of Theorem 3 is provided in the full version [18].

Acknowledgments. The second author thanks Dominik Scheder for very interesting discussions on Azuma's inequality, and Akshay Wadia for suggesting a simplification in the compiler. The same author thanks Ivan Visconti and Rafail Ostrovsky for valuable discussions.

References

1. Armknecht, F., Maes, R., Sadeghi, A.R., Standaert, F.X., Wachsmann, C.: A formalization of the security features of physical functions. In: IEEE Symposium on Security and Privacy, pp. 397–412. IEEE Computer Society (2011)
2. Armknecht, F., Maes, R., Sadeghi, A.R., Sunar, B., Tuyls, P.: Memory leakage-resilient encryption based on physically unclonable functions. In: Matsui, M. (ed.) ASIACRYPT 2009. LNCS, vol. 5912, pp. 685–702. Springer, Heidelberg (2009)
3. Barak, B., Canetti, R., Nielsen, J.B., Pass, R.: Universally composable protocols with relaxed set-up assumptions. In: Foundations of Computer Science (FOCS 2004), pp. 394–403 (2004)
4. Ben-Or, M., Goldwasser, S., Wigderson, A.: Completeness theorems for non-cryptographic fault-tolerant distributed computation (extended abstract). In: STOC, pp. 1–10 (1988)
5. Boit, C., Helfmeier, C., Nedospasaov, D., Seifert, J.P.: Cloning physically unclonable functions. In: IEEE HOST (2013)
6. Brassard, G., Chaum, D., Crépeau, C.: Minimum disclosure proofs of knowledge. J. Comput. Syst. Sci. 37(2), 156–189 (1988)
7. Brzuska, C., Fischlin, M., Schröder, H., Katzenbeisser, S.: Physically uncloneable functions in the universal composition framework. In: Rogaway, P. (ed.) CRYPTO 2011. LNCS, vol. 6841, pp. 51–70. Springer, Heidelberg (2011)
8. Brzuska, C., Fischlin, M., Schröder, H., Katzenbeisser, S.: Physically uncloneable functions in the universal composition framework. IACR Cryptology ePrint Archive 2011, 681 (2011)
9. Canetti, R.: Universally composable security: A new paradigm for cryptographic protocols. In: Foundations of Computer Science (FOCS 2001), pp. 136–145 (2001)
10. Canetti, R., Dodis, Y., Pass, R., Walfish, S.: Universally composable security with global setup. In: Vadhan, S.P. (ed.) TCC 2007. LNCS, vol. 4392, pp. 61–85. Springer, Heidelberg (2007)
11. Canetti, R., Kushilevitz, E., Lindell, Y.: On the limitations of universally composable two-party computation without set-up assumptions. In: Biham, E. (ed.) EUROCRYPT 2003. LNCS, vol. 2656, pp. 68–86. Springer, Heidelberg (2003)

12. Canetti, R., Lindell, Y., Ostrovsky, R., Sahai, A.: Universally composable two-party and multi-party secure computation. In: 34th Annual ACM Symposium on Theory of Computing, May 19-21, pp. 494–503. ACM Press, Montréal (2002)
13. Chandran, N., Goyal, V., Sahai, A.: New constructions for UC secure computation using tamper-proof hardware. In: Smart, N.P. (ed.) EUROCRYPT 2008. LNCS, vol. 4965, pp. 545–562. Springer, Heidelberg (2008)
14. Chaum, D., Crépeau, C., Damgård, I.: Multiparty unconditionally secure protocols (extended abstract). In: STOC. pp. 11–19 (1988)
15. Choi, S.G., Katz, J., Schröder, D., Yerukhimovich, A., Zhou, H.S.: (efficient) universally composable two-party computation using a minimal number of stateless tokens. IACR Cryptology ePrint Archive 2011, 689 (2011)
16. Crépeau, C., Kilian, J.: Achieving oblivious transfer using weakened security assumptions (extended abstract). In: FOCS, pp. 42–52. IEEE Computer Society (1988)
17. Damgård, I., Kilian, J., Salvail, L.: On the (Im)possibility of basing oblivious transfer and bit commitment on weakened security assumptions. In: Stern, J. (ed.) EUROCRYPT 1999. LNCS, vol. 1592, pp. 56–73. Springer, Heidelberg (1999)
18. Damgård, I., Scafuro, A.: Unconditionally secure and universally composable commitments from physical assumptions. IACR Cryptology ePrint Archive 2013, 108 (2013)
19. van Dijk, M., Rührmair, U.: Physical unclonable functions in cryptographic protocols: Security proofs and impossibility results. IACR Cryptology ePrint Archive 2012, 228 (2012)
20. Dodis, Y., Ostrovsky, R., Reyzin, L., Smith, A.: Fuzzy extractors: How to generate strong keys from biometrics and other noisy data. SIAM J. Comput. 38(1), 97–139 (2008)
21. Döttling, N., Kraschewski, D., Müller-Quade, J.: Unconditional and composable security using a single stateful tamper-proof hardware token. In: Ishai, Y. (ed.) TCC 2011. LNCS, vol. 6597, pp. 164–181. Springer, Heidelberg (2011)
22. Döttling, N., Kraschewski, D., Müller-Quade, J.: David & goliath oblivious affine function evaluation - asymptotically optimal building blocks for universally composable two-party computation from a single untrusted stateful tamper-proof hardware token. IACR Cryptology ePrint Archive 2012, 135 (2012)
23. Goyal, V., Ishai, Y., Mahmoody, M., Sahai, A.: Interactive locking, zero-knowledge pcps, and unconditional cryptography. In: Rabin, T. (ed.) CRYPTO 2010. LNCS, vol. 6223, pp. 173–190. Springer, Heidelberg (2010)
24. Goyal, V., Ishai, Y., Sahai, A., Venkatesan, R., Wadia, A.: Founding cryptography on tamper-proof hardware tokens. In: Micciancio, D. (ed.) TCC 2010. LNCS, vol. 5978, pp. 308–326. Springer, Heidelberg (2010)
25. Hofheinz, D.: Possibility and impossibility results for selective decommitments. J. Cryptology 24(3), 470–516 (2011)
26. Katz, J.: Universally composable multi-party computation using tamper-proof hardware. In: Naor, M. (ed.) EUROCRYPT 2007. LNCS, vol. 4515, pp. 115–128. Springer, Heidelberg (2007)
27. Katzenbeisser, S., KocabaNs, Ü., Rožić, V., Sadeghi, A.-R., Verbauwhede, I., Wachsmann, C.: PUFs: Myth, fact or busted? A security evaluation of physically unclonable functions (PUFs) cast in silicon. In: Prouff, E., Schaumont, P. (eds.) CHES 2012. LNCS, vol. 7428, pp. 283–301. Springer, Heidelberg (2012)

28. Kilian, J.: A note on efficient zero-knowledge proofs and arguments (extended abstract). In: Proceedings of the Twenty-Fourth Annual ACM Symposium on Theory of Computing, STOC 1992, pp. 723–732. ACM, New York (1992), http://doi.acm.org/10.1145/129712.129782

29. Kolesnikov, V.: Truly efficient string oblivious transfer using resettable tamper-proof tokens. In: Micciancio, D. (ed.) TCC 2010. LNCS, vol. 5978, pp. 327–342. Springer, Heidelberg (2010)

30. Maes, R., Verbauwhede, I.: Physically unclonable functions: A study on the state of the art and future research directions. In: Sadeghi, A.R., Naccache, D. (eds.) Towards Hardware-Intrinsic Security. Information Security and Cryptography, pp. 3–37. Springer, Heidelberg (2010)

31. Moran, T., Segev, G.: David and Goliath commitments: UC computation for asymmetric parties using tamper-proof hardware. In: Smart, N.P. (ed.) EUROCRYPT 2008. LNCS, vol. 4965, pp. 527–544. Springer, Heidelberg (2008)

32. Müller-Quade, J., Unruh, D.: Long-term security and universal composability. In: Vadhan, S.P. (ed.) TCC 2007. LNCS, vol. 4392, pp. 41–60. Springer, Heidelberg (2007)

33. Müller-Quade, J., Unruh, D.: Long-term security and universal composability. J. Cryptology 23(4), 594–671 (2010)

34. Naor, M.: Bit commitment using pseudo-randomness. In: Brassard, G. (ed.) CRYPTO 1989. LNCS, vol. 435, pp. 128–136. Springer, Heidelberg (1990)

35. Ostrovsky, R., Scafuro, A., Visconti, I., Wadia, A.: Universally composable secure computation with (malicious) physically uncloneable functions. IACR Cryptology ePrint Archive 2012, 143 (2012)

36. Ostrovsky, R., Scafuro, A., Visconti, I., Wadia, A.: Universally composable secure computation with (malicious) physically uncloneable functions. In: Johansson, T., Nguyen, P.Q. (eds.) EUROCRYPT 2013. LNCS, vol. 7881, pp. 702–718. Springer, Heidelberg (2013)

37. Pappu, R.S., Recht, B., Taylor, J., Gershenfeld, N.: Physical one-way functions. Science 297, 2026–2030 (2002)

38. Pappu, R.S.: Physical One-Way Functions. Ph.D. thesis, MIT (2001)

39. Pass, R., Wee, H.: Black-box constructions of two-party protocols from one-way functions. In: Reingold, O. (ed.) TCC 2009. LNCS, vol. 5444, pp. 403–418. Springer, Heidelberg (2009)

40. Rührmair, U.: Oblivious transfer based on physical unclonable functions. In: Acquisti, A., Smith, S.W., Sadeghi, A.R. (eds.) TRUST 2010. LNCS, vol. 6101, pp. 430–440. Springer, Heidelberg (2010)

41. Rührmair, U., van Dijk, M.: Pufs in security protocols: Attack models and security evaluations. In: IEEE Symposium on Security and Privacy (2013)

42. Sadeghi, A.R., Visconti, I., Wachsmann, C.: Enhancing rfid security and privacy by physically unclonable functions. In: Sadeghi, A.R., Naccache, D. (eds.) Towards Hardware-Intrinsic Security. Information Security and Cryptography, pp. 281–305. Springer, Heidelberg (2010)

Functional Encryption
from (Small) Hardware Tokens

Kai-Min Chung[1,*], Jonathan Katz[2,**], and Hong-Sheng Zhou[3,* * *]

[1] Academia Sinica
kmchung@iis.sinica.edu.tw
[2] University of Maryland
jkatz@cs.umd.edu
[3] Virginia Commonwealth University
hszhou@vcu.edu

Abstract. Functional encryption (FE) enables fine-grained access control of encrypted data while promising simplified key management. In the past few years substantial progress has been made on functional encryption and a weaker variant called predicate encryption. Unfortunately, fundamental impossibility results have been demonstrated for constructing FE schemes for general functions satisfying a simulation-based definition of security.

We show how to use *hardware tokens* to overcome these impossibility results. In our envisioned scenario, an authority gives a hardware token and some cryptographic information to each authorized user; the user combines these to decrypt received ciphertexts. Our schemes rely on *stateless* tokens that are *identical* for all users. (Requiring a different token for each user trivializes the problem, and would be a barrier to practical deployment.) The tokens can implement relatively "lightweight" computation relative to the functions supported by the scheme.

Our token-based approach can be extended to support hierarchal functional encryption, function privacy, and more.

1 Introduction

In traditional public-key encryption, a sender encrypts a message M with respect to the public key pk of a particular receiver, and only that receiver (i.e., the owner of the secret key associated with pk) can decrypt the resulting ciphertext and recover the underlying message. More recently, there has been an explosion of interest in encryption schemes that can provide greater flexibility and more refined access to encrypted data. Such schemes allow the sender to specify a *policy* at the time of encryption, and enable any user (decryptor) satisfying the policy (within the given system) to decrypt the resulting ciphertext.

---

* Work done while at Cornell University and supported by NSF awards CNS-1217821 and CCF-1214844, and the Sloan Research Fellowship of Rafael Pass.
** Work supported by NSF award #1223623.
* * * Work done while at the University of Maryland and supported by an NSF CI postdoctoral fellowship.

K. Sako and P. Sarkar (Eds.) ASIACRYPT 2013, Part II, LNCS 8270, pp. 120–139, 2013.

Work in this direction was spurred by constructions of identity-based encryption (IBE) [8], fuzzy IBE [43], and attribute-based encryption [29]. Each of these can be cast as special cases of *predicate encryption* [11,35], which is in turn a special case of the more powerful notion of *functional encryption* (FE) recently introduced by Boneh, Sahai, and Waters [10]. Roughly speaking, in an FE scheme a user's secret key SK_K is associated with a policy K. Given an encryption of some message M, a user in possession of the secret key SK_K associated with K can recover $F(K, M)$ for some function F fixed as part of the scheme itself. (In the most general case F might be a universal Turing machine, but weaker F are also interesting.)

Security of functional encryption, informally, guarantees that a group of users with secret keys $\mathsf{SK}_{K_1}, \ldots, \mathsf{SK}_{K_\ell}$ learn nothing from an encryption of M that is not implied by $F(K_1, M), \ldots, F(K_\ell, M)$ (plus the length of M). As far as formal definitions are concerned, early work on predicate encryption used an indistinguishability-based definition of security, but Boneh et al. [10] and O'Neill [41] independently showed that such a definitional approach is not, in general, sufficient for analyzing functional encryption. They suggest to use stronger, simulation-based definitions of security (similar in spirit to semantic security) instead.

In the past few years substantial progress has been made in this area [42,25,4,26,17,24,18,3,14,1]. Yet several open questions remain. First, it remains an unsolved problem to construct an FE scheme for *arbitrary* functions F with *unbounded* collusion resistance. Second, it is unknown how to realize the strongest simulation-based notion of security for functional encryption. In fact, Boneh et al. [10] and Agrawal et al. [1] showed fundamental limitations on achieving such definitions for FE schemes supporting arbitrary F.

Here we propose the use of (stateless) *hardware tokens* to solve both the above issues. In our envisioned usage scenario, an authority gives a hardware token along with a cryptographic key SK_K to each authorized user; the user combines these in order to decrypt received ciphertexts. We believe this would be a feasible approach for realizing functional encryption in small- or medium-size organizations where an authority could purchase hardware tokens, customize them as needed, and then give them directly to users in the system.

The idea of using physical devices to bypass cryptographic impossibility results has been investigated previously. Katz [34] showed that hardware tokens can be used for universally composable computation of arbitrary functions. His work motivated an extensive amount of follow-up work [12,40,13,27,28,15]. In the context of program obfuscation, several works [28,6,16] considered using hardware tokens to achieve program obfuscation, which is impossible in the plain model even for simple classes of programs [2].

A token-based approach should have the following properties:

1. The tokens used should be *universal*, in the sense that every user in the system is given an *identical* token. Having a single token used by everyone appears to be the only way to make a token-based approach viable.

2. In applications where the complexity of F is high, it is desirable that tokens be "lightweight" in the sense that the complexity of the token is smaller than the complexity of F.

In this work, we show token-based solutions that satisfy the above requirements. Additionally, our constructions satisfy a strong simulation-based notion of security and have succinct ciphertexts (of size independent of F). We provide the intuition behind our approach in the next section.

1.1   Our Results

Let pk, sk denote the public and secret keys for a (standard) public-key encryption scheme. Intuitively, a trivial construction of an FE scheme based on (stateless) tokens is to let pk be the master public key, and to give the user associated with key K a token which implements the functionality $\mathsf{token}_{sk,K}(C) = F(K, \mathsf{Dec}_{sk}(C))$. In this scheme the tokens are not *universal*, as each user must be given a token whose functionality depends on that user's key K. Perhaps more surprising is that this scheme is not secure: nothing prevents a user from modifying C before feeding the ciphertext to its token; if the encryption scheme scheme is *malleable* then a user might be able to use such a cheating strategy to learn disallowed information about the underlying message M. We will address both these issues in our solutions, described next.

Solution #1. We can address the universality issue by having the user provide K along with C as input to the token. (The token will then implement the functionality $\mathsf{token}_{sk}(K, C) = F(K, \mathsf{Dec}_{sk}(C))$.) Now we must prevent the user from changing either K or C. Modifications of the key are handled by *signing* K and hard-coding the verification key vk into the token; the token then verifies a signature on K before decrypting as before. We show that illegal modification of the ciphertext can be solved if the public-key encryption scheme is CCA2-secure; we give the details in Section 4.2.

Solution #2. In the previous solution, the complexity of the token was (at least) the complexity of computing F itself. We can use ideas from the area of verifiable outsource of computation [19] in order to obtain a solution in which the complexity of the token is independent of the complexity of F. The basic idea here is for the token to "outsource" most of the computation of F to the user. To do so, we now let the underlying public-key encryption scheme be *fully homomorphic* [21]. Given a ciphertext $C = \mathsf{Enc}_{pk}(M)$, the user can now compute the transformed ciphertext $\hat{C} = \mathsf{Enc}_{pk}(F(K, M))$ and feed this to the token for decryption. To enforce correct behavior with lightweight tokens, we let the user provide a *succinct non-interactive argument (SNARG)* [23,20,5] that the computation is done correctly.[1] The immediate problem with this approach is that any

---

[1] As a technical note, while SNARGs are constructed based on knowledge-type assumptions, we here rely on SNARGs for **P**, which can be formulated as a falsifiable assumption.

fully-homomorphic encryption scheme is completely malleable! We here instead rely on simulation-extractable non-interactive zero-knowledge proofs (NIZKs) to deal with the malleable issue, where we let the encryptor provide an NIZK proof that $C = \mathsf{Enc}_{pk}(M)$ is correctly encrypted. The computation done by the token now involves (1) verifying the signature on K as in previous solution, and (2) verifying the given SNARG and NIZK, (3) decrypting the given ciphertext, all of which have complexity independent of F. We give the details in Section 4.3.

While both of our schemes are simple, we show in Section 6 that both schemes satisfy a very strong notion of simulation-based security, where an adversary A gets full access to the scheme (in particular, A can make an arbitrary number of key queries and encryption queries in a fully adaptive way), yet cannot learn any information beyond what it should have learnt through the access. At a high level, our security proof crucially relies on the fact that in the simulation, the simulator gets to simulate token's answers to the queries made by the adversary, which bypasses the information-theoretic arguments underlying the impossibility results of Boneh et al. [10] and Agrawal et al. [1].

We remark that our constructions and the way we get around impossibility results share some similarities to the work of Bitansky et al [6] on achieving program obfuscation using stateless and universal hardware tokens, but the security notion of functional encryption and program obfuscation are different and the results from both contexts does not seem to imply each other. For example, one may think intuitively that by obfuscating the decryption circuit $\mathsf{Dec}_{sk,K}(C) = F(K, \mathsf{Dec}_{sk}(C))$, one obtains a "trivial" construction of functional encryption. However, such a construction cannot satisfy simulation-based security as it does not bypass the impossibility results of [10,1].

1.2   Extensions

Our approach can be extended in several ways; we sketch two extensions here.

Hierarchical functional encryption. Consider an encrypted database in a company where the top-level manager has the access control on the database that allows different first-level departments to access different part of the data; then any first level department, say, research department, allows different second level sub-department to run different analytic/learning algorithms over the encrypted data; this naturally induces a hierarchical access structure to the data. To support this natural variant of access control, we need *hierarchical functional encryption*, which generalizes many primitives considered in the literature, such as hierarchical IBE [33,22,7,44,37], hierarchical PE [36].

More precisely, to enable such a hierarchical structure, the global authority may delegate a first level user Alice (under some functionality key K_{Alice} with respect to functionality F_1) the ability to generate a second level secret key $\mathsf{SK}_{K_{Alice}:K_{Bob}}$ of functionality key K_{Bob} with respect to functionality F_2 for a second level user Bob. For a message M encrypted under global master public key, Bob should be able to decrypt $F_2(K_{Bob}, F_1(K_{Alice}, M))$ using $\mathsf{SK}_{K_{Alice}:K_{Bob}}$. Alice may further delegate Bob the ability to generate a third level secret key

$\mathsf{SK}_{K_{Alice}:K_{Bob}:K_{Carol}}$ of functionality key K_{Carol} with respect to functionality F_3, and so on.

Our solution #1 can be readily extended to the hierarchical setting using the idea of signature chains. Roughly, to delegate Alice such power, the global authority generates a key pair $(\mathsf{sk}_{Alice}, \mathsf{vk}_{Alice})$ of a digital signature scheme and "authorizes" it by signing $(K_{Alice}, \mathsf{vk}_{Alice})$; sk_{Alice} is given to Alice as a "delegation key" and $(K_{Alice}, \mathsf{vk}_{Alice})$ together with its signature are published. Alice can then generate $\mathsf{SK}_{K_{Alice}:K_{Bob}}$ by simply signing $K_{Alice} : K_{Bob}$ (using sk_{Alice}). To decrypt, Bob queries the token with the ciphertext together with the chain of signatures—including $(K_{Alice}, \mathsf{vk}_{Alice})$ together with its signature and $K_{Alice} : K_{Bob}$ together with its signature. The token returns $F_2(K_{Bob}, F_1(K_{Alice}, M))$ if the chain verifies. Alice can perform further delegation in a similar fashion.

The above solution has the drawback that all functionalities in the hierarchy need to be determined in the global setup and hard-wired in the token. We can further allow adaptively chosen functionalities by further "authorizing" the functionality as well, but at the price that the token needs to receive a description of the functionality (together with its authorization info) as its input, which results in long query length. This issue can be addressed in the framework of our solution #2, where the token only requires a succinct authorization information of the functionality (as such, the complexity of the token remains independent of the functionalities). We provide further details in the full version of this paper.

Function privacy. In a general FE scheme the secret key SK_K may leak K. Preventing such leakage is given as an interesting research direction in [10]. Very recently, Boneh et al. [9] studied the notion of *function privacy* for IBE, and gave several constructions. We can modify our token-based constructions to obtain function privacy in functional encryption: in the key generation, instead of obtaining a signature of K, the users obtain an encrypted version signature $\mathcal{E}(\sigma)$; the decryption key sk will be stored in token; at any moment when the users receive a ciphertext C, instead of providing (C, K, σ), the tuple $(C, K, \mathcal{E}(\sigma))$ will be given to token; the token would first decrypt $\mathcal{E}(\sigma)$ into σ and then verify that σ is a valid signature on K and, if so, return the result $F(K, \mathsf{Dec}_{sk}(C))$ as in basic functional encryption constructions.

2   Preliminaries

2.1   Fully Homomorphic Encryption

A fully homomorphic encryption scheme $\mathcal{FHE} = (\mathsf{FHE.Gen}, \mathsf{FHE.Enc}, \mathsf{FHE.Dec}, \mathsf{FHE.Eval})$ is a public-key encryption scheme that associates with an additional polynomial-time algorithm Eval, which takes as input a public key ek, a ciphertext $\mathsf{ct} = \mathsf{Enc}_{\mathsf{ek}}(m)$ and a circuit C, and outputs, a new ciphertext $\mathsf{ct}' = \mathsf{Eval}_{\mathsf{ek}}(\mathsf{ct}, C)$, such that $\mathsf{Dec}_{\mathsf{dk}}(\mathsf{ct}') = C(m)$, where dk is the secret key corresponding to the public key ek. It is required that the size of $\mathsf{ct}' = \mathsf{Eval}_{\mathsf{ek}}(\mathsf{Enc}_{\mathsf{ek}}(m), C)$ depends polynomially on the security parameter and the length of $C(m)$, but is otherwise independent of the size of the circuit C. We also require that Eval is deterministic, and the scheme has perfect correctness (i.e. it always holds that $\mathsf{Dec_{dk}}(\mathsf{Enc_{ek}}(m)) = m$ and that $\mathsf{Dec_{dk}}(\mathsf{FHE.Eval_{ek}}(\mathsf{Enc_{ek}}(m), C)) = C(m)$). Most known schemes satisfies these properties. For security, we simply require that $\mathcal{FHE}$ is semantically secure.

Since the breakthrough of Gentry [21], several fully homomorphic encryption schemes have been constructed with improved efficiency and based on more standard assumptions such as LWE (Learning With Error). In general, these constructions achieve leveled FHE, where the complexity of the schemes depend linearly on the depth of the circuits C that are allowed as inputs to Eval. However, under the additional assumption that these constructions are circular secure (i.e., remain secure even given an encryption of the secret key), the complexity of the schemes are independent of the allowed circuits, and the schemes can evaluate any polynomial-sized circuit.

2.2   Non-interactive Zero-Knowledge Arguments

Let R be a binary relation that is efficiently computable. Let L_R be the language defined by the relation R, that is, $L_R = \{x : \exists w \text{ s.t.}(x, w) \in R\}$. For any pair $(x, w) \in R$, we call x the statement and w the witness.

Definition 1 (NIZK). *A tuple of* PPT *algorithms* $\mathcal{NIZK}$ = (NIZK.Gen, NIZK.P, NIZK.V), *is a non-interactive zero-knowledge (NIZK) argument system for R if it has the following properties described below:*

Completeness. *For any* $(x, w) \in R$ *it holds that*

$$\Pr[\mathsf{crs} \leftarrow \mathsf{NIZK.Gen}(1^\kappa); \pi \leftarrow \mathsf{NIZK}.P(\mathsf{crs}, x, w) : \ \mathsf{NIZK}.V(\mathsf{crs}, x, \pi) = 1] = 1.$$

Soundness. *For any non-uniform* PPT $\mathcal{A}$, *it holds that*

$$\Pr[\mathsf{crs} \leftarrow \mathsf{NIZK.Gen}(1^\kappa); (x, \pi) \leftarrow \mathcal{A}(\mathsf{crs}) : \ \mathsf{NIZK}.V(\mathsf{crs}, x, \pi) = 1] \leq \mathrm{negl}(\kappa).$$

Zero-knowledge. *For any non-uniform* PPT $\mathcal{A}$, *there exists a* PPT $\mathcal{S} = (\mathcal{S}_1, \mathcal{S}_2)$ *such that it holds that* $|p_1 - p_2| \leq \mathrm{negl}(\kappa)$, *where*

$$p_1 = \Pr[\mathsf{crs} \leftarrow \mathsf{NIZK.Gen}(1^\kappa) : \mathcal{A}^{\mathsf{NIZK}.P(\mathsf{crs}, \cdot, \cdot)}(\mathsf{crs}) = 1]$$
$$p_2 = \Pr[(\mathsf{crs}, \tau, \xi) \leftarrow \mathcal{S}_1(1^\kappa) : \mathcal{A}^{Sim(\mathsf{crs}, \tau, \cdot, \cdot)}(\mathsf{crs}) = 1]$$

where $Sim(\mathsf{crs}, \tau, x, w) = \mathcal{S}_2(\mathsf{crs}, \tau, x)$ *for* $(x, w) \in R$. *Both oracles* NIZK.P() *and* Sim() *output* $\perp$ *if* $(x, w) \notin R$.

Next we define (unbounded) simulation-extractability of NIZK [30,32]. Intuitively, it says that even after seeing many simulated proofs, whenever the adversary makes a new proof we are able to extract a witness.

Definition 2 (Simulation-Extractability). *Let* $\mathcal{NIZK}$ = (NIZK.Gen, NIZK.P, NIZK.V) *be a NIZK argument system for R. We say $\mathcal{NIZK}$ is simulation-extractable if for all* PPT *adversaries $\mathcal{A}$, there exists a* PPT $\mathcal{S} = (\mathcal{S}_1, \mathcal{S}_2, \mathcal{S}_3)$ *so that*

$$\Pr\left[\begin{array}{c}(\mathsf{crs}, \tau, \xi) \leftarrow \mathcal{S}_1(1^\kappa); (x, \pi) \leftarrow \mathcal{A}^{\mathcal{S}_2(\mathsf{crs}, \tau, \cdot)}(\mathsf{crs}); w \leftarrow \mathcal{S}_3(\mathsf{crs}, \xi, x, \pi) : \\ \mathsf{NIZK}.V(\mathsf{crs}, x, \pi) = 1 \ \wedge \ (x, \pi) \notin Q \ \wedge \ (x, w) \notin R \end{array}\right] \leq \mathrm{negl}(\kappa)$$

where Q is the list of simulation queries and responses (x_i, π_i) that $\mathcal{A}$ makes to $\mathcal{S}_2()$.

2.3 SNARG

We present the definition of succinct non-interactive arguments (abbreviated SNARGs) [23,20,5]. A SNARG for a function class $\mathcal{F} = \{\mathcal{F}_\kappa\}_{\kappa \in \mathbb{N}}$ consists of a set of PPT algorithms $\mathcal{SNARG} = \text{SNARG}.\{\text{Gen}, P, V\}$: The generation algorithm Gen on input security parameter 1^κ and a function $F : \{0,1\}^{n(\kappa)} \to \{0,1\}^{m(\kappa)} \in \mathcal{F}_\kappa$ (represented as a circuit), outputs a reference string rs and a (short) verification state vrs.[2] The prover P on input rs and an input string $x \in \{0,1\}^n$, outputs an answer $y = F(x)$ together with a (short) proof ϖ. The verifier algorithm V on input vrs, x, y, and ϖ outputs a bit $b \in \{0,1\}$ represents whether V accepts or rejects. We require the following properties for a SNARG scheme.

- **Completeness:** For every $\kappa \in \mathbb{N}$, $F \in \mathcal{F}_\kappa$, $x \in \{0,1\}^n$, the probability that the verifier V rejects in the following experiment is negligible in κ: (i) $(\text{rs}, \text{vrs}) \leftarrow \text{Gen}(1^\kappa, F)$, (ii) $(y, \varpi) \leftarrow P(\text{rs}, x)$, and (iii) $b \leftarrow V(\text{vrs}, x, y, \varpi)$.
- **Soundness:** For every efficient adversary P^*, and every $\kappa \in \mathbb{N}$, the probability that P^* makes V accept an incorrect answer in the following experiment is negligible in κ: (i) P^* on input 1^κ outputs a function $F \in \mathcal{F}_\kappa$, (ii) $(\text{rs}, \text{vrs}) \leftarrow \text{Gen}(1^\kappa, F)$, (iii) P^* on input rs, outputs x, y, ϖ with $y \neq F(x)$, and (iv) $b \leftarrow V(\text{vrs}, x, y, \varpi)$.
- **Efficiency:** The running time of the verifier is $\text{poly}(\kappa, n + m, \log |F|)$ (which implies the succinctness of vrs and ϖ). The running time of the generation algorithm and the prover is $\text{poly}(\kappa, |F|)$.

We say $\mathcal{SNARG}$ is *publicly-verifiable* if the verification state vrs is part of the reference string rs.

We require a publicly-verifiable SNARG scheme $\mathcal{SNARG}$ for polynomial-size circuits. Such a SNARG scheme can be obtained by using Micali's CS proof [39] (with random oracle instantiated by some hash function heuristically), or provably secure based on publicly-verifiable succinct non-interactive arguments (SNARGs), which in turn can be constructed based on (non-falsifiable) q-PKE (q-power knowledge of exponent) and q-PDH (q-power Diffie-Hellman) assumptions on bilinear groups. Such SNARGs was first constructed implicitly in [31] and later improved by [38,20], where [20] explicitly constructs SNARGs. In the scheme of [20], the generation algorithm and the prover run in time quasi-linear in the size of F with rs length linear in $|F|$, and the verifier runs in linear time in the input and output length.

3   Definition of Functional Encryption

Functional encryption was recently introduced by Boneh, Sahai, and Waters [10]. Let $\mathcal{F} = \{\mathcal{F}_\kappa\}_{\kappa \in \mathbb{N}}$ where $\mathcal{F}_\kappa = \{F : \mathcal{K}_\kappa \times \mathcal{M}_\kappa \to \mathcal{M}_\kappa\}$ be an ensemble of

---

[2] We assume w.l.o.g. that rs contains a description of F.

functionality class indexed by a security parameter κ. A functional encryption scheme $\mathcal{FE}$ for a functionality class $\mathcal{F}$ consists of four PPT algorithms $\mathcal{FE} =$ FE.{Setup, Key, Enc, Dec} defined as follows.

- **Setup:** FE.Setup($1^\kappa, F$) is a PPT algorithm takes as input a security parameter 1^κ and a functionality $F \in \mathcal{F}_\kappa$ and outputs a pair of master public and secret keys (MPK, MSK).
- **Key Generation:** FE.Key(MSK, K) is a PPT algorithm that takes as input the master secret key MSK and a functionality key $K \in \mathcal{K}_\kappa$ and outputs a corresponding secret key SK_K.
- **Encryption:** FE.Enc(MPK, M) is a PPT algorithm that takes as input the master public key MPK and a message $M \in \mathcal{M}_\kappa$ and outputs a ciphertext CT.
- **Decryption:** FE.Dec(SK_K, CT) is a deterministic algorithm that takes as input the secret key SK_K and a ciphertext CT $=$ Enc(MPK, M) and outputs $F(K, M)$.

Definition 3 (Correctness). *A functional encryption scheme $\mathcal{FE}$ is correct if for every $\kappa \in \mathbb{N}$, $F \in \mathcal{F}_\kappa$, $K \in \mathcal{K}_\kappa$, and $M \in \mathcal{M}_\kappa$,*

$$\Pr\left[\begin{array}{l}(\mathsf{MPK}, \mathsf{MSK}) \leftarrow \mathsf{FE.Setup}(1^\kappa, F); \\ \mathsf{FE.Dec}(\mathsf{FE.Key}(\mathsf{MSK}, K), \mathsf{FE.Enc}(\mathsf{MPK}, M)) \neq F(K, M)\end{array}\right] = \mathsf{negl}(\kappa)$$

where the probability is taken over the coins of FE.Setup, FE.Key, *and* FE.Enc.

We next define a stronger simulation-based notion of security for functional encryption than the existing simulation-based security notions in the literature. We note that, while there are negative results [10,1] showing that even significantly weaker notions of security are impossible to achieve in the plain model, our token-based construction in Section 4 achieves our strong security notion in the token model.

Our definition is stronger in the sense that we allow the adversary $\mathcal{A}$ to take *full* control over the access of the encryption scheme, where $\mathcal{A}$ can choose the functionality F and request to see an arbitrary number of secret keys SK_K's and ciphertexts CT's in a *fully adaptive* fashion. Previous definitions either restrict the number of ciphertext queries and/or restrict the order of secret key and ciphertext queries (e.g., require $\mathcal{A}$ to ask for all challenge ciphertexts at once). Informally, the following definition says that even with full access to the encryption scheme, $\mathcal{A}$ still cannot learn any additional information than what it should have legally learnt from the received ciphertexts (using the received secret keys). This, as usual, is formalized by requiring that the ciphertexts can be simulated by an efficient simulator with only the "legal" information.

Definition 4 (Fully-Adaptive Simulation Security). *Let $\mathcal{FE}$ be a functional encryption scheme for a functionality class $\mathcal{F}$. For every PPT stateful adversary $\mathcal{A}$ and PPT stateful simulator* Sim, *consider the following two experiments.*

$\mathbf{Expt}^{real}_{\mathcal{FE},\mathcal{A}}(1^\kappa)$	$\mathbf{Expt}^{ideal}_{\mathcal{FE},\mathcal{A},\mathsf{Sim}}(1^\kappa)$
1: $F \leftarrow \mathcal{A}(1^\kappa)$;	1: $F \leftarrow \mathcal{A}(1^\kappa)$;
2: $(\mathsf{MPK}, \mathsf{MSK}) \leftarrow \mathsf{FE.Setup}(1^\kappa, F)$;	2: $\mathsf{MPK} \leftarrow \mathsf{Sim}(1^\kappa, F)$;
PROVIDE MPK TO $\mathcal{A}$;	PROVIDE MPK TO $\mathcal{A}$;
3: LET $i := 1$;	3: LET $i := 1$;
4: DO	4: DO
$\quad M_i \leftarrow \mathcal{A}^{\mathsf{FE.Key}(\mathsf{MSK},\cdot)}()$;	$\quad M_i \leftarrow \mathcal{A}^{\mathsf{Sim}^{\{F(\cdot,M_j)\}_{j<i}}}()$;
$\quad \mathsf{CT}_i \leftarrow \mathsf{FE.Enc}(\mathsf{MPK}, M_i)$;	$\quad \mathsf{CT}_i \leftarrow \mathsf{Sim}^{\{F(\cdot,M_j)\}_{j\leq i}}(1^{\lvert M_i\rvert})$;
$\quad$ PROVIDE CT_i TO $\mathcal{A}$;	$\quad$ PROVIDE CT_i TO $\mathcal{A}$;
$\quad i := i + 1$;	$\quad i := i + 1$;
$\quad$ UNTIL $\mathcal{A}$ BREAKS;	$\quad$ UNTIL $\mathcal{A}$ BREAKS;
5: $\alpha \leftarrow \mathcal{A}()$;	5: $\alpha \leftarrow \mathcal{A}()$;
6: OUTPUT $(\alpha, \{M_j\}_{j \in [i-1]})$;	6: OUTPUT $(\alpha, \{M_j\}_{j \in [i-1]})$;

In Step 4 of the ideal experiment, Sim *needs to provide answers to* $\mathsf{Key}(\mathsf{MSK}, \cdot)$ *queries of* $\mathcal{A}$. *During the execution of the ideal experiment, we say that* Sim*'s query K to oracles $\{F(\cdot, M_1), \ldots, F(\cdot, M_i)\}$ is* legal *if $\mathcal{A}$ already requested ciphertexts for $M_1, \ldots, M_i$, and made oracle query K to* $\mathsf{Key}(\mathsf{MSK}, \cdot)$. *We call a simulator algorithm* Sim admissible *if it only makes legal queries to its oracle throughout the execution.*

The functional encryption scheme $\mathcal{FE}$ *is said to be* fully-adaptive simulation-secure *if there is an admissible* PPT *stateful simulator* Sim *such that for every* PPT *stateful adversary* $\mathcal{A}$, *the following two distributions are computationally indistinguishable:*

$$\left\{ \mathbf{Expt}^{real}_{\mathcal{FE},\mathcal{A}}(1^\kappa) \right\}_\kappa \stackrel{c}{\approx} \left\{ \mathbf{Expt}^{ideal}_{\mathcal{FE},\mathcal{A},\mathsf{Sim}}(1^\kappa) \right\}_\kappa$$

4   Token Model and Constructions

4.1   Token-Based FE

Here we introduce a simple token model for encryption schemes and provide formal definitions of token-based functional encryption schemes. In our model, we consider *stateless* tokens that are initialized by the master authority in the setup stage, and are only used by users in decryption. Furthermore, we require token to be *universal* in the sense that tokens used by different users are identical. Thus, tokens are simply deterministic oracles that are generated by the Setup algorithm, and queried by the Dec algorithm.

Definition 5 (Token-based FE). *A token-based functional encryption scheme* $\mathcal{FE}$ *is defined identical to the definition of functional encryption scheme except for the following modifications.*

- **Setup:** *In addition to* MPK *and* MSK, *the algorithm* $\mathsf{FE.Setup}$ *also outputs a token* $\mathbf{T}$, *which is simply a deterministic oracle.*

- **Key Generation :** *In addition to the keys* SK_K, *the algorithm* $\mathsf{FE.Key}$ *also returns a copy of the token* $\mathbf{T}$ *to users.*
- **Decryption:** *The decryption algorithm* $\mathsf{FE.Dec}^{\mathbf{T}}$ *can query the* $\mathbf{T}$ *in order to decrypt.*

The correctness property extends straightforwardly. For security, we generalize fully-adaptive simulation security to the token model. As before, we allow the adversary $\mathcal{A}$ to take full control over the access of the encryption scheme; in particular, $\mathcal{A}$ is given the oracle access to token after setup. In the ideal world, the simulator is required to simulate answers to all queries made by $\mathcal{A}$, including the token queries, given only the "legal" information that $\mathcal{A}$ can learn from the received ciphertexts using the received secret keys.

Definition 6 (Fully-Adaptive Simulation Security for Token-Based FE). *Let* $\mathcal{FE}$ *be a token-basd functional encryption scheme for a functionality class* $\mathcal{F}$. *For every* PPT *stateful adversary* $\mathcal{A}$ *and* PPT *stateful simulator* Sim, *consider the following two experiments.*

$\mathbf{Expt}^{real}_{\mathcal{FE},\mathcal{A}}(1^\kappa)$	$\mathbf{Expt}^{ideal}_{\mathcal{FE},\mathcal{A},\mathsf{Sim}}(1^\kappa)$
1: $F \leftarrow \mathcal{A}(1^\kappa)$;	1: $F \leftarrow \mathcal{A}(1^\kappa)$;
2: $(\mathsf{MPK}, \mathsf{MSK}, \mathbf{T}) \leftarrow \mathsf{FE.Setup}(1^\kappa, F)$;	2: $\mathsf{MPK} \leftarrow \mathsf{Sim}(1^\kappa, F)$;
PROVIDE MPK TO $\mathcal{A}$;	PROVIDE MPK TO $\mathcal{A}$;
3: LET $i := 1$;	3: LET $i := 1$;
4: DO	4: DO
$\quad M_i \leftarrow \mathcal{A}^{\mathsf{FE.Key}(\mathsf{MSK},\cdot),\mathbf{T}(\cdot)}()$;	$\quad M_i \leftarrow \mathcal{A}^{\mathsf{Sim}^{\{F(\cdot,M_j)\}_{j<i}}}()$;
$\quad \mathsf{CT}_i \leftarrow \mathsf{FE.Enc}(\mathsf{MPK}, M_i)$;	$\quad \mathsf{CT}_i \leftarrow \mathsf{Sim}^{\{F(\cdot,M_j)\}_{j\leq i}}(1^{|M_i|})$;
PROVIDE CT_i TO $\mathcal{A}$;	PROVIDE CT_i TO $\mathcal{A}$;
$\quad i := i + 1$;	$\quad i := i + 1$;
UNTIL $\mathcal{A}$ BREAKS;	UNTIL $\mathcal{A}$ BREAKS;
5: $\alpha \leftarrow \mathcal{A}()$;	5: $\alpha \leftarrow \mathcal{A}()$;
6: OUTPUT $(\alpha, \{M_j\}_{j \in [i-1]})$;	6: OUTPUT $(\alpha, \{M_j\}_{j \in [i-1]})$;

In Step 4 of the ideal experiment, Sim *needs to provide answers to both* $\mathsf{Key}(\mathsf{MSK},\cdot)$ *and* $\mathbf{T}(\cdot)$ *queries of* $\mathcal{A}$. *During the execution of the ideal experiment, we say that* Sim*'s query* K *to oracles* $\{F(\cdot,M_1),\ldots,F(\cdot,M_i)\}$ *is* legal *if* $\mathcal{A}$ *already requested ciphertexts for* $M_1,\ldots,M_i$, *and made oracle query* K *to* $\mathsf{Key}(\mathsf{MSK},\cdot)$. *We call a simulator algorithm* Sim admissible *if it only makes legal queries to its oracle throughout the execution.*

The functional encryption scheme $\mathcal{FE}$ *is said to be* fully-adaptive simulation-secure *if there is an admissible* PPT *stateful simulator* Sim *such that for every* PPT *stateful adversary* $\mathcal{A}$, *the following two distributions are computationally indistinguishable:*

$$\left\{\mathbf{Expt}^{real}_{\mathcal{FE},\mathcal{A}}(1^\kappa)\right\}_\kappa \overset{c}{\approx} \left\{\mathbf{Expt}^{ideal}_{\mathcal{FE},\mathcal{A},\mathsf{Sim}}(1^\kappa)\right\}_\kappa$$

4.2 Token-Based FE Construction — Solution #1

Here we give the construction of a functional encryption scheme $\mathcal{FE} = $ FE.{Setup, Key, Enc, Dec} for a functionality F based on stateless and universal tokens. Our construction is based on a CCA2-secure public key encryption PKE.{Gen, Enc, Dec} and a strongly unforgeable signature scheme SIG.{Gen, Sign, Vrfy}. In the setup stage, the authority generates a key-pair (ek, dk) for encryption and a key-pair (vk, sk) for digital signature, and set MPK = (ek, vk) and MSK = sk. Additionally, the authority initializes the token $\mathbf{T}$ with the description of F, public keys ek, vk, and secret decryption key dk.

To encrypt a message M, one simply encrypts it using the underlying CCA2 public key ek; that is, the ciphertext is ct $\leftarrow$ PKE.Enc$_{ek}(M)$. The secret key SK$_K$ for a functionality key K is simply a signature of K; that is, SK$_K = \sigma_K \leftarrow$ SIG.Sign$_{sk}(K)$. To decrypt ct using secret key SK$_K$, the user queries its token $\mathbf{T}$ with (ct, K, σ_K). $\mathbf{T}$ verifies if σ_K is valid, and if so, $\mathbf{T}$ returns $F(K, \text{PKE.Dec}_{dk}(\text{ct}))$, and returns $\perp$ otherwise. A formal description of our scheme can be found in Figure 1.

Note that our scheme has succinct ciphertext size. Indeed, our ciphertext is simply a CCA2 encryption of the message, which is independent of the complexity of F. On the other hand, our token need to evaluate F to decrypt. Thus, our solution #1 is suitable for lower complexity functionalities (e.g., inner product functionality).

While our scheme is very simple, it satisfies the strong fully-adaptive simulation-security as defined in Definition 6. In fact, the security proof is rather straightforward: The simulator simply simulates Setup and Key queries honestly, and simulates encryption queries M_i by encryption of $0^{|M_i|}$. To answer a token query (ct, K, σ_K), when σ_K verifies, the simulator checks if ct is one of the simulated ciphertext (for some encryption query M_i). If so, the simulator queries its oracle and returns $F(K, M_i)$, and if not, it simulates the token honestly. Intuitively, the simulation works since by strong unforgeability, the simulator can learn correct answers for simulated ciphertexts from its oracle, and CCA2-security ensures that the simulation works for other ciphertexts.

We note that our security proof crucially relies on the fact that in the simulation, the simulator gets to simulate token's answers to the queries made by the adversary, which bypasses the information-theoretic arguments underlying the impossibility results of Boneh et al. [10] and Agrawal et al. [1].

Theorem 1. *If $\mathcal{SIG}$ is a strongly unforgeable signature scheme, $\mathcal{PKE}$ is a CCA2-secure public key encryption, then the above functional encryption construction $\mathcal{FE}$ is simulation-secure (Definition 6).*

Proof: We here prove that our scheme achieves the strong fully-adaptive simulation-security as defined in Definition 6. In order to prove the security, we need to construct a simulator Sim which interacts with an adversary $\mathcal{A}$. The ideal experiment $\mathbf{Expt}_{\mathcal{FE},\mathcal{A},\text{Sim}}^{ideal}(1^\kappa)$ is as follows:

- **Setup:** on input a security parameter 1^κ, a functionality $F \in \mathcal{F}_\kappa$, the setup algorithm Setup() performs the following steps to generate MPK, MSK, and a deterministic stateless token **T**.
    - Execute $(\mathsf{ek}, \mathsf{dk}) \leftarrow \mathsf{PKE.Gen}(1^\kappa)$, and $(\mathsf{vk}, \mathsf{sk}) \leftarrow \mathsf{SIG.Gen}(1^\kappa)$.
    - Initiate a token **T** with values $(\mathsf{dk}, \mathsf{ek}, \mathsf{vk}, F)$.
    - Output $\mathsf{MPK} = (\mathsf{ek}, \mathsf{vk})$, and $\mathsf{MSK} = (\mathsf{sk})$.
- **Key Generation:** on input a master secret key MSK and a functionality key K, the key generation algorithm Key() generates SK_K as follows.
    - Execute $\sigma_K \leftarrow \mathsf{SIG.Sign}_{\mathsf{sk}}(K)$. Output $\mathsf{SK}_K = (\sigma_K)$.
- **Encryption:** on input a master public key MPK and a message M, the encryption algorithm Enc() generates CT as follows.
    - Execute $\mathsf{ct} \leftarrow \mathsf{PKE.Enc}_{\mathsf{ek}}(M; \rho)$, where ρ is the randomness. Return $\mathsf{CT} = (\mathsf{ct})$.
- **Decryption:** on input $\mathsf{SK}_K = (\sigma_K)$ and a ciphertext $\mathsf{CT} = (\mathsf{ct})$ of a message M, with access to a token **T**, the decryption algorithm $\mathsf{Dec^T}()$ performs the following steps to decrypt $m = F(K, M)$:
    - Query the token $m \leftarrow \mathbf{T}(\mathsf{CT}, K, \mathsf{SK}_K)$. Output m.
- **Token Operations:** on query $(\mathsf{CT}, K, \mathsf{SK}_K)$, where $\mathsf{CT} = (\mathsf{ct})$ and $\mathsf{SK}_K = (\sigma_K)$, the token **T** carries out the following operations.
    - Execute $\mathsf{SIG.Vrfy}_{\mathsf{vk}}(K, \sigma_K)$.
    - If the above verification accepts, then compute $M \leftarrow \mathsf{PKE.Dec}_{\mathsf{dk}}(\mathsf{ct})$ and return $m = F(K, M)$. Otherwise, return $\perp$.

Fig. 1. Solution #1. Here PKE.{Gen, Enc, Dec} is a public-key encryption scheme, and SIG.{Gen, Sign, Vrfy} is a signature scheme.

- Upon obtaining functionality F from the adversary, the simulator runs $(\mathsf{vk}, \mathsf{sk}) \leftarrow \mathsf{SIG.Gen}()$ and $(\mathsf{ek}, \mathsf{dk}) \leftarrow \mathsf{PKE.Gen}()$, and set $\mathsf{MPK} = (\mathsf{ek}, \mathsf{vk})$, and give MPK to the adversary. From now on, oracle access to the token will be simulated for the adversary.
- In the key generation, upon receiving the request on K from the adversary, the simulator computes $\sigma_K \leftarrow \mathsf{SIG.Sign}_{\mathsf{sk}}(K)$, and returns σ_K to the adversary. Note that now the simulator records (K, σ_K) into history.
- At any point when the adversary provides message M, the simulator is allowed to see the length $|M|$ and it is granted an oracle $F(\cdot, M)$. The simulator then computes $\mathsf{ct} \leftarrow \mathsf{PKE.Enc}(0^{|M|}; \omega)$ where ω is randomly chosen, and sets ct as a pointer to the oracle $F(\cdot, M)$. The simulator records $(|M|, \mathsf{ct})$ into history, and returns ct to the adversary. If the same ct is recorded twice in the history, then the simulator returns Abort.
- At any point when the adversary queries the token with tuple $(\mathsf{ct}, K, \sigma_K)$, the simulator first checks if (K, σ_K) has been recorded in the history. If not, then it returns $\perp$. Else if the pair (K, σ_K) has been recorded, and ct has also been recorded in the history, then the simulator queries the corresponding oracle $F(\cdot, M)$, and learns $m = F(K, M)$. Then the simulator returns m to the adversary. Otherwise, if (K, σ_K) has been recorded but ct has not, the

simulator computes $M \leftarrow \mathsf{PKE.Dec_{dk}(ct)}$ and $m \leftarrow F(K, M)$, and returns m to the adversary.

- Let $M_1^*, \ldots, M_n^*$ be the messages that the adversary queried for ciphertexts. If the adversary finally outputs a value α, output $(\alpha, \{M_i^*\}_{i \in [n]})$.

From the above simulation, we can easily see that only valid users who participate in the key generation are able to use the token to decrypt ciphertexts. Furthermore, for a ciphertext $\mathsf{ct} = \mathsf{PKE.Enc}(M)$, the adversary cannot learn any extra information beyond $\{F(K_i, M)\}_i$ where $\{K_i\}_i$ have been registered in the key generation.

Next, we show that the ideal experiment is computationally close to the real experiment, by developing a sequence of hybrids between them.

Hybrid 0: This is the real experiment $\mathbf{Expt}^{real}_{\mathcal{FE}, \mathcal{A}}(1^\kappa)$. As described in construction $\mathcal{FE}$, upon obtaining functionality F, we first generate $(\mathsf{MPK}, \mathsf{MSK}, \mathbf{T}) \leftarrow \mathsf{FE.Setup}(1^\kappa, F)$ where $\mathsf{MPK} = (\mathsf{ek}, \mathsf{vk})$ and $\mathsf{MSK} = \mathsf{sk}$, and give MPK to the adversary. At any moment when the adversary queries $\mathsf{FE.Key}()$ with K, we return $\mathsf{SK}_K = \sigma_K$ where $\sigma_K \leftarrow \mathsf{SIG.Sign_{sk}}(K)$. At any point when the adversary outputs M_i^*, we return the adversary with $\mathsf{CT}_i^* = \mathsf{ct}_i^*$ where $\mathsf{ct}_i^* \leftarrow \mathsf{PKE.Enc_{ek}}(M_i^*; \omega_i^*)$. At any point when the adversary queries the token with tuple $(\mathsf{ct}, K, \sigma_K)$, the token will behave as follows: if the pair (K, σ_K) is not verified, then return $\bot$. Otherwise if the pair is verified, i.e., $\mathsf{SIG.Vrfy_{vk}}(K, \sigma_K) = 1$, then use dk to decrypt ct into $M \leftarrow \mathsf{PKE.Dec_{dk}(ct)}$, and return $m = F(K, M)$. We then return m to the adversary. Let $M_1^*, \ldots, M_n^*$ be the values that the adversary queried for ciphertexts. If the adversary finally outputs a value α, we output $(\alpha, \{M_i^*\}_{i \in [n]})$.

Hybrid 1: This hybrid is the same as Hybrid 0 except the following: In this hybrid, we change the token's responses to the adversary. At any point when the adversary queries the token with tuple $(\mathsf{ct}, K, \sigma_K)$, if $\mathsf{SIG.Vrfy_{vk}}(K, \sigma_K) = 1$ while the pair (K, σ_K) never appears in the queries to $\mathsf{FE.Key}()$, then the hybrid outputs $\mathtt{Abort}$.

Hybrid 1 and Hybrid 0 are the same except that $\mathtt{Abort}$ occurs. Based on the strong unforgeability of $\mathcal{SIG}$, we claim the event of $\mathtt{Abort}$ occurs with negligible probability. Therefore, Hybrid 1 and Hybrid 0 are computationally indistinguishable. Towards contradiction, assume there is a distinguisher $\mathcal{A}$ can distinguish Hybrid 0 from Hybrid 1. We next show an algorithm $\mathcal{B}$ that breaks the strong unforgeability of $\mathcal{SIG}$ as follows:

- Upon receiving the encryption key vk, $\mathcal{B}$ internally simulates $\mathcal{A}$. $\mathcal{B}$ works the same as in Hybrid 0 except the following: At any point when the adversary provides functionality F, $\mathcal{B}$ computes $(\mathsf{ek}, \mathsf{dk}) \leftarrow \mathsf{PKE.Gen}()$, and sets $\mathsf{MPK} := (\mathsf{ek}, \mathsf{vk})$. At any moment when the adversary queries $\mathsf{FE.Key}()$ with K, $\mathcal{B}$ queries its own signing oracle with K and receives σ_K, and then $\mathcal{B}$ returns σ_K to $\mathcal{A}$ as the response. At any point when the adversary queries the token with tuple $(\mathsf{ct}, K, \sigma_K)$, if $\mathsf{SIG.Vrfy_{vk}}(K, \sigma_K) = 1$, but (K, σ_K) never appears in the queries to $\mathsf{FE.Key}()$, then the event $\mathtt{Abort}$ occurs, $\mathcal{B}$ halts and output (K, σ_K) to its challenger as the forged pair.

We note that the view of the above simulated $\mathcal{A}$ is the same as that in Hybrid 1. We further note that as long as the event Abort does not occur, $\mathcal{A}$'s view is the same as that in Hybrid 0. Since $\mathcal{A}$ is able to distinguish the two hybrids, that means the event Abort will occur with non-negligible probability. That says, $\mathcal{B}$ is a successful unforgeability attacker against $\mathcal{SIG}$, which reaches a contradiction. Therefore, Hybrid 0 and Hybrid 1 are computationally indistinguishable.

Hybrid 2: This hybrid is the same as Hybrid 1 except the following: Whenever the adversary queries on M_i^*, we compute $\hat{\mathsf{ct}}_i^* \leftarrow \mathsf{PKE.Enc}_{\mathsf{ek}}(M_i^*; \omega_i^*)$, and record $(|M_i^*|, \hat{\mathsf{ct}}_i^*)$. Here we can easily simulate an oracle $F(\cdot, M_i^*)$ based on M_i^*, and we set the ciphertext ct^* as the pointer to the oracle. Furthermore, we change the token's responses to the adversary. At any point when the adversary queries the token with tuple $(\mathsf{ct}, K, \sigma_K)$ where the pair (K, σ_K) has been recorded, we carry out the following: if ct has been recorded then we based on it query the corresponding oracle $F(\cdot, M_i^*)$ with K and receive $m = F(K, M_i^*)$. Then we return m to the adversary.

We can easily see that the views of $\mathcal{A}$ are the same in Hybrid 1 and Hybrid 2.

Hybrid 3.j, where $j = 0, \ldots, n$: Here n is the total number of messages the adversary has queried for ciphertexts. This hybrid is the same as Hybrid 2 except the following:

When the adversary queries on $\{M_i^*\}_{i\in[n]}$, the messages $\{M_1^*, \ldots, M_j^*\}$ are blocked; instead, we are allowed to see the length of the messages, i.e, $|M_1^*|, \ldots, |M_j^*|$, and have oracle access to $F(\cdot, M_1^*), \ldots, F(\cdot, M_j^*)$. Note that we are now still allowed to see the messages $\{M_{j+1}^*, \ldots, M_n^*\}$, and therefore we can easily simulate the oracles $F(\cdot, M_{j+1}^*), \ldots, F(\cdot, M_n^*)$.

We change the response to the adversary's query on $\{M_i^*\}_{i=1,\ldots,n}$. We now return the adversary with $\mathsf{CT}_i^* = \hat{\mathsf{ct}}_i^*$ for all $i \in [n]$. Here $\hat{\mathsf{ct}}_i^* \leftarrow \mathsf{PKE.Enc}_{\mathsf{ek}}(0^{|M_i^*|}; \omega_i^*)$ for all $i \in [1, \ldots, j]$, and $\hat{\mathsf{ct}}_i^* \leftarrow \mathsf{PKE.Enc}_{\mathsf{ek}}(M_i^*; \omega_i^*)$ for all $i \in [j+1, \ldots, n]$.

Based on the CCA2-security of $\mathcal{PKE}$, we claim Hybrid 3.j and Hybrid 3.$(j+1)$ are computationally indistinguishable for all $j = 0, \ldots, n-1$. Towards contradiction, assume there is a distinguisher $\mathcal{A}$ who can distinguish Hybrid 3.j from Hybrid 3.$(j+1)$. We next show an algorithm $\mathcal{B}$ that breaks the CCA2-security of $\mathcal{PKE}$ as follows:

- Upon receiving the encryption key ek, $\mathcal{B}$ internally simulates $\mathcal{A}$. $\mathcal{B}$ works the same as in Hybrid 3.j except the following:
  - Upon $\mathcal{A}$'s query on M_{j+1}^*, $\mathcal{B}$ queries LR-oracle LR with $(M_{j+1}^*, 0^{|M_{j+1}^*|})$; in turn it gets back a ciphertext ct_{j+1}^* which is $\mathsf{PKE.Enc}$ $(0^{|M_{j+1}^*|})$ or $\mathsf{PKE.Enc}(M_{j+1}^*)$ from the LR-oracle.
  - Upon receiving $\mathcal{A}$'s query to the token with tuple $(\mathsf{ct}, K, \sigma_K)$ where (K, σ_K) has been recorded, if ct has been recorded then $\mathcal{B}$ simulates the corresponding oracle $F(\cdot, M_i^*)$ for K and provides $m = F(K, M_i^*)$ to $\mathcal{A}$. If ct has not been recorded, then $\mathcal{B}$ queries its decryption oracle to obtain the plaintext M of the ciphertext ct, and then return $m = F(K, M)$ to the adversary.

– Finally, $\mathcal{B}$ outputs whatever $\mathcal{A}$ outputs.

Let β be the hidden bit associated with the LR oracle. We note that when $\beta = 0$, the algorithm $\mathcal{B}$ exactly simulates the Hybrid $3.j$ to $\mathcal{A}$; when $\beta = 1$, $\mathcal{B}$ simulates exactly the Hybrid $3.(j+1)$ to $\mathcal{A}$. Under the assumption, since $\mathcal{A}$ is able to distinguish the two hybrids in non-negligible probability, that means the constructed $\mathcal{B}$ is successful CCA2 attacker against $\mathcal{PKE}$, which reaches a contradiction. Therefore Hybrid $3.j$ and Hybrid $3.(j+1)$ are computationally indistinguishable.

Furthermore, we note that Hybrid 3.0 is the same as Hybrid 2, and Hybrid $3.n$ is the ideal experiment. Based on the above argument we already see the real experiment and the ideal experiment are in distinguishable. This means the construction $\mathcal{FE}$ is simulation secure as defined in Definition 6. $\blacksquare$

4.3   Token-Based FE Construction — Solution #2

In our solution #1 presented in the previous section, the token size is linear of function F. Here we present our solution #2, a functional encryption scheme $\mathcal{FE} = \mathsf{FE}.\{\mathsf{Setup}, \mathsf{Key}, \mathsf{Enc}, \mathsf{Dec}\}$ in the token model where the complexity of token is independent of the complexity of F. We use the following tools: FHE, digital signature, publicly verifiable SNARG, and simulation-extractable NIZK. (Please refer to Section 2 for the definitions.)

In the setup stage, the authority generates key pairs $(\mathsf{ek}, \mathsf{dk})$ and $(\mathsf{vk}, \mathsf{sk})$ for FHE and for digital signature respectively. The authority also sets up the reference strings crs and $(\mathsf{rs}, \mathsf{vrs})$ for NIZK and for SNARG respectively. Note that the reference string vrs for SNARG verification is very short and it is independent of F. The authority sets $(\mathsf{ek}, \mathsf{vk}, \mathsf{crs}, \mathsf{rs}, \mathsf{vrs})$ as its master public key MPK, and sk as the master secret key MSK. In addition, the authority initializes the token $\mathbf{T}$ with the public information $(\mathsf{ek}, \mathsf{vk}, \mathsf{crs}, \mathsf{vrs})$, and the secret decryption key dk.

The key generation stage is the same as that in the previous solution; for each user associated with a key K, the authority uses the MSK to generate a signature σ_K on the key K; in addition the authority sends the user an identical copy of the token. The encryption algorithm is different from that in the previous solution: To encrypt a message M, one takes two steps: (1) encrypt it using the FHE public key ek; that is, the ciphertext is $\mathsf{ct} \leftarrow \mathsf{FHE}.\mathsf{Enc}_{\mathsf{ek}}(M)$; (2) generate an NIZK that the ciphertext ct is honestly generated. The ciphertext for message M is (ct, π).

The decryption algorithm is different from that in the previous solution as well. Our goal as stated before is to obtain a solution in which the complexity of the token is independent of the complexity of F. The idea is to let the token to "outsource" most of the computation of F to the user. Concretely, to decrypt a ciphertext (ct, π), the user who is associated with key K computes the transformed ciphertext $\tilde{\mathsf{ct}}$ by homomorphically evaluating ct with $F(K, \cdot)$; to be sure that the transformation is carried out correctly, the user also provides a SNARG ϖ. Then the user queries the token $\mathbf{T}$ with an input tuple $(\mathsf{ct}, \pi, K, \sigma_K, \tilde{\mathsf{ct}}, \varpi)$;

- **Setup:** on input a security parameter 1^κ, a functionality $F \in \mathcal{F}_\kappa$, the setup algorithm Setup() performs the following steps to generate MPK, MSK, and a deterministic stateless token **T**.
  - Execute $(\mathsf{ek}, \mathsf{dk}) \leftarrow \mathsf{FHE.Gen}(1^\kappa)$, $(\mathsf{vk}, \mathsf{sk}) \leftarrow \mathsf{SIG.Gen}(1^\kappa)$, and $\mathsf{crs} \leftarrow \mathsf{NIZK.Gen}(1^\kappa)$.
  - Define $\hat{F}(K, \mathsf{ct}) \triangleq \mathsf{FHE.Eval}_{\mathsf{ek}}(\mathsf{ct}, F(K, \cdot))$. Execute $(\mathsf{rs}, \mathsf{vrs}) \leftarrow \mathsf{SNARG.Gen}(1^\kappa, \hat{F})$.
  - Initiate a token **T** with values $(\mathsf{dk}, \mathsf{ek}, \mathsf{vk}, \mathsf{crs}, \mathsf{vrs})$.
  - Output $\mathsf{MPK} = (\mathsf{ek}, \mathsf{vk}, \mathsf{crs}, \mathsf{rs}, \mathsf{vrs})$, $\mathsf{MSK} = (\mathsf{sk})$.
- **Key Generation:** on input a master secret key MSK and a functionality key K, Key() generates SK_K as:
  - Execute $\sigma_K \leftarrow \mathsf{SIG.Sign}_{\mathsf{sk}}(K)$. Output $\mathsf{SK}_K = (\sigma_K)$.
- **Encryption:** on input a master public key MPK and a message M, the encryption algorithm Enc() generates CT as follows.
  - Execute $\mathsf{ct} \leftarrow \mathsf{FHE.Enc}_{\mathsf{ek}}(M; \omega)$, where ω is the randomness used in the encryption.
  - Execute $\pi \leftarrow \mathsf{NIZK}.P(\mathsf{crs}, (\mathsf{ek}, \mathsf{ct}), (M, \omega))$ with respect to the relation

  $$R_{\mathsf{FHE}} = \{((\mathsf{ek}, \mathsf{ct}), (M, \omega)) : \mathsf{FHE.Enc}_{\mathsf{ek}}(M; \omega) = \mathsf{ct}\}.$$

  - Output $\mathsf{CT} = (\mathsf{ct}, \pi)$
- **Decryption:** on input SK_K and a ciphertext $\mathsf{CT} = (\mathsf{ct}, \pi)$ of a message M, with access to a token **T**, the decryption algorithm $\mathsf{Dec}^{\mathbf{T}}()$ performs the following steps to decrypt $m = F(K, M)$:
  - Execute $(\tilde{\mathsf{ct}}, \varpi) \leftarrow \mathsf{SNARG}.P(\mathsf{rs}, (K, \mathsf{ct}))$. Here $\tilde{\mathsf{ct}} = \hat{F}(K, \mathsf{ct}) = \mathsf{FHE.Eval}_{\mathsf{ek}}(\mathsf{ct}, F(K, \cdot))$.
  - Query the token $m \leftarrow \mathbf{T}(\mathsf{CT}, K, \mathsf{SK}_K, \tilde{\mathsf{ct}}, \varpi)$. Output m.
- **Token Operations:** on query $(\mathsf{CT}, K, \mathsf{SK}_K, \tilde{\mathsf{ct}}, \varpi)$, where $\mathsf{CT} = (\mathsf{ct}, \pi)$ and $\mathsf{SK}_K = \sigma_K$, the token **T** carries out the following operations.
  - Execute $\mathsf{SIG.Vrfy}_{\mathsf{vk}}(K, \sigma_K)$, $\mathsf{NIZK}.V(\mathsf{crs}, (\mathsf{ek}, \mathsf{ct}), \pi)$, and $\mathsf{SNARG}.V(\mathsf{vrs}, (K, \mathsf{ct}), \tilde{\mathsf{ct}}, \varpi)$.
  - Return $\mathsf{FHE.Dec}_{\mathsf{dk}}(\tilde{\mathsf{ct}})$ if all above verifications accept, and return $\perp$ otherwise.

Fig. 2. Solution #2. Here $\mathsf{FHE}.\{\mathsf{Gen}, \mathsf{Enc}, \mathsf{Eval}, \mathsf{Dec}\}$ is a fully homomorphic encryption, $\mathsf{SIG}.\{\mathsf{Gen}, \mathsf{Sign}, \mathsf{Vrfy}\}$ is a signature scheme, $\mathsf{SNARG}.\{\mathsf{Gen}, P, V\}$ is a SNARG scheme, $\mathsf{NIZK}.\{\mathsf{Gen}, P, V\}$ is a NIZK scheme.

the token first verifies if signature, NIZK, SNARG are all valid; if so, the token decrypts the ciphertext $\tilde{ct}$ into message m and returns m, and it returns $\bot$ otherwise. A formal description of our scheme can be found in Figure 2.

Note that, similar to solution #1, our scheme here also has succinct ciphertext size. Our ciphertext consists of an FHE ciphertext and an NIZK, both of which are independent of the complexity of F. On the other hand, here our token does *not* need to evaluate F to decrypt, and thus the complexity of the token is independent of the complexity of F.

Our scheme here also satisfies the strong fully-adaptive simulation-security as defined in Definition 4. The proof idea is very similar to that in the previous section, which crucially relies on the fact that in the simulation, the simulator gets to simulate token's answers to the queries made by the adversary. Next, we briefly highlight the differences between the two solutions. In both constructions, the user can only provide authenticated inputs to the hardware token, and digital signature is used to authenticate K. But two different approaches are used to authenticate the ciphertext: in solution #1, the authentication is guaranteed by the CCA2 security of the encryption, while in solution #2, the authentication is provided by the simulation-extractability of the NIZK, and the soundness of the SNARG.

Theorem 2. *If $\mathcal{SNARG}$ is a publicly verifiable SNARG scheme, $\mathcal{NIZK}$ is a zero-knowledge and simulation-extractable NIZK scheme, $\mathcal{SIG}$ is a strong unforgeable signature scheme, $\mathcal{FHE}$ is a secure fully homomorphic encryption scheme, then the above construction $\mathcal{FE}$ is simulation-secure functional encryption scheme.*

Proof can be found in the full version.

Acknowledgments. We would like to thank the anonymous reviewers for helpful feedback.

References

1. Agrawal, S., Gorbunov, S., Vaikuntanathan, V., Wee, H.: Functional encryption: New perspectives and lower bounds. In: Canetti, R., Garay, J.A. (eds.) CRYPTO 2013, Part II. LNCS, vol. 8043, pp. 500–518. Springer, Heidelberg (2013)
2. Barak, B., Goldreich, O., Impagliazzo, R., Rudich, S., Sahai, A., Vadhan, S.P., Yang, K.: On the (im)possibility of obfuscating programs. J. ACM 59(2), 6 (2012)
3. Barbosa, M., Farshim, P.: On the semantic security of functional encryption schemes. In: Kurosawa, K., Hanaoka, G. (eds.) PKC 2013. LNCS, vol. 7778, pp. 143–161. Springer, Heidelberg (2013)
4. Bellare, M., O'Neill, A.: Semantically secure functional encryption: Possibility results, impossibility results, and the quest for a general definition. Cryptology ePrint Archive, Report 2012/515 (2012), http://eprint.iacr.org/
5. Bitansky, N., Canetti, R., Chiesa, A., Tromer, E.: Recursive composition and bootstrapping for snarks and proof-carrying data. In: STOC 2013 (2013)

6. Bitansky, N., Canetti, R., Goldwasser, S., Halevi, S., Kalai, Y.T., Rothblum, G.N.: Program obfuscation with leaky hardware. In: Lee, D.H., Wang, X. (eds.) ASIACRYPT 2011. LNCS, vol. 7073, pp. 722–739. Springer, Heidelberg (2011)

7. Boneh, D., Boyen, X., Goh, E.-J.: Hierarchical identity based encryption with constant size ciphertext. In: Cramer, R. (ed.) EUROCRYPT 2005. LNCS, vol. 3494, pp. 440–456. Springer, Heidelberg (2005)

8. Boneh, D., Franklin, M.: Identity-based encryption from the weil pairing. In: Kilian, J. (ed.) CRYPTO 2001. LNCS, vol. 2139, pp. 213–229. Springer, Heidelberg (2001)

9. Boneh, D., Raghunathan, A., Segev, G.: Function-private identity-based encryption: Hiding the function in functional encryption. In: Canetti, R., Garay, J.A. (eds.) CRYPTO 2013, Part II. LNCS, vol. 8043, pp. 461–478. Springer, Heidelberg (2013)

10. Boneh, D., Sahai, A., Waters, B.: Functional encryption: Definitions and challenges. In: Ishai, Y. (ed.) TCC 2011. LNCS, vol. 6597, pp. 253–273. Springer, Heidelberg (2011)

11. Boneh, D., Waters, B.: Conjunctive, subset, and range queries on encrypted data. In: Vadhan, S.P. (ed.) TCC 2007. LNCS, vol. 4392, pp. 535–554. Springer, Heidelberg (2007)

12. Chandran, N., Goyal, V., Sahai, A.: New constructions for UC secure computation using tamper-proof hardware. In: Smart, N.P. (ed.) EUROCRYPT 2008. LNCS, vol. 4965, pp. 545–562. Springer, Heidelberg (2008)

13. Damgård, I., Nielsen, J.B., Wichs, D.: Universally composable multiparty computation with partially isolated parties. In: Reingold, O. (ed.) TCC 2009. LNCS, vol. 5444, pp. 315–331. Springer, Heidelberg (2009)

14. De Caro, A., Iovino, V., Jain, A., O'Neill, A., Paneth, O., Persiano, G.: On the achievability of simulation-based security for functional encryption. In: Canetti, R., Garay, J.A. (eds.) CRYPTO 2013, Part II. LNCS, vol. 8043, pp. 519–535. Springer, Heidelberg (2013)

15. Döttling, N., Kraschewski, D., Müller-Quade, J.: Unconditional and composable security using a single stateful tamper-proof hardware token. In: Ishai, Y. (ed.) TCC 2011. LNCS, vol. 6597, pp. 164–181. Springer, Heidelberg (2011)

16. Döttling, N., Mie, T., Müller-Quade, J., Nilges, T.: Implementing resettable UC-functionalities with untrusted tamper-proof hardware-tokens. In: Sahai, A. (ed.) TCC 2013. LNCS, vol. 7785, pp. 642–661. Springer, Heidelberg (2013)

17. Garg, S., Gentry, C., Halevi, S., Sahai, A., Waters, B.: Attribute-based encryption for circuits from multilinear maps. In: Canetti, R., Garay, J.A. (eds.) CRYPTO 2013, Part II. LNCS, vol. 8043, pp. 479–499. Springer, Heidelberg (2013)

18. Garg, S., Gentry, C., Sahai, A., Waters, B.: Witness encryption and its applications. In: STOC 2013 (2013)

19. Gennaro, R., Gentry, C., Parno, B.: Non-interactive verifiable computing: Outsourcing computation to untrusted workers. In: Rabin, T. (ed.) CRYPTO 2010. LNCS, vol. 6223, pp. 465–482. Springer, Heidelberg (2010)

20. Gennaro, R., Gentry, C., Parno, B., Raykova, M.: Quadratic span programs and succinct NIZKs without PCPs. In: Johansson, T., Nguyen, P.Q. (eds.) EUROCRYPT 2013. LNCS, vol. 7881, pp. 626–645. Springer, Heidelberg (2013)

21. Gentry, C.: Fully homomorphic encryption using ideal lattices. In: 41st Annual ACM Symposium on Theory of Computing, pp. 169–178. ACM Press (2009)

22. Gentry, C., Silverberg, A.: Hierarchical ID-based cryptography. In: Zheng, Y. (ed.) ASIACRYPT 2002. LNCS, vol. 2501, pp. 548–566. Springer, Heidelberg (2002)

23. Gentry, C., Wichs, D.: Separating succinct non-interactive arguments from all falsifiable assumptions. In: 43rd Annual ACM Symposium on Theory of Computing, pp. 99–108. ACM Press (2011)
24. Goldwasser, S., Kalai, Y., Popa, R.A., Vaikuntanathan, V., Zeldovich, N.: Succinct functional encryption and applications: Reusable garbled circuits and beyond. In: STOC 2013 (2013)
25. Gorbunov, S., Vaikuntanathan, V., Wee, H.: Functional encryption with bounded collusions via multi-party computation. In: Safavi-Naini, R., Canetti, R. (eds.) CRYPTO 2012. LNCS, vol. 7417, pp. 162–179. Springer, Heidelberg (2012)
26. Gorbunov, S., Vaikuntanathan, V., Wee, H.: Attribute-based encryption for circuits. In: STOC 2013 (2013)
27. Goyal, V., Ishai, Y., Mahmoody, M., Sahai, A.: Interactive locking, zero-knowledge PCPs, and unconditional cryptography. In: Rabin, T. (ed.) CRYPTO 2010. LNCS, vol. 6223, pp. 173–190. Springer, Heidelberg (2010)
28. Goyal, V., Ishai, Y., Sahai, A., Venkatesan, R., Wadia, A.: Founding cryptography on tamper-proof hardware tokens. In: Micciancio, D. (ed.) TCC 2010. LNCS, vol. 5978, pp. 308–326. Springer, Heidelberg (2010)
29. Goyal, V., Pandey, O., Sahai, A., Waters, B.: Attribute-based encryption for fine-grained access control of encrypted data. In: ACM Conference on Computer and Communications Security, pp. 89–98. ACM Press (2006)
30. Groth, J.: Simulation-sound NIZK proofs for a practical language and constant size group signatures. In: Lai, X., Chen, K. (eds.) ASIACRYPT 2006. LNCS, vol. 4284, pp. 444–459. Springer, Heidelberg (2006)
31. Groth, J.: Short pairing-based non-interactive zero-knowledge arguments. In: Abe, M. (ed.) ASIACRYPT 2010. LNCS, vol. 6477, pp. 321–340. Springer, Heidelberg (2010)
32. Groth, J., Ostrovsky, R., Sahai, A.: Perfect non-interactive zero knowledge for NP. In: Vaudenay, S. (ed.) EUROCRYPT 2006. LNCS, vol. 4004, pp. 339–358. Springer, Heidelberg (2006)
33. Horwitz, J., Lynn, B.: Toward hierarchical identity-based encryption. In: Knudsen, L.R. (ed.) EUROCRYPT 2002. LNCS, vol. 2332, pp. 466–481. Springer, Heidelberg (2002)
34. Katz, J.: Universally composable multi-party computation using tamper-proof hardware. In: Naor, M. (ed.) EUROCRYPT 2007. LNCS, vol. 4515, pp. 115–128. Springer, Heidelberg (2007)
35. Katz, J., Sahai, A., Waters, B.: Predicate encryption supporting disjunctions, polynomial equations, and inner products. J. Cryptology 26(2), 191–224 (2013)
36. Lewko, A., Okamoto, T., Sahai, A., Takashima, K., Waters, B.: Fully secure functional encryption: Attribute-based encryption and (Hierarchical) inner product encryption. In: Gilbert, H. (ed.) EUROCRYPT 2010. LNCS, vol. 6110, pp. 62–91. Springer, Heidelberg (2010)
37. Lewko, A., Waters, B.: Unbounded HIBE and attribute-based encryption. In: Paterson, K.G. (ed.) EUROCRYPT 2011. LNCS, vol. 6632, pp. 547–567. Springer, Heidelberg (2011)
38. Lipmaa, H.: Progression-free sets and sublinear pairing-based non-interactive zero-knowledge arguments. In: Cramer, R. (ed.) TCC 2012. LNCS, vol. 7194, pp. 169–189. Springer, Heidelberg (2012)
39. Micali, S.: Computationally sound proofs. SIAM J. Computing 30(4), 1253–1298 (2000)

40. Moran, T., Segev, G.: David and Goliath commitments: UC computation for asymmetric parties using tamper-proof hardware. In: Smart, N.P. (ed.) EUROCRYPT 2008. LNCS, vol. 4965, pp. 527–544. Springer, Heidelberg (2008)
41. O'Neill, A.: Definitional issues in functional encryption. Cryptology ePrint Archive, Report 2010/556 (2010), `http://eprint.iacr.org/`
42. Sahai, A., Seyalioglu, H.: Worry-free encryption: functional encryption with public keys. In: ACM Conference on Computer and Communications Security, pp. 463–472. ACM Press (2010)
43. Sahai, A., Waters, B.: Fuzzy identity-based encryption. In: Cramer, R. (ed.) EUROCRYPT 2005. LNCS, vol. 3494, pp. 457–473. Springer, Heidelberg (2005)
44. Waters, B.: Dual system encryption: Realizing fully secure IBE and HIBE under simple assumptions. In: Halevi, S. (ed.) CRYPTO 2009. LNCS, vol. 5677, pp. 619–636. Springer, Heidelberg (2009)

Bounded Tamper Resilience:
How to Go beyond the Algebraic Barrier

Ivan Damgård[1], Sebastian Faust[2], Pratyay Mukherjee[1], and Daniele Venturi[1]

[1] Department of Computer Science, Aarhus University
[2] Security and Cryptography Laboratory, EPFL

Abstract. Related key attacks (RKAs) are powerful cryptanalytic attacks where an adversary can change the secret key and observe the effect of such changes at the output. The state of the art in RKA security protects against an a-priori unbounded number of certain algebraic induced key relations, e.g., affine functions or polynomials of bounded degree. In this work, we show that it is possible to go beyond the algebraic barrier and achieve security against *arbitrary* key relations, by restricting the number of tampering queries the adversary is allowed to ask for. The latter restriction is necessary in case of arbitrary key relations, as otherwise a generic attack of Gennaro *et al.* (TCC 2004) shows how to recover the key of almost any cryptographic primitive. We describe our contributions in more detail below.

1. We show that standard ID and signature schemes constructed from a large class of Σ-protocols (including the Okamoto scheme, for instance) are secure even if the adversary can *arbitrarily* tamper with the prover's state a *bounded* number of times and obtain some bounded amount of leakage. Interestingly, for the Okamoto scheme we can allow also independent tampering with the public parameters.

2. We show a *bounded* tamper and leakage resilient CCA secure public key cryptosystem based on the DDH assumption. We first define a weaker CPA-like security notion that we can instantiate based on DDH, and then we give a general compiler that yields CCA-security with tamper and leakage resilience. This requires a public tamper-proof common reference string.

3. Finally, we explain how to boost bounded tampering and leakage resilience (as in 1. and 2. above) to *continuous* tampering and leakage resilience, in the so-called *floppy model* where each user has a personal hardware token (containing leak- and tamper-free information) which can be used to refresh the secret key.

We believe that bounded tampering is a meaningful and interesting alternative to avoid known impossibility results and can provide important insights into the security of existing standard cryptographic schemes.

Keywords: related key security, bounded tamper resilience, public key encryption, identification schemes.

K. Sako and P. Sarkar (Eds.) ASIACRYPT 2013, Part II, LNCS 8270, pp. 140–160, 2013.

1   Introduction

Related key attacks (RKAs) are powerful cryptanalytic attacks against a cryptographic implementation that allow an adversary to change the key, and subsequently observe the effect of such modification on the output. In practice, such attacks can be carried out, e.g., by heating up the device or altering the internal power supply or clock [4,11], and may have severe consequences for the security of a cryptographic implementation. To illustrate such key tampering, consider a digital signature scheme Sign with public/secret key pair (pk, sk). The tampering adversary obtains pk and can replace sk with $T(sk)$ where T is some arbitrary tampering function. Then, the adversary gets access to an oracle $\mathsf{Sign}(T(sk), \cdot)$, i.e., to a signing oracle running with the tampered key $T(sk)$. As usual the adversary wins the game by outputting a valid forgery with respect to the original public key pk. Notice that T may be the identity function, in which case we get the standard security notion of digital signature schemes.

Bellare and Kohno [8] pioneered the formal security analysis of cryptographic schemes in the presence of related key attacks. In their setting an adversary tampers *continuously* with the key by applying functions T chosen from a set of *admissible* tampering functions $\mathcal{T}$. In the signature example from above, each signing query for message m would be accompanied with a tampering function $T \in \mathcal{T}$ and the adversary obtains $\mathsf{Sign}(T(sk), m)$. Clearly, a result in the RKA setting is stronger if the class of admissible functions $\mathcal{T}$ is larger, and hence several recent works have focussed on further broadening $\mathcal{T}$. The current state of the art (see discussion in Section 1.2) considers certain algebraic relations of the key, e.g., $\mathcal{T}$ is the set of all affine functions or all polynomials of bounded degree. A natural question that arises from these works is if we can further broaden the class of tampering functions — possibly showing security for *arbitrary* relations. In this work, we study this question and show that under certain assumptions security against arbitrary key relations can be achieved.

Is arbitrary key tampering possible? Unfortunately, the answer to the above question in its most general form is negative. As shown by Gennaro *et al.* [25], it is *impossible* to protect any cryptographic scheme against arbitrary key relations. In particular, there is an attack that allows to recover the secret key of most stateless cryptographic primitives after only a few number of tampering queries.[1] To prevent this attack the authors propose to use a *self-destruct* mechanism. That is, before each execution of the cryptographic scheme the key is checked for its validity. In case the key was changed the device self-destructs. In practice, such self-destruct can for instance be implemented by overwriting the secret key with the all-zero string, or by switching to a special mode in which the device outputs $\perp$.[2] In this work, we consider an alternative setting to avoid the

---

[1] The impossibility result of [25] leaves certain loopholes, which however seem very hard to exploit.

[2] We notice that the self-destruct has to be permanent as otherwise the attack of [25] may still apply.

impossibility results of [25], and assume that an adversary can only carry out a bounded number of (say t) tampering queries. To explain our setting consider again the example of a digital signature scheme. In our model, we give the adversary access to t tampered signing oracles $\mathsf{Sign}(T_i(sk), \cdot)$, where T_i can be an arbitrary adaptively chosen tampering function. Notice that of course each of these oracles can be queried a polynomial number of times, while t is typically linear in the security parameter.

Is security against bounded tampering useful? Besides from being a natural and non-trivial security notion, we believe that our adversarial model of *arbitrary, bounded* tampering is useful for a number of reasons:

1. It is a natural alternative to continuous restricted tampering: our security notion of *bounded, arbitrary* tampering is orthogonal to the traditional setting of RKA security where the adversary can tamper *continuously* but is *restricted* to certain classes of attacks. Most previous work in the RKA setting considers algebraic key relations that are tied to the scheme's algebra and may not reflect attacks in practice. For instance, it is not clear that heating up the device or shooting with a laser on the memory can be described by, e.g., an affine function — a class that is usually considered in the literature. We also notice that physical tampering may completely destroy the device, or may be detected by hardware countermeasures, and hence our model of bounded but arbitrary tampering may be sufficient in such settings.
2. It allows to analyze the security of *standard* cryptoschemes: as outlined above a common countermeasure to protect against arbitrary tampering is to implement a key validity check and self-destruct (or output a special failure symbol) in case such check fails. Unfortunately, most standard cryptographic implementations do not come with such a built-in procedure to check the validity of the key. Our notion of bounded tamper resilience allows to make formal security guarantees of *standard* cryptographic schemes where neither the construction, nor the implementation needs to be specially engineered.
3. It can be a useful as a building-block: even if the restriction of bounded tamper resilience may be too strong in some settings, it can be useful to achieve results in the stronger continuous tampering setting (we provide some first preliminary results on this in the full version [17]). Notice that this is similar to the setting of leakage resilient cryptography which also started mainly with "bounded leakage" that later turned out to be very useful to get results in the continuous leakage setting.

We believe that due to the above points the bounded tampering model is an interesting alternative to avoid known impossibility results for arbitrary tampering attacks.

1.1 Our Contribution

We initiate a general study of schemes resilient to both *bounded* tamper and leakage attacks. We call this model the *bounded leakage and tampering model*

Table 1. An overview of our results for bounded leakage and tamper resilience. All parameters $|\mathcal{X}|$, $|\mathcal{Y}|$ ℓ, p and n are a function of the security parameter k. For the case of Σ-protocol, the set $\mathcal{X}$ is the set of all possible witnesses and the set $\mathcal{Y}$ is the set of all possible statements for the language; we can achieve a better bound depending on the conditional average min-entropy of the witness given the statement (cf. Section 3).

Tampering Model	ID Schemes		IND-CCA PKE
	Σ-Protocols	Okamoto	BHHO
Secret Key	✓	✓	✓
Public Parameters	n.a.	✓	n.a.
Continuous Tampering iFloppy	✓	✓	✓
Key Length	$\log|\mathcal{X}|$	$\ell \log p$	$\ell \log p$
Tampering Queries	$\lfloor \log|\mathcal{X}|/\log|\mathcal{Y}| \rfloor - 2$	$\ell - 2$	$\ell - 3$

(BLT) model. While our general techniques use ideas from the leakage realm, we emphasize that bounded leakage resilience does *not* imply bounded tamper resilience. In fact, it is easy to find contrived schemes that are leakage resilient but completely break for a single tampering query. At a more technical level, we observe that a trivial strategy using leakage to simulate, e.g., faulty signatures, has to fail as the adversary can get any polynomial number of faulty signatures — which clearly cannot be simulated with bounded leakage only. Nevertheless, as we show in this work, we are able to identify certain classes of cryptoschemes for which a small amount of leakage is sufficient to simulate faulty outputs. We discuss this in more detail below.

Our concrete schemes are proven secure under standard assumptions (DL, factoring or DDH) and are efficient and simple. Moreover, we show that our schemes can easily be extended to the continual setting by putting an additional simple assumption on the hardware. We elaborate more on our main contributions in the following paragraphs (see also Table 1 for an overview of our results). Importantly, all our results allow arbitrary key tampering and do not need any kind of tamper detection mechanism.

Identification schemes. It is well known that the Generalized Okamoto identification scheme [34] provides security against bounded leakage from the secret key [3,30]. In Section 3, we show that additionally it provides strong security against tampering attacks. While in general the tampered view may contain a polynomial number of faulty transcripts that may potentially reveal a large amount of information about the secret key, we can show that fortunately this is not the case for the Generalized Okamaoto scheme. More concretely, we are able to identify a short amount of information that for each tampering query allows us to simulate any number of corresponding faulty transcripts. Hence, BLT security of the Generalized Okamoto scheme is implied by its leakage resilience.

Our results on the Okamoto identification can be further generalized to a large class of identification schemes (and signature schemes based on the Fiat-Shamir heuristic). More concretely, we show that Σ-protocols where the secret key is significantly longer than the public key are BLT secure. We can instantiate our result with the generalized Guillou-Quisquater ID scheme [27], and its variant based on factoring [24] yielding tamper resilient identification based on factoring. We give more details in Section 3.

Interestingly, for Okamoto identification security still holds in a stronger model where the adversary is allowed to tamper not only with the secret key of the prover, but also with the description of the public parameters (i.e., the generator g of a group $\mathbb{G}$ of prime order p). The only restrictions are: (i) tampering with the public parameters is independent from tampering with the secret key and (ii) the tampering with public parameters must map to its domain. We also show that the latter restrictions are necessary, by presenting explicit attacks when the adversary can tamper jointly with the secret key and the public parameters or he can tamper the public parameters to some particular range.

Public key encryption. We show how to construct IND-CCA secure public key encryption (PKE) in the BLT model. To this end, we first introduce a weaker CPA-like security notion, where an adversary is given access to a restricted (faulty) decryption oracle. Instead of decrypting adversarial chosen ciphertexts such an oracle accepts inputs (m, r), encrypts the message m using randomness r under the original public key, and returns the decryption using the faulty secret key. This notion already provides a basic level of tamper resilience for public key encryption schemes. Consider for instance a setting where the adversary can tamper with the decryption key, but has no control over the ciphertexts that are sent to the decryption oracle, e.g., the ciphertexts are sent over a secure authenticated channel.

Our notion allows the adversary to tamper adaptively with the secret key; intuitively this allows him to learn faulty decryptions of ciphertexts for which he already knows the corresponding plaintext (under the original public key) and the randomness. We show how to instantiate our basic tamper security notion under DDH. More concretely, we prove that the BHHO cryptosystem [12] is BLT and CPA secure. The proof uses similar ideas as in the proof of the Okamoto identification scheme.

We then show how to transform our extended CPA-like notion to CCA security in the BLT model. To this end, we follow the classical paradigm to transform IND-CPA security into IND-CCA security by adding an argument of "plaintext knowledge" π to the ciphertext. Our transformation requires a public tamper-proof common reference string similar to earlier work [29]. Intuitively, this works because the argument π enforces the adversary to submit to the faulty decryption oracle only ciphertexts for which he knows the corresponding plaintext (and the randomness used to encrypt it). The pairs (m, r) can then be extracted from the argument π, allowing to reduce IND-CCA BLT security to our extended IND-CPA security notion.

Updating the key in the iFloppy model. As mentioned earlier, if the key is not updated BLT security is the best we can hope for when we consider arbitrary tampering. To go beyond the bound of $|sk|$ tampering queries we may regularly update the secret key with fresh randomness, which renders information that the adversary has learned about earlier keys useless. The effectiveness of key updates in the context of tampering attacks has first been used in the important work of Kalai *et al.* [29]. We follow this idea but add an additional hardware assumption that allows for much simpler and more efficient key updates. More concretely, we propose the i*Floppy model* which is a variant of the floppy model proposed by Alwen *et al.* [3] and recently studied in depth by Agrawal *et al.* [2]. In the floppy model a user of a cryptodevice possesses a so-called *floppy* – a secure hardware token – that stores an update key.[3] The floppy is leakage and tamper proof and the update key that it holds is solely used to refresh the actual secret key kept on the cryptodevice. One may think of the floppy as a particularly secure device that the user keeps at home, while the cryptodevice, e.g., a smart-card, runs the actual cryptographic task and is used out in the wild prone to leakage and tampering attacks. We consider a variant called the i*Floppy model* (here "i" stands for individual). While in the floppy model of [2,3] all users can potentially possess an identical hardware token, in the i*Floppy model* we require that each user has an individual floppy storing some secret key related data. We note that from a practical point of view the i*Floppy model* is incomparable to the original floppy model. It may be more cumbersome to produce personalized hardware tokens, but on the other hand, in practice one would not want to distribute hardware tokens that all contain the same global update key as this constitutes a single point of failure.

We show in the i*Floppy model* a simple compiler that "boosts" any ID scheme with BLT security into a scheme with *continuous* leakage and tamper resilience (CLT security). Similarly, we show how to extend IND-CCA BLT security to the CLT setting for the BHHO cryptosystem (borrowing ideas from [2]). We emphasize that while the i*Floppy model* puts additional requirements on the way users must behave in order to guarantee security, it greatly simplifies cryptographic schemes, and allows us to base security on standard assumptions. Our results in the i*Floppy model* are mainly deferred to the full version [17].

Tampering with the computation via the BRM. Finally, we make a simple observation showing that if we instantiate the above ID compiler with an ID scheme that is secure in the bounded retrieval model [15,20,3] we can provide security in the i*Floppy model* even when the adversary can replace the original cryptoscheme with an arbitrary adversarial chosen functionality, i.e., we can allow arbitrary tampering with the computation (see the full version [17]). While easy to prove, we believe this is nevertheless noteworthy: it seems to us that results in the BRM naturally provide some form of tamper resilience and leave it as an open question for future research to explore this direction further.

---

[3] Notice that "floppy" is just terminology and we use it for consistency with earlier works.

1.2 Previous Work

Related key security. We already discussed the relation between BLT security and the traditional notion of RKA security above. Below we give further details on some important results in the RKA area. Bellare and Kohno [8] initiated the theoretical study of related-key-attacks. Their result mainly focused on symmetric key primitives (e.g. PRP, PRF). They proposed various block-cipher based constructions which are RKA-secure against certain restricted classes of tampering functions. Their constructions were further improved by [32,6]. Following these works other cryptographic primitives were constructed that are provably secure against certain classes of related key attacks. Most of these works consider rather restricted tampering functions that, e.g., can be described by a linear or affine function [8,32,6,5,35,37,10]. A few important exceptions are described below.

In [9] the authors show how to go beyond the linear barrier by extending the class of allowed tampering functions to the class of polynomials of bounded degree for a number of public-key primitives. Also, the work of Goyal, O'Neill and Rao [26] considers polynomial relations that are induced to the inputs of a hash function. Finally Bellare, Cash and Miller [7] develop a framework to transfer RKA security from a pseudorandom function to other primitives (including many public key primitives).

Tamper resilient encodings. A generic method for tamper protection has been put forward by Gennaro *et al.* [25]. The authors propose a general "compiler" that transforms any cryptographic device CS with secret state st, e.g., a block cipher, into a "transformed" cryptoscheme CS' running with state st' that is resilient to arbitrary tampering with st'. In their construction the original state is signed and the signature is checked before each usage. While the above works for any tampering function, it is limited to settings where CS does not change its state as it would need access to the secret signing key to authenticate the new state. This drawback is resolved by the concept of non-malleable codes pioneered by Dziembowski, Pietrzak and Wichs [21]. The original construction of [21] considers an adversary that can tamper independently with bits. This has been extended to small size blocks in [13], and recently to so-called split-state tampering [31,1]. While the above schemes provide surprisingly strong security guarantees, they all require certain assumptions on the hardware (e.g., the memory has to be split into two parts that cannot tampered with jointly), and require significant changes to the implementation for decoding, tamper detection and self-destruct.

Continuous tamper resilience via key updates. Kalai *et al.* [29] provide first feasibility results in the so-called *continuous leakage and tampering* model (CLT). Their constructions achieve strong security requirements where the adversary can arbitrarily tamper continuously with the state. This is achieved by updating the secret key after each usage. While the tampering adversary considered in [29] is clearly stronger (continuous as opposed to bounded tampering), the proposed schemes are non-standard, rather inefficient and rely on non-standard assumptions. Moreover, the approach of key updates requires a stateful device and large amounts of randomness which is costly in practice. The main focus of this work, are simple standard cryptosystems that neither require randomness for key updates nor need to keep state.

Tampering with computation. In all the above works (including ours) it is assumed that the circuitry that computes the cryptographic algorithm using the potentially tampered key runs correctly and is not subject to tampering attacks. An important line of works analyze to what extend we can guarantee security when the complete circuitry is prone to tampering attacks [28,22,16]. These works typically consider a restricted class of tampering attacks (e.g., individual bit tampering) and assume that large parts of the circuit (and memory) remain un-tampered.

2   Preliminaries

For space reasons, we defer some of the basic definitions to the full version [17].

Basic notation. We review the basic terminology used throughout the paper. For $n \in \mathbb{N}$, we write $[n] := \{1, \dots, n\}$. Given a set $\mathcal{S}$, we write $s \leftarrow \mathcal{S}$ to denote that element s is sampled uniformly from $\mathcal{S}$. If A is an algorithm, $y \leftarrow A(x)$ denotes an execution of A with input x and output y; if A is randomized, then y is a random variable. Vectors are denoted in bold. Given a vector $\mathbf{x} = (x_1, \dots, x_\ell)$ and some integer a, we write $a^{\mathbf{x}}$ for the vector $(a^{x_1}, \dots, a^{x_\ell})$.

We denote with k the security parameter. A function $\delta(k)$ is called *negligible* in k (or simply negligible) if it vanishes faster than the inverse of any polynomial in k. A machine A is called *probabilistic polynomial time* (PPT) if for any input $x \in \{0,1\}^*$ the computation of $A(x)$ terminates in at most $poly(|x|)$ steps and A is probabilistic (i.e., it uses randomness as part of its logic). Random variables are usually denoted by capital letters. We sometimes abuse notation and denote a distribution and the corresponding random variable with the same capital letter, say X.

Languages and relations. A *decision problem* related to a language $\mathfrak{L} \subseteq \{0,1\}^*$ requires to determine if a given string y is in $\mathfrak{L}$ or not. We can associate to any $\mathcal{NP}$-language $\mathfrak{L}$ a polynomial-time recognizable relation $\mathfrak{R} \subseteq \{0,1\}^* \times \{0,1\}^*$ defining $\mathfrak{L}$ itself, i.e. $\mathfrak{L} = \{y : \exists x \text{ s.t. } (y,x) \in \mathfrak{R}\}$ for $|x| \leq poly(|y|)$. The string x is called a *witness* for membership of $y \in \mathfrak{L}$.

Information theory. The min-entropy of a random variable X over a set $\mathcal{X}$ is defined as $\mathbf{H}_\infty(X) := -\log \max_x \Pr[X = x]$, and measures how X can be predicted by the best (unbounded) predictor. The conditional average min-entropy [19] of X given a random variable Z (over a set $\mathcal{Z}$) possibly dependent on X, is defined as $\widetilde{\mathbf{H}}_\infty(X|Z) := -\log \mathbb{E}_{z \leftarrow Z}[2^{-\mathbf{H}_\infty(X|Z=z)}]$. Following [3], we sometimes rephrase the notion of conditional min-entropy in terms of predictors A that are given some information Z (presumably correlated with X), so $\tilde{\mathbf{H}}_\infty(X|Z) = -\log(\max_A \Pr[A(Z) = X])$. The above notion of conditional min-entropy can be generalized to the case of interactive predictors A, which participate in some randomized experiment $\mathcal{E}$. An experiment is modeled as interaction between A and a challenger oracle $\mathcal{E}(\cdot)$ which can be randomized, stateful and interactive.

Definition 1 ([3]). *The conditional min-entropy of a random variable X, conditioned on the experiment $\mathcal{E}$ is $\tilde{\mathbf{H}}_\infty(X|\mathcal{E}) = -\log(\max_A \Pr[A^{\mathcal{E}(\cdot)}() = X])$. In the special case that $\mathcal{E}$ is a non-interactive experiment which simply outputs a random variable Z, then $\tilde{\mathbf{H}}_\infty(X|Z)$ can be written to denote $\tilde{\mathbf{H}}_\infty(X|\mathcal{E})$ abusing the notion.*

We will rely on the following basic properties (see [19, Lemma 2.2]).

Lemma 1. *For all random variables X, Z and Λ over sets $\mathcal{X}$, $\mathcal{Z}$ and $\{0,1\}^\lambda$ such that $\tilde{\mathbf{H}}_\infty(X|Z) \geq \alpha$, we have*

$$\tilde{\mathbf{H}}_\infty(X|Z, \Lambda) \geq \tilde{\mathbf{H}}_\infty(X|Z) - \lambda \geq \alpha - \lambda.$$

3   ID Schemes with BLT Security

In an identification scheme a prover tries to convince a verifier of its identity (corresponding to its public key pk). Formally, an identification scheme is a tuple of algorithms $\mathcal{ID} = (\mathsf{Setup}, \mathsf{Gen}, \mathsf{P}, \mathsf{V})$ defined as follows:

$pp \leftarrow \mathsf{Setup}(1^k)$: Algorithm Setup takes the security parameter as input and outputs public parameters pp. The set of all public parameter is denoted by $\mathcal{PP}$.

$(pk, sk) \leftarrow \mathsf{Gen}(1^k)$: Algorithm Gen outputs the public key and the secret key corresponding to the prover's identity. The set of all possible secret keys is denoted by $\mathcal{SK}$.

(P, V): We let $(\mathsf{P}(pp, sk) \rightleftarrows \mathsf{V}(pp))(pk)$ denote the interaction between prover P (holding sk and using public parameters pp) and verifier V on common input pk. Such interaction outputs a result in $\{accept, reject\}$, where $accept$ means P's identity is considered as valid.

Definition 2. *Let $\lambda = \lambda(k)$, $t = t(k)$ and $\delta = \delta(k)$ be parameters and let $\mathcal{T}$ be some set of functions such that $T \in \mathcal{T}$ has a type $T : \mathcal{SK} \times \mathcal{PP} \rightarrow \mathcal{SK} \times \mathcal{PP}$. We say that $\mathcal{ID}$ is (λ, t, δ)-bounded leakage and tamper secure (in short BLT-secure) against impersonation attacks with respect to $\mathcal{T}$ if the following properties are satisfied.*

(i) Correctness. For all $pp \leftarrow \mathsf{Setup}(1^k)$ and $(pk, sk) \leftarrow \mathsf{Gen}(1^k)$ we have that $(\mathsf{P}(pp, sk) \rightleftarrows \mathsf{V}(pp))(pk)$ outputs accept.

(ii) Security. For all PPT adversaries A we have that $\Pr[A \text{ wins}] \leq \delta(k)$ in the following game:

1. The challenger runs $pp \leftarrow \mathsf{Setup}(1^k)$ and $(pk, sk) \leftarrow \mathsf{Gen}(1^k)$, and gives (pp, pk) to A.
2. The adversary is given oracle access to $\mathsf{P}(pp, sk)$ that outputs polynomially many proof transcripts with respect to secret key sk.
3. The adversary may adaptively ask t tampering queries. During the ith query, A chooses a function $T_i \in \mathcal{T}$ and gets oracle access to $\mathsf{P}(\widetilde{pp}_i, \widetilde{sk}_i)$, where $(\widetilde{sk}_i, \widetilde{pp}_i) = T_i(sk, pp)$. The adversary can interact with the oracle $\mathsf{P}(\widetilde{pp}_i, \widetilde{sk}_i)$ a polynomially number of times, where it uses the tampered secret key $\widetilde{sk}_i$ and the public parameter $\widetilde{pp}_i$.
4. The adversary may adaptively ask leakage queries. In the jth query, A chooses a function $L_j : \{0,1\}^* \rightarrow \{0,1\}^{\lambda_j}$ and receives back the output of the function applied to sk.
5. The adversary loses access to all other oracles and interacts with an honest verifier V (holding pk). We say that A wins if $(\mathsf{A} \rightleftarrows \mathsf{V}(pp))(pk)$ outputs accept and $\sum_j \lambda_j \leq \lambda$.

Notice that in the above definition the leakage is from the original secret key sk. This is without loss of generality as our tampering functions are modeled as deterministic circuits.

In a slightly more general setting, one could allow A to leak on the original secret key also in the last phase where it has to convince the verifier. In the terminology of [3] this is reminiscent of so-called *anytime leakage* attacks. Our results can be generalized to this setting, however we stick to Definition 2 for simplicity.

The rest of this section is organized as follows. In Section 3.1 we prove that a large class of Σ-protocols are secure in the BLT model, where the tampering function is allowed to modify the secret state of the prover but not the public parameters. In Section 3.2 we look at a concrete instantiation based on the Okamoto ID scheme, and prove that this construction is secure in a stronger model where the tampering function can modify both the secret state of the prover and the public parameters (but independently). Finally, in Section 3.3 we illustrate that the latter assumption is necessary, as otherwise the Okamoto ID scheme can be broken by (albeit contrived) attacks.

3.1 Σ-protocols Are Tamper Resilient

We start by considering ID schemes based on Σ-protocols [14]. Σ-protocols are a special class of interactive proof systems for membership in a language $\mathfrak{L}$, where a prover $\mathsf{P} = (\mathsf{P}_0, \mathsf{P}_1)$ wants to convince a verifier $\mathsf{V} = (\mathsf{V}_0, \mathsf{V}_1)$ (both modelled as PPT algorithms) that a shared string y belongs to $\mathfrak{L}$. Denote with x the witness corresponding to y and let pp be public parameters. The protocol proceeds as follows: (1) The prover computes $a \leftarrow \mathsf{P}_0(pp)$ and sends it to the verifier; (2) The verifier chooses $c \leftarrow \mathsf{V}_0(pp, y)$ uniformly at random and sends it to the prover; (3) The prover answers with $z \leftarrow \mathsf{P}_1(pp, (a, c, x))$; (4) The verifier outputs a result $\mathsf{V}_1(pp, y, (a, c, z)) \in \{accept, reject\}$. We call this a *public-coin*

ID Scheme from Σ-Protocol

Let $((P_0, P_1), (V_0, V_1))$ be a Σ-protocol for a relation $\mathfrak{R}$.

Setup(1^k): Sample public parameters $pp \leftarrow \mathcal{PP}$ for the underlying relation $\mathfrak{R}$.

Gen(1^k): Output a pair $(y, x) \in \mathfrak{R}$, where $x \in \mathcal{X}$ and $y \in \mathcal{Y}$ and $|x|$ is polynomially bounded by $|y|$.

$(P(pp, x) \rightleftarrows V(pp))(y)$: The protocol works as follows.
1. The prover sends $a \leftarrow P_0(pp)$ to the verifier.
2. The verifier chooses a random challenge $c \leftarrow V_0(pp, y)$ and sends it to the prover.
3. The prover computes the answer $z \leftarrow P_1(pp, (a, c, x))$.
4. The verifier accepts iff $V_1(pp, y, (a, c, z))$ outputs *accept*.

Fig. 1. ID scheme based on Σ-protocol for relation $\mathfrak{R}$

three round interactive proof system. A formal definition of Σ-protocols can be found in the full version [17].

It is well known that Σ-protocols are a natural tool to design ID schemes. The construction is depicted in Figure 1. Consider now the class of tampering functions $\mathcal{T}_{sk} \subset \mathcal{T}$ such that $T \in \mathcal{T}_{sk}$ has the following form: $T = (T^{sk}, ID^{pp})$ where $T^{sk} : \mathcal{SK} \to \mathcal{SK}$ is an arbitrary polynomial time computable function and $ID^{pp} : \mathcal{PP} \to \mathcal{PP}$ is the identity function. This models tampering with the secret state of P without changing the public parameters (these must be hard-wired into the prover's code). The proof of the following theorem (which appears in the full version [17]) uses ideas of [3], but is carefully adjusted to incorporate tampering attacks.

Theorem 1. *Let $k \in \mathbb{N}$ be the security parameter and let* (P, V) *be a Σ-protocol for relation $\mathfrak{R}$ with $|\mathcal{Y}| = O(k^{\log k})$, such that the representation problem is hard for $\mathfrak{R}$. Assume that conditioned on the distribution of the public input $y \in \mathcal{Y}$, the witness $x \in \mathcal{X}$ has high average min entropy β, i.e., $\tilde{\mathbf{H}}_\infty(X|Y) \geq \beta$. Then, the ID scheme of Figure 1 is $(\lambda(k), t(k), negl(k))$-BLT secure against impersonation attacks with respect to $\mathcal{T}_{sk}$, where*

$$\lambda \leq \beta - t \log |\mathcal{Y}| - k \qquad and \qquad t \leq \left\lfloor \frac{\beta}{\log |\mathcal{Y}|} \right\rfloor - 1.$$

3.2 Concrete Instantiation with More Tampering

We extend the power of the adversary by allowing him to tamper not only with the witness, but also with the public parameters (used by the prover to generate the transcripts). However the tampering has to be independent on the two components. This is reminiscent of the so-called split-state model (considered

---

Generalized Okamoto ID Scheme

Let $\ell = \ell(k)$ be some function of the security parameter. Consider the following identification scheme.

Setup: Choose a group $\mathbb{G}$ of prime order p with generator g and a vector $\alpha \leftarrow \mathbb{Z}_p^\ell$, and output $pp = (\mathbb{G}, g, g^\alpha)$ where $g^\alpha = (g_1, \ldots, g_\ell)$.
Gen(1^k): Select a vector $\mathbf{x} \leftarrow \mathbb{Z}_p^\ell$ and set $y = pk = \prod_{i=1}^\ell g_i^{x_i}$ and $sk = \mathbf{x}$.
$(\mathsf{P}(pp, sk) \rightleftharpoons \mathsf{V}(pp))(pk)$: The protocol works as follows.
1. The prover chooses a random vector $\mathbf{r} \leftarrow \mathbb{Z}_p^\ell$ and sends $a = \prod_{i=1}^\ell g_i^{r_i}$ to the verifier.
2. The verifier chooses a random challenge $c \leftarrow \mathbb{Z}_p$ and sends it to the prover.
3. The prover computes the answer $\mathbf{z} = (r_1 + cx_1, \ldots, r_\ell + cx_\ell)$.
4. The verifier accepts if and only if $\prod_{i=1}^\ell g_i^{z_i} = a \cdot y^c$.

Fig. 2. Generalized Okamoto Identification Scheme

for instance in [31]), with the key difference that in our case the secret state does not need to be split into two parts.

We model this through the following class of tampering functions $\mathcal{T}_{\mathsf{split}}$: We say that $T \in \mathcal{T}_{\mathsf{split}}$ if we can write $T = (T^{sk}, T^{pp})$ where $T^{sk} : \mathcal{SK} \to \mathcal{SK}$ and $T^{pp} : \mathcal{PP} \to \mathcal{PP}$ are arbitrary polynomial time computable functions. Recall that the input/output domains of T^{sk}, T^{pp} are identical, hence the size of the witness and the public parameters cannot be changed. As we show in the next section, this restriction is necessary. Note also that $\mathcal{T}_{\mathsf{sk}} \subseteq \mathcal{T}_{\mathsf{split}} \subseteq \mathcal{T}$.

Generalized Okamoto. Consider the generalized version of the Okamoto ID scheme [34], depicted in Figure 2. The underlying hard relation here is the relation $\mathfrak{R}_{\mathsf{DL}}$ and the representation problem for $\mathfrak{R}_{\mathsf{DL}}$ is the ℓ-representation problem in a group $\mathbb{G}$. As proven in [3], this problem is equivalent to the Discrete Log problem in $\mathbb{G}$. The proof of the following corollary appears in the full version [17].

Corollary 1. *Let $k \in \mathbb{N}$ be the security parameter and assume the Discrete Log problem is hard in $\mathbb{G}$. Then, the generalized Okamoto ID scheme is $(\lambda(k), t(k),$ $\mathsf{negl}(k))$-BLT secure against impersonation attacks with respect to $\mathcal{T}_{\mathsf{split}}$, where*

$$\lambda \leq (\ell - 1 - t) \log(p) - k \qquad and \qquad t \leq \ell - 2.$$

3.3 Some Attacks

We show that for the Okamoto scheme it is hard to hope for BLT security beyond the class of tampering functions $\mathcal{T}_{\mathsf{split}}$. We illustrate this by concrete attacks which work in case one tries to extend the power of the adversary in two different ways: (1) Allowing A to tamper jointly with the witness and the public parameters; (2) Allowing A to tamper independently with the witness and with the public parameters but increase their size.

Tampering jointly with the public parameters. Consider the class of functions $\mathcal{T}$ introduced in Definition 2.

Claim. The generalized Okamoto ID scheme is *not* BLT-secure against impersonation attacks with respect to $\mathcal{T}$.

Proof. The attack uses a single tampering query. Define the tampering function $T(\mathbf{x}, pp) = (\widetilde{\mathbf{x}}, \widetilde{pp})$ to be as follows:

- The witness is unchanged, i.e., $\mathbf{x} = \widetilde{\mathbf{x}}$.
- The value $\widetilde{p}$ is some prime of size $|\widetilde{p}| \approx |p|$ such that the Discrete Log problem is easy in the corresponding group $\widetilde{\mathbb{G}}$. (This can be done efficiently by choosing $\widetilde{p} - 1$ to be the product of small prime (power) factors [36].)
- Let $\widetilde{g}$ be a generator of $\widetilde{\mathbb{G}}$ (which exists since $\widetilde{p}$ is a prime) and define the new generators as $\widetilde{g}_i = \widetilde{g}^{x_i} \bmod \widetilde{p}$.

Consider now a transcript $(a, c, \mathbf{z})$ produced by a run of $\mathsf{P}(\widetilde{pp}, \mathbf{x})$. We have $a = \widetilde{g}^{\sum_{i=1}^{\ell} x_i r_i} \bmod \widetilde{p}$ for random $r_i \in \mathbb{Z}_{\widetilde{p}}$. By computing the Discrete Log of a in base $\widetilde{g}$ (which is easy by our choice of $\widetilde{\mathbb{G}}$), we get one equation $\sum_{i=1}^{\ell} x_i r_i = \log_{\widetilde{g}}(a) \bmod \widetilde{p}$. Asking for polynomially many transcripts, yields ℓ linearly independent equations (with overwhelming probability) and thus allows to solve for $(x_1, \ldots, x_\ell)$. (Note here that with high probability $x_i \bmod p = x_i \bmod \widetilde{p}$ since $|p| \approx |\widetilde{p}|$.)

Tampering by "inflating" the prime p. Consider the following class of tampering functions $\mathcal{T}_{\mathsf{split}} \subseteq \mathcal{T}^*_{\mathsf{split}}$: We say that $T \in \mathcal{T}^*_{\mathsf{split}}$ if $T = (T^{sk}, T^{pp})$, where $T^{sk} : \mathcal{SK} \to \{0,1\}^*$ and $T^{pp} : \mathcal{PP} \to \{0,1\}^*$.

Claim. The generalized Okamoto ID scheme is *not* BLT-secure against impersonation attacks with respect to $\mathcal{T}^*_{\mathsf{split}}$.

Proof. The attack uses a single tampering query. Consider the following tampering function $T = (T^{sk}, T^{pp}) \in \mathcal{T}^*_{\mathsf{split}}$:

- Choose $\widetilde{p}$ to be a prime of size $|\widetilde{p}| = \Omega(\ell|p|)$, such that the Discrete Log problem is easy in $\widetilde{\mathbb{G}}$. (This can be done as in the proof of Claim 3.3.)
- Choose a generator $\widetilde{g}$ of $\widetilde{\mathbb{G}}$; define $\widetilde{g}_1 = \widetilde{g}$ and $\widetilde{g}_j = 1$ for all $j = 2, \ldots, \ell$.
- Define the witness to be $\widetilde{\mathbf{x}}$ such that $\widetilde{x}_1 = x_1 || \ldots || x_\ell$ and $\widetilde{x}_j = 0$ for all $j = 2, \ldots, \ell$.

Given a single transcript $(a, c, \mathbf{z})$ the adversary learns $a = \widetilde{g}^{r_1}$ for some $r_1 \in \mathbb{Z}_{\widetilde{p}}$. Since the Discrete Log is easy in this group, A can find r_1. Now the knowledge of c and $z_1 = r_1 + c\widetilde{x}_1$, allows to recover $\widetilde{x}_1 = (x_1, \ldots, x_\ell)$.

3.4 BLT-Secure Signatures

It is well known that every Σ-protocol can be turned into a signature scheme via the Fiat-Shamir heuristic [23]. By applying the Fiat-Shamir transformation to the protocol of Figure 1, we get efficient BLT-secure signatures in the random oracle model.

4   IND-CCA PKE with BLT Security

We start by defining IND-CCA public key encryption (PKE) with BLT security. A PKE scheme is a tuple of algorithms $\mathcal{PKE} = (\mathsf{Setup}, \mathsf{KGen}, \mathsf{Enc}, \mathsf{Dec})$ defined as follows. (1) Algorithm Setup takes as input the security parameter and outputs the description of public parameters pp; the set of all public parameters is denoted by $\mathcal{PP}$. (2) Algorithm KGen takes as input the security parameter and outputs a public/secret key pair (pk, sk); the set of all secret keys is denoted by $\mathcal{SK}$ and the set of all public keys by $\mathcal{PK}$. (3) The randomized algorithm Enc takes as input the public key pk, a message $m \in \mathcal{M}$ and randomness $r \in \mathcal{R}$ and outputs a ciphertext $c = \mathsf{Enc}(pk, m; r)$; the set of all ciphertexts is denoted by $\mathcal{C}$. (4) The deterministic algorithm Dec takes as input the secret key sk and a ciphertext $c \in \mathcal{C}$ and outputs $m = \mathsf{Dec}(sk, c)$ which is either equal to some message $m \in \mathcal{M}$ or to an error symbol $\perp$.

Definition 3. *Let $\lambda = \lambda(k)$, $t = t(k)$ and $\delta = \delta(k)$ be parameters and let $\mathcal{T}_{\mathsf{sk}}$ be some set of functions such that $T \in \mathcal{T}_{\mathsf{sk}}$ has a type $T : \mathcal{SK} \to \mathcal{SK}$. We say that $\mathcal{PKE}$ is IND-CCA $(\lambda(k), t(k), \delta(k))$-BLT secure with respect to $\mathcal{T}_{\mathsf{sk}}$ if the following properties are satisfied.*

(i) *Correctness. For all $pp \leftarrow \mathsf{Setup}(1^k)$, $(pk, sk) \leftarrow \mathsf{KGen}(1^k)$ we have that $\Pr[\mathsf{Dec}(sk, \mathsf{Enc}(pk, m)) = m] = 1$ (where the randomness is taken over the internal coin tosses of algorithm Enc).*

(ii) *Security. For all PPT adversaries A we have that $\Pr[\mathsf{A}\ wins] \leq \frac{1}{2} + \delta(k)$ in the following game:*

1. *The challenger runs $pp \leftarrow \mathsf{Setup}(1^k)$, $(pk, sk) \leftarrow \mathsf{KGen}(1^k)$ and gives (pp, pk) to A.*

2. *The adversary is given oracle access to $\mathsf{Dec}(sk, \cdot)$. This oracle outputs polynomially many decryptions of ciphertexts using secret key sk.*

3. *The adversary may adaptively ask t tampering queries. During the ith query, A chooses a function $T_i \in \mathcal{T}_{\mathsf{sk}}$ and gets oracle access to $\mathsf{Dec}(\widetilde{sk}_i, \cdot)$, where $\widetilde{sk}_i = T_i(sk)$. This oracle outputs polynomially many decryptions of ciphertexts using secret key $\widetilde{sk}_i$.*

4. *The adversary may adaptively ask polynomially many leakage queries. In the jth query, A chooses a function $L_j : \{0, 1\}^* \to \{0, 1\}^{\lambda_j}$ and receives back the output of the function applied to sk.*

5. *The adversary outputs two messages of the same length $m_0, m_1 \in \mathcal{M}$ and the challenger computes $c_b \leftarrow \mathsf{Enc}(pk, m_b)$ where b is a uniformly random bit.*

6. *The adversary keeps access to $\mathsf{Dec}(sk, \cdot)$ and outputs a bit b'. We say A wins if $b = b'$, $\sum_j \lambda_j \leq \lambda$ and c_b has not been queried for.*

In case $t = 0$ we get the notion of leakage resilient IND-CCA from [33] as a special case. Notice that A is not allowed to tamper with the secret key after seeing the challenge ciphertext. As we show in the full version [17], this restriction is necessary because otherwise A could overwrite the secret key depending on the plaintext encrypted in c_b, and thus gain some advantage in guessing the value of b by asking additional decryption queries.

We build an IND-CCA BLT-secure PKE scheme in two steps. In Section 4.1 we define a weaker notion which we call IND-CPA BLT security. In Section 4.2 we show a general transformation from IND-CPA BLT security to IND-CCA BLT security relying on tSE NIZK proofs [18] in the common reference string (CRS) model. The CRS is supposed to be tamper-free and must be hard-wired into the code of the encryption algorithm; however tampering and leakage can depend adaptively on the CRS and the public parameters. Finally, in Section 4.3, we prove that a variant of the BHHO encryption scheme [33] satisfies our notion of IND-CPA BLT security.

4.1 IND-CPA BLT Security

The main idea of our new security notion is as follows. Instead of giving A full access to a tampering oracle (as in Definition 3) we restrict his power by allowing him to see the output of the (tampered) decryption oracle only for ciphertexts c for which A already knows both the corresponding plaintext m and the randomness r used to generate c (via the real public key). Essentially this restricts A to submit to the tampering oracle only "well-formed" ciphertexts.

Definition 4. *Let $\lambda = \lambda(k)$, $t = t(k)$ and $\delta = \delta(k)$ be parameters and let $\mathcal{T}_{sk}$ be some set of functions such that $T \in \mathcal{T}_{sk}$ has a type $T : \mathcal{SK} \to \mathcal{SK}$. We say that $\mathcal{PKE}$ is IND-CPA $(\lambda(k), t(k), \delta(k))$-BLT secure with respect to $\mathcal{T}_{sk}$ if it satisfies property (i) of Definition 3 and property (ii) is modified as follows:*

(ii) *Security. For all PPT adversaries A we have that $\Pr[\text{A wins}] \leq \frac{1}{2} + \delta(k)$ in the following game:*

1. *The challenger runs $pp \leftarrow \text{Setup}(1^k)$, $(pk, sk) \leftarrow \text{KGen}(1^k)$ and gives (pp, pk) to A.*

2. *The adversary may adaptively ask t tampering queries. During the ith query, A chooses a function $T_i \in \mathcal{T}_{sk}$ and gets oracle access to $\text{Dec}^*(\widetilde{sk}_i, \cdot, \cdot)$, where $\widetilde{sk}_i = T_i(sk)$. This oracle answers polynomially many queries of the following form: Upon input a pair $(m, r) \in \mathcal{M} \times \mathcal{R}$, compute $c \leftarrow \text{Enc}(pk, m; r)$ and output a plaintext $\widetilde{m} = \text{Dec}(\widetilde{sk}_i, c)$ using the current tampered key.*

3. *The adversary may adaptively ask leakage queries. In the jth query, A chooses a function $L_j : \{0,1\}^* \to \{0,1\}^{\lambda_j}$ and receives back the output of the function applied to sk.*

4. *The adversary outputs two messages of the same length $m_0, m_1 \in \mathcal{M}$ and the challenger computes $c_b \leftarrow \text{Enc}(pk, m_b)$ where b is a uniformly random bit.*

5. *The adversary loses access to all oracles and outputs a bit b'. We say that A wins if $b = b'$ and $\sum_j \lambda_j \leq \lambda$.*

From IND-CPA BLT Security to IND-CCA BLT Security

Let $\mathcal{PKE} = (\mathsf{Setup}, \mathsf{KGen}, \mathsf{Enc}, \mathsf{Dec})$ be a PKE scheme and $(\mathsf{Gen}, \mathsf{Prove}, \mathsf{Verify})$ be a tSE NIZK argument system for the relation:

$$\mathfrak{R}_{\mathsf{PKE}} = \{(pk, c), (m, r) : \ c = \mathsf{Enc}(pk, m; r)\}.$$

Define the following PKE scheme $\mathcal{PKE}' = (\mathsf{Setup}', \mathsf{KGen}', \mathsf{Enc}', \mathsf{Dec}')$.

Setup': Sample $pp \leftarrow \mathsf{Setup}(1^k)$ and $(\omega, \mathsf{tk}, \mathsf{ek}) \leftarrow \mathsf{Gen}(1^k)$ and let $pp' = (pp, \omega)$.
KGen': Run $(pk, sk) \leftarrow \mathsf{KGen}(1^k)$ and set $pk' = pk$ and $sk' = sk$.
Enc': Sample $r \leftarrow \mathcal{R}$ and compute $c \leftarrow \mathsf{Enc}(pk, m; r)$. Output (c, π), where $\pi \leftarrow \mathsf{Prove}^\omega((pk, c), (m, r))$.
Dec': Check that $\mathsf{Verify}^\omega((pk, c), \pi) = 1$. If not output $\perp$; otherwise, output $m = \mathsf{Dec}(sk, c)$.

Fig. 3. How to transform IND-CPA BLT-secure PKE into IND-CCA BLT-secure PKE

4.2 A General Transformation

We compile an arbitrary IND-CPA BLT-secure encryption scheme into an IND-CCA BLT-secure one by appending to the ciphertext c an argument of "plaintext knowledge" π computed through a tSE NIZK argument system. The same construction has been already used by Dodis *et al.* [18] to go from IND-CPA security to IND-CCA security in the context of memory leakage.

The intuition why the transformation works is fairly simple: The argument π enforces the adversary to submit to the tampered decryption oracle only ciphertexts for which he knows the corresponding plaintext (and the randomness used to encrypt it). In the security proof the pair (m, r) can indeed be extracted from such argument, allowing to reduce IND-CCA BLT security to IND-CPA BLT security.

Theorem 2. *Let $k \in \mathbb{N}$ be the security parameter. Assume that $\mathcal{PKE}$ is an IND-CPA $(\lambda(k), t(k), \delta(k))$-BLT secure encryption scheme and that $(\mathsf{Gen}, \mathsf{Prove}, \mathsf{Verify})$ is a strong tSE NIZK argument system for relation $\mathfrak{R}_{\mathsf{PKE}}$. Then, the encryption scheme $\mathcal{PKE}'$ of Figure 3 is IND-CCA $(\lambda(k), t(k), \delta'(k))$-BLT secure for $\delta' \leq \delta + \mathsf{negl}(k)$.*

Proof. We prove the theorem by a series of games. All games are a variant of the IND-CCA BLT game and in all games the adversary gets correctly generated public parameters (pp, ω, pk). Leakage and tampering queries are answered using the corresponding secret key sk. The games will differ only in the way the challenge ciphertext is computed or in the way the decryption oracles work.

Game G_1. This is the IND-CCA BLT game of Definition 3 for the scheme $\mathcal{PKE}'$. Note in particular that all decryption oracles expect to receive as input a ciphertext of the form (c, π) and proceed to verify the proof π before decrypting the ciphertext (and output $\bot$ if such verification fails). The challenge ciphertext is a pair (c_b, π_b) such that $c_b = \mathsf{Enc}(pk, m_b; r)$ and $\pi_b \leftarrow \mathsf{Prove}^\omega((pk, c_b), (m_b, r))$, where $m_b \in \{m_0, m_1\}$ for a uniformly random bit b. By assumption we have that

$$\Pr[\mathsf{A} \text{ wins in } \mathsf{G}_1] \le \frac{1}{2} + \delta'(k).$$

Game G_2. In this game we change the way the challenge ciphertext is computed by replacing the argument π_b with a simulated argument $\pi_b \leftarrow \mathsf{S}((pk, c_b), \mathsf{tk})$. It follows from the composable NIZK property of the argument system that G_1 and G_2 are computationally close. In particular there exists a negligible function $\delta_1(k)$ such that $|\Pr[\mathsf{A} \text{ wins in } \mathsf{G}_1] - \Pr[\mathsf{A} \text{ wins in } \mathsf{G}_2]| \le \delta_1(k)$.

Game G_3. We change the way decryption queries are handled. Queries (c, π) to $\mathsf{Dec}(sk, \cdot)$ (such that π accepts) are answered by running the extractor Ext on π, yielding $(m, r) \leftarrow \mathsf{Ext}((pk, c), \pi, \mathsf{ek})$, and returning m.

Queries (c, π) to $\mathsf{Dec}(\widetilde{sk}_i, \cdot)$ (such that π accepts) are answered as follows. We first extract $(m, r) \leftarrow \mathsf{Ext}((pk, c), \pi, \mathsf{ek})$ as above. Then, instead of returning m, we recompute $c = \mathsf{Enc}(pk, m; r)$ and return $\widetilde{m} = \mathsf{Dec}(\widetilde{sk}_i, c)$.

It follows from true simulation extractability that G_2 and G_3 are computationally close. The reason for this is that A gets to see only a single simulated proof for a true statement (i.e., the pair (pk, c_b)) and thus cannot produce a pair $(c, \pi) \ne (c_b, \pi_b)$ such that the proof π accepts and Ext fails to extract the corresponding plaintext m. In particular there exists a negligible function $\delta_2(k)$ such that $|\Pr[\mathsf{A} \text{ wins in } \mathsf{G}_2] - \Pr[\mathsf{A} \text{ wins in } \mathsf{G}_3]| \le \delta_2(k)$.

Game G_4. In the last game we replace the ciphertext c_b in the challenge with an encryption of $0^{|m_b|}$, whereas we still compute the proof as $\pi_b \leftarrow \mathsf{S}((pk, c_b), \mathsf{tk})$. We claim that G_3 and G_4 are computationally close. This follows from IND-CPA BLT-security of $\mathcal{PKE}$. Assume there exists a distinguisher D between G_3 and G_4. We build an adversary B breaking IND-CPA BLT security for $\mathcal{PKE}$. The adversary B uses D as a black-box as follows.

Reduction B^D:
1. Receive (pp, pk) from the challenger, sample $(\omega, \mathsf{tk}, \mathsf{ek}) \leftarrow \mathsf{Gen}(1^k)$ and give $pp' = (pp, \omega)$ and $pk' = pk$ to A.
2. Upon input a normal decryption query (c, π) from A, run the extractor to compute $(m, r) \leftarrow \mathsf{Ext}((pk, c), \pi, \mathsf{ek})$ and return m.
3. Upon input a tampering query $T_i \in \mathcal{T}_{sk}$, forward T_i to the tampering oracle for $\mathcal{PKE}$. To answer a query (c, π), run the extractor to compute $(m, r) \leftarrow \mathsf{Ext}((pk, c), \pi, \mathsf{ek})$. Submit (m, r) to oracle $\mathsf{Dec}^*(\widetilde{sk}_i, \cdot, \cdot)$ and receive the answer $\widetilde{m}$. Return $\widetilde{m}$ to A.
4. Upon input a leakage query L_j, forward L_j to the leakage oracle for $\mathcal{PKE}$.
5. When A outputs $m_0, m_1 \in \mathcal{M}$, sample a random bit b' and output $(m_{b'}, 0^{|m_{b'}|})$. Let c_b be the corresponding challenge ciphertext. Compute $\pi_b \leftarrow \mathsf{S}((pk, c_b), \mathsf{tk})$ and forward (c_b, π_b) to A. Continue to answer normal decryption queries (c, π) from A as above.

6. Output whatever D does.

Notice that the reduction perfectly simulates the environment for A; in particular c_b is either the encryption of randomly chosen message among (m_0, m_1) (as in G_3) or an encryption of zero (as in G_4). Since $\mathcal{PKE}$ is (λ, t, δ)-BLT secure, it must be $|\Pr[\text{A wins in } G_3] - \Pr[\text{A wins in } G_4]| \leq \delta(k)$.

As clearly $\Pr[\text{A wins in } G_4] = 0$, we have obtained

$$
\begin{aligned}
\delta' &= |\Pr[\text{A wins in } G_1] - \Pr[\text{A wins in } G_4]| \\
&\leq |\Pr[\text{A wins in } G_1] - \Pr[\text{A wins in } G_2]| + |\Pr[\text{A wins in } G_2] \\
&\qquad - \Pr[\text{A wins in } G_3]| + |\Pr[\text{A wins in } G_3] - \Pr[\text{A wins in } G_4]| \\
&\leq \delta_1(k) + \delta_2(k) + \delta(k) = \delta(k) + negl(k).
\end{aligned}
$$

This concludes the proof.

4.3   Instantiation from BHHO

We show that the variant of the encryption scheme introduced by Boneh *et al.* [12] used in [33] is IND-CPA BLT-secure. The proof relies on the observation that one can simulate polynomially many decryption queries for a given tampered key by only leaking a bounded amount of information from the secret key. Hence, security follows from leakage resilience. The formal description of the scheme and the proof can be found in the full version [17].

5   Updating the Key in the *i*Floppy Model

We complement the results from the previous two sections by showing how to obtain security against an unbounded number of tampering queries in the floppy model of [3,2]. Recall that in this model we assume the existence of an external tamper-free and leakage-free storage (the floppy), which is needed to refresh the secret key on the tamperable device. An important difference between the floppy model considered in this paper and the model of [2] is that in our case the floppy can contain "user-specific" information, whereas in [2] it contains a *unique* master key which in principle could be equal for all users. To stress this difference, we refer to our model as the *i*Floppy model.

Clearly, the assumption of a unique master key makes production easier but it is also a single point of failure in the system since in case the content of the floppy is published (e.g., by a malicious user) the entire system needs to be re-initialized.[4] A solution for this is to assume that each floppy contains a different master key as is the case in the *i*Floppy model, resulting in a trade-off between security and production cost. Due to space restrictions, we defer the formal definitions and proofs to the full version [17].

---

[4] We note that in the schemes of [2] making the content of the floppy public does not constitute a total breach of security; however the security proof completely breaks down, leaving no security guarantee for the schemes at hand.

Acknowledgments. Ivan Damgård and Daniele Venturi acknowledge support from the Danish National Research Foundation, the National Science Foundation of China (under the grant 61061130540), the Danish Council for Independent Research (under the DFF Starting Grant 10-081612) and also from the CFEM research center within which part of this work was performed. Sebastian Faust was partially funded by the above grants. Pratyay Mukherjee's work at Aarhus University was supported by the above grants and a European Research Commission Starting Grant (no. 279447). Part of this work was done while this author was at the University of Warsaw and was supported by the WELCOME/2010-4/2 grant founded within the framework of the EU Innovative Economy Operational Programme.

References

1. Aggarwal, D., Dodis, Y., Lovett, S.: Non-malleable codes from additive combinatorics. IACR Cryptology ePrint Archive, 2013:201 (2013)

2. Agrawal, S., Dodis, Y., Vaikuntanathan, V., Wichs, D.: On continual leakage of discrete log representations. IACR Cryptology ePrint Archive, 2012:367 (2012)

3. Alwen, J., Dodis, Y., Wichs, D.: Leakage-resilient public-key cryptography in the bounded-retrieval model. In: Halevi, S. (ed.) CRYPTO 2009. LNCS, vol. 5677, pp. 36–54. Springer, Heidelberg (2009)

4. Anderson, R., Kuhn, M.: Tamper resistance: A cautionary note. In: WOEC 1996: Proceedings of the 2nd Conference on Proceedings of the Second USENIX Workshop on Electronic Commerce, p. 1. USENIX Association, Berkeley (1996)

5. Applebaum, B., Harnik, D., Ishai, Y.: Semantic security under related-key attacks and applications. In: ICS, pp. 45–60 (2011)

6. Bellare, M., Cash, D.: Pseudorandom functions and permutations provably secure against related-key attacks. In: Rabin, T. (ed.) CRYPTO 2010. LNCS, vol. 6223, pp. 666–684. Springer, Heidelberg (2010)

7. Bellare, M., Cash, D., Miller, R.: Cryptography secure against related-key attacks and tampering. In: Lee, D.H., Wang, X. (eds.) ASIACRYPT 2011. LNCS, vol. 7073, pp. 486–503. Springer, Heidelberg (2011)

8. Bellare, M., Kohno, T.: A theoretical treatment of related-key attacks: RKA-PRPs, RKA-PRFs, and applications. In: Biham, E. (ed.) EUROCRYPT 2003. LNCS, vol. 2656, pp. 491–506. Springer, Heidelberg (2003)

9. Bellare, M., Paterson, K.G., Thomson, S.: RKA security beyond the linear barrier: IBE, encryption and signatures. In: Wang, X., Sako, K. (eds.) ASIACRYPT 2012. LNCS, vol. 7658, pp. 331–348. Springer, Heidelberg (2012)

10. Bhattacharyya, R., Roy, A.: Secure message authentication against related key attack. In: FSE (2013)

11. Boneh, D., DeMillo, R.A., Lipton, R.J.: On the importance of eliminating errors in cryptographic computations. J. Cryptology 14(2), 101–119 (2001)

12. Boneh, D., Halevi, S., Hamburg, M., Ostrovsky, R.: Circular-secure encryption from decision diffie-hellman. In: Wagner, D. (ed.) CRYPTO 2008. LNCS, vol. 5157, pp. 108–125. Springer, Heidelberg (2008)

13. Choi, S.G., Kiayias, A., Malkin, T.: BiTR: Built-in tamper resilience. In: Lee, D.H., Wang, X. (eds.) ASIACRYPT 2011. LNCS, vol. 7073, pp. 740–758. Springer, Heidelberg (2011)

14. Cramer, R.: Modular Design of Secure yet Practical Cryptographic Protocols. PhD thesis, University of Amsterdam (November 1996)
15. Di Crescenzo, G., Lipton, R.J., Walfish, S.: Perfectly secure password protocols in the bounded retrieval model. In: Halevi, S., Rabin, T. (eds.) TCC 2006. LNCS, vol. 3876, pp. 225–244. Springer, Heidelberg (2006)
16. Dachman-Soled, D., Kalai, Y.T.: Securing circuits against constant-rate tampering. In: Safavi-Naini, R., Canetti, R. (eds.) CRYPTO 2012. LNCS, vol. 7417, pp. 533–551. Springer, Heidelberg (2012)
17. Damgård, I., Faust, S., Mukherjee, P., Venturi, D.: Bounded tamper resilience: How to go beyond the algebraic barrier. IACR Cryptology ePrint Archive (2013)
18. Dodis, Y., Haralambiev, K., López-Alt, A., Wichs, D.: Cryptography against continuous memory attacks. In: FOCS, pp. 511–520 (2010)
19. Dodis, Y., Ostrovsky, R., Reyzin, L., Smith, A.: Fuzzy extractors: How to generate strong keys from biometrics and other noisy data. SIAM J. Comput. 38(1), 97–139 (2008)
20. Dziembowski, S.: Intrusion-resilience via the bounded-storage model. In: Halevi, S., Rabin, T. (eds.) TCC 2006. LNCS, vol. 3876, pp. 207–224. Springer, Heidelberg (2006)
21. Dziembowski, S., Pietrzak, K., Wichs, D.: Non-malleable codes. In: ICS, pp. 434–452 (2010)
22. Faust, S., Pietrzak, K., Venturi, D.: Tamper-proof circuits: How to trade leakage for tamper-resilience. In: Aceto, L., Henzinger, M., Sgall, J. (eds.) ICALP 2011, Part I. LNCS, vol. 6755, pp. 391–402. Springer, Heidelberg (2011)
23. Fiat, A., Shamir, A.: How to prove yourself: Practical solutions to identification and signature problems. In: Odlyzko, A.M. (ed.) CRYPTO 1986. LNCS, vol. 263, pp. 186–194. Springer, Heidelberg (1987)
24. Fischlin, M., Fischlin, R.: The representation problem based on factoring. In: Preneel, B. (ed.) CT-RSA 2002. LNCS, vol. 2271, pp. 96–113. Springer, Heidelberg (2002)
25. Gennaro, R., Lysyanskaya, A., Malkin, T., Micali, S., Rabin, T.: Algorithmic tamper-proof (ATP) security: Theoretical foundations for security against hardware tampering. In: Naor, M. (ed.) TCC 2004. LNCS, vol. 2951, pp. 258–277. Springer, Heidelberg (2004)
26. Goyal, V., O'Neill, A., Rao, V.: Correlated-input secure hash functions. In: Ishai, Y. (ed.) TCC 2011. LNCS, vol. 6597, pp. 182–200. Springer, Heidelberg (2011)
27. Guillou, L.C., Quisquater, J.-J.: A "Paradoxical" identity-based signature scheme resulting from zero-knowledge. In: Goldwasser, S. (ed.) CRYPTO 1988. LNCS, vol. 403, pp. 216–231. Springer, Heidelberg (1990)
28. Ishai, Y., Prabhakaran, M., Sahai, A., Wagner, D.: Private circuits II: Keeping secrets in tamperable circuits. In: Vaudenay, S. (ed.) EUROCRYPT 2006. LNCS, vol. 4004, pp. 308–327. Springer, Heidelberg (2006)
29. Kalai, Y.T., Kanukurthi, B., Sahai, A.: Cryptography with tamperable and leaky memory. In: Rogaway, P. (ed.) CRYPTO 2011. LNCS, vol. 6841, pp. 373–390. Springer, Heidelberg (2011)
30. Katz, J., Vaikuntanathan, V.: Signature schemes with bounded leakage resilience. In: Matsui, M. (ed.) ASIACRYPT 2009. LNCS, vol. 5912, pp. 703–720. Springer, Heidelberg (2009)
31. Liu, F.-H., Lysyanskaya, A.: Tamper and leakage resilience in the split-state model. In: Safavi-Naini, R., Canetti, R. (eds.) CRYPTO 2012. LNCS, vol. 7417, pp. 517–532. Springer, Heidelberg (2012)

32. Lucks, S.: Ciphers secure against related-key attacks. In: Roy, B., Meier, W. (eds.) FSE 2004. LNCS, vol. 3017, pp. 359–370. Springer, Heidelberg (2004)
33. Naor, M., Segev, G.: Public-key cryptosystems resilient to key leakage. In: Halevi, S. (ed.) CRYPTO 2009. LNCS, vol. 5677, pp. 18–35. Springer, Heidelberg (2009)
34. Okamoto, T.: Provably secure and practical identification schemes and corresponding signature schemes. In: Brickell, E.F. (ed.) CRYPTO 1992. LNCS, vol. 740, pp. 31–53. Springer, Heidelberg (1993)
35. Pietrzak, K.: Subspace LWE. In: Cramer, R. (ed.) TCC 2012. LNCS, vol. 7194, pp. 548–563. Springer, Heidelberg (2012)
36. Pohlig, S., Hellman, M.: An improved algorithm for computing logarithms over and its cryptographic significance. IEEE Transactions on Information Theory 24(1), 106–110 (1978)
37. Wee, H.: Public key encryption against related key attacks. In: Fischlin, M., Buchmann, J., Manulis, M. (eds.) PKC 2012. LNCS, vol. 7293, pp. 262–279. Springer, Heidelberg (2012)

Tamper Resilient Circuits:
The Adversary at the Gates

Aggelos Kiayias and Yiannis Tselekounis

Department of Informatics and Telecommunications,
National and Kapodistrian University of Athens

Abstract. We initiate the investigation of *gate*-tampering attacks against cryptographic circuits. Our model is motivated by the plausibility of tampering directly with circuit gates and by the increasing use of *tamper resilient gates* among the known constructions that are shown to be resilient against *wire-tampering* adversaries. We prove that gate-tampering is *strictly* stronger than wire-tampering. On the one hand, we show that there is a gate-tampering strategy that perfectly simulates any given wire-tampering strategy. On the other, we construct families of circuits over which it is impossible for any wire-tampering attacker to simulate a certain gate-tampering attack (that we explicitly construct). We also provide a tamper resilience impossibility result that applies to both gate and wire tampering adversaries and relates the amount of tampering to the depth of the circuit. Finally, we show that defending against gate-tampering attacks is feasible by appropriately abstracting and analyzing the circuit compiler of Ishai et al. [18] in a manner which may be of independent interest. Specifically, we first introduce a class of compilers that, assuming certain well defined tamper resilience characteristics against a specific class of attackers, can be shown to produce tamper resilient circuits against that same class of attackers. Then, we describe a compiler in this class for which we prove that it possesses the necessary tamper-resilience characteristics against gate-tampering attackers.

Keywords: tamper resilient circuits, attack modeling.

1 Introduction

Traditionally, cryptographic algorithms are designed under the assumption that adversaries have *black box* access to the algorithms' implementation and private input. In this setting, the adversary chooses an input, supplies the algorithm with it, receives the corresponding output, and it is not allowed to alter the algorithm's internals during its execution. This mode of interaction is usually being modeled as a security game (e.g., chosen-ciphertext attack against an encryption scheme or chosen message attack against a digital signature) and the underlying cryptographic scheme is proven secure based on it. In reality though, besides observing the algorithms' input-output behaviour, an adversary may also land physical attacks on the algorithm's implementation. For instance, she may learn the secret key of an encryption scheme by measuring the power

K. Sako and P. Sarkar (Eds.) ASIACRYPT 2013, Part II, LNCS 8270, pp. 161–180, 2013.

consumed by the device during the encryption operation [23], or by measuring the time needed for the encryption to complete [22]. Besides *passive* attacks, the class of *active* attacks includes inducing faults to the computation [4,6,22], exposing the device to electromagnetic radiation [14,28,29], and several others [17,24,21,1,7,30]. Such attacks have proven to be a significant threat to the real-world security of cryptographic implementations.

1.1 Related Work and Motivation

The work of [18] followed by [11,9] undertook the difficult task of modeling and defending against adversaries that tamper directly with the implementation circuit. In this setting the adversary is given access to a circuit equipped with secret data stored in private memory; it is allowed to modify a bounded number of circuit wires and/or memory gates in each circuit invocation. The objective is to suitably modify the circuit operation so that tampering gives no (or -at least-bounded) advantage to the adversary.

In [18] the adversary is allowed to tamper with a bounded number of wires or memory gates in each computation, and for each component she may *set* its value to 1, *reset* it to 0, or *toggle* its value. The tampering effect can be persistent, i.e., if the value of a circuit wire or memory gate is modified during one run, it remains modified for all subsequent runs. Hence, the adversary can tamper with the entire circuit by persistently tampering with a bounded number of wires in each run. The proposed compiler, which is parameterized by $t \in \mathbb{N}$, transforms any boolean circuit C into C', where C' realizes the same functionality with C and is secure against adversaries who tamper with up to t of its wires in each computation, i.e., any adversary who tampers with up to t circuit wires of C' in one circuit invocation, cannot learn anything more about the circuit's private information than an adversary having *black-box* access to C. Formally, this notion is captured by the following *simulation-based* security definition: for every PPT adversary $\mathcal{A}$ tampering with C', there exists a simulator $\mathcal{S}$ having black-box access to C such that the output distribution of $\mathcal{A}$ and $\mathcal{S}$ are indistinguishable. The construction is based on a randomized secret sharing scheme which shares the bit-value of a wire in C among k wires, and then introduces redundancy by making $2kt$ copies of each wire, where k denotes the security parameter. The randomized encoding guarantees that any tampering with C' will produce an invalid encoding with high probability, triggering the circuit's self-destruction mechanism that erases the circuit's secret memory. Since this mechanism is also prone to tampering, the adversary could try to deactivate it so as to tamper with the rest of the circuit while keeping the secret state intact. In order to prevent such a scenario, C' incorporates an error-propagation mechanism which permeates the circuit and propagates errors induced by tampering attacks. The size of C' is larger than C by a factor of $O(t \cdot \log^3(\frac{1}{\epsilon}))$ where ϵ represents the simulation error.

In [11] the authors consider a different adversarial model, in which the adversary is allowed to tamper with every circuit wire, but each tampering attempt fails with probability $\delta \geq 0$ (*noisy* tampering). Moreover, they put forward a relaxed security definition in which the simulator does not have black-box access to the circuit, but requires logarithmically many bits of information about the circuit's secret memory. The resulting circuit is augmented by a $O(\delta^{-1}\log(\frac{1}{\epsilon}))$ factor for the simulation to fail with probability at most ϵ. Furthermore, it uses no randomness during execution and consists of subcircuits which perform computations over *Manchester encodings*, which encode a single bit into four bits. For each subcircuit, the compiler randomly encodes the 0,1-bits to elements in $\{0,1\}^4\backslash\{0000\}$, and employs *tamper-proof* gates that handle computations over these encodings. If the inputs are invalid, the gates output 0000 and the error propagates to the self-destruction mechanism, which is similar to the one employed in [18] but uses some additional tamper-proof gadgets. Besides error propagation and memory erasure, the self-destruction mechanism verifies that all subcircuits produce consistent outputs. Hence, in order to alter the computation effectively, an adversary needs to tamper with all k subcircuits in a way such that (i) all attacks produce valid, probably different due to randomization, encodings, and (ii) the encodings must produce the same decoded output. As it is proved in [11], this happens with negligible probability in k.

The adversarial model considered in [9] is similar to [18]. The main difference is that now persistent tampering is not allowed on circuit wires and, similarly, to [11] the simulator is allowed a logarithmic amount of leakage from the computation. Regarding the construction, [9] combines *error-correcting codes* and *probabilistically checkable proofs of proximity* (PCPP) in the following way: the circuit's secret state $\mathbf{s}$ and input $\mathbf{x}$ are encoded into $\mathbf{S}$ and $\mathbf{X}$, respectively. Then the transformed circuit computes[1] $y = C_{\mathbf{s}}(\mathbf{x})$ and a PCPP proof π for the validity of the tuple $(y, \mathbf{S} \circ \mathbf{X})$ with respect to the error-correcting code and C. The proof is verified by polynomially many verifiers who output 1 in case of validity, and 0 otherwise, and their output (i) is fed to a (tamper-proof) AND gate with *unbounded* fan-in and fan-out that erases the circuit's secret state if a verifier rejects π, and (ii) together with y they feed a (tamper-proof) AND gate with unbounded fan-in and one output wire, which is the circuit's output wire. If one of the verifiers outputs 0, then the circuit outputs the zero bit. The resulting circuit size is polynomial on the input circuit but a constant ratio of wire tampering can be tolerated.

Besides tampering against circuits' wires or memory gates, some works consider adversaries who tamper exclusively with memory gates [15,25,10,8]. In [15] the authors give an impossibility result on tamper resilience by showing that without using secure hardware an adversary can extract the circuit's private information, by sequentially setting or resetting the memory bits and observing the tampering effects on the circuit output. Apparently, [18,11,9] circumvent the impossibility result by erasing the private information in case of error detection. In [25] the authors consider adversaries who tamper and probe with the circuits' private memory and they give an impossibility result for circuits that do not have access to a source of random bits, with respect to both tampering and probing attacks. [10] introduces the notion of non-malleable codes. Such codes ensure that any adversary who tampers with a codeword, with respect to some specific

---

[1] Their construction [9] considers circuits that output one bit.

class of tampering functions, will either lead the decoding algorithm to output $\perp$, or output a codeword which is irrelevant to the original word. Moreover, they show how to construct non-malleable codes for specific classes of tampering functions. Finally, in [8] the authors introduce the notion of Built-in Tamper Resilience, which defines security for cryptographic protocols where some of the parties are implemented by hardware tokens that resist tampering attacks.

The above state of the art in tamper resilient circuits suggests a fundamental issue that is a source of theoretical motivation for our work. While tampering circuit wires seems to be a strong adversarial model, recent constructions do heavily exploit tamper-proof gates (e.g., the gate with unbounded high fan-in in [9]). This suggests that *tampering gates directly* might be an even stronger (and possibly in some cases even more plausible) adversarial model; how do wire and gate adversaries fare against each other? and is it possible to protect against both? what are the upper bounds in terms of amount of tampering that can be tolerated? Our work initiates the investigation of gate-tampering attacks and takes steps towards answering all those questions as explained below.

Besides its theoretical interest, our work is also motivated by practical attacks on circuit gates. For example, in [30] it is explained how illumination of a target transistor can cause it to conduct. Transistors are used to implement logical gates so such an optical probe attack will amount to gate tampering in a circuit (effectively changing the gate for another gate). Beyond that, fault injection in the SRAM of an FPGA also results to switching the computation of the FPGA circuit (because the program of the FPGA is stored in memory).

1.2   Our Contributions

Impossibility Results. Informally, the main idea behind our impossibility result (section 3) is the following: we define the notion of *non-triviality* of a cryptographic circuit which attempts to capture the essence of a meaningful cryptographic implementation. According to non-triviality, for every PPT algorithm $\mathcal{A}$ and circuit C with private memory $\mathbf{s}$, where C implements some cryptographic functionality, $\mathcal{A}$ should not be able to learn $\mathbf{s}$, while having black box access to C (observe that if $\mathcal{A}$ learns $\mathbf{s}$ then the implementation becomes obsolete as $\mathcal{A}$ can simulate it). Then we prove that any circuit C that satisfies non-triviality possesses necessarily a *weakly unpredictable bit*, i.e., there exists a secret state bit that cannot be extracted with probability very close to 1, while having black box access to C. Now, let d be the depth of C and assume it consists of gates with fan-in at most 2. If the adversary is allowed to tamper with up to d circuit components (we prove our result for either wires or gates), there exists a strategy that extracts the circuit's unpredictable bit with probability equal to 1. The impossibility result follows from this, since any simulator with black-box access to C only has no capability to predict the unpredictable bit Âǎ as good as the tampering adversary. The main observation here is that for any $d \in \mathbf{N}$, and every compiler T that receives a circuit C and produces a circuit C' of depth at most d, T cannot be secure against an adversary who tampers with d circuit gates or wires, *regardless of the size of C'*.

It is worth contrasting our result to that of [15], where the authors consider an adaptive adversary who is capable of tampering with private memory bits; by correlating circuit outputs to tampering set/reset operations within the secret-state they show that the whole secret state can be reconstructed. This suggests that the only way to attain tamper resilience of the secret-state is by employing internal integrity detection mechanisms and have the circuit self-destruct in any case of fault detection. With our impossibility result however we show that simulation will inevitably fail even in the presence of error-detection and self-destruction mechanisms in case we allow tampering with up to d circuit components (wires or gates), where d is the circuit's depth. This underlines the strength of tampering with circuit components vs. tampering the secret state.

Gate Adversaries Are Strictly Stronger than Wire Adversaries. We proceed to explore the relationship between gate and wire adversaries. In section 4 we first prove that any tampering attack on up to t circuit wires can be simulated by an adversary who tampers with up to t circuit gates, i.e., for every circuit $C_{\mathbf{s}}$ and any PPT adversary $\mathcal{A}$ who tampers with up to t wires of $C_{\mathbf{s}}$, there exists a PPT adversary $\mathcal{A}'$ who tampers with up to t circuit gates, such that the output of $\mathcal{A}$ and $\mathcal{A}'$ are exactly the same. Then we proceed to prove the other direction which is the most technically involved. Note that in the presence of unbounded fan-out (or fan-in) gates in a circuit it is clear that a gate adversary has an advantage since a wire adversary may be incapable of controlling sufficiently many wires to modify the behavior of the gate. However we demonstrate that even w.r.t. to bounded fan-in/fan-out circuits, gate adversary are strictly stronger. Specifically, we show that there exist a family of circuits $\tilde{C}_{\tilde{\mathbf{s}}}$ parameterized by n, t and a PPT adversary $\mathcal{A}$ who tampers with up to n circuit gates, such that for all PPT adversaries $\mathcal{A}'$ who tamper with up to t circuit wires (where t can be arbitrarily larger than n), $\mathcal{A}'$ fails to simulate $\mathcal{A}$. Intuitively, the idea behind our proof is the following. We construct a circuit that has a "critical area" comprised of n AND-gates. The input to the critical area is provided by a sub-circuit (referred to as C_1) that implements a PRF, a digital signature and a counter. The output of the critical area is fed to a second sub-circuit (referred to as C_2) that calculates a digital signature and a second counter. The key point is that a gate-adversary can transform all AND-gates of the critical area into XOR-gates. This enables the gate-attacker to produce a circuit output with a certain specific distribution that is verifiable in polynomial-time. The main technical difficulty is assembling the circuits C_1, C_2 suitably so that we can show that no matter what the wire-attacker does, it is incapable of simulating the distribution produced by the gate-attacker. Note that the wire-attacker is fully capable of controlling the input to the C_2 sub-circuit (by tampering with all the output wires of the critical area). Hence by running the circuit several times, the wire-attacker can attempt to learn the proper output distribution of the critical region and feed it to C_2. In our explicit construction, by appropriately assembling the main ingredients of each sub-circuit (PRF, digital signatures and counters) we show that there exists no efficient wire-tampering strategy that simulates the gate-tampering strategy assuming the security of the PRF and the digital signature. The circuit family we construct is executable an unbounded number of times (by either attacker). If one restricts the number of times the implementation can be executed by the tampering attackers (by having the implementation always self-destruct by design after one invocation) then the circuit family can be simplified (due to lack of space we provide this construction in the full version of the paper).

Tamper Resilience against Gate Adversaries. Given our separation result, the question that remains is whether it is possible to defend cryptographic implementations against gate adversaries. Towards that direction, we show (section 5) that gate attackers compromise the security of [18] by effectively eliminating the circuit's randomness, when it is produced by randomness gates, and then we prove that if we substitute those gates with pseudo-random generators, then [18] can be shown to be secure against gate adversaries. Based on [18], we give a general characterization (Definition 10) of a secure class of compilers and we use it in order to present our result in a self-contained way. The way we present our positive result may also be of independent interest: first, we define a class of compilers (Definition 10) that have the property that if they have certain tamper resilience characteristics against an arbitrary class of adversaries, then the circuits that they produce are tamper-resilient against that class of adversaries. Seen in this light, the result of [18] is a specific instance that belongs to this class of compilers that satisfies the basic tamper resilience characteristics of the class against appropriately bounded *wire* adversaries. We proceed analogously to prove that the circuit transformation that removes the randomness gates satisfies the necessary tamper resilience characteristics against appropriately bounded *gate* adversaries. The resulting compiler produces circuits of comparable size to those of [18] however the parameter t in our case reflects the bound on the gate-tampering adversary.

2   Preliminaries

Definition 1 (Circuit). *A Boolean circuit C, over a set of gates $\mathcal{G}$, with n input bits and q output bits, is a directed acyclic graph $C = (V, E)$, such that every $v \in V$ belongs to one of the following sets of nodes:*

- V_I *(**Input**): For all $v \in V_I$, the indegree is zero, the outdegree is greater than 1, and each v represents one input bit. We label these nodes by $i_1, \ldots, i_n$.*
- $V_{\mathcal{G}}$ *(**Gate**): Each $v \in V_{\mathcal{G}}$ represents a logic gate in $\mathcal{G}$, and its indegree is equal to the arity of the logic gate. The outdegree is at least 1.*
- V_O *(**Output**): For all $v \in V_O$, the indegree is one and the outdegree is zero, and each v represents one output bit. We label these nodes by $o_1, \ldots, o_q$.*

The cardinalities of the sets defined above are n, s and q respectively. Finally, the edges of the graph represent the wires of the circuit. The set of all circuits with respect to a set of gates $\mathcal{G}$, will be denoted by $\mathcal{C}_{\mathcal{G}}$.

The circuit's *private memory* is considered as a *peripheral* component and consists of m additional gates. The value m is the memory's storage capacity in bits.[2] Formally,

Definition 2 (Circuit with Private memory). *A graph C is a circuit with private memory provided that its set of vertices can be partitioned into two sets V, V' such that (i) the graph $C \setminus V'$ is a DAG conforming to definition 1, (ii) there are no edges between nodes in V', (iii) each vertex $v \in V'$ possesses at most one incoming and exactly one outgoing edge. The set V' represents the memory gates of C; in this case, we refer to C as a circuit with private memory V' and we also denote by E' the edges of C that are incident to V'. We denote $|V'| = m$.*

From now on we will use the terms *Circuit* and *Graph* interchangeably, as well as the terms wire/edge.

Definition 3 (Computation). *Let C be a circuit with private memory V' that contains $\mathbf{s} \in \{0,1\}^m$, and let $\mathbf{x} \in \{0,1\}^n$ be the circuit input. The computation $\mathbf{y} = C_{\mathbf{s}}(\mathbf{x})$ consists of the following steps:*

i. **Memory access:** *for each $v \in V'$, the value of v propagates to the outgoing edge.*
ii. **Breadth-first traversal of C:**
   - *Assign to each input node i_j, for $j \in [n]$, the value x_j and propagate x_j to its outgoing wires.*
   - *For $v \in V_G$ (gate node), apply the boolean function that corresponds to v on the incoming edges and propagate the result to the outgoing edges.*
   - *Every output node o_i, for $i \in [m]$, evaluates to the incoming value, and determines the value y_i.*
iii. **Memory update:** *every $v \in V'$ is updated according to its incoming value.*

Informally, in each *computation* the circuit receives an input $\mathbf{x}$, produces an output $\mathbf{y}$, and it may also update its private memory from $\mathbf{s}$ to some value $\mathbf{s}'$. From now on, a circuit with secret state $\mathbf{s}$ will be denoted by $C_{\mathbf{s}}$, or by C, if there is no need to refer to $\mathbf{s}$ explicitly.

In the following, we give a generalization of the above definition for multiple rounds and we formally define the adversarial models we consider.

Definition 4 (v-round computation). *Let C be a circuit and $\mathbf{Env} = (\mathbf{s}, \mathbf{v})$ a pair of random variables. A v-round circuit execution w.r.t. $\mathbf{Env}$ is a random variable $(\mathbf{v}, \mathcal{A}^{C(\cdot)}(\mathbf{v}))$ s.t. $\mathcal{A}$ is a polynomial-time algorithm that is allowed to submit v queries to the circuit which is initialized to state $\mathbf{s}$ and in each query it performs a calculation over its input as in Definition 3.*

---

[2] In many circumstances we refer to private memory using the terms *secret state* or *private state*.

In the above definition, $\mathbf{v}$ represents all public information related to private memory $\mathbf{s}$. For instance, if C implements the decryption algorithm of a public-key encryption scheme with secret key $\mathbf{s}$, then $\mathbf{v}$ should contain information such as the length of $\mathbf{s}$ and the corresponding public-key. Now we define the adversary of [18].

Definition 5 (t-wire tampered computation). *Let C be a circuit with private memory V' that contains $\mathbf{s} \in \{0,1\}^m$, and let $\mathbf{x} \in \{0,1\}^n$ be the circuit input. The t-wire tampered computation $\mathbf{y} = C_{\mathbf{s}}^{\mathcal{T}}(\mathbf{x})$ for some tampering strategy $\mathcal{T}$ is defined as follows.*

1. *$\mathcal{T}$ is a set of up to t triples of the form (α, e, p), where e is an edge of C, and α may be one of the following tampering attacks:*
   - *__Set__: set the value of e to 1,*
   - *__Reset__: set the value of e to 0,*
   - *__Toggle__: flip the value of e,*

   and if $p \in \{0,1\}$ is set to 1 then the attack is persistent, i.e., the modification made by α is preserved in all subsequent computations. For non-persistent attacks we write $(\alpha, e, 0)$ or just (α, e).
2. *The computation of the circuit is executed as in definition 3 taking into account the tampering instructions of $\mathcal{T}$.*

Next we define gate-tampered computation.

Definition 6 (t-gate tampered computation). *Let C be a circuit with private memory V' that contains $\mathbf{s} \in \{0,1\}^m$, and let $\mathbf{x} \in \{0,1\}^n$ be the circuit input. The t-gate tampered computation $\mathbf{y} = C_{\mathbf{s}}^{\mathcal{T}}(\mathbf{x})$ for some tampering strategy $\mathcal{T}$ is defined as follows.*

1. *$\mathcal{T}$ is a set of up to t triples of the form (f, g, p), where $g \in V_{\mathcal{G}} \cup V'$, f can be any function $f : \{0,1\}^l \rightarrow \{0,1\}$, where l is the arity of the gate represented by g, and p defines persistent (or not) attacks as in definition 5. The tampering substitutes the gate functionality of g to be the function f.*
2. *The computation of the circuit is executed as in definition 3 taking into account the tampering instructions of $\mathcal{T}$.*

Definition 7 (v-round wire (resp. gate) tampered computation). *Let C be a circuit and $\mathbf{Env} = (\mathbf{s}, \mathbf{v})$ a pair of random variables. A v-round tampered computation w.r.t. $\mathbf{Env}$ is a random variable $(\mathbf{v}, \mathcal{A}^{C^*(\cdot)}(\mathbf{v}))$ s.t. $\mathcal{A}$ is a polynomial-time algorithm that is allowed to submit v queries to the circuit which is initialized to state $\mathbf{s}$ and in the i-th query it performs a wire (resp. gate) tampered computation according to tampering instructions $\mathcal{T}_i$ specified by $\mathcal{A}$. Note that the computation respects persistent tampering as specified by $\mathcal{A}$.*

Notation. We will denote wire and gate adversaries respectively by $\mathcal{A}_\mathsf{w}$ and $\mathcal{A}_\mathsf{g}$. Moreover, $\mathcal{A}_\mathsf{w}^t$ (resp. $\mathcal{A}_\mathsf{g}^t$) denotes a wire (resp. gate) adversary who tampers with t circuit wires (resp. gates). The output of a single-round wire (resp. gate) adversary $\mathcal{A}_\mathsf{w}$ (resp. $\mathcal{A}_\mathsf{g}$) with strategy $\mathcal{T}_\mathsf{w}$ (resp. $\mathcal{T}_\mathsf{g}$) who performs a tampered computation on C is denoted by $\mathcal{A}_\mathsf{w}^{[C,\mathcal{T}_\mathsf{w}]}$ (resp. $\mathcal{A}_\mathsf{g}^{[C,\mathcal{T}_\mathsf{g}]}$). The output of a multi-round adversary A_g (resp. A_w) is denoted by $\mathcal{A}_\mathsf{g}^{C^*(\cdot)}$ (resp. $\mathcal{A}_\mathsf{w}^{C^*(\cdot)}$). For $n \in \mathbb{N}$, $[n]$ is the set $\{1, \ldots, n\}$. The statistical distance between two random variables X, Y, with range $\mathcal{D}$, is denoted by $\Delta(X, Y)$, i.e., $\Delta(X, Y) = \frac{1}{2} \sum_{u \in \mathcal{D}} |\Pr[X = u] - \Pr[Y = u]|$. Finally, if D is a distribution over $\mathcal{D}$, $X \sim D$ indicates that variable X follows distribution D.

3   Impossibility Results

In this section we prove that for any *non-trivial* cryptographic device implemented by some circuit $C \in \mathcal{C}_\mathcal{G}$ of depth d, where $\mathcal{G}$ contains boolean logic gates, tamper-resilience is impossible (i) when wire adversaries land $d(k-1)$ non-persistent tampering attacks on the wires of C, where k is the maximum fan-in of the elements in $\mathcal{G}$, and (ii) when gate adversaries non-persistently tamper with d gates of C. In order to do so, we define the notion of *non-triviality*, which characterizes meaningful implementations and then we prove that every non-trivial circuit C possesses a *weakly unpredictable bit* (Lemma 1). Then we define an adversarial strategy $\mathcal{T}$, such that for any $\mathbf{x} \in \{0,1\}^n$, $C^\mathcal{T}(\mathbf{x})$ is statistically far from the output of $\mathcal{S}^{C(\cdot)}$, for any PPT algorithm $\mathcal{S}$.

A non-trivial cryptographic device is one that contains a circuit for which no adversary can produce its entire secret-state in polynomial-time when allowed black-box access to it. Formally,

Definition 8 (non-triviality property). *Let* $\mathbf{Env} = (\mathbf{s}, \mathbf{v})$ *be a pair of random variables, and let* $C_\mathbf{s}$ *be a circuit with secret state* $\mathbf{s}$. *We say that* $C_\mathbf{s}$ *satisfies the non-triviality property w.r.t. environment* $\mathbf{Env}$ *if for every PPT algorithm* $\mathcal{A}$ *there exists a non-negligible function* $f(m)$ *such that*

$$\Pr[\mathcal{A}^{C_\mathbf{s}(\cdot)}(\mathbf{v}) = \mathbf{s}] < 1 - f(m).$$

The above definition is a necessary property from a cryptographic point of view, since its negation implies that the device can be replicated with only black-box access. Thus any attacker can render it redundant by recovering its secret state and instantiating it from scratch. We then focus on specific bits of the secret state. We define a bit to be weakly unpredictable if predicting its value always involves a non-negligible error given black-box access to the device.

Definition 9 (weakly unpredictable bit). *Let* $\mathbf{Env} = (\mathbf{s}, \mathbf{v})$ *be a pair of random variables, and* $C_\mathbf{s}$ *be a circuit with secret state* $\mathbf{s}$. *Then* $C_\mathbf{s}$ *possesses a weakly unpredictable bit w.r.t. environment* $\mathbf{Env}$ *if there exists an index* i, $1 \leq i \leq m$, *such that for every PPT algorithm* $\mathcal{A}$ *there exists a non-negligible function* $\delta(m)$ *such that*

$$\Pr[\mathcal{A}^{C_\mathbf{s}(\cdot)}(\mathbf{v}) = s_i] < 1 - \delta(m).$$

Armed with the above definitions we demonstrate that any non-trivial circuit possesses at least one weakly unpredictable bit.

Lemma 1. *Let* **Env** $= (\mathbf{s}, \mathbf{v})$ *be a pair of random variables. Then for every circuit* $C_{\mathbf{s}}$, *if* $C_{\mathbf{s}}$ *satisfies the non-triviality property w.r.t. enviroment* **Env**, *then* $C_{\mathbf{s}}$ *possesses a weakly unpredictable bit, again w.r.t.* **Env**.

The above lemma is proved[3] by contradiction: we consider a circuit $C_{\mathbf{s}}$ which satisfies the non-triviality property and none of its bits is a weakly unpredictable bit. Then we construct an algorithm which extracts $\mathbf{s}$ with probability $1 - f(m)$, for some non-negligible function $f(m)$.

We next give the impossibility result for circuits that consist of standard AND, NOT and OR gates, and for the case of wire adversaries. The impossibility is via a construction: we design a specific single round adversary $\mathcal{A}$ that is non-simulatable in polynomial time. The main idea behind the construction is exploiting the tampering instructions so that we correlate the output of the circuit with the weakly unpredictable bit contained in the secret state. Concretely, in the proof of theorem 1 we define a wire adversary $\mathcal{A}_{\mathsf{w}}$ who acts as follows: she targets the weakly unpredictable bit s_i and a path $\mathcal{P}$ from the i-th memory gate to one output gate, say y_j. Then she chooses a tampering strategy $\mathcal{T}_{\mathsf{w}}$ on wires which ensures that s_i remains unchanged during its pass through the circuit gates. For instance, if at some point the circuit computes $g(s_i, x)$, where x is an input or secret state bit, then (i) if g is an AND (resp. OR) gate then $\mathcal{A}_{\mathsf{w}}$ *sets* (resp. *resets*) the wire that carries x, and the circuit computes $\wedge(s_i, 1) = s_i$ (resp. $\vee(s_i, 0) = s_i$). After defining $\mathcal{T}_{\mathsf{w}}$, $\mathcal{A}_{\mathsf{w}}$ executes $C_{\mathbf{s}}^{\mathcal{T}_{\mathsf{w}}}(\mathbf{x})$ for some $\mathbf{x} \in \{0,1\}^n$ of her choice and outputs s_i. The challenge for $\mathcal{S}$ is to output the unpredictable bit with probability close to 1, while having only black box access to $C_{\mathbf{s}}$.

Theorem 1. *(Wire Adversaries - Binary Fan-in) Let* **Env** $= (\mathbf{s}, \mathbf{v})$ *be a pair of random variables. Then for every circuit* $C_{\mathbf{s}} \in \mathcal{C}_{\mathcal{G}}$ *of depth* $d \in \mathbb{N}$, *where* $\mathcal{G} = \{\wedge, \vee, \neg\}$ *and* $\mathbf{s} \in \{0,1\}^m$, *that satisfies the non-triviality property w.r.t.* **Env**, *there exists a single round adversary* $\mathcal{A}_{\mathsf{w}}$ *with strategy* $\mathcal{T}_{\mathsf{w}}$, *where* $|\mathcal{T}_{\mathsf{w}}| \leq d$, *such that for every PPT simulator* $\mathcal{S}$ *having black-box access to* $C_{\mathbf{s}}$, *it holds that* $\Delta(\mathcal{S}^{C_{\mathbf{s}}(\cdot)}(\mathbf{v}), \mathcal{A}_{\mathsf{w}}^{[C_{\mathbf{s}}, \mathcal{T}_{\mathsf{w}}]}(\mathbf{v})) \geq f(m)$, *for some non-negligible function* $f(m)$.

The above theorem also holds for circuits that contain NAND gates, when the adversary is allowed to tamper with $2d$ circuit wires. Concretely, the adversarial strategy against $g(s_i, x)$, where g is a NAND gate, is the following: $\mathcal{A}_{\mathsf{w}}$ *resets* the wire that carries x and *toggles* the wire that carries s_i. The next corollary generalizes the above theorem for circuits that consist of gates with fan-in greater than two. Consider for example an AND gate with fan-in k, which receives the weakly unpredictable bit, s_i, on some of its input wires. If the adversary *sets* the $k - 1$ remaining wires, then the gate outputs s_i.

Corollary 1. *(Wire Adversaries - Arbitrary Fan-in) Let* **Env** $= (\mathbf{s}, \mathbf{v})$ *be a pair of random variables. Then for every circuit* $C_{\mathbf{s}} \in \mathcal{C}_{\mathcal{G}}$ *of depth* $d \in \mathbb{N}$, *where*

---

[3] For detailed proofs see the paper's full version.

$\mathbf{s} \in \{0,1\}^m$ and $\mathcal{G} = \{\wedge, \vee, \neg\}$ with bounded fan-in k, that satisfies the non-triviality property, there exists a single round adversary $\mathcal{A}_w$ with strategy $\mathcal{T}_w$, where $|\mathcal{T}_w| \leq d(k-1)$, such that for every PPT simulator $\mathcal{S}$ having black-box access to $C_\mathbf{s}$, $\Delta(\mathcal{S}^{C_\mathbf{s}(\cdot)}(\mathbf{v}), \mathcal{A}_w^{[C_\mathbf{s}, \mathcal{T}_w]}(\mathbf{v})) \geq f(m)$, for some non-negligible function $f(m)$.

Finally, we give an impossibility result with respect to gate adversaries. The main idea behind $\mathcal{A}_g$'s strategy in the following theorem, e.g. against an AND gate with fan-in k that receives the weakly unpredictable bit s_i, is the following: $\mathcal{A}_g$ substitutes the AND gate with the function that projects the value of the incoming wire that carries s_i to all outgoing wires.

Theorem 2. *(Gate Adversaries - Arbitrary Fan-in) Let* $\mathbf{Env} = (\mathbf{s}, \mathbf{v})$ *be a pair of random variables. Then for every circuit* $C_\mathbf{s} \in \mathcal{C}_\mathcal{G}$ *of depth* $d \in \mathbb{N}$, *where* $\mathbf{s} \in \{0,1\}^m$ *and* $\mathcal{G} = \{\wedge, \vee, \neg\}$ *with bounded fan-in* k, *that satisfies the non-triviality property, there exists a single round adversary* $\mathcal{A}_g$ *with strategy* $\mathcal{T}_g$, *where* $|\mathcal{T}_g| \leq d$, *such that for every PPT simulator* $\mathcal{S}$ *having black-box access to* $C_\mathbf{s}$, $\Delta(\mathcal{S}^{C_\mathbf{s}(\cdot)}(\mathbf{v}), \mathcal{A}_g^{[C_\mathbf{s}, \mathcal{T}_g]}(\mathbf{v})) \geq f(m)$, *for some non-negligible function* $f(m)$.

Notice here that $|\mathcal{T}_g|$ does not depend on k.

4   Wire vs. Gate Adversaries

In this section we investigate the relation between wire and gate adversaries of Definitions 5 and 6, respectively. Concretely, we prove that for any boolean circuit $C_\mathbf{s}$ and PPT adversary $\mathcal{A}_w^t$ with strategy $\mathcal{T}_w$, $t \in \mathbb{N}$, there exists a PPT adversary $\mathcal{A}_g^t$ with strategy $\mathcal{T}_g$, such that for any tampering action in $\mathcal{T}_w$, there exists an action in $\mathcal{T}_g$ that produces the same tampering effect (Theorem 3). Then we show that the other direction does *not* hold, i.e., we prove that gate adversaries are strictly stronger than wire ones (Theorem 4).

Wire adversaries are subsumed by Gate adversaries. We show that any wire tampering strategy is possible to be simulated by a suitable gate tampering strategy.

Now we briefly discuss the main idea behind Theorem 3, i.e., we describe how the gate adversary simulates tampering attacks on circuit wires w.r.t. to an AND gate g with two input wires, x, y, and one output wire w. So, if $\mathcal{A}_w^t$, $t \geq 3$, *resets* x or y, then $\mathcal{A}_g^{t'}$, $t' \geq 1$, replaces g with the zero function, and if $\mathcal{A}_w^t$ sets both x and y, then the gate adversary replaces g with $f(x,y) = 1$. On the other hand, if $\mathcal{A}_w^t$ sets x (resp. y) then $\mathcal{A}_g^t$ substitutes g with $f(x,y) = y$ (resp. $f(x,y) = x$). The other cases consider more complex tampering combinations on input and output wires, as well as, tampering with memory gates, and they can be similarly dealt with.

Theorem 3. *Let* $\mathbf{Env} = (\mathbf{s}, \mathbf{v})$ *be a pair of random variables. For every circuit* C *with gates* $\mathcal{G} = \{\wedge, \vee, \neg\}$, $t \in \mathbb{N}$, *and any v-round PPT wire adversary* $\mathcal{A}_w^t$, *there exists a v-round gate adversary* $\mathcal{A}_g^l$, *for some* $l \in \mathbb{N}$, $l \leq t$, *such that* $\mathcal{A}_w^{C_\mathbf{s}^*(\cdot)}(\mathbf{v})$ *is identically distributed to* $\mathcal{A}_g^{C_\mathbf{s}^*(\cdot)}(\mathbf{v})$.

Gate adversaries are stronger than wire adversaries. Consider a PPT gate adversary $\mathcal{A}_{g}^{t}$, $t = 1$, who tampers with an AND or an OR gate g that consists of two input wires x, y, and a single output wire w, by replacing g with some g' that implements one of the 16 possible binary boolean functions[4] $f_i : \{0,1\}^2 \to \{0,1\}$, $i \in [16]$. For each f_i, $i \in [16]\backslash\{7,10\}$, we give a tampering strategy for $\mathcal{A}_{w}^{t}$, $t \leq 3$, that simulates the tampering effect of f_i, for both AND and OR gates. Here we abbreviate the attacks *set*, *reset* and *toggle* by S, R and T, respectively.

In the following, the variables x, y, w, will denote both circuit wires, as well as their bit-values.

Table 1. All boolean functions from $\{0,1\}^2$ to $\{0,1\}$ and $\mathcal{A}_{w}^{t}$'s tampering strategy

	(0, 0)	(0, 1)	(1, 0)	(1, 1)	Repr. 1	Repr. 2	$\mathcal{A}_{w}^{t}$'s strategy – AND gate	$\mathcal{A}_{w}^{t}$'s strategy – OR gate
f_1	1	1	1	1	$1 \wedge 1$	$1 \vee y$	$((S, x), (S, y))$	(S, x)
f_2	1	1	1	0	$\neg(x \wedge y)$	$(\neg x \vee \neg y)$	(T, w)	$((T, x), (T, y))$
f_3	1	1	0	1	$\neg(x \wedge \neg y)$	$\neg x \vee y$	$((T, y), (T, w))$	(T, x)
f_4	1	1	0	0	$\neg x \wedge 1$	$\neg x \vee 0$	$((T, x), (S, y))$	$((T, x), (R, y))$
f_5	1	0	1	1	$\neg(\neg x \wedge y)$	$x \vee \neg y$	$((T, x), (T, w))$	(T, y)
f_6	1	0	1	0	$\neg y \wedge 1$	$\neg y \vee 0$	$((T, y), (S, x))$	$((T, y), (R, x))$
f_7	1	0	0	1	$x == y$	$(x \wedge y) \vee (\neg x \wedge \neg y)$	—	—
f_8	1	0	0	0	$\neg x \wedge \neg y$	$\neg(x \vee y)$	$((T, x), (T, y))$	(T, w)
f_9	0	1	1	1	$\neg(\neg x \wedge \neg y)$	$x \vee y$	$((T, x), (T, y), (T, w))$	No action
f_{10}	0	1	1	0	$x \neq y$	$(x \wedge \neg y) \vee (\neg x \wedge y)$	—	—
f_{11}	0	1	0	1	$1 \wedge y$	$0 \vee y$	(S, x)	(R, x)
f_{12}	0	1	0	0	$\neg x \wedge y$	$\neg(x \vee \neg y)$	(T, x)	$((T, y), (T, w))$
f_{13}	0	0	1	1	$x \wedge 1$	$x \vee 0$	(S, y)	(R, y)
f_{14}	0	0	1	0	$x \wedge \neg y$	$\neg(\neg x \vee y)$	(T, y)	$((T, x), (T, w))$
f_{15}	0	0	0	1	$x \wedge y$	$\neg(\neg x \vee \neg y)$	No action	$((T, x), (T, y), (T, w))$
f_{16}	0	0	0	0	$0 \wedge y$	$0 \vee 0$	(R, x)	$((R, x), (R, y))$

We observe that there is no tampering strategy for $\mathcal{A}_{w}^{t}$ consisting of *set, reset* or *toggle* attacks on x, y and w, that simulates the tampering effect of $f_7(x, y) = (x == y)$ (NXOR) and $f_{10}(x, y) = (x \neq y)$ (XOR). We use this observation as a key idea behind theorem 4 which provides a "qualitative" separation between the two classes of adversaries.

In the following we prove that for any $n, l, k \in \mathbb{N}$, there exist a circuit $\tilde{C}$ whose size depends on n, l, k, and a PPT adversary $\mathcal{A}_{g}^{n}$, such that for all PPT adversaries $\mathcal{A}_{w}^{t}$, where $t \leq l$, $\mathcal{A}_{w}^{t}$ fails to simulate the view of $\mathcal{A}_{g}^{n}$ while interacting with $\tilde{C}$. Our construction for the counterexample circuit $\tilde{C}$ is presented in Figure 1. It consists of two subcircuits, C_1, C_2, which will be protected against adversaries who tamper with up to l of their wires (l-wire secure). C_1 is the secure transformation of some circuit C'_1 which implements a pseudorandom function $F_{\mathbf{s}}(x)$, together with a counter (Cr_a) and a signing algorithm $(\mathrm{Sign}_{sk'})$ of a digital signature scheme $\Pi =(\mathrm{Gen,Sign,Vrfy})$ with secret key sk', $|sk'| = 2n$. C_1 computes $F_{\mathbf{s}}(c)$ and produces two n-bit strings $\mathbf{s}'_a$ and $\mathbf{s}'_b$. Here, c denotes the current counter value and the computation is based on the secret $\mathbf{s}$. Afterwards, the computation $\sigma_1 =\mathrm{Sign}_{sk'}(c, \mathbf{s}'_a, \mathbf{s}'_b)$ takes place and $\mathbf{m}_1 = ((c, \mathbf{s}'_a, \mathbf{s}'_b), \sigma_1)$ is given as input to C_2, which is the l-wire secure transformation of a circuit C'_2. Furthermore, the two n-bit strings $\mathbf{s}'_a$ and $\mathbf{s}'_b$, are given as input to the AND gates

---

[4] For clarity, and besides f_7 and f_{10}, we give the functions' representation by logic formulas with respect to both $\wedge$ (Repr. 1) and $\vee$ (Repr. 2) operators.

which compute $\mathbf{s}'_a \wedge \mathbf{s}'_b$. The result $\mathbf{z}$ is given as input to C_2 which implements another instantiation of the signing algorithm on input $\mathbf{z}$ and the counter (Cr_2). Eventually C_2 computes $\sigma_2 = \mathsf{Sign}_{sk'}(c, \mathbf{z}, \mathbf{m}_1)$ and outputs $\mathbf{m}_2 = ((c, \mathbf{z}, \mathbf{m}_1), \sigma_2)$. Notice that Cr_1 and Cr_2 produce the same output in every round and their initial value is zero.

In order to construct the l-wire secure circuits C_1, C_2, we employ the compiler of [18]. This compiler receives the security parameter k, the maximum number of tampering attacks l, C'_1, and outputs C_1. In the same way we transform C'_2 to C_2. Since [18] considers *reversible* NOT gates, i.e., gates on which any tampering action on either side propagates to the other side as well, the NOT gates of C_1, C_2, are also reversible. The final circuit $\tilde{C}$ is the composition of C_1, C_2, as shown in Figure 1. Now, the area between C_1 and C_2 (we call this the *critical*

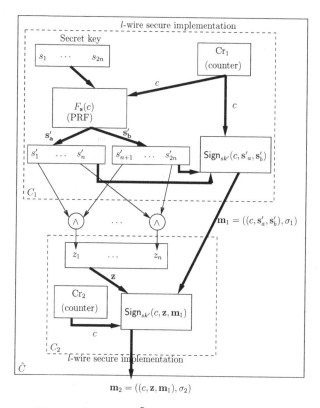

Fig. 1. The circuit $\tilde{C}$ for the separation theorem

area) is the part of C that the gate adversary will tamper with by substituting each AND gate with an XOR gate. This will be the main challenge for the wire adversary and its reason to fail the simulation. Specifically, in order to succeed in the simulation the wire adversary should produce two valid signatures σ_1 and

σ_2 on the messages $(c, \mathbf{s}'_a, \mathbf{s}'_b)$ and $(c, \mathbf{z}, \mathbf{m}_1)$ where c is an integer representing the number of rounds the circuit has been executed and $\mathbf{z} = \mathbf{s}'_a \oplus \mathbf{s}'_b$. Now observe that in normal execution the value $\mathbf{z}$ is defined as $\mathbf{s}'_a \wedge \mathbf{s}'_b$ and it is infeasible for the wire adversary to simulate XOR gates using wire tampering directly in the critical area. We emphasize that even by fully controlling the input $\mathbf{z}$ to the second circuit C_2 (and thus entirely circumventing the difficulty of manipulating the $\wedge$ gates) the wire adversary is insufficient since it will have to provide a valid signature in order to execute a proper C_2 evaluation and the only way such a string can come to its possession is via a previous round of circuit execution; this will make the counter found inside each of the two signatures of the final output to carry different values and thus be detectable as a failed simulation attempt.

Using the above logic we now proceed as follows. For the circuit that we have described we consider a simple one-round gate adversary $\mathcal{A}_{\mathbf{g}}^n$ that tampers with the gates in the critical area transforming them into XOR gates and then returns the output of the circuit. Then we show that there exists a polynomial-time distinguisher that given any wire-adversary operating on the same circuit for any polynomial number of rounds is capable of essentially always telling apart the output of the wire adversary from the output of the gate adversary. The impossibility result follows: the knowledge gained by the gate adversary from interacting with the circuit (just once!) is impossible to be derived by any wire adversary (no matter the number of rounds it is allowed to run the circuit).

In the following, the circuit defined above is called $\tilde{C}_{\tilde{\mathbf{s}}}$ with parameters n, k, $l \in \mathbb{N}$, where $\tilde{\mathbf{s}}$ denotes its secret state. Now we define a distinguisher D w.r.t. $\tilde{C}_{\tilde{\mathbf{s}}}$, which receives the public information $\mathbf{v}$ related to $\tilde{\mathbf{s}}$ and $\mathcal{A}^{\tilde{C}_{\tilde{\mathbf{s}}}^*(\cdot)}$, for some tampering adversary $\mathcal{A}$, and distinguishes the output of the gate adversary from the output of the wire adversary.

Distinguisher $D(\mathbf{v}, \mathbf{m}_2)$ w.r.t. $\tilde{C}_{\tilde{\mathbf{s}}}$:

Distinguisher precondition: The environment variable $\mathbf{Env} = (\tilde{\mathbf{s}}, \mathbf{v})$ where $\tilde{\mathbf{s}}$ determines the secret-state of $\tilde{C}$ is such that $\mathbf{v}$ consists of the public key pk of the digital signature Π and $\tilde{\mathbf{s}}$ contains two copies of the secret key of Π, sk', the secret-key of the PRF and the two counters initialized to 0.

Verification: On input $\mathbf{m}_2 = ((c', \mathbf{d}, \mathbf{m}_1), \sigma_2)$, where $\mathbf{m}_1 = ((c, \mathbf{d}_a, \mathbf{d}_b), \sigma_1)$:

$$\text{if } \mathsf{Vrfy}_{\mathsf{pk}}((c', \mathbf{d}, \mathbf{m}_1), \sigma_2) = 0, \text{ output } 0,$$
$$\text{else if } \mathsf{Vrfy}_{\mathsf{pk}}((c, \mathbf{d}_a, \mathbf{d}_b), \sigma_1)) = 0, \text{ output } 0,$$
$$\text{else if} \quad \mathbf{d} \neq \mathbf{d}_a \oplus \mathbf{d}_b, \quad \text{output } 0,$$
$$\text{else if} \quad c' \neq c, \quad \text{output } 0,$$
$$\text{else} \quad \text{output } 1.$$

Fig. 2. The distinguisher D

Theorem 4. *For all l, $k \in \mathbb{N}$, polynomial in n, for the circuit $\tilde{C}_{\bar{s}}$ of Figure 1 with parameters n, k, l and* **Env** *as in Figure 2, there exists 1-round gate adversary $\mathcal{A}_g^n$ such that for every (multi-round) PPT wire adversary $\mathcal{A}_w^l$ it holds for the distinguisher D defined in Figure 2:*

$$|\Pr[D(\mathbf{v}, \mathcal{A}_w^{\tilde{C}_{\bar{s}}^*(\cdot)}) = 1] - \Pr[D(\mathbf{v}, \mathcal{A}_g^{\tilde{C}_{\bar{s}}^*(\cdot)}) = 1]| = 1 - \mathsf{negl}(n).$$

In the above theorem, the circuit $\tilde{C}_{\bar{s}}$ which distinguishes wire and gate adversaries has a persistent private state which is a random cryptographic key and is operational for an unbounded number of invocations. If one accepts more restricted circuits to be used as counterexamples for separation, specifically circuits that self-destruct after one invocation, we can simplify the separation result via a much simpler circuit. For more details we refer the reader to the paper's full version.

5   Protecting against Gate Adversaries

5.1   Properties That Ensure Security

The following definition generalizes the properties of the compiler presented in [18] and formalizes the functionality for the main parts of the transformed circuit. Definition 10 is a versatile tool for providing tamper-resilient compilers that may be of independent interest. The logic is as follows: we define a (t, k)-secure circuit compiler to be a mapping that produces a circuit accompanied with certain distributions and gate encodings. Specifically the compiler substitutes each wire of the given circuit with a wire-bundle and each gate with a gate that operates over wire-bundles. Within each wire-bundle a specific probability distribution is supposed to exist that encodes probabilistically the 0's and 1's of the original circuit. We note that in the definition below we purposefully leave the exact nature of the class of tampering attacks undetermined.

Definition 10 $((t, k)$**-secure circuit compiler).** *For every t, $k \in \mathbb{N}$, the mapping T over circuits $C \in \mathcal{C}_\mathcal{G}$ with n input bits and m output bits where $\mathcal{G} = \{\wedge, \neg\}$ and n, $m \in \mathbb{N}$ and memory strings $\mathbf{s}$,*

$$(C, \mathbf{s}) \rightarrow \langle \mathcal{D}_0, \mathcal{D}_1, \mathcal{D}_\perp \rangle, \langle C_\wedge, C_\neg \rangle, \left\langle C_{\mathsf{enc}}, C_{\mathsf{dec}}, \hat{C}, \mathbf{s}', C_{\mathsf{cascade}} \right\rangle$$

is a (t, k)-secure circuit compiler if the circuit $C'_{\mathbf{s}'} = C_{\mathsf{dec}} \circ C_{\mathsf{cascade}} \circ \hat{C}_{\mathbf{s}'} \circ C_{\mathsf{enc}}$ realizes the same functionality with $C_{\mathbf{s}}$ and for any PPT adversary $\mathcal{A}$ with strategy $\mathcal{T}$, where $|\mathcal{T}| \leq t$, the following hold

1. *(**Encoding**) $\mathcal{D}_0$, $\mathcal{D}_1$ are distributions of strings in $\{0, 1\}^p$, which correspond to valid encodings of the bits 0 and 1, respectively. The length of the encoding, p, depends on the security parameter k and also on t. Moreover, let S_i be the support set of $\mathcal{D}_i$, for $i \in \{0, 1\}$. Then, the aforementioned distributions must satisfy the following properties:*

(a) $S_0 \cap S_1 = \emptyset$. The set of invalid encodings is $S_\perp = \{0,1\}^p \backslash (S_0 \cup S_1)$.

(b) Each tampering attack against a circuit component that affects a wire-bundle that contains either a sample from $\mathcal{D}_0$ or $\mathcal{D}_1$ may (i) leave the value unchanged, or (ii) produce an element in $S_\perp$. Moreover, there is an efficient way to predict the effect of the tampering (as a distribution over the two events (i) and (ii)).

2. (**Encoder-decoder**) The circuit C_{enc} for each input bit 0 (resp. input bit 1) samples $\mathcal{D}_0$ (resp. $\mathcal{D}_1$). Moreover, for any $\mathbf{x} \in \{0,1\}^n$ the distribution of $C_{\mathsf{enc}}^{\mathcal{T}}(\mathbf{x})$ is predictable given the tampering strategy $\mathcal{T}$ and $\mathbf{x}$. C_{dec} is a deterministic circuit which maps any element of S_0 to 0 and any element of S_1 to 1.

3. (**Circuit gates**) The secret state of C, $\mathbf{s}$, is substituted by $\mathbf{s}'$, where $\mathbf{s}'$ is the encoding of $\mathbf{s}$. Additionally, every gate in C with functionality $f \in \{\wedge, \neg\}$, n' input wires and m' output wires, is being substituted with the circuit C_f with pn' input wires and pm' output wires. Every wire of C is substituted by a bundle of wires $\mathbf{w}$, which carries an element in $S_0 \cup S_1$.
The resuling circuit is $\hat{C}$ and the following hold:

(a) (**Correctness**) For $i,j \in \{0,1\}$, if $\mathbf{x} \sim \mathcal{D}_i$, $\mathbf{y} \sim \mathcal{D}_j$ then it holds that $C_\wedge(\mathbf{x}, \mathbf{y}) \sim \mathcal{D}_{\wedge(i,j)}$ and $C_\neg(\mathbf{x}) \sim \mathcal{D}_{\neg i}$.

(b) (**Error propagation**) If $\mathbf{x} \in S_\perp$ or $\mathbf{y} \in S_\perp$, then $C_\wedge(\mathbf{x}, \mathbf{y}) \sim \mathcal{D}_\perp$ and $C_\wedge^{\mathcal{T}}(\mathbf{x}, \mathbf{y}) \in S_\perp$. The case for $C_\neg$ is similar.

(c) (**Simulatability**) For $i,j \in \{0,1\}$, $\mathbf{x} \sim \mathcal{D}_i$, $\mathbf{y} \sim \mathcal{D}_j$, one of the following must hold: (i) $C_\wedge^{\mathcal{T}}(\mathbf{x}, \mathbf{y}) = C_\wedge(\mathbf{x}, \mathbf{y})$ or (ii) $C_\wedge^{\mathcal{T}}(\mathbf{x}, \mathbf{y}) \in S_\perp$. Moreover, there is an efficient way to predict the effect of the tampering as a distribution over the events (i) and (ii), given $\mathcal{T}$. The case for $C_\neg$ is similar.

4. (**Error propagation & self destruction**) C_{cascade} is a circuit which receives $(\{0,1\}^p)^{m'}$ wires and returns output in $(\{0,1\}^p)^{m'}$, i.e., it receives m' wire-bundles and outputs m' wire-bundles. It is applied on the output wire-bundles of $\hat{C}$ as well as the wire-bundles of $\hat{C}$ that update the circuit's secret state (therefore $m \le m' \le m+q$). Its purpose is to propagate encoding errors and erase the circuit memory (if needed); it works as follows:

(a) If for all $i \in \{1, \ldots, m'\}$, $\mathbf{y}_i \in S_0 \cup S_1$, then (1) for all $i \in \{1, \ldots, m'\}$, the i-th output wire-bundle of $C_{\mathsf{cascade}}(\mathbf{y}_1, \ldots, \mathbf{y}_{m'})$ is equal to $\mathbf{y}_i$, and (2) the output distributions of $C_{\mathsf{cascade}}(\mathbf{y}_1, \ldots, \mathbf{y}_{m'})$ and $C_{\mathsf{cascade}}^{\mathcal{T}}(\mathbf{y}_1, \ldots, \mathbf{y}_{m'})$ are simulatable given $\mathcal{T}$ and $C_{\mathbf{s}}(\mathbf{x})$, where $\mathbf{x} \in \{0,1\}^n$ denotes the circuit input.

(b) If there exists $i \in \{1, \ldots, m'\}$, s.t. $\mathbf{y}_i \in S_\perp$, then, (1) all output wire-bundles of $C_{\mathsf{cascade}}(\mathbf{y}_1, \ldots, \mathbf{y}_{m'}))$ will be distributed according to $\mathcal{D}_\perp$, (2) all output wire-bundles of $C_{\mathsf{cascade}}^{\mathcal{T}}(\mathbf{y}_1, \ldots, \mathbf{y}_{m'})$ will be in $S_\perp$, and (3) the distribution of all output wire-bundles of $C_{\mathsf{cascade}}^{\mathcal{T}}(\mathbf{y}_1, \ldots, \mathbf{y}_{m'})$ will be simulatable given the tampering strategy $\mathcal{T}$ and $C_{\mathbf{s}}(\mathbf{x})$.

5.2   Tamper-Resilient Circuits against Gate Adversaries

Now we give a high level overview of [18] casted as an instance of Definition 10, and we define a gate adversary that compromises its security by attacking randomness gates. Then we prove that by substituting randomness gates with PRNGs, we receive a (t, k)-secure circuit compiler against gate adversaries who tamper with up to t circuit gates. Finally, we prove security for any compiler that satisfies the properties of Definition 10.

A high level description. In [18] the authors consider an encoding in which each input or secret state bit, say x, in the original circuit is encoded into a string of $2k^2t$ bits $(r_1^{2kt}||\ldots||r_k^{2kt})$, where each r_i is a random bit, $i \in [k-1]$, and $r_k = x \oplus r_1 \oplus \ldots \oplus r_{k-1}$. Here, k denotes the security parameter and t is the upper bound on the number of wires that the adversary may tamper with in each computation. The resulting encoding is handled by circuits that implement the functionality of the atomic AND and NOT gates, perform computations over encoded values and satisfy the properties of Definition 10 against wire adversaries. Concretely, let C be a circuit, x an input bit to C and s a secret state bit, and assume that some part of C computes $z = x \wedge s$. According to the aforementioned encoding, the transformed circuit C' encodes x to $(r_1^{2kt}||\ldots||r_k^{2kt})$, where r_i, $i \in [k-1]$, is the output of a randomness gate with fan-out $(2kt)$, [5] and computes $\mathbf{z} = C_\wedge(\mathbf{x}, \mathbf{e})$, using a subcircuit $C_\wedge$ that handles the encoded circuit values and "securely" implements the AND gate. Here, $\mathbf{e}$ and $\mathbf{z}$ denote the encoded version of s and z, respectively, and $\mathbf{z} = (z_1^{2kt}||\ldots||z_k^{2kt})$ is the output of $C_\wedge$ which satisfies $z_i = r_i s_i \oplus \bigoplus_{j \neq i} R_{i,j}$, for $1 \leq i < j \leq k$, $R_{i,j}$ is the output of a randomness gate with fan-out a multiple of $2kt$ and $R_{j,i} = (R_{i,j} \oplus r_i s_j) \oplus r_j b_i$. The number of randomness gates employed by $C_\wedge$ is $\frac{k(k-1)}{2}$. Observe that the value of each wire in the original circuit is shared among k wires and each one of them is replicated $2kt$ times, i.e., each "bundle" consists of k "subbundles" with $2kt$ wires each. The negation of an encoding $\mathbf{e}$ is computed by a circuit $C_\neg$ which consists of $2kt$ NOT gates that simply negate one of the subbundles of $\mathbf{e}$. The whole transformation is the composition of three compilers, and the above description refers to the the second compiler, say T_{rand}. The third compiler replaces randomness gates with circuits that generate pseudo-random bits.

Fact: The compiler of [18] conforms to Definition 10. Let $\mathcal{A}_w^t$ be a wire adversary for C', which is the t-secure transformation of C with respect to T_{rand}, and let s be a secret state bit of C. As we discussed above, s is encoded into $\mathbf{e} = (e_1^{2kt}||\ldots||e_k^{2kt})$, where each e_i, $i \in [k-1]$, is a random bit, and $e_k = s \oplus e_1 \oplus \ldots \oplus e_{k-1}$. Let us consider what happens if $\mathcal{A}_w^t$ tampers with up to t wires of C', where t can be greater than k, and moreover, assume that she tampers with at most $k-1$ different "subbundles" that carry randomized shares of the value s. In such a scenario, the size of each subbundle, which is $2kt$, and the randomization of the carrying values ensure that the adversary may leave the value of each subbundle unchanged or she may alter the value of up to t of its wires, in which case she instantly produces an invalid encoding. Moreover, the effect of the tampering is simulatable in the following way. The simulator simulates the output of the

---

[5] Besides the $2kt$ wires employed by the encoding, some extra copies of r_i^{2kt} are needed for computing r_k^{2kt}, $i \in [k-1]$.

randomness gates by producing her own randomness, and then she decides the effect of the tampering without touching the distribution of s. On the other hand, if the adversary tampers with all subbundles, and since the randomization on the circuit's signals ensures that each tampering attack produces a fault with constant probability, the simulator knows that the probability that none of the attacks produce an invalid encoding is exponentially small in k. Therefore, with all but negligible (in k) probability an error is induced and propagated by the following circuit components: the cascade phase (Property 4) and the circuits that implement the standard gates of the original circuit (Property 3). Since $C_\wedge$ and $C_\neg$ also produce randomized shares, a similar argument gives us simulatability against adversaries who tamper with such encodings.

Reversible gates. As we have already discussed in section 4, [18] assumes reversible NOT gates. As in [18], in this section we will consider *reversible tampering*, i.e., the adversary who tampers with a reversible NOT gate produces a tampering effect that propagates to the gate's incoming wire (note also that w.r.t. NOT gates the wire and gate adversaries are equivalent).

The compiler T_{rand} is insecure against gate tampering. Let x, s, z and $\mathbf{x}$, $\mathbf{e}$, $\mathbf{z}$ be the values defined above and consider an adversary who (*i*) sets to zero the $k-1$ randomness gates $R_{i,i+1}$, for $i \in [k-1]$, that lie on $C_\wedge$, (*ii*) sets to zero the $k-1$ randomness gates that lie on C_{enc} and produce the randomness which is used to encode an input bit x into $\mathbf{x} = (r_1^{2kt}||\dots||r_k^{2kt})$, and (*iii*) tampers with a gate that outputs z_k. Apparently, the $2(k-1)$ attacks on the randomness gates are fully simulatable. Nevertheless, we have $z_i = 0$, for $i \in [k-1]$ and $z_k = x \cdot s$. Hence, in order to simulate the attack on the gate that outputs z_k, the simulator has to make a "guess" on s and the simulation breaks. Notice, that since we consider persistent tampering, an adversary $\mathcal{A}_g^t$, with $t < k$, can land the aforementioned attack in $2\lceil \frac{k}{t} \rceil$ rounds by tampering with t circuit gates in each round. In general any persistent gate adversary may completely eliminate the circuit's randomness, and the second stage compiler T_{rand} of [18] collapses when subjected to this gate adversary attack. Now, we describe how to circumvent such attacks.

In the full version of the paper we describe how to substitute randomness gates with pseudo-random generators, and then we prove that the resulting compiler, named T_{comp}, satisfies the properties of Definition 10 against gate adversaries. Here we give the intuition on why this construction retains its properties against gate adversaries. The key idea is that eliminating randomness gates effectively removes the advantage of the gate adversary. This is the case because all other gates employed by T_{rand} even those whose fan-out is somehow big (and hence may be thought to be higher value targets for a gate attack), lead to different wire-subbundles. Therefore, a gate adversary that induces a fault will spread the fault to multiple circuit gates. The circuit's defense mechanisms of [18] will then be able to detect the invalid encodings with high probability.

Theorem 5. *For every t, $k \in \mathbb{N}$, the compiler T_{comp} is a (t,k)-secure circuit compiler per definition 10 w.r.t. the class of PPT gate attackers $\mathcal{A}_g^t$.*

The final theorem (which may be of independent interest) states that any compiler which satisfies the properties of definition 10, produces tamper resilient circuits with respect to the standard simulation based security definition.

Theorem 6. *Let C_s any boolean circuit, T_{comp} a (t,k)-secure circuit compiler, t, $k \in \mathbb{N}$, and let $C'_{s'}$ be the secure transformation of C_s w.r.t. T_{comp}. Then for every tampering adversary $\mathcal{A}$ for which definition 10 applies, there exists a simulator $\mathcal{S}$ such that $\Delta(\mathcal{S}^{C_s(\cdot)}(\mathbf{v}), \mathcal{A}^{C'^*_{s'}(\cdot)}(\mathbf{v}))$ is negligible in k.*

The proofs of the above theorems are given in the full version of this paper.

Acknowledgements. This research was supported by ERC project CODAM-ODA. The second author was also supported by EU FP7 project UaESMC.

References

1. Anderson, R., Kuhn, M.: Tamper resistance-a cautionary note. In: Proceedings of the Second Usenix Workshop on Electronic Commerce, vol. 2, pp. 1–11 (1996)
2. Barak, B., Goldreich, O., Impagliazzo, R., Rudich, S., Sahai, A., Vadhan, S., Yang, K.: On the (Im)possibility of obfuscating programs. In: Kilian, J. (ed.) CRYPTO 2001. LNCS, vol. 2139, pp. 1–18. Springer, Heidelberg (2001)
3. Ben-Sasson, E., Goldreich, O., Harsha, P., Sudan, M., Vadhan, S.: Robust pcps of proximity, shorter pcps and applications to coding. In: Proceedings of the Thirty-Sixth Annual ACM Symposium on Theory of Computing, pp. 1–10. ACM (2004)
4. Biham, E., Shamir, A.: Differential fault analysis of secret key cryptosystems. In: Kaliski Jr., B.S. (ed.) CRYPTO 1997. LNCS, vol. 1294, pp. 513–525. Springer, Heidelberg (1997)
5. Blömer, J., Seifert, J.-P.: Fault based cryptanalysis of the advanced encryption standard (AES). In: Wright, R.N. (ed.) FC 2003. LNCS, vol. 2742, pp. 162–181. Springer, Heidelberg (2003)
6. Boneh, D., DeMillo, R.A., Lipton, R.J.: On the importance of checking cryptographic protocols for faults. In: Fumy, W. (ed.) EUROCRYPT 1997. LNCS, vol. 1233, pp. 37–51. Springer, Heidelberg (1997)
7. Boneh, D., DeMillo, R.A., Lipton, R.J.: On the importance of eliminating errors in cryptographic computations. Journal of cryptology 14(2), 101–119 (2001)
8. Choi, S.G., Kiayias, A., Malkin, T.: BiTR: Built-in tamper resilience. In: Lee, D.H., Wang, X. (eds.) ASIACRYPT 2011. LNCS, vol. 7073, pp. 740–758. Springer, Heidelberg (2011)
9. Dachman-Soled, D., Kalai, Y.T.: Securing circuits against constant-rate tampering. In: Safavi-Naini, R., Canetti, R. (eds.) CRYPTO 2012. LNCS, vol. 7417, pp. 533–551. Springer, Heidelberg (2012)
10. Dziembowski, S., Pietrzak, K., Wichs, D.: Non-malleable codes. In: ICS, pp. 434–452. Tsinghua University Press (2010)
11. Faust, S., Pietrzak, K., Venturi, D.: Tamper-proof circuits: How to trade leakage for tamper-resilience. In: Aceto, L., Henzinger, M., Sgall, J. (eds.) ICALP 2011, Part I. LNCS, vol. 6755, pp. 391–402. Springer, Heidelberg (2011)
12. Gács, P., Gál, A.: Lower bounds for the complexity of reliable boolean circuits with noisy gates. IEEE Transactions on Information Theory 40(2), 579–583 (1994)

13. Gal, A., Szegedy, M.: Fault tolerant circuits and probabilistically checkable proofs. In: Proceedings of Tenth Annual IEEE Structure in Complexity Theory Conference, pp. 65–73. IEEE (1995)
14. Gandolfi, K., Mourtel, C., Olivier, F.: Electromagnetic analysis: Concrete results. In: Koç, Ç.K., Naccache, D., Paar, C. (eds.) CHES 2001. LNCS, vol. 2162, pp. 251–261. Springer, Heidelberg (2001)
15. Gennaro, R., Lysyanskaya, A., Malkin, T., Micali, S., Rabin, T.: Algorithmic tamper-proof (ATP) security: Theoretical foundations for security against hardware tampering. In: Naor, M. (ed.) TCC 2004. LNCS, vol. 2951, pp. 258–277. Springer, Heidelberg (2004)
16. Goldreich, O., Micali, S., Wigderson, A.: How to play any mental game. In: Proceedings of the Nineteenth Annual ACM Symposium on Theory of Computing, pp. 218–229. ACM (1987)
17. Govindavajhala, S., Appel, A.W.: Using memory errors to attack a virtual machine. In: Proceedings of the 2003 Symposium on Security and Privacy, pp. 154–165. IEEE (2003)
18. Ishai, Y., Prabhakaran, M., Sahai, A., Wagner, D.: Private circuits II: Keeping secrets in tamperable circuits. In: Vaudenay, S. (ed.) EUROCRYPT 2006. LNCS, vol. 4004, pp. 308–327. Springer, Heidelberg (2006)
19. Ishai, Y., Sahai, A., Wagner, D.: Private circuits: Securing hardware against probing attacks. In: Boneh, D. (ed.) CRYPTO 2003. LNCS, vol. 2729, pp. 463–481. Springer, Heidelberg (2003)
20. Katz, J., Lindell, Y.: Introduction to modern cryptography. Chapman & Hall (2008)
21. Kelsey, J., Schneier, B., Wagner, D., Hall, C.: Side channel cryptanalysis of product ciphers. In: Quisquater, J.-J., Deswarte, Y., Meadows, C., Gollmann, D. (eds.) ESORICS 1998. LNCS, vol. 1485, pp. 97–110. Springer, Heidelberg (1998)
22. Kocher, P., Jaffe, J., Jun, B.: Differential power analysis. In: Wiener, M. (ed.) CRYPTO 1999. LNCS, vol. 1666, pp. 388–397. Springer, Heidelberg (1999)
23. Kocher, P.C.: Timing attacks on implementations of Diffie-Hellman, RSA, DSS, and other systems. In: Koblitz, N. (ed.) CRYPTO 1996. LNCS, vol. 1109, pp. 104–113. Springer, Heidelberg (1996)
24. Kuhn, M.G., Anderson, R.J.: Soft tempest: Hidden data transmission using electromagnetic emanations. In: Aucsmith, D. (ed.) IH 1998. LNCS, vol. 1525, pp. 124–142. Springer, Heidelberg (1998)
25. Liu, F.-H., Lysyanskaya, A.: Algorithmic tamper-proof security under probing attacks. In: Garay, J.A., De Prisco, R. (eds.) SCN 2010. LNCS, vol. 6280, pp. 106–120. Springer, Heidelberg (2010)
26. Micali, S., Reyzin, L.: Physically observable cryptography. In: Naor, M. (ed.) TCC 2004. LNCS, vol. 2951, pp. 278–296. Springer, Heidelberg (2004)
27. Pippenger, N.: On networks of noisy gates. In: 26th Annual Symposium on Foundations of Computer Science, pp. 30–38. IEEE (1985)
28. Quisquater, J.-J., Samyde, D.: Electromagnetic analysis (EMA): Measures and counter-measures for smart cards. In: Attali, S., Jensen, T. (eds.) E-smart 2001. LNCS, vol. 2140, pp. 200–210. Springer, Heidelberg (2001)
29. Rao, J.R., Rohatgi, P.: Empowering side-channel attacks. IACR ePrint, 37 (2001)
30. Skorobogatov, S.P., Anderson, R.J.: Optical fault induction attacks. In: Kaliski Jr., B.S., Koç, Ç.K., Paar, C. (eds.) CHES 2002. LNCS, vol. 2523, pp. 2–12. Springer, Heidelberg (2003)

Efficient General-Adversary Multi-Party Computation

Martin Hirt and Daniel Tschudi*

ETH Zurich
{hirt,tschudid}@inf.ethz.ch

Abstract. Secure multi-party computation (MPC) allows a set $\mathcal{P}$ of n players to evaluate a function f in presence of an adversary who corrupts a subset of the players. In this paper we consider active, general adversaries, characterized by a so-called adversary structure $\mathcal{Z}$ which enumerates all possible subsets of corrupted players. In particular for small sets of players general adversaries better capture real-world requirements than classical threshold adversaries.

Protocols for general adversaries are "efficient" in the sense that they require $|\mathcal{Z}|^{\mathcal{O}(1)}$ bits of communication. However, as $|\mathcal{Z}|$ is usually very large (even exponential in n), the exact exponent is very relevant. In the setting with perfect security, the most efficient protocol known to date communicates $\mathcal{O}(|\mathcal{Z}|^3)$ bits; we present a protocol for this setting which communicates $\mathcal{O}(|\mathcal{Z}|^2)$ bits. In the setting with statistical security, $\mathcal{O}(|\mathcal{Z}|^3)$ bits of communication is needed in general (whereas for a very restricted subclass of adversary structures, a protocol with communication $\mathcal{O}(|\mathcal{Z}|^2)$ bits is known); we present a protocol for this setting (without limitations) which communicates $\mathcal{O}(|\mathcal{Z}|^1)$ bits.

Keywords: Secure Multiparty Computation, General Adversaries, Efficiency.

1 Introduction

Secure Multi-Party Computation. Secure Multi-Party Computation (MPC) allows a set $\mathcal{P}$ of n players to securely evaluate a function f even when a subset of the players is corrupted by a central adversary. MPC was introduced by Yao [Yao82]. A first solution (with computational security) was given by Goldreich, Micali, and Wigderson [GMW87]. Later solutions [BGW88, CCD88, RB89] provide statistical and even perfect security. All these protocols consider threshold adversaries (characterized by an upper bound t on the number of corrupted parties). This was generalized in [HM00] by considering so-called general adversaries, characterized by an adversary structure $\mathcal{Z} = \{Z_1, \ldots, Z_\ell\}$, which enumerates all possible subsets of corrupted players.

In the setting with perfect active security, MPC is achievable if and only if $t < \frac{n}{3}$, respectively $\mathcal{Q}^3(\mathcal{P}, \mathcal{Z})$ (the union of no *three* sets in $\mathcal{Z}$ covers $\mathcal{P}$). In the setting with statistical or cryptographic active security, MPC is achievable if and only if $t < \frac{n}{2}$, respectively $\mathcal{Q}^2(\mathcal{P}, \mathcal{Z})$ (the union of no *two* sets in $\mathcal{Z}$ covers $\mathcal{P}$).

---
* Research supported in part by the Swiss National Science Foundation (SNF), project no. 200020-132794.

K. Sako and P. Sarkar (Eds.) ASIACRYPT 2013, Part II, LNCS 8270, pp. 181–200, 2013.

Threshold vs. General Adversaries. Clearly, general adversaries are more flexible, which is relevant in particular when the set of players is not very large. However, general-adversary protocols are typically by orders of magnitude less efficient than threshold protocols; more specifically, threshold protocols usually communicate $\text{Poly}(n)$ bits per multiplication, whereas general-adversary protocols require $\text{Poly}(|Z|)$ bits. As typically $|Z|$ is exponential in n, this is a huge drawback. However, in some scenarios (e.g. with very different types of players), threshold protocols are not applicable, and general-adversary protocols must be used. In these settings, the concrete communication complexity of the general-adversary protocol is highly relevant: For example for $n = 25$, $|Z|$ is expected to be around one million, and a protocol communicating $|Z| \cdot \text{Poly}(n)$ might be acceptable, whereas a protocol communicating $|Z|^3 \cdot \text{Poly}(n)$ might be useless.

Contributions. In the statistically-secure model, one can tolerate at most adversary structures satisfying $\mathcal{Q}^2(\mathcal{P}, \mathcal{Z})$. The most efficient protocol known to date, which is also optimal in terms of resilience, requires $|\mathcal{Z}|^3 \cdot \text{Poly}(n, \kappa)$ bits of communication (where κ is the security parameter) [Mau02, HMZ08]. There exists a protocol with communication complexity of $|\mathcal{Z}|^2 \cdot \text{Poly}(n, \kappa)$ [PSR03]. But this results is non-optimal in terms of resilience, as it tolerates only adversaries satisfying $\mathcal{Q}^3$.

Using a new approach for multiplication, we construct a protocol communicating $|\mathcal{Z}| \cdot \text{Poly}(n, \kappa)$ bits and tolerating $\mathcal{Q}^2$ adversary structures. This protocol is optimal both in terms of overall efficiency and resilience. We stress that even with cryptographic security, $\mathcal{Q}^2$ is necessary and complexity linear in $|\mathcal{Z}|$ is required at least with respect to the computation (see [Hir01]).

Furthermore, we present a perfectly secure protocol (with no error probability) with communication complexity of $|\mathcal{Z}|^2 \cdot \text{Poly}(n)$. It is optimal in terms of resilience ($\mathcal{Q}^3$) and also the most efficient protocol up to date in the model with perfect security.

Table 1. Communication complexity of different protocols

Setting	Cond.	Bits / Mult.	Reference
passive perfect	$\mathcal{Q}^2$	$|\mathcal{Z}| \cdot \text{Poly}(n)$	[Mau02]
active perfect	$\mathcal{Q}^3$	$|\mathcal{Z}|^3 \cdot \text{Poly}(n)$	[Mau02]
active perfect	$\mathcal{Q}^3$	$|\mathcal{Z}|^2 \cdot \text{Poly}(n)$	our result
active unconditional	$\mathcal{Q}^2$	$|\mathcal{Z}|^3 \cdot \text{Poly}(n, \kappa)$	[Mau02]/[HMZ08]
active unconditional	$\mathcal{Q}^3$	$|\mathcal{Z}|^2 \cdot \text{Poly}(n, \kappa)$	[PSR03]
active unconditional	$\mathcal{Q}^2$	$|\mathcal{Z}| \cdot \text{Poly}(n, \kappa)$	our result

2   Preliminaries

Players and Computation. Let $\mathcal{P} = \{P_1, \ldots, P_n\}$ be a set of n players. The players in $\mathcal{P}$ want to compute a function f over some finite field $\mathbb{F}$. The function is specified by a circuit $\mathcal{C}$ consisting of input, output, random, addition, and multiplication gates.

In an ideal world, a trusted party does all the computation. In the real world, players are connected by a complete network of secure (private and authentic) synchronous channels. There exist authenticated broadcast channels (they can be simulated by the players, see e.g. [FM98] or [PW96]). In order to compute the function f, the players simulate the trusted party by using some MPC-protocol Π.

Adversary and Adversary Structure. Dishonesty is modeled in terms of a central adversary $\mathcal{A}$ who corrupts players. During the execution of the protocol the adversary can access the internal state of corrupted players and make them deviate from the protocol at will. We allow that the adversary is computationally unbounded. Before the execution of the protocol the adversary has to specify the players he wants to corrupt. His choice is limited by means of an adversary structure $\mathcal{Z} = \{Z_1, \ldots, Z_\ell\} \subseteq 2^{\mathcal{P}}$, i.e. all corrupted players must be part of an adversary set in $\mathcal{Z}$. We denote the chosen set by Z^*. Note that Z^* is not known to the honest players and is solely used for ease of notation. We say that $\mathcal{Z}$ satisfies the $\mathcal{Q}^k(\mathcal{P}, \mathcal{Z})$ property if $\mathcal{P} \not\subseteq Z_1 \cup \cdots \cup Z_k \ \forall Z_1, \ldots, Z_k \in \mathcal{Z}$.

Security. We say a protocol is $\mathcal{Z}$-secure if anything the adversary achieves during the execution of the protocol can be achieved in the ideal world as well. More precisely, for every adversary in the real world there exists an adversary in the ideal world such that both the information the adversary gets and the output of honest players are statistically indistinguishable for perfect security respectively statistically close for unconditional security. The main result from [HM97] states that $\mathcal{Q}^3(\mathcal{P}, \mathcal{Z})$ resp. $\mathcal{Q}^2(\mathcal{P}, \mathcal{Z})$ are the necessary and sufficient conditions for the existence of perfectly resp. unconditionally $\mathcal{Z}$-secure protocols considering active adversaries. For simplicity we assume that all messages sent during the execution of Π are from the right domain. If a player receives a message where this is not the case, he replaces it with an arbitrary element from the right domain. If a player receives an unexpected message, he ignores it.

3 Perfect Protocol

In this section we present a perfectly $\mathcal{Z}$-secure protocol for an arbitrary adversary structure $\mathcal{Z}$ satisfying the $\mathcal{Q}^3$ property. The communication complexity of the protocol is quadratic in $|\mathcal{Z}|$. The efficiency gain is due to an improved multiplication protocol. The sharing is (up to presentation) the same as in [Mau02].

3.1 Secret Sharing

Secret sharing allows a player to distribute a secret value among the player set, such that only qualified subsets of players are able reconstruct it. The secret sharing used for our protocol is based on the one from [Mau02] / [BFH+08]. It is characterized by a *sharing specification* $\mathbb{S} = (S_1, \ldots, S_h)$, which is a tuple of subsets of $\mathcal{P}$.

Definition 1. *A value s is* shared *with respect to sharing specification* $\mathbb{S} = (S_1, \ldots, S_h)$ *if the following holds:*

a) *There exist shares* $s_1, \ldots, s_h$ *such that* $s = \sum_{q=1}^{h} s_q$

b) *Each* s_q *is known to every (honest) player in* S_q

We denote the sharing of a value s by $[s]$ and use $[s]_q$ as notation for s_q, the q-th share. A sharing specification $\mathbb{S} = (S_1, \ldots, S_h)$ is called $\mathcal{Z}$-*private* if for every $Z \in \mathcal{Z}$ there is an $S \in \mathbb{S}$ such that $Z \cap S = \emptyset$. A sharing specification $\mathbb{S} = (S_1, \ldots, S_h)$ and an adversary structure $\mathcal{Z}$ satisfy $Q^k(\mathbb{S}, \mathcal{Z})$ if $S \not\subseteq Z_1 \cup \cdots \cup Z_k \quad \forall Z_1, \ldots, Z_k \in \mathcal{Z}$ $S \in \mathbb{S}$. If $\mathbb{S}$ is $\mathcal{Z}$-private, a sharing $[s]$ does not leak information to the adversary, as all shares known by the adversary are statistically independent of s. The players can compute a sharing of any linear combination of shared values (with respect to a sharing specification $\mathbb{S}$) by locally computing the linear combination of their shares. This property is called the linearity of the sharing. The following protocol Share allows a dealer P_D to correctly share value s among the players in $\mathcal{P}$.

Protocol Share$(\mathcal{P}, \mathcal{Z}, \mathbb{S}, P_D, s)$ [Mau02]

0: The dealer P_D takes s as input.

1: P_D splits s into random shares $s_1, \ldots, s_{|\mathbb{S}|}$ subject to $s = \sum_{q=1}^{|\mathbb{S}|} s_q$.

2: **for all** $q \in \{1, \ldots, |\mathbb{S}|\}$ **do**

3:     P_D sends s_q to every player in S_q.

4:     Each player in S_q forwards the received value to every player in S_q.

5:     Each player in S_q checks that the received values are all the same and broadcasts OK, or NOK accordingly.

6:     If a player in S_q broadcast NOK, the dealer broadcasts s_q and the players in S_q take this value (resp. some default value if the dealer does not broadcast) as share. Otherwise every player in S_q takes the value he received in Step 3 as share.

7: **end for**

8: The players in $\mathcal{P}$ collectively output $[s]$.

Lemma 1. *For any adversary structure* $\mathcal{Z}$ *the protocol* Share$(\mathcal{P}, \mathcal{Z}, \mathbb{S}, P_D, s)$ *securely computes a sharing* $[s']$. *For honest* P_D *it holds that* $s' = s$. *The protocol communicates at most* $|\mathbb{S}| (n^2 + n) \log |\mathbb{F}|$ *bits and broadcasts at most* $|\mathbb{S}| (\log |\mathbb{F}| + n)$ *bits.*

Proof. Correctness: For each s_q either all the honest players in S_q hold the same value after Step 3, or one of them complains and they receive a consistent value in Step 6. Hence the protocol outputs a (consistent) sharing $[s']$. If the dealer is honest he is able to ensure in Steps 3 and 6 that the honest players use the intended value for s_q such that $s = s'$. Privacy: Let the dealer be honest, as otherwise secrecy is trivially fulfilled. All a player learns beyond his designated output are values broadcast in Step 6. However this does not violate secrecy as these values are already known to the adversary (from Step 3). Complexity: For each share at most $n + n^2$ values are sent and at most $n + \log |\mathbb{F}|$ bits broadcast. $\quad\square$

For publicly known value s the players can invoke DefaultShare to get a sharing $[s]$ without having to communicate.

Protocol DefaultShare$(\mathcal{P}, \mathcal{Z}, \mathbb{S}, s)$

0: Every player takes s as input.

1: The share s_1 is set to s and all other shares are set to 0.

2: The players in $\mathcal{P}$ collectively output $[s]$.

Lemma 2. DefaultShare$(\mathcal{P}, \mathcal{Z}, \mathbb{S}, s)$ *securely computes a sharing $[s]$ where s is a publicly known value. The protocol does not communicate.*

Proof. Correctness: In Step 1 every honest player in S_q takes the same value for share s_q. As the sum of all shares is s, the protocol outputs a consistent sharing $[s]$. Privacy: During the protocol no communication occurs, hence the adversary does not obtain new information.   □

The protocol ReconstructShare allows the reconstruction of a share $[s]_q$ to the players in some set R. This implies that the players can reconstruct a shared value $[s]$ by invoking ReconstructShare for each share.

Protocol ReconstructShare$(\mathcal{P}, \mathcal{Z}, \mathbb{S}, [s]_q, R)$

0: The players in S_q take the share $[s]_q$ as input.

1: Every player P_i in S_q sends $[s]_q$ to every player in R.

2: For each player $P_j \in R$ let $v_{j,i}$ be the value received from P_i. Then P_j outputs some value v_j such that there exists a $Z \in \mathcal{Z}$ with $v_{j,i} = v_j$ for all $P_i \in S_q \setminus Z$.

Lemma 3. *If S_q and $\mathcal{Z}$ satisfy $Q^2(S_q, \mathcal{Z})$, the protocol* ReconstructShare *securely reconstructs the share $[s]_q$ to the players in R, such that every (honest) player outputs $[s]_q$. The protocol communicates at most $n^2 \log |\mathbb{F}|$ bits.*

Proof. Correctness: In Step 1 all honest player will send the same value $[s]_q$, which is a suitable choice for v_j for an (honest) player $P_j \in R$ in Step 2. For the sake of contradiction suppose there exist two values $v_1 \neq v_2$ with corresponding $Z_1, Z_2 \in \mathcal{Z}$ such that the condition of Step 2 holds for both of them. Hence $(S_q \setminus Z_1) \cap (S_q \setminus Z_2) = \emptyset$ and thus $S_q \subseteq Z_1 \cup Z_2$ which contradicts $Q^2(S_q, \mathcal{Z})$. Therefore every honest players outputs the value $[s]_q$. Privacy: The adversary learns at most $[s]_q$ (if a malicious player is part of R). Complexity: Each player in S_q sends his value to at most n players.   □

Protocol Reconstruct$(\mathcal{P}, \mathcal{Z}, \mathbb{S}, [s], R)$ [Mau02]

0: The players in $\mathcal{P}$ take collectively $[s]$ as input.

1: $\forall q \in \{1, \ldots, |\mathbb{S}|\}$ protocol ReconstructShare$(\mathcal{P}, \mathcal{Z}, \mathbb{S}, [s]_q, R)$ is invoked.

2: The players in R locally sum up the obtained shares and output the sum s.

Lemma 4. *If $\mathbb{S}$ and $\mathcal{Z}$ satisfy $Q^2(\mathbb{S}, \mathcal{Z})$ and $[s]$ is a sharing of the value s, then* Reconstruct$(\mathcal{P}, \mathcal{Z}, \mathbb{S}, [s], R)$ *securely reconstructs s to the players in R. The protocol communicates at most $|\mathbb{S}| n^2 \log |\mathbb{F}|$ bits.*

Proof. Correctness and privacy follow directly from Lemma 3. As ReconstructShare is invoked $|\mathbb{S}|$ times the complexity follows as well.   □

3.2 Multiplication

We present a protocol for the perfectly-secure computation of the (shared) product of two shared values $[a]$ and $[b]$ (with respect to a sharing specification $\mathbb{S}$). Along the lines of [Mau02] the fundamental idea of multiplication is to assign each local product $a_p b_q$ to a player in $S_p \cap S_q$, who computes and shares his designated products. The sum of all these sharings is a sharing of ab as long as no player actively cheated. So each player is mapped to a collection of local products, formalized by a function $I : [n] \to 2^{\{(p,q) \mid 1 \le p,q \le |\mathcal{Z}|\}}$ with the constraint that $\forall (p,q) \exists! i$ such that $(p,q) \in I(i)$. W.l.o.g let $I(i) := \{(p,q) \mid P_i = \min_P \{P \in S_p \cap S_q\}\}$. We first show an optimistic multiplication protocol which takes an additional parameter Z and computes the correct product if the actual adversary set Z^* is a subset of Z. In this protocol local products are assigned to players in $\mathcal{P} \setminus Z$ only. Clearly this is possible if and only if for each local product a player in $\mathcal{P} \setminus Z$ holds both involved shares, i.e. $\forall S_p, S_q \in \mathbb{S} : S_p \cap S_q \setminus Z \ne \emptyset$. So for each $Z \in \mathcal{Z}$ let I_Z be a mapping as above with the additional constraint that $\forall P_i \in Z\; I_Z(i) = \emptyset$. Without loss of generality, let $I_Z(i) := \{(p,q) \mid P_i = \min_P \{P \in S_p \cap S_q \setminus Z\}\}$.

Protocol OptimisticMult($\mathcal{P}, \mathcal{Z}, \mathbb{S}, [a], [b], Z$)

0: The players in $\mathcal{P}$ take collectively $[a], [b]$ and Z as input.

1:

    a) Each player $P_i \in \mathcal{P} \setminus Z$ (locally) computes his designated products and shares the sum $c_i = \sum_{(p,q) \in I_Z(i)} a_p b_q$.

    b) For each $P_i \in Z$ DefaultShare($\mathcal{P}, \mathcal{Z}, \mathbb{S}, 0$) is invoked to share $c_i = 0$.

2: The players collectively output $([c_1], \ldots, [c_n])$ and $[c] = \sum_{i=1}^n [c_i]$.

Lemma 5. *Let* $Z \subseteq \mathcal{P}$ *such that* $\forall S_p, S_q \in \mathbb{S} : S_p \cap S_q \setminus Z \ne \emptyset$. *Then the protocol* OptimisticMult *securely computes sharings* $[c], ([c_1], \ldots, [c_n])$. *If no player in* $\mathcal{P} \setminus Z$ *actively cheats (in particular, if* $Z^* \subseteq Z$), *then* $\forall i\; c_i = \sum_{(p,q) \in I_Z(i)} a_p b_q$ *and* $c = ab$. *The protocol communicates at most* $\mathcal{O}(|\mathbb{S}| \, n^3 \log |\mathbb{F}|)$ *bits and broadcasts at most* $\mathcal{O}(|\mathbb{S}| \, (n \log |\mathbb{F}| + n^2))$ *bits.*

Proof. Correctness: The properties of the sharing protocol guarantee that the outputs are valid sharings. If none of the players in $\mathcal{P} \setminus Z$ cheated actively, it holds for each P_i that $c_i = \sum_{(p,q) \in I_Z(i)} a_p b_q$. The condition $\forall S_p, S_q \in \mathbb{S} : S_p \cap S_q \setminus Z \ne \emptyset$ guarantees that $ab = \sum_{i=1}^n \sum_{(p,q) \in I_Z(i)} a_p b_q$. Hence it follows that $c = ab$. Privacy / Complexity: Follow directly from Lemmas 1 and 2. $\qquad\square$

As the players do not know the actual adversary set Z^*, they invoke OptimisticMult once for each set $Z \in \mathcal{Z}$ (Step 1 of the Multiplication protocol). This guarantees that at least one of the resulting sharings is correct. By comparing them the players can determine whether cheating occurred (Step 2 of the Multiplication protocol). If all sharings are equal, no cheating occurred and any of the sharings can serve as sharing of the product. Otherwise at least one player cheated. In this case the (honest) players can identify him and remove all sharings where he was involved in computation, as these sharings are potentially tampered (Step 3 of the Multiplication protocol). This checking and removing is repeated until all remaining sharing are equal (and hence correct). As the identification of cheaters does not reveal any information to the adversary, Multiplication allows the secure computation of the product of two shared secret values.

Protocol Multiplication($\mathcal{P}, \mathcal{Z}, \mathbb{S}, [a], [b]$)

0: Set $M = \emptyset$.

1: Invoke OptimisticMult($\mathcal{P}, \mathcal{Z}, \mathbb{S}, [a], [b], Z$) to compute $([c_1^{(Z)}], \dots, [c_n^{(Z)}])$ and $[c^{(Z)}]$ for each $Z \in \mathcal{Z}$.

2: Set $\mathcal{Z}_M := \{Z \in \mathcal{Z} \mid M \subseteq Z\}$, fix some $\widetilde{Z} \in \mathcal{Z}_M$ and reconstruct the differences $[c^{(\widetilde{Z})}] - [c^{(Z)}] \; \forall Z \in \mathcal{Z}_M$.

3: If all differences are zero, output $[c^{(\widetilde{Z})}]$ as sharing of the product.

Otherwise let $([d_1], \dots, [d_n]) := ([c_1^{(\widetilde{Z})}], \dots, [c_n^{(\widetilde{Z})}])$, $([e_1], \dots, [e_n]) := ([c_1^{(Z)}], \dots, [c_n^{(Z)}])$, $D := I_{\widetilde{Z}}$ and $E := I_Z$, where $[c^{(\widetilde{Z})}] - [c^{(Z)}] \neq 0$.

   a) Each P_i shares the $2n$ values $d_{i,j} = \sum_{(p,q) \in D(i) \cap E(j)} a_p b_q$ and $e_{i,j} = \sum_{(p,q) \in E(i) \cap D(j)} a_p b_q$

   b) For each player P_i reconstruct the differences $[d_i] - \sum_{j=1}^n [d_{i,j}]$ and $[e_i] - \sum_{j=1}^n [e_{i,j}]$. If one of them is non-zero set $M \leftarrow M \cup \{P_i\}$ and continue at Step 2.

   c) For each (ordered) pair (P_i, P_j) of players reconstruct the difference $[d_{i,j}] - [e_{j,i}]$. If it is non-zero, reconstruct $[d_{i,j}], [e_{j,i}]$ and all shares $\{a_p, b_q \mid (p,q) \in D(i) \cap E(j)\}$ to find the cheater $P \in \{P_i, P_j\}$. Set $M \leftarrow M \cup \{P\}$ and continue at Step 2.

Lemma 6. *If $\mathbb{S}$ and $\mathcal{Z}$ satisfy $Q^2(\mathbb{S}, \mathcal{Z})$ the protocol Multiplication yields a sharing $[c] = [ab]$. No information is leaked to the adversary. Multiplication communicates at most $\mathcal{O}(|\mathbb{S}|\,|\mathcal{Z}|\,n^3 \log |\mathbb{F}| + |\mathbb{S}|\,n^5 \log |\mathbb{F}|)$ bits and broadcasts at most $\mathcal{O}(|\mathbb{S}|\,|\mathcal{Z}|\,(n \log |\mathbb{F}| + n^2) + |\mathbb{S}|\,(n^3 \log |\mathbb{F}| + n^4))$ bits.*

Proof. Correctness: By invoking OptimisticMult for each $Z \in \mathcal{Z}$ it holds for Z^* that $[c^{(Z^*)}] = [ab]$ (due to $Q^2(\mathbb{S}, \mathcal{Z})$ $\forall S_p, S_q \in \mathbb{S} : S_p \cap S_q \setminus Z \neq \emptyset$ holds). If for every $Z \in \mathcal{Z}_M$ the difference in Step 2 is zero, then $[c^{(Z)}] = [ab]$ $\forall Z \in \mathcal{Z}_M$ ($M = \emptyset$ at the beginning). Hence the protocol terminates successfully outputting a sharing of ab. Otherwise there exists $[c^{(\widetilde{Z})}] - [c^{(Z)}] \neq 0$ and thus $\sum_{i=1}^n [d_i] \neq \sum_{i=1}^n [e_i]$. In Step 3a) each player is supposed to share a partition of his shares. Hence one of the following cases must occur: There exists a player P_i such that $[d_i] \neq \sum_{j=1}^n [d_{i,j}]$ or $[e_i] \neq \sum_{j=1}^n [e_{i,j}]$. Or there exists a pair of players (P_i, P_j) such that $[d_{i,j}] \neq [e_{j,i}]$. In the first case P_i will be detected as cheater in Step 3b). In the second case the cheater will be detected in Step 3c). In both cases $M \subseteq \mathcal{P}$ is strictly increased, hence the protocol will terminate after at most n iterations. It holds that $M \subseteq Z^*$ and thus $Z^* \in \mathcal{Z}_M$. Therefore the correct sharing $[c^{(Z^*)}]$ is always used in Step 2 and the protocol will output the correct result. *Privacy*: By the properties of the sharing scheme and Lemma 5 the invocation of Share, Reconstruct, OptimisticMult does not violate privacy. The adversary learns the differences reconstructed in Steps 2 and 3 of Multiplication, which are all zero unless the adversary cheats. In case of cheating the reconstructed values depends solely on the inputs of the adversary and are thus already known to him, thus privacy is not violated. All values further reconstructed in Step 3c) are known to the adversary before, as either P_i or P_j is corrupted. *Complexity*: Follows from Lemmas 1, 4 and 5 by counting the number of invocations of the corresponding sub-protocols. □

3.3 MPC Protocol

Combining Share, Reconstruct, and Multiplication the players can securely compute a circuit $\mathcal{C}$ over $\mathbb{F}$, where all intermediate values are shared according to Definition 1.

Protocol MPC$(\mathcal{P}, \mathcal{Z}, \mathcal{C})$

0: The players take $\mathbb{S} := \{\mathcal{P} \setminus Z | Z \in \mathcal{Z}\}$ as sharing specification.

1: For every gate of C being evaluated do the following:
  - *Input gate for P_D:* Share$(\mathcal{P}, \mathcal{Z}, \mathbb{S}, P_D, s)$ is invoked to share s, where P_D is the input-giving player.
  - *Linear gate:* The linear combination of the corresponding shares is computed locally using the linearity of the sharing.
  - *Random gate:* Each player shares a random value. The sum of these values is used as output of the gate.
  - *Multiplication gate:* Multiplication$(\mathcal{P}, \mathcal{Z}, \mathbb{S}, [a], [b])$ is used to multiply $[a]$ and $[b]$.
  - *Output gate:* The players invoke Reconstruct$(\mathcal{P}, \mathcal{Z}, \mathbb{S}, [s], R)$ to reconstruct the sharing $[s]$ to players in R.

Theorem 1. *Let $\mathcal{P}$ be a set of n players, $\mathcal{C}$ a circuit over $\mathbb{F}$ and $\mathcal{Z}$ an adversary structure satisfying $\mathcal{Q}^3(\mathcal{P}, \mathcal{Z})$, then $\mathsf{MPC}(\mathcal{P}, \mathcal{Z}, \mathcal{C})$ perfectly $\mathcal{Z}$-securely evaluates $\mathcal{C}$. It communicates $|\mathcal{C}| \, |\mathcal{Z}|^2 \cdot \mathrm{Poly}(n, \log |\mathbb{F}|)$ bits.*

Proof. It is easy to see that $\mathbb{S} := \{\mathcal{P} \setminus Z | Z \in \mathcal{Z}\}$ is a sharing specification satisfying $\mathcal{Q}^2(\mathbb{S}, \mathcal{Z})$. Hence by the properties of the sharing scheme and Lemma 6 the statement follows. The protocol communicates $\mathcal{O}(|\mathcal{C}| \, |\mathcal{Z}|^2 \, n^3 \log |\mathbb{F}| + |\mathcal{C}| \, |\mathcal{Z}| \, n^5 \log |\mathbb{F}|)$ bits and broadcasts $\mathcal{O}(|\mathcal{C}| \, |\mathcal{Z}|^2 \, (n \log |\mathbb{F}| + n^2) + |\mathcal{C}| \, |\mathcal{Z}| \, (n^3 \log |\mathbb{F}| + n^4))$ bits. Broadcast can be simulated with the protocol in [FM98], which communicates $\mathrm{Poly}(n)$ bits in order to broadcast one bit. This yields the claimed communication complexity. □

4   Unconditional Protocol

Our main result is an MPC protocol unconditionally $\mathcal{Z}$-secure for an $\mathcal{Q}^2$ adversary structure $\mathcal{Z}$. Its communication complexity is linear in $|\mathcal{Z}|$. This is the first protocol reaching the optimal lower bound of $\Omega(|\mathcal{Z}|)$ on the computational complexity (see Section 6).

4.1 Information Checking

In the perfect model, $\mathcal{Q}^3$ enables the honest players to securely reconstruct shares, as it assures that every share is held by enough honest players. Here, $\mathcal{Q}^2$ only ensures that each share is held by at least one honest player. Correctness is achieved by the use of information checking, a technique that prevents (malicious) players from announcing wrong values (see [RB89, Bea91a, CDD+99, HMZ08]). The following information-checking protocol is a slight variation of [CDD+99]. It is a three party protocol between a sender P_i, a recipient P_j and a verifier P_k. The sender P_i provides P_j with some authentication tag and P_k with some verification tag, such that P_j later can prove the authenticity of a value s to the verifier P_k. We assume that each pair P_i, P_k of players knows a fixed secret value $\alpha_{i,k} \in \mathbb{F} \setminus \{0, 1\}$.

Definition 2. *A vector (s, y, z, α) is 1-consistent if there exists a polynomial f of degree 1 over $\mathbb{F}$ such that $f(0) = s, f(1) = y, f(\alpha) = z$. We say a value s is (P_i, P_j, P_k)-authenticated if P_j knows s and some authentication tag y and P_k knows a verification tag z such that $(s, y, z, \alpha_{i,k})$ is 1-consistent. The vector $(y, z, \alpha_{i,k})$ is denoted by $A_{i,j,k}(s)$.*

Lemma 7. *A (P_i, P_j, P_k)-authenticated value s does not leak information to P_k.*

Proof. The verification tag z is statistically independent of the value s. $\square$

Lemma 8. *Let s be (P_i, P_j, P_k)-authenticated, i.e. $(s, y, z, \alpha_{i,k})$ is 1-consistent. Then for P_j being able to find an authentication tag y' for a value $s' \neq s$ such that $(s', y', z, \alpha_{i,k})$ is 1-consistent is equivalent to finding $\alpha_{i,k}$.*

Proof. If both $(s, y, z, \alpha_{i,k})$ and $(s', y', z, \alpha_{i,k})$ are 1-consistent, then also $(s - s', y - y', 0, \alpha_{i,k})$ is 1-consistent. The corresponding polynomial of degree 1 is not parallel to the x-axis, as $s - s' \neq 0$. Thus it has an unique root at $\alpha_{i,k} = \frac{s - s'}{s - s' - y + y'}$. $\square$

Lemma 9. *The players P_j and P_k can locally compute an authentication and a verification tag of any linear combination of (P_i, P_j, P_k)-authenticated values (for fixed P_i). This is called the linearity of the authentication.*

Proof. Let s_a and s_b be (P_i, P_j, P_k)-authenticated with authentication tags y_a, y_b and verification tags z_a, z_b and the (fixed) point $\alpha_{i,k}$ and let L be a linear function. Then $L(s_a, s_b)$ is (P_i, P_j, P_k)-authenticated with authentication tag $y = L(y_a, y_b)$ and verification tag $z = L(z_a, z_b)$. This works as the polynomials of degree 1 over $\mathbb{F}$ form a vector space, hence $(L(s_a, s_b), L(y_a, y_b), L(z_a, z_b), \alpha_{i,k})$ is 1-consistent. $\square$

Let s be a value known to P_j and P_k. Then these players can use the protocol Default Authenticate to (P_i, P_j, P_k)-authenticate s without communication for arbitrary P_i. Note that P_i does not play an (active) role in this protocol.

Protocol DefaultAuthenticate(P_i, P_j, P_k, s)

0: P_j, P_k take the value s as input.

1: P_j outputs authentication tag $y = s$. P_k outputs verification tag $z = s$.

Lemma 10. *If the value s is known to the honest players in $\{P_j, P_k\}$ protocol* Default Authenticate(P_i, P_j, P_k, s) *securely (P_i, P_j, P_k)-authenticates s without any communication.*

Proof. Correctness: $(s, s, s, \alpha_{i,k})$ is 1-consistent for any $\alpha_{i,k}$. *Privacy/Communication:* No communication occurs.   □

The non-robust protocol Authenticate allows to securely (P_i, P_j, P_k)-authenticate a (secret) value s.

Protocol Authenticate(P_i, P_j, P_k, s)

0: P_i and P_j take the value s as input.

1: P_i chooses random values $(y, z) \in \mathbb{F}$ such that $(s, y, z, \alpha_{i,k})$ is 1-consistent and random values $(s', y', z') \in \mathbb{F}$ such that $(s', y', z', \alpha_{i,k})$ is 1-consistent and sends (s', y, y') to P_j and (z, z') to P_k

2: P_k broadcasts random $r \in \mathbb{F}$.

3: P_i broadcasts $s'' = rs + s'$ and $y'' = ry + y'$.

4: P_j checks if $s'' = rs + s'$ and $y'' = ry + y'$ and broadcast OK or NOK accordingly. If NOK was broadcast the protocol is aborted.

5: P_k checks if $(s'', y'', rz + z', \alpha_{i,k})$ is 1-consistent. If yes P_k sends OK to P_j otherwise he sends $(\alpha_{i,k}, z)$ to P_j, who adjusts y such that $(s, y, z, \alpha_{i,k})$ is 1-consistent.

6: P_j outputs y and P_k outputs z.

Lemma 11. *If P_k is honest and s is known to the honest players in $\{P_i, P_j\}$. Then* Authenticate(P_i, P_j, P_k, s) *either securely (P_i, P_j, P_k)-authenticates s or aborts except with error probability of at most $\frac{1}{|\mathbb{F}|}$. In the case of an abort a player in $\{P_i, P_j\}$ is corrupted. The protocol communicates at most $7 \log |\mathbb{F}|$ bits and broadcasts at most $3 \log |\mathbb{F}| + 1$ bits.*

Proof. Correctness: If the protocol was aborted, either $s'' \neq rs + s'$ or $y'' \neq ry + y'$ meaning P_i is corrupted, or P_j misleadingly accused P_i. Otherwise, the players use some $(s, y, z, \alpha_{i,k})$ as authentication of s. The probability that $(s, y, z, \alpha_{i,k})$ is not 1-consistent is $|\mathbb{F}|^{-1}$, as for a fixed r there is exactly one way to choose y, z such that the inconsistency is not detected. *Privacy:* The verification tag z, the values s'' and y'' are statistically independent of the value s. Also $\alpha_{i,k}$ is sent only to P_j if either P_i or P_k is malicious. *Communication:* Seen by counting the number of messages sent or broadcast during the protocol.   □

Remark 1. If the (honest) players P_i and P_j do not know the same s the protocol will abort as well.

Assume that P_k knows a candidate s' for a (P_i, P_j, P_k)-authenticated value s. If P_j wants to prove the authenticity of s' (i.e. that $s' = s$) the players invoke the protocol Verify. If P_k accepts the proof he outputs s', otherwise he outputs $\perp$.

Protocol Verify$(P_i, P_j, P_k, s', A_{i,j,k}(s))$

0: Let $A_{i,j,k}(s) = (y, z, \alpha_{i,k})$. P_j takes y as input and P_k takes s', z as input.

1: P_j sends y to P_k

2: P_k outputs s' if $(s', y, z, \alpha_{i,k})$ is 1-consistent otherwise $\perp$.

Lemma 12. *Assume s is (P_i, P_j, P_k)-authenticated and let P_k be an honest player knowing s'. If P_j is honest and $s' = s$, P_k will output s in* Verify. *Otherwise P_k will output $\perp$ or s except with error probability of at most $\frac{1}{|\mathbb{F}|-2}$. The protocol communicates at most $\log |\mathbb{F}|$ bits.*

Proof. Correctness: Let P_k be an honest player, let $A_{i,j,k}(s) = (y, z, \alpha_{i,k})$ be consistent with s, i.e. $(s, y, z, \alpha_{i,k})$ is 1-consistent and assume that $s' = s$. If P_j sends the right y the vector $(s', y, z, \alpha_{i,k})$ is 1-consistent and P_k will output s. Otherwise P_k always outputs $\perp$. So assume $s' \neq s$. Then the probability of finding y' such that the vector $(s', y', z, \alpha_{i,k})$ is 1-consistent is at most $\frac{1}{|\mathbb{F}|-2}$, thus P_k outputs $\perp$ except with error probability of at most $\frac{1}{|\mathbb{F}|-2}$. *Privacy/Communication:* No information except y is sent. $\square$

4.2   Unconditional Secret Sharing

Starting from the secret sharing of Section 3.1 we construct a sharing scheme for the $\mathcal{Q}^2$ case using the information-checking scheme of the previous section.

Definition 3. *A value s is* shared *with respect to the sharing specification $\mathbb{S} = (S_1, \ldots, S_h)$, if the following holds:*

a) *There exist shares $s_1, \ldots, s_h$ such that $s = \sum_{q=1}^{h} s_q$*

b) *Each s_q is known to every (honest) player in S_q*

c) *$\forall P_i, P_j \in S_q$ $P_k \in \mathcal{P}$ s_q is (P_i, P_j, P_k)-authenticated.*

We denote the sharing of a value s by $[s]$. Let $[s]_q = (s_q, \{A_{i,j,k}(s_q)\})$, where s_q is the q-th share and $\{A_{i,j,k}(s_q)\}$ the set of all associated authentications. As the perfect sharing from Section 3.1 this sharing is linear and does not leak information to the adversary (for a $\mathcal{Z}$-private $\mathbb{S}$).

The following protocol allows a dealer P_D to securely share a secret value s.

Protocol Share$(\mathcal{P}, \mathcal{Z}, \mathbb{S}, P_D, s)$

0: The dealer P_D takes s as input.

1: P_D splits s into random shares $s_1, \ldots, s_{|\mathbb{S}|}$ subject to $s = \sum_{q=1}^{|\mathbb{S}|} s_q$.

2: **for all** $q \in \{1, \ldots, |\mathbb{S}|\}$ **do**

3:       P_D sends share s_q to every player in S_q.

4:       $\forall P_i, P_j \in S_q$ and $\forall P_k \in \mathcal{P}$ invoke Authenticate(P_i, P_j, P_k, s_q).
         If (for fixed q) any Authenticate(P_i, P_j, P_k, s_q) was aborted
         P_D broadcasts s_q, the players in S_q replace there share and
         DefaultAuthenticate(P_i, P_j, P_k, s_q) is invoked $\forall P_i, P_j \in S_q \forall P_k \in \mathcal{P}$.

5: **end for**

6: The players in $\mathcal{P}$ collectively output $[s]$.

Lemma 13. *For any adversary structure Z the protocol* Share$(\mathcal{P}, \mathcal{Z}, \mathbb{S}, P_D, s)$ *securely computes a sharing $[s']$ except with error probability of at most $\frac{1}{|\mathbb{F}|} n^3 |\mathbb{S}|$ and if P_D is honest $s' = s$. The protocol communicates at most $|\mathbb{S}| (7n^3 + n) \log |\mathbb{F}|$ bits and broadcasts at most $|\mathbb{S}| ((3n^3 + 1) \log |\mathbb{F}| + n^3)$ bits.*

Proof. Correctness: Assume that P_D does not send the same value to the (honest) players in S_q (Step 3). In this case at least one invocation of Authenticate will abort (see Remark 1) and P_D must broadcast the value. Otherwise all (honest) player use the same value s_q in Step 3. We have to show that every (honest) P_j gets his authentications $A_{i,j,k}(s_q)$. If all instances of Authenticate do not abort the statement follows from Lemma 11. Otherwise s_q is broadcast and the players use DefaultAuthenticate resulting in the proper sharing state (c.f. Lemma 10). Note that a single invocation of Authenticate has an error probability of at most $\frac{1}{|\mathbb{F}|}$, so the above upper bound on the error probability follows. *Privacy:* We only have to check that broadcasting s_q in Step 4 does not violate privacy. But s_q is only broadcast when at least one Authenticate was aborted. In this case either P_D or a player in S_q is malicious, hence s_q is known to the adversary before the broadcast (Lemma 11 and Remark 1). *Communication:* Follows directly by counting the numbers of messages sent or broadcast (c.f. Lemmas 11 and 10) □

If a value is publicly known the player can use DefaultShare to obtain a sharing of it.

Protocol DefaultShare$(\mathcal{P}, \mathcal{Z}, \mathbb{S}, s)$

0: Every player takes s as input.
1: The share s_1 is set to s and all other shares are set to 0.
2: DefaultAuthenticate(P_i, P_j, P_k, s_q) is invoked $\forall S_q \forall P_i, P_j \in S_q \ \forall P_k \in \mathcal{P}$.
3: The players in $\mathcal{P}$ collectively output $[s]$.

Lemma 14. DefaultShare$(\mathcal{P}, \mathcal{Z}, \mathbb{S}, s)$ *securely computes a sharing $[s]$ of s. The protocol does not communicate.*

Proof. The statement follows from Lemmas 2 and 10. □

The protocol ReconstructShare allows reconstruction of a share from some sharing $[s]$ to players in $R \subseteq \mathcal{P}$. Hence the players can reconstruct s by invoking protocol ReconstructShare for each share of $[s]$.

Protocol ReconstructShare$(\mathcal{P}, \mathcal{Z}, \mathbb{S}, [s]_q, R)$

0: The players in S_q take collectively $[s]_q = (s_q, \{A_{i,j,k}(s_q)\})$ as input.
1: Every player P_j in S_q sends s_q to every player in R.
2: **for all** $P_j \in S_q, P_k \in R$ **do**
3:    Invoke Verify$(P_i, P_j, P_k, s_q^{(j)}, A_{i,j,k}(s_q)) \ \forall P_i \in S_q$ where $s_q^{(j)}$ is the value received by P_k from P_j in Step 1. If P_k output $s_q^{(j)}$ in each invocation he accepts it as value for s_q.
4: **end for**
5: Each P_k outputs some value he accepted in Step 3 (or $\perp$ if never accepted a value).

Lemma 15. *Assume S_q and $\mathcal{Z}$ satisfy $\mathcal{Q}^1(S_q, \mathcal{Z})$ and let $[s]_q$ be a consistent share. Every honest player in R outputs s_q in* ReconstructShare *except with error probability of at most $\frac{1}{|\mathbb{F}|-2} n |S_q|$. The protocol communicates at most $(n^3 + n^2) \log |\mathbb{F}|$ bits and does not broadcast.*

Proof. Correctness: As S_q and $\mathcal{Z}$ satisfy $\mathcal{Q}^1(S_q, \mathcal{Z})$ there exists at least one honest player P_j in S_q, who sends the right value s_q to $P_k . \in R$ in Step 1. Hence every (honest) P_k will accept s_q in Step 3 from P_j, as P_j has a valid authentication for s_q from every player in S_q (c.f. Lemma 12). On the other hand a malicious player does not have a valid authentication for $s'_q \neq s_q$ from every player in S_q (one of them is honest!). So no honest player will accept $s'_q \neq s_q$ in Step 3 and thus P_k output s_q in the last step except with error probability of at most $\frac{1}{|\mathbb{F}|-2} |S_q|$ (c.f. Lemma 12). As there are at most n players in R the overall error probability follows. *Privacy:* Follows from Lemma 12. *Communication:* Follows directly by counting the numbers of messages sent (c.f. Lemma 12) □

Protocol Reconstruct$(\mathcal{P}, \mathcal{Z}, \mathbb{S}, [s], R)$
0: The players in $\mathcal{P}$ take collectively $[s]$ as input.
1: **for all** $q = 1, \ldots, |\mathbb{S}|$ **do**
2:     ReconstructShare$(\mathcal{P}, \mathcal{Z}, \mathbb{S}, [s]_q, R)$ is invoked.
3: **end for**
4: The players locally sum up the shares to obtain and output s.

Lemma 16. *Assume $\mathbb{S}$ and $\mathcal{Z}$ satisfy $\mathcal{Q}^1(\mathbb{S}, \mathcal{Z})$ and let $[s]$ be a sharing of the value s. Every honest player in R outputs s in* Reconstruct *except with error probability of at most $\frac{1}{|\mathbb{F}|-2} n^2 |\mathbb{S}|$. The protocol communicates at most $|\mathbb{S}| (n^3 + n^2) \log |\mathbb{F}|$ bits and does not broadcast.*

Proof. The statement follows directly from Lemma 15, as the players invoke the protocol ReconstructShare for each share. □

4.3 Multiplication

We present a protocol for the unconditionally-secure computation of the (shared) product of two shared values $[a]$ and $[b]$. The idea is, as in the perfect case, to use an optimistic multiplication. The protocol BasicMult takes a set M of (identified) malicious players as input and outputs the correct product given that no player in $\mathcal{P} \setminus M$ actively cheated. In a next step a probabilistic check is used to determine whether the product computed in BasicMult is correct. This allows us to detect malicious behaviour. If cheating occured, all involved sharings (from BasicMult) are reconstructed to identify a cheater in $\mathcal{P} \setminus M$. These reconstructions violate the privacy of the involved factors the protocol is not used directly in the actual circuit computation. Instead we use it to multiply two random values and make use of circuit randomization from [Bea91b] for actual multiplication gates.

Protocol BasicMult$(\mathcal{P}, \mathcal{Z}, \mathbb{S}, [a], [b], M)$

0: The players in $\mathcal{P}$ take collectively $[a]$, $[b]$ and M as input.

1: $\forall S_q : S_q \cap M \neq \emptyset$ invoke ReconstructShare to reconstruct a_q and b_q.

2:   *a)* Each player $P_i \in \mathcal{P} \setminus M$ (locally) computes his designated products and shares the sum $c_i = \sum_{(p,q) \in I(i)} a_p b_q$.

    b) For each $P_i \in M$ DefaultShare$(\mathcal{P}, \mathcal{Z}, \mathbb{S}, c_i)$ is invoked where $c_i = \sum_{(p,q) \in I(i)} a_p b_q$.

3: The players collectively output $([c_1], \ldots, [c_n])$ and $[c] = \sum_{i=1}^{n} [c_i]$.

Lemma 17. *Let $M \subseteq Z^*$ be a set of (identified) malicious players and assume that $\mathcal{Z}$ and $\mathbb{S}$ satisfy $Q^1(\mathbb{S}, \mathcal{Z})$. Then BasicMult$(\mathcal{P}, \mathcal{Z}, \mathbb{S}, [a], [b], M)$ securely computes sharings $[c], ([c_1], \ldots, [c_n])$ except with error probability of $\mathcal{O}(\frac{1}{|\mathbb{F}|} n^4 |\mathbb{S}|)$. If no player in $\mathcal{P} \setminus M$ actively cheats, then $\forall i \; c_i = \sum_{(p,q) \in I_Z(i)} a_p b_q$ and $c = ab$. The protocol communicates at most $\mathcal{O}(|\mathbb{S}| n^4 \log |\mathbb{F}|)$ bits and broadcasts at most $\mathcal{O}(|\mathbb{S}| n^4 \log |\mathbb{F}|)$ bits.*

Proof. Correctness: The properties of the sharing protocol guarantee that the outputs are valid sharings except with error probability of $\mathcal{O}(\frac{1}{|\mathbb{F}|} n^4 |\mathbb{S}|)$. The $Q^1(\mathbb{S}, \mathcal{Z})$ property allows the players to securely reconstruct shares and grants that there exists a proper assignment of players in $\mathcal{P}$ to the local products. If none of the players in $\mathcal{P} \setminus M$ cheated, it holds for each P_i that $c_i = \sum_{(p,q) \in I_Z(i)} a_p b_q$ (for players in M DefaultShare is used on reconstructed values). Privacy: All reconstructed shares a_q, b_q are known to players in M. Complexity: Follow directly from the properties of the sharing scheme (c.f. Lemmas 13, 14 and 15). □

Detectable Random Triple Generation. The following unconditionally secure protocol takes a set M of malicious players as an additional input and computes a random multiplication triple $([a], [b], [c])$ where $c = ab$ given that no player in $\mathcal{P} \setminus M$ actively cheats. Otherwise it outputs a set of malicious players M' such that $M \subsetneq M'$. This protocol uses a probabilistic check to detect cheating. First the players generate a shared random challenge $[r]$ and a blinding $[b']$. Then they use BasicMult to compute the sharings $[c] = [a][b]$, $[c'] = [a][b']$ and check whether $[a](r[b] + [b']) = (r[c] + [c'])$. If this is the case the multiplication triple $([a], [b], [c])$ is output. Otherwise the players identify (at least) one cheater in $\mathcal{P} \setminus M$ by reconstructing $[a], [b], [b'], [c], [c']$.

Lemma 18. *If $\mathbb{S}$ and $\mathcal{Z}$ satisfy $Q^1(\mathbb{S}, \mathcal{Z})$ and $M \subseteq Z^*$, the protocol RandomTriple outputs either a random multiplication triple $([a], [b], [c])$ or set $M' \subseteq Z^*$ where $M \subsetneq M'$ except with error probability of $\mathcal{O}(\frac{1}{|\mathbb{F}|} |\mathbb{S}| n^4) + \frac{1}{|\mathbb{F}|}$. No information is leaked to the adversary. RandomTriple communicates at most $\mathcal{O}(|\mathbb{S}| n^4 \log |\mathbb{F}|)$ bits and broadcasts at most $\mathcal{O}(|\mathbb{S}| n^4 \log |\mathbb{F}|)$ bits.*

Proof. Correctness: In Step 2, the players compute $[c]$ and $[c']$. Given that no player in $\mathcal{P} \setminus M$ actively cheated it holds that $c = ab$ and $c' = ab'$. In this case $[a](r[b] + [b']) - r[c] - [c']$, which is computed in Step 3, is zero for all r and the players reconstruct the random multiplication triple $([a], [b], [c])$. If $c \neq ab$ the difference $[a](r[b] + [b']) - r[c] -$

$[c']$ is non-zero except for at most one r and the players go to Step 5 with probability at least $(1 - \frac{1}{|\mathbb{F}|})$ (assuming that no errors happen in sharing and reconstruction of values). For at least one player $P_i \in \mathcal{P} \setminus M$ it must hold that $rc_i + c_i' \neq \sum_{(p,q) \in I(i)} r(a_p b_q) + (a_p b_q')$. By opening all involved sharing it is easy to find these players. Thus it holds that $M \subsetneq M'$ and $M' \subseteq Z^*$. The overall error probability is composed of the error probability of the sharing scheme and the one of the random challenge check in Step 3. *Privacy:* Neither the protocol BasicMult nor the sharing scheme do violate privacy (c.f. Lemma 17). The values e is statistically independent of $([a], [b], [c])$, as b' acts as blinding. If no cheating occurred the value d is always zero. If Step 5 is invoked, the reconstructed values are not used, and privacy is met. *Communication:* Follows from counting the number of messages sent (c.f. Lemmas 13, 16 and 17). □

Protocol RandomTriple$(\mathcal{P}, \mathcal{Z}, \mathbb{S}, M)$

0: The players take the set $M \subseteq \mathcal{P}$ as input.

1: The players generate random shared values $[a], [b], [b'], [r]$ by summing up shared random values (one from each player) for each value.

2: Invoke BasicMult$(\mathcal{P}, \mathcal{Z}, \mathbb{S}, [a], [b], M)$ to compute the sharing $[c]$ and the vector $([c_1], \ldots, [c_n])$ and invoke BasicMult$(\mathcal{P}, \mathcal{Z}, \mathbb{S}, [a], [b'], M)$ to compute the sharing $[c']$ and the vector $([c_1'], \ldots, [c_n'])$.

3: Reconstruct $[r]$ and (locally) compute $[e] = r[b] + [b']$ and reconstruct it to obtain e. Then $[d] = e[a] - r[c] - [c']$ is computed (locally) and reconstructed.

4: If the value d is zero the players output $([a], [b], [c])$.

5: Otherwise reconstruct the sharings $[a], [b], [b'], [c_1], \ldots, [c_n], [c_1'], \ldots, [c_n']$. The players output $M' = M \cup \{P_i : rc_i + c_i' \neq \sum_{(p,q) \in I(i)} r(a_p b_q) + (a_p b_q')\}$.

Multiplication with Circuit Randomization. The actual multiplication is based on circuit randomization [Bea91b]. It allows players to compute the product $[xy]$ of two shared values $[x]$ and $[y]$ at the cost of two reconstructions given a random multiplication triple $([a], [b], [c])$, where $ab = c$. The trick is to use that $xy = ((x - a) + a)((y - b) + b)$. By reconstructing $d = x - a$ and $e = y - b$ the players can compute $[xy]$ as $de + d[b] + [a]e + [c]$. This does not violate the secrecy of $[x]$ or $[y]$ as the random values $[a]$ and $[b]$ act as blinding.

Protocol Multiplication$(\mathcal{P}, \mathcal{Z}, \mathbb{S}, [x], [y])$

0: The players in $\mathcal{P}$ take collectively $[x], [y]$ as input and set $M := \emptyset$.

1: Invoke RandomTriple$(\mathcal{P}, \mathcal{Z}, \mathbb{S}, M)$. If the protocol outputs a set M', set $M \leftarrow M'$ and repeat Step 1. Otherwise use the output as random multiplication triple $([a], [b], [c])$.

2: Compute and reconstruct $[d_x] = [x] - [a]$ and $[d_y] = [y] - [b]$. Compute $d_x d_y + d_x [b] + d_y [a] + [c] = [xy]$ to obtain a sharing of xy.

Lemma 19. Multiplication($\mathcal{P}, \mathcal{Z}, \mathbb{S}, [x], [y]$) *is an unconditional secure multiplication protocol given that* $\mathbb{S}$ *and* $\mathcal{Z}$ *satisfy* $\mathcal{Q}^1(\mathbb{S}, \mathcal{Z})$. *The protocol has an error probability of* $\mathcal{O}(\frac{1}{|\mathbb{F}|} |\mathbb{S}| n^5)$, *communicates at most* $\mathcal{O}(|\mathbb{S}| n^5 \log |\mathbb{F}|)$ *bits and broadcasts at most* $\mathcal{O}(|\mathbb{S}| n^5 \log |\mathbb{F}|)$ *bits.*

Proof. Correctness: Assume that RandomTriple in Step 1 outputs a set M', then we have that $M \subsetneq M' \subseteq \mathcal{P}$. Hence this step is repeated less then n times and results in a random multiplication triple ($[a], [b], [c]$) (c.f. Lemma 18). The rest of the protocol is just the multiplication from [Bea91b]. *Privacy:* Follows from [Bea91b] and Lemma 18. *Communication:* Follows from counting the number of messages sent (c.f. Lemmas 16 and 18). $\qquad\square$

4.4 Unconditional MPC Protocol

The combination of Share, Reconstruct and Multiplication directly gives the following unconditionally secure MPC protocol.

Protocol MPC($\mathcal{P}, \mathcal{Z}, \mathcal{C}$)

0: The players take $\mathbb{S} := \{\mathcal{P} \setminus Z | Z \in \mathcal{Z}\}$ as sharing specification.

1: For every gate of $\mathcal{C}$ being evaluated do the following:
   - *Input gate for* P_D: Share($\mathcal{P}, \mathcal{Z}, \mathbb{S}, P_D, s$) is invoked to share s
   - *Linear gate:* The linear combination of the corresponding shares is computed locally using the linearity of the sharing.
   - *Random gate:* Each player shares a random value. The sum of these values is used as output of the gate.
   - *Multiplication gate:* Multiplication($\mathcal{P}, \mathcal{Z}, \mathbb{S}, [x], [y]$) is used to multiply $[x]$ and $[y]$.
   - *Output gate:* The players invoke Reconstruct($\mathcal{P}, \mathcal{Z}, \mathbb{S}, [s], R$) to reconstruct s for players in R.

Theorem 2. *Let* $\mathcal{C}$ *be a circuit over* $\mathbb{F}$, *where* $|F| \in \Omega(2^\kappa)$ *and* κ *is a security parameter, and let* $\mathcal{Z}$ *be an adversary structure satisfying* $\mathcal{Q}^2(\mathcal{P}, \mathcal{Z})$, *then* MPC($\mathcal{P}, \mathcal{Z}, \mathcal{C}$) $\mathcal{Z}$-*securely evaluates* $\mathcal{C}$ *with an error probability of* $2^{-\kappa} |\mathcal{C}| |\mathcal{Z}| \cdot \mathrm{Poly}(n, \kappa)$. *It communicates* $|\mathcal{C}| |\mathcal{Z}| \cdot \mathrm{Poly}(n, \kappa)$ *bits and broadcasts* $|\mathcal{C}| |\mathcal{Z}| \cdot \mathrm{Poly}(n, \kappa)$ *bits within* $\mathrm{Poly}(n, \kappa) \cdot d$ *rounds, where* d *denotes the multiplicative depth of* $\mathcal{C}$.

Proof. It is easy to see that $\mathbb{S} := \{\mathcal{P} \setminus Z | Z \in \mathcal{Z}\}$ is a sharing specification satisfying $\mathcal{Q}^1(\mathbb{S}, \mathcal{Z})$. Hence by the properties of the sharing scheme and Lemma 19 correctness and the bound on the error probability follow. The claimed communication and broadcast complexity follow directly from the used subprotocols. Inspection of the subprotocols also shows that it is possible to evaluate gates on the same multiplicative depth of $\mathcal{C}$ in parallel. As each subprotocol only requires $\mathrm{Poly}(n, \kappa)$ rounds, the total number of rounds follows. $\qquad\square$

Note that broadcast can be (unconditionally secure) simulated using the protocol from [PW96], which communicates $\text{Poly}(n, \kappa)$ bits in order to broadcast one bit (with error probability of $\mathcal{O}(2^{-\kappa})$). This results in an MPC protocol with the same efficiency and error probability as stated in Theorem 2.

The error probability of the presented protocol grows linearly in the size of the adversary structure $\mathcal{Z}$. As $|\mathcal{Z}|$ is typically exponential in n, the security parameter κ must be chosen accordingly (such that $|\mathcal{Z}| \in \text{Poly}(\kappa)$). This results in a huge security parameter and therefore in inefficient protocols. We therefore provide an extension of the previous protocol in which the error probability only depends on $\log|Z|$, and hence a reasonably large security parameter κ is sufficient.

5   Unconditional Protocol for Superpolynomial $|\mathcal{Z}|$

The protocol from the previous section has an error probability linear in $|\mathcal{Z}|$, which is problematic for large adversary structures $\mathcal{Z}$. In this section, we present modifications to the protocol that reduce the dependency to $\log|\mathcal{Z}|$, which is in $\text{Poly}(n)$.

The reason for the error probability being dependent on $|\mathcal{Z}|$ is twofold: Firstly, the protocol requires $\Omega(|\mathcal{Z}|)$ probabilistic checks, in each of them a cheating party might remain undetected with probability $2^{-\kappa}$. Secondly, the protocol requires $\Omega(|\mathcal{Z}|)$ broadcasts, each of them having a small probability of failure.

5.1   Information Checking

In each invocation of Authenticate / Verify, a cheating attempt of a malicious player P_i is not detected with probability of $\mathcal{O}(\frac{1}{|\mathbb{F}|})$ (c.f. Section 4.1). As these protocols are invoked $\Theta(|\mathcal{Z}|)$ times per sharing, the resulting error probability depends linearly on $|\mathcal{Z}|$. To avoid this we use local dispute control to deal with detected cheaters.

More formally, each player P_k locally maintains a list $\mathcal{L}_k$ of players whom he distrusts. At the beginning of the MPC protocol these lists are empty. Protocol Authenticate is modified, such that P_j puts P_i on his list $\mathcal{L}_j$ if the check in Step 4 fails. Once $P_i \in \mathcal{L}_j$, P_j behaves in all future invocations of the protocol as if the check in Step 4 failed independently whether this is the case or not. Similarly P_k puts P_i on his list $\mathcal{L}_k$ if the check in Step 5 fails. As soon as $P_i \in \mathcal{L}_k$, P_k behaves in Step 5 as if the corresponding check failed. Furthermore, in protocol Verify, P_k puts P_j on his list $\mathcal{L}_k$ if the check in Step 2 failed. Again P_k behaves for all $P_j \in \mathcal{L}_k$ as if the check failed independently whether this is the case or not.

In both protocols the adversary has a chance of $\mathcal{O}(\frac{1}{|\mathbb{F}|})$ to cheat successfully, but if he fails (with probability $\Omega(1 - \frac{1}{|\mathbb{F}|})$) one corrupted player P_i is put on the list $\mathcal{L}_k$ of an honest player P_k. From then on P_i is never able to cheat in instances of both protocols when P_k takes part (in the right position). This means that the adversary actually has at most n^2 attempts to cheat. Hence total error probability of arbitrary many instances of Verify and Authenticate is at most $\mathcal{O}(\frac{1}{|\mathbb{F}|}n^2)$ and no longer depends on $\mathcal{Z}$.

Note that the parallel invocation of Authenticate, as it is used in Share, requires special care. For example if in one of the parallel invocations of Authenticate (with P_i

and P_k) the consistency check fails P_k must assume that all other parallel checks failed. Analogous modifications are made in Verify and Multiplication.

Lemma 20. *The modified* Authenticate *and* Verify *protocols have a total error probability of* $\mathcal{O}(\frac{1}{|\mathbb{F}|}n^2)$ *independent of the number of invocations.*

5.2 Broadcast

Although broadcast is only needed in Share, the total number of broadcast calls is in $\Theta(|\mathcal{Z}|)$. If [PW96] is used, the resulting overall error probability depends linearly on $|\mathcal{Z}|$. To avoid this problem, the number of broadcast calls must be reduced.

To reach this goal we use the fact that the Share protocol only has constantly many rounds. In each round a player P_S must broadcast $\Theta(|\mathcal{Z}|)$ many messages of size $\mathcal{O}(\log|\mathbb{F}|)$. Instead of broadcasting these messages in parallel, P_S sends their concatenation to the other players, who then check that they received the same message. If an inconsistency is detected the protocol is repeated. To limit the number of repetitions we use the concept of dispute control from [BH06] which prevents the malicious players from repetitive cheating. Dispute control is realized by a publicly known *dispute set* $\Gamma \subseteq \mathcal{P} \times \mathcal{P}$, a set of unordered pairs of players. If $\{P_i, P_j\} \in \Gamma$ it means that there is a dispute between P_i and P_j and thus at least one of them is corrupted. Note that from P_i's view all player in $\{P_j | \{P_i, P_j\} \in \Gamma\}$ are malicious and thus he no longer trust them. At the beginning of the MPC protocol Γ is empty.

Protocol OptimisticBroadcast($\mathcal{P}, \mathcal{Z}, P_S, m$)

0: The player P_S takes $m \in \{0,1\}^w$ as input.

1: $\forall \{P_i, P_S\} \notin \Gamma$ P_S sends m as m_i to P_i.

2: $\forall \{P_i, P_j\} \notin \Gamma$ P_i sends m_i as m_{ij} to P_j.

3: $\forall P_i$ if all received values are the same P_i is happy, otherwise unhappy. P_i broadcasts using [PW96] his happy bit.

4: If all players are happy, each P_i outputs the value he holds. Otherwise, an unhappy player P_i (e.g. the one with the smallest index) broadcasts j, j', z, b where m_{ji} differs from $m_{j'i}$ at bit-position z and b is the bit of m_{ji} at position z. Then $P_S, P_j, P_{j'}$ broadcast their versions of the bit at position z. Using this information the players localize a dispute between two players of $\{P_i, P_S, P_j, P_{j'}\}$. Then the protocol is repeated with updated Γ.

Lemma 21. *The protocol* OptimisticBroadcast($\mathcal{P}, \mathcal{Z}, P_S, m$) *achieves the broadcast of a message* $m' \in \{0,1\}^w$. *The protocol communicates at most* $w \cdot \text{Poly}(n, \kappa)$ *bits and broadcasts at most* $\log w \cdot \text{Poly}(n, \kappa)$.

Proof. The properties of Γ guarantee that honest players will exchange in Step 2 their received messages from P_S. So if all honest player are happy they all will output the same message m'. For an honest P_S this also ensures that $m' = m$. If a player is unhappy, at least one player misbehaved. The actions taken in Step 4 then ensure that the honest players will find at least one dispute. The protocol will terminate, as the number of repetition is limited by $n(n-1)$. As the broadcast of z requires $\log w$ bits, the communication and broadcast complexities follow.   □

For a message of length $\Theta(|\mathcal{Z}|)$ the above protocol only needs to broadcast $\log |\mathcal{Z}| \cdot \mathrm{Poly}(n, \kappa)$ bits, hence the total number of broadcast calls per invocation of Share is reduced to $\log |\mathcal{Z}| \cdot \mathrm{Poly}(n, \kappa)$.

Lemma 22. *The modified* Share *protocol communicates* $|\mathcal{C}| |\mathcal{Z}| \cdot \mathrm{Poly}(n, \kappa)$ *bits and broadcasts* $|\mathcal{C}| \log |\mathcal{Z}| \cdot \mathrm{Poly}(n, \kappa)$ *bits.*

5.3 Summary

The combination of the above extension results in the following Lemma:

Lemma 23. *Let $\mathcal{C}$ be a circuit over $\mathbb{F}$, where $|\mathbb{F}| \in \Omega(2^\kappa)$ and κ is a security parameter, and let $\mathcal{Z}$ be an adversary structure satisfying $Q^2(\mathcal{P}, \mathcal{Z})$, then the modified protocol* $\mathrm{MPC}(\mathcal{P}, \mathcal{Z}, \mathcal{C})$ $\mathcal{Z}$-*securely evaluates $\mathcal{C}$ with an error probability of $2^{-\kappa} |\mathcal{C}| \cdot \mathrm{Poly}(n, \kappa)$. It communicates $|\mathcal{C}| |\mathcal{Z}| \cdot \mathrm{Poly}(n, \kappa)$ bits and broadcasts $|\mathcal{C}| \log |\mathcal{Z}| \cdot \mathrm{Poly}(n, \kappa)$ bits. The number of rounds is $\mathrm{Poly}(n, \kappa) \cdot d$, where d denotes the multiplicative depth of $\mathcal{C}$.*

Proof. Follows directly from Theorem 2 and Lemmas 20 and 22.   □

By replacing broadcast with the simulated one from [PW96], one gets for $|\mathcal{Z}| \in \mathcal{O}(2^n)$ and $|\mathcal{C}| \in \mathrm{Poly}(\kappa)$ the following theorem.

Theorem 3. *Let $\mathcal{C}$ be a circuit over $\mathbb{F}$, where $|\mathbb{F}| \in \Omega(2^\kappa)$ and κ is a security parameter, and let $\mathcal{Z}$ be an adversary structure satisfying $Q^2(\mathcal{P}, \mathcal{Z})$, then* $\mathrm{MPC}(\mathcal{P}, \mathcal{Z}, \mathcal{C})$ $\mathcal{Z}$-*securely evaluates $\mathcal{C}$ with an error probability of $2^{-\kappa} \cdot \mathrm{Poly}(n, \kappa)$. It communicates $|\mathcal{Z}| \cdot \mathrm{Poly}(n, \kappa)$ bits.*

6 Lower Bound on the Efficiency

The following theorem states that there exists a family of circuits and Q^2 adversary structures such that the length of unconditionally secure protocols tolerating these adversaries grows exponentially in the number of players. This implies that the computational complexity of our protocol from the previous section is optimal, as there exists no protocol with a computational complexity in $o(|\mathcal{Z}|)$.

Theorem 4. *[Hir01] Let $\mathcal{C}$ be the circuit which takes inputs from P_1 and P_2 and outputs the product to P_1. Then there exists a family $\mathcal{Z}_2, \mathcal{Z}_3, \ldots$ of Q^2 adversary structures for player sets $\mathcal{P}_2, \mathcal{P}_3, \ldots$ ($|\mathcal{P}_n| = n$) such that the length of the shortest unconditionally $\mathcal{Z}_n$-secure protocol for $\mathcal{C}$ grows exponentially in n.*

References

[Bea91a]   Beaver, D.: Secure multiparty protocols and zero-knowledge proof systems tolerating a faulty minority. Journal of Cryptology 4(2), 75–122 (1991)

[Bea91b]   Beaver, D.: Efficient multiparty protocols using circuit randomization. In: Feigenbaum, J. (ed.) CRYPTO 1991. LNCS, vol. 576, pp. 420–432. Springer, Heidelberg (1992)

[BFH+08]   Beerliová-Trubíniová, Z., Fitzi, M., Hirt, M., Maurer, U., Zikas, V.: MPC vs. SFE: Perfect security in a unified corruption model. In: Canetti, R. (ed.) TCC 2008. LNCS, vol. 4948, pp. 231–250. Springer, Heidelberg (2008)

[BGW88]   Ben-Or, M., Goldwasser, S., Wigderson, A.: Completeness theorems for non-cryptographic fault-tolerant distributed computation. In: STOC, pp. 1–10. ACM (1988)

[BH06]   Beerliová-Trubíniová, Z., Hirt, M.: Efficient multi-party computation with dispute control. In: Halevi, S., Rabin, T. (eds.) TCC 2006. LNCS, vol. 3876, pp. 305–328. Springer, Heidelberg (2006)

[CCD88]   Chaum, D., Crépeau, C., Damgard, I.: Multiparty unconditionally secure protocols. In: STOC, pp. 11–19. ACM (1988)

[CDD+99]   Cramer, R., Damgård, I.B., Dziembowski, S., Hirt, M., Rabin, T.: Efficient multiparty computations secure against an adaptive adversary. In: Stern, J. (ed.) EUROCRYPT 1999. LNCS, vol. 1592, p. 311. Springer, Heidelberg (1999)

[FM98]   Fitzi, M., Maurer, U.: Efficient byzantine agreement secure against general adversaries. In: Kutten, S. (ed.) DISC 1998. LNCS, vol. 1499, pp. 134–148. Springer, Heidelberg (1998)

[GMW87]   Goldwasser, S., Micali, S., Wigderson, A.: How to play any mental game, or a completeness theorem for protocols with an honest majority. In: STOC, vol. 87, pp. 218–229 (1987)

[Hir01]   Hirt, M.: Multi-Party Computation: Efficient Protocols, General Adversaries, and Voting. PhD thesis, ETH Zurich (September 2001), Reprint as ETH Series in Information Security and Cryptography, vol. 3. Hartung-Gorre Verlag, Konstanz (2001) ISBN 3-89649-747-2

[HM97]   Hirt, M., Maurer, U.: Complete characterization of adversaries tolerable in secure multi-party computation. In: PODC, pp. 25–34 (August 1997)

[HM00]   Hirt, M., Maurer, U.: Player simulation and general adversary structures in perfect multiparty computation. Journal of Cryptology 13(1), 31–60 (2000), Extended abstract in Proc. 16th of ACM PODC 1997 (1997)

[HMZ08]   Hirt, M., Maurer, U., Zikas, V.: MPC vs. SFE: Unconditional and computational security. In: Pieprzyk, J. (ed.) ASIACRYPT 2008. LNCS, vol. 5350, pp. 1–18. Springer, Heidelberg (2008)

[Mau02]   Maurer, U.: Secure multi-party computation made simple. In: Cimato, S., Galdi, C., Persiano, G. (eds.) SCN 2002. LNCS, vol. 2576, pp. 14–28. Springer, Heidelberg (2003)

[PSR03]   Prabhu, B., Srinathan, K., Pandu Rangan, C.: Trading players for efficiency in unconditional multiparty computation. In: Cimato, S., Galdi, C., Persiano, G. (eds.) SCN 2002. LNCS, vol. 2576, pp. 342–353. Springer, Heidelberg (2003)

[PW96]   Pfitzmann, B., Waidner, M.: Information-theoretic pseudosignatures and byzantine agreement for $t \geq n/3$. In Research report. IBM Research (1996)

[RB89]   Rabin, T., Ben-Or, M.: Verifiable secret sharing and multiparty protocols with honest majority. In: STOC, pp. 73–85. ACM (1989)

[Yao82]   Yao, A.C.: Protocols for secure computations. In: FOCS, pp. 160–164 (1982)

Fair and Efficient Secure Multiparty Computation with Reputation Systems*

Gilad Asharov, Yehuda Lindell, and Hila Zarosim**

Dept. of Computer Science, Bar-Ilan University, Israel
{asharog,zarosih}@cs.biu.ac.il, lindell@biu.ac.il

Abstract. A reputation system for a set of entities is essentially a list of scores that provides a measure of the reliability of each entity in the set. The score given to an entity can be interpreted (and in the reputation system literature it often is [12]) as the probability that an entity will behave honestly. In this paper, we ask whether or not it is possible to utilize reputation systems for carrying out secure multiparty computation. We provide formal definitions of secure computation in this setting, and carry out a theoretical study of feasibility. We present almost tight results showing when it is and is not possible to achieve *fair* secure computation in our model. We suggest applications for our model in settings where some information about the honesty of other parties is given. This can be preferable to the current situation where either an honest majority is arbitrarily assumed, or a protocol that is secure for a dishonest majority is used and the efficiency and security guarantees (including fairness) of an honest majority are not obtained.

Keywords: secure multiparty computation, reputation systems, new models.

1 Introduction

1.1 Background

In the setting of secure multiparty computation, a set of mutually distrustful parties $P_1, \ldots, P_m$ wish to compute a function of their inputs in the presence of adversarial behavior. The security requirements of such a computation are that nothing beyond the output should be learned (privacy), the output received must be correctly computed (correctness), the parties must choose their inputs independently of each other (independence of inputs), and either no parties receive output or all parties receive output (fairness). The formal definition of security requires that the result of a secure protocol be like the outcome of an

---

* This research was supported by the European Research Council under the European Union's Seventh Framework Programme (FP/2007-2013) / ERC Grant Agreement n. 239868.
** Hila Zarosim is grateful to the Azrieli Foundation for the award of an Azrieli Fellowship.

K. Sako and P. Sarkar (Eds.) ASIACRYPT 2013, Part II, LNCS 8270, pp. 201–220, 2013.

ideal execution where an incorruptible trusted party is used to compute the function for all the parties. We remark that if there is no honest majority, then it is impossible to achieve fairness in general [8].

Under the assumption that the majority of the parties are honest, there exist protocols with full security [3,7,15]. However, the security of known protocols totally collapses when this assumption does not hold; in particular, the adversary can learn the inputs of the honest parties. Based on this, it may seem prudent to use protocols that guarantee security *except for fairness* when any number of parties are corrupted [15]. Unfortunately, all known protocols of this type have the property that *just one corrupted party* can prevent the parties from terminating successfully, and can even breach fairness. Moreover, it is known that there exist no protocols that simultaneously achieve full security for the case of honest majority, and security-with-abort (i.e., without fairness) when there is no honest majority [18]. This leads to the following unfortunate situation: the parties need to make a decision in advance whether to run a protocol that is secure as long as there is an honest majority and thereby risk losing privacy if they are wrong, or to run a protocol that is secure for any number of corruptions and thereby give up on any hope of obtaining fairness. To make things worse, *this decision is essentially made with no concrete information.*

Reputation Systems. A reputation system is a system whose aim is to predict agents' behavior in future transactions. Such systems evaluate the data about agents' previous transactions and estimate the probability that an agent will behave honestly or dishonestly in future transactions [12]. Reputation systems are very popular today in the electronic commerce market and in peer to peer systems [2]. They are used in these contexts to choose which vendors are trustworthy, to determine the level of service obtained by a peer, and more. There is considerable work on how to construct reliable reputation systems, maintain them and so on. Such systems provide us with information regarding the honesty of parties, and therefore could be utilized.

Reputation Systems and Secure Computation. In this paper, we study the use of reputation systems in order to carry out secure multiparty computation. We consider a model where all parties are given a reputation vector $(r_1, \ldots, r_m)$ with the interpretation that the probability that party P_i is honest is r_i. Another possible interpretation of this model is that there exists an adversary who attempts to corrupt as many parties as possible. Then, r_i is the probability that party P_i remains uncorrupted, and can depend on the security measures employed by party P_i. The main question that we ask is:

Can reputation systems be utilized in order to achieve fair and efficient secure multiparty computation?

This model differs from the standard model of secure computation since all parties are given information about the honesty of the other parties and the level at which they can be trusted. Thus, there is hope that this can be used to achieve more than is possible in the standard model. For example, it may be possible to use protocols that require an honest majority (that are more efficient than those for a dishonest majority, and in addition also guarantee fairness), without just arbitrarily hoping that a majority of the parties are honest. It is important that this actually models a more general setting than just that of reputation systems; see below.

1.2 Our Results

Our main contributions are as follows. First, we suggest this novel model for secure computation and provide formal definitions of security in this model. Next, we study the problem of secure computation with reputation systems from a theoretical perspective. Specifically, we ask under what conditions on the reputation vector it is possible to achieve *fair* secure multiparty computation. We stress that our focus is on fairness since without this requirement one can just ignore the reputation system entirely and run a protocol like [15] that assumes no honest majority and guarantees security without fairness. We present both feasibility and impossibility results for this setting.

Regarding feasibility, we provide a criterion for when the reputations are such that there exists a subset of parties for which a majority are honest, except with negligible probability. Thus, when a reputation vector fulfills the criterion, it is possible to have this subset run a secure protocol that assumes an honest majority. Using the protocol of [10], this subset can be used to run the protocol for many other parties who just provide input and receive output (and it does not matter how many of these other parties are corrupted). Regarding impossibility, we present another criterion on the reputations and show that when this criterion is fulfilled it is impossible to securely toss a coin. This is proven by showing a reduction to the case of two-party coin tossing in the standard model of secure computation, for which the impossibility of fair coin tossing is well known [8]. Interestingly, we show that in the case of *constant* reputation values (that do not depend on the security parameter), our characterization is tight. That is, we prove the following very informally stated theorem:

Theorem 1.1 (Feasibility Characterization – Informal Statement). *Let* Rep *be a reputation system. Then, there exist protocols for securely computing every family of functionalities F with complete fairness with respect to* Rep *if and only if the number of parties with reputation greater than* $1/2$ *is superlogarithmic in the security parameter* n.

As we have mentioned, the positive result is obtained by showing that when the condition on the number of parties with reputation greater than $1/2$ is fulfilled then there exists a subset of parties within which there is an honest majority, except with negligible probability. Thus, standard protocols for secure computation with fairness can be run by this subset.

The main question that this leaves is whether or not it is possible to use a reputation system to achieve fairness in a different and more "interesting" way than just finding a subset for within which there is an honest majority. We show that in fact it is *impossible* to utilize reputations systems in any other way, and so our upper bound is almost tight.[1]

We also show how it is possible to use our feasibility result given a concrete reputation system; This is not immediate since our theoretical feasibility result is asymptotic also in the number of parties and requires finding a subset of parties whose reputation values fulfill a special property.

Reputation Systems with Correlations. The aforementioned basic model implicitly assumes independence between parties, since each party P_i is corrupted with probability r_i as given in the reputation vector. This therefore does not model the case that P_i and P_j are both corrupted if and only if some P_k is corrupted. We therefore also study the more general setting where the probabilities that parties are corrupted may be correlated. In this setting, we show that as long as the correlations are "limited" in the sense that each party is dependent on only ℓ other parties, then an honest majority exists (except with negligible probability) if the expected number of honest parties is "large enough". We formally define what it means for correlations to be limited to ℓ, and give a criterion on the required expected number of honest parties. We prove this using martingales.

We remark that although this extension allows a more general type of reputation system, real-world reputation systems work by providing a vector stating the individual probabilities that every party is corrupted. Thus, we view the basic model as our main model.

Covert Security. We observe that the model of security in the presence of covert adversaries [1], where the guarantee is that any cheating is detected by honest parties, is particularly suited to our setting where there is an existing reputation system. This is due to the fact that any cheating will go immediately punished by reporting such a cheating to the reputation system manager. In addition, we observe that it is possible to use the protocol of [9] that is only twice as expensive as a semi-honest information-theoretic protocol, and provides a deterrent of 1/4 (meaning that any cheating is detected with probability at least 1/4). Such protocols have been proven to be highly efficient.

Applications to other Settings. Our basic model for secure computation with reputations actually relates to any setting where additional information about the honesty of the parties is known. Two examples of such settings are as follows. First, consider an environment with an access control scheme where there is non-negligible probability of impersonation. Expressing this probability of cheating as a reputation system and using our protocols, it is possible to neutralize the threat from the impersonators. A second example relates to a set of servers where intrusion detection tools provide an indication as to whether or not a given server has been compromised. Rather than naively assuming

---

[1] We remark that in the general case that the parties' reputations may depend on the security parameter, our results are *not* completely tight; in the full version of this paper, we present a concrete example of reputation values for which neither our feasibility result nor impossibility result holds.

that a majority of the servers have not been compromised, it is possible to use the indicators of the intrusion detection system within our protocol in order to obtain a more robust solution. This setting fits in very nicely with numerous recent projects that offer a secure computation service, where a set of servers carry out the computation and security is guaranteed as long as a majority of them are not compromised [4,5].

2 Definitions – The Basic Model

2.1 Reputation Vectors and Secure Computation

Let f be an m-ary functionality and let π be an m-party protocol for computing f. A reputation vector r for the protocol π is a vector in $\mathbb{R}^m$ such that for every i, the value r_i indicates the probability that party P_i is honest in an execution of π. We assume that r is public information, and is obtained from an external authority that handles the reputation system. Our goal in this work is to study the advantages that such public information can provide in constructing secure protocols. We remark that reporting malicious behavior of individuals and maintaining the reputation system is out of this scope of this work. A huge amount of work deals with how to adapt the reputation of an individual in case it has been corrupted. Incorporating cryptography to this task and proving the system that indeed the entity behaves inappropriately, seems as an appealing future direction.

In the standard setting of secure computation, we are given a fixed m-ary function and our goal is to construct a secure m-party protocol π for computing f. Thus the functionality to be computed is fixed and hence its arity is fixed as well. However, in this paper we wish to work asymptotically in the number of parties as well, and this makes things more complicated. The reason that we work in this way is so that we can reason about the probability that some subset of parties of a given size is corrupted. In order to see this, consider a protocol that is secure as long as the majority of the parties are honest. Then, consider the case that all parties are honest with probability $3/4$ (and otherwise they are corrupted). Clearly, for a sufficiently large number of parties it is possible to apply the Chernoff bound in order to argue that the probability that there is no honest majority is negligible. However, this is only possible when we consider an asymptotic analysis over the number of parties. We stress that just like the use of a security parameter, in a real instantiation of a protocol one would set a concrete allowed error probability (e.g., 2^{-40}) and verify that for the given real number of parties and their reputation vector, the protocol error is below this allowed probability.

Toward this end, we consider a family of functionalities, each with a different arity, rather than considering a fixed functionality. We require the existence of a polynomial-time process that is given the requested arity m and security parameter n and outputs a circuit $C_{n,m}$ for computing the functionality f^m; this suits the natural case that f computes the same function for each m and the only difference is the number of inputs (e.g., statistics like median, majority and so on). Formally,

Definition 2.1. *Let $\mathcal{F} = \{f^m\}_{m \in \mathbb{N}}$ be an infinite family of functionalities, where f^m is an m-ary functionality. We say that $\mathcal{F}$ is a PPT family of functionalities if there exists a polynomial $p(\cdot)$ and a machine M that on input n and m outputs a circuit $C_{n,m}$ in time at most $p(n + m)$ such that for every $x_1, \ldots, x_m$, it holds that $C_{n,m}(x_1, \ldots, x_m) = f^{(m)}(1^n, x_1, \ldots, x_m)$.*

We define a family of protocols $\Pi(m, n)$ in the same way, and say that it is polynomial time if there exists a polynomial $p(\cdot)$ such that the running time of all parties is bounded by $p(m + n)$. We will consider the case where the number of parties $m = m(n)$ is bounded by a polynomial in the security parameter n. This makes sense since any given party cannot run more than poly(n) in any case, and so if the number of parties $m(n)$ is superpolynomial in n, then it will not be possible to even send a single message to all other parties.

Summary. We consider secure computation with $m = m(n)$ parties, where $m : \mathbb{N} \to \mathbb{N}$ is bounded by a polynomial in n.[2] The parties run a protocol $\Pi(m, n)$, which is an m-party protocol with security parameter n, that securely computes the functionality f^m in the class $\mathcal{F} = \{f^{m(n)}\}_{n \in \mathbb{N}}$. Finally, the parties have for auxiliary input a reputation vector r^m such that for every $i \in [m]$, the probability that party P_i is corrupted is r_i^m. As we will see, we will require that for all large enough values of n (which also determines $m = m(n)$), the protocol $\Pi(m, n)$ securely computes f^m with respect to the reputation vector r^m. Thus, we also need to consider a family of reputation vectors, one for each value of m; we denote the family of reputation vectors for every n by $\mathsf{Rep} = \{r^{m(n)}\}_{n \in \mathbb{N}}$.

2.2 Security with Respect to a Reputation Vector

We assume that the reader is familiar with the standard definition of security for secure computation (see [14,6]). We modify the definition to allow for a varying number of parties, that is, $m(n)$ for a given function $m : \mathbb{N} \to \mathbb{N}$.

Definition 2.2. *Let $m : \mathbb{N} \to \mathbb{N}$ be a function. We say that the protocol Π $t(\cdot)$-securely computes the functionality $\mathcal{F} = \{f^{m(n)}\}_{n \in \mathbb{N}}$ with respect to $m(\cdot)$, if for every PPT adversary $\mathcal{A}$, there exists a PPT simulator $\mathcal{S}$, such that for every PPT distinguisher D, there exists a negligible function $\mu(\cdot)$ such that for every $n \in \mathbb{N}$, every $I \subseteq [m(n)]$ with $|I| \leq t(m(n))$, every $x \in (\{0,1\}^*)^{m(n)}$ and $z \in \{0,1\}^*$, it holds that:*

$$\left| \Pr \left[D \left(\mathrm{IDEAL}_{\mathcal{F},\mathcal{S}(z),I} (n, m, x) \right) = 1 \right] \right.$$
$$\left. - \Pr \left[D \left(\mathrm{REAL}_{\Pi,\mathcal{A}(z),I} (n, m, x) \right) = 1 \right] \right| \leq \mu(n).$$

---

[2] We remark that the naive approach of taking m to be a parameter that is independent of n does not work, since security would also need to hold when m is superpolynomial in n. However, in such a case, one cannot rely on cryptographic hardness. In addition, bounding m by a polynomial in n is natural in the sense that parties cannot run in time that is superpolynomial in n in any case.

The protocol of GMW [15,14] satisfies this definition, and is secure even when the number of parties m is a function of n, as long as it is *polynomial*. Formally, let $\Pi(m, n)$ denote the GMW protocol with the following change. Let $\mathcal{F} = \{f^{m(n)}\}_{n \in \mathbb{N}}$ be a functionality. Then, upon input $1^n, 1^m$, each party runs the polynomial-time process to obtain the circuit $C_{n,m}$ for computing f^m. They then proceed to run the GMW protocol with m parties on this circuit. We are interested here in the version of GMW that assumes an honest majority and guarantees fairness. Thus, we have:

Fact 2.3. *Let $\mathcal{F} = \{f^{m(n)}\}_{n \in \mathbb{N}}$ be a functionality and let Π denote the GMW protocol as described above for $\mathcal{F}$. Then, for every polynomial $m(\cdot) : \mathbb{N} \to \mathbb{N}$, the protocol $\Pi(m, n) \frac{m(n)}{2}$-securely computes $\mathcal{F}$ with respect to $m(n)$.*

Having defined security with respect to a varying number of parties, and thus actually being asymptotic also in the number of parties, we proceed to include the reputation system as well. The definition is the same except that instead of quantifying over *all possible subsets of corrupted parties of a certain size*, the set of corrupted parties is chosen probabilistically according to the given reputation vector $r^m = (r_1^m, \dots, r_m^m)$. We denote by $I \leftarrow r^m$ the subset $I \subseteq [m]$ of parties chosen probabilistically where every $i \in I$ with probability $1 - r_i^m$ (independently of all $j \neq i$). We note that the output of IDEAL and REAL includes $1^n, 1^m, x, z$ and I. Thus the probabilistic choice of I is given to the distinguisher.

Definition 2.4 (Security with Respect to $(m(\cdot), \mathsf{Rep})$). *Let $m(\cdot)$, Rep, $\mathcal{F}$ and Π be as above. We say that Π securely computes $\mathcal{F}$ with respect to $(m(\cdot), \mathsf{Rep})$, if for every PPT adversary $\mathcal{A}$, there exists a PPT simulator $\mathcal{S}$, such that for every PPT distinguisher D, there exists a negligible function $\mu(\cdot)$ such that for every $n \in \mathbb{N}$, every $x \in (\{0,1\}^*)^{m(n)}$ and $z \in \{0,1\}^*$, it holds that:*

$$\left| \Pr_{I \leftarrow r^{m(n)}} \left[D \left(\mathrm{IDEAL}_{\mathcal{F}, \mathcal{S}(z), I} \left(n, m(n), x \right) \right) = 1 \right] \right.$$

$$\left. - \Pr_{I \leftarrow r^{m(n)}} \left[D \left(\mathrm{REAL}_{\Pi, \mathcal{A}(z), I} \left(n, m(n), x \right) \right) = 1 \right] \right| \leq \mu(n)$$

Observe that a protocol that is secure with respect to a reputation vector is allowed to always fail for a certain subset I of corrupted parties, if that specific corruption subset is only obtained with negligible probability with the reputation vectors in Rep.

3   A Theoretical Study

In this section we explore our model, and ask under what conditions on the reputation vector, security can be obtained. We first observe that when an honest majority (over all or just a subset of parties) can be guaranteed except with negligible probability, then it is possible to run protocols that are secure with an honest majority like [15,14] and [21]. We then present a simple condition on a family of reputation vectors that determines whether or not an honest majority on a subset exists. We also present a condition on the reputation vectors for which it is impossible to securely compute the coin tossing functionality. This is shown by reduction to the impossibility of computing two-party coin tossing with fairness [8]. Finally, we show that when the reputations are *constant*, then the above conditions are complementary. In the full version we give an example of a reputation vector with probabilities that depend on n for which neither of our conditions apply.

3.1   Feasibility

Reputation Vectors and Honest Majority. We begin by presenting a simple property for evaluating whether or not a family of reputation vectors guarantees an honest majority, except with negligible probability. It is clear that if all parties have reputation $r_i > \frac{1}{2} + \epsilon$ for some constant ϵ, then there will be an honest majority except with probability that is negligible in the number of parties; this can be seen by applying the Chernoff bound. Likewise, if at least two thirds of the parties have reputation $r_i > \frac{3}{4} + \epsilon$, then a similar calculation will yield an honest majority except with negligible probability. However, these calculations require a large subset of parties to have high reputation, and the use of Chernoff requires that we use the same probability for a large set. Thus, this does not address the case that $1/4$ of the parties have very high reputation (almost 1), another half have reputation $1/2$, and the remaining $1/4$ have low reputation. In order to consider this type of case, we use the Hoeffding Inequality [19]. This enables us to relate to the overall *sum* (or equivalently, *average*) of the reputations of all parties. Using this inequality, we obtain a very simple condition on reputation vectors. Namely, given a family $\mathsf{Rep} = \{r^{m(n)}\}_{n \in \mathbb{N}}$ and a polynomial $m = m(n)$, we simply require that for all sufficiently large n's, the average of the reputations is greater than: $1/2 + \omega\left(\sqrt{\frac{\log m}{m}}\right)$, or, equivalently, that the expected number of honest parties is greater than: $m/2 + \omega\left(\sqrt{m \cdot \log m}\right)$.

Before proceeding to the formal proof, we first state the Hoeffding Inequality [19] (see also [13, Sec. 1.2]). In our specific case, all of the random variables have values between 0 and 1, and we therefore write a simplified inequality for this case.

Lemma 3.1 (The Hoeffding Inequality). *Let $X_1, \ldots, X_m$ be m independent random variables, each ranging over the (real) interval $[0, 1]$, and let $\mu = \frac{1}{m} \cdot \mathrm{E}[\sum_{i=1}^m X_i]$ denote the expected value of the mean of these variables. Then, for every $\epsilon > 0$, $\Pr\left[\left|\frac{\sum_{i=1}^m X_i}{m} - \mu\right| \geq \epsilon\right] \leq 2 \cdot e^{-2\epsilon^2 \cdot m}$.*

Claim 3.2. *Let $m : \mathbb{N} \to \mathbb{N}$ be such that $O(\log m(n)) = O(\log n)$, let $\mathsf{Rep} = \{r^{m(n)}\}_{n \in \mathbb{N}}$ be a family of reputation vectors and let $m = m(n)$. If it holds that*

$$\sum_{i=1}^m r_i^m > \left\lfloor \frac{m}{2} \right\rfloor + \omega\left(\sqrt{m \cdot \log m}\right),$$

then there exists a negligible function $\mu(n)$ such that for every n,

$$\Pr_{I \leftarrow r^m}\left[|I| \geq \left\lfloor \frac{m}{2} \right\rfloor\right] < \mu(n) .$$

Proof: Fix n and let $m = m(n)$. For every $i \in [m]$, let X_i be a random variable that equals 1 when party P_i is honest, and 0 when it is corrupted. Thus, $\Pr[X_i = 1] = r_i$. Let $\bar{X} = \frac{\sum_{i=1}^m X_i}{m}$. Using linearity of expectations, we have that $\mathrm{E}[\bar{X}] = \frac{1}{m}\sum_{i=1}^m r_i$.

There is an honest majority when $|I| < \frac{m}{2}$; equivalently, when $\sum_{i=1}^m X_i \geq \lfloor m/2 \rfloor + 1$. Let $\Delta = (\sum_{i=1}^m r_i) - \lfloor m/2 \rfloor = m \cdot \mathrm{E}[\bar{X}] - \lfloor m/2 \rfloor$. By the Hoeffding inequality:

$$\Pr\left[\sum_{i=1}^m X_i \leq \left\lfloor \frac{m}{2} \right\rfloor\right] = \Pr\left[\sum_{i=1}^m X_i - m \cdot \mathrm{E}\left[\bar{X}\right] \leq \left\lfloor \frac{m}{2} \right\rfloor - m \cdot \mathrm{E}\left[\bar{X}\right]\right]$$

$$= \Pr\left[\sum_{i=1}^m X_i - m \cdot \mathrm{E}\left[\bar{X}\right] \leq -\Delta\right] = \Pr\left[\sum_{i=1}^m X_i - m \cdot \mathrm{E}\left[\bar{X}\right] \leq -m \cdot \frac{\Delta}{m}\right]$$

$$= \Pr\left[\frac{\sum_{i=1}^m X_i}{m} - \mathrm{E}\left[\bar{X}\right] \leq -\frac{\Delta}{m}\right] \leq 2e^{-\frac{2\Delta^2}{m}} .$$

The above holds for all n and $m = m(n)$. Asymptotically, by the assumption in the claim, $\Delta = \omega(\sqrt{m \cdot \log m})$ and thus $\frac{\Delta^2}{m} = \omega(\log m)$. Hence we have that, $\Pr\left[|I| \geq \lfloor \frac{m}{2} \rfloor\right] = \Pr\left[\sum_{i=1}^m X_i \leq \lfloor \frac{m}{2} \rfloor\right] \leq 2e^{-\frac{2\Delta^2}{m}} < 2e^{-\omega(\log m)}$ which is negligible in m. Since $m(\cdot)$ is a function such that $O(\log m(n)) = O(\log n)$, it holds that $e^{-\omega(\log m(n))}$ is a function that is negligible in n, as required. ∎

Intuitively, in order to use the above for secure computation, all the parties need to do is to run a protocol that is secure with an honest majority (like GMW [15]). Since there is guaranteed to be an honest majority except with negligible probability, then this is fine. We stress, however, that for this to work we need to use Fact 2.3 since here we refer to the version of GMW for which the number of parties m is a *parameter*, and is not fixed. We therefore conclude:

Theorem 3.3. *Let $\mathcal{F}$ be as above, and let $\Pi = \{\Pi(m,n)\}$ be the GMW protocol of Fact 2.3. Let $m(\cdot)$ be a function such that $O(\log m(n)) = O(\log n)$, let $m = m(n)$ and let Rep be as above. If*

$$\sum_{i=1}^m r_i^m > \left\lfloor \frac{m}{2} \right\rfloor + \omega(\sqrt{m \log m}) ,$$

then Π securely computes $\mathcal{F}$ with respect to $(m(\cdot), \mathsf{Rep})$.

The proof of this is immediate; if there is an honest majority then the real and ideal executions are indistinguishable, and there is an honest majority except with negligible probability.

Subset Honest Majority. In order to achieve secure computation with complete fairness, it suffices to have a subset of parties for which there is guaranteed to be an honest majority except with negligible probability [10]. This works by having the subset carry out the actual computation for all other parties. Specifically, all the parties send shares of their inputs to the subset, who compute shares of the output and return them. In more detail, the protocol of [10] works as follows. Let $T \subseteq [m(n)]$ be the subset of parties that carry out the actual computation; these are called the *servers*. All the parties distribute their inputs to the set of servers T using VSS (verifiable secret sharing) with threshold $|T|/2+1$. The servers then compute shares of the outputs by computing the circuit gate by gate. At the output phase, the servers send the appropriate shares of the outputs to each party, who then reconstructs the output. See [10] for details, and for a proof that the protocol is secure as long as a majority of the parties in T are honest. Thus, as long as there exists a subset of parties $T \subseteq [m(n)]$ with honest majority except with negligible probability, there exists a protocol for this $m(\cdot)$ and family of reputation vectors. Thus, we have:

Claim 3.4. *Let $\mathcal{F}$, $m(\cdot)$ and Rep be as above. If there exists a negligible function $\mu(\cdot)$, such that for every n there exists a subset $T_n \subset [m(n)]$ for which $\Pr_{I \leftarrow r^{m(n)}} \left[|T_n \cap I| \leq \frac{|T_n|}{2} \right] \leq \mu(n)$, then there exists a (non-uniform) protocol Π that securely computes $\mathcal{F}$ with respect to $(m(\cdot), \text{Rep})$.*

The proof of this claim is the same as the proof of Theorem 3.3: if there is a subset with an honest majority then the security of [10] holds, and the probability that there is not an honest majority is negligible. There is one subtle point here, which is that the protocol as described is *non-uniform* since the subset T_n may be different for every n. Nevertheless, as we will see below, our criteria for the existence of such a subset is such that an appropriate subset T_n can always be efficiently computed from the reputation vector $r^{m(n)}$ (assuming its existence).

Criteria on the Reputation Vector. Our next goal is to analyze when a family of reputation vectors guarantees a subset of parties with an honest majority, except with negligible probability. We first present the technical criteria, and then explain its significance below.

Claim 3.5. *Let $m(\cdot)$, and Rep be as above. For every n and subset $T_n \subseteq [m(n)]$, let $\Delta_{T_n} \stackrel{\text{def}}{=} \sum_{i \in T_n} r_i^{m(n)} - \frac{|T_n|}{2}$. If there exists a series of subsets $\{T_n\}_{n \in \mathbb{N}}$ (each $T_n \subseteq [m(n)]$) such that $\frac{(\Delta_{T_n})^2}{|T_n|} = \omega(\log n)$, then there exists a negligible function $\mu(\cdot)$ such that for every n, $\Pr_{I \leftarrow r^{m(n)}} \left[|I \cap T_n| > \frac{|T_n|}{2} \right] \leq \mu(n)$.*

Proof: The proof of this claim is very similar to that of Claim 3.2, and uses the Hoeffding inequality. Let $\{T_n\}_{n \in \mathbb{N}}$ be a series of subsets as in the claim. Fix n and $T = T_n$. Then, for every $i \in T$, let X_i be a random variable that equals 1 when the party is honest and 0 when it is corrupted. An identical calculation to that carried out in the proof of Claim 3.2 yields that for every n,

$$\Pr_{I \leftarrow r^{m(n)}} \left[|I \cap T_n| > \frac{|T_n|}{2} \right] = \Pr \left[\sum_{i \in T_n} X_i \leq \frac{|T_n|}{2} \right] \leq 2e^{-\frac{2(\Delta_{T_n})^2}{|T_n|}}. \quad (1)$$

Since $\frac{(\Delta_{T_n})^2}{|T_n|} = \omega(\log n)$, we conclude that $\Pr_{I \leftarrow r^{m(n)}} \left[|I \cap T_n| > \frac{|T_n|}{2} \right] \leq 2e^{-\omega(\log n)}$ which is negligible in n, as required. ∎

Combining Claim 3.5 with Claim 3.4 we conclude that:

Corollary 3.6. *Let $\mathcal{F}$, $m(\cdot)$ and Rep be as above. For every n and subset $T_n \subseteq [m(n)]$, let $\Delta_{T_n} \overset{\text{def}}{=} \sum_{i \in T_n} r_i^{m(n)} - \frac{|T_n|}{2}$. If there exists a series of subsets $\{T_n\}_{n \in \mathbb{N}}$ (each $T_n \subseteq [m(n)]$) such that $\frac{(\Delta_{T_n})^2}{|T_n|} = \omega(\log n)$, then there exists a (non-uniform) protocol Π that securely computes $\mathcal{F}$ with respect to $(m(\cdot), \text{Rep})$.*

Efficiently Finding the Subset T_n. The non-uniformity of the protocol is due to the fact that in general, the subset T_n may not be efficiently computable from the reputation vector. Nevertheless, we show now that assuming the existence of a subset T_n fulfilling the condition, it is easy to find a subset T_n' (not necessarily equal to T_n) that also fulfills the condition.

In order to see this, first note that for any size t, the largest value of Δ (over all subsets $T_n \subseteq [m(n)]$ of size t) is obtained by taking the t indices i for which r_i is largest. This follows since $\Delta_{T_n} = (\sum_{i \in T_n} r_i) - \frac{|T_n|}{2}$ and so replacing an r_i in the sum with a larger r_j always gives a larger Δ_{T_n}. This gives the following algorithm for finding an appropriate subset:

1. Given r^m, sort the values in decreasing order; let $r_{i_1}, \ldots, r_{i_m}$ be the sorted values.
2. For every $j = 1, \ldots, m$, compute $\Delta_j = \left(\sum_{k=1}^{j} r_{i_k} \right) - \frac{j}{2}$.
3. Let j be the index for which $\frac{(\Delta_j)^2}{j}$ is maximum over all j's. Then, output the subset $T = \{i_1, \ldots, i_j\}$.

In order to see that this fulfills the requirement, observe that by the above observation, the maximum value of Δ_{T_n} for *all possible* subsets T_n is one of the values of $\Delta_1, \ldots, \Delta_m$. Therefore, if there exists a subset that fulfills the requirement, the subset output by the algorithm also fulfills the requirement.

A Protocol for the Concrete Setting. Our protocols above are proven secure under the *assumption* that there exists a subset T_n fulfilling the required *asymptotic* property. However, concretely, how can a set of parties know that there exists such a subset, and which subset to take? This turns out to be very easy since the Hoeffding inequality is exact, and not asymptotic. Thus, for a given error parameter δ (e.g., $\delta = 2^{-40}$ or $\delta = 2^{-80}$) and subset T_n, it is possible to simply compute Δ_{T_n} and then check if $2e^{-\frac{2(\Delta_{T_n})^2}{|T_n|}} < \delta$ (this bound is obtained similarly to the bound in the proof of Claim 3.2; see Eq. (1) for more details). By the algorithm given above for finding the subset, it is possible to efficiently find an appropriate T_n with an error below the allowed bound, if one exists. (If one does not exist, then the parties know that they cannot run the protocol.) We remark that for efficiency, it is best to take the smallest subset that gives a value below the allowed error parameter, since this means that the protocol run by the parties has less participants, and so is more efficient.

Inaccurate Reputation System. Sometimes, the reputation system may be inaccurate, where the true reputation value of the party is ϵ-far from its public value. This error may arise in both directions, that is, sometimes the public reputation might be lower than the true one, and sometimes it might be higher. However, it is easy to generalize our results to deal with this case as well, while taking the reputation as the minimum guaranteed value and considering the worst case scenario.

3.2   Impossibility

We now turn to study under what conditions on the family of reputation vectors it is not possible to achieve (general) secure computation. We stress that we focus on the question of fairness here since one can always ignore the reputation vector and run a general protocol for secure computation *with abort* that is resilient to any number of corrupted parties. We therefore consider the problem of coin tossing, since it is impossible to fairly toss a coin without an honest majority [8] (or, more accurately, with only two parties).

Let $m(\cdot)$ be a function and let $\mathsf{Rep} = \{r^{m(n)}\}_{n\in\mathbb{N}}$ be a family of reputation vectors. For every n, we denote by $\mathsf{H}_n^{1/2}$ the set of all indices i of parties P_i such that $\frac{1}{2} < r_i^{m(n)} < 1$. Recall that $m(\cdot)$ denotes the number of parties and so the size of the set $\mathsf{H}_n^{1/2}$ is bounded by $m(n)$. We denote by $\mathcal{F} = \{f_{CT}^{m(n)}\}_{n\in\mathbb{N}}$ the coin-tossing functionality: $f_{CT}^{m(n)}(1^n, \ldots, 1^n) = (U_1, \ldots, U_1)$, where U_1 denotes a uniform random bit; i.e., the output of the function is the same random bit for all parties.

The idea behind our proof of impossibility is as follows. Consider first for simplicity the case that all the reputations are at most $1/2$, and thus $\mathsf{H}_n^{1/2}$ is empty. This means that the expected number of corrupted parties is at least half and thus, intuitively, any protocol that is secure with respect to such a reputation vector must be secure in the presence of a dishonest majority. We show that this implies the existence of a two party protocol for fair coin tossing. We also prove impossibility when $\mathsf{H}_n^{1/2}$ is not empty but the probability of all parties in $\mathsf{H}_n^{1/2}$ being corrupted is non-negligible. In this case, we show that since security must hold even when all parties in $\mathsf{H}_n^{1/2}$ are corrupted, we can reduce to fair coin tossing even here.

Theorem 3.7. *Let $m(\cdot)$ be polynomially bounded, and let Rep be a family of reputation vectors. If there exists a polynomial $p(\cdot)$ such that for infinitely many*

n's it holds that the probability that all parties in $\mathsf{H}_n^{1/2}$ are corrupted is at least $\frac{1}{p(n)}$, then there does not exist a protocol Π that securely computes the multiparty coin-tossing functionality $\mathcal{F} = \{f_{CT}^{m(n)}\}_{n\in\mathbb{N}}$ with respect to $(m(\cdot), \mathsf{Rep})$.

Proof Sketch: Assume the existence of a protocol Π that securely computes the family $\mathcal{F} = \{f_{CT}^{m(n)}\}_{n\in\mathbb{N}}$ with respect to a polynomial $m(\cdot)$ and a family of reputation vectors Rep, and that there exists a polynomial $p(\cdot)$ such that for infinitely many n's, $\Pr_{I\leftarrow \boldsymbol{r}^{m(n)}}\left[\mathsf{H}_n^{1/2} \subseteq I\right] \geq \frac{1}{p(n)}$. We show that this implies the existence of an infinitely-often[3] non-uniform two-party protocol $\pi' = \langle P_0', P_1'\rangle$ for the coin-tossing functionality that is secure in the presence of malicious adversaries, in contradiction to the fact that fair coin tossing cannot be achieved [8].[4]

We start with an informal description of our transformation and we provide an informal explanation of why it works; we then describe in Protocol 3.8 the construction of π', the formal proof appears in the full version of this paper. We begin with the simpler case where Rep is such that for infinitely many n's, $\mathsf{H}_n^{1/2}$ is empty; that is, each party is honest with probability at most $1/2$. We use this to construct a two-party protocol π' in which on security parameter n, the two parties P_0' and P_1' emulate an execution of the $m = m(n)$-party protocol $\Pi(m, n)$ by randomly choosing which of the m parties in $\Pi(m, n)$ is under the control of P_0' and which of the parties in Π is under the control of P_1'. This can be done by tossing m coins, and giving the control of each (virtual) party for which the coin is 0 to P_0', and the control of each (virtual) party for which the coin is 1 to P_1'. The two parties then emulate an execution of the m-party protocol Π for the coin-tossing functionality f_{CT}^m, and determine the resulting bit according to the outputs of the (virtual) parties under their control.

Loosely speaking, we claim that the security of Π implies that this emulation is secure as well. To see this, note that in an execution of π', the m-party protocol Π is invoked when each of the parties of Π is under the control of P_0' with probability $1/2$ and under the control of P_1' with probability $1/2$. Thus, for every adversary controlling one of the parties in π', we expect to have about half of the parties in Π under its control (since each party in Π is under the adversary's control with probability $1/2$). Since Π is secure with respect to a family of reputation vectors Rep such that $\mathsf{H}_n^{1/2}$ is empty (and so, $r_i^m \leq \frac{1}{2}$ for every i), Π is secure when the expected number of the corrupted parties is $\sum_i (1 - r_i^m) \geq \frac{m}{2}$, and thus can handle the number of corruptions in the emulation carried out by the two party protocol π'.

So far, we have ignored two issues in the construction of π'. First, the coin tossing we use to decide which of the parties is controlled by P_0' and which is controlled by P_1' is only secure with abort, and so an adversary controlling P_0' or P_1' might abort before the other party sees the output of the coin tossing. However, we show that in this case the honest party can simply output a random bit and this adds no bias to the output. Intuitively, the reason that this works

---

[3] This means that the security of the protocol holds for infinitely many n's.

[4] We note that the proof of impossibility of two-party coin tossing of [8] holds also for infinitely-often non-uniform protocols (the proof of [8] holds for every fixed n).

is that if a party aborts before beginning the emulation of Π, then it has no meaningful information and so cannot bias the outcome. Thus, the other party may just output a random bit. Second, after the emulation ends, each of the parties P_0' and P_1' should determine their outputs. Recall that each of the two parties has a subset of the m parties in Π under its control and hence at the end of the emulation of Π, each of P_0' and P_1' sees a set of outputs (as each party in Π has an output). However, we expect that when the parties play honestly, all parties in Π output the same output. Moreover, even if some of the parties in Π are corrupted, by the security of Π, the output of the honest parties should be the same (except with a negligible probability). Therefore, intuitively it seems that P_0' (resp. P_1') can determine its output by considering the set of all outputs of the parties under its control. If all those parties have the same output, then P_0' outputs the common bit. Since we expect the event of not all parties having the same output happen only with a negligible probability, in this case P_0' (resp. P_1') can just output a $\perp$ symbol. However, when trying to formalize this idea, a technical problem arises because the expected number of honest outputs in π' may be larger than the expected number of honest outputs in Π (recall that in π' the expected number of honest parties is $\frac{m}{2}$ whereas in a real execution of Π the expected number of honest parties is $\sum_i r_i^m \leq \frac{m}{2}$). We overcome this by having the honest party in π' not consider the set of *all* outputs of the parties under its control, but rather choose a subset of the outputs that is expected to be of size $\sum_i r_i^m$. To do this, the parties in π' must know the vector $\boldsymbol{r}^m$ and hence the protocol we construct is non-uniform.

So far we only discussed the case that for infinitely many n's, $\mathsf{H}_n^{1/2}$ is empty. Now, assume that this does not hold and $\mathsf{H}_n^{1/2}$ is non-empty. In this case, the construction of π' fails because in the execution simulated by π', each party is corrupted with probability $\frac{1}{2}$ whereas in the real execution of Π, we have parties whose probabilities to be corrupted are strictly less than $\frac{1}{2}$. For example, assume that a certain party P_i in $\Pi(m, n)$ is honest with probability 0.9 (and hence - corrupted with probability 0.1). However, by the way π' is defined, this party will be under the control of the adversary with probability $\frac{1}{2}$. In this case, it might be that π' is not secure even though Π is secure, simply because the party P_i is more likely to be corrupted in π' than in $\Pi(m, n)$. However, we show that if for infinitely many n's, the probability of all parties in $\mathsf{H}_n^{1/2}$ being corrupted is polynomial in n, then Π must remain (infinitely often) secure even conditioned on the event that the parties in $\mathsf{H}_n^{1/2}$ are corrupted. This will imply that we can slightly modify the construction of π' such that one of the parties always controls the parties in $\mathsf{H}_n^{1/2}$, and obtain that even though these parties are more likely to be corrupted when π' simulates Π than in real executions of Π, the simulation carried out by π' still remains secure.

A formal construction of π' is given in Protocol 3.8 and the full proof that π' securely computes the two-party coin-tossing functionality with fairness appears in the full version of this paper.   ■

PROTOCOL 3.8 (Protocol $\pi' = \langle P_0', P_1' \rangle$ for two-party coin-tossing)

- (Non-uniform) auxiliary input: 1^n, $r^{m(n)}$.
- The Protocol:
  1. *Set up subsets for emulation:* Parties P_0' and P_1' invoke $m = m(n)$ executions of the f_{CT}^2 functionality with security-with-abort in order to obtain m coins; let $\boldsymbol{b} \in \{0,1\}^m$ be the resulting coins. If one of the parties receives $\bot$ (i.e., abort) for output, then it outputs a random bit and halts.
     Otherwise, the parties define $I_0 = \{i \mid b_i = 0\} \cup \mathsf{H}_n^{1/2}$, and $I_1 = [m] \setminus I_0$.
  2. *Emulate Π:* The parties P_0' and P_1' emulate an execution $\Pi = \Pi(m(n), n)$ for computing f_{CT}^m where P_0' controls the parties in I_0 and P_1' controls the parties in I_1; all parties use input 1^n.
  3. *Determine outputs:*
     (a) Party P_0' selects a subset $S^0 \subseteq I_0$ of the (virtual) parties under its control as follows. For every $i \in \mathsf{H}_n^{1/2}$, P_i is added to S^0 with probability r_i^m; for every $i \in I_0 \setminus \mathsf{H}_n^{1/2}$, P_i is added to S^0 with probability $2r_i^m$ (note that since $i \notin \mathsf{H}_n^{1/2}$, it holds that $r_i^m \leq \frac{1}{2}$ and hence $2r_i^m$ is a valid probability).
     P_0' outputs the bit $b \in \{0,1\}$ if *all* the virtual parties in S^0 output b in Π. Otherwise, it outputs $\bot$.
     (b) Party P_1' selects a subset $S^1 \subseteq I_1$ of the (virtual) parties under its control by adding each P_i (for $i \in I_1$) with probability $2r_i^m$ (as before, $i \notin \mathsf{H}_n^{1/2}$ and hence $2r_i^m$ is a valid probability).
     P_1' outputs the bit $b \in \{0,1\}$ if *all* the virtual parties in S^1 output b in Π. Otherwise, it outputs $\bot$.

3.3 Tightness of the Feasibility and Impossibility Results

Our feasibility result states that if there exists a series of subsets $\{T_n\}_{n \in \mathbb{N}}$ (each $T_n \subseteq [m(n)]$) such that $\frac{(\Delta_{T_n})^2}{|T_n|} = \omega(\log n)$, then there exists a secure protocol. In contrast, our impossibility result states that if for infinitely many n's, the probability that all parties in $\mathsf{H}_n^{1/2}$ are corrupted is $1/p(n)$, then there exists no protocol. In this section, we clarify the relation between these two results.

Constant Reputations. We consider the case that all reputations are constant. This is somewhat tricky to define since the reputation vectors are modeled asymptotically themselves (each vector of length $m(n)$ can have different values and thus can depend on n). We therefore define "constant" by saying that there exists a finite set $\mathcal{R} = \{r_1, \ldots, r_L\}$ such that all reputation values (in all vectors) in Rep are in $\mathcal{R}$, and $1 \notin \mathcal{R}$ (if $1 \in \mathcal{R}$ then this is an uncorruptible trusted party and secure computation is trivial). In this case, we say that Rep has constant reputations. We have the following theorem:

Theorem 3.9. *Let $m(\cdot)$ be a polynomial, and let Rep be a family of constant reputations. Then, there exist protocols for securely computing every PPT family of functionalities $\mathcal{F}$ with respect to $(m(\cdot), \mathsf{Rep})$, if and only if it holds that $|\mathsf{H}_n^{1/2}| = \omega(\log n)$.*

Proof: The existence of protocols for every family of functionalities $\mathcal{F}$ when $|H_n^{1/2}| = \omega(\log n)$ can be seen as follows. Let r be the smallest value greater than $1/2$ in $\mathcal{R}$. This implies that all reputations in $H_n^{1/2}$ for all n are at least r. Thus, $\Delta_{H_n^{1/2}} = \sum_{i \in H_n^{1/2}} \left(r_i^{m(n)} - \frac{1}{2} \right) \geq \sum_{i \in H_n^{1/2}} \left(r - \frac{1}{2} \right) = |H_n^{1/2}| \cdot \left(r - \frac{1}{2} \right)$. Now, take $T_n = H_n^{1/2}$ and we have that $\frac{(\Delta_{T_n})^2}{|T_n|} \geq \frac{|T_n|^2 \cdot (r - \frac{1}{2})^2}{|T_n|} = |T_n| \cdot \left(r - \frac{1}{2} \right)^2 = \omega(\log n)$, where the last equality holds because $(r - 1/2)^2$ is constant, and by the assumption $|T_n| = |H_n^{1/2}| = \omega(\log n)$. Thus, by Corollary 3.6, there exists a protocol for every family $\mathcal{F}$.

For the other direction, assume that it is not the case that $|H_n^{1/2}| = \omega(\log n)$, for every n. This implies that there exists a constant c such that for infinitely many n's, $|H_n^{1/2}| \leq c \cdot \log n$. Now, let r' be the highest value in $\mathcal{R}$. It follows that for every $i \in H_n^{1/2}$, $r_i \leq r'$. Thus, for infinitely many n's the probability that all parties in $H_n^{1/2}$ are corrupted is at least $(1 - r')^{-c \cdot \log n}$. Since $(1 - r')$ is constant, $(1 - r')^{-c \cdot \log n}$ is $1/\text{poly}(n)$. Thus, by Theorem 3.7, there exists no protocol for coin-tossing (and so it does not hold that all functionalities can be securely computed). ∎

We conclude that in the case of constant reputations, our results are *tight*. In the full version we give an example of a family of reputation vectors with non-constant reputations, for which neither our upper bound nor lower bound applies.

4   Secure Computation with Correlated Reputations

Until now, we have considered reputation systems where the probability that each party is corrupted is independent of the probability that all other parties are corrupted. This follows directly from how we define reputation systems (and, indeed, the way these systems are typically defined in other fields). A natural question that arises is what happens when the probabilities that parties are corrupted are *not* independent; that is, when there is a correlation between the probability that some party P_i is corrupted and the probability that some P_j (or some subset of other parties) is corrupted. In this section we take a first step towards exploring the feasibility of fair secure computation with correlated reputation systems.

We begin by extending Definition 2.4 to this more general case. First, observe that a reputation system can no longer be represented as a vector of probabilities, since this inherently assumes independence. (Specifically, no vector can represent a corruption situation where with probability $1/2$ both parties P_1 and P_2 are corrupted and with probability $1/2$ they are both honest.) Thus, we represent a reputation system with $m = m(n)$ parties where $m : \mathbb{N} \to \mathbb{N}$ is bounded by a polynomial in n by a probabilistic polynomial-time sampling machine M that receives the security parameter $n \in \mathbb{N}$ and outputs a set $I \subseteq [m(n)]$, such that P_i is corrupted if $i \in I$. We call (m, M) the **reputation system**.

The definition of security is the natural extension of Definition 2.4. Specifically, we say that a protocol Π securely computes $\mathcal{F}$ with respect to a reputation system

(m, M) if all is the same as in Definition 2.4 except that the set I of corrupted parties is chosen by running $M(1^n)$.

We remark that unlike the case of *reputation vectors*, it does not suffice to look at the expected number of honest parties here. In order to see this, consider the case of m parties such that with probability $1/100$ all the parties but one are corrupted, and with probability $99/100$ all the parties are honest. According to this, the expected number of honest parties is $1 \cdot 1/100 + 99/100 \cdot m$ which is greater than $0.99m$. Nevertheless, the probability that there is a dishonest majority is $1/100$ which is clearly non-negligible. Thus, in the setting of correlated reputations where there is dependency between parties, a high expected number of honest parties does not imply that there is an honest majority except with negligible probability.

We show that when the "amount of dependence" between the parties is limited, then it is possible to obtain fair secure computation. In a nutshell, we show that if each party is dependent on at most $\ell(m)$ other parties, where $\ell(m)$ is some function, and the expected number of honest parties in a sampling by M is $\frac{m}{2} + \omega(\ell(m) \cdot \sqrt{m \log m})$, then there is an honest majority except with negligible probability. Given this fact, it is possible to run any multiparty computation protocol that is secure with an honest majority. Thus, in contrast to the above example, we conclude that when dependence is limited, it is possible to use the expected number of honest parties in order to bound the probability of a dishonest majority. *This is a direct generalization of Theorem 3.3, where the bound in Theorem 3.3 is obtained by setting $\ell(m) = 1$.* This result is proven by defining a martingale based on the random variables indicating the corruption status of each party, and then applying Azuma's inequality. We also consider a generalization of our model, where a party can be correlated also with all other $m - \ell(m)$ parties, but only to a very small extent. In addition, as with the case of reputation vectors, we also show how to extend this result to the case of a large enough subset with high enough expectation.

Another way to interpret the aforementioned result is as follows. In practical applications the reputation system is usually represented as a vector, although dependency between the parties may exist. Thus, the parties do not have all the information about the honesty of the parties. However, our analysis shows that the vector of reputations alone may still be useful. By linearity of expectation, the expected number of honest parties is the sum of reputations even if those are somehow dependent. Therefore, if this sum is large enough, honest majority is guaranteed except with negligible probability, even if there exists some correlation between the parties.

We now provide our definition for "limited correlations" and a formal statement of the main result. The full proof, together with additional results appear in the full version of this paper.

Defining Limited Dependence. Let $X_1, \dots, X_m$ be Boolean random variables such that $X_i = 1$ if and only if P_i is honest, or equivalently if and only if $i \notin I$. We begin by defining what it means for two parties to be dependent on each other. It is important to note that in our context the naive approach of saying that

P_i and P_j are dependent if the random variables X_i and X_j are dependent does *not* suffice. In order to see this, assume that there are three parties P_1, P_2 and P_3, such that P_3 is honest if and only if only one of P_1 and P_2 is honest, and that P_1 and P_2 are honest with probability $1/2$ (independently of each other). Clearly, by the standard notion of dependence P_1 and P_2 are independent. However, the honesty or lack thereof of P_1 and P_2 is influenced by P_3; stated differently, the random variables X_1 and X_2 are *not independent* when conditioning on X_3. Since we need to consider the global context of who is honest and who is corrupted, in such a case we should define that P_1 and P_2 *are* correlated.

Based on the above discussion, we define a new notion of dependency that we call *correlation amongst* $\mathcal{P}$, where $\mathcal{P} = \{P_1, \ldots, P_m\}$ is the set of all parties. Intuitively, we say that a pair of parties P_i and P_j are correlated amongst $\mathcal{P}$ if there exists a subset of parties in $\mathcal{P}$ such that the random variables X_i and X_j are not independent when conditioning on the corruption status of the parties in the subset. We stress that P_i and P_j are correlated as soon as the above holds for any subset. We believe that this is quite a natural definition that captures the intuitive meaning of dependence where the probability that a party is corrupted can depend on coalitions amongst other parties and whether or not they are corrupted. Formally:

Definition 4.1 (Correlated Amongst $\mathcal{P}$). *Let (m, M) be a reputation system. We say that parties P_i and P_j are* correlated amongst $\mathcal{P}$ *if there exists a subset $S \subseteq [m]$, and Boolean values $b_i, b_j, \{b_k\}_{k \in S}$ such that*

$$\Pr\left[X_i = b_i \wedge X_j = b_j \mid \{X_k = b_k\}_{k \in S}\right]$$
$$\neq \Pr\left[X_i = b_i \mid \{X_k = b_k\}_{k \in S}\right] \cdot \Pr\left[X_j = b_j \mid \{X_k = b_k\}_{k \in S}\right].$$

Let $\mathcal{D}(i)$ be the set of all parties P_j for which P_i and P_j are correlated amongst $\mathcal{P}$. Intuitively, we say that a reputation system has an ℓ-limited correlation if for every i it holds that the size of $\mathcal{D}(i)$ is at most ℓ; that is the number of parties with which any party P_i is correlated is at most ℓ.

Definition 4.2 (ℓ-Limited Correlation). *Let (m, M) be a reputation system and $\ell = \ell(m)$. We say that (m, M) has an ℓ-limited correlation if for every $i \in [m]$, it holds that $|\mathcal{D}(i)| \leq \ell(m)$.*

An Honest Majority in ℓ-limited Correlated Reputation Systems. We show that if the expected number of honest parties in an ℓ-limited correlated reputation systems is large enough, as a function of m and ℓ, then an honest majority is guaranteed except with negligible probability. The proof of this fact uses martingales and Azuma's inequality, and appears in the full version of this paper. Recall that X_i is a random variable that equals 1 when P_i is honest; thus $\sum_{i=1}^{m} X_i$ gives the number of honest parties. We show that the probability that this sum is less than $m/2$ is negligible.

Theorem 4.3. *Let $m : \mathbb{N} \to \mathbb{N}$ be a function such that $O(\log m(n)) = O(\log n)$, let $(m(n), M)$ be a family of reputation systems, and let $\ell = \ell(m)$ where $m = m(n)$. If $(m(n), M)$ has an ℓ-limited correlation and*

$$E\left[\sum_{i=1}^{m} X_i\right] \geq \frac{m}{2} + \omega\left(\ell(m) \cdot \sqrt{m \log m}\right),$$

then there exists a negligible function $\mu(\cdot)$ such that for every n,

$$\Pr\left[\sum_{i=1}^{m} X_i \leq \frac{m}{2}\right] < \mu(n).$$

The full proof and additional results appear in the full version of the paper.

5   Reputation and Covert Security

In the model of secure computation in the presence of covert adversaries [1], the security guarantee is that if a party cheats then it will be detected cheating with some probability ϵ (this probability is called the deterrent). The deterrent parameter ϵ can be tailored depending on the requirements. For $\epsilon = 1/2$, the cost of computing is between 2 and 4 times the cost of protocols that are secure for semi-honest adversaries. Thus, this is much more efficient than security in the presence of malicious adversaries. The model of covert adversaries is *particularly suited to a setting with reputation systems* since if cheating is detected, then an immediate "punishment" can be incurred via a report to the reputation system manager. Thus, the use of protocols that are secure for covert adversaries makes real sense here.

In addition to the above, we observe that great efficiency can be obtained by using the protocol of [9]. This protocol assumes an honest majority and obtains security in the presence of covert adversaries with deterrent $\frac{1}{4}$, at just twice the cost of obtaining information-theoretic security in the semi-honest model. Thus, this protocol is extremely efficient. Combining this with the fact that it is possible to run the protocol on the smallest subset that yields an honest majority except with probability below the allowed error δ (see Section 3.1), we have that large-scale computation involving many thousands of participants can be efficiently computed by taking a much smaller subset to run the protocol of [9].

References

1. Aumann, Y., Lindell, Y.: Security Against Covert Adversaries: Efficient Protocols for Realistic Adversaries. In: Vadhan, S.P. (ed.) TCC 2007. LNCS, vol. 4392, pp. 137–156. Springer, Heidelberg (2007)
2. Babaioff, M., Chuang, J., Feldman, M.: Incentives in Peer-to-Peer Systems. In: Algorithmic Game Theory, ch. 23. Cambridge University Press (2007)

3. Ben-Or, M., Goldwasser, S., Wigderson, A.: Completeness Theorems for Non-Cryptographic Fault-Tolerant Distributed Computation. In: 20th STOC, pp. 1–10 (1988)

4. Bogetoft, P., et al.: Secure Multiparty Computation Goes Live. In: Dingledine, R., Golle, P. (eds.) FC 2009. LNCS, vol. 5628, pp. 325–343. Springer, Heidelberg (2009)

5. Bogdanov, D., Laur, S., Willemson, J.: Sharemind: A Framework for Fast Privacy-Preserving Computations. In: Jajodia, S., Lopez, J. (eds.) ESORICS 2008. LNCS, vol. 5283, pp. 192–206. Springer, Heidelberg (2008)

6. Canetti, R.: Security and Composition of Multiparty Cryptographic Protocols. Journal of Cryptology 13(1), 143–202 (2000)

7. Chaum, D., Crépeau, C., Damgård, I.: Multi-party Unconditionally Secure Protocols. In: 20th STOC, pp. 11–19 (1988)

8. Cleve, R.: Limits on the Security of Coin Flips when Half the Processors are Faulty. In: 18th STOC, pp. 364–369 (1986)

9. Damgård, I., Geisler, M., Nielsen, J.B.: From Passive to Covert Security at Low Cost. In: Micciancio, D. (ed.) TCC 2010. LNCS, vol. 5978, pp. 128–145. Springer, Heidelberg (2010)

10. Damgård, I., Ishai, Y.: Constant-Round Multiparty Computation Using a Black-Box Pseudorandom Generator. In: Shoup, V. (ed.) CRYPTO 2005. LNCS, vol. 3621, pp. 378–394. Springer, Heidelberg (2005)

11. Dubhashi, D., Panconesi, A.: Concentration of Measure for the Analysis of Randomized Algorithms, 1st edn. Cambridge University Press, New York (2009)

12. Friedman, E., Resnick, P., Sami, R.: Manipulation-Resistant Reputation Systems. In: Algorithmic Game Theory, ch. 27. Cambridge University Press (2007)

13. Goldreich, O.: Foundations of Cryptography: Volume 1 – Basic Tools. Cambridge University Press (2001)

14. Goldreich, O.: Foundations of Cryptography: Volume 2 – Basic Applications. Cambridge University Press (2004)

15. Goldreich, O., Micali, S., Wigderson, A.: How to Play any Mental Game – A Completeness Theorem for Protocols with Honest Majority. In: 19th STOC, pp. 218–229 (1987), For details see [14]

16. Goyal, V., Mohassel, P., Smith, A.: Efficient Two Party and Multi Party Computation Against Covert Adversaries. In: Smart, N.P. (ed.) EUROCRYPT 2008. LNCS, vol. 4965, pp. 289–306. Springer, Heidelberg (2008)

17. Halevi, S., Lindell, Y., Pinkas, B.: Secure Computation on the Web: Computing without Simultaneous Interaction. In: Rogaway, P. (ed.) CRYPTO 2011. LNCS, vol. 6841, pp. 132–150. Springer, Heidelberg (2011)

18. Ishai, Y., Katz, J., Kushilevitz, E., Lindell, Y., Petrank, E.: On Achieving the "Best of Both Worlds" in Secure Multiparty Computation. SIAM J. Comput. 40(1), 122–141 (2011)

19. Hoeffding, W.: Probability Inequalities for Sums of Bounded Random Variables. Journal of the American Statistical Association 58(301), 13–30 (1963)

20. Mitzenmacher, M., Upfal, E.: Probability and Computing: Randomized Algorithms and Probabilistic Analysis. Cambridge University Press, New York (2005)

21. Rabin, T., Ben-Or, M.: Verifiable Secret Sharing and Multi-party Protocols with Honest Majority. In: 21st STOC, pp. 73–85 (1989)

Between a Rock and a Hard Place: Interpolating between MPC and FHE

Ashish Choudhury, Jake Loftus, Emmanuela Orsini, Arpita Patra, and Nigel P. Smart

Dept. Computer Science,
University of Bristol,
United Kingdom
{Ashish.Choudhary,Emmanuela.Orsini,Arpita.Patra}@bristol.ac.uk,
{loftus,nigel}@cs.bris.ac.uk

Abstract. We present a computationally secure MPC protocol for threshold adversaries which is parametrized by a value L. When $L = 2$ we obtain a classical form of MPC protocol in which interaction is required for multiplications, as L increases interaction is reduced, in that one requires interaction only after computing a higher degree function. When L approaches infinity one obtains the FHE based protocol of Gentry, which requires no interaction. Thus one can trade communication for computation in a simple way. Our protocol is based on an interactive protocol for "bootstrapping" a somewhat homomorphic encryption (SHE) scheme. The key contribution is that our presented protocol is highly communication efficient enabling us to obtain reduced communication when compared to traditional MPC protocols for relatively small values of L.

1 Introduction

In the last few years computing on encrypted data via either Fully Homomorphic Encryption (FHE) or Multi-Party Computation (MPC) has been subject to a remarkable number of improvements. Firstly, FHE was shown to be possible [23]; and this was quickly followed by a variety of applications and performance improvements [6,9,8,24,25,29,30]. Secondly, whilst MPC has been around for over thirty years, only in the last few years we have seen an increased emphasis on practical instantiations; with some very impressive results [5,18,28].

We focus on MPC where n parties wish to compute a function on their respective inputs. Whilst the computational overhead of MPC protocols, compared to computing "in the clear", is relatively small (for example in practical protocols such as [20,28] a small constant multiple of the "in the clear" cost), the main restriction on practical deployment of MPC is the communication cost. Even for protocols in the preprocessing model, evaluating arithmetic circuits over a field $\mathbb{F}_p$, the communication cost in terms of number of bits per multiplication gate and per party is a constant multiple of the bit length, $\log p$, of the data being manipulated for a typically large value of the constant. This is a major drawback of MPC protocols since communication is generally more expensive than computation. Theoretical results like [15] (for the computational case) and [16] (for the information theoretic case) bring down the per gate per party communication cost to a very small quantity; essentially $\mathcal{O}(\frac{\log n}{n} \cdot \log |C| \cdot \log p)$ bits for a

K. Sako and P. Sarkar (Eds.) ASIACRYPT 2013, Part II, LNCS 8270, pp. 221–240, 2013.

circuit C of size $|C|$. While these results suggest that the communication cost can be asymptotically brought down to a constant for large n, the constants are known to be large for any practical purpose. Our interest lies in constructing efficient MPC protocols where the efficiency is measured in terms of *exact* complexity rather than the *asymptotic* complexity.

In his thesis, Gentry [22] showed how FHE can be used to reduce the communication cost of MPC down to virtually zero for any number of parties. In Gentry's MPC protocol all parties send to each other the encryptions of their inputs under a shared FHE public key. They then compute the function homomorphically, and at the end perform a shared decryption. This implies an MPC protocol whose communication is limited to a function of the input and output sizes, and not to the complexity of the circuit. However, this reduction in communication complexity comes at a cost, namely the huge expense of evaluating homomorphically the function. With current understanding of FHE technology, this solution is completely infeasible in practice.

A variant of Gentry's protocol was presented by Asharov et al. in [1] where the parties outsource their computation to a server and only interact via a distributed decryption. The key innovation in [1] was that independently generated (FHE) keys can be combined into a "global" FHE key with distributed decryption capability. We do not assume such a functionality of the keys (but one can easily extend our results to accommodate this); instead we focus on using distributed decryption to enable *efficient* multi-party bootstrapping. In addition the work of [1], in requiring an FHE scheme, as opposed to the somewhat homomorphic encryption (SHE) scheme of our work, requires the assumption of circular security of the underlying FHE scheme (and hence more assumptions). Although in our instantiation, for efficiency reasons, we use a scheme which assumes circular security; this is not however theoretically necessary.

In [20], following on the work in [4], the authors propose an MPC protocol which uses an SHE scheme as an "optimization". Based in the preprocessing model, the authors utilize an SHE scheme which can evaluate circuits of multiplicative depth one to optimize the preprocessing step of an essentially standard MPC protocol. The optimizations, and use of SHE, in [20] are focused on the case of computational improvements. In this work we invert the use of SHE in [20], by using it for the online phase of the MPC protocol, so as to optimize the communication efficiency for any number of parties.

In essence we interpolate between the two extremes of traditional MPC protocols (with high communication but low computational costs) and Gentry's FHE based solution (with high computation but low communication costs). Our interpolation is dependent on a parameter, which we label as L, where $L \geq 2$. At one extreme, for $L = 2$ our protocol resembles traditional MPC protocols, whilst at the other extreme, for $L = \infty$ our protocol is exactly that of Gentry's FHE based solution. We emphasize that our construction is general in that *any* SHE can be used which supports homomorphic computation of depth *two* circuits and threshold decryption. Thus the requirements on the underlying SHE scheme are much weaker than the previous SHE (FHE) based MPC protocols, such as the one by Asharov et al. [1], which relies on the specifics of LWE (learning with errors) based SHE i.e. *key-homomorphism* and demands homomorphic computation of depth L circuits for big enough L to bootstrap.

The solution we present is in the preprocessing model; in which we allow a preprocessing phase which can compute data which is neither input, nor function, dependent. This preprocessed data is then consumed in the online phase. As usual in such a model our goal is for efficiency in the online phase only. We present our basic protocol and efficiency analysis for the case of passive threshold adversaries only; i.e. we can tolerate up to t passive corruptions where $t < n$. We then note that security against t active adversaries with $t < n/3$ can be achieved for no extra cost in the online phase. For the active security case, essentially the same communication costs can be achieved even when $t < n/2$, bar some extra work (which is *independent* of $|C|$) to eliminate the cheating parties when they are detected. The security of our protocols are proven in the standard universal composability (UC) framework [10].

Finally we note that our results on communication complexity, both in a practical and in an asymptotic sense, in the computational setting are comparable (if not better) than the best known results in the information theoretic and computational settings. Namely the best known optimally resilient statistically secure MPC protocol with $t < n/2$ has (asymptotic) communication complexity of $\mathcal{O}(n)$ per multiplication [3], whereas ours is $\mathcal{O}(n/L)$ (see Section 6 for the analysis of our protocol). With near optimal resiliency of $t < (\frac{1}{3} - \epsilon)n$, the best known perfectly secure MPC protocol has (asymptotic) communication complexity of $\mathcal{O}(\text{polylog } n)$ per multiplication [16], but a huge constant is hiding under the $\mathcal{O}$. In the computational settings, with near optimal resiliency of $t < (\frac{1}{2} - \epsilon)n$, the best known MPC protocol has (asymptotic) communication complexity of $\mathcal{O}(\text{polylog } n)$ per multiplication [15], but again a huge constant is hiding under the $\mathcal{O}$. All these protocols can not win over ours when *exact* communication complexity is compared for even small values of L.

Overview: Our protocol is intuitively simple. We first take an L-levelled SHE scheme (strictly it has $L + 1$ levels, but can evaluate circuits with L levels of multiplications) which possesses a distributed decryption protocol for the specific access structure required by our MPC protocol. We assume that the SHE scheme is implemented over a ring which supports N embeddings of the underlying finite field $\mathbb{F}_p$ into the message space of the SHE scheme. Almost all known SHE schemes support such packing of the finite field into the plaintext slots in an SIMD manner [24,30]; and such packing has been crucial in the implementation of SHE in various applications [17,20,25].

Clearly with such a setup we can implement Gentry's MPC solution for circuits of multiplicative depth L. All that remains is how to "bootstrap" from circuits with multiplicative depth L to arbitrary circuits. The standard solution would be to bootstrap the FHE scheme directly, following the blueprint outlined in Gentry's thesis. However, in the case of applications to MPC we could instead utilize a protocol to perform the bootstrapping. In a nutshell that is exactly what we propose.

The main issue then is show how to efficiently perform the bootstrapping in a distributed manner; where efficiency is measured in terms of computational and communication performance. Naively performing an MPC protocol to execute the bootstrapping phase will lead to a large communication overhead, due to the inherent overhead in dealing with homomorphic encryptions. But on its own this is enough to obtain our asymptotic interpolation between FHE and MPC; we however aim to provide an efficient and practical interpolation. That is one which is efficient for small values of L.

It turns out that a special case of a suitable bootstrapping protocol can be found as a subprocedure of the MPC protocol in [20]. We extract the required protocol, generalise it, and then apply it to our MPC situation.

To ease exposition we will not utilize the packing from [24] to perform evaluations of the depth L sub-circuits; we see this as a computational optimization which is orthogonal to the issues we will explore in this paper. In any practical instantiation of the protocol of this paper such a packing could be used, as described in [24], in evaluating the circuit of multiplicative depth L. However, we will use this packing to perform the bootstrapping in a communication efficient manner.

The bootstrapping protocol runs in two phases. In the first (offline) phase we repeatedly generate sets of ciphertexts, one set for each party, such that all parties learn the ciphertexts but only the given party learns their underlying messages (which are assumed to be packed). The offline phase can be run in either a passive, covert or active security model, irrespective of the underlying access structure of the MPC protocol following ideas from [18]. In the second (online) phase the data to be bootstrapped is packed together, a random mask is added (computed from the offline phase data), a distributed decryption protocol is executed to obtain the masked data which is then re-encrypted, the mask is subtracted and then the data is unpacked. All these steps are relatively efficient, with communication only being required for the distributed decryption.

To apply our interactive bootstrapping method efficiently we need to make a mild assumption on the circuit being evaluated; this is similar to the assumptions used in [15,16,21]. The assumption can be intuitively seen as saying that the circuit is relatively wide enough to enable packing of enough values which need to be bootstrapped at each respective level. We expect that most circuits in practice will satisfy our assumption, and we will call the circuits which satisfy our requirement "well formed".

We pause to note that the ability to open data within the MPC protocol enables one to perform more than a simple evaluation of an arithmetic circuit. This observation is well known in the MPC community, where it has been used to obtain efficient protocols for higher level functions [11,14]. Thus enabling a distributed bootstrapping also enables one to produce more efficient protocols than purely FHE based ones.

We instantiate our protocol with the BGV scheme [7] and obtain sufficient parameter sizes following the methodology in [18,25]. Due to the way we utilize the BGV scheme we need to restrict to MPC protocols for arithmetic circuits over a finite field $\mathbb{F}_p$, with $p \equiv 1 \pmod{m}$ with $m = 2 \cdot N$ and $N = 2^r$ for some r. The distributed decryption method uses a "smudging" technique (see the full version of the paper) which means that the modulus used in the BGV scheme needs to be larger than what one would need to perform just the homomorphic operations. Removing this smudging technique, and hence obtaining an efficient protocol for distributed decryption, for *any* SHE scheme is an interesting open problem; with many potential applications including that described in this paper.

We show that even for a very small value of L, in particular $L = 5$, we can achieve better communication efficiency than many practical MPC protocols in the preprocessing model. Most practical MPC protocols such as [5,20,28] require the transmission of at least two finite field elements per multiplication gate between each pair of parties.

In [20] a technique is presented which can reduce this to the transmission of an average of three field elements per multiplication gate per party (and not per pair of parties). Note the models in [5] (three party, one passive adversary) and [20,28] (n party, dishonest majority, active security) are different from ours (we assume honest majority, active security); but even mapping these protocols to our setting of n party honest majority would result in the same communication characteristics. We show that for relatively small values of L, i.e. $L > 8$, one can obtain a communication efficiency of less than one field element per gate and party (details available in Section 6).

Clearly, by setting L appropriately one can obtain a communication efficiency which improves upon that in [15,16]; albeit we are only interested in communication in the online phase of a protocol in the preprocessing model whilst [15,16] discuss total communication cost over all phases. But we stress this is not in itself interesting, as Gentry's FHE based protocol can beat the communication efficiency of [15,16] in any case. What is interesting is that we can beat the communication efficiency of the online phase of practical MPC protocols, with very small values of L indeed. Thus the protocol in this paper may provide a practical tradeoff between existing MPC protocols (which consume high bandwidth) and FHE based protocols (which require huge computation).

Our protocol therefore enables the following use-case: it is known that SHE schemes only become prohibitively computationally expensive for large L; indeed one of the reasons why the protocols in [18,20] are so efficient is that they restrict to evaluating homomorphically circuits of multiplicative depth one. With our protocol parties can a priori decide the value of L, for a value which enables them to produce a computationally efficient SHE scheme. Then they can execute an MPC protocol with communication costs reduced by effectively a factor of L. Over time as SHE technology improves the value of L can be increased and we can obtain Gentry's original protocol. Thus our methodology enables us to interpolate between the case of standard MPC and the eventual goal of MPC with almost zero communication costs.

2   Well Formed Circuits

In this section we define what we mean by well formed circuits, and the pre-processing which we require on our circuits. We take as given an arithmetic circuit C defined over a finite field $\mathbb{F}_p$. In particular the circuit C is a directed acyclic graph consisting of edges made up of n_I input wires, n_O output wires, and n_W internal wires, plus a set of nodes being given by a set of gates $\mathbb{G}$. The gates are divided into sets of Add gates $\mathbb{G}_A$ and Mult gates $\mathbb{G}_M$, with $\mathbb{G} = \mathbb{G}_A \cup \mathbb{G}_M$, with each Add/Mult gate taking two wires (or a constant value in $\mathbb{F}_p$) as input and producing one wire as output. The circuit is such that all input wires are open on their input ends, and all output wires are open on their output ends, with the internal wires being connected on both ends. We let the depth of the circuit d be the length of the maximum path from an input wire to an output wire. Our definition of a well formed circuit is parametrized by two positive integer values N and L.

We now associate inductively to each wire in the circuit an integer valued label as follows. The input wires are given the label one; then all other wires are given a label as follows (where we assume a constant input to a gate has label L):

$$\text{Label of output wire of Add gate} = \min(\text{Label of input wires}),$$
$$\text{Label of output wire of Mult gate} = \min(\text{Label of input wires}) - 1.$$

Thus the minimum value of a label is $1 - d$ (which is negative for a general d). Looking ahead, the reason for starting with an input label of one is when we match this up with our MPC protocol this will result in low communication complexity for the input stage of the computation.

We now augment the circuit, to produce a new circuit C^{aug} which will have labels in the range $[1, \ldots, L]$, by adding in some special gates which we will call Refresh gates; the set of such gates are denoted as $\mathbb{G}_R$. A Refresh gate takes as input a maximum of N wires, and produces as output an exact copy of the specified input wires. The input requirement is that the input wires must have label in the range $[1, \ldots, L]$, and all that the Refresh gate does is relabel the labels of the gate's input wires to be L. At the end of the augmentation process we require the invariant that all wire labels in C^{aug} are then in the range $[1, \ldots, L]$, and the circuit is now essentially a collection of "sub-circuits" of multiplicative depth at most $L - 1$ glued together using Refresh gates. However, we require that this is done with as small a number of Refresh gates as possible.

Definition 1 (Well Formed Circuit). *A circuit C will be called well formed if the number of Refresh gates in the associated augmented circuit C^{aug} is at most $\frac{2 \cdot |\mathbb{G}_M|}{L \cdot N}$.*

We expect that "most" circuits will be well formed due to the following argument: We first note that the only gates which concern us are multiplication gates; so without loss of generality we consider a circuit C consisting only of multiplication gates. The circuit has d layers, and let the width of C (i.e. the number of gates) at layer i be w_i. Consider the algorithm to produce C^{aug} which considers each layer in turn, from $i = 1$ to d and adds Refresh gates where needed. When reaching level i in our algorithm to produce C^{aug} we can therefore assume (by induction) that all input wires at this layer have labels in the range $[1, \ldots, L]$. To maintain the invariant we only need to apply a Refresh operation to those input wires which have label one. Let p_i denote the proportion of wires at layer i which have label one when we perform this process. It is clear that the number of required Refresh gates which we will add into C^{aug} at level i will be at most $\lceil 2 \cdot p_i \cdot w_i/N \rceil$, where the factor of two comes from the fact that each multiplication gate has two input wires.

Assuming a large enough circuit we can assume for most layers that this proportion p_i will be approximately $1/L$, since wires will be refreshed after their values have passed through L multiplication gates. So summing up over all levels, the expected number of Refresh gates in C^{aug} will be:

$$\sum_{i=1}^{d} \left\lceil \frac{2 \cdot w_i}{L \cdot N} \right\rceil \approx \frac{2}{L \cdot N} \cdot \sum_{i=1}^{d} w_i = \frac{2 \cdot |\mathbb{G}_M|}{L \cdot N}.$$

Note, we would expect that for most circuits this upper bound on the number of Refresh gates could be easily met. For example our above rough analysis did not take into account the presence of gates with fan-out greater than one (meaning there are less wires to Refresh than we estimated above), nor did it take into account utilizing unused slots in the Refresh gates to refresh wires with labels not equal to one.

Determining an optimum algorithm for moving from C to a suitable C^{aug}, with a minimal number of Refresh gates, is an interesting optimization problem which we leave as an open problem; however clearly the above outlined greedy algorithm will work for most circuits.

3 Threshold L-Levelled Packed Somewhat Homomorphic Encryption (SHE)

In this section, we present a detailed explanation of the syntax and requirements for our Threshold L-Levelled Packed Somewhat Homomorphic Encryption Scheme. The scheme will be parametrized by a number of values; namely the security parameter κ, the number of levels L, the amount of packing of plaintext elements which can be made into one ciphertext N, a statistical security parameter sec (for the security of the distributed decryption) and a pair (t, n) which defines the threshold properties of our scheme. In practice the parameter N will be a function of L and κ. The message space of the SHE scheme is defined to be $\mathcal{M} = \mathbb{F}_p^N$, and we embed the finite field $\mathbb{F}_p$ into $\mathcal{M}$ via a map $\chi : \mathbb{F}_p \longrightarrow \mathcal{M}$.

Let $\mathcal{C}(L)$ denote the family of circuits consisting of addition and multiplication gates whose labels follow the conventions in Section 2; except that input wires have label L and whose minimum wire label is zero. Thus $\mathcal{C}(L)$ is the family of standard arithmetic circuits of multiplicative depth at most L which consist of 2-input addition and multiplication gates over $\mathbb{F}_p$, whose wire labels lie in the range $[0, \ldots, L]$. Informally, a threshold L-levelled SHE scheme supports homomorphic evaluation of any circuit in the family $\mathcal{C}(L)$ with the provision for distributed (threshold) decryption, where the input wire values v_i are mapped to ciphertexts (at level L) by encrypting $\chi(v_i)$.

As remarked in the introduction we could also, as in [24], extend the circuit family $\mathcal{C}(L)$ to include gates which process N input values at once as

$$N\text{-Add}\left(\langle u_1, \ldots, u_N \rangle, \langle v_1, \ldots, v_N \rangle\right) := \langle u_1 + v_1, \ldots, u_N + v_N \rangle,$$
$$N\text{-Mult}\left(\langle u_1, \ldots, u_N \rangle, \langle v_1, \ldots, v_N \rangle\right) := \langle u_1 \times v_1, \ldots, u_N \times v_N \rangle.$$

But such an optimization of the underlying circuit is orthogonal to our consideration. However, the underlying L-levelled packed SHE scheme supports such operations on its underlying plaintext (we will just not consider these operations in our circuits being evaluated).

We can evaluate subcircuits in $\mathcal{C}(L)$; and this is how we will describe the homomorphic evaluation below (this will later help us to argue the correctness property of our general MPC protocol). In particular if $C \in \mathcal{C}(L)$, we can deal with sub-circuits C^{sub} of C whose input wires have labels $\mathfrak{l}_1^{in}, \ldots, \mathfrak{l}_{\ell_{in}}^{in}$, and whose output wires have labels $\mathfrak{l}_1^{out}, \ldots, \mathfrak{l}_{\ell_{out}}^{out}$, where $\mathfrak{l}_i^{in}, \mathfrak{l}_i^{out} \in [0, \ldots, L]$. Then given ciphertexts $c_1, \ldots, c_{\ell_{in}}$ encrypting the messages $\mathbf{m}_1, \ldots, \mathbf{m}_{\ell_{in}}$, for which the ciphertexts are at level $\mathfrak{l}_1^{in}, \ldots, \mathfrak{l}_{\ell_{in}}^{in}$, the homomorphic evaluation function will produce ciphertexts $\hat{c}_1, \ldots, \hat{c}_{\ell_{out}}$, at levels $\mathfrak{l}_1^{out}, \ldots, \mathfrak{l}_{\ell_{out}}^{out}$, which encrypt the messages corresponding to evaluating C^{sub} on the components of the vectors $\mathbf{m}_1, \ldots, \mathbf{m}_{\ell_{in}}$ in a SIMD manner. More formally:

Definition 2 (Threshold L-levelled Packed SHE). *An L-levelled public key packed somewhat homomorphic encryption (SHE) scheme with the underlying message space $\mathcal{M} = \mathbb{F}_p^N$, public key space $\mathcal{PK}$, secret key space $\mathcal{SK}$, evaluation key space $\mathcal{EK}$, ciphertext space $\mathcal{CT}$ and distributed decryption key space $\mathcal{DK}_i$ for $i \in [1, \ldots, n]$ is a collection of the following PPT algorithms, parametrized by a computational security parameter κ and a statistical security parameter sec:*

1. $\mathsf{SHE.KeyGen}(1^\kappa, 1^{\mathsf{sec}}, n, t) \to (\mathsf{pk}, \mathsf{ek}, \mathsf{sk}, \mathsf{dk}_1, \ldots, \mathsf{dk}_n)$: *The key generation algorithm outputs a public key $\mathsf{pk} \in \mathcal{PK}$, a public evaluation key $\mathsf{ek} \in \mathcal{EK}$, a secret key $\mathsf{sk} \in \mathcal{SK}$ and n keys $(\mathsf{dk}_1, \ldots, \mathsf{dk}_n)$ for the distributed decryption, with $\mathsf{dk}_i \in \mathcal{DK}_i$.*

2. $\mathsf{SHE.Enc_{pk}}(\mathbf{m}, r) \to (\mathfrak{c}, L)$: *The encryption algorithm computes a ciphertext $\mathfrak{c} \in \mathcal{CT}$, which encrypts a plaintext vector $\mathbf{m} \in \mathcal{M}$ under the public key pk using the randomness[1] r and outputs $(\mathfrak{c}, L)$ to indicate that the associated level of the ciphertext is L.*

3. $\mathsf{SHE.Dec_{sk}}(\mathfrak{c}, \mathfrak{l}) \to \mathbf{m}'$: *The decryption algorithm decrypts a ciphertext $\mathfrak{c} \in \mathcal{CT}$ of associated level $\mathfrak{l}$ where $\mathfrak{l} \in [0, \ldots, L]$ using the decryption key sk and outputs a plaintext $\mathbf{m}' \in \mathcal{M}$. We say that $\mathbf{m}'$ is the plaintext associated with $\mathfrak{c}$.*

4. $\mathsf{SHE.ShareDec_{dk_i}}(\mathfrak{c}, \mathfrak{l}) \to \bar{\mu}_i$: *The share decryption algorithm takes a ciphertext $\mathfrak{c}$ with associated level $\mathfrak{l} \in [0, \ldots, L]$, a key dk_i for the distributed decryption, and computes a decryption share $\bar{\mu}_i$ of $\mathfrak{c}$.*

5. $\mathsf{SHE.ShareCombine}((\mathfrak{c}, \mathfrak{l}), \{\bar{\mu}_i\}_{i \in [1, \ldots, n]}) \to \mathbf{m}'$: *The share combine algorithm takes a ciphertext $\mathfrak{c}$ with associated level $\mathfrak{l} \in [0, \ldots, L]$ and a set of n decryption shares and outputs a plaintext $\mathbf{m}' \in \mathcal{M}$.*

6. $\mathsf{SHE.Eval_{ek}}(C^{\mathsf{sub}}, (\mathfrak{c}_1, \mathfrak{l}_1^{in}), \ldots, (\mathfrak{c}_{\ell_{in}}, \mathfrak{l}_{\ell_{in}}^{in})) \to (\hat{\mathfrak{c}}_1, \mathfrak{l}_1^{out}), \ldots, (\hat{\mathfrak{c}}_{\ell_{out}}, \mathfrak{l}_{\ell_{out}}^{out})$: *The homomorphic evaluation algorithm is a deterministic polynomial time algorithm (polynomial in L, ℓ_{in}, ℓ_{out} and κ) that takes as input the evaluation key ek, a subcircuit C^{sub} of a circuit $C \in \mathcal{C}(L)$ with ℓ_{in} input gates and ℓ_{out} output gates as well as a set of ℓ_{in} ciphertexts $\mathfrak{c}_1, \ldots, \mathfrak{c}_{\ell_{in}}$, with associated level $\mathfrak{l}_1^{in}, \ldots, \mathfrak{l}_{\ell_{in}}^{in}$, and outputs ℓ_{out} ciphertexts $\hat{\mathfrak{c}}_1, \ldots, \hat{\mathfrak{c}}_{\ell_{out}}$, with associated levels $\mathfrak{l}_1^{out}, \ldots, \mathfrak{l}_{\ell_{out}}^{out}$ respectively, where each $\mathfrak{l}_i^{in}, \mathfrak{l}_i^{out} \in [0, \ldots, L]$ is the label associated to the given input/output wire in C^{sub}.*
   Algorithm $\mathsf{SHE.Eval}$ associates the input ciphertexts with the input gates of C^{sub} and homomorphically evaluates C^{sub} gate by gate in an SIMD manner on the components of the input messages. For this, $\mathsf{SHE.Eval}$ consists of separate algorithms $\mathsf{SHE.Add}$ and $\mathsf{SHE.Mult}$ for homomorphically evaluating addition and multiplication gates respectively. More specifically, given two ciphertexts $(\mathfrak{c}_1, \mathfrak{l}_1)$ and $(\mathfrak{c}_2, \mathfrak{l}_2)$ with associated levels $\mathfrak{l}_1$ and $\mathfrak{l}_2$ respectively where $\mathfrak{l}_1, \mathfrak{l}_2 \in [0, \ldots, L]$ then[2]:

   - $\mathsf{SHE.Add_{ek}}((\mathfrak{c}_1, \mathfrak{l}_1), (\mathfrak{c}_2, \mathfrak{l}_2)) \to (\mathfrak{c}_{\mathsf{Add}}, \min(\mathfrak{l}_1, \mathfrak{l}_2))$: *The deterministic polynomial time addition algorithm takes as input $(\mathfrak{c}_1, \mathfrak{l}_1), (\mathfrak{c}_2, \mathfrak{l}_2)$ and outputs a ciphertext $\mathfrak{c}_{\mathsf{Add}}$ with associated level $\min(\mathfrak{l}_1, \mathfrak{l}_2)$.*

---

[1] In the paper, unless it is explicitly specified, we assume that some randomness has been used for encryption.

[2] Without loss of generality we assume that we can perform homomorphic operations on ciphertexts of different levels, since we can always deterministically downgrade the ciphertext level of any ciphertext to any value between zero and its current value using $\mathsf{SHE.LowerLevel_{ek}}$.

- SHE.Mult$_{ek}$$((c_1, l_1), (c_2, l_2)) \rightarrow (c_{Mult}, \min(l_1, l_2) - 1)$: *The deterministic polynomial time multiplication algorithm takes as input* $(c_1, l_1), (c_2, l_2)$ *and outputs a ciphertext* c_{Mult} *with associated level* $\min(l_1, l_2) - 1$.
- SHE.ScalarMult$_{ek}$$((c_1, l_1), a) \rightarrow (c_{Scalar}, l_1)$: *The deterministic polynomial time scalar multiplication algorithm takes as input* (c_1, l_1) *and a plaintext* $a \in \mathcal{M}$ *and outputs a ciphertext* c_{Scalar} *with associated level* l_1.

7. SHE.Pack$_{ek}$$((c_1, l_1), \ldots, (c_N, l_N)) \rightarrow (c, \min(l_1, \ldots, l_N))$: *If* c_i *is a ciphertext with associated plaintext* $\chi(m_i)$, *then this procedure produces a ciphertext* $(c, \min(l_1, \ldots, l_N))$ *with associated plaintext* $\mathbf{m} = (m_1, \ldots, m_N)$.

8. SHE.Unpack$_{ek}$$(c, l) \rightarrow ((c_1, l), \ldots, (c_N, l))$: *If* c *is a ciphertext with associated plaintext* $\mathbf{m} = (m_1, \ldots, m_N)$, *then this procedure produces* N *ciphertexts* (c_1, l), $\ldots, (c_N, l)$ *such that* c_i *has associated plaintext* $\chi(m_i)$.

9. SHE.LowerLevel$_{ek}$$((c, l), l') \rightarrow (c', l')$: *This procedure, for* $l' < l$, *produces a ciphertext* c' *with the same associated plaintext as* c, *but at level* l'.   □

We require the following homomorphic property to be satisfied:

- *Somewhat Homomorphic SIMD Property*: Let $C^{sub} : \mathbb{F}_p^{\ell_{in}} \rightarrow \mathbb{F}_p^{\ell_{out}}$ be any sub-circuit of a circuit C in the family $\mathcal{C}(L)$ with respective inputs $\mathbf{m}_1, \ldots, \mathbf{m}_{\ell_{in}} \in \mathcal{M}$, such that C^{sub} when evaluated N times in an SIMD fashion on the N components of the vectors $\mathbf{m}_1, \ldots, \mathbf{m}_{\ell_{in}}$, produces N sets of ℓ_{out} output values $\hat{\mathbf{m}}_1, \ldots, \hat{\mathbf{m}}_{\ell_{out}} \in \mathcal{M}$. Moreover, for $i \in [1, \ldots, \ell_{in}]$ let c_i be a ciphertext of level l_i^{in} with associated plaintext vector $\mathbf{m}_i$ and let $(\hat{c}_1, l_1^{out}), \ldots, (\hat{c}_{\ell_{out}}, l_{\ell_{out}}^{out}) = $ SHE.Eval$_{ek}$$(C^{sub}$, $(c_1, l_1^{in}), \ldots, (c_{\ell_{in}}, l_{\ell_{in}}^{in}))$. Then the following holds with probability one for each $i \in [1, \ldots, \ell_{out}]$:

$$\text{SHE.Dec}_{sk}(\hat{c}_i, l_i^{out}) = \hat{\mathbf{m}}_i.$$

We also require the following security properties:

- *Key Generation Security*: Let S and D_i be the random variables which denote the probability distribution with which the secret key sk and the ith key dk$_i$ for the distributed decryption is selected from $\mathcal{SK}$ and $\mathcal{DK}_i$ by SHE.KeyGen for $i = 1, \ldots, n$. Moreover, for a set $I \subseteq \{1, \ldots, n\}$, let D_I denote the random variable which denote the probability distribution with which the set of keys for the distributed decryption, belonging to the indices in I, are selected from the corresponding $\mathcal{DK}_i$s by SHE.KeyGen. Then the following two properties hold:
  - *Correctness*: For any set $I \subseteq \{1, \ldots, n\}$ with $|I| \geq t+1$, $H(S|D_I) = 0$. Here $H(X|Y)$ denotes the conditional entropy of a random variable X with respect to a random variable Y [13].
  - *Privacy*: For any set $I \subset \{1, \ldots, n\}$ with $|I| \leq t$, $H(S|D_I) = H(S)$.
- *Semantic Security*: For every set $I \subset \{1, \ldots, n\}$ with $|I| \leq t$ and all PPT adversaries $\mathcal{A}$, the advantage of $\mathcal{A}$ in the following game is negligible in κ:
  - *Key Generation*: The challenger runs SHE.KeyGen$(1^\kappa, 1^{sec}, n, t)$ to obtain (pk, ek, sk, dk$_1$, ..., dk$_n$) and sends pk, ek and $\{dk_i\}_{i \in I}$ to $\mathcal{A}$.
  - *Challenge*: $\mathcal{A}$ sends plaintexts $\mathbf{m}_0, \mathbf{m}_1 \in \mathcal{M}$ to the challenger, who randomly selects $b \in \{0, 1\}$ and sends $(c, L) = $ SHE.Enc$_{pk}$$(\mathbf{m}_b, r)$ for some randomness r to $\mathcal{A}$.

- *Output*: $\mathcal{A}$ outputs b'.

The advantage of $\mathcal{A}$ in the above game is defined to be $|\frac{1}{2} - \Pr[b' = b]|$.

- *Correct Share Decryption*: For any $(\mathsf{pk}, \mathsf{ek}, \mathsf{sk}, \mathsf{dk}_1, \dots, \mathsf{dk}_n)$ obtained as the output of SHE.KeyGen, the following should hold for any ciphertext $(\mathfrak{c}, \mathfrak{l})$ with associated level $\mathfrak{l} \in [0, \dots, L]$:

$$\mathsf{SHE.Dec}_{\mathsf{sk}}(\mathfrak{c}, \mathfrak{l}) = \mathsf{SHE.ShareCombine}((\mathfrak{c}, \mathfrak{l}), \{\mathsf{SHE.ShareDec}_{\mathsf{dk}_i}(\mathfrak{c}, \mathfrak{l})\}_{i \in [1, \dots, n]}).$$

- *Share Simulation Indistinguishability*: There exists a PPT simulator SHE.ShareSim, which on input a subset $I \subset \{1, \dots, n\}$ of size at most t, a ciphertext $(\mathfrak{c}, \mathfrak{l})$ of level $\mathfrak{l} \in [0, \dots, L]$, a plaintext $\mathbf{m}$ and $|I|$ decryption shares $\{\bar{\mu}_i\}_{i \in I}$ outputs $n - |I|$ simulated decryption shares $\{\bar{\mu}_j^*\}_{j \in \bar{I}}$ with the following property: For any $(\mathsf{pk}, \mathsf{ek}, \mathsf{sk}, \mathsf{dk}_1, \dots, \mathsf{dk}_n)$ obtained as the output of SHE.KeyGen, any subset $I \subset \{1, \dots, n\}$ of size at most t, any $\mathbf{m} \in \mathcal{M}$ and any $(\mathfrak{c}, \mathfrak{l})$ where $\mathbf{m} = \mathsf{SHE.Dec}_{\mathsf{sk}}(\mathfrak{c}, \mathfrak{l})$, the following distributions are statistically indistinguishable:

$$(\{\bar{\mu}_i\}_{i \in I}, \mathsf{SHE.ShareSim}((\mathfrak{c}, \mathfrak{l}), \mathbf{m}, \{\bar{\mu}_i\}_{i \in I})) \overset{s}{\approx} \left(\{\bar{\mu}_i\}_{i \in I}, \{\bar{\mu}_j\}_{j \in \bar{I}}\right),$$

where for all $i \in [1, \dots, n]$, $\bar{\mu}_i = \mathsf{SHE.ShareDec}_{\mathsf{dk}_i}(\mathfrak{c}, \mathfrak{l})$. We require in particular that the statistical distance between the two distributions is bounded by $2^{-\mathsf{sec}}$. Moreover

$$\mathsf{SHE.ShareCombine}((\mathfrak{c}, \mathfrak{l}), \{\bar{\mu}_i\}_{i \in I} \cup \mathsf{SHE.ShareSim}((\mathfrak{c}, \mathfrak{l}), \mathbf{m}, \{\bar{\mu}_i\}_{i \in I}))$$

outputs the result $\mathbf{m}$. Here $\bar{I}$ denotes the complement of the set I; i.e. $\bar{I} = \{1, \dots, n\} \setminus I$.

In the full version of the paper we instantiate the abstract syntax with a threshold SHE scheme based on the BGV scheme [7]. We pause to note the difference between our underlying SHE, which is just an SHE scheme which supports distributed decryption, and that of [1] which requires a special key homomorphic FHE scheme.

4   MPC from SHE – The Semi-Honest Settings

In this section we present our generic MPC protocol for the computation of any arbitrary depth d circuit using an abstract threshold L-levelled SHE scheme. For the ease of exposition we first concentrate on the case of semi-honest security, and then we deal with active security in Section 5.

Without loss of generality we make the simplifying assumption that the function f to be computed takes a single input from each party and has a single output; specifically $f : \mathbb{F}_p^n \to \mathbb{F}_p$. The ideal functionality $\mathcal{F}_f$ presented in Figure 1 computes such a given function f, represented by a well formed circuit C. We will present a protocol to realise the ideal functionality $\mathcal{F}_f$ in a hybrid model in which we are given access to an ideal functionality $\mathcal{F}_{\mathsf{SETUPGEN}}$ which implements a distributed key generation for the underlying SHE scheme. In particular the $\mathcal{F}_{\mathsf{SETUPGEN}}$ functionality presented in Figure 2 computes the public key, secret key, evaluation key and the keys for the distributed decryption of an L-levelled SHE scheme, distributes the public key and the evaluation key

---

Functionality $\mathcal{F}_f$

$\mathcal{F}_f$ interacts with the parties $P_1, \ldots, P_n$ and the adversary $\mathcal{S}$ and is parametrized by an n-input function $f : \mathbb{F}_p^n \to \mathbb{F}_p$.

- Upon receiving (sid, i, x_i) from the party P_i for every $i \in [1, \ldots, n]$ where $x_i \in \mathbb{F}_p$, compute $y = C(x_1, \ldots, x_n)$, send (sid, y) to all the parties and the adversary $\mathcal{S}$ and halt. Here C denotes the (publicly known) well formed arithmetic circuit over $\mathbb{F}_p$ representing the function f.

Fig. 1. The Ideal Functionality for Computing a Given Function

to all the parties and sends the ith key dk_i (for the distributed decryption) to the party P_i for each $i \in [1, \ldots, n]$. In addition, the functionality also computes a random encryption c_1 with associated plaintext $\mathbf{1} = (1, \ldots, 1) \in \mathcal{M}$ and sends it to all the parties. Looking ahead, c_1 will be required while proving the security of our MPC protocol. The ciphertext c_1 is at level one, as we only need it to pre-multiply the ciphertexts which are going to be decrypted via the distributed decryption protocol; thus the output of a multiplication by c_1 need only be at level zero. Looking ahead, this ensures that (with respect to our instantiation of SHE) the noise is kept to a minimum at this stage of the protocol.

Functionality $\mathcal{F}_{\text{SETUPGEN}}$

$\mathcal{F}_{\text{SETUPGEN}}$ interacts with the parties $P_1, \ldots, P_n$ and the adversary $\mathcal{S}$ and is parametrized by an L-levelled SHE scheme.

- Upon receiving (sid, i) from the party P_i for every $i \in [1, \ldots, n]$, compute $(pk, ek, sk, dk_1, \ldots, dk_n) = \text{SHE.KeyGen}(1^\kappa, 1^{sec}, n, t)$ and $(c_1, 1) = \text{SHE.LowerLevel}_{ek}((\text{SHE.Enc}_{pk}(\mathbf{1}, r), 1))$ for $\mathbf{1} = (1, \ldots, 1) \in \mathcal{M}$ and some randomness r. Finally send $(sid, pk, ek, dk_i, (c_1, 1))$ to the party P_i for every $i \in [1, \ldots, n]$ and halt.

Fig. 2. The Ideal Functionality for Key Generation

4.1 The MPC Protocol in the $\mathcal{F}_{\text{SETUPGEN}}$-Hybrid Model

Here we present our MPC protocol Π_f^{SH} in the $\mathcal{F}_{\text{SETUPGEN}}$-hybrid model. Let C be the (well formed) arithmetic circuit representing the function f and C^{aug} be the associated augmented circuit (which includes the necessary Refresh gates). The protocol Π_f^{SH} (see Figure 3) runs in two phases: offline and online. The computation performed in the offline phase is completely independent of the circuit and (private) inputs of the parties and therefore can be carried out well ahead of the time (namely the online phase) when the function and inputs are known. If the parties have more than one input/output then one can apply packing/unpacking at the input/output stages of the protocol; we leave this minor modification to the reader.

In the offline phase, the parties interact with $\mathcal{F}_{\text{SETUPGEN}}$ to obtain the public key, evaluation key and their respective keys for performing distributed decryption, corresponding to a threshold L-levelled SHE scheme. Next each party sends encryptions of ζ random elements and then additively combines them (by applying the homomorphic addition to the ciphertexts encrypting the random elements) to generate ζ ciphertexts at level L of truly random elements (unknown to the adversary). Here ζ is assumed to be large enough, so that for a typical circuit it is more than the number of refresh gates in the circuit, i.e. $\zeta > \mathbb{G}_R$. Looking ahead, these random ciphertexts created in the offline phase are used in the online phase to evaluate refresh gates by (homomorphically) masking the messages associated with the input wires of a refresh gate.

During the online phase, the parties encrypt their private inputs and distribute the corresponding ciphertexts to all other parties. These ciphertexts are transmitted at level one, thus consuming low bandwidth, and are then elevated to level L by the use of a following Refresh gate (which would have been inserted by the circuit augmentation process). Note that the inputs of the parties are in $\mathbb{F}_p$ and so the parties first apply the mapping χ (embedding $\mathbb{F}_p$ into the message space $\mathcal{M}$ of SHE) before encrypting their private inputs.

The input stage is followed by the homomorphic evaluation of C^{aug} as follows: The addition and multiplication gates are evaluated locally using the addition and multiplication algorithm of the SHE. For each refresh gate, the parties execute the following protocol to enable a "multiparty bootstrapping" of the input ciphertexts: the parties pick one of the random ciphertext created in the offline phase (for each refresh gate a different ciphertext is used) and perform the following computation to refresh N ciphertexts with levels in the range $[1, \ldots, L]$ and obtain N fresh level L ciphertexts, with the associated messages unperturbed:

- Let $(\mathfrak{c}_1, \mathfrak{l}_1), \ldots, (\mathfrak{c}_N, \mathfrak{l}_N)$ be the N ciphertexts with associated plaintexts $\chi(z_1), \ldots, \chi(z_N)$ with every $z_i \in \mathbb{F}_p$, that need to be refreshed (i.e. they are the inputs of a refresh gate).
- The N ciphertexts are then (locally) packed into a single ciphertext $\mathfrak{c}$, which is then homomorphically masked with a random ciphertext from the offline phase.
- The resulting masked ciphertext is then publicly opened via distributed decryption. This allows for the creation of a fresh encryption of the opened value at level L.
- The resulting fresh encryption is then homomorphically unmasked so that its associated plaintext is the same as original plaintext prior to the original masking.
- This fresh (unmasked) ciphertext is then unpacked to obtain N fresh ciphertexts, having the same associated plaintexts as the original N ciphertexts $\mathfrak{c}_i$ but at level L.

By packing the ciphertexts together we only need to invoke distributed decryption once, instead of N times. This leads to a more communication efficient online phase, since the distributed decryption is the only operation that demands communication. Without

---

Protocol Π_f^{SH}

Let C^{aug} denote an augmented circuit for a well formed circuit C over $\mathbb{F}_p$ representing f and let SHE be a threshold L-levelled SHE. Moreover, let $\mathcal{P} = \{P_1, \ldots, P_n\}$ be the set of n parties. For the session ID sid the parties do the following:

Offline Computation: Every party $P_i \in \mathcal{P}$ does the following:
- Call $\mathcal{F}_{\text{SETUPGEN}}$ with (sid, i) and receive $(\text{sid}, \text{pk}, \text{ek}, \text{dk}_i, (\mathfrak{c}_1, 1))$.
- Randomly select ζ plaintexts $\mathbf{m}_{i,1}, \ldots, \mathbf{m}_{i,\zeta} \in \mathcal{M}$, and compute $(\mathfrak{c}_{\mathbf{m}_{i,k}}, L) = \text{SHE.Enc}_{\text{pk}}(\mathbf{m}_{i,k}, r_{i,k})$. Send $(\text{sid}, i, (\mathfrak{c}_{\mathbf{m}_{i,1}}, L), \ldots, (\mathfrak{c}_{\mathbf{m}_{i,\zeta}}, L))$ to all parties in $\mathcal{P}$.
- Upon receiving $(\text{sid}, j, (\mathfrak{c}_{\mathbf{m}_{j,1}}, L), \ldots, (\mathfrak{c}_{\mathbf{m}_{j,\zeta}}, L))$ from all parties $P_j \in \mathcal{P}$, apply SHE.Add for $1 \leq k \leq \zeta$, on $(\mathfrak{c}_{\mathbf{m}_{1,k}}, L), \ldots, (\mathfrak{c}_{\mathbf{m}_{n,k}}, L)$, set the resultant ciphertext as the kth offline ciphertext $\mathfrak{c}_{\mathbf{m}_k}$ with the (unknown) associated plaintext $\mathbf{m}_k = \mathbf{m}_{1,k} + \cdots + \mathbf{m}_{n,k}$.

Online Computation: Every party $P_i \in \mathcal{P}$ does the following:
- **Input Stage:** On having input $x_i \in \mathbb{F}_p$, compute $(\mathfrak{c}_{\mathbf{x}_i}, 1) = \text{SHE.LowerLevel}_{\text{ek}}(\text{SHE.Enc}_{\text{pk}}(\chi(x_i), r_i), 1)$ with randomness r_i and send $(\text{sid}, i, (\mathfrak{c}_{\mathbf{x}_i}, 1))$ to each party. Receive $(\text{sid}, j, (\mathfrak{c}_{\mathbf{x}_j}, 1))$ from each party $P_j \in \mathcal{P}$.
- **Computation Stage:** Associate the received ciphertexts with the corresponding input wires of C^{aug} and then homomorphically evaluate the circuit C^{aug} gate by gate as follows:
  - **Addition Gate and Multiplication Gate:** Given $(\mathfrak{c}_1, \mathfrak{l}_1)$ and $(\mathfrak{c}_2, \mathfrak{l}_2)$ associated with the input wires of the gate where $\mathfrak{l}_1, \mathfrak{l}_2 \in [1, \ldots, L]$, locally compute $(\mathfrak{c}, \mathfrak{l}) = \text{SHE.Add}_{\text{ek}}((\mathfrak{c}_1, \mathfrak{l}_1), (\mathfrak{c}_2, \mathfrak{l}_2))$ with $\mathfrak{l} = \min(\mathfrak{l}_1, \mathfrak{l}_2)$ for an addition gate and $(\mathfrak{c}, \mathfrak{l}) = \text{SHE.Mult}_{\text{ek}}((\mathfrak{c}_1, \mathfrak{l}_1), (\mathfrak{c}_2, \mathfrak{l}_2))$ with $\mathfrak{l} = \min(\mathfrak{l}_1, \mathfrak{l}_2) - 1$ for a multiplication gate; for the multiplication gate, $\mathfrak{l}_1, \mathfrak{l}_2 \in [2, \ldots, L]$, instead of $[1, \ldots, L]$. Associate $(\mathfrak{c}, \mathfrak{l})$ with the output wire of the gate.
  - **Refresh Gate:** For the kth refresh gate in the circuit, the kth offline ciphertext $(\mathfrak{c}_{\mathbf{m}_k}, L)$ is used. Let $(\mathfrak{c}_1, \mathfrak{l}_1), \ldots, (\mathfrak{c}_N, \mathfrak{l}_N)$ be the ciphertexts associated with the input wires of the refresh gate where $\mathfrak{l}_1, \ldots, \mathfrak{l}_N \in [1, \ldots, L]$:
    * **Packing:** Locally compute $(\mathfrak{c}_{\mathbf{z}}, \mathfrak{l}) = \text{SHE.Pack}_{\text{ek}}(\{(\mathfrak{c}_i, \mathfrak{l}_i)\}_{i \in [1, \ldots, N]})$ where $\mathfrak{l} = \min(\mathfrak{l}_1, \ldots, \mathfrak{l}_N)$.
    * **Masking:** Locally compute $(\mathfrak{c}_{\mathbf{z}+\mathbf{m}_k}, 0) = \text{SHE.Add}_{\text{ek}}(\text{SHE.Mult}_{\text{ek}}((\mathfrak{c}_{\mathbf{z}}, \mathfrak{l}), (\mathfrak{c}_1, 1)), (\mathfrak{c}_{\mathbf{m}_k}, L))$
    * **Decrypting:** Locally compute the decryption share $\bar{\mu}_i = \text{SHE.ShareDec}_{\text{dk}_i}(\mathfrak{c}_{\mathbf{z}+\mathbf{m}_k}, 0)$ and send $(\text{sid}, i, \bar{\mu}_i)$ to every other party. On receiving $(\text{sid}, j, \bar{\mu}_j)$ from every $P_j \in \mathcal{P}$, compute the plaintext $\mathbf{z} + \mathbf{m}_k = \text{SHE.ShareCombine}((\mathfrak{c}_{\mathbf{z}+\mathbf{m}_k}, 0), \{\bar{\mu}_j\}_{j \in [1, \ldots, n]})$.
    * **Re-encrypting:** Locally re-encrypt $\mathbf{z} + \mathbf{m}_k$ by computing $(\hat{\mathfrak{c}}_{\mathbf{z}+\mathbf{m}_k}, L) = \text{SHE.Enc}_{\text{pk}}(\mathbf{z} + \mathbf{m}_k, r)$ using a publicly known (common) randomness r, (This can simply be the zero string for our BGV instantiation, we only need to map the known plaintext into a ciphertext element).
    * **Unmasking:** Locally subtract $(\mathfrak{c}_{\mathbf{m}_k}, L)$ from $(\hat{\mathfrak{c}}_{\mathbf{z}+\mathbf{m}_k}, L)$ to obtain $(\hat{\mathfrak{c}}_{\mathbf{z}}, L)$.
    * **Unpacking:** Locally compute $(\hat{\mathfrak{c}}_1, L), \ldots, (\hat{\mathfrak{c}}_N, L) = \text{SHE.Unpack}_{\text{ek}}(\hat{\mathfrak{c}}_{\mathbf{z}}, L)$ and associate $(\hat{\mathfrak{c}}_1, L), \ldots, (\hat{\mathfrak{c}}_N, L)$ with the output wires of the refresh gate.
- **Output Stage:** Let $(\mathfrak{c}, \mathfrak{l})$ be the ciphertext associated with the output wire of C^{aug} where $\mathfrak{l} \in [1, \ldots, L]$.
  - **Randomization:** Compute a random encryption $(\mathfrak{c}_i, L) = \text{SHE.Enc}_{\text{pk}}(\mathbf{0}, r_i')$ of $\mathbf{0} = (0, \ldots, 0)$ and send $(\text{sid}, i, (\mathfrak{c}_i, L))$ to every other party. On receiving $(\text{sid}, j, (\mathfrak{c}_j, L))$ from every $P_j \in \mathcal{P}$, apply SHE.Add on $\{(\mathfrak{c}_j, L)\}_{j \in [1, \ldots, n]}$ to obtain $(\mathfrak{c}_{\mathbf{0}}, L)$. Compute $(\hat{\mathfrak{c}}, 0) = \text{SHE.Add}_{\text{ek}}(\text{SHE.Mult}_{\text{ek}}((\mathfrak{c}, \mathfrak{l}), (\mathfrak{c}_1, 1)), (\mathfrak{c}_{\mathbf{0}}, L))$.
  - **Output Decryption:** Compute $\bar{\gamma}_i = \text{SHE.ShareDec}_{\text{dk}_i}(\hat{\mathfrak{c}}, 0)$ and send $(\text{sid}, i, \bar{\gamma}_i)$ to every party. On receiving $(\text{sid}, j, \bar{\gamma}_j)$ from every $P_j \in \mathcal{P}$, compute $\mathbf{y} = \text{SHE.ShareCombine}((\hat{\mathfrak{c}}, 0), \{\bar{\gamma}_j\}_{j \in [1, \ldots, n]})$, output y and halt, where $y = \chi^{-1}(\mathbf{y})$.

Fig. 3. The Protocol for Realizing $\mathcal{F}_f$ against a Semi-Honest Adversary in the $\mathcal{F}_{\text{SETUPGEN}}$-hybrid Model

affecting the correctness of the above technique, but to ensure security, we add an additional step while doing the masking: the parties homomorphically pre-multiply the ciphertext c with c_1 before masking. Recall that c_1 is an encryption of $\underline{1} \in \mathcal{M}$ generated by $\mathcal{F}_{\text{SETUPGEN}}$ and so by doing the above operation, the plaintext associated with c remains the same. During the simulation in the security proof, this step allows the simulator to set the decrypted value to the random mask (irrespective of the circuit inputs), by playing the role of $\mathcal{F}_{\text{SETUPGEN}}$ and replacing c_1 with c_0, a random encryption of $\underline{0} = (0, \ldots, 0)$. Furthermore, this step explains the reason why we made provision for an extra multiplication during circuit augmentation by insisting that the refresh gates take inputs with labels in $[1, \ldots, L]$, instead of $[0, \ldots, L]$; the details are available in the simulation proof of security of our MPC protocol.

Finally, the function output y is obtained by another distributed decryption of the output ciphertext. However, this step is also not secure unless the ciphertext is randomized again by pre-multiplication by c_1 and adding n encryptions of $\underline{0}$ where each party contributes one encryption. In the simulation, the simulator gives encryption of $\chi(y)$ on behalf of one honest party and replaces c_1 by c_0, letting the output ciphertext correspond to the actual output y, even though the circuit is evaluated with zero as the inputs of the honest parties during the simulation (the simulator will not know the real inputs of the honest parties and thus will simulate them with zero). A similar idea was also used in [19]; details can be found in the security proof.

Intuitively, privacy follows because at any stage of the computation, the keys of the honest parties for the distributed decryption are not revealed and so the adversary will not be able to decrypt any intermediate ciphertext. Correctness follows from the properties of the SHE and the fact that the level of each ciphertext in the protocol remains in the range $[1, \ldots, L]$, thanks to the refresh gates. So even though the circuit C may have any arbitrary depth $d > L$, we can homomorphically evaluate C using an L-levelled SHE.

Theorem 1. *Let $f : \mathbb{F}_p^n \to \mathbb{F}_p$ be a function over $\mathbb{F}_p$ represented by a well formed arithmetic circuit C of depth d over $\mathbb{F}_p$. Let $\mathcal{F}_f$ (presented in Figure 1) be the ideal functionality computing f and let SHE be a threshold L-levelled SHE scheme. Then the protocol Π_f^{SH} UC-secure realizes $\mathcal{F}_f$ against a static, semi-honest adversary $\mathcal{A}$, corrupting upto $t < n$ parties in the $\mathcal{F}_{\text{SETUPGEN}}$-hybrid Model.*

The proof is given in the full version of the paper.

5 MPC from SHE – The Active Setting

The functionalities from Section 4 are in the passive corruption model. In the presence of an active adversary, the functionalities will be modified as follows: the respective functionality considers the input received from the majority of the parties and performs the task it is supposed to do on those inputs. For example, in the case of $\mathcal{F}_f$, the functionality considers for the computation those x_is, corresponding to the P_is from which the functionality has received the message (sid, i, x_i); on the behalf of the remaining P_is, the functionality substitutes 0 as the default input for the computation. Similarly for $\mathcal{F}_{\text{SETUPGEN}}$, the functionality performs its task if it receives the message (sid, i) from the majority of the parties. These are the standard notions of defining ideal functionalities for various corruption scenarios and we refer [26] for the complete formal details; we will not present separately the ideal functionality $\mathcal{F}_f$ and $\mathcal{F}_{\text{SETUPGEN}}$ for the malicious setting.

A closer look at Π_f^{SH} shows that we can "compile" it into an actively secure MPC protocol tolerating t active corruptions if we ensure that every corrupted party "proves" in a zero knowledge (ZK) fashion that it constructed the following correctly: **(1)** The ciphertexts in the offline phase; **(2)** The ciphertexts during the input stage and **(3)** The randomizing ciphertexts during the output stage.

Apart from the above three requirements, we also require a "robust" version of the SHE.ShareCombine method which works correctly even if up to t input decryption shares are incorrect. In the full version we show that for our specific SHE scheme, the SHE.ShareCombine algorithm (based on the standard error-correction) is indeed robust, provided $t < n/3$. For the case of $t < n/2$ we also show that by including additional steps and zero-knowledge proofs (namely proof of correct decryption), one can also obtain a robust output. Interestingly the MPC protocol requires the transmission of at most $\mathcal{O}(n^3)$ such additional zero-knowledge proofs; i.e. the communication needed to obtain robustness is *independent* of the circuit. We stress that $t < n/2$ is the *optimal resilience* for computationally secure MPC against active corruptions (with robustness and fairness) [12,27]. To keep the protocol presentation and its proof simple, we assume a robust SHE.ShareCombine (i.e. for the case of $t < n/3$), which applies error correction for the correct decryption.

Functionality $\mathcal{F}_{\text{ZK}}^R$

$\mathcal{F}_{\text{ZK}}^R$ interacts with a prover $P_i \in \{P_1, \ldots, P_n\}$ and the set of n verifiers $\mathcal{P} = \{P_1, \ldots, P_n\}$ and the adversary $\mathcal{S}$.

- Upon receiving $(\text{sid}, i, (x, w))$ from the prover $P_i \in \{P_1, \ldots, P_n\}$, the functionality sends (sid, i, x) to all the verifiers in $\mathcal{P}$ and $\mathcal{S}$ if $R(x, w)$ is true. Else it sends $(\text{sid}, i, \bot)$ and halts.

Fig. 4. The Ideal Functionality for ZK

The actively secure MPC protocol is given in Figure 4, it uses an ideal ZK functionality $\mathcal{F}_{\text{ZK}}^R$, parametrized with an NP-relation R. We apply this ZK functionality to the following relations to obtain the functionalities $\mathcal{F}_{\text{ZK}}^{R_{enc}}$ and $\mathcal{F}_{\text{ZK}}^{R_{zeroenc}}$. We note that UC-secure realizations of $\mathcal{F}_{\text{ZK}}^{R_{enc}}$ and $\mathcal{F}_{\text{ZK}}^{R_{zeroenc}}$ can be obtained in the CRS model, similar techniques to these are used in [2]. Finally we do not worry about the instantiation of $\mathcal{F}_{\text{SETUPGEN}}$ as we consider it a one time set-up, which can be done via standard techniques (such as running an MPC protocol).

- $R_{enc} = \{((\mathfrak{c}, \mathfrak{l}), (\mathbf{x}, r)) \mid (\mathfrak{c}, \mathfrak{l}) = \text{SHE.Enc}_{\text{pk}}(\mathbf{x}, r) \text{ if } \mathfrak{l} = L \lor (\mathfrak{c}, \mathfrak{l}) = \text{SHE.LowerLevel}_{\text{ek}}(\text{SHE.Enc}_{\text{pk}}(\mathbf{x}, r), 1) \text{ if } \mathfrak{l} = 1\}$: we require this relation to hold

---

Protocol Π_f^{MAL}

Let C be the well formed arithmetic circuit over $\mathbb{F}_p$ representing the function f, let C^{aug} denote an augmented circuit associated with C, and let SHE be a threshold L-levelled SHE scheme. For session ID sid the parties in $\mathcal{P} = \{P_1, \ldots, P_n\}$ do the following:

Offline Computation: Every party $P_i \in \mathcal{P}$ does the following:

- Call $\mathcal{F}_{\text{SETUPGEN}}$ with (sid, i) and receive $(\text{sid}, \text{pk}, \text{ek}, \text{dk}_i, (\mathfrak{c}_{\underline{1}}, 1))$.
- Do the same as in the offline phase of Π_f^{SH}, except that for every message $\mathbf{m}_{ik}$ for $k \in [1, \ldots, \varsigma]$ and the corresponding ciphertext $(\mathfrak{c}_{\mathbf{m}_{ik}}, L) = \text{SHE.Enc}_{\text{pk}}(\mathbf{m}_{ik}, r_{ik})$, call $\mathcal{F}_{\text{ZK}}^{R_{enc}}$ with $(\text{sid}, i, ((\mathfrak{c}_{\mathbf{m}_{ik}}, L), (\mathbf{m}_{ik}, r_{ik})))$. Receive $(\text{sid}, j, (\mathfrak{c}_{\mathbf{m}_{jk}}, L))$ from $\mathcal{F}_{\text{ZK}}^{R_{enc}}$ for $k \in [1, \ldots, \varsigma]$ corresponding to each $P_j \in \mathcal{P}$. If $(\text{sid}, j, \perp)$ is received from $\mathcal{F}_{\text{ZK}}^{R_{enc}}$ for some $P_j \in \mathcal{P}$, then consider ς publicly known level L encryptions of random values from $\mathcal{M}$ as $(\mathfrak{c}_{\mathbf{m}_{jk}}, L)$ for $k \in [1, \ldots, \varsigma]$.

Online Computation: Every party $P_i \in \mathcal{P}$ does the following:

- **Input Stage:** On having input $x_i \in \mathbb{F}_p$, compute level L ciphertext $(\mathfrak{c}_{\mathbf{x}_i}, 1) = \text{SHE.LowerLevel}_{\text{ek}}(\text{SHE.Enc}_{\text{pk}}(\chi(x_i), r_i), 1)$ with randomness r_i and call $\mathcal{F}_{\text{ZK}}^{R_{enc}}$ with the message $(\text{sid}, i, ((\mathfrak{c}_{\mathbf{x}_i}, 1), (\chi(x_i), r_i)))$. Receive $(\text{sid}, j, (\mathfrak{c}_{\mathbf{x}_j}, 1))$ from $\mathcal{F}_{\text{ZK}}^{R_{enc}}$ corresponding to each $P_j \in \mathcal{P}$. If $(\text{sid}, j, \perp)$ is received from $\mathcal{F}_{\text{ZK}}^{R_{enc}}$ for some $P_j \in \mathcal{P}$, then consider a publicly known level 1 encryption of $\chi(0)$ as $(\mathfrak{c}_{\mathbf{x}_j}, 1)$ for such a P_j.
- **Computation Stage:** Same as Π_f^{SH}, except that now the robust SHE.ShareCombine is used.
- **Output Stage:** Let $(\mathfrak{c}, \mathfrak{l})$ be the ciphertext associated with the output wire of C^{aug} where $\mathfrak{l} \in [1, \ldots, L]$.
  - **Randomization:** Compute a random encryption $(c_i, L) = \text{SHE.Enc}_{\text{pk}}(\underline{0}, r_i')$ of $\underline{0} = (0, \ldots, 0)$ and call $\mathcal{F}_{\text{ZK}}^{R_{zeroenc}}$ with the message $(\text{sid}, i, ((c_i, L), (\underline{0}, r_i')))$. Receive $(\text{sid}, j, (c_j, L))$ from $\mathcal{F}_{\text{ZK}}^{R_{zeroenc}}$ corresponding to each $P_j \in \mathcal{P}$. If $(\text{sid}, j, \perp)$ is received from $\mathcal{F}_{\text{ZK}}^{R_{zeroenc}}$ for some $P_j \in \mathcal{P}$, then consider a publicly known level L encryption of $\underline{0}$ as (c_j, L) for such a P_j.
  - The rest of the steps are same as in Π_f^{SH}, except that now the robust SHE.ShareCombine is used.

Fig. 5. The Protocol for Realizing $\mathcal{F}_f$ against an Active Adversary in the $(\mathcal{F}_{\text{SETUPGEN}}, \mathcal{F}_{\text{ZK}}^{R_{enc}}, \mathcal{F}_{\text{ZK}}^{R_{zeroenc}})$-hybrid Model

for the offline stage ciphertexts (where $\mathfrak{l} = L$) and for the input stage ciphertexts (where $\mathfrak{l} = 1$).

- $R_{zeroenc} = \{((\mathfrak{c}, L), (\mathbf{x}, r)) \mid (\mathfrak{c}, L) = \mathsf{SHE.Enc}_{\mathsf{pk}}(\mathbf{x}, r) \wedge \mathbf{x} = \underline{\mathbf{0}}\}$: we require this relation to hold for the randomizing ciphertexts during the output stage.

We are now ready to present the protocol Π_f^{MAL} (see Figure 5) in the ($\mathcal{F}_{\mathrm{SETUPGEN}}$, $\mathcal{F}_{\mathrm{ZK}}^{R_{enc}}$, $\mathcal{F}_{\mathrm{ZK}}^{R_{zeroenc}}$)-hybrid model and assuming a robust SHE.ShareCombine based on error-correction (i.e. for the case $t < n/3$).

Theorem 2. *Let $f : \mathbb{F}_p^n \to \mathbb{F}_p$ be a function represented by a well-formed arithmetic circuit C over $\mathbb{F}_p$. Let $\mathcal{F}_f$ (presented in Figure 1) be the ideal functionality computing f and let SHE be a threshold L-levelled SHE scheme such that SHE.ShareCombine is robust. Then the protocol Π_f^{MAL} UC-secure realises $\mathcal{F}_f$ in the ($\mathcal{F}_{\mathrm{SETUPGEN}}, \mathcal{F}_{\mathrm{ZK}}^{R_{enc}}$, $\mathcal{F}_{\mathrm{ZK}}^{R_{zeroenc}}$)-hybrid Model against a static, active adversary $\mathcal{A}$ corrupting t parties.*

See the full version for a proof of this theorem.

6 Estimating the Consumed Bandwidth

In the full version we determine the parameters for the instantiation of our SHE scheme using BGV by adapting the analysis from [18,25]. In this section we use this parameter estimation to show that our MPC protocol can in fact give improved communication complexity compared to the standard MPC protocols, for relatively small values of the parameter L. We are interested in the communication cost of our online stage computation. To ease our exposition we will focus on the passively secure case from Section 4; the analysis for the active security case with $t < n/3$ is exactly the same (bar the additional cost of the exchange of zero-knowledge proofs for the input stage and the output stage). For the case of active security with $t < n/2$ we also need to add in the communication related to the dispute control strategy outlined in the full version for attaining robust SHE.ShareCombine with $t < n/2$; but this is a cost which is proportional to $\mathcal{O}(n^3)$.

To get a feel for the parameters we now specialise the BGV instantiation from the full version of this paper to the case of finite fields of size $p \approx 2^{64}$, statistical security parameter sec of 40, and for various values of the computational security level κ. We estimate in Table 1 the value of N, assuming a small value for n (we need to restrict to small n to ensure a large enough range in the PRF needed in the distributed decryption protocol; see the full version).

Since a Refresh gate requires the transmission of $n-1$ elements (namely the decryption shares) in the ring R_{q_0} from party P_i to the other parties, the total communication in our protocol (in bits) is

$$|\mathbb{G}_R| \cdot n \cdot (n-1) \cdot |R_{q_0}|,$$

where $|R_{q_0}|$ is the number of bits needed to transmit an element in R_{q_0}, i.e. $N \cdot \log_2 p_0$. Assuming the circuit meets our requirement of being well formed, this implies that total communication cost for our protocol is

$$\frac{2 \cdot |\mathbb{G}_M| \cdot n \cdot (n-1) \cdot N \cdot \log_2 p_0}{L \cdot N} = \frac{2 \cdot n \cdot (n-1) \cdot |\mathbb{G}_M|}{L} \cdot \log_2(309 \cdot 2^{\mathrm{sec}} \cdot p \cdot \sqrt{N}).$$

Table 1. The value of N for various values of κ and L

L	$\kappa = 80$	$\kappa = 128$	$\kappa = 256$
2	16384	16384	32768
3	16384	16384	32768
4	16384	32768	32768
5	32768	32768	65536
6	32768	32768	65536
7	32768	32768	65536
8	32768	65536	65536
9	32768	65536	65536
10	65536	65536	65536

Using the batch distributed decryption technique (of efficiently and parallely evaluating $t + 1$ independent Refresh gates simultaneously) from the full version this can be reduced to

$$\mathsf{Cost} = \frac{4 \cdot n \cdot (n - 1) \cdot |\mathbb{G}_M|}{L \cdot (t + 1)} \cdot \log_2(309 \cdot 2^{\mathsf{sec}} \cdot p \cdot \sqrt{N}).$$

We are interested in the *overhead per multiplication gate*, in terms of equivalent numbers of finite field elements in $\mathbb{F}_p$, which is given by $\mathsf{Cost}/(|\mathbb{G}_M| \cdot \log_2 p)$, and the cost per party is $\mathsf{Cost}/(|\mathbb{G}_M| \cdot n \cdot \log_2 p)$.

At the 128 bit security level, with $p \approx 2^{64}$, and $\mathsf{sec} = 40$ (along with the above estimated values of N), this means for $n = 3$ parties, and at most $t = 1$ corruption, we obtain the following cost estimates:

L		2	3	4	5	6	7	8	9	10
Total Cost	$\mathsf{Cost}/(|\mathbb{G}_M| \cdot \log_2 p)$	12.49	8.33	6.31	5.05	4.21	3.61	3.19	2.84	2.55
Per party Cost	$\mathsf{Cost}/(|\mathbb{G}_M| \cdot n \cdot \log_2 p)$	4.16	2.77	2.10	1.68	1.40	1.20	1.06	0.94	0.85

Note for $L = 2$ our protocol becomes the one which requires interaction after every multiplication, for $L = 3$ we require interaction only after every two multiplications and so on. Note that most practical MPC protocols in the preprocessing model have a per gate per party communication cost of at least 2 finite field elements, e.g. [20]. Thus, even when $L = 5$, we obtain better communication efficiency in the online phase than traditional practical protocols in the preprocessing model with these parameters.

Acknowledgements. This work has been supported in part by ERC Advanced Grant ERC-2010-AdG-267188-CRIPTO, by EPSRC via grant EP/I03126X, and by Defense Advanced Research Projects Agency (DARPA) and the Air Force Research Laboratory (AFRL) under agreement number FA8750-11-2-0079[3]. The second author was

---

[3] The US Government is authorized to reproduce and distribute reprints for Government purposes notwithstanding any copyright notation thereon. The views and conclusions contained herein are those of the authors and should not be interpreted as necessarily representing the official policies or endorsements, either expressed or implied, of Defense Advanced Research Projects Agency (DARPA) or the U.S. Government.

supported by an Trend Micro Ltd, and the fifth author was supported by in part by a Royal Society Wolfson Merit Award.

References

1. Asharov, G., Jain, A., López-Alt, A., Tromer, E., Vaikuntanathan, V., Wichs, D.: Multiparty computation with low communication, computation and interaction via threshold FHE. In: Pointcheval, D., Johansson, T. (eds.) EUROCRYPT 2012. LNCS, vol. 7237, pp. 483–501. Springer, Heidelberg (2012)
2. Asharov, G., Jain, A., Wichs, D.: Multiparty computation with low communication, computation and interaction via threshold FHE. IACR Cryptology ePrint Archive, 2011:613 (2011)
3. Ben-Sasson, E., Fehr, S., Ostrovsky, R.: Near-linear unconditionally-secure multiparty computation with a dishonest minority. In: Safavi-Naini, R., Canetti, R. (eds.) CRYPTO 2012. LNCS, vol. 7417, pp. 663–680. Springer, Heidelberg (2012)
4. Bendlin, R., Damgård, I., Orlandi, C., Zakarias, S.: Semi-homomorphic encryption and multiparty computation. In: Paterson, K.G. (ed.) EUROCRYPT 2011. LNCS, vol. 6632, pp. 169–188. Springer, Heidelberg (2011)
5. Bogdanov, D., Laur, S., Willemson, J.: Sharemind: A framework for fast privacy-preserving computations. In: Jajodia, S., Lopez, J. (eds.) ESORICS 2008. LNCS, vol. 5283, pp. 192–206. Springer, Heidelberg (2008)
6. Brakerski, Z.: Fully homomorphic encryption without modulus switching from classical GapSVP. In: Safavi-Naini, R., Canetti, R. (eds.) CRYPTO 2012. LNCS, vol. 7417, pp. 868–886. Springer, Heidelberg (2012)
7. Brakerski, Z., Gentry, C., Vaikuntanathan, V.: (Leveled) fully homomorphic encryption without bootstrapping. In: ITCS, pp. 309–325. ACM (2012)
8. Brakerski, Z., Vaikuntanathan, V.: Efficient fully homomorphic encryption from (standard) LWE. In: FOCS, pp. 97–106. IEEE (2011)
9. Brakerski, Z., Vaikuntanathan, V.: Fully homomorphic encryption from ring-LWE and security for key dependent messages. In: Rogaway, P. (ed.) CRYPTO 2011. LNCS, vol. 6841, pp. 505–524. Springer, Heidelberg (2011)
10. Canetti, R.: Universally composable security: A new paradigm for cryptographic protocols. In: FOCS, pp. 136–145 (2001)
11. Catrina, O., Saxena, A.: Secure computation with fixed-point numbers. In: Sion, R. (ed.) FC 2010. LNCS, vol. 6052, pp. 35–50. Springer, Heidelberg (2010)
12. Cleve, R.: Limits on the security of coin flips when half the processors are faulty (Extended abstract). In: STOC, pp. 364–369. ACM (1986)
13. Cover, T.M., Thomas, J.A.: Elements of Information theory. Wiley (2006)
14. Damgård, I., Fitzi, M., Kiltz, E., Nielsen, J.B., Toft, T.: Unconditionally secure constant-rounds multi-party computation for equality, comparison, bits and exponentiation. In: Halevi, S., Rabin, T. (eds.) TCC 2006. LNCS, vol. 3876, pp. 285–304. Springer, Heidelberg (2006)
15. Damgård, I., Ishai, Y., Krøigaard, M., Nielsen, J.B., Smith, A.: Scalable multiparty computation with nearly optimal work and resilience. In: Wagner, D. (ed.) CRYPTO 2008. LNCS, vol. 5157, pp. 241–261. Springer, Heidelberg (2008)
16. Damgård, I., Ishai, Y., Krøigaard, M.: Perfectly secure multiparty computation and the computational overhead of cryptography. In: Gilbert, H. (ed.) EUROCRYPT 2010. LNCS, vol. 6110, pp. 445–465. Springer, Heidelberg (2010)
17. Damgård, I., Keller, M., Larraia, E., Miles, C., Smart, N.P.: Implementing AES via an actively/Covertly secure dishonest-majority MPC protocol. In: Visconti, I., De Prisco, R. (eds.) SCN 2012. LNCS, vol. 7485, pp. 241–263. Springer, Heidelberg (2012)

18. Damgård, I., Keller, M., Larraia, E., Pastro, V., Scholl, P., Smart, N.P.: Practical covertly secure MPC for dishonest majority – or: Breaking the SPDZ limits. In: Crampton, J., Jajodia, S., Mayes, K. (eds.) ESORICS 2013. LNCS, vol. 8134, pp. 1–18. Springer, Heidelberg (2013)

19. Damgård, I., Nielsen, J.B.: Universally composable efficient multiparty computation from threshold homomorphic encryption. In: Boneh, D. (ed.) CRYPTO 2003. LNCS, vol. 2729, pp. 247–264. Springer, Heidelberg (2003)

20. Damgård, I., Pastro, V., Smart, N., Zakarias, S.: Multiparty computation from somewhat homomorphic encryption. In: Safavi-Naini, R., Canetti, R. (eds.) CRYPTO 2012. LNCS, vol. 7417, pp. 643–662. Springer, Heidelberg (2012)

21. Damgård, I., Zakarias, S.: Constant-overhead secure computation of boolean circuits using preprocessing. In: Sahai, A. (ed.) TCC 2013. LNCS, vol. 7785, pp. 621–641. Springer, Heidelberg (2013)

22. Gentry, C.: A fully homomorphic encryption scheme. PhD thesis, Stanford University (2009), http://crypto.stanford.edu/craig

23. Gentry, C.: Fully homomorphic encryption using ideal lattices. In: STOC, pp. 169–178. ACM (2009)

24. Gentry, C., Halevi, S., Smart, N.P.: Fully homomorphic encryption with polylog overhead. In: Pointcheval, D., Johansson, T. (eds.) EUROCRYPT 2012. LNCS, vol. 7237, pp. 465–482. Springer, Heidelberg (2012)

25. Gentry, C., Halevi, S., Smart, N.P.: Homomorphic evaluation of the AES circuit. In: Safavi-Naini, R., Canetti, R. (eds.) CRYPTO 2012. LNCS, vol. 7417, pp. 850–867. Springer, Heidelberg (2012)

26. Goldreich, O.: The Foundations of Cryptography - Volume 2, Basic Applications. Cambridge University Press (2004)

27. Hirt, M., Nielsen, J.B.: Robust multiparty computation with linear communication complexity. In: Dwork, C. (ed.) CRYPTO 2006. LNCS, vol. 4117, pp. 463–482. Springer, Heidelberg (2006)

28. Nielsen, J.B., Nordholt, P.S., Orlandi, C., Burra, S.S.: A new approach to practical active-secure two-party computation. In: Safavi-Naini, R., Canetti, R. (eds.) CRYPTO 2012. LNCS, vol. 7417, pp. 681–700. Springer, Heidelberg (2012)

29. Smart, N.P., Vercauteren, F.: Fully homomorphic encryption with relatively small key and ciphertext sizes. In: Nguyen, P.Q., Pointcheval, D. (eds.) PKC 2010. LNCS, vol. 6056, pp. 420–443. Springer, Heidelberg (2010)

30. Smart, N.P., Vercauteren, F.: Fully homomorphic SIMD operations. To Appear in Designs, Codes and Cryptography (2012)

Building Lossy Trapdoor Functions
from Lossy Encryption

Brett Hemenway[1] and Rafail Ostrovsky[2]

[1] University of Pennsylvania
fbrett@cis.upenn.edu
[2] UCLA
rafail@cs.ucla.edu

Abstract. Injective one-way trapdoor functions are one of the most fundamental cryptographic primitives. In this work we show how to derandomize lossy encryption (with long messages) to obtain lossy trapdoor functions, and hence injective one-way trapdoor functions.

Bellare, Halevi, Sahai and Vadhan (CRYPTO '98) showed that if Enc is an IND-CPA secure cryptosystem, and H is a random oracle, then $x \mapsto \mathsf{Enc}(x, H(x))$ is an injective trapdoor function. In this work, we show that if Enc is a lossy encryption with messages at least 1-bit longer than randomness, and h is a pairwise independent hash function, then $x \mapsto \mathsf{Enc}(x, h(x))$ is a lossy trapdoor function, and hence also an injective trapdoor function.

The works of Peikert, Vaikuntanathan and Waters and Hemenway, Libert, Ostrovsky and Vergnaud showed that statistically-hiding 2-round Oblivious Transfer (OT) is equivalent to Lossy Encryption. In their construction, if the sender randomness is shorter than the message in the OT, it will also be shorter than the message in the lossy encryption. This gives an alternate interpretation of our main result. In this language, we show that *any* 2-message statistically sender-private semi-honest oblivious transfer (OT) for strings longer than the sender randomness implies the existence of *injective* one-way trapdoor functions. This is in contrast to the black box separation of injective trapdoor functions from many common cryptographic protocols, e.g. IND-CCA encryption.

Keywords: public-key cryptography, derandomization, injective trapdoor functions, oblivious transfer, lossy trapdoor functions.

1 Introduction

One-way functions are one of the most basic cryptographic primitives, and their existence is necessary for much of modern cryptography. Despite their immense value in cryptography, one-way functions are not sufficient for many useful cryptographic applications [IR89,RTV04], and in many situations a *trapdoor* is needed.

Constructing injective one-way trapdoor functions (a deterministic primitive) from a secure protocol, e.g. public-key encryption or oblivious transfer

K. Sako and P. Sarkar (Eds.) ASIACRYPT 2013, Part II, LNCS 8270, pp. 241–260, 2013.

(randomized primitives) has received much attention over the years with little success. One step in this direction was given by Bellare, Halevi, Sahai and Vadhan [BHSV98], who showed that in the *Random Oracle Model* IND-CPA secure encryption implies injective one-way trapdoor functions. Since it is known ([GKM+00]) 2-message OT implies IND-CPA encryption, the results of Bellare et al. can be viewed as a construction of injective one-way trapdoor functions from 2-message oblivious transfer in the *random oracle model*.

Our main contribution is to give a simple construction of injective trapdoor functions from lossy encryption (with long messages). In contrast to the results of [BHSV98], our results are in the *standard model*, and do not rely on random oracles. Our results can also be viewed as a derandomization of the basic cryptographic primitive Oblivious Transfer (OT) [Rab81,EGL85].

Lossy Encryption [KN08,PVW08,BHY09], is a public-key encryption protocol with two indistinguishable types of public keys, injective keys and lossy keys. Ciphertexts created under injective keys can be decrypted, while ciphertexts created under lossy keys are statistically independent of the underlying plaintext. The security of the encryption is then guaranteed by the indistinguishability of the two types of keys.

Building on the construction of [PW08], in [PVW08], Peikert, Vaikuntanathan and Waters showed that lossy encryption implies statistically sender-private 2-message oblivious transfer. In [HLOV11], Hemenway, Libert, Ostrovsky and Vergnaud showed that the two primitives are, in fact, equivalent.[1] Throughout this work, we will use the terminology of lossy encryption because it makes the constructions more transparent.

If $\mathcal{PKE} = (\mathsf{Gen}, \mathsf{Enc}, \mathsf{Dec})$ is a lossy encryption scheme our construction has a simple description: we choose as our function candidate,

$$F_{pk,h}(x) = \mathsf{Enc}(pk, x, h(x))$$

where h is some 2-wise independent hash function. Our main theorem says that if messages in $\mathcal{PKE}$ are at least one-bit longer than the encryption randomness, then $F_{pk,h}(\cdot)$ is a family of injective one-way trapdoor functions. In this setting, we are able to prove that this is secure even though the randomness is *dependent on the message*. In [BBO07], Bellare et al. used a similar construction, and they showed that $\mathsf{Enc}(pk, x, h(pk\|m))$ is one-way (in fact a deterministic encryption) if h is a *Random Oracle*. In their results the random oracle serves to break the dependence between the message and the randomness. In this work, we do not rely on random oracles, instead we use the lossiness of the encryption scheme to break this dependence. This is interesting given how difficult it has been to realize other forms of circular security, e.g. Key Dependent Message (KDM) security [CL01,BRS03,BHHO08].

The primary limitation of our construction is the requirement that the message space be larger than the randomness space. The lossy encryption protocols

---

[1] Their construction of lossy encryption from OT also preserves the randomness and message lengths, so if the OT uses sender randomness shorter than the messages so does the lossy encryption.

based on the Paillier cryptosystem [RS08,FGK⁺10], satisfy this requirement, and in Appendix B we give constructions of lossy encryption with short randomness based on the DDH, DCR and QR assumptions. These do not lead to significantly new constructions of LTDFs, however, as direct constructions of lossy trapdoor functions are already known from these assumptions. It is an intriguing question whether this restriction can be removed. In addition to increasing the applicability of our construction, a removal of the restriction on message length would imply a black-box separation between OT and statistically sender-private OT by the results of [GKM⁺00].

Although our construction does not provide more efficient ways of constructing lossy trapdoor functions, it provides an interesting theoretical connection between lossy trapdoor functions and lossy encryption. Constructing injective trapdoor functions from randomized primitives such as public-key encryption or oblivious transfer has proven to be an elusive goal, and our results provide one of the few positive examples in this area without relying on random oracles.

The notion of RDM security has also been studied by Birrell, Chung, Pass and Telang [BCPT13], who show that full RDM security (where the message can be an arbitrary function of the randomness) is not possible. Birrell et al. go on to show that any encryption scheme where the randomness is longer than the message can be transformed into a bounded-RDM secure cryptosystem. Their work, which requires the message to be shorter than the encryption randomness, nicely complements this work where we insist the opposite, that the message is longer than the encryption randomness.[2] While Birrell et al. focus on the goal of building RDM secure encryption for general circuits, in this work, we use RDM encryption as a stepping stone for building injective trapdoor functions from lossy encryption. Birrell et al. provide more general constructions of RDM encryption than we do, but their constructions do not immediately imply injective trapdoor functions.

1.1   Previous Work

Injective one-way trapdoor functions were one of the first abstract cryptographic primitives to be defined, and their value is well recognized. In [Yao82], Yao showed that injective trapdoor functions imply IND-CPA secure encryption, and Gertner, Malkin and Reingold [GMR01] showed a black-box separation between injective (also poly-to-one) trapdoor functions and public-key encryption schemes. Gertner, Kannan, Malkin, Reingold, and Viswanathan [GKM⁺00] showed a black-box separation between 2-message oblivious transfer (OT) and injective trapdoor functions, in both directions.

In this work, we show that statistically sender-private OT for long strings implies injective one-way trapdoor functions. Combining our results with the separation results of [GKM⁺00] gives a separation between standard OT and statistically sender-private OT for long strings.

---

[2] We also require the initial cryptosystem to be lossy, while their construction works with any semantically secure cryptosystem with short messages.

This work actually constructs lossy trapdoor functions (LTDFs), a stronger primitive than injective one-way trapdoor functions. Peikert and Waters introduced LTDFs in [PW08]. Lossy trapdoor functions imply injective trapdoor functions [PW08,MY10], but appear to be a strictly stronger primitive, as they cannot be constructed in a black-box manner from even one-way trapdoor permutations as shown by Rosen and Segev [RS09]. This separation was later extended by Vahlis in [Vah10]. A family of lossy trapdoor functions contains two computationally indistinguishable types of functions: injective functions with a trapdoor, and lossy functions, which are functions that statistically lose information about their input. The indistinguishability of the two types of functions shows that the injective functions are, in fact, one-way.

A similar property can be defined for cryptosystems [GOS06,PVW08,KN08,BHY09]. A cryptosystem is called lossy encryption, if there are two indistinguishable types of public keys, injective keys which behave normally, and lossy keys, which have the property that ciphertexts created under a lossy key are statistically independent of the plaintext. It was shown in Bellare, Hofheinz and Yilek [BHY09] that just as injective trapdoor functions imply IND-CPA secure encryption, LTDFs imply lossy encryption. One interpretation of our main theorem is as a partial converse of that result.

Although LTDFs immediately imply injective one-way trapdoor functions, Rosen and Segev [RS09] showed that LTDFs cannot be constructed from one-way trapdoor permutations in a black-box manner, and currently the only known generic construction of LTDFs is from homomorphic smooth hash proof systems [HO12]. In this work, we construct lossy trapdoor functions, and hence injective one-way trapdoor functions from lossy encryption with long plaintexts.

While lossy trapdoor functions were created as a building block for IND-CCA secure encryption, lossy encryption was initially created to help prove security against an active adversary in the Multiparty Computation Setting. Lossy encryption has gone by many names. Groth, Ostrovsky and Sahai called it "parameter-switching" in the context of perfect non-interactive zero knowledge proofs [GOS06]. In [KN08], Kol and Naor called it "Meaningful/Meaningless" encryption, in [PVW08], Peikert, Vaikuntanathan and Waters called it "Dual-Mode Encryption", and in [BHY09] Bellare, Hofheinz and Yilek called it "Lossy Encryption". In this work, we use the name Lossy Encryption to highlight its connection to Lossy Trapdoor Functions.. Despite the apparent utility of lossy encryption, it has proven rather easy to construct, and in [HLOV11], Hemenway, Libert, Ostrovsky and Vergnaud give constructions of lossy encryption from, rerandomizable encryption, statistically-hiding oblivious transfer, universal hash proofs, private information retrieval schemes and homomorphic encryption. Combining the results of [PVW08] and [HLOV11], shows that lossy encryption with short randomness can be viewed exactly as a statistically sender private $\binom{2}{1}$-oblivious transfer with short randomness. Thus, throughout this work, we will use the terminology of lossy encryption because it preserves the intuition of our construction, but it should be noted that lossy encryption can be substituted throughout the paper by 2-message statistically sender-private $\binom{2}{1}$-oblivious transfer and all of our results remain valid.

1.2 Our Contributions

One of the most fundamental techniques in modern cryptography is the use of randomization in protocols to achieve higher levels of security. On the other hand, because deterministic protocols are more versatile, a significant body of research has explored the question of where deterministic primitives can be created from their randomized counterparts. One (negative) example of this type was the results of Gertner, Malkin and Reingold showing that IND-CPA secure encryption cannot be used in a black-box way to construct injective one-way trapdoor functions. Our work is perhaps best viewed in this light. We show that lossy encryption, a randomized primitive, which is a strengthening of the standard IND-CPA secure encryption, can be used to construct lossy trapdoor functions, a deterministic primitive, which is the analogous strengthening of injective one-way trapdoor functions. Since we construct a deterministic primitive from the analogous randomized one, it is natural to think of these results as a "derandomization" of lossy encryption[3].

Our main result is to give a black-box construction of LTDFs (and hence injective one-way trapdoor functions, and IND-CCA secure encryption) from any lossy encryption over a plaintext space which is (at least 1-bit) larger than its randomness space. This is an interesting connection because many generic constructions of lossy encryption exist, while injective one-way trapdoor functions have proven difficult to construct and are black-box separated from many common primitives ([Rud89,IR89,GKM+00,GMR01]).

Main Theorem. Suppose $\mathcal{PKE} = (\mathsf{Gen}, \mathsf{Enc}, \mathsf{Dec})$ is a lossy encryption scheme over message space $\mathcal{M}$, randomness space $\mathcal{R}$ and ciphertext space $\mathcal{C}$. If $|\mathcal{M}| > 2|\mathcal{R}|$, i.e. messages are at least one bit longer than the randomness, and $\mathcal{H}$ is a 2-wise independent hash family, with $h : \mathcal{M} \to \mathcal{R}$, for $h \in \mathcal{H}$, then the function

$$F_{pk,h} : \mathcal{M} \to \mathcal{C}$$
$$x \mapsto \mathsf{Enc}(pk, x, h(x))$$

is a slightly lossy trapdoor function.

While these functions are fairly simple to describe, the circular nature of the construction makes the proof very delicate. Applying the results of Mol and Yilek [MY10], we have the following corollaries:

Corollary. If there exists a lossy encryption scheme with messages at least one bit longer than the encryption randomness, then there exist Correlated Product secure functions.

---

[3] It is important to note, however, that any notion of one-wayness depends inherently on the fact that the inputs are randomized. While one-way functions must have "random" inputs to provide any one-wayness guarantees they do not require auxiliary random inputs as public-key encryption does.

Corollary. If there exists a lossy encryption scheme with messages at least one bit longer than the encryption randomness, then there exists IND-CCA secure encryption.

Applying the recent results of Kiltz, Mohassel and O'Neill [KMO10], we have

Corollary. If there exists a lossy encryption scheme with messages at least one bit longer than the encryption randomness, then there exist adaptive trapdoor functions.

2   Preliminaries

2.1   Notation

If $f : X \to Y$ is a function, for any $Z \subset X$, we let $f(Z) = \{f(x) : x \in Z\}$. If A is a PPT machine, then we use $a \xleftarrow{\$} A$ to denote running the machine A and obtaining an output, where a is distributed according to the internal randomness of A. If R is a set, and no distribution is specified, we use $r \xleftarrow{\$} R$ to denote sampling from the uniform distribution on R.

If X and Y are families of distributions indexed by a security parameter λ, we say that X is statistically close to Y, (written $X \approx_s Y$) to mean that for all polynomials p and sufficiently large λ, we have $\sum_x |\Pr[X = x] - \Pr[Y = x]| < \frac{1}{p(\lambda)}$.

We say that X and Y are computationally close (written $X \approx_c Y$) to mean that for all PPT adversaries A, for all polynomials p, and for all sufficiently large λ, we have $|\Pr[A^X = 1] - \Pr[A^Y = 1]| < 1/p(\lambda)$.

2.2   Lossy Trapdoor Functions

We briefly review the notion of *Lossy Trapdoor Functions* (LTDFs) as described in [PW08]. Intuitively, a family of Lossy Trapdoor Functions is a family of functions which have two modes, or branches, injective mode, which has a trapdoor, and lossy mode which is guaranteed to have a small image size. This implies that with high probability the preimage of an element in the image will be a large set. Formally we have:

Definition 1. *A tuple* $(S_{ltdf}, F_{\mathrm{ltdf}}, F_{\mathrm{ltdf}}^{-1})$ *of PPT algorithms is called a family of* (n, k)-*Lossy Trapdoor Functions if the following properties hold:*

- *Sampling Injective Functions: $S_{ltdf}(1^\lambda, 1)$ outputs s, t where s is a function index, and t its trapdoor. We require that $F_{\mathrm{ltdf}}(s, \cdot)$ is an injective deterministic function on $\{0, 1\}^n$, and $F_{\mathrm{ltdf}}^{-1}(t, F_{\mathrm{ltdf}}(s, x)) = x$ for all x.*
- *Sampling Lossy Functions: $S_{ltdf}(1^\lambda, 0)$ outputs $(s, \perp)$ where s is a function index and $F_{\mathrm{ltdf}}(s, \cdot)$ is a function on $\{0, 1\}^n$, where the image of $F_{\mathrm{ltdf}}(s, \cdot)$ has size at most 2^{n-k}.*
- *Indistinguishability: The first outputs of $S_{ltdf}(1^\lambda, 0)$ and $S_{ltdf}(1^\lambda, 1)$ are computationally indistinguishable.*

We recall a basic result about Lossy Trapdoor Functions from [PW08].

Lemma 1. *(From [PW08]) Let λ be a security parameter. If $(S_{ltdf}, F_{\text{ltdf}}, F_{\text{ltdf}}^{-1})$ is a family of (n, k) Lossy Trapdoor Functions, and $k = \omega(\log(\lambda))$, then the injective branches form a family of injective one-way trapdoor functions.*

In [MY10], Mol and Yilek observed that if f is an (n, k)-LTDF, then defining $\boldsymbol{f}(x_1, \ldots, x_t) = (f(x_1), \ldots, f(x_t))$, is a (tn, tk)-LTDF. Thus if $k > 1/\text{poly}$, t can be chosen such that $tk = \omega(\log(\lambda))$, and hence $\boldsymbol{f}$ is a injective one-way trapdoor function by Lemma 1. Mol and Yilek went on to show how to construct correlated product secure functions, and hence IND-CCA secure cryptosystems from these *slightly lossy trapdoor functions.*

2.3   Lossy Encryption

Peikert and Waters introduced LTDFs as a means of constructing IND-CCA secure cryptosystems. In their original work, however, they also observed that LTDFs can be used to create a simple IND-CPA secure cryptosystem, $\mathsf{Enc}(m, r) = (F_{\text{ltdf}}(r), h(r) + m)$. This simple cryptosystem has powerful lossiness properties. The lossiness of this cryptosystem was further developed and explored in [PVW08] where Peikert, Vaikuntanathan and Waters defined Dual-Mode Encryption, as a means of constructing efficient and composable oblivious transfer. Dual-Mode encryption is a type of cryptosystem with two types public-keys, injective keys on which the cryptosystem behaves normally and "lossy" or "messy" keys on which the system loses information about the plaintext. In particular they require that the encryptions of any two plaintexts under a lossy key yield distributions that are statistically close, yet injective and lossy keys remain computationally indistinguishable. Groth, Ostrovsky and Sahai [GOS06] previously used a similar notion in the context of non-interactive zero knowledge. With the goal of creating selective opening secure cryptosystems, in [BHY09] Bellare, Hofheinz and Yilek defined *Lossy Encryption*, extending the definitions of Dual-Mode Encryption in [PVW08], Meaningful/Meaningless Encryption in [KN08] and Parameter-Switching [GOS06]. We review the definition of Lossy Encryption here:

Definition 2. *Formally, a* lossy public-key encryption scheme *is a tuple $\mathcal{PKE} = (\mathsf{Gen}, \mathsf{Enc}, \mathsf{Dec})$ of polynomial-time algorithms such that*

- $\mathsf{Gen}(1^\lambda, inj)$ *outputs keys (pk, sk), keys generated by $\mathsf{Gen}(1^\lambda, inj)$ are called* injective keys.
- $\mathsf{Gen}(1^\lambda, lossy)$ *outputs keys $(pk_{\mathsf{lossy}}, \perp)$, keys generated by $\mathsf{Gen}(1^\lambda, lossy)$ are called* lossy keys.
- $\mathsf{Enc}(pk, \cdot, \cdot) : \mathcal{M} \times \mathcal{R} \to \mathcal{C}.$

Additionally, the algorithms must satisfy the following properties:

1. *Correctness on injective keys. For all $x \in \mathcal{M}$,*

$$\Pr\left[(pk, sk) \xleftarrow{\$} \mathsf{Gen}(1^\lambda, inj); r \xleftarrow{\$} \mathcal{R} : \mathsf{Dec}(sk, \mathsf{Enc}(pk, x, r)) = x\right] = 1.$$

2. Indistinguishability of keys. *We require that the public key, pk, in lossy mode and injective mode are computationally indistinguishable. Specifically, if* proj : $(pk, sk) \mapsto pk$ *is the projection map,*

$$\{\text{proj}(\text{Gen}(1^\lambda, inj))\} \approx_c \{\text{proj}(\text{Gen}(1^\lambda, lossy))\}$$

3. Lossiness of lossy keys. *For all* $(pk_{\text{lossy}}, sk_{\text{lossy}}) \overset{\$}{\leftarrow} \text{Gen}(1^\lambda, lossy)$, *and all* $x_0, x_1 \in \mathcal{M}$, *the two distributions* $\{r \overset{\$}{\leftarrow} \mathcal{R} : (pk_{\text{lossy}}, \text{Enc}(pk_{\text{lossy}}, x_0, r))\}$ *and* $\{r \overset{\$}{\leftarrow} \mathcal{R} : (pk_{\text{lossy}}, \text{Enc}(pk_{\text{lossy}}, x_1, r))\}$ *are statistically close, i.e. the statistical distance is negligible in* λ.

We call a cryptosystem ν-lossy if for all $(pk_{\text{lossy}}, sk_{\text{lossy}}) \overset{\$}{\leftarrow} \text{Gen}(1^\lambda, lossy)$ we have

$$\max_{x_0, x_1 \in \mathcal{M}} \Delta(\{r \overset{\$}{\leftarrow} \mathcal{R} : (pk_{\text{lossy}}, \text{Enc}(pk_{\text{lossy}}, x_0, r))\}, \{r \overset{\$}{\leftarrow} \mathcal{R} : (pk_{\text{lossy}}, \text{Enc}(pk_{\text{lossy}}, x_1, r))\}) < \nu.$$

We call a cryptosystem *perfectly lossy* if the distributions are identical. The works of [PW08,PVW08,HLOV11], show that lossy encryption is identical to statistically sender private $\binom{2}{1}$-OT.

3 Constructing Slightly Lossy Trapdoor Functions

In this section we describe our main result: a generic construction of a slightly lossy trapdoor functions from lossy encryption. Let $\mathcal{PKE} = (\text{Gen}, \text{Enc}, \text{Dec})$ be a Lossy Encryption, with $\text{Enc}(pk, \cdot, \cdot) : \mathcal{M} \times \mathcal{R} \to \mathcal{C}_{pk}$. Let $\mathcal{H}$ be a family of pairwise independent hash functions, with $h : \mathcal{M} \to \mathcal{R}$, for all $h \in \mathcal{H}$. The construction is described in Figure 1.

The injectivity, and correctness of inversion of the functions described in Figure 1 is clear, it remains only to show that the lossy branch of $F_{pk,h}$ is slightly lossy.

Sampling an Injective Function: **Evaluation:**

   $(pk, sk) \overset{\$}{\leftarrow} \text{Gen}(1^\lambda, inj)$ $F_{pk,h} : \mathcal{M} \to \mathcal{C}$,

   $h \overset{\$}{\leftarrow} \mathcal{H}$ $F_{pk,h}(x) = \text{Enc}(pk, x, h(x))$

Sampling a Slightly Lossy Function: **Trapdoor:**

   $(pk, \perp) \overset{\$}{\leftarrow} \text{Gen}(1^\lambda, lossy)$ $F_{pk,h}^{-1}(c) = \text{Dec}(sk, c)$

   $h \overset{\$}{\leftarrow} \mathcal{H}$

Fig. 1. Slightly Lossy Trapdoor Functions from Lossy Encryption

4 Proof of Security

In this section we prove that the function family defined in Figure 1 is slightly lossy. To build intuition, we begin by considering the case when the encryption scheme $\mathcal{PKE} = (\mathsf{Gen}, \mathsf{Enc}, \mathsf{Dec})$ is *perfectly* lossy, i.e. for a lossy key pk, the distributions $\mathsf{Enc}(pk, x)$ and $\mathsf{Enc}(pk, y)$ are identical for any $x, y \in \mathcal{M}$.

4.1 The Perfectly Lossy Case

Lemma 2. *If* $\mathcal{PKE} = (\mathsf{Gen}, \mathsf{Enc}, \mathsf{Dec})$, *be a* perfectly *lossy encryption scheme, then for all* $pk \xleftarrow{\$} \mathsf{Gen}(1^\lambda, lossy)$, *the sets* $\mathsf{Enc}(pk, \mathcal{M}, \mathcal{R})$ *and* $\mathsf{Enc}(pk, 0, \mathcal{R})$ *are equal.*

Proof. The perfect lossiness property says that

$$\Pr[r \xleftarrow{\$} \mathcal{R} : \mathsf{Enc}(pk, x) = c] = \Pr[r \xleftarrow{\$} \mathcal{R} : \mathsf{Enc}(pk, y) = c],$$

for all $x, y \in \mathcal{M}$ and all $c \in \mathcal{C}$, thus we have that *as sets* $\mathsf{Enc}(pk, x, \mathcal{R}) = \mathsf{Enc}(pk, y, \mathcal{R})$. Since $\mathsf{Enc}(pk, \mathcal{M}, \mathcal{R}) = \bigcup_{x \in \mathcal{M}} \mathsf{Enc}(pk, x, \mathcal{R})$, the claim follows.

Lemma 3. *If* $\mathcal{PKE} = (\mathsf{Gen}, \mathsf{Enc}, \mathsf{Dec})$, *is a* perfectly *lossy encryption scheme, and* h *is any function from* $\mathcal{M}$ *to* $\mathcal{R}$, *then the function defined in Figure 1 is a* $(\log |\mathcal{M}|, \log |\mathcal{M}| - \log |\mathcal{R}|)$*-LTDF.*

Proof. The indistinguishability of injective and lossy modes follows from the indistinguishability of injective and lossy keys for $\mathcal{PKE}$. The trapdoor follows from the correctness of decryption for $\mathcal{PKE}$.

Notice that for any function h, the image of $F_{pk,h}$ is a subset of the ciphertext space $\mathcal{C} = \mathsf{Enc}(pk, \mathcal{M}, \mathcal{R})$. In lossy mode, from Lemma 2 we have that the set $\mathsf{Enc}(pk, \mathcal{M}, \mathcal{R})$ is equal to the set $\mathsf{Enc}(pk, 0, \mathcal{R})$, but $|\mathsf{Enc}(pk, 0, \mathcal{R})| \leq |\mathcal{R}|$, so if pk is a lossy key, the image size of $F_{pk,h}$ is at most $|\mathcal{R}|$, and the result follows.

Notice that the specific form of the function h was never used in the proof of Lemma 3. For example, we could choose h to be a constant function, and the result would still hold! In particular, if the hypotheses of Lemma 3 hold and $|\mathcal{M}| > |\mathcal{R}|$, the function $F_{pk,h}(x) = \mathsf{Enc}(pk, x, 0)$ is one-way. It is instructive to examine this a little further. For most ordinary encryption schemes, the function $F_{pk,h}(x) = \mathsf{Enc}(pk, x, 0)$, i.e. encrypting the message x using some fixed randomness (in this case the zero string), will not be a one-way function. To see this, we can take any IND-CPA secure encryption scheme and modify it so that if the zero string is used for the randomness, the encryption algorithm simply outputs the message in the clear. This will not affect the CPA security of the encryption scheme, but it will mean the function $F_{pk,h}$ defined in this way will be the identity function, and hence trivially invertible. On the other hand, if $\mathcal{PKE}$ is a perfectly lossy encryption, and $|\mathcal{M}| > |\mathcal{R}|$, then this modification will break the perfect lossiness of $\mathcal{PKE}$.

It is tempting to conclude that if $\mathcal{PKE}$ were only statistically lossy, then Lemma 3 would still hold. To see that this is not the case, notice that the counterexample given in the previous paragraph applies even when $\mathcal{PKE}$ is statistically lossy. In the next section, we will construct a lossy trapdoor function from statistically lossy encryption, but significantly more machinery is needed.

As remarked above, one reason why this argument *does not* trivially extend to the statistically-lossy case is that although the distributions $\{r \xleftarrow{\$} \mathcal{R} : \mathrm{Enc}(pk, x, r)\}$ and $\{r \xleftarrow{\$} \mathcal{R} : \mathrm{Enc}(pk, y, r)\}$, will be statistically close for any $x, y \in \mathcal{M}$, we are not choosing the randomness uniformly. In our situation, the randomness is uniquely defined by the message, so new techniques are needed. See Appendix C for further discussion.

4.2 The Statistically Lossy Case

In the preceding section, we examined the perfectly lossy case. There, we were free to choose the function h arbitrarily, even a constant function sufficed to prove security! In the statistical setting the proofs are significantly more delicate, and we will need to make use of the fact that h is a pairwise independent hash function.

For the following, consider a fixed (lossy) public key pk. Let C_0 be the set of encryptions of 0, i.e. $C_0 = \mathrm{Enc}(pk, 0, \mathcal{R})$. This immediately implies that $|C_0| \leq |\mathcal{R}|$. For $x \in \mathcal{M}$, define A_x to be the event (over the random choice of $h \xleftarrow{\$} \mathcal{H}$) that $F_{pk,h}(x) \notin C_0$. Let $d_x = \Pr[A_x] = \mathbb{E}(1_{A_x})$. Let $d = \frac{1}{|\mathcal{M}|} \sum_{x \in \mathcal{M}} d_x$. Thus Cauchy-Schwarz says that $\sum_{x \in X} d_x^2 \geq |\mathcal{M}| d^2$. Let Z be the random variable denoting the number of elements in the domain that map outside of C_0, so $Z = \sum_{x \in \mathcal{M}} 1_{A_x} = \sum_{x \in \mathcal{M}} 1_{F_{pk,h}(x) \notin C_0}$. Thus the image of $F_{pk,h}$ has size bounded by $|C_0| + Z$.

To show that $F_{pk,h}$ is a lossy trapdoor function, we must show that with high probability (over the choice of h), the image of $F_{pk,h}$ is small (relative to the domain $\mathcal{M}$). We begin with the easy observation:

$$\mathbb{E}(Z) = \mathbb{E}\left(\sum_{x \in \mathcal{M}} 1_{A_x}\right) = \sum_{x \in \mathcal{M}} d_x = |\mathcal{M}| d. \tag{1}$$

Notice as well, that since h pairwise independent, it is 1-universal and hence $\Pr[h \xleftarrow{\$} \mathcal{H} : F_{pk,h}(x) = c] = \Pr[r \leftarrow \mathcal{R} : \mathrm{Enc}(pk, x, r) = c]$ for all $x \in \mathcal{M}, c \in \mathcal{C}$. We will use this fact to show that d is small. In fact, it's not hard to see that d is bounded by the lossiness of $\mathcal{PKE}$.

This shows that the expected image size is small, but we wish to show that with high probability the image size of $F_{pk,h}$ is small. To do this we examine the variance of Z. Since $Z = \sum_{x \in \mathcal{M}} 1_{A_x}$, where the variables 1_{A_x} are bernoulli random variables with parameter d_x. The variables 1_{A_x} are pairwise independent (because h is pairwise independent), thus we have

$$\mathrm{Var}(Z) = \sum_{x \in \mathcal{M}} \mathrm{Var}(1_{A_x}) = \sum_{x \in \mathcal{M}} d_x(1 - d_x) = |\mathcal{M}|d - \sum_{x \in \mathcal{M}} d_x^2$$

Thus by Cauchy-Schwarz, we arrive at the upper bound

$$\mathrm{Var}(Z) \le |\mathcal{M}|d - |\mathcal{M}|d^2 = |\mathcal{M}|(d - d^2). \tag{2}$$

On the other hand, we have

$$\mathrm{Var}(Z) = \sum_{z=0}^{|\mathcal{M}|}(z - \mathbb{E}(Z))^2 \Pr[Z = z] = \sum_{z=0}^{|\mathcal{M}|}(z - |\mathcal{M}|d)^2 \Pr[Z = z]$$

$$\ge \sum_{z=(1-\epsilon)|\mathcal{M}|}^{|\mathcal{M}|}(z - |\mathcal{M}|d)^2 \Pr[Z = z] \ge \sum_{z=(1-\epsilon)|\mathcal{M}|}^{|\mathcal{M}|}((1-\epsilon)|\mathcal{M}| - |\mathcal{M}|d)^2 \Pr[Z = z]$$

$$= (1 - \epsilon - d)^2 |\mathcal{M}|^2 \sum_{z=(1-\epsilon)|\mathcal{M}|}^{|\mathcal{M}|} \Pr[Z = z]$$

For any ϵ with $0 < \epsilon < 1$, and $1 - \epsilon > d$. Since the parameter ϵ is under our control, we can always ensure that this is the case. This will not be a stringent restriction, however, since d (the proportion of inputs that map outside of C_0) will always negligible by the statistical lossiness of $\mathcal{PKE}$. In the proof of the following, we will find another restriction on ϵ, namely to achieve a useful degree of lossiness, ϵ must be chosen so that $\epsilon > \frac{|\mathcal{R}|}{|\mathcal{M}|}$.

Rearranging, we have

$$\sum_{z=(1-\epsilon)|\mathcal{M}|}^{|\mathcal{M}|} \Pr[Z = z] \le \frac{\mathrm{Var}(Z)}{(1 - \epsilon - d)^2 |\mathcal{M}|^2}.$$

Applying the bound on the variance obtained in Equation 2, we have

$$\sum_{z=(1-\epsilon)|\mathcal{M}|}^{|\mathcal{M}|} \Pr[Z = z] \le \frac{|\mathcal{M}|(d - d^2)}{(1 - \epsilon - d)^2 |\mathcal{M}|^2} \le \frac{d(1 - d)}{(1 - \epsilon - d)^2 |\mathcal{M}|}. \tag{3}$$

This upper bound on the probability that Z is large can be extended to show:

Lemma 4. *If $\mathcal{PKE} = (\mathsf{Gen}, \mathsf{Enc}, \mathsf{Dec})$ is a ν-lossy encryption, and if $|\mathcal{M}| = t|\mathcal{R}|$, for some $t > 1$, then for any $0 < \epsilon < 1$ such that $1 - \epsilon$ is noticeable, and $\epsilon > \frac{1}{t}$, with all but negligible probability over the choice of h, the function $F_{pk,h}$ is a $(\log |\mathcal{M}|, -\log(1 - \epsilon + \frac{1}{t}))$-LTDF family.*

Proof. Suppose $\mathcal{PKE}$ is ν-Lossy, i.e. $\Delta(\{r \xleftarrow{\$} \mathcal{R} : \mathsf{Enc}(pk, x, r)\}, \{r \xleftarrow{\$} \mathcal{R} : \mathsf{Enc}(pk, y, r)\}) < \nu$. Then by the pairwise independence of h, $\Delta(\{h \xleftarrow{\$} \mathcal{H} : F_{pk,h}(0)\}, \{h \xleftarrow{\$} \mathcal{H} : F_{pk,h}(x)\}) < \nu$ for all $x \in \mathcal{M}$. In particular, $d_x = \Pr(A_x) < \nu$ for all d_x, so $d = \frac{1}{|\mathcal{M}|} \sum_{x \in \mathcal{M}} d_x < \nu$. Because the random variable Z represents the number of $x \in \mathcal{M}$ such that $F_{pk,h}(x) \notin C_0$, we have $|F_{pk,h}(\mathcal{M})| \leq |C_0| + Z$. Since $|C_0| \leq |\mathcal{R}| = \frac{1}{t}|\mathcal{M}|$, by Equation 3, we have

$$\Pr[|F_{pk,h}(\mathcal{M})| > (1 - \epsilon + \frac{1}{t})|\mathcal{M}|] < \frac{(\nu - \nu^2)}{(1 - \epsilon - \nu)^2 |\mathcal{M}|}.$$

We would like to choose ϵ as close to 1 as possible but subject to the constraint that $\frac{\nu - \nu^2}{(1-\epsilon-\nu)^2|\mathcal{M}|}$ is negligible. Since ν is negligible, and $\frac{1}{|\mathcal{M}|}$ is negligible, the right hand side will certainly be negligible if $1 - \epsilon - \nu$ is non-negligible. But this holds because ν is negligible, and $1 - \epsilon$ is non-negligible. Thus with all but negligible probability, the residual leakage is $\log((1 - \epsilon + \frac{1}{t})|\mathcal{M}|)$, so the lossiness is $\log(|\mathcal{M}|) - \log((1 - \epsilon + \frac{1}{t})|\mathcal{M}|) = -\log(1 - \epsilon + \frac{1}{t})$.

From Lemma 4, we see that if $1 - \frac{1}{t}$ is non-negligible, such an ϵ will exist. This immediately implies the result:

Theorem 1 (Main Theorem). *If $\mathcal{PKE}$ is a ν-Lossy Encryption with $|\mathcal{M}| = t|\mathcal{R}|$, for some $t > 1$ with $1 - \frac{1}{t}$ non-negligible, then the functions described in Figure 1 form a family of lossy trapdoor functions.*

Proof. From the proof of Lemma 4, it suffices to find an ϵ such that $1 - \epsilon$ is noticeable, and $\epsilon - \frac{1}{t}$ is noticeable.

In this case, we can take $\epsilon = \frac{1}{2} + \frac{1}{2t}$. In this case $1 - \epsilon = \epsilon - \frac{1}{t} = \frac{1 - \frac{1}{t}}{2}$ which is noticeable since $1 - \frac{1}{t}$ was assumed to be noticeable. In this case, the lossiness of the function will be $-\log(1 - \epsilon + \frac{1}{t}) = \sum_{j=1}^{\infty} \frac{(\epsilon - \frac{1}{t})^j}{j} \geq \epsilon - \frac{1}{t} = \frac{1}{2}(1 - \frac{1}{t})$, which is noticeable.

Taking $t = 2$, and applying the results of [MY10], we have

Corollary 1. *If there exists Lossy Encryption with $|\mathcal{M}| > 2|\mathcal{R}|$, and there is an efficiently computable family of 2-wise independent hash functions from $\mathcal{M}$ to $\mathcal{R}$, then there exist injective one-way trapdoor functions, Correlated Product secure functions and IND-CCA2 secure encryption.*

Although Theorem 1 provides lossy trapdoor functions and hence IND-CCA secure encryption [MY10], we would like to see exactly how lossy the functions can be. This is captured in Corollary 2.

Corollary 2. *If $|\mathcal{M}| = t|\mathcal{R}|$, and $\frac{1}{t}$ is negligible, i.e. the messages are $\omega(\log \lambda)$ bits longer than the randomness, then the functions described in Figure 1 form a family of injective one-way trapdoor functions.*

Proof. From Equation 3, we have

$$\Pr[|F_{pk,h}(\mathcal{M})| > (1 - \epsilon + \frac{1}{t})|\mathcal{M}|] < \frac{(\nu - \nu^2)}{(1 - \epsilon - \nu)^2|\mathcal{M}|}.$$

If we set $\epsilon = 1 - \nu - \frac{1}{\sqrt{|\mathcal{M}|}}$, then the right hand side becomes $\nu - \nu^2$, which

is negligible. The lossiness is then $-\log\left(1 - \epsilon + \frac{1}{t}\right) = -\log\left(\nu + \frac{1}{t} + \frac{1}{\sqrt{|\mathcal{M}|}}\right) >$
$-\log(\nu + \frac{1}{t} + |\mathcal{M}|^{-1/2})$. Since both ν and $\frac{1}{t}$ were assumed to be negligible, and
since $|\mathcal{M}| > |\mathcal{R}|$, the sum $\nu + \frac{1}{t} + |\mathcal{M}|^{-1/2}$ is also negligible. But this means that
$-\log(\nu + \frac{1}{t} + |\mathcal{M}|^{-1/2}) \in \omega(\log \lambda)$. Thus we can apply Lemma 1 to conclude
that $F_{pk,h}$ is a family of injective one-way trapdoor functions.

Finally, we observe that applying the results of [KMO10], we can construct
adaptive trapdoor functions from lossy encryption with messages one bit longer
than the randomness.

Corollary 3. *If there exists lossy encryption with messages at least one bit
longer than the encryption randomness then there exist adaptive trapdoor
functions.*

5   Conclusion

The results of Gertner, Malkin and Reingold [GMR01] show that injective one-
way trapdoor functions cannot be constructed in a black-box manner from IND-
CPA secure encryption. Our results show that when the cryptosystem is indistinguishable from a one which loses information about the plaintext (i.e. lossy
encryption), then we can construct injective trapdoor functions from it which are
indistinguishable from functions that statistically lose information about their
inputs (i.e. lossy trapdoor functions). The only requirement we have is that the
plaintext space of the cryptosystem be larger than its randomness space.

An interesting feature of this result is that it does not parallel the standard
(non-lossy) case. This result somewhat surprising as well given the number of
generic primitives that imply lossy encryption, and the lack of constructions of
injective one-way trapdoor functions from general assumptions. Our proof relies
crucially on showing that lossy encryption with long plaintexts remains one-
way even when encrypting with *randomness that is dependent on the message*.
The notion of security in the presence of randomness dependent messages is an
interesting one, and we hope it will prove useful in other constructions.

Applying the results of [MY10] to our constructions immediately gives a construction of IND-CCA secure encryption from lossy encryption with long plaintexts. Applying the results of [KMO10] to our constructions gives a construction
of adaptive trapdoor functions from lossy encryption with long plaintexts.

The primary limitation of our results is the requirement that the message
space be larger than the randomness space. Whether this restriction can be
removed is an important open question.

References

ACPS09. Applebaum, B., Cash, D., Peikert, C., Sahai, A.: Fast cryptographic primitives and circular-secure encryption based on hard learning problems. In: Halevi, S. (ed.) CRYPTO 2009. LNCS, vol. 5677, pp. 595–618. Springer, Heidelberg (2009)

BBO07. Bellare, M., Boldyreva, A., O'Neill, A.: Deterministic and efficiently searchable encryption. In: Menezes, A. (ed.) CRYPTO 2007. LNCS, vol. 4622, pp. 535–552. Springer, Heidelberg (2007)

BCPT13. Birrell, E., Chung, K.-M., Pass, R., Telang, S.: Randomness-Dependent Message Security. In: Sahai, A. (ed.) TCC 2013. LNCS, vol. 7785, pp. 700–720. Springer, Heidelberg (2013)

BHHO08. Boneh, D., Halevi, S., Hamburg, M., Ostrovsky, R.: Circular-secure encryption from decision diffie-hellman. In: Wagner, D. (ed.) CRYPTO 2008. LNCS, vol. 5157, pp. 108–125. Springer, Heidelberg (2008)

BHSV98. Bellare, M., Halevi, S., Sahai, A., Vadhan, S.P.: Many-to-one trapdoor functions and their relation to public-key cryptosystems. In: Krawczyk, H. (ed.) CRYPTO 1998. LNCS, vol. 1462, pp. 283–298. Springer, Heidelberg (1998)

BHY09. Bellare, M., Hofheinz, D., Yilek, S.: Possibility and impossibility results for encryption and commitment secure under selective opening. In: Joux, A. (ed.) EUROCRYPT 2009. LNCS, vol. 5479, pp. 1–35. Springer, Heidelberg (2009)

BRS03. Black, J., Rogaway, P., Shrimpton, T.: Encryption-scheme security in the presence of key-dependent messages. In: Nyberg, K., Heys, H.M. (eds.) SAC 2002. LNCS, vol. 2595, pp. 62–75. Springer, Heidelberg (2003)

CL01. Camenisch, J., Lysyanskaya, A.: An efficient system for non-transferable anonymous credentials with optional anonymity revocation. In: Pfitzmann, B. (ed.) EUROCRYPT 2001. LNCS, vol. 2045, pp. 93–118. Springer, Heidelberg (2001)

EGL85. Even, S., Goldreich, O., Lempel, A.: A randomized protocol for signing contracts. Communications of the ACM 28(6), 637–647 (1985)

FGK+10. Freeman, D.M., Goldreich, O., Kiltz, E., Rosen, A., Segev, G.: More constructions of lossy and correlation-secure trapdoor functions. In: Nguyen, P.Q., Pointcheval, D. (eds.) PKC 2010. LNCS, vol. 6056, pp. 279–295. Springer, Heidelberg (2010)

GKM+00. Gertner, Y., Kannan, S., Malkin, T., Reingold, O., Viswanathan, M.: The relationship between public key encryption and oblivious transfer. In: FOCS 2000, p. 325. IEEE Computer Society Press, Washington, DC (2000)

GMR01. Gertner, Y., Malkin, T., Reingold, O.: On the impossibility of basing trapdoor functions on trapdoor predicates. In: FOCS 2001, p. 126. IEEE Computer Society Press, Washington, DC (2001)

GOS06. Groth, J., Ostrovsky, R., Sahai, A.: Perfect non-interactive zero knowledge for NP. In: Vaudenay, S. (ed.) EUROCRYPT 2006. LNCS, vol. 4004, pp. 339–358. Springer, Heidelberg (2006)

HLOV11. Hemenway, B., Libert, B., Ostrovsky, R., Vergnaud, D.: Lossy encryption: Constructions from general assumptions and efficient selective opening chosen ciphertext security. In: Lee, D.H., Wang, X. (eds.) ASIACRYPT 2011. LNCS, vol. 7073, pp. 70–88. Springer, Heidelberg (2011)

HO12. Hemenway, B., Ostrovsky, R.: Extended-DDH and Lossy Trapdoor Functions. In: Fischlin, M., Buchmann, J., Manulis, M. (eds.) PKC 2012. LNCS, vol. 7293, pp. 627–643. Springer, Heidelberg (2012)

HU08.   Hofheinz, D., Unruh, D.: Towards key-dependent message security in the standard model. In: Smart, N.P. (ed.) EUROCRYPT 2008. LNCS, vol. 4965, pp. 108–126. Springer, Heidelberg (2008)

IR89.   Impagliazzo, R., Rudich, S.: Limits on the provable consequences of one-way permutations. In: STOC 1989, pp. 44–61. ACM (1989)

KMO10.   Kiltz, E., Mohassel, P., O'Neill, A.: Adaptive trapdoor functions and chosen-ciphertext security. In: Gilbert, H. (ed.) EUROCRYPT 2010. LNCS, vol. 6110, pp. 673–692. Springer, Heidelberg (2010)

KN08.   Kol, G., Naor, M.: Cryptography and game theory: Designing protocols for exchanging information. In: Canetti, R. (ed.) TCC 2008. LNCS, vol. 4948, pp. 320–339. Springer, Heidelberg (2008)

MY10.   Mol, P., Yilek, S.: Chosen-ciphertext security from slightly lossy trapdoor functions. In: Nguyen, P.Q., Pointcheval, D. (eds.) PKC 2010. LNCS, vol. 6056, pp. 296–311. Springer, Heidelberg (2010), http://eprint.iacr.org/2009/524/

PVW08.   Peikert, C., Vaikuntanathan, V., Waters, B.: A Framework for Efficient and Composable Oblivious Transfer. In: Wagner, D. (ed.) CRYPTO 2008. LNCS, vol. 5157, pp. 554–571. Springer, Heidelberg (2008)

PW08.   Peikert, C., Waters, B.: Lossy trapdoor functions and their applications. In: STOC 2008: Proceedings of the 40th Annual ACM Symposium on Theory of Computing, pp. 187–196. ACM, New York (2008)

Rab81.   Rabin, M.: How to exchange secrets by oblivious transfer. Technical Report TR-81, Harvard University (1981)

RS08.   Rosen, A., Segev, G.: Efficient lossy trapdoor functions based on the composite residuosity assumption. Cryptology ePrint Archive, Report 2008/134 (2008)

RS09.   Rosen, A., Segev, G.: Chosen-Ciphertext Security via Correlated Products. In: Reingold, O. (ed.) TCC 2009. LNCS, vol. 5444, pp. 419–436. Springer, Heidelberg (2009)

RTV04.   Reingold, O., Trevisan, L., Vadhan, S.: Notions of Reducibility between Cryptographic Primitives. In: Naor, M. (ed.) TCC 2004. LNCS, vol. 2951, pp. 1–20. Springer, Heidelberg (2004)

Rud89.   Rudich, S.: Limits on the Provable Consequences of One-way Permutations. PhD thesis, University of California, Berkeley (1989)

Vah10.   Vahlis, Y.: Two is a crowd? a black-box separation of one-wayness and security under correlated inputs. In: Micciancio, D. (ed.) TCC 2010. LNCS, vol. 5978, pp. 165–182. Springer, Heidelberg (2010)

Yao82.   Yao, A.: Theory and applications of trapdoor functions. In: FOCS 1982, pp. 82–91. IEEE Computer Society (1982)

Appendix

A   Randomness Dependent Message (RDM) Security

It is well-established that the semantic security of a public-key cryptosystem may not hold when the messages being encrypted cannot be efficiently computed by an adversary given access to the public key alone. Previous work has explored the notion of security when the plaintext is allowed to depend on the secret key (dubbed key dependent message (KDM) security) [BRS03,BHHO08,HU08,ACPS09]. In this work we consider new notions of security when the plaintext may depend on the encryption randomness. While the need for KDM security arises naturally in practical applications, the notion of Randomness Dependent Message (RDM) security arises naturally in cryptographic *constructions*.

Definition 3 (Strong RDM Security). *We consider two experiments:*

RDM (Real)	RDM (Ideal)
$pk \xleftarrow{\$} \mathsf{Gen}(1^\lambda)$	$pk \xleftarrow{\$} \mathsf{Gen}(1^\lambda)$
$r = (r_1, \ldots, r_n) \xleftarrow{\$} \mathsf{coins}(\mathsf{Enc})$	$r = (r_1, \ldots, r_n) \xleftarrow{\$} \mathsf{coins}(\mathsf{Enc})$
$(f_1, \ldots, f_n) \xleftarrow{\$} \mathcal{A}_1(pk)$	$(f_1, \ldots, f_n) \xleftarrow{\$} \mathcal{A}_1(pk))$
$c = (\mathsf{Enc}(pk, f_1(r), r_1), \ldots, \mathsf{Enc}(pk, f_n(r), r_n))$	$c = (\mathsf{Enc}(pk, 0, r_1), \ldots, \mathsf{Enc}(pk, 0, r_n))$
$b \leftarrow A_2(c)$	$b \xleftarrow{\$} A_2(c).$

Fig. 2. RDM security

A cryptosystem $\mathcal{PKE} = (\mathsf{Gen}, \mathsf{Enc}, \mathsf{Dec})$ *is called* Strongly RDM Secure *with respect to* $\mathcal{F}$ *if for all polynomials* $n = n(\lambda)$, *and all PPT adversaries* $A = (A_1, A_2)$ *for which* A_1 *only outputs* $f_i \in \mathcal{F}$, *we have*

$$\left| \Pr[A^{RDM_{\mathsf{real}}} = 1] - \Pr[A^{RDM_{\mathsf{ideal}}} = 1] \right| < \nu$$

for some negligible function $\nu = \nu(\lambda)$.

It is natural as well to consider a weakened notion of RDM security, called RDM One-wayness.

Definition 4 (RDM One-Way). *Let* $\mathcal{PKE} = (\mathsf{Gen}, \mathsf{Enc}, \mathsf{Dec})$ *be a public key cryptosystem. Consider the following experiment*

A cryptosystem $\mathcal{PKE} = (\mathsf{Gen}, \mathsf{Enc}, \mathsf{Dec})$ *is called* RDM One-Way *with respect to family* $\mathcal{F}$ *if for all polynomials* $n = n(\lambda)$, *and all PPT adversaries* $A = (A_1, A_2)$ *for which* A_1 *only outputs* $f_i \in \mathcal{F}$, *we have* $\Pr[r = r'] < \nu$ *for some negligible function* $\nu = \nu(\lambda)$.

A special case of RDM one-wayness, is the encryption of a randomness cycle. As before we can consider both the decision and the search variants.

RDM One-Way

$pk \xleftarrow{\$} \mathsf{Gen}(1^\lambda)$

$\boldsymbol{r} = (r_1, \ldots, r_n) \xleftarrow{\$} \mathcal{R}$

$(f_1, \ldots, f_n) \xleftarrow{\$} \mathcal{A}_1(pk)$

$\boldsymbol{c} = (\mathsf{Enc}(pk, f_1(\boldsymbol{r}), r_1), \ldots, \mathsf{Enc}(pk, f_n(\boldsymbol{r}), r_n))$

$\boldsymbol{r}' \leftarrow \mathcal{A}_2(\boldsymbol{c})$

Fig. 3. RDM One-Way

Definition 5 (RCIRC Security). *A cryptosystem* $\mathcal{PKE} = (\mathsf{Gen}, \mathsf{Enc}, \mathsf{Dec})$ *will be called* randomness circular secure *(RCIRC secure) if we have*

$$\{pk, \mathsf{Enc}(pk, r_2, r_1), \mathsf{Enc}(pk, r_3, r_2), \ldots, \mathsf{Enc}(pk, r_n, r_{n-1}), \mathsf{Enc}(pk, r_1, r_n)\} \approx_c$$

$$\{pk, \mathsf{Enc}(pk, 0, r_1), \ldots, \mathsf{Enc}(pk, 0, r_n)\},$$

where $pk \xleftarrow{\$} \mathsf{Gen}(1^\lambda)$, *and* $r_i \xleftarrow{\$} \mathsf{coins}(\mathsf{Enc})$ *for* $i = 1, \ldots, n$.

When using a cryptosystem as a building block in a more complicated proto-col, it is sometimes desirable to encrypt messages that are correlated with the randomness. Similar to the notion of circular security ([CL01,BRS03,BHHO08]), which talks about security when encrypting *key cycles*, we define a notion of se-curity related to encrypting *randomness cycles*. We call this property RCIRC One-Wayness.

Definition 6 (RCIRC One-wayness). *We say that a cryptosystem is RCIRC One-Way if the family of functions, parametrized by* pk

$$F_{pk} : \mathsf{coins}(\mathsf{Enc})^n \to \mathcal{C}^n$$

$$(r_1, \ldots, r_n) \mapsto (\mathsf{Enc}(pk, r_2, r_1), \ldots, \mathsf{Enc}(pk, r_1, r_n)),$$

is one-way.

It is not hard to see that a cryptosystem that is RCIRC One-Way gives rise to an injective one-way trapdoor function.

An immediate corollary of Theorem 1 is that if the functions described in Figure 1 are a family of injective one-way trapdoor functions, that means that the underlying cryptosystem, is RCIRC One-Way

Corollary 4. *If* $\mathcal{PKE} = (\mathsf{Gen}, \mathsf{Enc}, \mathsf{Dec})$ *is a lossy encryption, and if* $|\mathcal{M}| = t|\mathcal{R}|$, *and* $\frac{1}{t}$ *is negligible, if we define* $\widetilde{\mathcal{PKE}} = (\widetilde{\mathsf{Gen}}, \widetilde{\mathsf{Enc}}, \widetilde{\mathsf{Dec}})$, *with*

- $\widetilde{\mathsf{Gen}}(1^\lambda)$, *generates* $(pk, sk) \xleftarrow{\$} \mathsf{Gen}(1^\lambda)$, *and* $h \xleftarrow{\$} \mathcal{H}$ *and sets* $\widetilde{pk} = (pk, h)$, $\widetilde{sk} = sk$.
- $\widetilde{\mathsf{Enc}}(\widetilde{pk}, m, r) = \mathsf{Enc}(pk, m, h(r))$.
- $\widetilde{\mathsf{Dec}}(\widetilde{sk}, c) = \mathsf{Dec}(sk, c)$.

Then $\widetilde{\mathcal{PKE}}$ *is RCIRC One-Way.*

We remark that the construction outlined above is RCIRC-OW for one input. A straightforward modification of the above arguments shows that if h is a $2k$-wise independent hash family, then $\widetilde{\mathcal{PKE}}$ is RCIRC-OW for k inputs.

B   Constructing Lossy Encryption with Long Plaintexts

In [HLOV11], Hemenway et al. showed that lossy encryption can be constructed from statistically rerandomizable encryption and from statistically sender private $\binom{2}{1}$-oblivious transfer. This immediately yields constructions of lossy encryption from homomorphic encryption and smooth universal hash proof systems. Using the generic transformation from re-randomizable encryption to lossy encryption given in [HLOV11], we have efficient Lossy Encryption from the Damgård-Jurik cryptosystem.

Recall, that with a standard IND-CPA secure cryptosystem $\mathcal{PKE} = (\mathsf{Gen}, \mathsf{Enc}, \mathsf{Dec})$ we can arbitrarily extend the plaintext space by expanding the randomness with a pseudorandom generator. Specifically, if PRG is pseudorandom generator, such that $\mathsf{PRG} : \mathcal{R} \to \mathcal{R}^k$, we can define a new cryptosystem, with encryption of $(m_1, \ldots, m_k)$ under randomness r given by setting $r_1, \ldots, r_k = \mathsf{PRG}(r)$, and setting the ciphertext as $\mathsf{Enc}(m_1, r_1), \ldots, \mathsf{Enc}(m_k, r_k)$. It is important to notice that applying this construction to a lossy encryption scheme, will yield an IND-CPA secure scheme, but not necessarily a lossy encryption scheme.

Below, we describe lossy encryption protocols that have plaintexts that can be made much longer than the encryption randomness. These schemes are based on the Extended Decisional Diffie Hellman (EDDH) assumption. The EDDH assumption is a slight generalization of the DDH assumption. The EDDH assumption has semantics that are very similar to the DDH assumption but the EDDH assumption is implied by the DCR, DDH and QR assumptions, so by framing our cryptosystems in this language we achieve unified constructions based on different hardness assumptions.

B.1   The EDDH Assumption

Hemenway and Ostrovsky [HO12] introduced the Extended Decisional Diffie-Hellman (EDDH) assumption

Definition 7 (The EDDH Assumption). *For a group $\mathbb{G}$, and a (samplable) subgroup $\mathbb{H} \lhd \mathbb{G}$, with samplable subsets $G \subset \mathbb{G}$, and $K \subset \mathbb{Z}$ the extended decisional diffie hellman (EDDH) assumption posits that the following two distributions are computationally indistinguishable:*

$$\{(g, g^a, g^b, g^{ab}) : g \xleftarrow{\$} G, a, b \xleftarrow{\$} K\} \approx_c \{(g, g^a, g^b, g^{ab}h) : g \xleftarrow{\$} G, a, b \xleftarrow{\$} K, h \xleftarrow{\$} \mathbb{H}\}$$

It follows immediately that if $K = \{1, \ldots, |\mathbb{G}|\}$, and $\mathbb{H} = \mathbb{G}$, then the EDDH assumption is just the DDH assumption in the group $\mathbb{G}$. A straightforward argument shows:

Lemma 5. *If the EDDH assumption holds in a group $\mathbb{G}$, then for any fixed $h^* \in \mathbb{H}$, the distributions*

$$\{(g, g^a, g^b, g^{ab}) : g \xleftarrow{\$} G, a, b \xleftarrow{\$} K\} \approx_c \{(g, g^a, g^b, g^{ab}h^*) : g \xleftarrow{\$} G, a, b \xleftarrow{\$} K\}$$

are computationally indistinguishable.

Lemma 6. *If the EDDH assumption holds in a group* $\mathbb{G}$, *then for any* $m \in \{0,1\}^n$, *and any* $h \in \mathbb{H}$, *the distributions*

$$\Lambda = \{(h, g, g^a, g^{b_1}, \ldots, g^{b_n}, g^{ab_1}, \ldots, g^{ab_n}) : g \xleftarrow{\$} G, a, b_1, \ldots, b_n \xleftarrow{\$} K\},$$

$$\Lambda_m = \{(h, g, g^a, g^{b_1}, \ldots, g^{b_n}, g^{ab_1} h^{m_1}, \ldots, g^{ab_n} h^{m_n}) : g \xleftarrow{\$} G, a, b_1, \ldots, b_n \xleftarrow{\$} K, h \xleftarrow{\$} \mathbb{H}\}$$

are computationally indistinguishable.

Proof. Let e_i denote the ith standard basis vector, i.e. e_i has a one in the ith position and zeros elsewhere. By a standard hybrid argument, it is enough to show that the distributions $\Lambda_m \approx \Lambda_{m+e_i}$.

Given an EDDH challenge $(g, g_1, g_2, g_3) = (g, g^a, g^b, g_3)$, we sample $b_1, \ldots, b_{i-1}, b_{i+1}, \ldots, b_n \xleftarrow{\$} K$ and create the vector

$$v = (h, g, g_1, g^{b_1}, \ldots, g^{b_{i-1}}, g_2, g^{b_{i+1}}, \ldots, g^{b_n}, g_1^{b_1} h^{m_1}, \ldots, g_1^{b_{i-1}} h^{m_{i-1}},$$
$$g_3, g_1^{b_{i+1}} h^{m_{i+1}}, \ldots, g_1^{b_n} h^{m_n})$$

The vector v will be in Λ_m or Λ_{m+e_i} depending on whether $g_3 = g^{ab} h$ or $g_3 = g^{ab}$.

B.2   Lossy Encryption from EDDH

In this section, we describe a simple lossy encryption scheme based on the EDDH assumption.

- **Public Parameters:**
  A group $\mathbb{G}$ under which the EDDH assumption holds. A generator $g \xleftarrow{\$} G$, an element $1 \neq h \in \mathbb{H}$.
- **Lossy Key Generation:**
  Sample $a_0, a_1, \ldots, a_n, b_1, \ldots, b_n \xleftarrow{\$} K$. Set $\mathbf{v} = (g^{a_0 b_1}, \ldots, g^{a_0 b_n})$
  $$\mathbf{v}_1 = (g^{a_1 b_1} h, g^{a_1 b_2}, \ldots, g^{a_1 b_n})$$
  $$\vdots$$
  $$\mathbf{v}_n = (g^{a_n b_1}, g^{a_n b_2}, \ldots, g^{a_n b_{n-1}}, g^{a_n b_n} h)$$

  and set $g_i = g^{a_i}$ for $i = 0, \ldots, n$. The public key will be $(g_0, \ldots, g_n, \mathbf{v}, \mathbf{v}_1, \ldots, \mathbf{v}_n)$. The secret key will be $b_1, \ldots, b_n$.
- **Injective Key Generation:**
  Sample $a, a_1, \ldots, a_n, b_1, \ldots, b_n \xleftarrow{\$} K$. Set $\mathbf{v} = (g^{ab_1}, \ldots, g^{ab_n})$
  $$\mathbf{v}_1 = (g^{a_1 b_1}, g^{a_1 b_2}, \ldots, g^{a_1 b_n})$$
  $$\vdots$$
  $$\mathbf{v}_n = (g^{a_n b_1}, g^{a_n b_2}, \ldots, g^{a_n b_{n-1}}, g^{a_n b_n})$$

  and set $g_i = g^{a_i}$ for $i = 0, \ldots, n$. The public key will be $(g_0, \ldots, g_n, \mathbf{v}, \mathbf{v}_1, \ldots, \mathbf{v}_n)$.

- **Encryption:**
  To encrypt a message $\mathbf{m} \in \{0,1\}^n$, choose an element $r \xleftarrow{\$} K$, and set

  $$\mathbf{c} = \mathbf{v}^r \mathbf{v}_1^{m_1} \cdots \mathbf{v}^{m_n} \in \mathbb{G}^n$$

  where all the operations are done coordinate-wise (the natural group action in the cartesian product group). and $c_0 = g_0^r \prod_{i=1}^n g_i^{m_i}$.
- **Decryption:**
  Given $(c_0, \mathbf{c})$, calculate $(\mathbf{c}_1 c_0^{-b_1}, \ldots, \mathbf{c}_n c_0^{-b_n}) = (h^{m_1}, \ldots, h^{m_n})$ and the m_i can be recovered by inspection.

The injective and lossy modes are indistinguishable by Lemma 6. In lossy mode, the ciphertext space has size bounded by the order of g. By choosing n large enough so that 2^n is much greater than the order of g we can achieve any degree of lossiness. The encryption randomness is a single element $r \xleftarrow{\$} K$, so choosing $n > K$, makes the plaintexts longer than the encryption randomness.

C   Perfectly Lossy Encryption

The perfect lossiness property discussed in Section 4.1 is so strong that we can actually extend Lemma 3.

Lemma 7. *If $\mathcal{PKE} = (\mathsf{Gen}, \mathsf{Enc}, \mathsf{Dec})$, is a perfectly lossy encryption scheme, $0 < t \in \mathbb{Z}$, and $h_1, \ldots, h_t$ are any functions from $\mathcal{M}^t$ to $\mathcal{R}$, then the function*

$$F_{pk,h} : \mathcal{M}^t \to \mathcal{C}^t$$
$$(x_1, \ldots, x_t) \mapsto (\mathsf{Enc}(pk, x_1, h_1(x_1, \ldots, x_t)), \ldots, \mathsf{Enc}(pk, x_t, h_t(x_1, \ldots, x_t))),$$

is a $(t \log |\mathcal{M}|, t(\log |\mathcal{M}| - \log |\mathcal{R}|))$-LTDF.

The proof is essentially identical to the proof of Lemma 3. One simple consequence of Lemma 7 is

Lemma 8. *If $\mathcal{PKE} = (\mathsf{Gen}, \mathsf{Enc}, \mathsf{Dec})$, is a perfectly lossy encryption scheme, $0 < t \in \mathbb{Z}$, and $\log(|\mathcal{M}|/|\mathcal{R}|) = \omega(\log(\lambda))$, then for any map $h : \mathcal{M} \to \mathcal{R}$, the encryption $\widehat{\mathsf{Enc}}(pk, x, y) = \mathsf{Enc}(pk, x, h(y))$ is strongly t-RCIRC-One Way.*

Pseudorandom Generators from Regular One-Way Functions: New Constructions with Improved Parameters

Yu Yu[1,2], Xiangxue Li[2], and Jian Weng[3]

[1] Institute for Theoretical Computer Science, Institute for Interdisciplinary Information Sciences, Tsinghua University, P.R. China
[2] Department of Computer Science, East China Normal University, P.R. China
[3] Department of Computer Science, Jinan University, Guangzhou 510632, P.R. China

Abstract. We revisit the problem of basing pseudorandom generators on regular one-way functions, and present the following constructions:

- For any known-regular one-way function (on n-bit inputs) that is known to be ε-hard to invert, we give a neat (and tighter) proof for the folklore construction of pseudorandom generator of seed length $\Theta(n)$ by making a single call to the underlying one-way function.
- For any unknown-regular one-way function with known ε-hardness, we give a new construction with seed length $\Theta(n)$ and $O(n/\log(1/\varepsilon))$ calls. Here the number of calls is also optimal by matching the lower bounds of Holenstein and Sinha (FOCS 2012).

Both constructions require the knowledge about ε, but the dependency can be removed while keeping nearly the same parameters. In the latter case, we get a construction of pseudo-random generator from any unknown-regular one-way function using seed length $\tilde{O}(n)$ and $\tilde{O}(n/\log n)$ calls, where $\tilde{O}$ omits a factor that can be made arbitrarily close to constant (e.g. $\log\log\log n$ or even less). This improves the *randomized iterate* approach by Haitner, Harnik and Reingold (CRYPTO 2006) which requires seed length $O(n\cdot\log n)$ and $O(n/\log n)$ calls.

1 Introduction

The seminal work of Håstad, Impagliazzo, Levin and Luby (HILL) [14] that one-way functions (OWFs) imply pseudorandom generators (PRGs) constitutes one of the centerpieces of modern cryptography. Technical tools and concepts (e.g. pseudo-entropy, leftover hash lemma) developed and introduced in [14] were found useful in many other contexts (such as leakage-resilient cryptography). Nevertheless, a major drawback of [14] is that the construction is quite involved and too inefficient to be of any practical use, namely, to obtain a PRG with comparable security to the underlying OWF on security parameter n, one needs a seed of length $O(n^8)^1$. Research efforts (see [15,13,23], just to name a few) have been followed up towards simplifying and improving the constructions, and

---

[1] More precisely, the main construction of [14] requires seed length $O(n^{10})$, but [14] also sketches another construction of seed length $O(n^8)$, which was proven in [15].

K. Sako and P. Sarkar (Eds.) ASIACRYPT 2013, Part II, LNCS 8270, pp. 261–279, 2013.

the current state-of-the-art construction [23] requires seed length $O(n^3)$. Let us mention all aforementioned approaches are characterized by a parallel construction, namely, they run sufficiently many independent copies of the underlying OWFs (rather than running a single trail and feeding its output back to the input iteratively) and there seems an inherent lower bound on the number of copies needed. This is recently formalized by Holenstein and Sinha [16], in particular, they showed that any black-box construction of a PRG from an arbitrary OWF f requires $\Omega(n/\log n)$ calls to f in general.[2]

PRGs FROM SPECIAL OWFs. Another line of research focuses on OWFs with special structures that give rise to more efficient PRGs. Blum, Micali [3] and Yao [26] independently introduced the notion of PRGs, and observed that PRGs can be efficiently constructed from one-way permutations (OWPs). That is, given a OWP f on input x and its hardcore function h_c (e.g. by Goldreich and Levin [10]), a single invocation of f already implies a PRG $g(x) = (f(x), h_c(x))$ with a stretch[3] of $\Omega(\log n)$ bits and it extends to arbitrary stretch by repeated iterations (seen by a hybrid argument):

$$g^\ell(x) \;=\; (\; h_c(x), h_c(f^1(x)), \ldots, h_c(f^\ell(x)), \ldots)$$

where $f^i(x)\overset{\text{def}}{=}f(f^{i-1}(x))$ and $f^1(x)\overset{\text{def}}{=}f(x)$. The above PRG, often referred to as the BMY generator, enjoys many advantages such as simplicity, optimal seed length, and minimal number of calls. Levin [19] observed that f is not necessarily a OWP, but it suffices to be one-way on its own iterate. Unfortunately, an arbitrary OWF doesn't have this property. Goldreich, Krawczyk, and Luby [9] assumed known-regular[4] OWFs and gave a construction of seed length $O(n^3)$ by iterating the underlying OWFs and applying k-wise independent hashing in between every two iterations. Later Goldreich showed a more efficient (and nearly optimal) construction from known-regular OWFs in his textbook [7], where in the concrete security setting the construction does only a single call to the underlying OWF (or $\omega(1)$ calls in general). The construction was also implicit in many HILL-style constructions (e.g. [15,13]). Haitner, Harnik and Reingold [12] refined the technique used in [9] (which they called the *randomized iterate*) and adapted the construction to unknown regular OWFs with reduced seed length $O(n \cdot \log\ n)$. Informally, the randomized iterate follows the route of [9] and applies a random pairwise independent hash function h_i in between every two applications of f, i.e.

$$f^1(x)\overset{\text{def}}{=}f(x); \text{for } i{\geq}2 \text{ let } f^i(x; h_1, \ldots, h_{i-1})\overset{\text{def}}{=}f(h_{i-1}(f^{i-1}(x; h_1, \cdots, h_{i-2})))$$

The key observation is *"the last iterate is hard-to-invert"*[11], more precisely, function f, when applied to $h_{i-1}(f^{i-1}; h_1, \cdots, h_{i-2})$, is hard-to-invert even if

---

[2] The lower bound of [16] also holds in the concrete security setting, namely, $\Omega(n/\log(1/\varepsilon))$ calls from any ε-hard OWF.

[3] The stretch of a PRG refers to the difference between output and input lengths.

[4] A function $f(x)$ is regular if the every image has the same number (say α) of preimages, and it is known- (resp., unknown-) regular if α is efficiently computable (resp., inefficient to approximate) from the security parameter.

$h_1, \ldots, h_{i-1}$ are made public. The generator follows by running the iterate $O(n/\log n)$ times, and outputting $\Omega(\log n)$ hardcore bits per iteration, which requires seed length $O(n^2/\log n)$ and can be further pushed to $O(n \cdot \log n)$ using derandomization techniques (e.g., Nisan's bounded-space generator [20]). The randomized iterate matches the lower bound on the number of OWF calls[5], but it remains open if any efficient construction can achieve linear seed length and $O(n/\log n)$ OWF calls simultaneously.

SUMMARY OF CONTRIBUTIONS. We contribute an alternative proof for the folklore construction of PRGs from known-regular OWFs via the notion of unpredictability pseudo-entropy, which significantly simplifies and tightens the proofs in [7]. We also give a new construction from any unknown-regular one-way function using seed length $\tilde{O}(n)$ and making $\tilde{O}(n/\log n)$ calls, where both parameters are optimal in the concrete security setting and nearly optimal in general (up to an arbitrarily close to constant factor), and this improves the randomized iterate [11]. We sketch both constructions as follows.

ENTROPY OBSERVATION. We start by assuming a (t,ε)-OWF f (see Definition 2) with known regularity 2^k (i.e., every image has 2^k preimages under f). The key observation is that for uniform X (over $\{0,1\}^n$) we have X given $f(X)$ has $k + \log(1/\varepsilon)$ bits of pseudo-entropy (defined by the game below and formally in Definition 5). That is, no adversary A of running time t can win the following game against the challenger C with probability greater than $(2^{-k} \cdot \varepsilon)$. The rationale is that conditioned on any $f(X) = y$ random variable X is uni-

Challenger C		Adversary A
$x \leftarrow U_n;\ y := f(x)$	$\xrightarrow{\quad y \quad}$	
	$\xleftarrow{\quad x' \quad}$	$x' := A(y)$
A wins iff $x' = x$		

Fig. 1. The interactive game between A and C that defines unpredictability pseudo-entropy, where $x \leftarrow U_n$ denotes sampling a random $x \in \{0,1\}^n$

formly distributed on set $f^{-1}(y) \stackrel{\text{def}}{=} \{x : f(x) = y\}$ of size 2^k, and thus even if any deterministic (or probabilistic) A recovers a $x' \in f^{-1}(y)$, the probability that $X = x'$ is only 2^{-k}.

PRGs FROM KNOWN-REGULAR OWFs. Given the above observation, we immediately obtain the following folklore construction using three extractions along with a three-line proof.

---

[5] As explicitly stated in [16], the lower bound of $\Omega(n/\log n)$ calls also applies to unknown regular OWFs.

- RANDOMNESS EXTRACTION FROM F(X). $f(X)$ has min-entropy $n - k$, and thus we can extract nearly $n - k$ statistically random bits.

- RANDOMNESS EXTRACTION FROM X. X has min-entropy k given any $y = f(X)$, so we can extract another k statistically random bits.

- PSEUDORANDOMNESS EXTRACTION FROM X. The second extraction reduces the unpredictability pseudo-entropy of X given $f(X)$ by no more than k (i.e., $\log(1/\varepsilon)$ bits remaining by the entropy chain rule), and hence we use Goldreich-Levin hardcore functions [10] to extract another $O(\log(1/\varepsilon))$ bits.

NOVELTY OF OUR ANALYSIS. While the construction was already known in literature (explicit in [7] and implicit in HILL-style generators [14,13,23]), its full proof was only seen in [7] and we further simplify the analysis via the use of unpredictability pseudo-entropy. In addition to simplicity, our technique can also be used to refine and tighten the proof given in [7] (see Section 3.2 and Remark 2 for details). We mention that our proofs sketched above are not implied by the recent work of [13,23]. In particular, the construction of [13] applies a dedicated universal hash function h of description length $O(n^2)$ to $f(X)$ such that the first k output bits are statistically random and the next $O(\log n)$ bits are computationally random, and this holds even if k is unknown (which is desired for a general OWF f whose preimage size may vary for different images). However, in our context k is known and it is crucial that the description length of the hash functions is linear, for which we do two extractions from $f(X)$ using h_2 and h_c respectively. We also stress that our observation that "X given $f(X)$ has unpredictability pseudo-entropy $k + \log(1/\varepsilon)$" is incomparable with the counterpart in [23, Thm 1.5], which was informally stated as "$(f(X), X)$ has next-bit pseudo-entropy $n + \Omega(\log n)$". Firstly, our proof enjoys simplicity and tightness whereas theirs employs the uniform version of Min-Max Theorem which is much involved and interesting in its own right. Secondly, next-bit pseudo-entropy was a newly introduced notion [13] and whether it implies (or is implied by) unpredictability pseudo-entropy is unknown to our knowledge, and the ways of extraction from these two sources are different. See Section 3.2 and Remark 3 for details.

CONCRETE VS. ASYMPTOTIC SECURITY. The above construction is optimal (in seed length and the number of OWF calls), but requires the knowledge about parameter ε, more precisely, we need ε to decide entropy loss d such that the first extraction outputs $n - k - d$ bits with statistical error bounded by $2^{-d/2}$ (by the Leftover Hash Lemma [14]) and let the third extraction output more than d bits to achieve a positive stretch. It is unknown how to remove the dependency on ε for free (see also the discussions in [7]). Fortunately, there is a known repetition trick to solve the problem using seed length $\tilde{O}(n)$ and $\tilde{O}(1)$ OWF calls, where notation $\tilde{O}$ omits a factor of $q \in \omega(1)$ (i.e. q can be any factor arbitrarily close to constant such as $\log \log \log n$).

PRGS FROM UNKNOWN-REGULAR OWFS. We also give a new construction oblivious of the regularity of f. The construction follows the steps below.

- CONVERT TO KNOWN REGULARITY. The key idea is to transform any un-known regular OWF into another known regular OWF (over a special do-main). That is, for a (length-preserving) unknown-regular (t, ε)-OWF $f : \{0,1\}^n \rightarrow \mathcal{Y}$ where $\mathcal{Y} \subseteq \{0,1\}^n$ is the range of f, define function $\bar{f} : \mathcal{Y} \times \{0,1\}^n \rightarrow \mathcal{Y}$ as $\bar{f}(y, r) \stackrel{\text{def}}{=} f(y \oplus r)$ where "$\oplus$" denotes bitwise XOR. It is not hard to see that $\bar{f}$ has regularity 2^n (regardless of the regularity of f) and it preserves the hardness of f.

- CONSTRUCT A Z-SEEDED PRG. Similar to that observed in the 1st con-struction, $\bar{f}(Y, R)$ hides $n + \log(1/\varepsilon)$ bits of pseudo-entropy about (Y, R), and thus we can extract $n + O(\log(1/\varepsilon))$ pseudorandom bits, namely, we get a special PRG $\bar{g}$ that maps random elements over $\mathcal{Y} \times \{0,1\}^n$ to pseudorandom ones over $\mathcal{Y} \times \{0,1\}^{n+O(\log(1/\varepsilon))}$. This PRG is known as the "Z-seeded PRG" [23], one that given input distribution Z outputs (Z', U_s) which is computa-tionally indistinguishable from (Z, U_s), where in the above case $Z = (Y, R)$ and stretch $s = O(\log(1/\varepsilon))$. Note that if Z were U_{2n} then this would be a standard PRG.

- ITERATIVE COMPOSITION OF Z-SEEDED PRG. Nevertheless, to use the above Z-seeded PRG $\bar{g}$ we need to efficiently sample from $Y = f(U_n)$ (i.e. uniform distribution over $\mathcal{Y}$), which costs n random bits despite that the entropy of Y may be far less than n. Quite naturally (following [23,3]), the construction invests n bits (to sample a random $y \leftarrow f(U_n)$) at initialization, runs $\bar{g}$ in iterations, and outputs $O(\log(1/\varepsilon))$ bits per iteration. The stretch becomes positive after $O(n/\log(1/\varepsilon))$ iterations, which matches the lower bounds of [16]. The seed length remains of order $\Theta(n)$ by reusing the coins for universal hash and G-L functions at every iteration, thanks to the hybrid argument.

- REMOVE DEPENDENCY ON ε (OPTIONAL). Similarly, in case that ε is un-known, we pay a penalty factor $\tilde{O}(1)$ for using the repetition trick. That is, we construct a PRG from any unknown-regular OWF using seed length $\tilde{O}(n)$ and $\tilde{O}(n/\log n)$ OWF calls.

2 Preliminaries

NOTATIONS AND DEFINITIONS. We use capital letters (e.g. X, Y, A) for random variables, standard letters (e.g. x, y, a) for values, and calligraphic letters (e.g. $\mathcal{X}, \mathcal{Y}, \mathcal{S}$) for sets. $|\mathcal{S}|$ denotes the cardinality of set $\mathcal{S}$. For function f, we let $f(\mathcal{X}) \stackrel{\text{def}}{=} \{f(x) : x \in \mathcal{X}\}$ be the set of images that are mapped from $\mathcal{X}$ under f, and denote by $f^{-1}(y)$ the set of y's preimages under f, i.e. $f^{-1}(y) \stackrel{\text{def}}{=} \{x : f(x) = y\}$. We say that distribution X is flat if it is uniformly distributed over some set $\mathcal{X}$. We use $s \leftarrow S$ to denote sampling an element s according to distribution S, and let $s \leftarrow \mathcal{S}$ denote sampling s uniformly from set $\mathcal{S}$, and $y := f(x)$ denote value assignment. We use U_n to denote the flat distribution over $\{0,1\}^n$ independent of the rest random variables in consideration, and let

$f(U_n)$ be the distribution induced by applying function f to U_n. We use $\mathsf{CP}(X)$ to denote the collision probability of X, i.e., $\mathsf{CP}(X) \stackrel{\text{def}}{=} \sum_x \Pr[X = x]^2$, and collision entropy $\mathbf{H}_2(X) \stackrel{\text{def}}{=} -\log \mathsf{CP}(X) \geq \mathbf{H}_\infty(X)$. We also define average (aka conditional) collision entropy and average min-entropy of a random variable X conditioned on another random variable Z by

$$\mathbf{H}_2(X|Z) \stackrel{\text{def}}{=} -\log \left(\mathbb{E}_{z \leftarrow Z} \left[\sum_x \Pr[X = x|Z = z]^2 \right] \right)$$
$$\mathbf{H}_\infty(X|Z) \stackrel{\text{def}}{=} -\log \left(\mathbb{E}_{z \leftarrow Z} \left[\max_x \Pr[X = x|Z = z] \right] \right)$$

An entropy source refers to a random variable that has some non-trivial amount of entropy. A function $\mu : \mathbb{N} \rightarrow [0, 1]$ is negligible if for every polynomial poly we have $\mu(n) < 1/\text{poly}(n)$ holds for all sufficiently large n's. We define the *computational distance* between distribution ensembles $X \stackrel{\text{def}}{=} \{X_n\}_{n \in \mathbb{N}}$ and $Y \stackrel{\text{def}}{=} \{Y_n\}_{n \in \mathbb{N}}$ as follows: we say that X and Y are $(t(n), \varepsilon(n))$-close, denoted by $\mathsf{CD}_{t(n)}(X, Y) \leq \varepsilon(n)$, if for every probabilistic distinguisher D of running time up to $t(n)$ it holds that

$$\left| \Pr[\mathsf{D}(1^n, X_n) = 1] - \Pr[\mathsf{D}(1^n, Y_n) = 1] \right| \leq \varepsilon(n) .$$

The *statistical distance* between X and Y, denoted by $\mathsf{SD}(X, Y)$, is defined by

$$\mathsf{SD}(X, Y) \stackrel{\text{def}}{=} \frac{1}{2} \sum_x |\Pr[X = x] - \Pr[Y = x]| = \mathsf{CD}_\infty(X, Y)$$

We use $\mathsf{SD}(X, Y|Z)$ (resp. $\mathsf{CD}_t(X, Y|Z)$) as shorthand for $\mathsf{SD}((X, Z), (Y, Z))$ (resp. $\mathsf{CD}_t((X, Z), (Y, Z))$).

SIMPLIFYING ASSUMPTIONS AND NOTATIONS. To simplify the presentation, we make the following assumptions without loss of generality. It is folklore that one-way functions can be assumed to be length-preserving (see [12] for formal proofs). Throughout, most parameters are functions of the security parameter n (e.g., $t(n)$, $\varepsilon(n)$, $\alpha(n)$) and we often omit n when clear from the context (e.g., t, ε, α). Parameters (e.g. ε, α) are said to be known if they are known to be polynomial-time computable from n. By notation $f : \{0, 1\}^n \rightarrow \{0, 1\}^l$ we refer to the ensemble of functions $\{f_n : \{0, 1\}^n \rightarrow \{0, 1\}^{l(n)}\}_{n \in \mathbb{N}}$. As slight abuse of notion, poly might be referring to the set of all polynomials or a certain polynomial, and h might be either a function or its description, which will be clear from the context.

Definition 1 (universal hash functions [4]). *A family of functions* $\mathcal{H} \stackrel{\text{def}}{=} \{h : \{0, 1\}^n \rightarrow \{0, 1\}^l\}$ *is called a* universal hash family, *if for any* $x_1 \neq x_2 \in \{0, 1\}^n$ *we have* $\Pr_{h \leftarrow \mathcal{H}}[h(x_1) = h(x_2)] \leq 2^{-l}$.

Definition 2 (one-way functions). *A function* $f : \{0, 1\}^n \rightarrow \{0, 1\}^{l(n)}$ *is* $(t(n), \varepsilon(n))$-one-way *if f is polynomial-time computable and for any probabilistic algorithm* A *of running time* $t(n)$

$$\Pr_{y \leftarrow f(U_n)} [\mathsf{A}(1^n, y) \in f^{-1}(y)] \leq \varepsilon(n).$$

For $\varepsilon(n) = 1/t(n)$, we simply say that f is $\varepsilon(n)$-hard. f is a one-way function if it is $\varepsilon(n)$-hard for some negligible function $\varepsilon(n)$.

Definition 3 (regular functions). *A function f is α-regular if there exists an integer function α, called the regularity function, such that for every $n \in \mathbb{N}$ and $x \in \{0,1\}^n$ we have*

$$|f^{-1}(f(x))| = \alpha(n).$$

In particular, f is known-regular if α is polynomial-time computable, or called unknown-regular otherwise. Further, f is a (known-/unknown-) regular OWF if f is a OWF with (known/unknown) regularity.

Definition 4 (pseudorandom generators [3,26]). *A function $g : \{0,1\}^n \to \{0,1\}^{l(n)}$ $(l(n) > n)$ is a $(t(n),\varepsilon(n))$-secure PRG if g is polynomial-time computable and*

$$\mathsf{CD}_{t(n)}(\ g(1^n, U_n)\ ,\ U_{l(n)}\ ) \ \leq\ \varepsilon(n).$$

where $(l(n) - n)$ is the stretch of g, and we often omit 1^n (security parameter in unary) from g's parameter list. We say that g is a pseudorandom generator if both $1/t(n)$ and $\varepsilon(n)$ are negligible.

Definition 5 (unpredictability pseudo-entropy[2,17]). *For distribution ensemble $(X, Z) \stackrel{\text{def}}{=} \{(X_n, Z_n)\}_{n\in\mathbb{N}}$, we say that X has $k(n)$ bits of pseudo-entropy conditioned on Z for all $t(n)$-time adversaries, denoted by $\mathbf{H}_{t(n)}(X|Z) \geq k(n)$, if for any $n \in \mathbb{N}$ and any probabilistic adversary A of running time $t(n)$*

$$\Pr_{(x,z)\leftarrow(X_n,Z_n)}[\mathsf{A}(1^n, z) = x] \ \leq\ 2^{-k(n)}$$

Alternatively, we say that X is $2^{-k(n)}$-hard to predict given Z for all $t(n)$-time adversaries.

Unpredictability pseudo-entropy can be seen as a relaxed form of min-entropy by weakening adversary's running time from unbounded to parameter $t(n)$, which (presumably) characterizes the class of practical adversaries we care about. Note that the notion seems only meaningful in its conditional form as otherwise (when Z is empty) non-uniform attackers can simply hardwire the best guess about X, and thus $\mathbf{H}_{t(n)}$ collapses to $\mathbf{H}_\infty$. Let us mention the unpredictability pseudo-entropy is different from (and in fact, strictly weaker than [2,17]) the HILL pseudo-entropy [14], which is another relaxed notion of min-entropy by considering its computationally indistinguishable analogues.

3   Pseudorandom Generators from Regular One-Way Functions

3.1   Technical Tools

The first technical tool we use is the leftover hash lemma. Informally, it states that when applying a random universal hash function to min-entropy (or Rényi entropy) source, one obtain random strings that are statistical close to uniform even conditioned on the description of hash function. The objects were later formalized as randomness extractors [21]. Universal hash functions are also good condensers (whose outputs have nearly maximal entropy) for a wider range of parameters than extractors.

Lemma 1 (leftover hash lemma [14]). *For any integers $d<k\leq n$, there exists a (efficiently computable) universal hash function family $\mathcal{H} \overset{def}{=} \{h : \{0,1\}^n \to \{0,1\}^{k-d}\}$ such that for any joint distribution (X,Z) where $X \in \{0,1\}^n$ and $\mathbf{H}_2(X|Z) \geq k$, we have*

$$\mathsf{SD}(H(X),\ U_{k-d} \mid H, Z) \leq 2^{-\frac{d}{2}}$$

where H is uniformly distributed over the members of $\mathcal{H}$, the description size of H is called seed length, and d is called entropy loss, i.e., the difference between the entropy of X (given Z) and the number of bits that were extracted from X.

Lemma 2 (condensers from hash functions). *Let $\mathcal{H} \overset{def}{=} \{h : \{0,1\}^n \to \{0,1\}^k\}$ be any universal hash function family and let (X,Z) be any random variable with $X \in \{0,1\}^n$ and $\mathbf{H}_2(X|Z) \geq k$. Then, for H uniform distributed over $\mathcal{H}$ we have $\mathbf{H}_2(H(X) \mid H, Z) \geq k - 1$.*

Proof. Let X_1 and X_2 be i.i.d. to $X \mid Z = z$ (i.e. X conditioned on $Z = z$).

$$2^{-\mathbf{H}_2(H(X)|H,Z)} = \mathbb{E}_{h\leftarrow H, z\leftarrow Z}\left[\Pr_{x_1\leftarrow X_1, x_2\leftarrow X_2}[H(x_1) = H(x_2)|H = h, Z = z]\right]$$

$$\leq \mathbb{E}_{z\leftarrow Z}\left[\Pr_{x_1\leftarrow X_1, x_2\leftarrow X_2}[x_1 = x_2|Z = z]\right]$$

$$+ \ \mathbb{E}_{z\leftarrow Z}\left[\Pr_{h\leftarrow H}[\ h(x_1) = h(x_2) \mid x_1\neq x_2, Z = z]\right]$$

$$\leq 2^{-k} + 2^{-k} = 2^{-(k-1)}\ .$$

We refer to [22,6,18] for extremely efficient constructions of universal hash functions with short description (of length $\Theta(n)$), such as multiplications between matrices and vectors, or over finite fields.

RECONSTRUCTIVE EXTRACTORS. We will also need objects that extract pseudorandomness from unpredictability pseudo-entropy sources. Unfortunately, the leftover hash lemma (and randomness extractors [21] in general) does not serve the purpose. Goldreich and Levin [10] showed that the inner product function is a reconstructive bit-extractor for unpredictability pseudo-entropy sources. Further, there are two ways to extend the inner product to multiple-bit extractors: (1) multiplication with a random matrix of length $O(n^2)$ and extracts almost all entropy (by a hybrid argument); (2) multiplication with a random Toeplitz matrix of length $\Theta(n)$ and extracts $O(\log(1/\varepsilon))$ bits (due to Vazirani's XOR lemma [25,10]). We will use the latter multi-bit variant (as stated below) to keep the seed length linear. Interestingly, the Toeplitz matrix based functions also constitute pairwise independent and universal hash function families.

Theorem 1 (Goldreich-Levin [10]). *For distribution ensemble* $(X,Y) \in \{0,1\}^n \times \{0,1\}^*$, *and for any integer* $m \leq n$, *there exists*[6] *a function family* $\mathcal{H}_C \stackrel{\text{def}}{=} \{h_c : \{0,1\}^n \to \{0,1\}^m\}$ *of description size* $\Theta(n)$, *such that*

- *If* $Y = f(X)$ *for any* (t,ε)-*OWF* f *and* X *uniform over* $\{0,1\}^n$, *then we have*

$$\mathsf{CD}_{t'}(\ H_C(X)\ ,\ U_m\ |\ Y, H_C) \in O(2^m \cdot \varepsilon)\ . \tag{1}$$

- *If* X *is* ε-*hard to predict given* Y *for all* t-*time adversaries, namely, entropy condition satisfies* $\mathbf{H}_t(X|Y) \geq \log(1/\varepsilon)$, *then we have*

$$\mathsf{CD}_{t'}(\ H_C(X)\ ,\ U_m\ |\ Y, H_C) \in O(2^m \cdot (n \cdot \varepsilon)^{\frac{1}{3}})\ . \tag{2}$$

where $t' = t \cdot (\varepsilon/n)^{O(1)}$ *and function* H_C *is uniformly distributed over the members of* $\mathcal{H}_C$.

Remark 1 (unpredictability vs. one-wayness). To see the difference between the two versions above, consider the interactive game in Figure 1, where by unpredictability A's prediction is successful only if $x = x'$, but in contrast A inverts OWF f as long as he finds any x' satisfying $f(x') = y$. Recall that the proof of the theorem can be seen as an efficient local list decoding procedure for the Hadamard code, where in the former case the decoder returns a random member from the candidate list while in the latter case it goes through all candidates and outputs the one x' satisfying $f(x') = y$ (if exists). We refer to Goldreich's exposition [8] for further details.

We recall two folklore facts below, namely the chain rule of unpredictability (pseudo-)entropy and the replacement inequality. Intuitively, any leakage $Y \in \{0,1\}^l$ decreases the unpredictability about secret X by a factor of no more than 2^l, which can be seen by a simple reduction (e.g., by replacing Y with a random string). The replacement inequality states that any information that is (efficiently) computable from the knowledge of the adversary does not help further reduce the unpredictability (pseudo-)entropy of the secret in consideration.

Fact 1 (chain rule of entropies). *For any joint distribution* (X,Y,Z) *where* $Y \in \{0,1\}^l$, *we have*

$$\mathbf{H}_\infty(X|Y,Z) \geq \mathbf{H}_\infty(X|Z) - l\ ,$$

$$\mathbf{H}_{t'}(X|Y,Z) \geq \mathbf{H}_t(X|Z) - l\ ,$$

where $t' \approx t$.

Fact 2 (replacement inequalities). *For any joint distribution* (X,Y,Z) *and any* t_h-*time computable function* $h : \mathcal{Y} \to \{0,1\}^*$, *we have*

$$\mathbf{H}_\infty(X|h(Y), h, Z) \geq \mathbf{H}_\infty(X|Y,Z)\ ,$$

$$\mathbf{H}_{t-t_h}(X|h(Y), h, Z) \geq \mathbf{H}_t(X|Y,Z)\ .$$

---

[6] For example (see [10]), we can use an $m \times n$ Toeplitz matrix $a_{m,n}$ to describe the family of functions, i.e., $\mathcal{H}_C \stackrel{\text{def}}{=} \{h_c(x) \stackrel{\text{def}}{=} a_{m,n} \cdot x$, where $x \in \{0,1\}^n$, $a_{m,n} \in \{0,1\}^{m+n-1}\}$.

3.2 PRGs from OWFs with Known Regularity and Hardness

We state our motivating observation as the lemma below.

Lemma 3 (regular OWFs imply unpredictability pseudo-entropy). *Let*
$f : \mathcal{X} \to \mathcal{Y}$ *be a 2^k-regular (t,ε)-OWF. Then, we have*

$$\mathbf{H}_t(X \mid f(X)) \geq k + \log(1/\varepsilon) \ , \tag{3}$$

where X is uniform over $\mathcal{X}$.

Proof. The (t,ε)-one-wayness of f guarantees that for any deterministic adversary A of running time t

$$\Pr_{x \leftarrow \mathcal{X}, y := f(x)} [\ \mathsf{A}(y) \in f^{-1}(y)\ ] \leq \varepsilon$$

which in turn implies (as conditioned on $f(X) = y$, X is uniform over $f^{-1}(y)$ of size 2^k):

$$\Pr_{x \leftarrow \mathcal{X}, y := f(x)} [\ \mathsf{A}(y) = x\ ] \leq 2^{-k} \cdot \varepsilon$$

which is essentially Equation (3) by taking a negative logarithm. Note that the above argument extends to probabilistic t-time A as well, by considering $\mathsf{A}(y; r)$ on every fixing of his random coin r.

THE CONSTRUCTION FOR KNOWN α AND ε. As sketched in introduction, our first construction essentially extracts from the joint distribution $(X, f(X))$ three times, namely, use universal hash function h_1 to extract nearly (up to entropy loss) $n - k$ bits from $f(X)$, and then apply h_2 and h_c to extract k statistical random bits and another $\Theta(\log(1/\varepsilon n))$ pseudo-random bits from X respectively. For convenience, we assume without loss of generality that the regularity is a power of two, i.e., $\alpha = 2^k$.

Theorem 2 (preliminary construction based on known regularity and hardness). *Let $f : \{0,1\}^n \to \{0,1\}^n$ be a known 2^k-regular length-preserving (t,ε)-OWF, let d, s be any integer functions satisfying $9d + 6s = 2\log(1/\varepsilon n)$, let $\mathcal{H}_1 \overset{\text{def}}{=} \{h_1 : \{0,1\}^n \to \{0,1\}^{n-k-d}\}$, $\mathcal{H}_2 \overset{\text{def}}{=} \{h_2 : \{0,1\}^n \to \{0,1\}^k\}$ be universal hash function families, let $\mathcal{H}_C \overset{\text{def}}{=} \{h_c : \{0,1\}^n \to \{0,1\}^{d+s}\}$ be a Goldreich-Levin function family, and let g be*

$$g : \{0,1\}^n \times \mathcal{H}_1 \times \mathcal{H}_2 \times \mathcal{H}_C \to \{0,1\}^{(n-k-d)+k+(d+s)} \times \mathcal{H}_1 \times \mathcal{H}_2 \times \mathcal{H}_C$$
$$(x, h_1, h_2, h_c) \mapsto (h_1(f(x)), h_2(x), h_c(x), h_1, h_2, h_c)$$

where $x \in \{0,1\}^n$, $h_1 \in \mathcal{H}_2$, $h_2 \in \mathcal{H}_2$, $h_c \in \mathcal{H}_C$. Then, g is a $(\ t \cdot (\varepsilon/n)^{O(1)}, O((2^{3s} \cdot \varepsilon \cdot n)^{\frac{1}{9}})\ )$-secure PRG with stretch s.
We deal with the situation where $n - k - d \leq 0$ by letting h_1 output nothing. Another special case $k = 0$ (i.e., f is a OWP) is handled by letting h_1 and h_2 output the identity and empty strings respectively.

Proof. The entropy conditions for the (pseudo)-randomness extractions are guaranteed by Lemma 4. We have by Equation (4), Equation (5) and the leftover hash lemma that the first $n - d$ bits extracted are statistically random, namely,

$$\mathsf{SD}(\ (H_1(f(X)), H_2(X)),\ U_{n-d}\ |\ H_1, H_2)$$
$$\leq \mathsf{SD}(\ H_1(f(X)),\ U_{n-k-d}\ |\ H_1) + \mathsf{SD}(\ H_2(X),\ U_k\ |\ H_1(f(X)), H_1, H_2)$$
$$\leq 2 \cdot 2^{-\frac{d}{2}} = 2 \cdot 2^{\frac{s}{3} + \frac{1}{9}\log(\varepsilon n)} = O((2^{3s} \cdot \varepsilon \cdot n)^{\frac{1}{9}})$$

Next, as stated in Equation (6), conditioned on the prefix of $n - d$ random bits (and the seeds used), X remains $(t - n^{O(1)}, \varepsilon)$-hard to predict, and thus by Goldreich-Levin (Theorem 1)

$$\mathsf{CD}_{t'}(\ H_C(X),\ U_{d+s}\ |\ H_1(f(X)), H_2(X), H_1, H_2, H_C) = O(2^{d+s} \cdot (n \cdot \varepsilon)^{\frac{1}{3}})$$
$$=\ O(2^{-\frac{d}{2}}) = O((2^{3s} \cdot \varepsilon \cdot n)^{\frac{1}{9}})$$

holds for $t' = t \cdot (\varepsilon/n)^{O(1)}$. The conclusion follows by a triangle inequality.

Lemma 4 (entropy conditions). *Let $f, \mathcal{H}_1, \mathcal{H}_2$ be defined as in Theorem 2, we have*

$$\mathbf{H}_\infty(f(X)) = n - k\ , \tag{4}$$

$$\mathbf{H}_\infty(X \mid h_1(f(X)), h_1) \geq \mathbf{H}_\infty(X) - (n - k - d)\ =\ k + d\ , \tag{5}$$

$$\mathbf{H}_{t-n^{O(1)}}(X \mid h_1(f(X)), h_2(X), h_1, h_2) \geq \mathbf{H}_t(X \mid f(X), h_2(X), h_2) \geq\ \log(1/\varepsilon) \tag{6}$$

hold for every $h_1 \in \mathcal{H}_1$, $h_2 \in \mathcal{H}_2$, and X uniform over $\{0,1\}^n$.

Proof. Equation (4) follows from the regularity of f, i.e., every $y = f(x)$ has 2^k preimages, and thus $f(X)$ is uniformly distributed over a set of size 2^{n-k}. Equation (5) is due to the chain rule of min-entropy (see Fact 1). The first inequality of Equation (6) is the replacement inequality (see Fact 2), and the second one is obtained by applying the chain rule of unpredictability entropy to Equation (3), i.e., $\mathbf{H}_t(X \mid f(X), h_2(X), h_2) \geq \mathbf{H}_t(X|f(X)) - k = \log(1/\varepsilon)$.

Therefore, we already complete the proof for the PRG with linear seed length by doing a single call to any 2^k-regular ε-hard OWF provided that ε and k are known. We provide an alternative (and simpler) proof to that given by Goldreich [7] for essentially the same construction via unpredictability pseudo-entropy.

ON TIGHTENING SECURITY BOUNDS. Concretely, if the underlying OWF is $n^{-\log n}$- (resp., $2^{-\frac{n}{3}}$-) hard, then the outputs of the resulting PRG will be nearly $n^{-\frac{\log n}{9}}$- (resp., $2^{-\frac{n}{27}}$-) close to uniform (with respect to reasonably weakened adversaries than counterparts of the OWF). The main lossy step in the reduction is that we considered function $f'(x, h_2) \overset{\text{def}}{=} (f(x), h_2(x), h_2)$, where by Equation (6) X is (ε, t)-hard to predict given $f'(X)$ and thus we directly applied Equation (2) to get the inferior bounds. However, a closer look at f' suggests that it is almost 1-to-1, which implies that f' is a OWF (stated as in Lemma 5), which allows us to use the tight version of Goldreich-Levin Theorem (see Equation (1)). This is actually the approach taken by [7], where however f' was only shown to be roughly $\varepsilon^{1/5}$-hard (by checking the proof of [7, Prop 3.5.9]). We give a refined analysis below to get the tighter $\sqrt{\varepsilon}$-hardness of f', and this eventually leads to the improved construction as in Theorem 3.

Lemma 5 (unpredictability and almost 1-to-1 imply one-wayness). *Let f and $\mathcal{H}_2$ be as defined in Theorem 2, then function $f'(x, h_2) \stackrel{\text{def}}{=} (f(x), h_2(x), h_2)$ is a $(t, 3\sqrt{\varepsilon})$- one-way function.*

Proof. Suppose for contradiction there exists A of running time t such that

$$\Pr\left[\mathsf{A}(f'(X, H_2)) \in f'^{-1}(f'(X, H_2))\right] > 3\sqrt{\varepsilon}$$

Recall that $f(X)$ has min-entropy $n - k$ and conditioned on any $y = f(X)$ X has min-entropy k, and thus by the condensing property of universal hashing (see Lemma 2, setting $Z = f(X)$ and $H = H_2$) $\mathbf{H}_2(H_2(X)|H_2, f(X)) \geq k - 1$, which implies that $\mathsf{CP}(f(X), H_2(X) \mid H_2) \leq 2^{-(n-k)} \cdot 2^{-(k-1)} = 2^{-(n-1)}$. It follows from Lemma 6 (setting $a = 2^{-n}/\sqrt{\varepsilon}$, $X_1 = (f(X), H_2(X))$, $Z_1 = H_2$) that $f'(X, H_2)$ hits set $\mathcal{S}$ (defined below) with negligible probability, i.e., $\Pr[f'(X, H_2) \in \mathcal{S}] \leq 2\sqrt{\varepsilon}$ where

$$\mathcal{S} \stackrel{\text{def}}{=} \{(y, w, h_2) : \Pr[(f(X), h_2(X)) = (y, w) \mid H_2 = h_2] \geq 2^{-n}/\sqrt{\varepsilon}\}$$
$$= \{(y, w, h_2) : |f'^{-1}(y, w, h_2)| \geq 1/\sqrt{\varepsilon}\} \ .$$

Then, let $\mathcal{E}$ be the event that A inverts f' on any image whose preimage size is bounded by $1/\sqrt{\varepsilon}$, i.e., $\mathcal{E} \stackrel{\text{def}}{=} \mathsf{A}(f'(X, H_2)) \in f'^{-1}(f'(X, H_2)) \ \wedge \ f'(X, H_2) \notin \mathcal{S}$

$$\Pr\left[\mathsf{A}(f'(X, H_2)) = X\right] \geq \Pr\left[\mathcal{E}\right] \cdot \Pr[\mathsf{A}(f'(X, H_2)) = X \mid \mathcal{E}]$$
$$> (3\sqrt{\varepsilon} - 2\sqrt{\varepsilon}) \cdot \left(\frac{1}{1/\sqrt{\varepsilon}}\right) = \varepsilon \ ,$$

where the probability of hard-to-invertness is related to unpredictability by the maximal preimage size. The conclusion follows by reaching a contradiction to the (t, ε)-unpredictability of X given $f'(X, H_2)$ (as stated in Equation (6)).

Lemma 6 ($\mathbf{H}_2$ implies $\mathbf{H}_\infty$ with small slackness). *Let (X_1, Z_1) be a random variable, for $a > 0$ define $\mathcal{S}_a \stackrel{\text{def}}{=} \{(x, z) : \Pr[X_1 = x | Z_1 = z] \geq a\}$, it holds that $\Pr[(X_1, Z_1) \in \mathcal{S}_a] \leq \mathsf{CP}(X_1|Z_1)/a$.*

Proof. The proof is a typical Markov type argument.

$$\mathsf{CP}(X_1|Z_1) = \mathbb{E}_{z \leftarrow Z_1}\left[\sum_x \Pr[X_1 = x | Z_1 = z]^2\right]$$

$$= \sum_{(x,z)} \Pr[(X_1, Z_1) = (x, z)] \cdot \Pr[X_1 = x | Z_1 = z]$$

$$\geq \sum_{(x,z) \in \mathcal{S}_a} \Pr[(X_1, Z_1) = (x, z)] \cdot \Pr[X_1 = x | Z_1 = z]$$

$$\geq a \cdot \Pr[(X_1, Z_1) \in \mathcal{S}_a] \ .$$

Theorem 3 (improved construction based on known regularity and hardness). *For the same f, g, $\mathcal{H}_1$, $\mathcal{H}_2$, $\mathcal{H}_C$ as assumed in Theorem 2 except that d and s satisfy $3d + 2s = \log(1/\varepsilon)$, we have that g is a ($t \cdot (\varepsilon/n)^{O(1)}$, $O((2^{2s} \cdot \varepsilon)^{1/6})$)-secure PRG with stretch s.*

Proof sketch. The proof is similar to Theorem 2. The first $n - d$ bits extracted are $2^{-d/2}$-statistically random, conditioned on which the next $d + s$ bits are $O(2^{d+s}\sqrt{\varepsilon})$-computationally random. It follows that the bound is $2^{-d/2} + O(2^{d+s}\sqrt{\varepsilon}) = O(2^{-d/2}) = O((2^{2s} \cdot \varepsilon)^{1/6})$. $\square$

Remark 2 (a comparison with [7]). We provide an alternative (and much simplified) proof to the counterpart in [7]. Both approaches start by considering function $f'(x, h_2) = (f(x), h_2(x), h_2)$ and observing that distribution $f'(X, H_2)$ is nearly of full entropy (i.e. the amount of random bits used to sample $f'(X, H_2)$). The analysis of [7] then gets somewhat involved to show that f' is a $(t, O(\varepsilon^{1/5}))$-OWF, and we simply apply the chain rule to get that the unpredictability pseudo-entropy about X given $f'(X, H)$ is at least $\log(1/\varepsilon)$ (see Lemma 4). Therefore, by Goldreich-Levin one can extract more bits from X to make a PRG. Combined with another idea that f' is nearly 1-to-1 and thus unpredictability implies one-wayness, our proof also implies a tighter version of [7], namely, f' is a $(t, 3\sqrt{\varepsilon})$-OWF.

Remark 3 (next-bit vs. unpredictability pseudo-entropy). We mention that our observation that "X given $f(X)$ has unpredictability pseudo-entropy $k+\log(1/\varepsilon)$" is incomparable with the counterpart[7] in [23] that "$(f(X),X)$ has next-bit pseudo-entropy $n+\Omega(\log n)$". First, the proof of [23] is fundamentally different via the uniform version of Min-Max Theorem which is technically involved and useful in much broader contexts [24]. Secondly, there are no known reductions in relating unpredictability pseudo-entropy to next-bit pseudo-entropy from either directions, and in the former case one needs special extractors (that support reconstruction) while for the latter one needs to concatenate many copies of next-bit entropy sources and to extract many times (see [23, Figure 1]).

THREE EXTRACTIONS ARE NECESSARY. We argue that three extractions (using h_1, h_2 and h_c) seem necessary. One might think that the first two extractions (using h_1 and h_2) can be merged using a single universal hash function (that applies to the source $(X, f(X))$ and outputs $n - d$ bits). However, by doing so we cannot ensure the entropy condition (see Equation (6)) for the third extraction (using h_c). From another perspective, the merge would remove the dependency on the regularity and thus result in a generic construction that does a single call to any unknown regular OWFs, which is a contradiction to [16]. Furthermore, it seems necessary to extract from X at least twice, namely, using h_2 and h_c to get statistically and computationally random bits respectively.

---

[7] In fact, this was first observed in [13] via the application of a special universal hash function of description length $O(n^2)$, and the work of [23] shows that the use of the hash function is not necessary.

3.3   PRGs from Any Known Regular OWFs: Removing the Dependency on ε

The parameterization of the aforementioned construction depends on ε, but sometimes ε is unknown or not polynomial-time computable. It is thus more desirable to have a construction based on any known-regular OWF regardless of parameter ε (as long as it is negligible). We observe that by setting entropy loss to zero (in which case hash functions are condensers) and letting G-L functions extract $O(\log n)$ bits the resulting generator is a generic (i.e. without relying on ε) pseudo-entropy generator (PEG) with a (collision) entropy stretch of $O(\log n)$ bits. Note however the output of the PEG is not indistinguishable from uniform but from some high collision entropy sources (with small constant entropy deficiency), which implies a PRG by running $q \in \omega(1)$ copies of the PEG and doing a single extraction from the concatenated outputs.

Definition 6 (pseudo-entropy generators). *Function $g : \{0,1\}^n \to \{0,1\}^{l+e}$ ($l > n$) is a (t,ε) $\mathbf{H}_2$-pseudo-entropy generator (PEG) if g is polynomial-time computable and there exists a random variable $Y \in \{0,1\}^{l+e}$ with $\mathbf{H}_2(Y) \geq l$*

$$\mathsf{CD}_t(\ g(U_n)\ ,\ Y\ ) \ \leq\ \varepsilon.$$

where $(l - n)$ is the stretch of g, and e is the entropy deficiency. We say that g is an $\mathbf{H}_2$-pseudo-entropy generator if $1/\varepsilon$ and t are both super-polynomial.

Theorem 4 (PEGs from any known-regular OWFs). *For the same f, g, $\mathcal{H}_1$, $\mathcal{H}_2$, $\mathcal{H}_C$ as assumed in Theorem 2 except that $d = 0$ and $s = 2\log n + 2$, we have that if f is a known-regular one-way function then g is a $\mathbf{H}_2$-pseudo-entropy generator with stretch $2\log n$ and entropy deficiency 2.*

Proof sketch. It is not hard to see (using Lemma 2) that for $d = 0$ we have

$$2^{-\mathbf{H}_2(H_1(f(X)),H_2(X)\ |\ H_1,H_2)} = \mathsf{CP}(H_1(f(X)),H_2(X)\ |\ H_1,H_2)$$

$$\leq \Pr_{x_1,x_2 \leftarrow U_n, h_1 \leftarrow H_1}[h_1(f(x_1)) = h_1(f(x_2))]$$

$$\times \Pr_{x_1 \leftarrow X_1, x_2 \leftarrow X_2, h_2 \leftarrow H_2}[h_2(x_1) = h_2(x_2)\ |\ f(X_1) = f(X_2)\ ]$$

$$\leq 2^{-(n-k-1)} \cdot 2^{-(k-1)} = 2^{-(n-2)}\ .$$

And we have by Lemma 5 and Goldreich-Levin the $2\log n + 2$ hardcore bits are pseudo-random given $H_1(f(X))$ and $H_2(X)$, which completes the proof.   □

Theorem 5 (PRGs from any known-regular OWFs). *For any known k, there exists a generic construction of pseudo-random generator with seed length $\tilde{O}(n)$ by making $\tilde{O}(1)$ calls to any (length-preserving) 2^k-regular one-way function.*

Proof sketch. The idea is to run $q \in \omega(1)$ independent copies of the PEGs as in Theorem 4 to get an entropy stretch of $2q \log n$ followed by a single randomness extraction with entropy loss $q \log n$. This yields a PRG with stretch $q \log n$ that is roughly $O(q \cdot n^2 \sqrt{\varepsilon} + n^{-q})$ computationally indistinguishable from uniform randomness, where n^{-q} is negligible for any $q \in \omega(1)$. $\square$

3.4 PRGs from Any Unknown Regular OWFs

THE FIRST ATTEMPT: A PARALLEL CONSTRUCTION. A straightforward way to adapt the construction to unknown regular OWFs is to pay a factor of $n/\log n$. That is, it is not hard to see the construction for known regularity $\alpha = 2^k$ remains secure even by using an approximated value $\tilde{\alpha} = 2^{\tilde{k}}$ with accuracy $|\tilde{k} - k| \leq \log n$. This immediately implies a parallel construction by running $n/\log n$ independent copies of our aforementioned construction, where each ith copy assumes regularity $2^{i \cdot \log n}$. Therefore, at least one (unknown) copy will be a PRG and thus we simply XOR the outputs of all copies and produce it as the output. Unfortunately, similar to the HILL approach, the parallelism turns out an inherent barrier to linear seed length. We will avoid this route by giving a sequential construction.

STEP 1: CONVERT TO KNOWN REGULARITY. Now we present the construction from any (length-preserving) unknown-regular OWF. We first transform it into a hardness-preserving equivalent with known regularity 2^n, as stated in Lemma 7.

Lemma 7 (unknown to known regularity). *For any length-preserving unknown regular (t,ε)-OWF $f : \{0,1\}^n \to \{0,1\}^n$, define*

$$\bar{f} : \mathcal{Y} \times \{0,1\}^n \to \mathcal{Y}$$
$$\bar{f}(y,r) \overset{\text{def}}{=} f(y \oplus r) \tag{7}$$

where $\mathcal{Y} \overset{\text{def}}{=} f(\{0,1\}^n) \subseteq \{0,1\}^n$, "$\oplus$" denotes bit-wise XOR. Then, $\bar{f}$ is a 2^n-regular $(t - O(n),\varepsilon)$-OWF.

Proof. On uniform (y,r) over $\mathcal{Y} \times \{0,1\}^n$, $y \oplus r$ is uniform over $\{0,1\}^n$. Thus, any algorithm inverts $\bar{f}$ to produce (y, r) with probability ε implies another algorithm that inverts f with the same probability by outputting $y \oplus r$. Let us assume that f is α-regular. Then, for any $y_1 = \bar{f}(y,r) = f(y \oplus r)$ we have $|f^{-1}(y_1)| = \alpha$, and for any $x \in f^{-1}(y_1)$ we have $|\{(y,r) \in \mathcal{Y} \times \{0,1\}^n : y \oplus r = x\}| = |\mathcal{Y}| = 2^n/\alpha$, which implies $|\bar{f}^{-1}(y_1)\}| = \alpha \cdot (2^n/\alpha) = 2^n$.

STEP 2: $\mathcal{Z}$-SEEDED PRG. Similarly to the known regular case, we first assume ε is known and then eliminate the dependency. Intuitively, the output of $\bar{f}$ hides n bits of min-entropy about its input (by the 2^n-regularity) plus another $\log(1/\varepsilon)$ bits of pseudo-entropy (due to the one-wayness), and thus one can extract $n + O(\log(1/\varepsilon))$ pseudorandom bits. This is formalized in Lemma 8, where we build a $\mathcal{Z}$-seeded PRG $\bar{g}$ that expands random elements over $\mathcal{Y} \times \{0,1\}^n$ into pseudorandom ones over $\mathcal{Y} \times \{0,1\}^{n+O(\log(1/\varepsilon))}$. The proof of Lemma 8 is similar to that of Theorem 2, and we defer it to the appendix.

Definition 7 (Z-seeded PRG [23]). *A function $g^z : \mathcal{Z} \to \mathcal{Z} \times \{0,1\}^s$ is a (t,ε)-secure Z-seeded PRG with stretch s if g^z is polynomial-time computable and $\mathrm{CD}_t(\, g^z(Z)\,,\,(Z,U_s)\,) \leq \varepsilon$.*

Lemma 8 (construct Z-seeded PRG). *Let f, $\bar{f}$ be defined as in Lemma 7, for any integers d, s satisfying $7d + 6s = 2\log(1/\varepsilon n)$, let $\mathcal{H} \overset{\text{def}}{=} \{h : \{0,1\}^{2n} \to \{0,1\}^{n-d}\}$ be a universal hash function family, let $\mathcal{H}_C \overset{\text{def}}{=} \{h_c : \{0,1\}^{2n} \to \{0,1\}^{d+s}\}$ be a G-L function family, define $\bar{g}$ as*

$$\bar{g} : \mathcal{Y} \times \{0,1\}^n \times \mathcal{H} \times \mathcal{H}_C \to \mathcal{Y} \times \{0,1\}^{n+s} \times \mathcal{H} \times \mathcal{H}_C \tag{8}$$
$$\bar{g}(y,r,h,h_c) \overset{\text{def}}{=} (\, \bar{f}(y,r),\ (h(y,r),h_c(y,r)),\ h,\ h_c)$$

Then, we have that $\bar{g}$ is a $(t \cdot (\varepsilon/n)^{O(1)}, O((2^{3s} \cdot \varepsilon \cdot n)^{\frac{1}{7}}))$-secure Z-seeded PRG for $Z = (Y, R, H, H_C)$, where (Y, R) is identically distributed to $(f(U_n^1), U_n^2)$, and H, H_C are uniform over $\mathcal{H}$, $\mathcal{H}_C$ respectively.

STEP 3: SEQUENTIAL COMPOSITION. Notice, however, $\bar{g}$ is NOT a standard PRG with positive stretch as the only black-box way to sample distribution Y is to compute $f(U_n)$, which costs n random bits (despite that $\mathbf{H}_\infty(Y)$ might be far less than n). Quite naturally and thanks to the sequential composition, the construction simply iterates $\bar{g}$, reuses the random seeds (in each iteration), and outputs $s = O(\log(1/\varepsilon))$ bits per iteration.

Lemma 9 (sequential composition [23,3]). *Let $g^z : \mathcal{Z} \to \mathcal{Z} \times \{0,1\}^s$ be a (t,ε)-secure Z-seeded PRG, for $1 \leq i \leq \ell$ iteratively compute $(z_i, w_i) := g^z(z_{i-1})$, and define $g^{z,\ell}(z_0) \overset{\text{def}}{=} (z_\ell, w_1, \ldots, w_\ell)$. Then, we have that $g^{z,\ell}$ is a $(t - \ell \cdot n^{O(1)}, \ell \cdot \varepsilon)$-secure Z-seeded PRG with stretch $\ell \cdot s$.*

Proof. The proof is seen by a hybrid argument.

Theorem 6 (PRGs from any unknown-regular OWFs with known hardness). *Let function $f : \{0,1\}^n \to \{0,1\}^n$ be any (possibly unknown) regular length-preserving (t,ε)-OWF, define $\bar{f}, \bar{g}, \mathcal{H}, \mathcal{H}_C, s$ as in Lemma 8, and define*

$$g : \{0,1\}^n \times \{0,1\}^n \times \mathcal{H} \times \mathcal{H}_C \to (\{0,1\}^s)^\ell \times \{0,1\}^n \times \mathcal{H} \times \mathcal{H}_C$$
$$g(x, r_0, h, h_c) \overset{\text{def}}{=} (w_1, w_2, \ldots, w_\ell, r_\ell, h, h_c)$$

where g computes $y = f(x)$, and sequentially composes (as in Lemma 9) the Z-seeded PRG $\bar{g}$ (on input $z_0 = (y, r_0, h, h_c)$) ℓ times to produce output $w_1, w_2, \ldots, w_\ell, r_\ell, h, h_c$. Then, for any $s \leq \log(1/\varepsilon n)/3$, we have that function g is a $(t \cdot (\varepsilon/n)^{O(1)} - \ell \cdot n^{O(1)}, O(\ell \cdot (2^{3s} \cdot \varepsilon \cdot n)^{\frac{1}{7}}))$-secure PRG with stretch $\ell \cdot s - n$.

Proof. We can almost complete the proof by Lemma 8 and Lemma 9 except that the stretch of g (as a standard PRG) is $\ell \cdot s - n$ instead of $\ell \cdot s$. This is because we need to take into account that n bits are used to sample y at initialization.

CONCRETE PARAMETERS. Therefore, for any unknown-regular OWF with known hardness, we obtain a PRG with linear seed length, and by letting $s \in \Theta(\log(\frac{1}{\varepsilon n}))$ the number of calls $\ell \in \Theta(n/s) = \Theta(n/\log(1/\varepsilon n))$ matches the lower bound of [16]. This extends to the general case (where the hardness parameter is unknown) by repetition.

Theorem 7 (PRGs from any unknown-regular OWFs). *There exists a generic construction of pseudo-random generator with seed length $\tilde{O}(n)$ by making $\tilde{O}(n/\log n)$ calls to any unknown-regular one-way function.*

Proof sketch. For any unknown-regular OWF f, define $\bar{g}$ as in Lemma 8 except setting $d = 0$ and $s = 2\log n + 1$. It is not hard to see that the resulting $\bar{g}$ is a $\mathbf{H}_2$-pseudo-entropy generator with stretch $2\log n$ and entropy deficiency 1 (proof similar to that in Theorem 4). We then use the repetition trick (similar to Theorem 5), namely, for any $q \in \omega(1)$ run q independent copies of $\bar{g}$ and do a single extraction on the concatenated output with entropy loss set to $q\log n$. This gives us a Z'-seeded PRG $\bar{g}'$ for $Z' = (Y, U_n, H, H_c)^q$ with stretch $q \cdot \log n$. Again, sequential composing $\bar{g}'$ for $\ell' = \lceil (qn+1)/q\log n \rceil \in O(n/\log n)$ iterations yields a standard PRG

$$g' : \{0,1\}^{2qn} \times \mathcal{H}^q \times \mathcal{H}_C^q \to \{0,1\}^{2qn+s'} \times \mathcal{H}^q \times \mathcal{H}_C^q$$

where the stretch $s' = (q \cdot \log n) \cdot \ell' - q \cdot n \geq 1$. This completes the proof.   □

Acknowledgements. We thank Guang Yang, Colin Jia Zheng and Yunlei Zhao for useful comments. We are also grateful to Thomas Holenstein for clarifying his lower bound results [16] at the very early stage of this work. We thank the anonymous reviewers of ASIACRYPT 2013 for very helpful comments and suggestions that significantly improve the presentations of the PRG constructions. Finally, Yu Yu thanks Leonid Reyzin for interesting discussions about "saving private randomness [5]" at the Warsaw workshop [1], and thank Stefan Dziembowski for making the meeting possible. This research work was supported by the National Basic Research Program of China Grant 2011CBA00300, 2011CBA00301, the National Natural Science Foundation of China Grant 61033001, 61172085, 61061130540, 61073174, 61103221, 61070249, 60703031, 11061130539, 61021004 and 61133014. This work was also supported by the National Science Foundation of China under Grant Nos. 61272413 and 61133014, the Fok Ying Tung Education Foundation under Grant No. 131066, and the Program for New Century Excellent Talents in University under Grant No. NCET-12-0680.

References

1. Workshop on leakage, tampering and viruses (June 2013),
   http://www.crypto.edu.pl/events/workshop2013
2. Barak, B., Shaltiel, R., Wigderson, A.: Computational analogues of entropy. In: Arora, S., Jansen, K., Rolim, J.D.P., Sahai, A. (eds.) RANDOM 2003 and AP-PROX 2003. LNCS, vol. 2764, pp. 200–215. Springer, Heidelberg (2003)

3. Blum, M., Micali, S.: How to generate cryptographically strong sequences of pseudo random bits. In: Proceedings of the 23rd IEEE Symposium on Foundation of Computer Science, pp. 112–117 (1982)

4. Carter, J.L., Wegman, M.N.: Universal classes of hash functions. Journal of Computer and System Sciences 18, 143–154 (1979)

5. Dedić, N., Harnik, D., Reyzin, L.: Saving private randomness in one-way functions and pseudorandom generators. In: Canetti, R. (ed.) TCC 2008. LNCS, vol. 4948, pp. 607–625. Springer, Heidelberg (2008)

6. Dodis, Y., Elbaz, A., Oliveira, R., Raz, R.: Improved randomness extraction from two independent sources. In: APPROX-RANDOM, pp. 334–344 (2005)

7. Goldreich, O.: Foundations of Cryptography: Basic Tools. Cambridge University Press (2001)

8. Goldreich, O.: Three XOR-lemmas — an exposition. In: Goldreich, O. (ed.) Studies in Complexity and Cryptography. LNCS, vol. 6650, pp. 248–272. Springer, Heidelberg (2011)

9. Goldreich, O., Krawczyk, H., Luby, M.: On the existence of pseudorandom generators. SIAM Journal on Computing 22(6), 1163–1175 (1993)

10. Goldreich, O., Levin, L.A.: A hard-core predicate for all one-way functions. In: Johnson, D.S. (ed.) Proceedings of the Twenty First Annual ACM Symposium on Theory of Computing, Seattle, Washington, May 15-17, pp. 25–32 (1989)

11. Haitner, I., Harnik, D., Reingold, O.: On the power of the randomized iterate. In: Dwork, C. (ed.) CRYPTO 2006. LNCS, vol. 4117, pp. 22–40. Springer, Heidelberg (2006)

12. Haitner, I., Harnik, D., Reingold, O.: On the power of the randomized iterate. SIAM Journal on Computing 40(6), 1486–1528 (2011)

13. Haitner, I., Reingold, O., Vadhan, S.P.: Efficiency improvements in constructing pseudorandom generators from one-way functions. In: Proceedings of the 42nd ACM Symposium on the Theory of Computing, pp. 437–446 (2010)

14. Håstad, J., Impagliazzo, R., Levin, L.A., Luby, M.: Construction of pseudorandom generator from any one-way function. SIAM Journal on Computing 28(4), 1364–1396 (1999)

15. Holenstein, T.: Pseudorandom generators from one-way functions: A simple construction for any hardness. In: Halevi, S., Rabin, T. (eds.) TCC 2006. LNCS, vol. 3876, pp. 443–461. Springer, Heidelberg (2006)

16. Holenstein, T., Sinha, M.: Constructing a pseudorandom generator requires an almost linear Number of calls. In: Proceedings of the 53rd IEEE Symposium on Foundation of Computer Science, pp. 698–707 (2012)

17. Hsiao, C.-Y., Lu, C.-J., Reyzin, L.: Conditional computational entropy, or toward separating pseudoentropy from compressibility. In: Naor, M. (ed.) EUROCRYPT 2007. LNCS, vol. 4515, pp. 169–186. Springer, Heidelberg (2007)

18. Lee, C.-J., Lu, C.-J., Tsai, S.-C., Tzeng, W.-G.: Extracting randomness from multiple independent sources. IEEE Transactions on Information Theory 51(6), 2224–2227 (2005)

19. Levin, L.A.: One-way functions and pseudorandom generators. Combinatorica 7(4), 357–363 (1987)

20. Nisan, N.: Pseudorandom generators for space-bounded computation. Combinatorica 12(4), 449–461 (1992)

21. Nisan, N., Zuckerman, D.: Randomness is linear in space. Journal of Computer and System Sciences 52(1), 43–53 (1996)

22. Stinson, D.R.: Universal hash families and the leftover hash lemma, and applications to cryptography and computing. Journal of Combinatorial Mathematics and Combinatorial Computing 42, 3–31 (2002),
    `http://www.cacr.math.uwaterloo.ca/~dstinson/publist.html`
23. Vadhan, S.P., Zheng, C.J.: Characterizing pseudoentropy and simplifying pseudorandom generator constructions. In: Proceedings of the 44th ACM Symposium on the Theory of Computing, pp. 817–836 (2012)
24. Vadhan, S.P., Zheng, C.J.: A uniform min-max theorem with applications in cryptography. In: Canetti, R., Garay, J.A. (eds.) CRYPTO 2013, Part I. LNCS, vol. 8042, pp. 93–110. Springer, Heidelberg (2013)
25. Vazirani, U.V., Vazirani, V.V.: Efficient and secure pseudo-random number generation (extended abstract). In: Proceedings of the 25th IEEE Symposium on Foundation of Computer Science, pp. 458–463 (1984)
26. Yao, A.C.-C.: Theory and applications of trapdoor functions (extended abstract). In: Proceedings of the 23rd IEEE Symposium on Foundation of Computer Science, pp. 80–91 (1982)

A   Proofs Omitted

Proof of Lemma 8. Note that $\bar{f}(Y, R)$ is identically distributed to Y, so it is equivalent to show

$$\mathsf{CD}_{t \cdot (\varepsilon/n)^{O(1)}} \left(\left(H(Y, R), H_C(Y, R) \right), U_{n+s} \mid \bar{f}(Y, R), H, H_C \right) = O((2^{3s} \cdot \varepsilon \cdot n)^{\frac{1}{7}}) .$$

It follows from the $(t - O(n), \varepsilon)$-one-way-ness of $\bar{f}$ (see Lemma 7) and Lemma 3 that

$$\mathbf{H}_{t-O(n)}((Y, R) \mid \bar{f}(Y, R)) \geq n + \log(1/\varepsilon) . \tag{9}$$

Then, similar to Lemma 4, we have the following entropy conditions

$$\mathbf{H}_\infty((Y, R) \mid \bar{f}(Y, R)) = n ,$$

$$\mathbf{H}_{t-O(n)}((Y, R) \mid \bar{f}(Y, R), h(Y, R), h) \geq \mathbf{H}_{t-O(n)}((Y, R) \mid \bar{f}(Y, R)) - (n - d)$$
$$\geq d + \log(1/\varepsilon) ,$$

hold for any $h \in \mathcal{H}$, where the second inequality is by applying the chain rule to Equation (9). Therefore,

$$\mathsf{CD}_{t \cdot (\varepsilon/n)^{O(1)}} \left(\left(H(Y, R), H_C(Y, R) \right), U_{n+s} \mid \bar{f}(Y, R), H, H_C \right)$$
$$\leq \mathsf{SD}(H(Y, R), U_{n-d} \mid \bar{f}(Y, R), H)$$
$$+ \mathsf{CD}_{t \cdot (\varepsilon/n)^{O(1)}}(H_C(Y, R), U_{d+s} \mid \bar{f}(Y, R), H(Y, R), H, H_C)$$
$$\leq 2^{-\frac{d}{2}} + O(2^{d+s} \cdot (n \cdot \varepsilon \cdot 2^{-d})^{\frac{1}{3}}) = 2^{-\frac{d}{2}} + O(2^{d+s} \cdot (2^{\frac{-(7d+6s)}{2}} \cdot 2^{-d})^{\frac{1}{3}})$$
$$= O(2^{-\frac{d}{2}}) = O(2^{\frac{3s+\log(\varepsilon \cdot n)}{7}}) = O((2^{3s} \cdot \varepsilon \cdot n)^{\frac{1}{7}})$$

where the first inequality is triangle, the statistical distance is due to the leftover hash lemma and the computational distance of the second inequality is by the Goldreich-Levin Theorem.   □

Constrained Pseudorandom Functions
and Their Applications[*]

Dan Boneh[1] and Brent Waters[2]

[1] Stanford University
dabo@cs.stanford.edu
[2] U.T. Austin
bwaters@cs.utexas.edu

Abstract. We put forward a new notion of pseudorandom functions (PRFs) we call constrained PRFs. In a standard PRF there is a master key k that enables one to evaluate the function at all points in the domain of the function. In a constrained PRF it is possible to derive constrained keys k_s from the master key k. A constrained key k_s enables the evaluation of the PRF at a certain subset S of the domain and nowhere else. We present a formal framework for this concept and show that constrained PRFs can be used to construct powerful primitives such as identity-based key exchange and a broadcast encryption system with optimal ciphertext size. We then construct constrained PRFs for several natural set systems needed for these applications. We conclude with several open problems relating to this new concept.

1 Introduction

Pseudorandom functions(PRF) [20] are a fundamental concept in modern cryptography. A PRF is a function $F : \mathcal{K} \times \mathcal{X} \to \mathcal{Y}$ that can be computed by a deterministic polynomial time algorithm: on input $(k, x) \in \mathcal{K} \times \mathcal{X}$ the algorithm outputs $F(k, x) \in \mathcal{Y}$. Note that given the key $k \in \mathcal{K}$, the function $F(k, \cdot)$ can be efficiently evaluated at *all* points $x \in \mathcal{X}$.

In this paper we put forward a new notion of PRFs we call *constrained PRFs*. Consider a PRF $F : \mathcal{K} \times \mathcal{X} \to \mathcal{Y}$ and let $k_0 \in \mathcal{K}$ be some key for F. In a constrained PRF one can derive constrained keys k_s from the master PRF key k_0. Each constrained key k_s corresponds to some subset $S \subseteq X$ and enables one to evaluate the function $F(k_0, x)$ for $x \in S$, but at no other points in the domain $\mathcal{X}$. A constrained PRF is secure if given several constrained keys for sets $S_1, \ldots, S_q$ of the adversary's choice, the adversary cannot distinguish the PRF from random for points x outside these sets, namely for $x \notin \cup_{i=1}^q S_i$. We give precise definitions in Section 3.

While constrained PRFs are a natural extension of the standard concept of PRFs, they have surprisingly powerful applications beyond what is possible with standard PRFs. We list a few examples here and present more applications in Section 6:

---

[*] The full version is available as Cryptology ePrint Archive, Report 2013/352.

K. Sako and P. Sarkar (Eds.) ASIACRYPT 2013, Part II, LNCS 8270, pp. 280–300, 2013.
© International Association for Cryptologic Research 2013

- **Left-Right PRFs:** Let $F : \mathcal{K} \times \mathcal{X}^2 \to \mathcal{Y}$ be a secure PRF. Its domain is $\mathcal{X} \times \mathcal{X}$. Now, suppose that for every $w \in \mathcal{X}$ there are two constrained keys $k_{w,\text{left}}$ and $k_{w,\text{right}}$. The key $k_{w,\text{left}}$ enables the evaluation of $F(k_0, \cdot)$ at the subset of points $\{(w, y) : y \in \mathcal{X}\}$ (i.e. at all points where the left side is w). The key $k_{w,\text{right}}$ enables the evaluation of $F(k_0, \cdot)$ at the subset of points $\{(x, w) : x \in \mathcal{X}\}$ (i.e. at all points where the right side is w). We show that such a constrained PRF can be used to construct an identity-based non-interactive key exchange (ID-NIKE) system [31,14,27,16].
- **Bit-Fixing PRFs:** Let $\mathcal{X} = \{0,1\}^n$ be the domain of the PRF. For a vector $v \in \{0, 1, ?\}^n$ let $S_v \subseteq \mathcal{X}$ be the set of n-bit strings that match v at all the coordinates where v is not '?'. We say that S_v is bit-fixed to v. For example, the set containing all n-bit strings starting with 00 and ending in 11 is bit-fixed to $v = 00? \ldots ?11$.

  Now, suppose that for every bit-fixed subset S of $\{0,1\}^n$ there is a constrained key k_S that enables the evaluation of $F(k_0, x)$ at $x \in S$ and nowhere else. We show that such a constrained PRF can be used to construct an *optimal* secret-key[1] broadcast encryption system [15]. In particular, the length of the private key and the broadcast ciphertext are all *independent* of the number of users. We compare these constructions to existing broadcast systems in Section 6.1.
- **Circuit PRFs:** Let $F : \mathcal{K} \times \{0,1\}^n \to \mathcal{Y}$ be a secure PRF. Suppose that for every polynomial size circuit C there is a constrained key k_C that enables the evaluation of $F(k_0, x)$ at all points $x \in \{0,1\}^n$ such that $C(x) = 1$. We show that such a constrained PRF gives rise to a non-interactive policy-based key exchange mechanism: a group of users identified by a complex policy (encoded as a circuit) can non-interactively setup a secret group key that they can then use for secure communications among group members. A related concept was studied by Gorantla et al. [21], but the schemes presented are interactive, analyzed in the generic group model, and only apply to policies represented as polynomial size formulas.

In the coming sections we present constructions for all the constrained PRFs discussed above as well as several others. Some of our constructions use bilinear maps while others require κ-linear maps [7,17,11] for $\kappa > 2$. It would be quite interesting and useful to develop constructions for these constrained PRFs from other assumptions such as Learning With Errors (LWE) [28]. This will give new key exchange and broadcast encryption systems from the LWE problem.

In defining security for a constrained PRF in Section 3 we allow the adversary to adaptively request constrained keys of his choice. The adversary's goal is to distinguish the PRF from a random function at input points where he cannot compute the PRF using the constrained keys at his disposal. The definition of security allows the adversary to *adaptively* choose the challenge point at which he tries to distinguish the PRF from random. However, to prove security of our constructions we require that the attacker commit to the challenge point ahead

---

[1] Secret-key broadcast encryption refers to the fact that the broadcaster's key is known only to the broadcaster.

of time thereby only proving a weaker notion of security called selective security. A standard argument called *complexity leveraging* (see e.g. [4, Sec. 7.1]) shows that selective security implies adaptive security via a non-polynomial time reduction. Therefore, to obtain adaptive security we must increase the parameters of our schemes so that security is maintained under the complexity leveraging reduction. A fascinating open problem is to construct standard model constrained PRFs that are adaptively secure under a polynomial time reduction.

Related work. Concurrently with this paper, similar notions to constrained PRFs were recently proposed by Kiayias et al. [24] where they were called delegatable PRFs and Boyle et al. [9] where they were called functional PRFs. Both papers give constructions for prefix constraints discussed in Section 3.3. A related concept applied to digital signatures was explored by Bellare and Fuchsbauer [1] where it was called policy-based signatures and by Boyle et al. [9] where it was called functional signatures.

2   Preliminaries: Bilinear and κ-Linear Maps

Recently, Garg, Gentry, and Halevi [17] proposed candidate constructions for leveled multilinear forms. Building on their work Coron, Lepoint, and Tibouchi [11] gave a second candidate. We will present some of our constructions using the abstraction of leveled multilinear groups.

The candidate constructions of [17,11] implement an abstraction called graded encodings which is similar, but slightly different from multilinear groups. In the full version [8] we show how to map our constructions to the language of graded encodings.

Leveled multilinear groups. We assume the existence of a group generator $\mathcal{G}$, which takes as input a security parameter 1^λ and a positive integer κ to indicate the number of levels. $\mathcal{G}(1^\lambda, \kappa)$ outputs a sequence of groups $\boldsymbol{G} = (\mathbb{G}_1, \ldots, \mathbb{G}_\kappa)$ each of large prime order $p > 2^\lambda$. In addition, we let g_i be a canonical generator of $\mathbb{G}_i$ that is known from the group's description. We let $g = g_1$.

We assume the existence of a set of bilinear maps $\{e_{i,j} : G_i \times G_j \to G_{i+j} \mid i, j \geq 1; \ i + j \leq \kappa\}$. The map $e_{i,j}$ satisfies the following relation:

$$e_{i,j}\left(g_i^a, g_j^b\right) = g_{i+j}^{ab} \ : \ \forall a, b \in \mathbb{Z}_p$$

We observe that one consequence of this is that $e_{i,j}(g_i, g_j) = g_{i+j}$ for each valid i, j. When the context is obvious, we will sometimes drop the subscripts i, j. For example, we may simply write:

$$e\left(g_i^a, g_j^b\right) = g_{i+j}^{ab}.$$

We define the κ-Multilinear Decisional Diffie-Hellman (κ-MDDH) assumption as follows:

Assumption 1 (κ-Multilinear Decisional Diffie-Hellman: κ-MDDH)
The κ-Multilinear Decisional Diffie-Hellman (κ-MDDH) problem states the following: A challenger runs $\mathcal{G}(1^\lambda, \kappa)$ to generate groups and generators of order p. Then it picks random $c_1, \ldots, c_{\kappa+1} \in \mathbb{Z}_p$.

The assumption then states that given $g = g_1, g^{c_1}, \ldots, g^{c_{\kappa+1}}$ it is hard to distinguish the element $T = g_\kappa^{\prod_{j \in [1, \kappa+1]} c_j} \in \mathbb{G}_\kappa$ from a random group element in $\mathbb{G}_\kappa$, with better than negligible advantage in the security parameter λ.

3   Constrained Pseudorandom Functions

We now give a precise definition of constrained Pseudorandom Functions. We begin with the syntax of the constrained PRF primitive and then define the security requirement.

3.1   The Constrained PRF Framework

Recall that a pseudorandom function (PRF) [20] is defined over a key space $\mathcal{K}$, a domain $\mathcal{X}$, and a range $\mathcal{Y}$ (and these sets may be parameterized by the security parameter λ). The PRF itself is a function $F : \mathcal{K} \times \mathcal{X} \to \mathcal{Y}$ that can be computed by a deterministic polynomial time algorithm: on input $(k, x) \in \mathcal{K} \times \mathcal{X}$ the algorithm outputs $F(k, x) \in \mathcal{Y}$. A PRF can include a setup algorithm $F.\mathsf{setup}(1^\lambda)$ that takes a security parameter λ as input and outputs a random secret key $k \in \mathcal{K}$.

A PRF $F : \mathcal{K} \times \mathcal{X} \to \mathcal{Y}$ is said to be *constrained* with respect to a set system $\mathcal{S} \subseteq 2^\mathcal{X}$ if there is an additional key space $\mathcal{K}_c$ and two additional algorithms $F.\mathsf{constrain}$ and $F.\mathsf{eval}$ as follows:

- $F.\mathsf{constrain}(k, S)$ is a randomized polynomial-time algorithm that takes as input a PRF key $k \in \mathcal{K}$ and the description of a set $S \in \mathcal{S}$ (so that $S \subseteq \mathcal{X}$). The algorithm outputs a constrained key $k_S \in \mathcal{K}_c$. This key k_S enables the evaluation of $F(k, x)$ for all $x \in S$ and no other x.
- $F.\mathsf{eval}(k_S, x)$ is a deterministic polynomial-time algorithm (in λ) that takes as input a constrained key $k_s \in \mathcal{K}_c$ and an $x \in \mathcal{X}$. If k_S is the output of $F.\mathsf{constrain}(k, S)$ for some PRF key $k \in \mathcal{K}$ then $F.\mathsf{eval}(k_S, x)$ outputs

$$F.\mathsf{eval}(k_S, x) = \begin{cases} F(k, x) & \text{if } x \in S \\ \bot & \text{otherwise} \end{cases}$$

where $\bot \notin \mathcal{Y}$. As shorthand we will occasionally write $F(k_S, x)$ for $F.\mathsf{eval}(k_S, x)$.

Note that while in general deciding if $x \in S$ may not be a poly-time problem, our formulation of $F.\mathsf{eval}$ effectively avoids this complication by requiring that all $S \in \mathcal{S}$ are poly-time decidable by the algorithm $F.\mathsf{eval}(k_S, \cdot)$. This poly-time algorithm outputs non-$\perp$ when $x \in S$ and $\perp$ otherwise thereby deciding S in polynomial time.

Occasionally it will be convenient to treat the set system $S \subseteq 2^{\mathcal{X}}$ as a family of predicates $\text{PP} = \{p : \mathcal{X} \to \{0,1\}\}$. For a predicate $p \in \text{PP}$ we have $F.\text{eval}(k_p, x) = F(k, x)$ whenever $p(x) = 1$ and $\perp$ otherwise. In this case we say that the PRF F is constrained with respect to the family of predicates PP.

The trivial constrained PRF. All PRFs $F : \mathcal{K} \times \mathcal{X} \to \mathcal{Y}$ are constrained with respect to the set system S consisting of all singleton sets: $S = \{\{x\} : x \in \mathcal{X}\}$. To see why, fix some PRF key $k \in \mathcal{K}$. Then the constrained key $k_{\{x\}}$ for the singleton set $\{x\}$ is simply $k_{\{x\}} = F(k, x)$. Given this key $k_{\{x\}}$, clearly anyone can evaluate $F(k, x)$ at the point x. This shows that we may assume without loss of generality that set systems S used to define a constrained PRF contain all singleton sets. More generally, we may also assume that S contains all *polynomial size* sets (polynomial in the security parameter λ). The constrained key k_S for a polynomial size set $S \subseteq \mathcal{X}$ is simply the set of values $F(k, x)$ for all $x \in S$. This construction fails for super-polynomial size sets since the constrained key k_S for such sets is too large.

3.2   Security of Constrained Pseudorandom Functions

Next, we define the security properties of constrained PRFs. The definition captures the property that given several constrained keys as well as several function values at points of the attacker's choosing, the function looks random at all points that the attacker cannot compute himself.

Let $F : \mathcal{K} \times \mathcal{X} \to \mathcal{Y}$ be a constrained PRF with respect to a set system $S \subseteq 2^{\mathcal{X}}$. We define constrained security using the following two experiments denoted $\text{EXP}(0)$ and $\text{EXP}(1)$ with an adversary $\mathcal{A}$. For $b = 0, 1$ experiment $\text{EXP}(b)$ proceeds as follows:

First, a random key $k \in \mathcal{K}$ is selected and two helper sets $C, V \subseteq \mathcal{X}$ are initialized to $\emptyset$. The set $V \subseteq \mathcal{X}$ will keep track of all the points at which the adversary can evaluate $F(k, \cdot)$. The set $C \subseteq \mathcal{X}$ will keep track of the points where the adversary has been challenged. The sets C and V will ensure that the adversary cannot trivially decide whether challenge values are random or pseudorandom. In particular, the experiments maintain the invariant that $C \cap V = \emptyset$.

The adversary $\mathcal{A}$ is then presented with three oracles as follows:
- F.eval: given $x \in \mathcal{X}$ from $\mathcal{A}$, if $x \notin C$ the oracle returns $F(k, x)$ and otherwise returns $\perp$. The set V is updated as $V \leftarrow V \cup \{x\}$.
- F.constrain: given a set $S \in S$ from $\mathcal{A}$, if $S \cap C = \emptyset$ the oracle returns a key $F.\text{constrain}(k, S)$ and otherwise returns $\perp$. The set V is updated as $V \leftarrow V \cup S$.
- Challenge: given $x \in \mathcal{X}$ from $\mathcal{A}$ where $x \notin V$, if $b = 0$ the adversary is given $F(k, x)$; otherwise the adversary is given a random (consistent) $y \in \mathcal{Y}$. The set C is updated as $C \leftarrow C \cup \{x\}$.

Once the adversary $\mathcal{A}$ is done interrogating the oracles it outputs $b' \in \{0,1\}$.
For $b = 0, 1$ let W_b be the event that $b' = 1$ in $\text{EXP}(b)$. We define the adversary's advantage as $\text{AdvPRF}_{\mathcal{A},F}(\lambda) = |\Pr[W_0] - \Pr[W_1]|$.

Definition 1. *The PRF F is a secure constrained PRF with respect to $\mathcal{S}$ if for all probabilistic polynomial time adversaries $\mathcal{A}$ the function $AdvPRF_{\mathcal{A},F}(\lambda)$ is negligible.*

When constructing constrained functions it will be more convenient to work with a definition that slightly restricts the adversary's power, but is equivalent to Definition 1. In particular, we only allow the adversary to issue a *single* challenge query (but multiple queries to the other two oracles). A standard hybrid argument shows that a PRF secure under this restricted definition is also secure under Definition 1.

3.3   Example Predicate Families

Next we introduce some notation to capture the predicate families described in the introduction.

Bit-Fixing Predicates. Let $F : \mathcal{K} \times \{0,1\}^n \to \mathcal{Y}$ be a PRF. We wish to support constrained keys $k_{\mathbf{v}}$ that enable the evaluation of $F(k, x)$ at all points x that match a particular bit pattern. To do so define for a vector $\mathbf{v} \in \{0, 1, ?\}^n$ the predicate $p_{\mathbf{v}}^{(\text{BF})} : \{0,1\}^n \to \{0,1\}$ as

$$p_{\mathbf{v}}^{(\text{BF})}(x) = 1 \quad \Longleftrightarrow \quad (\mathbf{v}_i = x_i \text{ or } \mathbf{v}_i =?) \text{ for all } i = 1, \ldots, n .$$

We say that $F : \mathcal{K} \times \{0,1\}^n \to \mathcal{Y}$ supports bit fixing if it is constrained with respect to the set of predicates

$$\mathcal{P}_{\text{BF}} = \{p_{\mathbf{v}}^{(\text{BF})} : \mathbf{v} \in \{0, 1, ?\}\}$$

Prefix Predicates. Prefix predicates are a special case of bit fixing predicates in which only the prefix is fixed. More precisely, we say that $F : \mathcal{K} \times \{0,1\}^n \to \mathcal{Y}$ supports prefix fixing if it is constrained with respect to the set of predicates

$$\mathcal{P}_{\text{PRE}} = \{p_{\mathbf{v}}^{(\text{BF})} : \mathbf{v} \in \{0,1\}^\ell \: ?^{n-\ell}, \: \ell \in [n]\}$$

Secure PRFs that are constrained with respect to $\mathcal{P}_{\text{PRE}}$ can be constructed directly from the GGM PRF construction [20]. For a prefix $\mathbf{v} \in \{0,1\}^\ell$ the constrained key $k_{\mathbf{v}}$ is simply the secret key in the GGM tree computed at the internal node associated with the string $\mathbf{v}$. Clearly this key enables the evaluation of $F(k, \mathbf{v}\|x)$ for any $x \in \{0,1\}^{n-|\mathbf{v}|}$. A similar construction, in a very different context, was used by Fiat and Naor [15] and later by Naor, Naor, and Lotspiech [25] to construct combinatorial broadcast encryption systems. The security proof for this GGM-based prefix constrained PRF is straight forward if the adversary commits to his challenge point ahead of time (a.k.a selective security). Full security can be achieved, for example, using standard complexity leveraging by guessing the adversary's challenge point ahead of time as in [4, Sec. 7.1].

Left/Right Predicates. Let $F : \mathcal{K} \times \mathcal{X}^2 \to \mathcal{Y}$ be a PRF. For all $w \in \mathcal{X}$ we wish to support constrained keys $k_{w,\text{LEFT}}$ that enable the evaluation of $F(k,\ (x,y))$ at all points $(w, y) \in \mathcal{X}^2$, that is, at all points in which the left side is fixed to w. In addition, we want constrained keys $k_{w,\text{RIGHT}}$ that fix the right hand side of (x, y) to w. More precisely, for an element $w \in \mathcal{X}$ define the two predicates $p_w^{(\text{L})}, p_w^{(\text{R})} : \mathcal{X}^2 \to \{0,1\}$ as

$$p_w^{(\text{L})}(x, y) = 1 \iff x = w \qquad \text{and} \qquad p_w^{(\text{R})}(x, y) = 1 \iff y = w$$

We say that F supports left/right fixing if it is constrained with respect to the set of predicates

$$\mathcal{P}_{LR} = \{p_w^{(\text{L})},\ p_w^{(\text{R})}\ :\ w \in \mathcal{X}\}$$

Constructing left/right constrained PRFs. We next show that secure PRFs that are constrained with respect to $\mathcal{P}_{LR}$ can be constructed straightforwardly in the random oracle model [3]. Constructing left/right constrained PRFs *without* random oracles is a far more challenging problem. We do so, and more, in the next section.

To construct a left/right constrained PRF in the random oracle model let $e : \mathbb{G} \times \mathbb{G} \to \mathbb{G}_T$ be a bilinear map where $\mathbb{G}$ and $\mathbb{G}_T$ are groups of prime order p. Let $H_1, H_2 : \mathcal{X} \to \mathbb{G}$ be two hash functions that will be modeled as random oracles. The setup algorithm will choose such a group and a random key $k \in \mathbb{Z}_p$. Define the following PRF:

$$F(k,\ (x, y)) = e(H_1(x), H_2(y))^k . \tag{1}$$

For $(x^*, y^*) \in \mathcal{X}^2$ the constrained keys for the predicates $p_{x^*}^{(\text{L})}$ and $p_{y^*}^{(\text{R})}$ are

$$k_{x^*} = H_1(x^*)^k \qquad \text{and} \qquad k_{y^*} = H_2(y^*)^k$$

respectively. Clearly k_{x^*} is sufficient for evaluating $f(k, y) = F(k,\ (x^*, y))$ and k_{y^*} is sufficient for evaluating $g(k, x) = F(k,\ (x, y^*))$, as required. We note the structural similarities between the above construction and the Boneh-Franklin [5] IBE system and the Sakai-Ohgishi-Kasahara [31] non-interactive key exchange system.

Theorem 2. *The PRF F defined in Eq. (1) is a secure constrained PRF with respect to $\mathcal{P}_{LR}$ assuming the decision bilinear Diffie-Hellman assumption (DBDH) holds for $(\mathbb{G}, \mathbb{G}_T, e)$ and the functions H_1, H_2 are modeled as random oracles.*

Due to space constraints the proof, which uses a standard argument, is given in the full version of the paper [8].

Circuit Predicates. Let $F : \mathcal{K} \times \{0,1\}^n \to \mathcal{Y}$ be a PRF. For a boolean circuit c on n inputs we wish to support a constrained key k_c that enable the evaluation of $F(k, x)$ at all points $x \in \mathcal{X}$ for which $c(x) = 1$.

Let $\mathcal{C}$ be the set of polynomial size circuits. We say that F supports circuit predicates if it is constrained with respect to the set of predicates

$$\mathcal{P}_{\text{circ}} = \{c\ :\ c \in \mathcal{C}\}$$

4   A Bit-Fixing Construction

We now describe our bit-fixing constrained PRF. We will present our construction in terms of three algorithms which include a setup algorithm F.setup in addition to F.constrain and F.eval. Our construction builds on the Naor-Reingold DDH-based PRF [26].

4.1   Construction

F.setup($1^\lambda, 1^n$):
The setup algorithm takes as input the security parameter λ and the bit length, n, of PRF inputs. The algorithm runs $\mathcal{G}(1^\lambda, \kappa = n + 1)$ and outputs a sequence of groups $\mathbb{G} = (\mathbb{G}_1, \ldots, \mathbb{G}_\kappa)$ of prime order p, with canonical generators $g_1, \ldots, g_\kappa$, where we let $g = g_1$. It then chooses random exponents $\alpha \in \mathbb{Z}_p$ and $(d_{1,0}, d_{1,1}), \ldots, (d_{n,0}, d_{n,1}) \in \mathbb{Z}_p{}^2$ and computes $D_{i,\beta} = g^{d_{i,\beta}}$ for $i \in [1, n]$ and $\beta \in \{0, 1\}$. The PRF master key k consists of the group sequence $(\mathbb{G}_1, \ldots, \mathbb{G}_\kappa)$ along with $\alpha, d_{i,\beta}$ and $D_{i,\beta}$ for $i \in [1, n]$ and $\beta \in \{0, 1\}$.

The domain $\mathcal{X}$ is $\{0, 1\}^n$ and the range of the function is $\mathbb{G}_\kappa$.[2] Letting x_i denote the i-th bit of $x \in \{0, 1\}^n$, the keyed function is defined as

$$F(k, x) = g_\kappa^{\alpha \prod_{i \in [1,n]} d_{i,x_i}} \in \mathbb{G}_\kappa .$$

F.constrain($k, \mathbf{v}$):
The constrain algorithm takes as input the master key k and a vector $\mathbf{v} \in \{0, 1, ?\}^n$. (Here we use the vector $\mathbf{v}$ to represent the set for which we want to allow evaluation.) Let V be the set of indices $i \in [1, n]$ such that $\mathbf{v}_i \neq ?$. That is the the indices for which the bit is fixed to 0 or 1.

The first component of the constrained key is computed as

$$k'_{\mathbf{v}} = (g_{1+|V|})^{\alpha \prod_{i \in V} d_{i,\mathbf{v}_i}}$$

Note if V is the empty set we interpret the product to be 1. The constrained key $k_{\mathbf{v}}$ consists of $k'_{\mathbf{v}}$ along with $D_{i,\beta} \; \forall i \notin V, \beta \in \{0, 1\}$.

F.eval($k_{\mathbf{v}}, x$):
Again let V be the set of indices $i \in [1, n]$ such that $\mathbf{v}_i \neq ?$. If $\exists i \in V$ such that $x_i \neq \mathbf{v}_i$ the algorithm aborts. If $|V| = n$ then all bits are fixed and the output of the function is $k_{\mathbf{v}}$. Otherwise, using repeated application of the pairing and $D_{i,\beta} \; \forall i \notin V, \beta \in \{0, 1\}$ the algorithm can compute the intermediate value

$$T = (g_{n-|V|})^{\prod_{i \in [1,n] \setminus V} (d_{i,x_i})} .$$

Finally, it computes $e(T, k'_{\mathbf{v}}) = g_\kappa^{\alpha \prod_{i \in [1,n]} d_{i,x_i}} = F(k, x)$.

---

[2] In practice one can use an extractor on the output to produce a bit string.

A few notes. We note that the values $D_{i,\beta} = g^{d_{i,\beta}}$ for $i \in [1, n]$ and $\beta \in \{0, 1\}$ could either be computed in setup and stored or computed as needed during the F.constrain function. As an alternative system one might save storage by utilizing a trusted common setup and make the group description plus the $D_{i,\beta}$ values public. These values would be shared and only the α parameter would be chosen per key. Our proof though will focus solely on the base system described above.

In the full version [8] we show how to map the construction above stated using multilinear maps to the language of graded encodings for which [17,11] provide a candidate instantiation.

4.2   Proof of Security

To show that our bit-fixing construction is secure we show that for an n-bit domain, if the $\kappa = n + 1$-Multilinear Decisional Diffie-Hellman assumption holds then our construction is secure for appropriate choice of the group generator security parameter.

As stated in Section 3 a standard hybrid argument allows us to prove security in a definition where the attacker is allowed a single query x^* to the challenge oracle. Our proof will use the standard complexity leveraging technique of guessing the challenge x^* technique to prove adaptive security. The guess will cause a loss of $1/2^n$ factor in the reduction. An interesting problem is to prove security with only a polynomial factors. The reduction will program all values of $D_{i,\beta}$ to be g_i^c if $x_i = \beta$ and g^{z_i} otherwise for known z_i.

Theorem 3. *If there exists a poly-time attack algorithm $\mathcal{A}$ that breaks our bit-fixing construction n-bit input with advantage $\epsilon(\lambda)$ there there exists a poly-time algorithm $\mathcal{B}$ that breaks the $\kappa = n + 1$-Multilinear Decisional Diffie-Hellman assumption with advantage $\epsilon(\lambda)/2^n$.*

Proof. We show how to construct $\mathcal{B}$. The algorithm $\mathcal{B}$ first receives an $\kappa = n + 1$-MDDH challenge consisting of the group sequence description G and $g = g_1, g^{c_1}, \ldots, g^{c_{\kappa+1}}$ along with T where T is either $g_k^{\prod_{j \in [1, k+1]} c_j}$ or a random group element in $\mathbb{G}_\kappa$. It then chooses a value $x^* \in \{0, 1\}^n$ uniformly at random. Next, it chooses random $z_1, \ldots, z_n$ (internally) sets

$$D_{i,\beta} = \begin{cases} g^{c_i} & \text{if } x_i^* = \beta \\ g^{z_i} & \text{if } x_i^* \neq \beta \end{cases}$$

for $i \in [1, n], \beta \in \{0, 1\}$. This corresponds to setting $d_{i,\beta} = c_i$ if $x_i^* = \beta$ and z_i otherwise. We observe this is distributed identically to the real scheme. In addition, it will internally view $\alpha = c_k \cdot c_{k+1}$.

Constrain Oracle We now describe how the algorithm responds to the key query oracle. Suppose a query is made for a secret key for $\mathbf{v} \in \{0, 1, ?\}^n$. Let V be the set of indices $i \in [1, n]$ such that $\mathbf{v}_i \neq ?$. That is the the indices for which the bit is fixed to 0 or 1. $\mathcal{B}$ identifies an arbitrary $i \in V$ such that $\mathbf{v}_i \neq x_i^*$. If no such i exists this means that the key cannot be produced since it could be used to evaluate $F(k, x^*)$. In this case abort and output a random guess for $\delta' \in \{0, 1\}$.

If the query did not cause an abort, $\mathcal{B}$ first computes $g_2^\alpha = e(g^{ck}, g^{ck+1})$. It then gathers all $D_{j,\mathbf{v}_j}$ for $j \in V/i$. It uses repeated application of the pairing with these values to compute $(g_{1+|V|})^{\alpha \prod_{j \in V/i} d_{j,\mathbf{v}_j}}$. (Recall, our previous assignments to d_j, β.) Finally, it raises this value to $d_{i,\mathbf{v}_I} = z_i$ which is known to the attacker to get. $k'_{vv} = (g_{1+|V|})^{\alpha \prod_{j \in V/i} d_{j,\mathbf{v}_j}}$. The rest of the key is simply the $D_{j,\beta}$ values for $j \notin V, \beta \in \{0, 1\}$.

Evaluate Oracle To handle the evaluation oracle, we observe that the output of $F(k, x)$ for $x \in \{0, 1\}$ is identical to asking a key for $k_{\mathbf{v}=x}$ (a key with no ? symbols. Therefore, queries to this oracle can be handled as secret key queries described above.

Challenge Finally, the attacker can query a challenge oracle once. If the query to this oracle is not equal to x^* then $\mathcal{B}$ randomly guesses $\delta' \in \{0, 1\}$. Otherwise, it outputs T as a response to the oracle query.

The attack algorithm will eventually output a guess b'. If $\mathcal{B}$ has not aborted, it will simply output $\delta' = b'$.

We now analyze the probability that $\mathcal{B}$'s guess $\delta' = \delta$, where δ indicates if T was an MDDH tuple. We have

$$\Pr[\delta' = \delta] = \Pr[\delta' = \delta | \text{abort}] \cdot \Pr[\text{abort}] + \Pr[\delta' = \delta | \overline{\text{abort}}] \cdot \Pr[\overline{\text{abort}}]$$
$$= \frac{1}{2}(1 - 2^{-n}) + \Pr[\delta' = \delta | \overline{\text{abort}}] \cdot (2^{-n})$$
$$= \frac{1}{2}(1 - 2^{-n}) + (\frac{1}{2} + \epsilon(\lambda)) \cdot (2^{-n})$$
$$= \frac{1}{2} + \epsilon(\lambda) \cdot (2^{-n})$$

The set of equations shows that the advantage of $\mathcal{B}$ is $\epsilon(\lambda)2^{-n}$. The second equation is derived since the probability of $\mathcal{B}$ not aborting is 2^{-n}. The third equation comes from the fact that the probability of the attacker winning given a conditioned on not aborting is the same as the original probability of the attacker winning. The reason is that the attacker's success is independent of whether $\mathcal{B}$ guessed x^*. This concludes the proof.

5   Constrained PRFs for Circuit Predicates

Next we build constrained PRFs where the accepting set for a key can be described by a polynomial size circuit. Our construction utilizes the structure used in a recent Attribute-Based Encryption scheme due to Garg, Gentry, Halevi, Sahai, and Waters [18].

We present our circuit construction for constrained PRFs in terms of three algorithms which include a setup algorithm F.setup in addition to F.constrain and F.eval. The setup algorithm will take an additional input ℓ which is the maximum depth of circuits allowed. For simplicity we assume all circuits are depth ℓ and are leveled. We use the same notation for circuits as in [18]. We include the notation in Appendix A for completeness. In addition, like [18] we also build our construction for monotone circuits (limiting ourselves to AND and OR gates); however, we make the standard observation that by pushing NOT gates to the input wires using De Morgan's law we obtain the same result for general circuits.

5.1   Construction

F.setup$(1^\lambda, 1^n, 1^\ell)$:
The setup algorithm takes as input the security parameter λ and the bit length, n, of inputs to the PRF and ℓ the maximum depth of the circuit. The algorithm runs $\mathcal{G}(1^\lambda, \kappa = n + \ell)$ and outputs a sequence of groups $\mathbb{G} = (\mathbb{G}_1, \ldots, \mathbb{G}_\kappa)$ of prime order p, with canonical generators $g_1, \ldots, g_\kappa$, where we let $g = g_1$. It then chooses random exponents $\alpha \in \mathbb{Z}_p$ and $(d_{1,0}, d_{1,1}), \ldots, (d_{n,0}, d_{n,1}) \in \mathbb{Z}_p{}^2$ and computes $D_{i,\beta} = g^{d_{i,\beta}}$ for $i \in [1, n]$ and $\beta \in \{0, 1\}$. The key k consists group sequence $(\mathbb{G}_1, \ldots, \mathbb{G}_\kappa)$ along with $\alpha, d_{i,\beta}$ and $D_{i,\beta}$ for $i \in [1, n]$ and $\beta \in \{0, 1\}$.

The domain $\mathcal{X}$ is $\{0, 1\}^n$ and the range of the function is $\mathbb{G}_\kappa$. Letting x_i denote the i-th bit of $x \in \{0, 1\}^n$, the keyed function is defined as

$$F(k, x) = g_\kappa^{\alpha \prod_{i \in [1, n]} d_{i, x_i}} \in \mathbb{G}_\kappa \quad .$$

F.constrain$(k, \ f = (n, q, A, B, \texttt{GateType}))$:
The constrain algorithm takes as input the key and a circuit description f. The circuit has $n + q$ wires with n input wires, q gates and the wire $n + q$ designated as the output wire.

To generate a constrained key k_f the key generation algorithm chooses random $r_1, \ldots, r_{n+q-1} \in \mathbb{Z}_p$, where we think of the random value r_w as being associated with wire w. It sets $r_{n+q} = \alpha$. The first part of the constrained key is given out as simply all $D_{i,\beta}$ for $i \in [1, n]$ and $\beta \in \{0, 1\}$.

Next, the algorithm generates key components for every wire w. The structure of the key components depends upon if w is an input wire, an OR gate, or an AND gate. We describe how it generates components for each case.

– *Input wire*
  By our convention if $w \in [1, n]$ then it corresponds to the w-th input. The key component is:
  $$K_w = g_2^{r_w d_{w,1}}$$

– *OR gate*
  Suppose that wire $w \in$ Gates and that $\texttt{GateType}(w) = $ OR. In addition, let $j = \texttt{depth}(w)$ be the depth of wire w. The algorithm will choose random $a_w, b_w \in \mathbb{Z}_p$. Then the algorithm creates key components:

  $$K_{w,1} = g^{a_w}, \ K_{w,2} = g^{b_w}, \ K_{w,3} = g_j^{r_w - a_w \cdot r_{A(w)}}, \ K_{w,4} = g_j^{r_w - b_w \cdot r_{B(w)}}$$

- *AND gate*

  Suppose that wire $w \in$ Gates and that $\texttt{GateType}(w) = \text{AND}$. In addition, let $j = \texttt{depth}(w)$ be the depth of wire w. The algorithm will choose random $a_w, b_w \in \mathbb{Z}_p$.

  $$K_{w,1} = g^{a_w}, \quad K_{w,2} = g^{b_w}, \quad K_{w,3} = g_j^{r_w - a_w \cdot r_{A(w)} - b_w \cdot r_{B(w)}}$$

  The constrained key k_f consists of all these $n + q$ key components along with $\{D_{i,\beta}\}$ for $i \in [1,n]$ and $\beta \in \{0,1\}$.

F.eval(k_f, x):

The evaluation algorithm takes as input k_f for circuit $f = (n, q, A, B, \texttt{GateType})$ and an input x. The algorithm first checks that $f(x) = 1$; it not it aborts.

The goal of the algorithm is to compute $F(k, x) = (g_{\kappa = n + \ell})^{\alpha \prod_{i \in [1,n]} d_{i,x_i}}$. We will evaluate the circuit from the bottom up. Consider wire w at depth j; if $f_w(x) = 1$ then, our algorithm will compute $E_w = (g_{j+n})^{r_w \prod_i d_{i,x_i}}$. (If $f_w(x) = 0$ nothing needs to be computed for that wire.) Our decryption algorithm proceeds iteratively starting with computing E_1 and proceeds in order to finally compute E_{n+q}. Computing these values in order ensures that the computation on a depth $j - 1$ wire (that evaluates to 1) will be defined before computing for a depth j wire. Since $r_{n+q} = \alpha$, $E_{n+q} = F(k, x)$.

We show how to compute E_w for all w where $f_w(x) = 1$, again breaking the cases according to whether the wire is an input, AND or OR gate.

- *Input wire*

  By our convention if $w \in [1, n]$ then it corresponds to the w-th input. Suppose that $x_w = f_w(x) = 1$. The algorithm computes $E_w = g_{n+1}^{r_w \prod_i d_{i,x_i}}$. Using the pairing operation successively it can compute $g_{n-1}^{\prod_{i \neq w} d_{i,x_i}}$ from the values $D_{x_i,\beta}$ for $i \in [1,n] \neq w$. It then computes

  $$E_w = e(K_w, g_{n-1}^{\prod_{i \neq w} d_{i,x_i}}) = e(g_2^{r_w d_{w,1}}, g_{n-1}^{\prod_{i \neq w} d_{i,x_i}}) = g_{n+1}^{r_w \prod_i d_{i,x_i}}$$

- *OR gate*

  Consider a wire $w \in$ Gates and that $\texttt{GateType}(w) = \text{OR}$. In addition, let $j = \texttt{depth}(w)$ be the depth of wire w. For exposition we define $D(x) = g_n^{\prod_i d_{i,x_i}}$. This is computable via the pairing operation from $D_{x_i,\beta}$ for $i \in [1,n]$. The computation is performed if $f_w(x) = 1$. If $f_{A(w)}(x) = 1$ (the first input evaluated to 1) then we compute:

  $$E_w = e(E_{A(w)}, K_{w,1}) \cdot e(K_{w,3}, D(x)) =$$

  $$= e((g_{j+n-1})^{r_{A(w)}} \prod_i d_{i,x_i}, g^{a_w}) \cdot e(g_j^{r_w - a_w \cdot r_{A(w)}}, g_n^{\prod_i d_{i,x_i}}) = (g_{j+n})^{r_w g_n^{\prod_i d_{i,x_i}}}$$

  Otherwise, if $f_{A(w)}(x) = 0$, but $f_{B(w)}(x) = 1$, then we compute:

  $$E_w = e(E_{B(w)}, K_{w,2}) \cdot e(K_{w,4}, D(x)) =$$

  $$= e((g_{j+n-1})^{r_{B(w)}} \prod_i d_{i,x_i}, g^{b_w}) \cdot e(g_j^{r_w - b_w \cdot r_{B(w)}}, g_n^{\prod_i d_{i,x_i}}) = (g_{j+n})^{r_w g_n^{\prod_i d_{i,x_i}}}$$

– *AND gate*

Consider a wire $w \in$ Gates and that GateType$(w) =$ AND. In addition, let $j = $ depth(w) be the depth of wire w. Suppose that $f_w(x) = 1$. Then $f_{A(w)}(x) = f_{B(w)}(x) = 1$ and we compute:

$$
\begin{aligned}
E_w &= e(E_{A(w)}, K_{w,1}) \cdot e(E_{B(w)}, K_{w,2}) \cdot e(K_{w,3}, D(x)) \\
&= e\big((g_{j+n-1})^{r_{A(w)} \prod_i d_{i,x_i}}, g^{a_w}\big) \cdot e\big((g_{j+n-1})^{r_{B(w)} \prod_i d_{i,x_i}}, g^{b_w}\big) \cdot \\
&\qquad \cdot e\big(g_j^{r_w - a_w \cdot r_{A(w)} - c_w \cdot r_{B(w)}}, g_n^{\prod_i d_{i,x_i}}\big) \\
&= (g_{j+n})^{r_w \prod_i d_{i,x_i}}
\end{aligned}
$$

The procedures above are evaluated in order for all w for which $f_w(x) = 1$. The final output gives $E_{n+q} = F(k, x)$.

5.2   Proof of Security

We now prove security of the circuit constrained construction. We show that for an n-bit domain and circuits of depth ℓ, if the $\kappa = n + \ell$-Multilinear Decisional Diffie-Hellman assumption holds then our construction is secure for appropriate choice of the group generator security parameter.

Our proof begins as in the bit-fixing proof where a where we use the standard complexity leveraging technique of guessing the challenge x^* ahead of time to prove adaptive security. The guess will cause a loss of $1/2^n$ factor in the reduction. The delegate oracle queries, however, are handled quite differently.

Theorem 4. *If there exists a poly-time attack algorithm $\mathcal{A}$ that breaks our circuit constrained construction n-bit input and circuits of depth ℓ with advantage $\epsilon(\lambda)$ there there exists a poly-time algorithm $\mathcal{B}$ that breaks the $\kappa = n + \ell$-Multilinear Decisional Diffie-Hellman assumption with advantage $\epsilon(\lambda)/2^n$.*

Due to space constraints the proof appears in the full version of the paper [8].

6   Applications

Having constructed constrained PRFs for several predicate families we now explore a number of remarkable applications for these concepts. Our primary goal is to demonstrate the versatility and general utility of constrained PRFs.

6.1   Broadcast Encryption with Optimal Ciphertext Length

We start by showing that a bit-fixing constrained PRF leads a broadcast encryption system with *optimal* ciphertext size. Recall that a broadcast encryption system [15] is made up of three randomized algorithms:

Setup(λ, n). Takes as input the security parameter λ and the number of receivers n. It outputs n private keys $d_1, \ldots, d_n$ and a broadcaster key bk. For $i = 1, \ldots, n$, recipient number i is given the private key d_i.

***Encrypt*(bk, S).** Takes as input a subset $S \subseteq \{1, \ldots, n\}$, and the broadcaster's key bk. It outputs a pair (hdr, k) where hdr is called the header and $k \in \mathcal{K}$ is a message encryption key chosen from the key space $\mathcal{K}$. We will often refer to hdr as the broadcast ciphertext.

Let m be a message to be broadcast that should be decipherable precisely by the receivers in S. Let c_m be the encryption of m under the symmetric key k. The broadcast data consists of (S, hdr, c_m). The pair (S, hdr) is often called the full header and c_m is often called the broadcast body.

***Decrypt*($i, d_i, S,$ hdr).** Takes as input a subset $S \subseteq \{1, \ldots, n\}$, a user id $i \in \{1, \ldots, n\}$ and the private key d_i for user i, and a header hdr. If $i \in S$ the algorithm outputs a message encryption key $k \in \mathcal{K}$. Intuitively, user i can then use k to decrypt the broadcast body c_m and obtain the message m.

In what follows the broadcaster's key bk is a secret key known only to the broadcaster and hence our system is a secret-key broadcast encryption.

The **length efficiency** of a broadcast encryption system is measured in the length of the header hdr. The shorter the header the more efficient the system. Remarkably, some systems such as [6,13,12,7,30] achieve a fixed size header that depends only on the security parameter and is independent of the size of the recipient set S.

As usual, we require that the system be correct, namely that for all subsets $S \subseteq \{1, \ldots, n\}$ and all $i \in S$ if $(\mathsf{bk}, (d_1, \ldots, d_n)) \xleftarrow{R} Setup(n)$ and $(\mathsf{hdr}, k) \xleftarrow{R} Encrypt(\mathsf{bk}, S)$ then $Decrypt(i, d_i, S, \mathsf{hdr}) = k$.

A broadcast encryption system is said to be semantically secure if an adaptive adversary $\mathcal{A}$ that obtains recipient keys d_i for $i \in S$ of its choice, cannot break the semantic security of a broadcast ciphertext intended for a recipient set S^* in the complement of S, namely $S^* \subseteq [n] \setminus S$. More precisely, security is defined using the following experiment, denoted $\mathrm{EXP}(b)$, parameterized by the total number of recipients n and by a bit $b \in \{0, 1\}$:

$(\mathsf{bk}, (d_1, \ldots, d_n)) \xleftarrow{R} Setup(\lambda, n)$

$b' \leftarrow \mathcal{A}^{\mathsf{RK}(\cdot), \mathsf{SK}(\cdot), \mathsf{RoR}(b, \cdot, \cdot)}(\lambda, n)$

where

$\quad$ $\mathsf{RK}(i)$ is a recipient key oracle that takes as input $i \in [n]$ and returns d_i,

$\quad$ $\mathsf{SK}(S)$ takes as input $S \subseteq [n]$ and returns $Encrypt(\mathsf{bk}, S)$, and

$\quad$ $\mathsf{RoR}(b, S^*)$ is a real-or-random oracle: it takes as input $b \in \{0, 1\}$ and $\quad\quad S^* \subseteq [n]$, computes $(\mathsf{hdr}, k_0) \xleftarrow{R} Encrypt(\mathsf{bk}, S^*)$ and $k_1 \xleftarrow{R} \mathcal{K}$, $\quad\quad$ and returns (hdr, k_b).

We require that all sets S^* given as input to oracle RoR are distinct from all sets S given as input to SK and that S^* does not contain any index i given as input to RK. For $b = 0, 1$ let W_b be the event that $b' = 1$ in $\mathrm{EXP}(b)$ and as usual define $\mathrm{AdvBE}_{\mathcal{A}}(\lambda) = |\Pr[W_0] - \Pr[W_1]|$.

Definition 2. *We say that a broadcast encryption is semantically secure if for all probabilistic polynomial time adversaries $\mathcal{A}$ the function $AdvBE_{\mathcal{A}}(\lambda)$ is negligible.*

An length-optimal broadcast encryption construction. A bit-fixing PRF such as the one constructed in Section 4 gives a broadcast encryption system with optimal ciphertext length. Specifically, the header size is always 0 for all recipient sets $S \subseteq [n]$. The system, denoted BE_F works as follows:

Setup(λ, n): Let $F : \mathcal{K} \times \{0,1\}^n \rightarrow \mathcal{Y}$ be a secure bit-fixing constrained PRF. Choose a random key $\mathsf{bk} \xleftarrow{R} \mathcal{K}$ and for $i = 1, \ldots, n$ compute

$$d_i \leftarrow F.\mathsf{constrain}(\mathsf{bk},\ p_i)$$

where $p_i : \{0,1\}^n \rightarrow \{0,1\}$ is the bit-fixing predicate satisfying $p_i(x) = 1$ iff $x_i = 1$. Thus, the key d_i enables the evaluation of $F(\mathsf{bk}, x)$ at any point $x \in \{0,1\}^n$ for which $x_i = 1$. Output $(\mathsf{bk}, (d_1, \ldots, d_n))$.

Encrypt(bk, S): Let $x \in \{0,1\}^n$ be the characteristic vector of S and compute $k \leftarrow F(\mathsf{bk}, x)$. Output the pair (hdr, k) where $\mathsf{hdr} = \epsilon$. That is, the output header is simply the empty string.

Decrypt(i, d_i, S, hdr): Let $x \in \{0,1\}^n$ be the characteristic vector of S. If $i \in S$ then the bit-fixing predicate p_i satisfies $p_i(x) = 1$. Therefore, d_i can be used to compute $F(\mathsf{bk}, x)$, as required.

Theorem 5. *BE_F is a semantically secure broadcast encryption system against adaptive adversaries assuming that the underlying constrained bit-fixing PRF is secure.*

Proof. Security follows immediately from the security of the bit-fixing PRF. Specifically, oracle RK in the broadcast encryption experiment is implemented using oracle $F.\mathsf{constrain}$ in the constrained security game (Section 3.2). Oracle SK is implemented using oracle $F.\mathsf{eval}$ in the constrained security game. Finally, the broadcast encryption real-or-random oracle is the same as the $\mathsf{Challenge}$ oracle in the constrained security game. Therefore, an attacker who succeeds in breaking semantic security of the broadcast encryption system will break security of the bit-fixing PRF.

Comparison to existing fully collusion resistant schemes. While our primary goal is to illustrate applications of abstract constrained PRFs, it is instructive to examine the specific broadcast system that results from instantiating the system above with the bit-fixing PRF in Section 4. We briefly compare this system to existing broadcast encryption systems such as [6,13,12,30]. These existing systems are built from bilinear maps, they allow the broadcaster's key to be public, and the broadcast header contains a constant number of group elements. The benefit of the instantiated system above is that the header length is smaller: its length is zero. However, the system uses multi-linear maps and the broadcaster's key is secret. The system is closely related to the multilinear-based broadcast system of Boneh and Silverberg [7] which has similar parameters. To re-iterate, our goal is to show the general utility of constrained PRFs. Nevertheless, we hope that future constrained PRFs will lead to new families of broadcast systems.

6.2 Identity-Based Key Exchange

Next, we show that a left/right constrained PRF directly implies an identity-based non-interactive key exchange (ID-NIKE) system [31,14,27,16]. Recall that such a system is made up of three algorithms:

- $Setup(\lambda)$ outputs public parameters pp and a master secret msk,
- $Extract(\mathsf{msk}, \mathsf{id})$ generates a secret key sk_id for identity id, and
- $KeyGen(\mathsf{pp}, \mathsf{sk}_\mathsf{id}, \mathsf{id}')$ outputs a shared key $k_{\mathsf{id},\mathsf{id}'}$.

For correctness we require that $KeyGen(\mathsf{pp}, \mathsf{sk}_\mathsf{id}, \mathsf{id}') = KeyGen(\mathsf{pp}, \mathsf{sk}_{\mathsf{id}'}, \mathsf{id})$ for all $\mathsf{id} \neq \mathsf{id}'$ and pp generated by $Setup$.

Briefly, the security requirement, defined by Dupont and Enge [14] and further refined by Paterson and Srinivasan [27], is that an adversary $\mathcal{A}$ who obtains secret keys sk_id for all identities $\mathsf{id} \in S$ for a set S of his choice, cannot distinguish the shared key $k_{\mathsf{id}_*, \mathsf{id}'_*}$ from random for identities $\mathsf{id}_*, \mathsf{id}'_* \notin S$ of his choice. The adversary may also ask to reveal the shared key $k_{\mathsf{id}, \mathsf{id}'}$ for any pair of identities $(\mathsf{id}, \mathsf{id}') \neq (\mathsf{id}_*, \mathsf{id}'_*)$.

Identity-based key exchange from left/right constrained PRFs. The system works as follows:

- $Setup(\lambda)$: let $F : \mathcal{K} \times \mathcal{X}^2 \to \mathcal{Y}$ be a secure left/right constrained PRF. Choose a random $\mathsf{msk} \xleftarrow{R} \mathcal{K}$ and output msk. The public parameters pp are the (optional) public parameters of the PRF.
- $Extract(\mathsf{msk}, \mathsf{id})$: compute $d_\mathrm{L} = F.\mathsf{constrain}(\mathsf{msk}, p_\mathsf{id}^{(\mathrm{L})})$ and $d_\mathrm{R} = F.\mathsf{constrain}(\mathsf{msk}, p_\mathsf{id}^{(\mathrm{R})})$. Output $\mathsf{sk}_\mathsf{id} = (d_\mathrm{L}, d_\mathrm{R})$.
- $KeyGen(\mathsf{sk}_\mathsf{id}, \mathsf{id}')$: We assume that the identity strings are lexicographically ordered. Output $k_{\mathsf{id},\mathsf{id}'} = F(\mathsf{msk}, (\mathsf{id}, \mathsf{id}'))$ if $\mathsf{id} < \mathsf{id}'$ using d_L. Output $k_{\mathsf{id},\mathsf{id}'} = F(\mathsf{msk}, (\mathsf{id}', \mathsf{id}))$ if $\mathsf{id} > \mathsf{id}'$ using d_R. By definition of a left/right constrained PRF, both values can be computed just given sk_id.

Correctness of the system follows directly from the correctness of the constrained PRF and lexicographic convention. Security again follows directly from the security definition of a constrained PRF. Oracle $F.\mathsf{constrain}$ in the constrained security game (Section 3.2) enables the adversary $\mathcal{A}$ to request the secret keys for any set of identities S of her choice. Oracle $F.\mathsf{eval}$ enables the adversary $\mathcal{A}$ to reveal the shared key $k_{\mathsf{id},\mathsf{id}'}$ for any pair of identities $(\mathsf{id}, \mathsf{id}')$. If $\mathcal{A}$ could then distinguish $F(\mathsf{msk}, (\mathsf{id}_*, \mathsf{id}'_*))$ from random for some $\mathsf{id}_*, \mathsf{id}'_* \notin S$ and for which reveal was not called then she would solve the challenge in the constrained security game.

Comparison to existing ID-NIKE. While our primary goal is to explore applications of general constrained PRFs, it is instructive to examine the specific ID-NIKE systems obtained by instantiating the ID-NIKE above with our specific PRFs. The first concrete ID-NIKE is obtained from the left/right constrained PRF in Eq. (1). This ID-NIKE is identical to the Sakai-Ohgishi-Kasahara [31] ID-NIKE which was analyzed in [14,27]. A second ID-NIKE is obtained by using the bit-fixing constrained PRF in Section 4 as a left/right constrained PRF. The resulting ID-NIKE is related to a recent ID-NIKE due to Freire et al. [16] which is the first ID-NIKE proven secure in the standard model. While *KeyGen* in our instantiated ID-NIKE uses fewer group elements than [16], we achieve adaptive security via complexity leveraging which forces our multilinear groups to be substantially larger. This likely results in an overall less efficient ID-NIKE when compared to [16].

As stated above, our primary goal here is to explore the power of constrained PRFs. We hope that future constrained PRFs, especially ones built from the learning with errors (LWE) assumption, will give new ID-NIKE systems.

6.3   Policy-Based Key Distribution

More generally, our constrained PRF construction for circuit predicates (Section 5) gives rise to a powerful non-interactive group key distribution mechanism.

Suppose each user in the system is identified by a vector id $\in \{0,1\}^n$ that encodes a set of attributes for that user. Our goal is that for any predicate $p : \{0,1\}^n \to \{0,1\}$, users whose id satisfies $p(\text{id}) = 1$ will be able to compute a shared key k_p. However, a coalition of users for which $p(\text{id}) = 0$ for all members of the coalition learns nothing about k_p. We call this mechanism *non-interactive policy-based key exchange* (PB-NIKE) since only those users whose set of attributes satisfies the policy p are able to compute the shared key k_p.

For example, consider the policy p that is true for users who are members of the IACR and have a driver's license. All such users will be able to derive the policy shared key k_p, but to all other users the key k_p will be indistinguishable from random. This k_p can then be used for secure communication among the group members. This functionality is related to the concept of Attribute-Based Encryption [29,22].

We implement policy-based key agreement using a constrained PRF $F : \mathcal{K} \times \{0,1\}^m \to \mathcal{Y}$ for circuit predicates. To do so, let $U(\cdot, \cdot)$ denote a universal circuit that takes two inputs: an identity id $\in \{0,1\}^n$ and an m-bit description of a circuit for a predicate $p : \{0,1\}^n \to \{0,1\}$. The universal circuit $U(\text{id}, p)$ is defined as:

$$U(\text{id}, p) = p(\text{id}) \in \{0,1\}$$

We define the secret key sk_{id} given to user id to be the constrained PRF key that lets user id evaluate $F(\mathsf{msk}, p)$ for all p for which $U(\text{id}, p) = p(\text{id}) = 1$. Thus, users whose set of attributes id satisfies $p(\text{id}) = 1$ can compute the policy key $k_p = F(\mathsf{msk}, p)$ using their secret key sk_{id}. All other users cannot.

In more detail, the system works as follows:

- *Setup*(λ) : let $F : \mathcal{K} \times \{0,1\}^m \to \mathcal{Y}$ be a secure constrained PRF for circuit predicates. The master secret msk is chosen as a random key in $\mathcal{K}$.
- *Extract*$(\mathsf{msk}, \mathsf{id})$: output $\mathsf{sk}_{\mathsf{id}} = F.\mathsf{constrain}\big(\mathsf{msk}, \ U(\mathsf{id}, \cdot) \ \big)$. By definition, this key $\mathsf{sk}_{\mathsf{id}}$ enables the evaluation of $F(\mathsf{msk}, p)$ at all p such that $U(\mathsf{id}, p) = p(\mathsf{id}) = 1$, as required.

The properties of F imply that for any predicate p (whose description is at most m bits), the group key $k_p = F(\mathsf{msk}, p)$ can be computed by any user whose id satisfies $p(\mathsf{id}) = 1$. Moreover, the security property for constrained PRFs implies that a coalition of users for which $p(\mathsf{id}) = 0$ for all members of the coalition cannot distinguish k_p from random.

7 Extensions and Open Problems

We constructed constrained PRFs for several natural predicate families and showed applications for all these constructions. Here we point out a few possible directions for future research.

First, it would be interesting to generalize the constrained concept to allow for multiple levels of delegation. That is, the master key for the PRF can be used to derive a constrained key k_S for some set $S \subset \mathcal{X}$. That key k_S can be used in turn to derive a further constrained key k'_S for some subset $S' \subset S$, and so on. This concept is in similar spirit to Hierarchical IBE [23,19,10] or delegation in ABE [22]. For the GGM prefix system, this is straightforward. Some of our constructions, such as the bit fixing PRF, extend naturally to support more than one level of delegation while others do not.

Second, for the most interesting predicate families our constructions are based on multilinear maps. It would be quite useful to provide constructions based on other assumptions such as Learning With Errors (LWE) or simple bilinear maps.

Acknowledgments. Dan Boneh is supported by NSF, the DARPA PROCEED program, an AFO SR MURI award, a grant from ONR, an IARPA project provided via DoI/NBC, and by a Google faculty research award.

Brent Waters is supported by NSF CNS-0915361, CNS-0952692, CNS-1228599, DARPA N00014-11-1-0382, DARPA N11AP20006, Google Faculty Research award, the Alfred P. Sloan Fellowship, Microsoft Faculty Fellowship, and Packard Foundation Fellowship.

Opinions, findings and conclusions or recommendations expressed in this material are those of the author(s) and do not necessarily reflect the views of DARPA or IARPA.

References

1. Bellare, M., Fuchsbauer, G.: Policy-based signatures. Cryptology ePrint Archive, Report 2013/413 (2013)

2. Bellare, M., Hoang, V.T., Rogaway, P.: Foundations of garbled circuits. In: ACM Conference on Computer and Communications Security, pp. 784–796 (2012)
3. Bellare, M., Rogaway, P.: Random oracles are practical: A paradigm for designing efficient protocols. In: ACM Conference on Computer and Communications Security, pp. 62–73 (1993)
4. Boneh, D., Boyen, X.: Efficient selective-ID secure identity-based encryption without random oracles. In: Cachin, C., Camenisch, J.L. (eds.) EUROCRYPT 2004. LNCS, vol. 3027, pp. 223–238. Springer, Heidelberg (2004)
5. Boneh, D., Franklin, M.K.: Identity-based encryption from the weil pairing. SIAM J. Comput. 32(3), 586–615 (2001); Extended abstract in Kilian, J. (ed.) CRYPTO 2001. LNCS, vol. 2139, p. 213–229. Springer, Heidelberg (2001)
6. Boneh, D., Gentry, C., Waters, B.: Collusion resistant broadcast encryption with short ciphertexts and private keys. In: Shoup, V. (ed.) CRYPTO 2005. LNCS, vol. 3621, pp. 258–275. Springer, Heidelberg (2005)
7. Boneh, D., Silverberg, A.: Applications of multilinear forms to cryptography. Contemporary Mathematics 324, 71–90 (2003)
8. Boneh, D., Waters, B.: Constrained pseudorandom functions and their applications. Cryptology ePrint Archive, Report 2013/352 (2013)
9. Boyle, E., Goldwasser, S., Ivan, I.: Functional signatures and pseudorandom functions. Cryptology ePrint Archive, Report 2013/401 (2013)
10. Canetti, R., Halevi, S., Katz, J.: A forward-secure public-key encryption scheme. In: Biham, E. (ed.) EUROCRYPT 2003. LNCS, vol. 2656, pp. 255–271. Springer, Heidelberg (2003)
11. Coron, J.-S., Lepoint, T., Tibouchi, M.: Practical multilinear maps over the integers. Cryptology ePrint Archive, Report 2013/183 (2013)
12. Delerablée, C.: Identity-based broadcast encryption with constant size ciphertexts and private keys. In: Kurosawa, K. (ed.) ASIACRYPT 2007. LNCS, vol. 4833, pp. 200–215. Springer, Heidelberg (2007)
13. Delerablée, C., Paillier, P., Pointcheval, D.: Fully collusion secure dynamic broadcast encryption with constant-size ciphertexts or decryption keys. In: Takagi, T., Okamoto, T., Okamoto, E., Okamoto, T. (eds.) Pairing 2007. LNCS, vol. 4575, pp. 39–59. Springer, Heidelberg (2007)
14. Dupont, R., Enge, A.: Provably secure non-interactive key distribution based on pairings. Discrete Applied Mathematics 154(2), 270–276 (2006)
15. Fiat, A., Naor, M.: Broadcast encryption. In: Stinson, D.R. (ed.) CRYPTO 1993. LNCS, vol. 773, pp. 480–491. Springer, Heidelberg (1994)
16. Freire, E.S.V., Hofheinz, D., Paterson, K.G., Striecks, C.: Programmable hash functions in the multilinear setting. In: Canetti, R., Garay, J.A. (eds.) CRYPTO 2013, Part I. LNCS, vol. 8042, pp. 513–530. Springer, Heidelberg (2013)
17. Garg, S., Gentry, C., Halevi, S.: Candidate multilinear maps from ideal lattices. In: Johansson, T., Nguyen, P.Q. (eds.) EUROCRYPT 2013. LNCS, vol. 7881, pp. 1–17. Springer, Heidelberg (2013)
18. Garg, S., Gentry, C., Halevi, S., Sahai, A., Waters, B.: Attribute-based encryption for circuits from multilinear maps. In: Canetti, R., Garay, J.A. (eds.) CRYPTO 2013, Part II. LNCS, vol. 8043, pp. 479–499. Springer, Heidelberg (2013)
19. Gentry, C., Silverberg, A.: Hierarchical ID-based cryptography. In: Zheng, Y. (ed.) ASIACRYPT 2002. LNCS, vol. 2501, pp. 548–566. Springer, Heidelberg (2002)
20. Goldreich, O., Goldwasser, S., Micali, S.: How to construct random functions. J. ACM 34(4), 792–807 (1986)

21. Choudary Gorantla, M., Boyd, C., Nieto, J.M.G.: Attribute-based authenticated key exchange. In: Steinfeld, R., Hawkes, P. (eds.) ACISP 2010. LNCS, vol. 6168, pp. 300–317. Springer, Heidelberg (2010)
22. Goyal, V., Pandey, O., Sahai, A., Waters, B.: Attribute-based encryption for fine-grained access control of encrypted data. In: ACM Conference on Computer and Communications Security, pp. 89–98 (2006)
23. Horwitz, J., Lynn, B.: Toward hierarchical identity-based encryption. In: Knudsen, L.R. (ed.) EUROCRYPT 2002. LNCS, vol. 2332, pp. 466–481. Springer, Heidelberg (2002)
24. Kiayias, A., Papadopoulos, S., Triandopoulos, N., Zacharias, T.: Delegatable pseudorandom functions and applications. In: Proceedings ACM CCS (2013)
25. Naor, D., Naor, M., Lotspiech, J.: Revocation and tracing schemes for stateless receivers. In: Kilian, J. (ed.) CRYPTO 2001. LNCS, vol. 2139, pp. 41–62. Springer, Heidelberg (2001)
26. Naor, M., Reingold, O.: Number-theoretic constructions of efficient pseudo-random functions. In: FOCS 1997, pp. 458–467 (1997)
27. Paterson, K., Srinivasan, S.: On the relations between non-interactive key distribution, identity-based encryption and trapdoor discrete log groups. Des. Codes Cryptography 52(2), 219–241 (2009)
28. Regev, O.: On lattices, learning with errors, random linear codes, and cryptography. In: Proc. of STOC 2005, pp. 84–93 (2005)
29. Sahai, A., Waters, B.: Fuzzy identity-based encryption. In: Cramer, R. (ed.) EUROCRYPT 2005. LNCS, vol. 3494, pp. 457–473. Springer, Heidelberg (2005)
30. Sakai, R., Furukawa, J.: Identity-based broadcast encryption. Cryptology ePrint Archive, Report 2007/217 (2007)
31. Sakai, R., Ohgishi, K., Kasahara, M.: Cryptosystems based on pairing. In: SCIS (2000)

A   Circuit Notation

We now define our notation for circuits that adapts the model and notation of Bellare, Hoang, and Rogaway [2] (Section 2.3). For our application we restrict our consideration to certain classes of boolean circuits. First, our circuits will have a single output gate. Next, we will consider layered circuits. In a layered circuit a gate at depth j will receive both of its inputs from wires at depth $j - 1$. Finally, we will restrict ourselves to monotonic circuits where gates are either AND or OR gates of two inputs. [3]

Our circuits will be a five tuple $f = (n, q, A, B, \texttt{GateType})$. We let n be the number of inputs and q be the number of gates. We define inputs $= \{1, \ldots, n\}$, Wires $= \{1, \ldots, n + q\}$, and Gates $= \{n + 1, \ldots, n + q\}$. The wire $n + q$ is the designated output wire. $A :$ Gates $\rightarrow$ Wires/outputwire is a function where $A(w)$ identifies w's first incoming wire and $B :$ Gates $\rightarrow$ Wires/outputwire is a function where $B(w)$ identifies w's second incoming wire. Finally, $\texttt{GateType} :$

---

[3] These restrictions are mostly useful for exposition and do not impact functionality. General circuits can be built from non-monotonic circuits. In addition, given a circuit an equivalent layered exists that is larger by at most a polynomial factor.

Gates $\to$ {AND, OR} is a function that identifies a gate as either an AND or OR gate.

We require that $w > B(w) > A(w)$. We also define a function $\mathtt{depth}(w)$ where if $w \in$ inputs $\mathtt{depth}(w) = 1$ and in general $\mathtt{depth}(w)$ of wire w is equal to the shortest path to an input wire plus 1. Since our circuit is layered we require that for all $w \in$ Gates that if $\mathtt{depth}(w) = j$ then $\mathtt{depth}(A(w)) = \mathtt{depth}(B(w)) = j-1$.

We will abuse notation and let $f(x)$ be the evaluation of the circuit f on input $x \in \{0,1\}^n$. In addition, we let $f_w(x)$ be the value of wire w of the circuit on input x.

Fully Homomorphic Message Authenticators

Rosario Gennaro[1,*] and Daniel Wichs[2,**]

[1] City College, CUNY
[2] Northeastern University

Abstract. We define and construct a new primitive called a *fully homomorphic message authenticator*. With such scheme, anybody can perform arbitrary computations over authenticated data and produce a *short* tag that authenticates the result of the computation (without knowing the secret key). This tag can be verified using the secret key to ensure that the claimed result is indeed the correct output of the specified computation over previously authenticated data (without knowing the underlying data). For example, Alice can upload authenticated data to "the cloud", which then performs some specified computations over this data and sends the output to Bob, along with a short tag that convinces Bob of correctness. Alice and Bob only share a secret key, and Bob never needs to know Alice's underlying data. Our construction relies on fully homomorphic encryption to build fully homomorphic message authenticators.

1 Introduction

The rise of the *cloud computing* paradigm requires that users can securely outsource their data to a remote service provider while allowing it to reliably perform computations over the data. The recent ground-breaking development of *fully homomorphic encryption* [24] allows us to maintain *confidentiality/privacy* of outsourced data in this setting. In this work, we look at the analogous but orthogonal question of providing *integrity/authenticity* for computations over outsourced data. In particular, if a remote server claims that the execution of some program $\mathcal{P}$ over the user's outsourced data results in an output y, how can the user be sure that this is indeed the case?

More generally, we can consider a group of mutually-trusting users that share a secret key – each user can authenticate various data items at various times (without keeping state) and upload the authenticated data to an untrusted cloud. The cloud should be able to perform a joint computation over various data of several users and convince any user in the group of the validity of the result.

Toward this goal, we define and instantiate a new primitive, called a *fully homomorphic message authenticator*. This primitive can be seen as a *symmetric-key* version of fully homomorphic *signatures*, which were defined by Boneh and Freeman [10], but whose construction remains an open problem. We will

---

* rosario@cs.ccny.cuny.edu. Work done while at the IBM Research, T.J.Watson.
** wichs@ccs.neu.edu. Work done while at IBM Research, T.J.Watson.

K. Sako and P. Sarkar (Eds.) ASIACRYPT 2013, Part II, LNCS 8270, pp. 301–320, 2013.

return to survey the related work on partially homomorphic signatures and authenticators, as well as related work on *delegating memory and computation*, in Section 1.3. First, we describe our notion of fully homomorphic message authenticators, which will be the focus of this work.

1.1   What Are Homomorphic Message Authenticators?

Simplified Description. In a homomorphic message-authenticator scheme, Alice can authenticate some large data D using her secret key sk. Later, anybody can homomorphically execute an arbitrary program $\mathcal{P}$ over the authenticated data to produce a short tag ψ (without knowing sk), which certifies the value $y = \mathcal{P}(D)$ as the output of $\mathcal{P}$. It is important that ψ does *not* simply authenticate y out of context; it only certifies y as the output of a specific program $\mathcal{P}$. Another user Bob, who shares the secret key sk with Alice, can verify the triple $(y, \mathcal{P}, \psi)$ to ensure that y is indeed the output of the program $\mathcal{P}$ evaluated on Alice's previously authenticated data D (without knowing D). The tag ψ should be *succinct*, meaning that its size is independent of the size of the data D or the complexity of the program $\mathcal{P}$. In other words, homomorphic message authenticators allow anyone to certify the output of a complex computation over a large authenticated data with only a short tag.

Labeled Data and Programs. The above high-level description considers a restricted scenario where a single user Alice authenticates a single large data D in one shot. We actually consider a more general setting where many users, who share a secret key, can authenticate various data-items (say, many different files) at different times without keeping any local or joint state. In this setting, we need to establish some syntax for specifying *which* data is being authenticated and which data a program $\mathcal{P}$ should be evaluated on. For this purpose, we rely on the notion of *labeled* data and programs.

Whenever the user wants to authenticate some data-item D, she chooses a *label* τ for it, and the authentication algorithm authenticates the data D with respect to the label τ. For example, the label τ could be a file name. For the greatest level of granularity, we will assume that the user authenticates individual *bits* of data separately. Each bit b is authenticated with respect to its own label τ via a secretly-keyed authentication algorithm $\sigma \leftarrow \mathsf{Auth}_{sk}(b, \tau)$. For example, to authenticate a long file named "salaries" containing the data-bits $D = (b_1, \ldots, b_t)$, the user can create separate labels $\tau_i = (\text{"salaries"}, i)$ to denote the ith bit of the file, and then authenticate each bit b_i of the file under the label τ_i. Our scheme is oblivious to how the labels for each bit are chosen and whether they have any meaningful semantics.

Correspondingly, we consider *labeled programs* $\mathcal{P}$, where each input bit of the program has an associated label τ_i indicating which data it should be evaluated on. For example, a labeled-program $\mathcal{P}$ meant to compute the *median* of the "salaries" data would have its input bits labeled by $\tau_i = (\text{"salaries"}, i)$. In general, the description of the labeled program $\mathcal{P}$ could be as long as, or even longer than, the input data itself. However, as in the above example, we envision the typical use-case to be one where $\mathcal{P}$ has some succinct description, such as computing the "median of the salaries data". We note that a labeled program can compute over data authenticated by different users at different times, as long as it was authenticated with the same shared secret key,

Homomorphic authenticators allow us to certify the output of a labeled program, given authentication tags for correspondingly labeled input data. In particular, there is a public homomorphic evaluation algorithm $\psi = \mathsf{Eval}(\mathcal{P}, \sigma_1, \ldots, \sigma_t)$ that takes as input tags σ_i authenticating some data-bits b_i with respect to some labels τ_i, and a labeled program $\mathcal{P}$ with matching input labels $\tau_1, \ldots, \tau_t$. It outputs a tag ψ that certifies the value $y = \mathcal{P}(b_1, \ldots, b_t)$ as the correct output of the program $\mathcal{P}$. The verification algorithm $\mathsf{Ver}_{sk}(y, \mathcal{P}, \psi)$ uses the secret key sk to verify that y is indeed the output of the labeled program $\mathcal{P}$ on previously authenticated labeled input-data, without needing to know the original data.

Composition. Our homomorphic authenticators are also *composable* so that we can incrementally combine authenticated outputs of partial computations to derive an authenticated output of a larger computation. In particular, if the tags $\psi_1, \ldots, \psi_t$ authenticate some bits $b_1, \ldots, b_t$ as the outputs of some labeled programs $\mathcal{P}_1, \ldots, \mathcal{P}_t$ respectively, then $\psi^* = \mathsf{Eval}(\mathcal{P}^*, \psi_1, \ldots, \psi_t)$ should authenticate the bit $b^* = \mathcal{P}^*(b_1, \ldots, b_t)$ as the output of the *composed program*, which first evaluates $\mathcal{P}_1, \ldots, \mathcal{P}_t$ on the appropriately labeled authenticated data, and then runs $\mathcal{P}^*$ on the outputs.

Succinct Tags vs. Efficient Verification. The main requirement that makes our definition of homomorphic authenticators interesting is that the tags should be *succinct*. Otherwise, there is a trivial solution where we can authenticate the output of a computation $\mathcal{P}$ by simply providing all of its input bits and their authentication tags. The succinctness requirement ensures that we can certify the output of a computation $\mathcal{P}$ over authenticated data with much smaller *communication* than that of simply transmitting the input data.[1] Therefore, this primitive is especially useful when verifying computations that read a lot of input data but have a short output (e.g., computing the median in a large database).

However, we note that the verification algorithm in a homomorphic authenticator schemes is allowed to have a large *computational complexity*, proportional to the complexity of the computation $\mathcal{P}$ being verified. Therefore, although homomorphic authenticators allow us to save on *communication*, they do not necessarily save on the *computational complexity* of verifying computations over outsourced data. We believe that communication-efficient solutions are already interesting, and may be useful, even without the additional constraint of computational efficiency. In Section 4, we explore how to combine our communication-efficient homomorphic authenticators with techniques from *delegating computation* to also achieve computationally efficient verification.

---

[1] Note that we do not count the cost of transmitting the labeled-program $\mathcal{P}$ itself. As a previous note explains, we envision that in the typical use-case such programs should have a succinct description.

1.2   Overview of Our Construction

Our construction of fully homomorphic authenticators relies on the use of fully homomorphic encryption (FHE). Let n denote the *security parameter*, and define $[n] \stackrel{\text{def}}{=} \{1, \ldots, n\}$. The secret key of the authenticator consists of: a random *verification set* $S \subseteq [n]$ of size $|S| = n/2$, a key-pair (pk, sk) for a fully homomorphic encryption (FHE) scheme and a pseudo-random function (PRF) $f_K(\cdot)$.

To authenticate a bit b under a label τ, Alice creates n ciphertexts $c_1, \ldots, c_n$ as follows. For $i \in [n] \setminus S$, she chooses the ciphertexts $c_i \leftarrow \mathsf{Enc}_{pk}(b)$ as random encryptions of the actual bit b being authenticated. For $i \in S$, she computes $c_i = \mathsf{Enc}_{pk}(0; f_K((i, \tau)))$ as *pseudorandom encryptions of 0*, where the random coins are derived using the PRF. Notice that for the indices $i \in S$ in the verification set, the pseudorandom ciphertexts c_i can be easily re-computed from Alice's secret key and the label τ alone, *without knowing the data* bit b. Alice outputs the *authentication tag* $\sigma = (c_1, \ldots, c_n)$, consisting of the n ciphertexts.

Given some program $\mathcal{P}$ with t input-labels and t authentication tags $\{\sigma_j = (c_{1,j}, \ldots, c_{n,j})\}_{j \in [t]}$ for the correspondingly labeled data, we can homomorphically derive an authentication tag $\psi = (c_1^*, \ldots, c_n^*)$ for the output by setting $c_i^* = \mathsf{Eval}(\mathcal{P}, c_{i,1}, \ldots, c_{i,t})$, where Eval is the homomorphic evaluation of the FHE scheme. In other words, for each position $i \in [n]$, we perform a homomorphic evaluations of the program $\mathcal{P}$ over the t FHE ciphertexts that lie in position i. We assume (without loss of generality) that the evaluation procedure for the FHE scheme is deterministic so that the results are reproducible.

Alice can verify the triple $(y, \mathcal{P}, \psi)$, where the tag $\psi = (\hat{c}_1, \ldots, \hat{c}_n)$ is supposed to certify that y is the output of the labeled program $\mathcal{P}$. Let $\tau_1, \ldots, \tau_t$ be the input labels of $\mathcal{P}$. For the indices $i \in S$ in the verification set, Alice can re-compute the pseudo-random input ciphertexts $\{c_{i,j} = \mathsf{Enc}_{pk}(0; f_K((i, \tau_j)))\}_{j \in [t]}$ using the PRF, without knowing the actual input bits. She then computes $c_i^* = \mathsf{Eval}(\mathcal{P}, c_{i,1}, \ldots, c_{i,t})$ and checks that the ciphertexts in the tag ψ were computed correctly with $\hat{c}_i \stackrel{?}{=} c_i^*$ for indices $i \in S$. [2] If this is the case, and all of the other ciphertexts $\hat{c}_i$ for $i \in [n] \setminus S$ decrypt to the claimed bit y, then Alice accepts.

Intuitively, the only way that an attacker can lie about the output of some program $\mathcal{P}$ is by producing a tag $\psi = (\hat{c}_1, \ldots, \hat{c}_t)$ where the ciphertexts $\hat{c}_i$ for indices in the verification set $i \in S$ are computed correctly but for $i \in [n] \setminus S$ they are all modified so as to encrypt the wrong bit. But this is impossible since the security of the FHE should ensure that the attacker cannot distinguish encryptions of 0 from those of the authenticated bits, and hence cannot learn anything about the set S. In particular, the FHE hides the difference between data-independent pseudorandom ciphertexts $c_i : i \in S$, which allow Alice to check that the computation was performed correctly, and data-containing ciphertexts $c_i : i \in [n] \setminus S$, which allow Alice to check that the output bit y is the correct one. Note that the authentication tags ψ in our scheme always consist of n (= security parameter) ciphertexts, no matter how many inputs the program $\mathcal{P}$ takes and what its complexity is. Therefore, we satisfy the succinctness requirement.

---

[2] In this step, Alice has to perform work comparable to that of computing $\mathcal{P}$.

We remark that several recent schemes for *delegating computation* [21,16,2] (see Section 1.3 on related work) also use FHE in a similar manner to check that a computation is performed correctly by a remote server. In particular, the work of Chung, Kalai and Vadhan [16] relies on a similar idea, where the output of the homomorphic evaluation for some "data-independent ciphertexts" is known in advance and used to check that the computation was done correctly for the relevant "data-containing" ciphertexts. However, in all these previous works, the technique is used to verify computations over a short known input, whereas in our case we use it to verify computations over unknown authenticated data. The main novelty in our use of the technique is to notice that the "data-independent ciphertexts" can be made *pseudorandom*, so that they can be re-derived in the future given only a short secret PRF key without needing to sacrifice any significant storage.

Security and Verification Queries. We show that our construction is secure in the setting where the attacker can adaptively make arbitrarily many *authentication queries* for various labels, but cannot make *verification queries* to test if a maliciously constructed tag verifies correctly. In practice, this means that the user needs to abort and completely stop using the scheme whenever she gets the first tag that doesn't verify correctly. It is easy to allow for some fixed a-priori bounded number of verification queries q, just by increasing the number of ciphertexts contained in an authentication tag from n to $n + q$.

The difficulty of allowing arbitrarily many verification queries also comes up in most prior schemes for delegating computation in the "pre-processing" model [21,16,2] (see Section 1.3 on related work), and remains an important open problem in both areas.

Fast Verification. One of the limitations of our solution above is that the verification algorithm is *no more efficient* than running the computation $\mathcal{P}$. Therefore, although it saves tremendously on the *communication complexity* of verifying computations over outsourced data, it does not save on user's computational complexity. In Section 4, we explore the option of using schemes for *delegating computation* to also offload the computational cost of the verification procedure to the remote server. As one of our contributions, we show how to achieve fully homomorphic MACs with fast verification using our initial construction and *succinct non-interactive arguments* for polynomial-time computation: **P-SNARGs**. In contrast, a simple solution that bypasses fully-homomorphic MACs would require succinct non-interactive arguments *of knowledge* for all *non-deterministic polynomial-time* computation: **NP-SNARKs**.

1.3   Related Work

Homomorphic Signatures and MACs. Many prior works consider the question of homomorphic message authentication (private verification) and signatures (public verification) for restricted homomorphisms, and almost exclusively for *linear functions*. Perhaps the first work to propose this problem is that of Johnson et al. [29]. Since then, many works have considered this notion in the context of *network coding*, yielding a long line of positive results [1,9,23,10,5,11,15,20]. Another line of works considered this notion in the context of *proofs of data possession and retrievability* [3,33,18,4].

The only work that considers a larger class of homomorphisms *beyond linear functions* is that of Boneh and Freeman [10], who show how to get homomorphic signatures for *bounded (constant) degree polynomials*. In that work, they also present a general definition along the same lines as the definition we use in this work, and pose the question of constructing *fully* homomorphic signatures for *arbitrary* functions.[3] Although the question of fully homomorphic *publicly verifiable* signatures under standard assumptions still remains open, our work provides the first positive result for the case of private verification.

Succinct Arguments of Knowledge. One method that would allow us to construct fully homomorphic (publicly verifiable) signatures is to rely on CS-Proofs [30] or, more generally, any *succinct non-interactive argument of knowledge for all of* **NP** (**NP**-SNARK) [7]. This primitive allows us to create a short "argument" π for any **NP** statement, to prove "knowledge" of the corresponding witness. The length of π is independent of the statement/witness size, and the complexity of verifying π only depends on the size size of the statement.

Using SNARKs, we can authenticate the output y of a labeled program $\mathcal{P}$, by creating a short argument π that proves the knowledge of some "labeled input data D along with valid signatures authenticating D under the appropriate labels, such that $\mathcal{P}(D) = y$". Since this is an argument of *knowledge*, a forged signature for the output of some program $\mathcal{P}$ would allow us to extract out a forged signature for the underlying input data, breaking the security of signatures.

Unfortunately, constructing succinct non-interactive arguments for **NP** is known to require the use of *non-standard* assumptions [25]. Current constructions either rely on the *random-oracle model* [30] or on various "knowledge" assumptions (see, e.g., [28,7,8,22]).

Delegating Computation. Several prior works consider the problem of *delegating computation* to a remote server while maintaining the ability to efficiently verify the result [27,21,16,2,6,31]. In this scenario, the server needs to convince the user that $\mathcal{P}(x) = y$, where the user knows the program $\mathcal{P}$, the input x and the output y, but does not want to do the *work* of computing $\mathcal{P}(x)$. In contrast, in our scenario the verifier only knows $\mathcal{P}, y$, but does *not* know the previously authenticated inputs that $\mathcal{P}$ should have been executed on. On the other hand, we are not trying to minimize work, just communication.

Despite these differences, some of the results on delegating computation in the "pre-processing" model [21,16,2], can also be (re-)interpreted for our setting. In this model, the user "pre-processes" a circuit C and stores some value σ on the server. Later, the user can ask the server to compute $C(x)$ for various inputs

---

[3] See [10,20] for an explanation of how this definition generalizes that of prior works on network coding.

x, and the server uses σ to derive a short and efficiently verifiable proof ψ that certifies correctness of the computation. One caveat is that, in all these schemes, the user first needs to send some challenge $c = chall(x)$ to the server, and the proof ψ is computed as a response to c.

We can apply these results to our setting of outsourced data as follows. Consider outsourcing the "universal circuit" $C_D(\cdot)$ that has the data D "hard-coded", gets as input a program $\mathcal{P}$, and outputs $C_D(\mathcal{P}) = \mathcal{P}(D)$. Then, we can think of the pre-processing of C_D as creating an authentication tag σ for the data D. Later, the user can take a program $\mathcal{P}$, create a challenge $c = chall(\mathcal{P})$, and get back a short tag ψ that authenticates $y = C_D(\mathcal{P}) = \mathcal{P}(D)$.[4] One advantage of this approach is that the tag ψ can be verified efficiently, with less work than that of computing $\mathcal{P}(D)$. However, there are several important disadvantages of this approach as compared to our notion of homomorphic authenticators:

1. *Interaction:* Homomorphic authenticators allow anybody to evaluate a chosen program $\mathcal{P}$ over authenticated data and *non-interactively* authenticate the output. The above delegation-based schemes require a *round of interaction*; the user first creates a challenge $chall(\mathcal{P})$ for the program $\mathcal{P}$, and only then can the server authenticate the output $\mathcal{P}(D)$ with respect to this challenge.
2. *Single Use:* Homomorphic authenticators allow several users to authenticate various labeled data *"on the fly"* (without any state) and verify arbitrary computations over all of the data in the future using a fixed secret key. The above delegation-based schemes require that a single user outsources all of the data D in *one shot*, and stores some small secret state associated with the data to verify computations over only this data in the future.
3. *Bounded Size:* The above delegation-based schemes require that the circuit-size of the computations $\mathcal{P}$ is *a-priori bounded* by some fixed polynomial chosen during authentication. Furthermore, the complexity of authentication is proportional to this polynomial. Our fully homomorphic authenticators have no such restriction.
4. *No composition:* The above delegation-based schemes does not support the *composition* of several partial authenticated computations.

Memory Delegation. The work of Chung et al. [17] on *memory delegation* explicitly considers the problem of outsourcing a large amount of data while maintaining the ability to efficiently verify later computations over it. The main *advantages* of memory delegation over our work are that: (I) the verification is more efficient than the computation, (II) the data can be efficiently updated by the user in the future, (III) it does not suffer from the verification problem. However, memory delegation suffers from many of the same disadvantaged outlined above for delegation-based schemes. In particular, the general memory delegation scheme of [17] is interactive, requiring 4 rounds of interaction during verification. The paper also provides a non-interactive solution where the size of

---

[4] Here, we assume that $\mathcal{P}$ has a short uniform description so that reading /transmitting $\mathcal{P}$ is much more efficient than evaluating $\mathcal{P}$.

the tag grows with the *depth* of the computation-circuit (and therefore does not satisfy our succinctness property). Furthermore, memory delegation considers the setting where a single user outsources a single data item D in *one shot*, and store some small secret state associated with the data to verify computations over it in the future. In particular, it does not provide a method where various users can authenticate various (small) pieces of data independently, and verify joint computations over all of the data in the future. Lastly, the memory-delegation solutions do not support composition.

Follow-Up Work. Following our work, Catalano and Fiore [14] (Eurocrypt '13) give a very efficient and simple construction of homomorphic-message authenticators for low-depth arithmetic circuits – in particular, the computation/verification time and the tag size depend polynomially on the *degree* of the circuit, which can be exponential in the depth of the circuit. Their construction only relies on one-way functions. In contrast, our construction here is significantly more general (works for any polynomial-size boolean circuit) but relies on the "heavier machinery" of fully homomorphic encryption.

2   Definitions

2.1   Homomorphic Authenticators

Labeled Programs. We begin by defining the concept of a labeled program, where the labels denote *which* data the program should be evaluated on. Formally, a *labeled-program* $\mathcal{P} = (f, \tau_1, \ldots, \tau_k)$ consists of a circuit $f : \{0,1\}^k \to \{0,1\}$ along with a distinct *input label* $\tau_i \in \{0,1\}^*$ for each input wire $i \in [k]$. [5]

Given some labeled programs $\mathcal{P}_1, \ldots, \mathcal{P}_t$ and a circuit $g : \{0,1\}^t \to \{0,1\}$, we can define the *composed program*, denoted by $\mathcal{P}^* = g(\mathcal{P}_1, \ldots, \mathcal{P}_t)$, which corresponds to evaluating g on the outputs of $\mathcal{P}_1, \ldots, \mathcal{P}_t$. The labeled inputs of the composed program $\mathcal{P}^*$ are just all the *distinct* labeled inputs of $\mathcal{P}_1, \ldots, \mathcal{P}_t$, meaning that we collect all the input wires with the same label and convert them into a single input wire.

We define the *identity program with label* τ as $\mathcal{I}_\tau := (g_{id}, \tau)$ where g_{id} is the *canonical identity circuit* and $\tau \in \{0,1\}^*$ is some label. Notice that any program $\mathcal{P} = (f, \tau_1, \ldots, \tau_k)$ can be written as a composition of identity programs $\mathcal{P} = f(\mathcal{I}_{\tau_1}, \ldots, \mathcal{I}_{\tau_k})$.

Syntax. A *homomorphic authenticator scheme* consists of the probabilistic-polynomial time algorithms (KeyGen, Auth, Ver, Eval) with the following syntax:

---

[5] Although the above description of $\mathcal{P}$ is long (proportional to its input size and complexity), in many scenarios it is possible that $\mathcal{P}$ may also have an alternative succinct description. For example, $\mathcal{P}$ may compute the *median* value in a large file called "salaries" and its input labels are simply $\tau_i = (\text{"salaries"}, i)$ for each bit i. Therefore, although we are concerned with succinctness, we will ignore the cost of communicating the program $\mathcal{P}$ from future consideration.

- KeyGen$(1^n) \to (evk, sk)$: Outputs the secret key sk and an evaluation key evk.
- Auth$_{sk}(b, \tau) \to \sigma$: Creates a *tag* σ that authenticates the bit $b \in \{0, 1\}$ under the label $\tau \in \{0, 1\}^*$. (Equivalently, we say that σ authenticates b as the output of the identity program $\mathcal{I}_\tau$.)
- Eval$_{evk}(f, \boldsymbol{\sigma}) \to \psi$: The deterministic evaluation procedure takes a vector of tags $\boldsymbol{\sigma} = (\sigma_1, \ldots, \sigma_k)$ and a circuit $f : \{0, 1\}^k \to \{0, 1\}$. It outputs a tag ψ. If each σ_i authenticates a bit b_i as the output of some labeled-program $\mathcal{P}_i$ (possibly the identity program), then ψ should authenticate $b^* = f(b_1, \ldots, b_k)$ as the output of the composed program $\mathcal{P}^* = f(\mathcal{P}_1, \ldots, \mathcal{P}_k)$.
- Ver$_{sk}(e, \mathcal{P}, \psi) \to \{\texttt{accept}, \texttt{reject}\}$: The deterministic verification procedure uses the tag ψ to check that $e \in \{0, 1\}$ is the output of the program $\mathcal{P}$ on previously authenticated labeled data.

We require that the scheme satisfies the following properties, defined below: *authentication correctness, evaluation correctness, succinctness* and *authenticator security*.

Authentication Correctness. We require that for any $b \in \{0, 1\}$ and any label $\tau \in \{0, 1\}^*$, we have:

$$\Pr\left[\mathsf{Ver}_{sk}(b, \mathcal{I}_\tau, \sigma) = \texttt{accept} \,\middle|\, (evk, sk) \leftarrow \mathsf{KeyGen}(1^n), \sigma \leftarrow \mathsf{Auth}_{sk}(b, \tau)\right] = 1$$

where $\mathcal{I}_\tau$ is the identity program with label τ. In other words, the tag $\sigma = \mathsf{Auth}_{sk}(b, \tau)$ correctly authenticates b under the label τ, which is equivalent to saying that it authenticates b as the output of the identity program $\mathcal{I}_\tau$.

Evaluation Correctness. Fix any (evk, sk) in the support of $\mathsf{KeyGen}(1^n)$. Fix any circuit $g : \{0, 1\}^t \to \{0, 1\}$ and any set of program/bit/tag triples $\{(\mathcal{P}_i, b_i, \psi_i)\}_{i=1}^t$ such that $\mathsf{Ver}_{sk}(b_i, \mathcal{P}_i, \psi_i) = \texttt{accept}$. Set:

$$b^* := g(b_1, \ldots, b_t), \quad \mathcal{P}^* := g(\mathcal{P}_1, \ldots, \mathcal{P}_t), \quad \psi^* := \mathsf{Eval}_{evk}(g, (\psi_1, \ldots, \psi_t)).$$

Then we require $\mathsf{Ver}_{sk}(b^*, \mathcal{P}^*, \psi^*) = \texttt{accept}$.

In words, assume that each tag ψ_i certifies that the output of the labeled program $\mathcal{P}_i$ is b_i. Then the tag ψ^* certifies that b^* is the output of the composed program $\mathcal{P}^*$. If all of the programs $\mathcal{P}_i = \mathcal{I}_{\tau_i}$ are identity programs, then the above says that as long as the tags ψ_i authenticate bits b_i under the labels τ_i, the tag ψ^* authenticates b^* as the output of $\mathcal{P}^* = (g, \tau_1, \ldots, \tau_t)$. Therefore, the above definition captures the basic correctness of computing over freshly authenticated data, as well as the *composability* of computing over the authenticated outputs of prior computations.

Succinctness. We require that the tag-size is always bounded by some fixed polynomial in the security parameter n, and is independent of the size of the evaluated circuit or the number of inputs it takes. That is, there exists some polynomial $p(\cdot)$ such that, for any (evk, sk) in the support of $\mathsf{KeyGen}(1^n)$, the output-size of $\mathsf{Auth}_{sk}(\cdot, \cdot)$ and of $\mathsf{Eval}_{evk}(\cdot, \cdot)$ is bounded by $p(n)$ for any choice of their input.

Authenticator Security. Consider the following game $\mathsf{ForgeGame}^{\mathcal{A}}(1^n)$ between an attacker $\mathcal{A}(1^n)$ and a challenger:

1. The challenger chooses $(evk, sk) \leftarrow \mathsf{KeyGen}(1^n)$ and gives evk to $\mathcal{A}$. It initializes $T := \emptyset$.
2. The attacker $\mathcal{A}$ can adaptively submit arbitrarily many *authentication queries* of the form (b, τ) to the challenger. On each such query, if there is some $(\tau, \cdot) \in T$ (i.e. the label τ is not fresh) then the challenger ignores it. Else it updates $T := T \cup \{(\tau, b)\}$, associating the label τ with the authenticated bit b, and replies with $\sigma \leftarrow \mathsf{Auth}_{sk}(b, \tau)$.
3. Finally, the attacker outputs some *forgery* $(e^*, \mathcal{P}^* = (f^*, \tau_1^*, \ldots, \tau_k^*), \psi^*)$. The output of the game is 1 iff $\mathsf{Ver}_{sk}(e^*, \mathcal{P}^*, \psi^*) = \mathtt{accept}$ *and* one of the following two conditions holds:
   - *Type I Forgery:* There is some $i \in [k]$ such that the label $(\tau_i^*, \cdot)$ does not appear in T. (*i.e., No bit was ever authenticated under the label τ_i^* involved in the forgery.*)
   - *Type II Forgery:* The set T contains tuples $(\tau_1^*, b_1), \ldots, (\tau_k^*, b_k)$, for some bits $b_1, \ldots, b_k \in \{0, 1\}$ such that $f^*(b_1, \ldots, b_k) \neq e^*$. (*i.e., The labeled program $\mathcal{P}^*$ does not output e^* when executed on previously authenticated labeled data $b_1, \ldots, b_k$.*)

We say that a homomorphic authenticator scheme is *secure (without verification queries)* if, for any probabilistic polynomial-time $\mathcal{A}$, we have $\Pr[\mathsf{ForgeGame}^{\mathcal{A}}(1^n) = 1] \leq \mathsf{negl}(n)$. We can also define a stronger variant, called *security with verification queries*, where we insist that the above probability holds for a modified version of the game, in which the attacker can also adaptively make arbitrarily many verification queries of the form $(e, \mathcal{P}, \psi)$, and the challenger replies with $\mathsf{Ver}_{sk}(e, \mathcal{P}, \psi)$.

2.2 Homomorphic Encryption

A *fully homomorphic (public-key) encryption* (FHE) scheme is a quadruple of PPT algorithms $\mathsf{HE} = (\mathsf{HE.KeyGen}, \mathsf{HE.Enc}, \mathsf{HE.Dec}, \mathsf{HE.Eval})$ defined as follows.

- $\mathsf{HE.KeyGen}(1^n) \rightarrow (pk, evk, sk)$: Outputs a public encryption key pk, a public evaluation key evk and a secret decryption key sk.
- $\mathsf{HE.Enc}_{pk}(b) \rightarrow c$: Encrypts a bit $b \in \{0, 1\}$ under public key pk. Outputs ciphertext c.
- $\mathsf{HE.Dec}_{sk}(c) \rightarrow b$: Decrypts ciphertext c using sk to a plaintext bit $b \in \{0, 1\}$.
- $\mathsf{HE.Eval}_{evk}(g, c_1, \ldots, c_t) \rightarrow c^*$: The *deterministic evaluation algorithm* takes the evaluation key evk, a boolean circuit $g : \{0, 1\}^t \rightarrow \{0, 1\}$, and a set of t ciphertexts $c_1, \ldots, c_t$. It outputs the result ciphertext c^*.

An FHE should also satisfy the following properties.

Encryption Correctness. For all $b \in \{0,1\}$ we have:

$$\Pr\left[\mathsf{HE.Dec}_{sk}(\mathsf{HE.Enc}_{pk}(b)) = b \mid (pk, evk, sk) \leftarrow \mathsf{HE.KeyGen}(1^n)\right] = 1$$

Evaluation Correctness. For any (pk, evk, sk) in the support of $\mathsf{HE.KeyGen}(1^n)$, any ciphertexts $c_1, \ldots, c_t$ such that $\mathsf{HE.Dec}_{sk}(c_i) = b_i \in \{0,1\}$, and any circuit $g : \{0,1\}^t \to \{0,1\}$ we have

$$\mathsf{HE.Dec}_{sk}(\mathsf{HE.Eval}_{evk}(g, c_1, \ldots, c_t)) = g(b_1, \ldots, b_t).$$

Succinctness. We require that the ciphertext-size is always bounded by some fixed polynomial in the security parameter, and is independent of the size of the evaluated circuit or the number of inputs it takes. That is, there exists some polynomial $p(\cdot)$ such that, for any (pk, evk, sk) in the support of $\mathsf{HE.KeyGen}(1^n)$, the output-size of $\mathsf{HE.Enc}_{pk}(\cdot)$ and of $\mathsf{Eval}_{evk}(\cdot)$ is bounded by $p(n)$, for any choice of their inputs.

Semantic Security. Lastly, an FHE should satisfy the standard notion of *semantic security* for public-key encryption, where we consider the evaluation key evk as a part of the public key. That is, for any PPT attacker $\mathcal{A}$ we have:

$$\mid \Pr\left[\mathcal{A}(1^n, pk, evk, c_0) = 1\right] - \Pr\left[\mathcal{A}(1^n, pk, evk, c_1) = 1\right] \mid \leq \mathsf{negl}(n)$$

where the probability is over $(pk, evk, sk) \leftarrow \mathsf{KeyGen}(1^n), \{c_b \leftarrow \mathsf{HE.Enc}_{pk}(b)\}_{b \in \{0,1\}}$, and the coins of $\mathcal{A}$.

Canonical FHE. We can take any FHE scheme and make it *canonical*, meaning that the $\mathsf{HE.Eval}$ procedure just evaluates the circuit recursively, level-by-level and gate-by-gate. In particular, for any circuit $g : \{0,1\}^k \to \{0,1\}$ taking input bits $b_1, \ldots, b_k$, if the top gate of g is $h : \{0,1\}^t \to \{0,1\}$ and the inputs to h are computed by sub-circuits $f_1(b_{i_{1,1}}, \ldots b_{i_{1,k_1}}) \ldots, f_t(b_{i_{t,1}}, \ldots b_{i_{t,k_t}})$ then

$$\mathsf{HE.Eval}_{evk}(g, c_1, \ldots, c_k) = \mathsf{HE.Eval}_{evk}(h, c_1^*, \ldots, c_t^*)$$

where $c_j^* = \mathsf{HE.Eval}_{evk}(f_j, c_{i_{j,1}}, \ldots, c_{i_{j,k_j}})$ for $j \in [t]$. We also assume that, if g_{id} is the canonical identity circuit on one input, then $\mathsf{HE.Eval}_{evk}(g_{id}, c) = c$. Making the FHE scheme canonical will be important when we want to reason about *composition*, since it will ensure that the evaluation procedure outputs the exact same ciphertext when we homomorphically evaluate the entire circuit in one shot as when we first homomorphically evaluate some sub-circuits and then combine the results via additional independent homomorphic evaluations.

See, e.g., the works of [24,34,13,12] for constructions of fully homomorphic encryption.

3   Constructing Homomorphic Authenticators

We now describe our construction of homomorphic authenticators. Although it closely follows our high-level description in the introduction, there are some differences. Most notably, the simple scheme in the introduction does not appropriately protect against *type I* forgeries. For example, it is possible that for the function g_0 which always outputs 0, the FHE scheme always outputs $\mathsf{HE.Eval}_{evk}(g_0, c) = C_0$ where C_0 is some fixed and known ciphertext encrypting 0. In that case, the attacker can always authenticate the output of the labeled-program $\mathcal{P}_0 = (g_0, \tau)$ for any label τ, even if he never saw any previously authenticated data under the label τ. This would qualify as a type I forgery, breaking our definition. To fix this, we add an extra component to our tags that ensures that the attacker must have seen authentication tags for all of the underlying input labels. We describe this component below.

Hash Tree of a Circuit. If $g : \{0,1\}^k \to \{0,1\}$ is a circuit and $H : \{0,1\}^* \to \{0,1\}^m$ is some hash function, we define the *hash-tree of g*, denoted g^H, as a *Merkle-Tree* that has the same structure as the circuit g, but replaces all internal gates with the hash function H. More precisely, the hash-tree $g^H : (\{0,1\}^*)^k \to \{0,1\}^m$ is a function which takes as input strings $\nu_i \in \{0,1\}^*$ for each input wire of g. For every wire w in the circuit g, we define the *value of $g^H(\nu_1, \ldots, \nu_k)$ at w* inductively as:

- $val(w) = H(\nu_i)$ if w is the ith input wire of g.
- $val(w) = H(val(w_1), \ldots, val(w_t))$ if w is the output wire of some gate with input wires $w_1, \ldots, w_t$.

We define the output of the function $g^H(\nu_1, \ldots, \nu_k)$ to be its its value at the output wire of g.

Construction. Let $\mathsf{HE} = (\mathsf{HE.KeyGen}, \mathsf{HE.Enc}, \mathsf{HE.Dec}, \mathsf{HE.Eval})$ be a canonical fully homomorphic encryption scheme, where the encryption algorithm uses $r = r(n) = \omega(\log(n))$ random bits. Let $\left\{ f_K : \{0,1\}^* \to \{0,1\}^{r(n)} \right\}_{K \in \{0,1\}^n}$ be a (variable-input-length) *pseudo-random function* PRF family. Let $\mathcal{H}$ be a family of (variable-length) *collision-resistant hash functions (CRHF)* $H : \{0,1\}^* \to \{0,1\}^{m(n)}$. We define the authenticator scheme $\Pi = (\mathsf{KeyGen}, \mathsf{Auth}, \mathsf{Eval}, \mathsf{Ver})$ as follows:

$\mathsf{KeyGen}(1^n)$: Choose a PRF key $K \leftarrow \{0,1\}^n$ and a CRHF $H \leftarrow \mathcal{H}$. Choose an encryption key $(pk, evk', sk') \leftarrow \mathsf{HE.KeyGen}(1^n)$. Select a subset $S \subseteq [n]$ by choosing whether to add each index $i \in [n]$ to the set S independently with probability $\frac{1}{2}$. Output $evk = (evk', H)$, $sk = (pk, evk', H, sk', K, S)$.

$\mathsf{Auth}_{sk}(b, \tau)$: Given $b \in \{0,1\}$ and $\tau \in \{0,1\}^*$ do the following:
1. Choose random coins $rand_1, \ldots, rand_n$ by setting $rand_i = f_K((\tau, i))$. Set $\nu := f_K(\tau)$.

2. Create n ciphertexts $c_1, \ldots, c_n$ as follows. For $i \in [n] \setminus S$, choose $c_i = \mathsf{HE.Enc}_{pk}(b; rand_i)$ as encryptions of the bit b. For $i \in S$, choose $c_i = \mathsf{HE.Enc}_{pk}(0; rand_i)$ as encryption of 0.

3. Output $\sigma = (c_1, \ldots, c_n, \nu)$.

$\mathsf{Eval}_{evk}(g, \boldsymbol{\sigma})$: Given $\boldsymbol{\sigma} = (\sigma_1, \ldots, \sigma_t)$, parse each $\sigma_j = (c_{1,j}, \ldots, c_{n,j}, \nu_j)$.

   – For each $i \in [n]$, compute $c_i^* = \mathsf{HE.Eval}_{evk'}(g, c_{i,1}, \ldots, c_{i,t})$.

   – Compute $\nu^* = g^H(\nu_1, \ldots, \nu_t)$ to be the output of the hash-tree of g evaluated at $\nu_1, \ldots \nu_t$.

   Output $\psi = (c_1^*, \ldots, c_n^*, \nu^*)$.

$\mathsf{Ver}_{sk}(e, \mathcal{P}, \psi)$: Parse $\mathcal{P} = (g, \tau_1, \ldots, \tau_t)$ and $\psi = (c_1^*, \ldots, c_n^*, \nu^*)$.

1. Compute $\nu_1 := f_K(\tau_1), \ldots, \nu_t = f_K(\tau_t)$ and $\nu' := g^H(\nu_1, \ldots, \nu_t)$. If $\nu' \neq \nu^*$, output reject.

2. For $i \in S$, $j \in [t]$, compute $rand_{i,j} := f_K((\tau_j, i))$ and set $c_{i,j} := \mathsf{HE.Enc}_{pk}(0; rand_{i,j})$.
   For each $i \in S$, evaluate $c_i' := \mathsf{HE.Eval}_{evk'}(g, c_{i,1}, \ldots, c_{i,t})$ and if $c_i' \neq c_i^*$ output reject.

3. For each $i \in [n] \setminus S$, decrypt $e_i := \mathsf{HE.Dec}_{sk'}(c_i^*)$ and if $e \neq e_i$ output reject.

If the above doesn't reject, output accept.

Theorem 1. *If $\{f_K\}$ is a PRF family, $\mathcal{H}$ is a CRHF family and HE is a semantically secure canonical FHE, then the homomorphic authenticator scheme Π is secure without verification queries.*

Proof. It is easy to verify that the *authentication correctness* of Π just follows from the encryption correctness of HE, and the *evaluation correctness* of Π follows from that of HE, along with the fact that HE is *canonical*.

We now prove the security of Π (without verification queries). Let $\mathcal{A}$ be some PPT attacker and let $\mu(n) = \Pr[\mathsf{ForgeGame}^{\mathcal{A}}(1^n) = 1]$. We use a series of hybrid games modifying $\mathsf{ForgeGame}$ to prove that $\mu(n)$ must be negligible.

Game_1: We modify $\mathsf{ForgeGame}$ so as to replace the PRF outputs with truly random consistent values. That is, the challenger replaces all calls to f_K needed to answer authentication queries and to check the winning condition $\mathsf{Ver}_{sk}(e^*, \mathcal{P}^*, \psi^*) \overset{?}{=} \mathsf{accept}$, with calls to a completely random function $F : \{0,1\}^* \to \{0,1\}^{r(n)}$, whose outputs it chooses efficiently "on the fly". By the pseudo-randomness of $\{f_K\}$, we must have $\Pr[\mathsf{Game}_1^{\mathcal{A}}(n) = 1] \geq \mu(n) - \mathsf{negl}(n)$.

Game_2: We now define Game_2 by modifying the winning condition, so that the attacker only wins (the game outputs 1) if the attacker outputs a valid *type (II) forgery* (and the game outputs 0 on a type I forgery). Let E be the event that attacker wins with a type I forgery in Game_1. Then we claim that, under the collision-resistance of $\mathcal{H}$, we have $\Pr[E] = \mathsf{negl}(n)$ and therefore:

$$\Pr[\mathsf{Game}_2^{\mathcal{A}}(n) = 1] \geq \Pr[\mathsf{Game}_1^{\mathcal{A}}(n) = 1] - \Pr[E] \geq \mu(n) - \mathsf{negl}(n)$$

Assume otherwise that $\Pr[E]$ is non-negligible. Recall that the event E only occurs when the attacker submits a forgery

$$e^*, \mathcal{P}^* = (g, \tau_1^*, \ldots, \tau_t^*), \psi^* = (c_1^*, \ldots, c_n^*, \nu^*)$$

such that the attacker never asked any authentication query containing one of the labels τ_j^* for some $j \in [t]$, and verification accepts. During the computation of $\mathsf{Ver}_{sk}(e^*, \mathcal{P}^*, \psi^*)$, when checking that verification accepts in Game_1, the challenger chooses the value $\nu_j = F(\tau_j^*) \leftarrow \{0,1\}^{r(n)}$ freshly at random, since the label τ_j^* was never queried before. If we rewind and re-sample $\nu_j' \leftarrow \{0,1\}^{r(n)}$ freshly an independently at random again, then the probability that verification accepts both times, which we denote by the event E^2, is at least $\Pr[E^2] \geq \Pr[E]^2$. Let C be the event that E^2 occurs and the values $\nu_j \neq \nu_j'$ are distinct, so that $\Pr[C] \geq \Pr[E]^2 - 2^{-r(n)} = \Pr[E]^2 - \mathsf{negl}(n)$ is non-negligible. When the event C occurs then we must have $\nu^* = g^H(\nu_1, \ldots, \nu_j, \ldots, \nu_t) = g^H(\nu_1, \ldots, \nu_j', \ldots, \nu_t)$ which immediately gives us some collision on H at some level of the hash tree g^H. Therefore, $\mathcal{A}$ can be used to efficiently find collisions on $H \xleftarrow{\$} \mathcal{H}$ with non-negligible probability, which gives us a contradiction.

Game_3: In Game_3 we modify the winning condition yet again. When answering authentication queries, the challenger now also remembers the tag σ that it uses, storing (τ, b, σ) in T. If the attacker outputs a type II forgery

$$e^*, \mathcal{P}^* = (g, \tau_1^*, \ldots, \tau_t^*), \psi^* = (c_1^*, \ldots, c_n^*, \nu^*),$$

we modify how the challenger checks $\mathsf{Ver}_{sk}(e^*, \mathcal{P}^*, \psi^*) \stackrel{?}{=} \mathsf{accept}$. Recall that for a type II forgery, the tags τ_i^* were previously used in authentication queries, so that T must contain some tuples

$$((\tau_1^*, b_1, \sigma_1 = (c_{1,1}, \ldots, c_{n,1}, \nu_1)), \ldots, (\tau_t^*, b_t, \sigma_t = (c_{1,t}, \ldots, c_{n,t}, \nu_t)).$$

Let $\hat{c}_i := \mathsf{HE.Eval}_{evk}(g, c_{i,1}, \ldots, c_{i,t})$ for $i \in [n]$ be the *"honest ciphertexts"* that would be included in an honestly generated tag ψ for the program $\mathcal{P}^*$. In Game_3, we replace steps (2), (3) of the verification procedure as follows:
2'. For each $i \in S$: if $\hat{c}_i \neq c_i^*$ then output reject.
3'. For each $i \in [n] \setminus S$: if $\hat{c}_i = c_i^*$ then output reject.
Notice that step (2') is actually the same as the original step (2) used in Game_2, since in both cases we just check the forgery ciphertexts c_i^* against the honest ciphertexts $c_i' = \hat{c}_i$. The only difference is that previously we re-computed c_i' from scratch by re-encrypting $c_{i,j}$, and now we compute $\hat{c}_i$ using the stored ciphertexts $c_{i,j}$ in T (but the values are equivalent). Step (3'), in Game_3 is different from the original step (3) in Game_2. In the original step (3), we decrypted the forgery ciphertexts for $i \in [n] \setminus S$ and checked that they decrypt to the claimed output $e^* \stackrel{?}{=} \mathsf{Dec}_{sk'}(c_i^*)$. Let $e = g(b_1, \ldots, b_t)$ be the correct output of g on previously authenticated data. In an accepting type II forgery, we must have $e^* \neq e$ but the decryption of the

"honest ciphertexts" will satisfy $\mathsf{HE.Dec}_{sk'}(\hat{c}_i) = e$. So it must be the case that $c_i^* \neq \hat{c}_i$ for all $i \in [n] \setminus S$ for any accepting type II forgery in Game_2. Therefore, any type II forgery that's accepting in Game_2 is also accepting in Game_3 and hence: $\Pr[\mathsf{Game}_3^{\mathcal{A}}(n) = 1] \geq \Pr[\mathsf{Game}_2^{\mathcal{A}}(n) = 1] \geq \mu(n) - \mathsf{negl}(n)$.

Game_4: We modify Game_3 so that, when answering *authentication queries*, the challenger computes all of the ciphertexts c_i (even for $i \in S$) as encryptions of the correct bit b in step (2) of the authentication procedure. In particular, the choice of S is ignored when answering authentication queries. We claim that:

$$\Pr[\mathsf{Game}_4^{\mathcal{A}}(n) = 1] \geq \Pr[\mathsf{Game}_3^{\mathcal{A}}(n) = 1] - \mathsf{negl}(n) \geq \mu(n) - \mathsf{negl}(n). \quad (1)$$

This simply follows by the semantic security of the encryption scheme HE. Given challenge ciphertexts which either encrypt the attacker's bits b or 0, we can embed these into the authentication procedures for positions $i \in S$ and either simulate Game_3 or Game_4. We can efficiently determine if the output of the game is 1, since the decryption secret key sk' is never used in these games. Therefore, if the above didn't hold, the attacker $\mathcal{A}$ would break semantic security.

Negligible Advantage. We now claim that, information theoretically, $\Pr[\mathsf{Game}_4^{\mathcal{A}}(n) = 1] \leq 2^{-n}$. Together with equation (1), this shows that $\mu(n) \leq 2^{-n} + \mathsf{negl}(n) = \mathsf{negl}(n)$, as we wanted to show. In Game_4, the choice of the set $S \subseteq [n]$ is not used at all when answering authentication queries and so we can think of the challenger as only picking the set S during verification. For any type II forgery $e^*, \mathcal{P}^*, \psi^* = (c_1^*, \ldots, c_n^*, \nu^*)$, let $c_1', \ldots, c_n'$ be the *"honest ciphertexts"* that would be included in an honestly generated tag ψ for the output of $\mathcal{P}^*$ (see description of Game_3). Let $S' := \{i \in [n] : c_i^* = c_i'\}$ be the indices on which the forged and honest ciphertexts match. The attacker only wins if steps (2'), (3') of verification pass, which only occurs if $S = S'$. But this only occurs with probability 2^{-n} over the random choice of S.

3.1 Fully Homomorphic Authenticator-Encryption

We can also extend homomorphic message authenticators to homomorphic *authenticator-encryption*. Given the secret key sk, it should be possible to *decrypt* the correct bit b from the tag authenticating it, but given only the evaluation key evk, the tags should not reveal any information about the authenticated bits.

We can allow decryption generically. Take any homomorphic authenticator scheme $(\mathsf{KeyGen}, \mathsf{Auth}, \mathsf{Eval}, \mathsf{Ver})$ and define $\mathsf{VerDec}(\mathcal{P}, \psi) \to \{0, 1, \mathtt{reject}\}$ as follows: run $\mathsf{Ver}(e, \mathcal{P}, \psi)$ for both choices of $e \in \{0, 1\}$ and, if exactly one of the runs is accepting, return the corresponding e, else return $\mathtt{reject}$.

We notice that our specific construction of homomorphic authenticators already provides *chosen-plaintext-attack (CPA)* security for the above authenticator-encryption scheme. Even if the attacker gets evk and access to the authentication oracle $\mathsf{Auth}_{sk}(\cdot, \cdot)$, if he later sees a tag $\sigma \leftarrow \mathsf{Auth}_{sk}(b, \tau)$ for a fresh label τ, he cannot distinguish between the cases $b = 0$ and $b = 1$.

3.2   Security with Verification Queries?

The scheme presented in Section 3 only provides security *without* verification queries, and it remains an interesting open problem to construct fully homomorphic authenticators that allow for the stronger notion of security *with* verification queries. We make several observations here.

An Efficient Attack. We note that there is an efficient attack against our basic scheme from Section 3, in the setting of security *with* verification queries.

The attacker gets evk and makes a single authentication query to get a tag $\sigma \leftarrow \mathsf{Auth}_{sk}(1, \tau)$, authenticating the bit $b = 1$ under some arbitrary label τ. We parse $\sigma = (c_1, \ldots, c_n, \nu)$.

The attacker then makes several verification queries whose aim is to learn the secret set $S \subseteq [n]$. It does so as follows: for each $i \in [n]$ he computes c_i' by adding an encryption of 0 to c_i (or performing any homomorphic operation that changes the ciphertext while preserving the plaintext) and sets the modified tag σ_i to be the same as σ, but with c_i replaced by c_i'. Then, for each $i \in [n]$, the attacker makes a verification query $\mathsf{Ver}_{sk}(1, \mathcal{I}_\tau, \sigma_i)$ to test if the modified tag σ_i is valid. If the query rejects then the attacker guesses that $i \in S$ and otherwise guesses $i \notin S$. With overwhelming probability the attacker correctly recovers the entire set S.

Now the attacker can construct a type II forgery for the identity program $\mathcal{I}_\tau$, claiming that its value is 0 (recall, that we previously authenticated $b = 1$ under the label τ). To do so, the attacker takes the tag σ and, for $i \notin S$, replaces the ciphertexts c_i with fresh encryptions of 0. Let's call the resulting modified tag σ^*. Then it's easy to see that $(0, \mathcal{I}_\tau, \sigma^*)$ is a valid type II forgery.

Bounded Verification Queries. The above attack requires n verification queries to break the scheme. It is relatively easy to show that the scheme is secure against $O(\log(n))$ verification queries. In particular, the attacker only gets $O(\log(n))$ bits of information about the set S, which is not enough to break the scheme. Similarly, for any a-priori bound q, we can modify our scheme so that the tags contain $n + q$ ciphertexts to get security against q verification queries.

Computation with Long Output. Our basic scheme considers homomorphic authentication for a program $\mathcal{P}$ with a 1-bit output. Of course, we can extend the scheme to authenticate a program with longer output, by simply authenticating each bit of the output separately. However, this means that the tag is proportional to the output size of the computation. A simple trick allows us to authenticate a long output of some program $\mathcal{P}$ with only a short tag which is independent of the output size. Instead of homomorphically authenticating the output of the program $\mathcal{P}$, we just authenticate the output of a program $H(\mathcal{P})$ which first computes the output y of $\mathcal{P}$ and then outputs $H(y)$ where H is a collision-resistant hash function. Since $H(\mathcal{P})$ has a short output even when $\mathcal{P}$ has a long output, the resulting tag is short. Moreover, by getting the short tag ψ computed as above and the long output y, the verifier can check $\mathsf{Ver}_{sk}(H(y), H(\mathcal{P}), \psi) = \mathsf{accept}$ to ensure that y is the output of the computation $\mathcal{P}$ over previously authenticated data.

4   Improving Verification Complexity

One of the main limitations of our homomorphic authenticator scheme from the previous section is that the complexity of the verification algorithm is no better than that of executing the program $\mathcal{P}$. Therefore, although the scheme saves on the *communication complexity* of transmitting the input data, it does not save on the *computation complexity* of executing $\mathcal{P}$. As we discussed in the introduction, works in the area of *delegating computation* obtain efficient verification whose complexity is independent (or at least much smaller than) the computation of $\mathcal{P}$, but require that the user knows the entire input x. We explore the idea of "marrying" these two techniques by *delegating* the computation required to *verify* the authentication tag in our homomorphic authenticator scheme.

Firstly, we notice that the verification procedure $\mathsf{Ver}_{sk}(e, \mathcal{P}, \psi)$ of our scheme as described in Section 3 has special structure. The only "expensive" computation (proportional to the complexity of the the program $\mathcal{P}$) is *independent* of the tag ψ. In particular, this computation uses the secret key sk and the program $\mathcal{P}$ to compute the output of the "hash-tree" ν' and the ciphertexts c_i' for $i \in S$ derived by evaluating $\mathcal{P}$ over the pseudorandom encryptions of 0. We call this computation $\mathtt{Expensive}(\mathcal{P}, sk)$. Given the outputs of $\mathtt{Expensive}(\mathcal{P}, sk)$, the rest of the verification procedure $\mathsf{Ver}_{sk}(e, \mathcal{P}, \psi)$ consists of simple comparisons and is incredibly efficient (independent of the complexity of $\mathcal{P}$). Therefore, we can delegate the computation of $\mathtt{Expensive}(\mathcal{P}, sk)$ prior to knowing the tag ψ that needs to be verified. One issue is that the computation does depend on the secret key sk, which needs to be kept private. We note that we can always (generically) keep the input of a delegated computation private by encrypting it under an FHE scheme. In our context, we can encrypt the value sk under an independently chosen FHE key, and publish this ciphertext C_{sk} in the evaluation key. We can then delegate the computation $\mathtt{Expensive}'(\mathcal{P}, C_{sk})$ which takes $\mathcal{P}$ as an input and homomorphically executed $\mathtt{Expensive}(\mathcal{P}, sk)$.

We now explore the advantages of using the above approach with some concrete delegation schemes.

Using SNARGs. We can use succinct non-interactive arguments for polynomial-time computation (**P**-SNARGs). This primitive allows anyone to provide a short proof π certifying the correctness of an arbitrary polynomial-time computation $y = f(x)$, where f is a Turing Machine. The tuple (f, y, x, π) can be verified in some fixed polynomial-time $p(|x|, |y|, |f|)$, that only depends on the *description-length* of the machine f, but is independent of the *running time* of f. Given **P**-SNARGs, we get a completely non-interactive delegation of computation (assuming the computation has a short uniform description). Therefore, using the above approach, we get homomorphic authenticators satisfying our original non-interactive syntax and security, but also allowing efficient verification for programs $\mathcal{P}$ having a short uniform description. During *evaluation*, we simply also have the server compute $\mathtt{Expensive}'(\mathcal{P}, C_{sk})$ and provide a SNARG proof π that it was done correctly. Recall that, in the introduction, we described a significantly simpler solution to the problem of efficiently verifiable homomorphic authenticators (and signatures) using succinct non-interactive argument of knowledge for all of **NP**, or **NP**-SNARKs. Therefore, the main advantage of the above technique is that now we only require **P**-SNARGs, which is a much weaker primitive. For example, if we instantiate the random-oracle CS-Proofs of Micali [30] with some cryptographic hash function, it may be more reasonable to assume that we get a **P**-SNARG, than it is to assume that we get an **NP**-SNARK. In particular, the former *is* a falsifiable assumption whereas the latter cannot be proved under any falsifiable assumption (see [25]).

Using Delegation with Pre-Processing. Alternatively, we can use the delegation techniques in the "pre-processing" model [21,16,2] to outsource the computation of $\texttt{Expensive}'(\mathcal{P}, C_{sk})$ where $\mathcal{P}$ is given as an input. This scheme will have many of the same advantages and disadvantages as the approach of using delegation with "pre-processing" directly to outsource the data (see a description of this latter approach in Section 1.3). In particular, in both approaches, the verification will be efficient but the scheme will now require one round of interaction, where the user needs to create a challenge $chall(\mathcal{P})$ for the computation $\mathcal{P}$ that she wants to verify. The main advantages of our approach, combining delegation with homomorphic authenticators, over a direct delegation-based scheme is the following. When using delegation directly, the user needs to outsource all of the data in one shot and remember some short partial information about it; now the user can arbitrarily authenticate fresh labeled data "on the fly" and verify computations over all of it using a single independent secret key.

5   Conclusions

In this work we give the first solution to fully homomorphic message authenticators, allowing a user to verify computations over previously authenticated data. The authentication tag is short, independent of the size of the authenticated input to the computation. Our work leaves many interesting open questions. Perhaps the most ambitious one is to construct fully homomorphic signatures with *public verification*. Less ambitiously, construct fully homomorphic authenticators that allow an unbounded number of verification queries. Lastly, it would be interesting to improve the verification efficiency of our construction. One pressing question is to make the verification complexity independent of the complexity of the program $\mathcal{P}$ while maintaining all of the advantages of our scheme (standard assumptions, no interaction). But a less ambitious, still interesting question is to just reduce the tag size from $O(n)$ ciphertexts to something smaller, say a single ciphertext.

Acknowledgement. We thank Craig Gentry for his valuable comments. In particular, a prior version of this work included a speculative suggestion for achieving security with verification queries via "randomness-homomorphic encryption"; Craig pointed out that this latter primitive cannot exist.

References

1. Agrawal, S., Boneh, D.: Homomorphic MACs: MAC-based integrity for network coding. In: Abdalla, M., Pointcheval, D., Fouque, P.-A., Vergnaud, D. (eds.) ACNS 2009. LNCS, vol. 5536, pp. 292–305. Springer, Heidelberg (2009)
2. Applebaum, B., Ishai, Y., Kushilevitz, E.: From secrecy to soundness: Efficient verification via secure computation. In: Abramsky, S., Gavoille, C., Kirchner, C., Meyer auf der Heide, F., Spirakis, P.G. (eds.) ICALP 2010. LNCS, vol. 6198, pp. 152–163. Springer, Heidelberg (2010)
3. Ateniese, G., Burns, R.C., Curtmola, R., Herring, J., Kissner, L., Peterson, Z.N.J., Song, D.: Provable data possession at untrusted stores. In: Ning, P., di Vimercati, S.D.C., Syverson, P.F. (eds.) ACM CCS 2007, pp. 598–609. ACM Press (October 2007)
4. Ateniese, G., Kamara, S., Katz, J.: Proofs of storage from homomorphic identification protocols. In: Matsui, M. (ed.) ASIACRYPT 2009. LNCS, vol. 5912, pp. 319–333. Springer, Heidelberg (2009)
5. Attrapadung, N., Libert, B.: Homomorphic network coding signatures in the standard model. In: Catalano, D., Fazio, N., Gennaro, R., Nicolosi, A. (eds.) PKC 2011. LNCS, vol. 6571, pp. 17–34. Springer, Heidelberg (2011)
6. Benabbas, S., Gennaro, R., Vahlis, Y.: Verifiable delegation of computation over large datasets. In: Rogaway (ed.) [32], pp. 111–131
7. Bitansky, N., Canetti, R., Chiesa, A., Tromer, E.: From extractable collision resistance to succinct non-interactive arguments of knowledge, and back again. In: Goldwasser (ed.) [26], pp. 326–349
8. Bitansky, N., Canetti, R., Chiesa, A., Tromer, E.: Recursive composition and bootstrapping for snarks and proof-carrying data. Cryptology ePrint Archive, Report 2012/095 (2012), http://eprint.iacr.org/
9. Boneh, D., Freeman, D., Katz, J., Waters, B.: Signing a linear subspace: Signature schemes for network coding. In: Jarecki, S., Tsudik, G. (eds.) PKC 2009. LNCS, vol. 5443, pp. 68–87. Springer, Heidelberg (2009)
10. Boneh, D., Freeman, D.M.: Homomorphic signatures for polynomial functions. In: Paterson, K.G. (ed.) EUROCRYPT 2011. LNCS, vol. 6632, pp. 149–168. Springer, Heidelberg (2011)
11. Boneh, D., Freeman, D.M.: Linearly homomorphic signatures over binary fields and new tools for lattice-based signatures. In: Catalano, D., Fazio, N., Gennaro, R., Nicolosi, A. (eds.) PKC 2011. LNCS, vol. 6571, pp. 1–16. Springer, Heidelberg (2011)
12. Brakerski, Z., Gentry, C., Vaikuntanathan, V.: (leveled) fully homomorphic encryption without bootstrapping. In: Goldwasser (ed.) [26], pp. 309–325
13. Brakerski, Z., Vaikuntanathan, V.: Efficient fully homomorphic encryption from (standard) lwe. In: Ostrovsky, R. (ed.) FOCS, pp. 97–106. IEEE (2011)
14. Catalano, D., Fiore, D.: Practical homomorphic MACs for arithmetic circuits. In: Johansson, T., Nguyen, P.Q. (eds.) EUROCRYPT 2013. LNCS, vol. 7881, pp. 336–352. Springer, Heidelberg (2013)
15. Catalano, D., Fiore, D., Warinschi, B.: Efficient network coding signatures in the standard model. In: Fischlin, et al. (eds.) [19], pp. 680–696
16. Chung, K.-M., Kalai, Y., Vadhan, S.: Improved delegation of computation using fully homomorphic encryption. In: Rabin, T. (ed.) CRYPTO 2010. LNCS, vol. 6223, pp. 483–501. Springer, Heidelberg (2010)

17. Chung, K.-M., Kalai, Y.T., Liu, F.-H., Raz, R.: Memory delegation. In: Rogaway (ed.) [32], pp. 151–168
18. Dodis, Y., Vadhan, S., Wichs, D.: Proofs of retrievability via hardness amplification. In: Reingold, O. (ed.) TCC 2009. LNCS, vol. 5444, pp. 109–127. Springer, Heidelberg (2009)
19. Fischlin, M., Buchmann, J., Manulis, M. (eds.): PKC 2012. LNCS, vol. 7293. Springer, Heidelberg (2012)
20. Freeman, D.M.: Improved security for linearly homomorphic signatures: A generic framework. In: Fischlin, et al. (eds.) [19], pp. 697–714
21. Gennaro, R., Gentry, C., Parno, B.: Non-interactive verifiable computing: Outsourcing computation to untrusted workers. In: Rabin, T. (ed.) CRYPTO 2010. LNCS, vol. 6223, pp. 465–482. Springer, Heidelberg (2010)
22. Gennaro, R., Gentry, C., Parno, B., Raykova, M.: Quadratic span programs and succinct nizks without pcps. Cryptology ePrint Archive, Report 2012/215 (2012), http://eprint.iacr.org/
23. Gennaro, R., Katz, J., Krawczyk, H., Rabin, T.: Secure network coding over the integers. In: Nguyen, P.Q., Pointcheval, D. (eds.) PKC 2010. LNCS, vol. 6056, pp. 142–160. Springer, Heidelberg (2010)
24. Gentry, C.: Fully homomorphic encryption using ideal lattices. In: Mitzenmacher, M. (ed.) 41st ACM STOC, pp. 169–178. ACM Press (May/June 2009)
25. Gentry, C., Wichs, D.: Separating succinct non-interactive arguments from all falsifiable assumptions. In: Fortnow, L., Vadhan, S.P. (eds.) 43rd ACM STOC, pp. 99–108. ACM Press (June 2011)
26. Goldwasser, S. (ed.): Innovations in Theoretical Computer Science 2012, Cambridge, MA, USA, January 8-10. ACM (2012)
27. Goldwasser, S., Kalai, Y.T., Rothblum, G.N.: Delegating computation: interactive proofs for muggles. In: Ladner, R.E., Dwork, C. (eds.) ACM STOC, pp. 113–122. ACM Press (May 2008)
28. Groth, J.: Short pairing-based non-interactive zero-knowledge arguments. In: Abe, M. (ed.) ASIACRYPT 2010. LNCS, vol. 6477, pp. 321–340. Springer, Heidelberg (2010)
29. Johnson, R., Molnar, D., Song, D., Wagner, D.: Homomorphic signature schemes. In: Preneel, B. (ed.) CT-RSA 2002. LNCS, vol. 2271, pp. 244–262. Springer, Heidelberg (2002)
30. Micali, S.: CS proofs (extended abstracts). In: FOCS, pp. 436–453. IEEE Computer Society (1994)
31. Papamanthou, C., Shi, E., Tamassia, R.: Signatures of correct computation. In: Sahai, A. (ed.) TCC 2013. LNCS, vol. 7785, pp. 222–242. Springer, Heidelberg (2013)
32. Rogaway, P. (ed.): CRYPTO 2011. LNCS, vol. 6841. Springer, Heidelberg (2011)
33. Shacham, H., Waters, B.: Compact proofs of retrievability. In: Pieprzyk, J. (ed.) ASIACRYPT 2008. LNCS, vol. 5350, pp. 90–107. Springer, Heidelberg (2008)
34. van Dijk, M., Gentry, C., Halevi, S., Vaikuntanathan, V.: Fully homomorphic encryption over the integers. In: Gilbert, H. (ed.) EUROCRYPT 2010. LNCS, vol. 6110, pp. 24–43. Springer, Heidelberg (2010)

Non-uniform Cracks in the Concrete:
The Power of Free Precomputation

Daniel J. Bernstein[1,2] and Tanja Lange[2]

[1] Department of Computer Science
University of Illinois at Chicago, Chicago, IL 60607–7053, USA
djb@cr.yp.to
[2] Department of Mathematics and Computer Science
Technische Universiteit Eindhoven, P.O. Box 513, 5600 MB Eindhoven,
The Netherlands
tanja@hyperelliptic.org

Abstract. AES-128, the NIST P-256 elliptic curve, DSA-3072, RSA-3072, and various higher-level protocols are frequently conjectured to provide a security level of 2^{128}. Extensive cryptanalysis of these primitives appears to have stabilized sufficiently to support such conjectures.

In the literature on provable concrete security it is standard to define 2^b security as the nonexistence of high-probability attack algorithms taking time $\leq 2^b$. However, this paper provides overwhelming evidence for the existence of high-probability attack algorithms against AES-128, NIST P-256, DSA-3072, and RSA-3072 taking time considerably below 2^{128}, contradicting the standard security conjectures.

These attack algorithms are not realistic; do not indicate any actual security problem; do not indicate any risk to cryptographic users; and do not indicate any failure in previous cryptanalysis. Any actual use of these attack algorithms would be much more expensive than the conventional 2^{128} attack algorithms. However, this expense is not visible to the standard definitions of security. Consequently the standard definitions of security fail to accurately model actual security.

The underlying problem is that the standard set of algorithms, namely the set of algorithms taking time $\leq 2^b$, fails to accurately model the set of algorithms that an attacker can carry out. This paper analyzes this failure in detail, and analyzes several ideas for fixing the security definitions.

Keywords: provable security, concrete security, algorithm cost metrics, non-uniform algorithms, non-constructive algorithms.

Full version. See http://cr.yp.to/nonuniform.html for the full version of this paper. Appendices appear only in the full version.

This work was supported by the National Science Foundation under grant 1018836, by the Netherlands Organisation for Scientific Research (NWO) under grant 639.073.005, and by the European Commission under Contract ICT-2007-216676 ECRYPT II. Permanent ID of this document: 7e044f2408c599254414615c72b3adbf. Date: 2013.09.10.

K. Sako and P. Sarkar (Eds.) ASIACRYPT 2013, Part II, LNCS 8270, pp. 321–340, 2013.

1   Introduction

> *The Basic Principles of Modern Cryptography ...*
>
> *Principle 1—Formulation of Exact Definitions*
>
> *One of the key intellectual contributions of modern cryptography has been the realization that formal definitions of security are* essential *pre-requisites for the design, usage, or study of any cryptographic primitive or protocol.*   —Katz and Lindell [53]

> *In this paper we will show that CBC MAC construction is secure if the underlying block cipher is secure. To make this statement meaningful we need first to discuss what we mean by security in each case.*
>   —Bellare, Kilian, and Rogaway [12, Section 1.2]

Why do we believe that AES-CBC-MAC is secure? More precisely: Why do we believe that an attacker limited to 2^{100} bit operations, and 2^{50} message blocks, cannot break AES-CBC-MAC with probability more than 2^{-20}?

The standard answer to this question has three parts. The first part is a concrete definition of what it means for a cipher or a MAC to be secure. We quote from the classic paper [12, Section 1.3] by Bellare, Kilian, and Rogaway: the PRP-"insecurity" of a cipher such as AES (denoted "$\mathbf{Adv}^{\mathrm{prp}}_{\mathrm{AES}}(q',t')$") is defined as the "maximum, over all adversaries restricted to q' input-output examples and execution time t', of the 'advantage' that the adversary has in the game of distinguishing [the cipher for a secret key] from a random permutation." The PRF-insecurity of m-block AES-CBC-MAC (denoted "$\mathbf{Adv}^{\mathrm{prf}}_{\mathrm{CBC}^m\text{-}\mathrm{AES}}(q,t)$") is defined similarly, using a uniform random function rather than a uniform random permutation.

The second part of the answer is a concrete security theorem bounding the insecurity of AES-CBC-MAC in terms of the insecurity of AES, or more generally the insecurity of F-CBC-MAC in terms of the insecurity of F for any ℓ-bit block cipher F. Specifically, here is the main theorem of [12]: "for any integers $q, t, m \geq 1$,

$$\mathbf{Adv}^{\mathrm{prf}}_{\mathrm{CBC}^m\text{-}F}(q,t) \leq \mathbf{Adv}^{\mathrm{prp}}_F(q',t') + \frac{q^2 m^2}{2^{l-1}}$$

where $q' = mq$ and $t' = t + O(mql)$." One can object that the O constant is unspecified, making this theorem meaningless as stated for any specific q, t, m values; but it is easy to imagine a truly concrete theorem replacing $O(mql)$ with the time for mql specified operations.

The third part of the answer is a concrete conjecture regarding the security of AES. NIST's call for AES submissions [66, Section 4] identified "the extent to which the algorithm output is indistinguishable from [the output of] a [uniform] random permutation" as one of the "most important" factors in evaluating candidates; cryptanalysts have extensively studied AES without finding any worrisome PRP-attacks; it seems reasonable to conjecture that no dramatically better attacks exist. Of course, this part of the story depends on the details of

AES; analogous conjectures regarding, e.g., DES would have to be much weaker. For example, Bellare and Rogaway in [16, Section 3.6] wrote the following:

"For example we might conjecture something like:

$$\mathbf{Adv}_{\mathrm{DES}}^{\mathrm{prp\text{-}cpa}}(A_{t,q}) \leq c_1 \cdot \frac{t/T_{\mathrm{DES}}}{2^{55}} + c_2 \cdot \frac{q}{2^{40}}$$

... In other words, we are conjecturing that the best attacks are either exhaustive key search or linear cryptanalysis. We might be bolder with regard to AES and conjecture something like

$$\mathbf{Adv}_{\mathrm{AES}}^{\mathrm{prp\text{-}cpa}}(B_{t,q}) \leq c_1 \cdot \frac{t/T_{\mathrm{AES}}}{2^{128}} + c_2 \cdot \frac{q}{2^{128}}."$$

One can again object that the c_1 and c_2 are unspecified here, making these conjectures non-concrete and unfalsifiable as stated. A proper concrete conjecture would specify, e.g., $c_1 = c_2 = 3$. One can also quibble that the T_{DES} and T_{AES} factors do not properly account for inner-loop speedups in exhaustive key search (see, e.g., [27]), that $q/2^{40}$ is a rather crude model of the success probability of linear cryptanalysis, etc., but aside from such minor algorithm-analysis details the conjectures seem quite reasonable.

This AES security conjecture (with small specified c_1 and c_2) says, in particular, that the attacker cannot PRP-break AES with probability more than 2^{-21} after 2^{50} cipher outputs and 2^{100} bit operations. The CBC-MAC security theorem (with small specified O) then says that the same attacker cannot PRF-break AES-CBC-MAC with probability more than 2^{-20}.

Of course, this answer does not *prove* that AES-CBC-MAC is secure; it relies on a conjecture regarding AES security. Why not simply conjecture that AES-CBC-MAC is secure? The answer is scalability. It is reasonable to ask cryptanalysts to intensively study AES, eventually providing confidence in the security of AES, while it is much less reasonable to ask cryptanalysts to intensively study AES-CBC-MAC, AES-OMAC, AES-CCM, AES-GCM, AES-OCB, and hundreds of other AES-based protocols. Partitioning the AES-CBC-MAC security conjecture into an AES security conjecture and a CBC-MAC security proof drastically simplifies the cryptanalyst's job.

The same three-part pattern has, as illustrated by Appendix L (in the full version), become completely standard throughout the literature on concrete "provable security". First part: The insecurity of X — where X is a primitive such as AES or RSA, or a higher-level protocol such as AES-CBC-MAC or RSA-PSS — is defined as the maximum, over all algorithms A ("attacks") that cost at most C, of the probability (or advantage in probability) that A succeeds in breaking X. This insecurity is explicitly a function of the cost limit C; typically C is separated into (1) a time limit t and (2) a limit q on the number of oracle queries. Note that this function depends implicitly on how the "cost" of an algorithm is defined.

Often "the (q,t)-insecurity of X is at most ϵ" is abbreviated "X is (q,t,ϵ)-secure". Many papers prefer the more concise notation and do not even mention the insecurity function. We emphasize, however, that this is merely a superficial change in notation, and that both of the quotes in this paragraph refer to exactly the same situation: namely, the nonexistence of algorithms that cost at most (q, t) and that break X with probability more than ϵ.

Second part: Concrete "provable security" theorems state that the insecurity (or security) of a complicated object is bounded in terms of the insecurity (or security) of a simpler object. Often these theorems require restrictions on the types of attacks allowed against the complicated object: for example, Bellare and Rogaway in [14] showed that RSA-OAEP has similar security to RSA against generic-hash attacks (attacks in the "random-oracle model").

Third part: The insecurity of a well-studied primitive such as AES or RSA-1024 is conjectured to match the success probability of the best attack known. For example, Bellare and Rogaway, evaluating the concrete security of RSA-FDH and RSA-PSS, hypothesized that "it takes time $Ce^{1.923(\log N)^{1/3}(\log \log N)^{2/3}}$ to invert RSA"; Bellare, evaluating the concrete security of NMAC-h and HMAC-h, hypothesized that "the best attack against h as a PRF is exhaustive key search". See [15, Section 1.4] and [7, Section 3.2]. These conjectures seem to precisely capture the idea that cryptanalysts will not make significant further progress in attacking these primitives.

1.1. Primary Contribution of This Paper.

Our primary goal in this paper is to convincingly undermine all of the standard security conjectures reviewed above. Specifically, Sections 2, 3, 4, and 5 show — assuming standard, amply tested heuristics — that there *exist* high-probability attacks against AES, the NIST P-256 elliptic curve, DSA-3072, and RSA-3072 taking considerably less than 2^{128} time. In other words, the insecurity of AES, NIST P-256, DSA-3072, and RSA-3072, according to the standard concrete-security definitions, reaches essentially 100% for a time bound considerably below 2^{128}. The conjectures by Bellare and Rogaway in [15, Section 1.4], [16, Section 3.6], [7, Section 3.2], etc. are false for every reasonable assignment of the unspecified constants.

The same ideas show that there *exist* high-probability attacks against AES-CBC-MAC, RSA-3072-PSS, RSA-3072-OAEP, and thousands of other "provably secure" protocols, in each case taking considerably less than 2^{128} time. It is not clear that similar attacks exist against *every* such protocol in the literature, since in some cases the security reductions are unidirectional, but undermining these conjectures also means undermining all of the security arguments that have those conjectures as hypotheses.

We do not claim that this reflects any actual security problem with AES, NIST P-256, DSA-3072, and RSA-3072, or with higher-level protocols built from these primitives. On the contrary! Our constructions of these attacks are very slow; we conjecture that any *fast* construction of these attacks has negligible probability of success. Users have nothing to worry about.

However, the standard metrics count only the cost of running the attack, not the cost of finding the attack in the first place. This means that there is a very large gap between the actual insecurity of these primitives and their insecurity according to the standard metrics.

This gap is not consistent across primitives. We identify different gaps for different primitives (for example, the asymptotic exponents for high-probability attacks drop by a factor of 1.5 for ECC and a factor of only 1.16 for RSA), and we expect that analyzing more primitives and protocols in the same way will show even more diversity. In principle a single attack is enough to illustrate that the standard definitions of security do not accurately model actual security, but the quantitative variations from one attack to another are helpful in analyzing the merits of ideas for fixing the definitions. It is of course also possible that the gaps for the primitives we discuss will have to be reevaluated in light of even better attacks.

1.2. Secondary Contribution of This Paper (in the full version). Our secondary goal in this paper is to propose a rescue strategy: a new way to define security — a definition that restores, to the maximum extent possible, the attractive three-part security arguments described above.

All of the gaps considered in this paper come from errors in quantifying feasibility. Each of the high-probability attacks presented in this paper (1) has a cost t according to the standard definitions, but (2) is obviously infeasible, even for an attacker able to carry out a "reasonable" algorithm that costs t according to the same definitions. The formalization challenge is to say exactly what "reasonable" means. Our core objective here is to give a new definition that accurately captures what is actually feasible for attackers.

This accuracy has two sides. First, the formally defined set of algorithms must be large enough. Security according to the definition does not imply actual security if the definition ignores algorithms that are actually feasible. Second, the formally defined set of algorithms must be small enough. One cannot conjecture security on the basis of cryptanalysis if infeasible attacks ignored by cryptanalysts are misdeclared to be feasible by the security definition.

We actually analyze four different ideas for modifying the notion of feasibility inside existing definitions:

- Appendix B.2: switching the definitions from the RAM metric used in [12] to the NAND metric, an "alternative" mentioned in [12];
- Appendix B.3: switching instead to the AT metric, a standard hardware-design metric formally defined by Brent and Kung in [29] in 1981;
- Appendix B.4: adding constructivity to the definitions, by a simple trick that we have not seen before (with a surprising spinoff, namely progress towards formalizing collision resistance); and
- Appendix B.5: adding uniformity (families) to the definitions.

Readers unfamiliar with the RAM, NAND, and AT metrics should see Appendix A (in the full version) for a summary and pointers to the literature.

The general idea of modifying security definitions, to improve the accuracy with which those definitions model actual security, is not new. A notable example is the change from the algorithm cost metric used in [11], the original Crypto '94 version of [12], to a more complicated algorithm cost metric used in subsequent definitions of security; readers unfamiliar with the details should see Appendix A for a review. The attacks in this paper show that this modification was not enough, so we push the same general idea further, analyzing the merits of the four modifications listed above. It is conceivable that this general idea is not the best approach, so we also analyze the merits of two incompatible approaches: (Appendix B.1) preserving the existing definitions of security; (Appendix B.7) trying to build an alternate form of "provable security" *without* definitions of security.

Ultimately we recommend the second and third modifications (*AT* and constructivity) as producing much more accurate models of actual feasibility. We also recommend refactoring theorems (see Appendix B.6) to simplify further changes, whether those changes are for even better accuracy or for other reasons. We recommend against the first and fourth modifications (NAND and uniformity). Full details of our analysis appear in Appendix B; the NAND and *AT* analyses for individual algorithms appear in Sections 2, 3, 4, and 5. Appendix Q (in the full version) is a frequently-asked-questions list, serving a role for this paper comparable to the role that a traditional index serves for a book.

Our recommended modifications have several positive consequences. Incorrect conjectures in the literature regarding the concrete security of primitives such as AES can be replaced by quite plausible conjectures using the new definitions. Our impression is that *most* of the proof ideas in the literature are compatible with the new definitions, modulo quantitative changes, so *most* concrete-security theorems in the literature can be replaced by meaningful concrete-security theorems using the new definitions. The conjectures and theorems together will then produce reasonable conclusions regarding the concrete security of protocols such as AES-CBC-MAC.

We do not claim that *all* proofs can be rescued, and it is even possible that some theorems will have to be abandoned entirely. Some troublesome examples have been pointed out by Koblitz and Menezes in [55] and [56]. Our experience indicates, however, that such examples are unusual. For example, there is nothing troublesome about the CBC-MAC proof or the FDH proof; these proofs simply need to be placed in a proper framework of meaningful definitions, conjectures, and theorem statements.

1.3. Priority Dates; Credits; New Analyses. On 20 March 2012 we publicly announced the trouble with the standard AES conjectures; on 17 April 2012 we publicly announced the trouble with the standard NIST P-256, DSA-3072, and RSA-3072 conjectures. The low-probability case of the AES trouble was observed independently by Koblitz and Menezes and announced earlier in March 2012; further credits to Koblitz and Menezes appear below. We are not aware of previous publications disputing the standard concrete-security conjectures.

Our attacks on AES, NIST P-256, DSA-3072, and RSA-3072 use many standard cryptanalytic techniques cited in Sections 2, 3, 4, and 5. We introduce new cost analyses in all four sections, and new algorithm improvements in Sections 3, 4, and 5; our improvements are critical for beating 2^{128} in Section 5. In Sections 2, 3, and 4 the standard techniques were already adequate to (heuristically) disprove the standard 2^{128} concrete-security conjectures, but as far as we know we were the first to point out these contradictions. We do not think the contradictions were obvious; in many cases the standard techniques were published decades *before* the conjectures!

This paper was triggered by a 23 February 2012 paper [55], in which Koblitz and Menezes objected to the non-constructive nature of Bellare's security proof [7] for NMAC. Bellare's security theorem states a quantitative relationship between the standard-definition-insecurity of NMAC-h and the standard-definition-insecurity of h: the *existence* of a fast attack on NMAC-h implies the *existence* of a fast attack on h. The objection is that the proof does not reveal a fast method to compute the second attack from the first: the proof left open the possibility that the fastest algorithm that can be *found* to attack NMAC-h is much faster than the fastest algorithm that can be *found* to attack h.

An early-March update of [55] added weight to this objection by pointing out the (heuristic) existence of a never-to-be-found fast algorithm to attack any 128-bit function h. The success probability of the algorithm was only about 2^{-64}, but this was still enough to disprove Bellare's security conjectures. Koblitz and Menezes commented on "how difficult it is to appreciate all the security implications of assuming that a function has prf-security even against unconstructible adversaries".

Compared to [55], we analyze a much wider range of attacks, including higher-probability PRF attacks and attacks against various public-key systems, showing that the difficulties here go far beyond PRF security. We also show quantitative variations of the difficulties between one algorithm cost metric and another, and we raise the possibility of eliminating the difficulties by carefully selecting a cost metric.

Readers who find these topics interesting may also be interested in the followup paper [56] by Koblitz and Menezes, especially the detailed discussion in [56, Section 2] of "two examples where the non-uniform model led researchers astray". See also Appendices Q.13, Q.14, and Q.15 of our paper for further comments on the concept of non-uniformity.

2   Breaking AES

This section analyzes the cost of various attacks against AES. All of the attacks readily generalize to other block ciphers; none of the attacks exploit any particular weakness of AES. We focus on AES because of its relevance in practice and to have concrete numbers to illustrate the attacks.

All of the (single-target) attacks here are "PRP" attacks: i.e., attacks that distinguish the cipher outputs for a uniform random key (on attacker-selected inputs) from outputs of a uniform random permutation. Some of the attacks go further, recovering the cipher key, but this is not a requirement for a distinguishing attack.

2.1. Breaking AES with MD5. We begin with an attack that does not use any precomputations. This attack is feasible, and in fact quite efficient; its success probability is low, but not nearly as low as one might initially expect. This is a warmup for the higher-success-probability attack of Section 2.2.

Let P be a uniform random permutation of the set $\{0,1\}^{128}$; we label elements of this set in little-endian form as integers $0,1,2,\ldots$ without further comment. The pair $(P(0),P(1))$ is nearly a uniform random 256-bit string: it avoids 2^{128} strings of the form (x,x) but is uniformly distributed among the remaining $2^{256}-2^{128}$ strings.

If k is a uniform random 128-bit string then the pair $(\mathrm{AES}_k(0),\mathrm{AES}_k(1))$ is a highly nonuniform random 256-bit string, obviously incapable of covering more than 2^{128} possibilities. One can reasonably guess that an easy way to distinguish this string from $(P(0),P(1))$ is to feed it through MD5 and output the first bit of the result. The success probability p of this attack — the absolute difference between the attack's average output for input $(\mathrm{AES}_k(0),\mathrm{AES}_k(1))$ and the attack's average output for input $(P(0),P(1))$ — is far below 1, but it is almost certainly above 2^{-80}, and therefore many orders of magnitude above 2^{-128}. See Appendix V for relevant computer experiments.

To understand why p is so large, imagine replacing the first bit of MD5 with a uniform random function from $\{0,1\}^{256}$ to $\{0,1\}$, and assume for simplicity that the 2^{128} keys k produce 2^{128} distinct strings $(\mathrm{AES}_k(0),\mathrm{AES}_k(1))$. Each key k then has a 50% chance of choosing 0 and a 50% chance of choosing 1, and these choices are independent, so the probability that $2^{127}+\delta$ keys k choose 1 is exactly $\binom{2^{128}}{2^{127}+\delta}/2^{2^{128}}$; the probability that $at\ least\ 2^{127}+\delta$ keys k choose 1 is exactly $\sum_{i\geq\delta}\binom{2^{128}}{2^{127}+i}/2^{2^{128}}$; the probability that $at\ most\ 2^{127}-\delta$ keys k choose 1 is the same. The other $2^{256}-2^{129}$ possibilities for $(P(0),P(1))$ are practically guaranteed to have far smaller bias. Consequently p is at least $\approx\delta/2^{128}$ with probability approximately $2\sum_{i\geq\delta}\binom{2^{128}}{2^{127}+i}/2^{2^{128}}\approx 1-\mathrm{erf}(\delta/\sqrt{2^{127}})\approx\exp(-\delta^2/2^{127})$, where erf is the standard error function. For example, p is at least $\approx 2^{-65}$ with probability above 30%, and is at least $\approx 2^{-80}$ with probability above 99.997%.

Of course, MD5 is not actually a uniform random function, but it would be astonishing for MD5 to interact with AES in such a way as to spoil this attack. More likely is that there are some collisions in $k\mapsto(\mathrm{AES}_k(0),\mathrm{AES}_k(1))$; but such collisions are rare unless AES is deeply flawed, and in any event will tend to push δ away from 0, helping the attack.

2.2. Precomputing Larger Success Probabilities.

The same analysis applies to a modified attack D_s that appends a short string s to the AES outputs $(\mathrm{AES}_k(0),\mathrm{AES}_k(1))$ before hashing them: with probability $\approx\exp(-\delta^2/2^{127})$ the attack D_s has success probability at least $\approx\delta/2^{128}$. If s is long enough to push the hash inputs beyond one block of MD5 input then the iterated structure of MD5 seems likely to spoil the attack, so we define D_s using "capacity-1024 Keccak" rather than MD5.

Consider, for example, $\delta=2^{67}$: with probability $\approx 1-\mathrm{erf}(2^{3.5})\approx 2^{-189}$ the attack D_s has success probability at least $\approx 2^{-61}$. There are 2^{192} choices of 192-bit strings s, so presumably at least one of them will have D_s having success probability at least $\approx 2^{-61}$. Of course, actually $finding$ such an s would require inconceivable amounts of computation by the best methods known (searching 2^{189} choices of s, and computing 2^{128} hashes for each choice); but this is not relevant to the definition of insecurity, which considers only the time taken by D_s.

More generally, for any $n \in \{0, 1, 2, \dots, 64\}$ and any s, with probability $\approx 1 - \operatorname{erf}(2^{n+0.5}) \approx \exp(-2^{2n+1})$, the attack D_s has success probability at least $\approx 2^{n-64}$. There are $2^{3 \cdot 2^{2n}}$ choices of $(3 \cdot 2^{2n})$-bit strings s, and $2^{3 \cdot 2^{2n}}$ is considerably larger than $\exp(2^{2n+1})$, so presumably at least one of these values of s will have D_s having success probability at least $\approx 2^{n-64}$.

Similar comments apply to essentially any short-key cipher. There almost certainly *exists* a $(3 \cdot 2^{2n})$-bit string s such that the following simple attack achieves success probability $\approx 2^{n-K/2}$, where K is the number of bits in the cipher key: query $2K$ bits of cipher output, append s, and hash the result to 1 bit. Later we will write p for the success probability; note that the string length is close to $2^K p^2$.

As n increases, the cost of hashing $3 \cdot 2^{2n} + 2K$ bits grows almost linearly with 2^{2n} in the RAM metric and the NAND metric. It grows more quickly in the AT metric: storing the $3 \cdot 2^{2n}$ bits of s uses area at least $3 \cdot 2^{2n}$, and even a heavily parallelizable hash function will take time proportional to 2^n simply to communicate across this area, for a total cost proportional to 2^{3n}. In each metric there are also lower-order terms reflecting the cost of hashing per bit; we suppress these lower-order terms since our concern is with much larger gaps.

2.3. Iteration (Hellman etc.). Large success probabilities are more efficiently achieved by a different type of attack that iterates, e.g., the function $f_7 : \{0, 1\}^{128} \to \{0, 1\}^{128}$ defined by $f_7(k) = \operatorname{AES}_k(0) \oplus 7$.

Choose an attack parameter n. Starting from $f_7(k)$, compute the sequence of iterates $f_7(k), f_7^2(k), f_7^3(k), \dots, f_7^{2^n}(k)$. Look up each of these iterates in a table containing the precomputed quantities $f_7^{2^n}(0), f_7^{2^n}(1), \dots, f_7^{2^n}(2^n - 1)$. If $f_7^j(k)$ matches $f_7^{2^n}(i)$, recompute $f_7^{2^n - j}(i)$ as a guess for k, and verify this guess by checking $\operatorname{AES}_k(1)$.

This computation finds the target key k if k matches any of the following keys: $0, f_7(0), \dots, f_7^{2^n - 1}(0)$; $1, f_7(1), \dots, f_7^{2^n - 1}(1)$; etc. If n is not too large (see the next paragraph) then there are close to 2^{2n} different keys here. The computation involves $\leq 2^n$ initial iterations; 2^n table lookups; and, in case of a match, $\leq 2^n$ iterations to recompute $f_7^{2^n - j}(i)$. The *precomputation* performs many more iterations, but this precomputation is only the cost of *finding* the algorithm, not the cost of *running* the algorithm.

This heuristic analysis begins to break down as $3n$ approaches the key size K. The central problem is that a chain $f_7(i), f_7^2(i), \dots$ could collide with one of the other $2^n - 1$ chains; this occurs with probability $\approx 2^{3n}/2^K$, since there are 2^n keys in this chain and almost 2^{2n} keys in the other chains. The colliding chains will then merge, reducing the coverage of keys and at the same time requiring extra iterations to check more than one value of i. This phenomenon loses a small constant factor in the algorithm performance for $n \approx K/3$ and much more for larger n.

Assume from now on that n is chosen to be close to $K/3$. The algorithm then has success chance $\approx 2^{-K/3}$. The algorithm cost is on the scale of $2^{K/3}$ in both the RAM metric and the NAND metric; for the NAND metric one computes the 2^n independent table lookups by sorting and merging.

This attack might not sound better (in the RAM metric) than the earlier attack D_s, which achieves success chance $\approx 2^{-K/3}$ for some string s with $\approx 2^{K/3}$ bits. The critical feature of this attack is that it recognizes its successes. If the attack fails to find k then one can change 7 to another number and try again, almost doubling the success chance of the algorithm at the expense of doubling its cost; for comparison, doubling the success chance of D_s requires quadrupling its cost. Repeating this attack $2^{K/3}$ times reaches success chance ≈ 1 at cost $2^{2K/3}$.

In the AT metric this attack is much more expensive. The table of precomputed quantities $f_7^{2^n}(0), f_7^{2^n}(1), \ldots, f_7^{2^n}(2^n - 1)$ uses area on the scale of 2^n, and computing $f_7^{2^n}(k)$ takes time on the scale of 2^n, for a total cost on the scale of 2^{2n} for an attack that finds $\approx 2^{2n}$ keys. One can *compute* $f_7^{2^n}(0), f_7^{2^n}(1), \ldots, f_7^{2^n}(2^n - 1)$ in parallel within essentially the same bounds on time and area, replacing each precomputed key with a small circuit that computes the key from scratch; precomputation does not change the exponent of the attack. One can, more straightforwardly, compute any reasonable sequence of 2^{2n} guesses for k within essentially the same cost bound. Achieving success probability p costs essentially $2^K p$.

2.4. Multiple Targets. Iteration becomes more efficient when there are multiple targets: U cipher outputs $\mathrm{AES}_{k_1}(0), \mathrm{AES}_{k_2}(0), \ldots, \mathrm{AES}_{k_U}(0)$ for U independent uniform random keys $k_1, \ldots, k_U$. Assume for simplicity that U is much smaller than 2^K; the hypothesis $U \le 2^{K/4}$ suffices for all heuristics used below.

Compute the iterates $f_7(k_1), f_7^2(k_1), \ldots, f_7^{2^n}(k_1)$, and similarly for each of $k_2, \ldots, k_U$; this takes $2^n U$ iterations. Look up each iterate in a table of $2^n U$ precomputed keys. Handle any match as above.

In the RAM metric or the NAND metric this attack has cost on the scale of $2^n U$, just like applying the previous attack to the U keys separately. The benefit of this attack is that it uses a larger table, producing a larger success probability for each key: the precomputation covers $2^{2n} U$ keys instead of just 2^{2n} keys. To avoid excessive chain collisions one must limit 2^n to $2^{K/3} U^{-1/3}$ so that $2^{3n} U$ does not grow past 2^K; the attack then finds each key with probability $2^{2n} U / 2^K = 2^{-K/3} U^{1/3}$, with a cost of $2^n = 2^{K/3} U^{-1/3}$ per key, a factor of $U^{2/3}$ better than handling each key separately. Finding each key with high probability costs $2^{2K/3} U^{-2/3}$ per key.

As before, the AT metric assigns a much larger cost than the RAM and NAND metrics. The computation of $f_7^{2^n}(k_1), f_7^{2^n}(k_2), \ldots, f_7^{2^n}(k_U)$ is trivially parallelized, taking time on the scale of 2^n, but the $2^n U$ precomputed keys occupy area $2^n U$, for a total cost on the scale of $2^{2n} U$, i.e., 2^{2n} per key, for success probability $2^{2n} U / 2^K$ per key. Note that one can carry out the precomputation using essentially the same area and time. There is a large benefit from handling U keys together — finding all U keys costs essentially 2^K, i.e., $2^K/U$ per key — but this benefit exists whether or not precomputation costs are taken into account.

2.5. Comparison. We summarize the insecurity established by the best attacks presented above. Achieving success probability p against U keys costs

- RAM metric: $\approx 2^K p^2$ for $p \leq 2^{-K/3} U^{-2/3}$; $\approx (2^{2K/3} U^{-2/3}) p$ for larger p.
- NAND metric: same.
- AT metric: $\approx 2^{3K/2} p^3$ for $p \leq 2^{-K/4} U^{-1/2}$; $\approx 2^K U^{-1} p$ for larger p.

Figure G.1 graphs these approximations for $U = 1$, along with the cost of exhaustive search.

2.6. Previous Work. All of the attacks described here have appeared before. In fact, when the conjectures in [16, Section 3.6] and [7, Section 3.2] were made, they were already inconsistent with known attacks.

The iteration idea was introduced by Hellman in [45] for the special case $U = 1$. Many subsequent papers (see, e.g., [25] and [49]) have explored variants and refinements of Hellman's attack, including the easy generalization to larger U. Hellman's goal was to attack many keys for a lower RAM cost than attacking each key separately; Hellman advertised a "cost per solution" of $2^{2K/3}$ using a precomputed table of size $2^{2K/3}$. The generalization to larger U achieves the same goal at lower cost, but the special case $U = 1$ remains of interest as a non-uniform single-key attack.

Koblitz and Menezes in [55] recently considered a family of attacks analogous to D_s. They explained that there should be a short string s where D_s has success probability at least $\approx 2^{-K/2}$, and analyzed some consequences for provable concrete secret-key security. However, they did not analyze higher levels of insecurity.

Replacing D_s with a more structured family of attacks, namely linear cryptanalysis, can be *proven* to achieve insecurity $2^{-K/2}$ at low cost. (See, for example, [39, Section 7], which says that this is "well known in complexity theory".) De, Trevisan, and Tulsiani in [36] proved cost $\approx 2^K p^2$, for both the RAM metric and the NAND metric, for any insecurity level p. A lucid discussion of the gap between these attacks and exhaustive search appears in [36, Section 1], but without any analysis of the resulting trouble for the literature on provable concrete secret-key security, and without any analysis of possible fixes.

Biham, Goren, and Ishai in [23, Section 1.1] pointed out that Hellman's attack causes problems for defining strong one-way functions. The only solution that they proposed was adding uniformity. Note that this solution abandons the goal of giving a definition for, e.g., the strength of AES as a one-way function, or the strength of protocols built on top of AES. We analyze this solution in detail in Appendix B.5.

Our AT analysis appears to be new. In particular, we are not aware of previous literature concluding that switching to the AT metric removes essentially all of the benefit of precomputation for large p, specifically $p > 2^{-K/4} U^{-1/2}$.

3   Breaking the NIST P-256 Elliptic Curve

This section analyzes the cost of an attack against NIST P-256 [67], an elliptic curve of 256-bit prime order ℓ over a 256-bit prime field $\mathbf{F}_p$. The attack computes discrete logarithms on this curve, recovering the secret key from the public key and thus completely breaking typical protocols that use NIST P-256.

The attack does not exploit any particular weakness of NIST P-256. Switching from NIST P-256 to another group of the same size (another curve over the same field, a curve over another field, a hyperelliptic curve, a torus, etc.) does not stop the attack. We focus on NIST P-256 for both concreteness and practical relevance, as in the previous section.

3.1. The Standard Attack without Precomputation. Let P be the specified base point on the NIST P-256 curve. The discrete-logarithm problem on this curve is to find, given another point Q on this curve, the unique integer k modulo ℓ such that $Q = kP$. The standard attack against the discrete-logarithm problem is the parallelization by van Oorschot and Wiener [72] of Pollard's rho method [73], described in the following paragraphs.

This attack uses a pseudorandom walk on the curve points. To obtain the $(i + 1)$-st point P_{i+1}, apply a hash function $h : \mathbf{F}_p \to I$ to the x-coordinate of P_i, select a step $S_{h(x(P_i))}$ from a sequence of precomputed steps $S_j = r_j P$ (with random scalars r_j for $j \in I$), and compute $P_{i+1} = P_i + S_{h(x(P_i))}$. The size of I is chosen large enough to have the walk simulate a uniform random walk; a common choice, recommended in [87], is $|I| = 20$. The walk continues until it hits a distinguished point: a point P_i where the last t bits of $x(P_i)$ are equal to zero. Here t is an attack parameter.

The starting point of the bth walk is of the form $aP + bQ$ where a is chosen randomly. Each step increases the multiple of P, so the distinguished point has the form $a'P + bQ$ for known a', b. The triple $(a'P + bQ, a', b)$ is stored and a new walk is started from a different starting point. If two walks hit the same distinguished point then $a'P + bQ = c'P + dQ$ which gives $(a' - c')P = (d - b)Q$; by construction $d \not\equiv b \bmod \ell$, revealing $k \equiv (a' - c')/(d - b) \bmod \ell$.

After $\sqrt{\ell} \approx 2^{128}$ additions (in approximately 2^{128-t} walks, using storage 2^{128-t}), there is a high chance that the same point has been obtained in two different walks. This collision is recognized from a repeated distinguished point within approximately 2^t additional steps.

3.2. Precomputed Distinguished Points. To use precomputations in this attack, build a database of triples of the form $(a'P, a', 0)$, i.e., starting each walk at a multiple of P. The attack algorithm takes this database and starts a new walk at $aP + bQ$ for random a and b. If this walk ends in a distinguished point present in the database, the DLP is solved. If the walk continues for more than 2^{t+1} steps (perhaps because it is in a cycle) or reaches a distinguished point not present in the database, the attack starts again from a new pair (a, b).

The parameter t is critical for RAM cost here, whereas it did not significantly affect RAM cost in Section 3.1. Choose t as $\lceil (\log_2 \ell)/3 \rceil$. One can see from the following analysis that significantly smaller values of t are much less effective, and that significantly larger values of t are much more expensive without being much more effective.

Construct the database to have exactly 2^t distinct triples, each obtained from a walk of length at least 2^t, representing a total of at least 2^{2t} (and almost certainly $O(2^{2t})$) points. Achieving this requires searching for starting points in the precomputation (and optionally also varying the steps S_j and the hash function) as follows. A point that enters a cycle without reaching a distinguished point is discarded. A point that reaches a distinguished point in fewer than 2^t steps is discarded; each point survives this with probability approximately $(1 - 1/2^t)^{2^t} \approx 1/e$. A point that produces a distinguished point already in the database is discarded; to see that a point survives this with essentially constant probability (independent of ℓ), observe that each new step has chance 2^{-t} of reaching a distinguished point, and chance $O(2^{2t}/\ell) = O(2^{-t})$ of reaching one of the previous $O(2^{2t})$ points represented by the database. Computer experiments that we reported in [22], as a followup to this paper, show that all the O constants here are reasonably close to 1.

Now consider a walk starting from $aP + bQ$. This walk has chance approximately $1/e$ of continuing for at least 2^t steps. If this occurs then those 2^t steps have chance approximately $1 - (1 - 2^{2t}/\ell)^{2^t} \approx 1 - \exp(-2^{3t}/\ell) \geq 1 - 1/e$ of reaching one of the 2^{2t} points in the precomputed walks that were within 2^t of the distinguished points in the database. If this occurs then the walk is guaranteed to reach a distinguished point in the database within a total of 2^{t+1} steps. The algorithm thus succeeds (in this way) with probability at least $(1 - 1/e)/e \approx 0.23$. This is actually an underestimate, since the algorithm can also succeed with an early distinguished point or a late collision.

To summarize, the attack uses a database of approximately $\sqrt[3]{\ell}$ distinguished points; one run of the attack uses approximately $2\sqrt[3]{\ell}$ curve additions and succeeds with considerable probability. The overall attack cost in the RAM metric is a small constant times $\sqrt[3]{\ell}$. The security of NIST P-256 in this metric has thus dropped to approximately 2^{86}. Note that the precomputation here is on the scale of 2^{170}, much larger than the precomputation in Section 2.3 but much smaller than the precomputation in Section 2.2.

In the NAND metric it is simplest to run each walk for exactly 2^{t+1} steps, keeping track of the first distinguished point found by that walk and then comparing that distinguished point to the 2^t points in the database. The overall attack cost is still on the scale of $\sqrt[3]{\ell}$.

In the AT metric the attack cost is proportional to $\sqrt[3]{\ell^2}$, larger than the standard $\sqrt{\ell}$. In this metric one does better by running many walks in parallel: if Z points are precomputed, one should run approximately Z walks in parallel with inputs depending on Q. The precomputation then covers $2^t Z$ points, and the computations involving Q cover approximately $2^t Z$ points, leading to a high probability of success when $2^t Z$ reaches $\sqrt{\ell}$. The AT cost is also $2^t Z$. This attack has the same cost as the standard Pollard rho method, except for small constants; there is no benefit in the precomputations.

3.3. Comparison. We summarize the insecurity established by the best attacks presented above. Achieving success probability p costs

- RAM metric: $\approx (p\ell)^{1/3}$.
- NAND metric: same.
- AT metric: $\approx (p\ell)^{1/2}$.

Figure G.2 graphs these approximations.

3.4. Related Work. Kuhn and Struik in [58] and Hitchcock, Montague, Carter, and Dawson in [46] considered the problem of solving multiple DLPs at once. They obtain a speedup of $\sqrt{U}$ per DLP for solving U DLPs at once. Their algorithm reuses the distinguished points found in the attack on Q_1 to attack Q_2, reuses the distinguished points found for Q_1 and Q_2 to attack Q_3, etc. However, their results do not seem to imply our $\sqrt[3]{\ell}$ result: they do not change the average walk length and distinguished-point probabilities, and they explicitly limit U to $c\sqrt[4]{\ell}$ with $c < 1$. See also the recent paper [61] by Lee, Cheon, and Hong, which considered solving DLPs with massive precomputation for trapdoor DL-groups. None of these papers noticed any implications for provable security, and none of them went beyond the RAM metric.

Our followup paper [22] experimentally verified the algorithm stated above, improved it to $1.77 \cdot \sqrt[3]{\ell}$ additions using $\sqrt[3]{\ell}$ distinguished points, extended it to DLPs in intervals (using slightly more additions), and showed constructive applications in various protocols.

4   Breaking DSA-3072

This section briefly analyzes the cost of an attack against the DSA-3072 signature system. The attack computes discrete logarithms in the DSA-3072 group, completely breaking the signature system.

DSA uses the unique order-q subgroup of the multiplicative group $\mathbf{F}_p^*$, where p and q are primes with q (and not q^2) dividing $p - 1$. DSA-3072 uses a 3072-bit prime p and is claimed to achieve 2^{128} security. The standard parameter choices for DSA-3072 specify a 256-bit prime q, allowing the 2^{86} attack explained in Section 3, but this section assumes that the user has stopped this attack by increasing q to 384 bits (at a performance penalty).

4.1. The Attack. Take $y = 2^{110}$, and precompute $\log_g x^{(p-1)/q}$ for every prime number $x \le y$, where g is the specified subgroup generator. There are almost exactly $y/\log y \approx 2^{103.75}$ such primes, and each $\log_g x^{(p-1)/q}$ fits into 48 bytes, for a total of $2^{109.33}$ bytes.

To compute $\log_g h$, first try to write h as a quotient h_1/h_2 in $\mathbf{F}_p^*$ with $h_2 \in \{1, 2, 3, \ldots, 2^{1535}\}$, $h_1 \in \{-2^{1535}, \ldots, 0, 1, \ldots, 2^{1535}\}$, and $\gcd\{h_1, h_2\} = 1$; and then try to factor h_1, h_2 into primes $\le y$. If this succeeds then $\log_g h^{(p-1)/q}$ is a known combination of known quantities $\log_g x^{(p-1)/q}$, revealing $\log_g h$. If this fails, try again with hg, hg^2, etc.

One can write h as h_1/h_2 with high probability, approximately $(6/\pi^2)2^{3071}/p$, since there are approximately $(6/\pi^2)2^{3071}$ pairs (h_1, h_2) and two distinct such pairs have distinct quotients. Finding the decomposition of h as h_1/h_2 is a very fast extended-Euclid computation.

The probability that h_1 is y-smooth (i.e., has no prime divisors larger than y) is very close to $u^{-u} \approx 2^{-53.06}$ where $u = 1535/110$. The same is true for h_2; overall the attack requires between $2^{107.85}$ and $2^{108.85}$ iterations, depending on $2^{3071}/p$. Batch trial division, analyzed in detail in Section 5, finds the y-smooth values among many choices of h_1 at very low cost in both the RAM metric and the NAND metric. This attack is much slower in the AT metric.

4.2. Previous Work. Standard attacks against DSA-3072 do not rely on pre-computation and cost more than 2^{128} in the RAM metric. These attacks have two stages: the first stage computes discrete logarithms of all primes $\leq y$, and the second stage computes $\log_g h$. Normally y is chosen to minimize the cost of the first stage, whereas we replace the first stage by precomputation and choose y to minimize the cost of the second stage.

The simple algorithm reviewed here is not the state-of-the-art algorithm for the second stage; see, e.g., the "special-q descent" algorithms in [51] and [32]. The gap between known algorithms and existing algorithms is thus even larger than indicated in this section. We expect that reoptimizing these algorithms to minimize the cost of the second stage will produce even better results. We emphasize, however, that none of the algorithms perform well in the AT metric.

5   Breaking RSA-3072

This section analyzes the cost of an attack against RSA-3072. The attack completely breaks RSA-3072, factoring any given 3072-bit public key into its prime factors, so it also breaks protocols such as RSA-3072-FDH and RSA-3072-OAEP.

This section begins by stating a generalization of the attack to any RSA key size, and analyzing the asymptotic cost exponents of the generalized attack. It then analyzes the cost more precisely for 3072-bit keys.

5.1. NFS with Precomputation. This attack is a variant of NFS, the standard attack against RSA. For simplicity this description omits several NFS optimizations. See [30] for an introduction to NFS.

The attack is determined by four parameters: a "polynomial degree" d; a "radix" m; a "height bound" H; and a "smoothness bound" y. Each of these parameters is a positive integer. The attack also includes a precomputed "factory"

$$F = \left\{ (a, b) \in \mathbf{Z} \times \mathbf{Z} : \begin{array}{l} -H \leq a \leq H; \ 0 < b \leq H; \\ \gcd\{a, b\} = 1; \text{ and } a - bm \text{ is } y\text{-smooth} \end{array} \right\}.$$

The standard estimate (see [30]) is that F has $(12/\pi^2)H^2/u^u$ elements where $u = (\log Hm)/\log y$. This estimate combines three approximations: first, there are about $12H^2/\pi^2$ pairs $(a, b) \in \mathbf{Z} \times \mathbf{Z}$ such that $-H \leq a \leq H$, $0 < b \leq H$, and $\gcd\{a, b\} = 1$; second, $a - bm$ has approximately the same smoothness chance as a uniform random integer in $[1, Hm]$; third, the latter chance is approximately $1/u^u$.

The integers N factored by the attack will be between m^d and m^{d+1}. For example, with parameters $m = 2^{256}$, $d = 7$, $H = 2^{55}$, and $y = 2^{50}$, the attack factors integers between 2^{1792} and 2^{2048}. Parameter selection is analyzed later in more detail. The following three paragraphs explain how the attack handles N.

Write N in radix m: i.e., find $n_0, n_1, \ldots, n_d \in \{0, 1, \ldots, m-1\}$ such that $N = n_d m^d + n_{d-1} m^{d-1} + \cdots + n_0$. Compute the "set of relations"

$$R = \{(a, b) \in F : n_d a^d + n_{d-1} a^{d-1} b + \cdots + n_0 b^d \text{ is } y\text{-smooth}\}$$

using Bernstein's batch trial-division algorithm [19]. The standard estimate is that R has $(12/\pi^2)H^2/(u^u v^v)$ elements where $v = (\log((d+1)H^d m))/\log y$.

We pause the attack description to emphasize two important ways that this attack differs from conventional NFS: first, conventional NFS chooses m as a function of N, while this attack does not; second, conventional NFS computes R by sieving all pairs (a, b) with $-H \leq a \leq H$ and $0 < b \leq H$ to detect smoothness of $a - bm$ and $n_d a^d + \cdots + n_0 b^d$ simultaneously, while this attack computes R by batch trial division of $n_d a^d + \cdots + n_0 b^d$ for the limited set of pairs $(a, b) \in F$.

The rest of the attack proceeds in the same way as conventional NFS. There is a standard construction of a sparse vector modulo 2 for each $(a, b) \in R$, and there is a standard way to convert several linear dependencies between the vectors into several congruences of squares modulo N, producing the complete prime factorization of N; see [30] for details. The number of components of each vector is approximately $2y/\log y$, and standard sparse-matrix techniques find linear dependencies using about $4y/\log y$ simple operations on dense vectors of length $2y/\log y$. If the number of elements of R is larger than the number of components of each vector then linear dependencies are guaranteed to exist.

5.2. Asymptotic Exponents. Write $L = \exp((\log N)^{1/3}(\log \log N)^{2/3})$. For the RAM metric it is best to choose

$$d \in (1.1047\ldots + o(1))(\log N)^{1/3}(\log \log N)^{-1/3},$$

$$\log m \in (0.9051\ldots + o(1))(\log N)^{2/3}(\log \log N)^{1/3},$$

$$\log y \in (0.8193\ldots + o(1))(\log N)^{1/3}(\log \log N)^{2/3} = (0.8193\ldots + o(1))\log L,$$

$$\log H \in (1.0034\ldots + o(1))(\log N)^{1/3}(\log \log N)^{2/3} = (1.0034\ldots + o(1))\log L.$$

so that

$$u \in (1.1047\ldots + o(1))(\log N)^{1/3}(\log \log N)^{-1/3},$$

$$u \log u \in (0.3682\ldots + o(1))(\log N)^{1/3}(\log \log N)^{2/3} = (0.3682\ldots + o(1))\log L,$$

$$d \log H \in (1.1085\ldots + o(1))(\log N)^{2/3}(\log \log N)^{1/3},$$

$$v \in (2.4578\ldots + o(1))(\log N)^{1/3}(\log \log N)^{-1/3},$$

$$v \log v \in (0.8193\ldots + o(1))(\log N)^{1/3}(\log \log N)^{2/3} = (0.8193\ldots + o(1))\log L.$$

Out of the $L^{2.0068...+o(1)}$ pairs (a, b) with $-H \le a \le H$ and $0 < b \le H$, there are $L^{1.6385...+o(1)}$ pairs in the factory F, and $L^{0.8193...+o(1)}$ relations in R, just enough to produce linear dependencies if the $o(1)$ terms are chosen appropriately. Linear algebra uses $y^{2+o(1)} = L^{1.6385...+o(1)}$ bit operations.

The total RAM cost of this factorization algorithm is thus $L^{1.6385...+o(1)}$. For comparison, factorization is normally claimed to cost $L^{1.9018...+o(1)}$ (in the RAM metric) with state-of-the-art variants of NFS. Similar comments apply to the NAND metric.

This algorithm runs into trouble in the AT metric. The algorithm needs space to store all the elements of F, and can compute R in time $L^{o(1)}$ using a chip of that size (applying ECM to each input in parallel rather than using batch trial division), but even the most heavily parallelized sparse-matrix techniques need much more than $L^{o(1)}$ time, raising the AT cost of the algorithm far above the size of F. A quantitative analysis shows that one obtains a better cost exponent by skipping the precomputation of F and instead computing the elements of F one by one on a smaller circuit, for AT cost $L^{1.9760...+o(1)}$.

5.3. RAM Cost for RSA-3072.
This attack breaks RSA-3072 with RAM cost considerably below the 2^{128} security level usually claimed for RSA-3072. Of course, justifying this estimate requires replacing the above $o(1)$ terms with more precise cost analyses.

For concreteness, assume that the RAM supports 128-bit pointers, unit-cost 256-bit vector operations, and unit-cost 256-bit floating-point multiplications. As justification for these assumptions, observe that real computers ten years ago supported 32-bit pointers, unit-cost 64-bit vector operations, and unit-cost 64-bit floating-point multiplications; that the RAM model requires operations to scale logarithmically with the machine size; and that previous NFS cost analyses implicitly make similar assumptions.

Take $m = 2^{384}$, $d = 7$, $H = 2^{62} + 2^{61} + 2^{57}$, and $y = 2^{66} + 2^{65}$. There are about $12H^2/\pi^2 \approx 2^{125.51}$ pairs (a, b) with $-H \le a \le H, 0 < b \le H$, and $\gcd\{a, b\} = 1$, and the integers $a - bm$ have smoothness chance approximately $u^{-u} \approx 2^{-18.42}$ where $u = (\log Hm)/\log y \approx 6.707$, so there are about $2^{107.09}$ pairs in the factory F. Each pair in F is small, easily encoded as just 16 bytes.

The quantities $n_d a^d + n_{d-1} a^{d-1} b + \cdots + n_0 b^d$ are bounded by $(d+1)mH^d \approx 2^{825.3}$. If they were uniformly distributed up to this bound then they would have smoothness chance approximately $v^{-v} \approx 2^{-45.01}$ where $v = (\log((d+1)mH^d))/\log y \approx 12.395$, so there would be approximately $(12H^2/\pi^2)u^{-u}v^{-v} \approx 2^{62.08}$ relations, safely above $2y/\log y \approx 2^{62.06}$. The quantities $n_d a^d + n_{d-1} a^{d-1} b + \cdots + n_0 b^d$ are actually biased towards smaller values and thus have larger smoothness chance, but this refinement is unnecessary here.

Batch trial division checks smoothness of 2^{58} of these quantities simultaneously; here 2^{58} is chosen so that the product of those quantities is larger (about $2^{67.69}$ bits) than the product of all the primes $\le y$ (about $2^{67.11}$ bits). The main steps in batch trial division are computing a product tree of these quantities and then computing a scaled remainder tree. Bernstein's cost analysis in [20, Section 3] shows that the overall cost of these two steps, for T inputs having a B-bit product, is approximately $(5/6) \log_2 T$ times the cost of a single multiplication of two $(B/2)$-bit integers. For us $T = 2^{58}$ and $B \approx 2^{67.69}$, and the cost of batch trial division is approximately $2^{5.59}$ times the cost of multiplying two $(B/2)$-bit integers; the total cost of smoothness detection for all $(a, b) \in F$ is approximately $2^{54.68}$ times the cost of multiplying two $(B/2)$-bit integers.

It is easiest to follow a standard floating-point multiplication strategy, dividing each $(B/2)$-bit input into $B/(2w)$ words for some word size $w \in \Omega(\log_2 B)$ and then performing three real floating-point FFTs of length B/w. Each FFT uses approximately $(17/9)(B/w) \log_2(B/w)$ arithmetic operations (additions, subtractions, and multiplications) on words of slightly more than $2w$ bits, for a total of $(17/3)(B/w) \log_2(B/w)$ arithmetic operations. A classic observation of Schönhage [82] is that the RAM metric allows constant-time multiplication of $\Theta(\log_2 B)$-bit integers in this context even if the machine model is not assumed to be equipped with a multiplier, since one can afford to build large multiplication tables; but it is simpler to take advantage of the hypothesized 256-bit multiplier, which comfortably allows $w = 69$ and $B/w < 2^{61} + 2^{60}$, for a total multiplication cost of $2^{70.03}$. Computing R then costs approximately $2^{124.71}$.

Linear algebra involves $2^{63.06}$ simple operations on vectors of length $2^{62.06}$. Each operation produces each output bit by xoring together a small number of input bits, on average fewer than 32 bits. A standard block-Wiedemann computation merges 256 xors of bits into a single 256-bit xor with negligible overhead, for a total linear-algebra cost of $2^{122.12}$. All other steps in the algorithm have negligible cost, so the final factorization cost is $2^{124.93}$.

5.4. Previous Work. There are two frequently quoted cost exponents for NFS without precomputation. Buhler, Lenstra, and Pomerance in [30] obtained RAM cost $L^{1.9229...+o(1)}$. Coppersmith in [33] introduced a "multiple number fields" tweak and obtained RAM cost $L^{1.9018...+o(1)}$.

Coppersmith also introduced NFS with precomputation in [33], using ECM for smoothness detection. Coppersmith called his algorithm a "factorization factory", emphasizing the distinction between precomputation time (building the factory) and computation time (running the factory). Coppersmith computed the same RAM exponent 1.6385... shown above for the cost of one factorization using the factory.

We save a subexponential factor in the RAM cost of Coppersmith's algorithm by switching from ECM to batch trial division. This is not visible in the asymptotic exponent 1.6385... but is important for RSA-3072. Our concrete analysis of RSA-3072 security is new, and as far as we know is the *first concrete analysis of Coppersmith's algorithm*.

Bernstein in [18] obtained AT exponent 1.9760... for NFS without precomputation, and emphasized the gap between this exponent and the RAM exponent 1.9018.... Our AT analysis of NFS with precomputation, and in particular our conclusion that this precomputation increases the AT cost of NFS, appears to be new.

References (see full paper for more references, appendices)

[7] Bellare, M.: New proofs for NMAC and HMAC: security without collision-resistance. In: Crypto 2006 [40], pp. 602–619 (2006) Cited in §1, §1.1, §1.3, §2.6

[11] Bellare, M., Kilian, J., Rogaway, P.: The security of cipher block chaining. In: Crypto 1994 [38], pp. 341–358 (1994); see also newer version [12]. Cited in §1.2

[12] Bellare, M., Kilian, J., Rogaway, P.: The security of the cipher block chaining message authentication code. Journal of Computer and System Sciences 61, 362–399 (2000); see also older version [11]. Cited in §1, §1, §1, §1.2, §1.2, §1.2

[14] Bellare, M., Rogaway, P.: Optimal asymmetric encryption — how to encrypt with RSA. In: Eurocrypt 1994 [37], pp. 92–111 (1995) Cited in §1

[15] Bellare, M., Rogaway, P.: The exact security of digital signatures: how to sign with RSA and Rabin. In: Eurocrypt 1996 [64], pp. 399–416 (1996) Cited in §1, §1.1

[16] Bellare, M., Rogaway, P.: Introduction to modern cryptography (2005), http://cseweb.ucsd.edu/~mihir/cse207/classnotes.html. Cited in §1, §1.1, §2.6

[18] Bernstein, D.J.: Circuits for integer factorization: a proposal (2001), http://cr.yp.to/papers.html#nfscircuit. Cited in §5.4

[19] Bernstein, D.J.: How to find smooth parts of integers (2004), http://cr.yp.to/papers.html#smoothparts. Cited in §5.1

[20] Bernstein, D.J.: Scaled remainder trees (2004), http://cr.yp.to/papers.html#scaledmod. Cited in §5.3

[22] Bernstein, D.J., Lange, T.: Computing small discrete logarithms faster. In: Indocrypt 2012 [41], pp. 317–338 (2012) Cited in §3.2, §3.4

[23] Biham, E., Goren, Y.J., Ishai, Y.: Basing weak public-key cryptography on strong one-way functions. In: TCC 2008 [31], pp. 55–72 (2008) Cited in §2.6

[25] Biryukov, A., Shamir, A.: Cryptanalytic time/memory/data tradeoffs for stream ciphers. In: Asiacrypt 2000 [70], pp. 1–13 (2000) Cited in §2.6

[27] Bogdanov, A., Khovratovich, D., Rechberger, C.: Biclique cryptanalysis of the full AES. In: Asiacrypt 2011 [60], pp. 344–371 (2011) Cited in §1

[29] Brent, R.P., Kung, H.T.: The area-time complexity of binary multiplication. Journal of the ACM 28, 521–534 (1981) Cited in §1.2

[30] Buhler, J.P., Lenstra Jr., H.W., Pomerance, C.: Factoring integers with the number field sieve. In: [63], pp. 50–94 (1993) Cited in §5.1, §5.1, §5.1, §5.4

[31] Canetti, R. (ed.): TCC 2008. LNCS, vol. 4948. Springer (2008). See [23]

[32] Commeine, A., Semaev, I.: An algorithm to solve the discrete logarithm problem with the number field sieve. In: PKC 2006 [91], pp. 174–190 (2006) Cited in §4.2

[33] Coppersmith, D.: Modifications to the number field sieve. Journal of Cryptology 6, 169–180 (1993) Cited in §5.4, §5.4

[35] De, A., Trevisan, L., Tulsiani, M.: Non-uniform attacks against one-way functions and PRGs. Electronic Colloquium on Computational Complexity 113 (2009); see also newer version [36]

[36] De, A., Trevisan, L., Tulsiani, M.: Time space tradeoffs for attacks against one-way functions and PRGs. In: Crypto 2010 [75], pp. 649–665 (2010); see also older version [35]. Cited in §2.6, §2.6

[37] De Santis, A. (ed.): Eurocrypt 1994. LNCS, vol. 950. Springer (1995). See [14]

[38] Desmedt, Y. (ed.): Crypto 1994. LNCS, vol. 839. Springer (1994). See [11]

[39] Dodis, Y., Steinberger, J.: Message authentication codes from unpredictable block ciphers. In: Crypto 2009 [44], pp. 267–285 (2009) Cited in §2.6

[40] Dwork, C. (ed.): Crypto 2006. LNCS, vol. 4117. Springer (2006). See [7]

[41] Galbraith, S., Nandi, M. (eds.): Indocrypt 2012. LNCS, vol. 7668. Springer (2012). See [22]

[44] Halevi, S. (ed.): Crypto 2009. LNCS, vol. 5677. Springer (2009). See [39]

[45] Hellman, M.E.: A cryptanalytic time-memory tradeoff. IEEE Transactions on Information Theory 26, 401–406 (1980) Cited in §2.6

[46] Hitchcock, Y., Montague, P., Carter, G., Dawson, E.: The efficiency of solving multiple discrete logarithm problems and the implications for the security of fixed elliptic curves. International Journal of Information Security 3, 86–98 (2004) Cited in §3.4

[49] Hong, J., Sarkar, P.: New applications of time memory data tradeoffs. In: Asiacrypt 2005 [78], pp. 353–372 (2005) Cited in §2.6

[51] Joux, A., Lercier, R.: Improvements to the general number field sieve for discrete logarithms in prime fields. A comparison with the Gaussian integer method. Mathematics of Computation 72, 953–967 (2003) Cited in §4.2

[53] Katz, J., Lindell, Y.: Introduction to modern cryptography: principles and protocols. Chapman & Hall/CRC (2007) Cited in §1

[55] Koblitz, N., Menezes, A.: Another look at HMAC (2012), http://eprint.iacr.org/2012/074. Cited in §1.2, §1.3, §1.3, §1.3, §2.6

[56] Koblitz, N., Menezes, A.: Another look at non-uniformity (2012), http://eprint.iacr.org/2012/359. Cited in §1.2, §1.3, §1.3

[58] Kuhn, F., Struik, R.: Random walks revisited: extensions of Pollard's rho algorithm for computing multiple discrete logarithms. In: SAC 2001 [89], pp. 212–229 (2001) Cited in §3.4

[60] Lee, D.H., Wang, X. (eds.): Asiacrypt 2011. LNCS, vol. 7073. Springer (2011). See [27]

[61] Lee, H.T., Cheon, J.H., Hong, J.: Accelerating ID-based encryption based on trapdoor DL using pre-computation, 11 January 2012 (2012), http://eprint.iacr.org/2011/187. Cited in §3.4

[63] Lenstra, A.K., Lenstra Jr., H.W. (eds.): The development of the number field sieve. LNM, vol. 1554. Springer (1993). See [30]

[64] Maurer, U.M. (ed.): Eurocrypt 1996. LNCS, vol. 1070. Springer (1996). See [15]

[66] NIST: Announcing request for candidate algorithm nominations for the Advanced Encryption Standard (AES) (1997), http://www.gpo.gov/fdsys/pkg/FR-1997-09-12/pdf/97-24214.pdf. Cited in §1

[67] NIST: Digital signature standard, Federal Information Processing Standards Publication 186-2 (2000), http://csrc.nist.gov. Cited in §3

[70] Okamoto, T. (ed.): Asiacrypt 2000. LNCS, vol. 1976. Springer (2000). See [25]

[72] van Oorschot, P.C., Wiener, M.: Parallel collision search with cryptanalytic applications. Journal of Cryptology 12, 1–28 (1999) Cited in §3.1

[73] Pollard, J.M.: Monte Carlo methods for index computation mod p. Mathematics of Computation 32, 918–924 (1978) Cited in §3.1

[75] Rabin, T. (ed.): Crypto 2010. LNCS, vol. 6223. Springer (2010). See [36]

[78] Roy, B. (ed.): Asiacrypt 2005. LNCS, vol. 3788. Springer (2005). See [49]

[82] Schönhage, A.: Storage modification machines. SIAM Journal on Computing 9, 490–508 (1980) Cited in §5.3

[87] Teske, E.: On random walks for Pollard's rho method. Mathematics of Computation 70, 809–825 (2001) Cited in §3.1

[89] Vaudenay, S., Youssef, A.M. (eds.): SAC 2001. LNCS, vol. 2259. Springer (2001). See [58]

[91] Yung, M., Dodis, Y., Kiayias, A., Malkin, T. (eds.): PKC 2006. LNCS, vol. 3958. Springer (2006). See [32]

Factoring RSA Keys from Certified Smart Cards: Coppersmith in the Wild

Daniel J. Bernstein[1,2], Yun-An Chang[3], Chen-Mou Cheng[3], Li-Ping Chou[4], Nadia Heninger[5], Tanja Lange[2], and Nicko van Someren[6]

[1] Department of Computer Science, University of Illinois at Chicago, USA
djb@cr.yp.to
[2] Department of Mathematics and Computer Science
Technische Universiteit Eindhoven, The Netherlands
tanja@hyperelliptic.org
[3] Research Center for Information Technology Innovation
Academia Sinica, Taipei, Taiwan
{ghfjdksl,doug}@crypto.tw
[4] Department of Computer Science and Information Engineering
Chinese Culture University, Taipei, Taiwan
randomalg@gmail.com
[5] Department of Computer and Information Science, University of Pennsylvania
nadiah@cis.upenn.edu
[6] Good Technology Inc.
nicko@good.com

Abstract. This paper explains how an attacker can efficiently factor 184 distinct RSA keys out of more than two million 1024-bit RSA keys downloaded from Taiwan's national "Citizen Digital Certificate" database. These keys were generated by government-issued smart cards that have built-in hardware random-number generators and that are advertised as having passed FIPS 140-2 Level 2 certification.

These 184 keys include 103 keys that share primes and that are efficiently factored by a batch-GCD computation. This is the same type of computation that was used last year by two independent teams (USENIX Security 2012: Heninger, Durumeric, Wustrow, Halderman; Crypto 2012: Lenstra, Hughes, Augier, Bos, Kleinjung, Wachter) to factor tens of thousands of cryptographic keys on the Internet.

The remaining 81 keys do not share primes. Factoring these 81 keys requires taking deeper advantage of randomness-generation failures: first using the shared primes as a springboard to characterize the failures, and then using Coppersmith-type partial-key-recovery attacks. This is the first successful public application of Coppersmith-type attacks to keys found in the wild.

Keywords: RSA, smart cards, factorization, Coppersmith, lattices.

This work was supported by NSF (U.S.) under grant 1018836, by NWO (Netherlands) under grants 639.073.005 and 040.09.003, and by NSC (Taiwan) under grant 101-2915-I-001-019. Cheng worked on this project while at Technische Universität Darmstadt under the support of Alexander von Humboldt-Stiftung. Heninger worked on this project while at Microsoft Research New England. Permanent ID of this document: 278505a8b16015f4fd8acae818080edd. Date: 2013.09.10.

K. Sako and P. Sarkar (Eds.) ASIACRYPT 2013, Part II, LNCS 8270, pp. 341–360, 2013.

1   Introduction

In 2003, Taiwan introduced an e-government initiative to provide a national public-key infrastructure for all citizens. This national certificate service allows citizens to use "smart" ID cards to digitally authenticate themselves to government services, such as filing income taxes and modifying car registrations online, as well as to a growing number of non-government services. RSA keys are generated by the cards, digitally signed by a government authority, and placed into an online repository of "Citizen Digital Certificates".

On some of these smart cards, unfortunately, the random-number generators used for key generation are fatally flawed, and have generated real certificates containing keys that provide no security whatsoever. This paper explains how we have computed the secret keys for 184 different certificates.

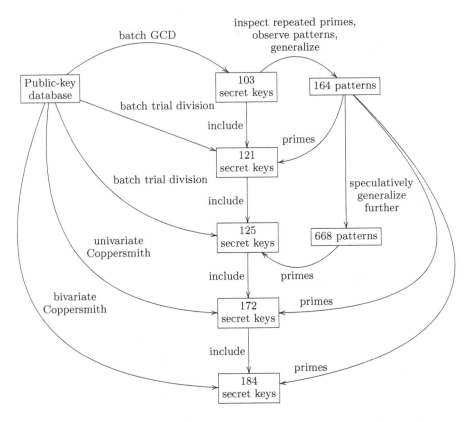

Fig. 1. Retrospective summary of the data flow leading to successful factorizations. After successfully factoring keys using a batch GCD algorithm, we characterized the failures, and used trial division to check for broader classes of specified primes (input on the right) as exact divisors. We then extended the attack and applied Coppersmith's method to check for the specified primes as approximate divisors.

1.1   Factorization Techniques

Bad randomness is not new. Last year two independent research teams [13,17] exploited bad randomness to break tens of thousands of keys of SSL certificates on the Internet, a similar number of SSH host keys, and a few PGP keys.

Our starting point in this work is the same basic attack used in those papers against poorly generated RSA keys, namely scanning for pairs of distinct keys that share a common divisor (see Section 3). The basic GCD attack, applied to the entire database of Citizen Digital Certificates, shows that 103 keys factor into 119 different primes.

We go beyond this attack in several ways. First, the shared primes provide enough data to build a model of the prime-generation procedure. It is surprising to see *visible* patterns of non-randomness in the primes generated by these smart cards, much more blatant non-randomness than the SSL key-generation failures identified by [13,17]. One expects smart cards to be controlled environments with built-in random-number generators, typically certified to meet various standards and practically guaranteed to avoid such obvious patterns. For comparison, the SSL keys factored last year were typically keys generated by low-power networked devices such as routers and firewalls running the Linux operating system while providing none of the sources of random input that Linux expects.

The next step is extrapolation from these prime factors: we hypothesize a particular model of randomness-generation failures consistent with 18 of the common divisors. The same model is actually capable of generating 164 different primes, and testing all of those primes using batch trial division successfully factors further keys. One might also speculate that the cards can generate primes fitting a somewhat broader model; this speculation turns out to be correct, factoring a few additional keys and bringing the total to 125. See Section 4 for a description of the patterns in these primes.

There are also several prime factors that are similar to the 164 patterns but that contain sporadic errors: some bits flipped here and there, or short sequences of altered bits. We therefore mount several Coppersmith-style lattice-based partial-key-recovery attacks to efficiently find prime divisors close to the patterns. The univariate attacks (Section 5) allow an arbitrary stretch of errors covering the bottom 40% of the bits of the prime. The bivariate attacks (Section 6) allow two separate stretches of errors. The internal structure of the patterns makes them particularly susceptible to these attacks. These attacks produce dozens of additional factorizations, raising the total to 184.

In the end nearly half of the keys that we factored did *not* share any common divisors with other keys; most of these were factored by the Coppersmith-style attacks. This is, to our knowledge, the first publicly described instance of a Coppersmith-style attack breaking keys in the wild.

1.2   Certification

The flawed keys were generated by government-issued smart cards that both the certification authority and manufacturer advertise as having passed stringent standards certifications. See Section 2.1.

It is clear from their externally visible behavior, as shown in this paper, that the random-number generators used to generate the vulnerable keys actually fall far short of these standards. This demonstrates a failure of the underlying hardware and the card's operating system, both of which are covered by certification.

1.3   Response to Vulnerabilities

When we reported the common-divisor vulnerabilities to government authorities, their response was to revoke exactly the certificates sharing common factors and to issue new cards only to those users. See Section 7 for more details.

Our further factorizations demonstrate how dangerous this type of response is. Randomness-generation failures sometimes manifest themselves as primes appearing twice, but sometimes manifest themselves as primes that appear only once, such as the primes that we found by Coppersmith-type attacks. Both cases are vulnerable to attackers with adequate models of the randomness-generation process, while only the first case is caught by central testing for repeated primes.

We endorse the idea of centrally testing RSA moduli for common divisors as a mechanism to detect some types of randomness-generation failures. We emphasize that finding repeated primes is much more than an indication that those particular RSA keys are vulnerable: it shows that the underlying randomness-generation system is malfunctioning. The correct response is not merely to eliminate those RSA keys but to revoke all keys generated with that generation of hardware and throw away the entire randomness-generation system, replacing it with a properly engineered system.

We also emphasize that an absence of common divisors is not an indication of security. If the primes generated by these smart cards had been modified to include a card serial number as their top bits then the keys would have avoided common divisors but the primes would still have been reasonably predictable to attackers. Our work illustrates several methods of translating different types of malfunctioning behavior into concrete vulnerabilities. There are many potential vulnerabilities resulting from bad randomness; it is important to thoroughly test every component of a random-number generator, not merely to look for certain types of extreme failures.

2   Background

2.1   The Taiwan Citizen Digital Certificate Program

Taiwan's Citizen Digital Certificates (CDCs) are a standard means of authentication whenever Taiwanese citizens want to do business over the Internet with the government and an increasing number of private companies.

CDCs are issued by the Ministry of Interior Certificate Authority (MOICA), a level 1 subordinate CA of the Taiwanese governmental PKI. Since the program's launch in 2003, more than 3.5 million CDCs have been issued, providing public key certificate and attribute certificate services. These digital certificates form a basis for the Taiwanese government's plan to migrate to electronic certificates from existing paper certificates for a range of applications including national and other identification cards, driver's licenses, and various professional technician licenses.

Today, Taiwanese citizens can already use the CDC to authenticate themselves over the Internet in a number of important government applications, e.g., to file personal income taxes, update car registration, and make transactions with government agencies such as property registries, national labor insurance, public safety, and immigration. In addition, the CDC is accepted as a means of authentication by a variety of organizations such as the National Science Council, several local governments, and recently some private companies such as Chunghwa Telecom. Overall, the CDC program appears quite successful as a two-sided network, as it has attracted an increasing number of both applications and subscribers.

Certificate registration: In order to generate CDCs, citizens bring their (paper) ID cards to a government registration office. A government official places the (smart) ID card into a registration device. The device prompts the card to generate a new cryptographic key, and the public key is incorporated into a certificate to be signed by MOICA. The certificate is made available in a database online for authentication purposes. In general, an individual will have two certificates: one for signing, and one for encryption, each with distinct keys.

Standards certifications: MOICA states that these cards are "high security", and "have been accredited to FIPS 140-1 level 2", and also that "A private key is created inside and the private key can't export from IC card after key created". (See [20] or search for "FIPS" on MOICA's website http://moica.nat.gov.tw/html/en/index.htm.) For comparison, the SSL keys factored last year were generated by software-hardware combinations that had never claimed to be evaluated for cryptographic security, such as Linux running on a home router.

2.2   Collecting Certificates

In March 2012, inspired by the results of [13] and [17], we retrieved 3002273 CDCs from the MOICA LDAP directory at ldap://moica.nat.gov.tw. Out of these CDCs, 2257569 have 1024-bit RSA keys, while the remaining, newer 744704 have 2048-bit RSA keys, as in 2010 MOICA migrated to 2048-bit RSA and stopped issuing certificates of 1024-bit RSA keys.

The 1024-bit CDCs contain 2086177 distinct moduli, of which 171366 moduli appear more than once. The repeated moduli appear to all be due to expired certificates still contained in the database, which contain the same keys as renewal certificates issued to the same individuals.

2.3   Random Number Generation

While generating high-quality random numbers is critical to the security of cryptographic systems, it is also notoriously difficult to do. Non-deterministic behavior is considered to be a fault in almost every other component of a computer, but it is a crucial component of generating random numbers that an attacker cannot predict. Several national and international standards for random number generation [22,1,11] specify correct behavior for these types of systems. In general, software pseudo-random number generators require a significant amount of entropy before their output is useful for cryptographic purposes.

As we will see later in the paper, the smart cards used in the PKI we examined fail to follow many well-known best practices and standards in hardware random number generation: they appear to utilize a source of randomness that is prone to failing, they fail to perform any run-time testing before generating keys, and they clearly do not apply any post-processing to the randomness stream. The lack of testing or post-processing causes the initial randomness-generation failure to be much more damaging than it would have been otherwise.

Analog RNG circuits: An analog circuit is the standard choice when hardware designers have the luxury of designing dedicated circuits for random-number generation. An analog circuit allows the designer to obtain randomness from simple quantum effects. While the use of radioactive decay is rare in commercial products, the quantum noise exhibited by a current through a suitably biased diode can be amplified and sampled to deliver a high-quality entropy source.

On-chip RNG circuits: Mixing analog and digital circuits on the same die is costly, so chip designers often seek other sources of unpredictability. These sources can include variation in gate propagation delays or gate metastability, which exhibit inherent randomness. Designers can explicitly harness gate-delay variation by building sets of free-running ring oscillators and sampling the behavior at hopefully uncorrelated intervals. To take advantage of randomness in gate metastability, designers build circuits that output bits based on the time it takes for the circuit to settle to a steady state, a variable which should be hard to predict. These designs are often tricky to get right, as the chip fabrication process can reduce or eliminate these variations, and subtle on-chip effects such as inductive coupling or charge coupling between components can cause free-running oscillators to settle into synchronised patterns and metastable circuits to predictably land one way or the other depending on other components nearby on the chip.

Handling entropy sources: Even with a perfectly unpredictable source of randomness, care needs to be taken to convert the raw signal into usable random numbers. Generally, designers characterize circuits in advance to understand the entropy density, test the signal from the entropy source at run time, and run the output through a compression function such as a cryptographically secure hash function. These practices are required by a number of security standards such as FIPS 140 [21].

3   Batch GCD

This section reviews the approach of [13,17] for detecting common factors in a collection of RSA keys, and reports the results of this approach applied to the collection of Citizen Digital Certificates.

If there are two distinct RSA moduli $N_1 = pq_1$ and $N_2 = pq_2$ sharing exactly one prime factor p, then the greatest common divisor of N_1 and N_2 will be p. Computing this GCD is fast, and dividing it out of N_1 and N_2 produces the other factors q_1 and q_2.

Of course, this type of vulnerability should never arise for properly generated RSA keys. However, since [13,17] had observed weak random-number generators producing keys with repeated factors in the wild, we began by checking whether there were repeated factors among the Citizen Digital Certificates.

Instead of the naive quadratic-time method of doing this computation (checking each N_1 against each N_2), we used a faster batch-GCD algorithm using product and remainder trees described in [2,13]. We used the C implementation available at `https://factorable.net/resources.html`.

We ran this implementation on the 3192962 distinct RSA moduli and found that 103 moduli were factored due to nontrivial common factors. This computation, parallelized across four cores of a 3.1GHz AMD FX-8120, finished in just 45 minutes.

4   Attacking Patterned Factors

A properly functioning random number generator would never generate identical 512-bit primes, so the discovery of repeated prime factors described in the previous section immediately indicates that the random-number-generation process producing these keys is broken. This section analyzes the structure of the repeated factors generated by the flawed random-number generator and designs a targeted attack against this structure.

The 103 moduli with repeated factors show a remarkable distribution of the shared factors; see Figure 2. The complete list of factors found using the GCD approach is given in Appendix A.

One prime factor, p110, appears a total of 46 times with different second primes. The hexadecimal representation of this factor is

```
0xc0000000000000000000000000000000000000000000000000000000000000000
000000000000000000000000000000000000000000000000000000000000002f9
```

which is the next prime after $2^{511} + 2^{510}$.

The next most common factor, repeated 7 times, is

```
0xc92424922492924992494924492424922492924992494924492424922492924
99249492449242492249292499249492449242492249292499249492449242 4e5
```

which displays a remarkable periodic structure. The binary representation of this integer, excluding a few most and least significant bits, is a repeated sequence of the string 001 with a "hiccup" every 16 bits.

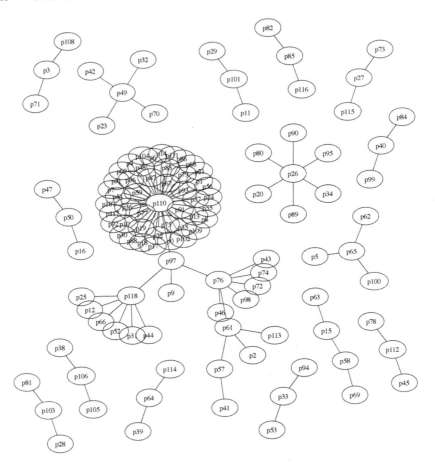

Fig. 2. Relationships between keys with shared factors. Each ellipse represents a prime; edges connect prime factors dividing the same modulus.

Nearly all of the shared prime factors had a similar and immediately apparent periodic structure. We hypothesized that nearly every repeated prime factor had been generated using the following process:

1. Choose a bit pattern of length 1, 3, 5, or 7 bits, repeat it to cover more than 512 bits, and truncate to exactly 512 bits.
2. For every 32-bit word, swap the lower and upper 16 bits.
3. Fix the most significant two bits to 11.
4. Find the next prime greater than or equal to this number.

We generated the 164 distinct primes of this form corresponding to all patterns of length 1, 3, 5, and 7 and tested divisibility with each modulus. This factored a total of 105 moduli, including 18 previously unfactored moduli, for a total of 121.

None of the repeated primes exhibit a (minimal) period of length 9 or larger. On the other hand, the data for period lengths 1, 3, 5, 7 shows that patterns with longer periods typically appear in fewer keys than patterns with shorter periods, and are thus less likely to appear as divisors of two or more keys, raising the question of whether there are primes with larger periods that appear in only one key and that are thus not found by the batch-GCD computation. We therefore extended this test to include length-9 periods and length-11 periods. The length-9 periods factored 4 more keys but the length-11 periods did not factor any new keys, leading us to speculate that 3, 5, and 7 are the only factors of the period length. We then ran a complete test on all length-15 patterns but did not find any further factors. The total number of certificates broken by these divisibility tests, together with the initial batch-GCD computation, is 125.

Sporadic errors: The handful of shared prime factors in our sample of GCD-factored keys that did not match the above form were differing from patterns in very few positions. We experimented with finding more factors using brute-force search starting from 0xc0...0 and found a few new factors, but these factors are more systematically and efficiently found using LLL in Coppersmith's method, as described in the next section.

We also experimented with searching for sporadic errors in factors using the techniques of Heninger and Shacham [14] and Paterson, Polychroniadou, and Sibborn [23]. The main idea is to assume that both of the factors of a weak modulus share nearly all bits in common with a known pattern, with only sporadic errors in each. It is then possible to recover the primes by enumerating, bit by bit, a search tree of all possible prime factors, and using depth- or breadth-first search with pruning to find a low-Hamming weight path through the search tree of all solutions.

Unfortunately, there are a few difficulties in applying this idea to the case at hand. The first is that because the primes are generated by incrementing to the next prime, a single sporadic error is likely to cause the least significant 9 bits of each prime to appear random (except for the least significant bit which is set to 1), so generating a solution tree from the least significant bits necessarily begins with that much brute forcing. Second, there is only a single constraint on the solutions (the fact that $pq = N$), instead of four constraints, which results in a lower probability of an incorrect solution being pruned than in the examples considered by [14,23]. And finally, in order to apply the algorithms, we must guess the underlying pattern, which in our case requires applying the algorithm to 164^2 possibilities for each modulus.

Applying this algorithm using only the all-zeros pattern for both factors to the 45 moduli with 20 bits of consecutive zeros took 13 minutes and factored 5 moduli. All of these moduli were also factored by the GCD method or Coppersmith methods described in the next section.

5   Univariate Coppersmith

Several of the factors computed via the GCD algorithm in Section 3 follow the
bit patterns described in Section 4, but are interrupted by what appear to be
sporadic errors. Coppersmith's method [6,7] factors RSA moduli if the top bits
of the primes are known, which matches our situation if the errors appear in the
bottom few bits of a factor. The method uses lattice basis reduction to factor in
polynomial time if at least half of the most significant bits of a prime factor are
known; however, since the running time scales very poorly as one approaches this
bound, we will be more interested in less optimal parameters that are efficient
enough to apply speculatively to millions of keys.

This section presents this method following Howgrave-Graham [16] for lattices
of dimension 3 and 5 and gives an outlook of how more keys could be factored
using larger dimensions. The idea is as follows: we assume that some prime factor
p of N is of the form

$$p = a + r$$

where a is a known 512-bit integer (one of the bit patterns described in the
previous section) and r is a small integer error to account for a sequence of bit
errors (and incrementing to next prime) among the least significant bits of p.

In the Coppersmith/Howgrave-Graham method, we can write a polynomial

$$f(x) = a + x$$

and we would like to find a root r of f modulo a large divisor of N (of size
approximately $N^{1/2} \approx p$). Let X be the bound on the size of the root we are
searching for. We will use lattice basis reduction to construct a new polynomial
$g(x)$ where $g(r) = 0$ over the integers, and thus we can factor g to discover r.

Let L be the lattice generated by the rows of the basis matrix

$$\begin{bmatrix} X^2 & Xa & 0 \\ 0 & X & a \\ 0 & 0 & N \end{bmatrix}$$

corresponding to the coefficients of the polynomials $Xxf(Xx), f(Xx), N$. Any
vector in L can be written as an integer combination of basis vectors, and,
after dividing by the appropriate power of X, corresponds to the coefficients
of a polynomial $g(x)$ which is an integer combination of f and N, and is thus
divisible by p by construction. A prime p is found by this method if we can find
g such that $g(r_i) \equiv 0 \bmod p$ holds not only modulo p but over the integers. The
latter is ensured if the coefficients of g are sufficiently small, which corresponds
to finding a short vector in L.

To find such a short vector, we apply the LLL lattice basis reduction algorithm [18]. To finish the algorithm, we regard the shortest vector in the reduced
basis as the coefficients of a polynomial $g(Xx)$, compute the roots r_i of $g(x)$,
and check if $a + r_i$ divides N. If so, we have factored N.

The shortest vector v_1 found by LLL is of length

$$|v_1| \leq 2^{(\dim L - 1)/4} (\det L)^{1/\dim L},$$

which must be smaller than p for the attack to succeed.

In our situation this translates to

$$2^{1/2} \left(X^3 N\right)^{1/3} < N^{1/2} \Leftrightarrow X < 2^{-1/2} N^{1/6},$$

so for $N \approx 2^{1024}$ we can choose X as large as 2^{170}, meaning that for a fast attack using dimension-3 lattices up to the bottom third of a prime can deviate from the pattern a. In the following we ignore the factor $2^{(\dim L - 1)/4}$ since all lattices we deal with are of small dimension and the contribution compared to N is negligible.

5.1   Experimental Results

A straightforward implementation using Sage 5.8 took about one hour on one CPU core to apply this method for one of the 164 patterns identified in Section 4. Running it for all 164 patterns factored 160 keys, obviously including all 105 keys derived from the patterns without error, and found 39 previously unfactored keys.

It is worth noting that the 160 keys included all but 2 of the 103 keys factored with the GCD method, showing that most of the weak primes are based on the patterns we identified and that errors predominantly appeared in the bottom third of the bits. The missing 2 keys are those divisible by 0xe0000...0f. Including 0xd0000...0, 0xe0000...0, 0xf0000...0 as additional bit patterns did not reveal any factors beyond the known ones, ruling out the hypothesis that the prime generation might set the top 4 bits rather than just 2. Instead this prime must have received a bit error in the top part.

5.2   Handling More Errors

Coppersmith's method can find primes with errors in up to $1/2$ of their bits using lattices of higher dimension. Getting close to this bound is prohibitively expensive, but trying somewhat larger dimensions than 3 is possible. For dimension 5 we used basis

$$\{N^2, Nf(xX), f^2(xX), xXf^2(xX), (xX)^2 f^2(xX)\}$$

which up to LLL constants handles $X < N^{1/5}$, i.e. up to 204 erroneous bottom bits in p for N of 1024 bits. The computation took about 2 hours per pattern and revealed 6 more factors.

We did not use higher dimensions because the "error" patterns we observed are very sparse making it more profitable to explore multivariate attacks (see Section 6).

5.3   Errors in the Top Bits

The factor 0xe000...f ($2^{511}+2^{510}+2^{509}+15$) appeared as a common factor after taking GCDs but was not found by the lattice attacks described in this section applied to the basic patterns described in Section 4. We can apply Coppersmith's attack to search for errors in higher bits of p by defining the polynomial f as $f(x) = a + 2^t x$. Here t is a bit offset giving the location of the errors we hope to learn. The method and bounds described in this section apply as well to this case.

However, since we hypothesize that the prime factors are generated by incrementing to the next prime after a sequence of bits output by the flawed RNG, we will not know the least significant bits of a because they have been modified in the prime generation process. This problem might speculatively be overcome by brute forcing the m least significant bits of each pattern: for each application of the algorithm to a single pattern a, we would apply the algorithm to the 2^{m-1} patterns generated by fixing a and varying the bottom m bits, with the least significant bit always fixed to 1. This will find factors if finding the next prime from the base string with errors did not require incrementing by more than those bottom m bits.

The following rough analysis suggests that for this attack to have a 50% chance of success, we need to apply the algorithm to 128 new patterns for every old pattern. Recall that the chance that a number around z is prime is approximately $1/\log z$, where log is the natural logarithm. In particular, each number around 2^{512} has about a 1/355 chance of being prime. Since $1 - (1 - 1/355)^{256} \approx 0.5$, trying 128 patterns for the bottom eight bits for odd patterns has a 50% chance of covering a sufficiently large interval to find a prime. See [12] for more precise estimates. Applying this to our 164 base patterns, our implementation would require 20992 core hours, or close to 2.5 core years. It is fairly likely that more factors would be found with this search but the method presented in the following section is more efficient at handling errors in top and bottom positions unless a very large portion of the top bits are erroneous.

6   Bivariate Coppersmith

The lattice attacks described in the previous section let us factor keys with unpredictable bits occurring in the least significant bits of one of the factors, with all of the remaining bits of the factor following a predictable pattern. In this section, we describe how we extended this attack to factor keys with unpredictable bits among the middle or most significant bits of one of the factors, without resorting to brute-forcing the bottom bits.

In the basic setup of the problem, we assume that one of the factors p of N has the form

$$p = a + 2^t s + r$$

where a is a 512-bit integer with a predictable bit pattern (as described in Section 4), t is a bit offset where a sequence of bit errors s deviating from the predictable pattern in a occurred during key generation, and r is an error at the least significant bits to account for the implementation incrementing to the next prime.

To apply Coppersmith's method, we can define an equation $f(x, y) = a + 2^t x + y$ and try to use lattice basis reduction to find new polynomials $Q_i(x, y)$ with the property that if $f(s, r)$ vanishes modulo a large unknown divisor p of N and s and r are reasonably small, then $Q_i(s, r) = 0$ over the integers. In that case, we can attempt to find appropriate zeros of Q_i. The most common method to do this is to look at multiple distinct polynomials Q_i and hope that their common solution set is not too large.

These types of bivariate Coppersmith attacks have many cryptanalytic applications, perhaps most prominently Boneh and Durfee's attack against RSA private key $d < N^{0.29}$ [3]. Our approach is very similar to that described by Herrmann and May for factoring RSA moduli with bits known [15], although for the application we describe here, we are less interested in optimal parameters, and more in speed: we wish to find the keys most likely to be factored using very low dimensional lattices.

Algebraic independence: Nearly all applications of multivariate Coppersmith methods require a heuristic assumption that the attacker can obtain two (or several) algebraically independent polynomial equations determined by the short vectors in a LLL-reduced lattice; this allows the attacker to compute a finite (polynomially-sized) set of common solutions. Most theorem statements in these papers include this heuristic assumption of algebraic independence as a matter of course, and note briefly (if at all) that it appears to be backed up experimentally.

Notably, in our experiments, this assumption *did not* hold in general. That is, most of the time the equations we obtained after lattice basis reduction were not algebraically independent, and in particular, the algebraic dependencies arose because all of the short vectors in the lattice were polynomial multiples of a single bivariate linear equation. This linear equation did in fact vanish at the desired solution, but without further information, there are an infinite number of additional solutions that we could not rule out. However, we were often able to find the solution using a simple method that we describe below.

Herrmann and May [15] describe one case where the assumption of algebraic independence did not hold in their experiments, namely when X and Y were significantly larger than the values of s and r. Similar to our case they observed that the polynomials of small norm shared a common factor but unlike in our case this factor was the original polynomial f. Note that the linear polynomial in our case vanishes over the integers at (s, r) while f vanishes only modulo p.

We experimented with running smaller dimensional lattice attacks in order to generate this sublattice more directly. The attack worked with smaller degree equations than theoretically required to obtain a result, but when we experimented with lattices generated from linear equations, this sublattice did not appear. Note that we specify a slightly different basis for the lattice, in terms of monomial powers rather than powers of f, which may have an effect on the output of the algorithm compared to the examples in [15] and might explain why we find a useful linear equation in the sublattice instead of the useless factor f.

6.1 Implementation Details

Lattice construction: Let X and Y be bounds on the size of the roots at x and y we wish to find. Our lattice is constructed using polynomial multiples of $f(x, y) = a + 2^t x X + yY$ and N up to degree k vanishing to degree 1 modulo p. Our lattice basis consists of the coefficient vectors of the set of polynomials

$$\{(Yy)^h (Xx)^i f^j N^\ell \mid j + \ell = 1, 0 \leq h + i + j \leq k\}$$
$$= \{N, xXN, f, (xX)^2 N, (xX)f, \ldots, (yY)^{k-2}(xX)f, (yY)^{k-1}f\},$$

using coefficients of the monomials $\{1, x, y, x^2, \ldots, y^{k-1}x, y^k\}$. The determinant of this lattice is

$$\det L = N^{k+1}(XY)^{\binom{k+2}{3}}.$$

and the dimension is $\binom{k+2}{2}$. Omitting the approximation factor of LLL, we want to ensure that

$$(\det L)^{1/\dim L} < p$$
$$\left(N^{k+1}(XY)^{\binom{k+2}{3}}\right)^{1/\binom{k+2}{2}} < N^{1/2}.$$

So for $N \approx 2^{1024}$, setting $k = 3$ should let us find $XY < 2^{102}$ and $k = 4$ should let us find $XY < 2^{128}$. The parameter choice $k = 2$ results in a theoretical bound $XY < 1$, but we also experimented with this choice; see below.

Solving for solutions: After running LLL on our lattice, we needed to solve the system of equations it generated over the integers to find our desired roots. The usual method of doing this in bivariate Coppersmith applications is to hope that the two shortest vectors in the reduced basis correspond to algebraically independent polynomials, and use resultants or Gröbner bases to compute the set of solutions. Unfortunately, in nearly all of our experiments, this condition did not hold, and thus there were an infinite number of possible solutions.

However, a simple method sufficed to compute these solutions in our experiments. In general, the algebraic dependencies arose because the short vectors in the reduced basis corresponded to a sublattice of multiples of the same degree-one equation, with seemingly random coefficients, which vanished at the desired roots. (The coefficient vectors were linearly independent, but the underlying polynomials were not algebraically independent.) The other polynomial factors of these short polynomials did not vanish at these roots. This linear equation has an infinite number of solutions, but in our experiments our desired roots corresponded to the smallest integer solution, which we could obtain by rounding. Let

$$ux + vy - w = 0$$

be an equation we want to solve for x and y. If u and v are relatively prime, then we can write $c_1 u + c_2 v = 1$, and parametrize an integer family of solutions

$$x = c_1 w + vz$$
$$y = c_2 w - uz$$

with $z = c_2 x - c_1 y$.

In experiments with the already-factored moduli, we observed that the solution was often the minimum integer value of x or y among the solution family. So we searched for z among the rounded values of $-c_1 w/v$ and $c_2 w/u$. This solution successfully factored the moduli in our dataset whenever the shortest-vector polynomial returned by lattice basis reduction was not irreducible.

For the handful of cases where the lattice did result in independent equations, we computed the solutions using a Gröbner basis generated by the two shortest vectors.

6.2 Experimental Results

We ran our experiments using Sage 5.8 [24] parallelized across eight cores on a 3.1GHz AMD FX-8120 processor. We used fpLLL [4] for lattice basis reduction, and Singular [8] to factor polynomials and compute Gröbner bases. For each lattice, we attempted to solve the system of equations either by factoring the polynomial into linear factors and looking for small solutions of the linear equations as described above or using Gröbner bases.

We attempted to factor each of the 2,086,171 1024-bit moduli using several different parameter settings. For $k = 3$, we had 10-dimensional lattices, and attempted to factor each modulus with the base pattern $a = 0$ using $Y = 2^{30}$, $X = 2^{70}$, and $t = 442$. We then experimented with $k = 4$, $Y = 2^{28}$, and $X = 2^{100}$, which gave us 15-dimensional lattices, and experimented with a base pattern $a = 2^{511} + 2^{510}$ and five different error offsets: $t = 0$ with $Y = 2^{128}$ and $X = 1$, and $t = 128, t = 228, t = 328$, and $t = 428$ with $Y = 2^{28}$ and $X = 2^{100}$. Finally, we experimented with the choice $k = 2$, $X = 4$, $Y = 4$ and the choices of t and a used in the $k = 4$ experiments, which used 6-dimensional lattices and theoretically should not have produced output, but in fact turned out to produce nearly all of the same factorizations as the choices above. We ran one very large experiment, using $k = 2$, $t = 1$, $Y = 2^{28}$, $X = 2^{74}$, $t = 438$, and running against all 164 patterns, which produced 155 factored keys, including two previously undiscovered factorizations. The choice $k = 1$ with the same parameter choices as $k = 2$ did not produce results.

6.3 Handling More Errors

From these experimental settings, it seems likely that many more keys could be factored by different choices of parameters and initial pattern values; one is limited merely by time and computational resources. We experimented with iterating over all patterns, but the computation quickly becomes very expensive.

Table 1. Experimental results from factoring keys using a bivariate Coppersmith approach, using the parameters listed in the text. Where we collected data, we noted the very small number of cases where the lattice produced algebraically independent polynomials; all of the other cases were solved via the heuristic methods described above.

k	$\log_2(XY)$	# t	# patterns	# factored keys	# alg. indep. eqns.	running time
2	4	5	1	104	3	4.3 hours
2	4	1	164	154	21	195 hours
3	100	1	1	112	-	2 hours
4	128	5	1	108	4	20 hours

Patterned factors: Mysteriously, using the base patterns $a = 0$ and $a = 2^{511} + 2^{510}$, the algorithm produced factorizations of keys with other patterned factorizations. This is because the product of the bit pattern of the relevant factor multiplied with a small factor produced an integer of the form we searched for, but we are as yet unable to characterize this behavior in general.

Higher powers of p: Similar to the univariate case we can construct higher-dimensional lattices in which each vector is divisible by higher powers of p, e.g. using multiples of N^2, Nf, and f^2 for divisibility by p^2. However, this approach is successful in covering larger ranges of XY only for lattices of dimension at least 28, which would incur a significantly greater computational cost to run over the entire data set of millions of keys.

More variables: More isolated errors can be handled by writing $p = a + \sum_{i=1}^{c} 2^{t_i} s_i$ with appropriate bounds on the $s_i < X_i$ so that the intervals do not overlap. The asymptotically optimal case is described in [15] and reaches similar bounds for $\prod_{i=1}^{c} X_i$ as in the univariate and bivariate case. However, the lattice dimension increases significantly with c. For $c = 3$, i.e. two patches of errors together with changed bottom bits to generate a prime, the condition $(\det L)^{\dim 1/L} < p$ holds only for lattices of dimension at least 35 at which point $X_1 X_2 X_3 < N^{1/14}$ can be found. A lattice of dimension 20 leads to the condition $X_1 X_2 X_3 < 1$. A sufficiently motivated attacker can run LLL on lattices of these dimensions but we decided that factors found thus far were sufficient to prove our point that the smart cards are fatally flawed.

6.4   Extension to Implicit Factoring

Ritzenhofen and May [19] and Faugère, Marinier, and Renault [9] give algorithms to factor RSA moduli when it is known that two or more moduli have prime factors that share large numbers of bits in common. Unfortunately, these results seem to apply only when the moduli have prime factors of unbalanced size, whereas in our case, both prime factors have 512 bits.

7   Hardware Details, Disclosure, and Response

Around 2006–2007, MOICA switched card platforms from their initial supplier and began to use Chunghwa Telecom's HiCOS PKI smart cards, specifically Chunghwa Telecom HD65145C1 cards (see [5]), using the Renesas AE45C1 smart card microcontroller (see [10]). We have confirmed these details with MOICA.

Unfortunately, the hardware random-number generator on the AE45C1 smart card microcontroller sometimes fails, as demonstrated by our results. These failures are so extreme that they should have been caught by standard health tests, and in fact the AE45C1 does offer such tests. However, as our results show, those tests were not enabled on some cards. This has now also been confirmed by MOICA. MOICA's estimate is that about 10000 cards were issued without these tests, and that subsequent cards used a "FIPS mode" (see below) that enabled these tests.

The random numbers generated by the batch of problematic cards obviously do not meet even minimal standards for collecting and processing entropy. This is a fatal flaw, and it can be expected to continue causing problems until all of the vulnerable cards are replaced.

The AE45C1 chip was certified conformant with Protection Profile BSI-PP-0002-2001 at CC assurance level EAL4+ [10]. The HD65145C1 card and HICOS operating system were accredited to FIPS 140-2 Level 2 [5]. The CC certification stated "The TOE software for random number postprocessing shall be implemented by the embedded software developer", and the FIPS certification was limited to "FIPS mode" (see http://www.cryptsoft.com/fips140/out/cert/614.html). However, neither certification prevented the same card from also offering a non-FIPS mode, and neither certification caught the underlying RNG failures. We recommend that industry move to stronger certifications that prohibit error-prone APIs and that include assessments of RNG quality.

In April 2012 we shared with MOICA our preliminary list of 103 certificates compromised by GCD. We announced these results in a talk in Taiwan in July 2012. We provided an extended list of compromised certificates to MOICA and Chunghwa Telecom in June 2013, along with an early draft of this paper. MOICA and Chunghwa Telecom subsequently confirmed our results; asked the cardholders to come in for replacement cards; revoked the compromised certificates; and initiated the task of contacting 408 registration offices across Taiwan to manually trace and replace all of the vulnerable cards from the same batch.

Acknowledgements. We thank J. Alex Halderman and Mark Wooding for discussion during this project. We also thank Kenny Paterson and the anonymous reviewers for helpful comments and suggestions, and in particular the encouragement to experiment with sparse key recovery methods.

References

1. ANSI. ANSI X9.31:1998: Digital Signatures Using Reversible Public Key Cryptography for the Financial Services Industry (rDSA). American National Standards Institute (1998)
2. Bernstein, D.J.: How to find the smooth parts of integers (May 2004), http://cr.yp.to/papers.html#smoothparts
3. Boneh, D., Durfee, G.: Cryptanalysis of RSA with private key d less than $n^{0.292}$. In: Stern, J. (ed.) EUROCRYPT. LNCS, vol. 1592, pp. 1–11. Springer (1999)
4. Cadé, D., Pujol, X., Stehlé, D.: fpLLL (2013), http://perso.ens-lyon.fr/damien.stehle/fplll/
5. Ltd. Chunghwa Telecom Co. Hicos pki smart card security policy (2006), http://www.cryptsoft.com/fips140/vendors/140sp614.pdf
6. Coppersmith, D.: Finding a small root of a bivariate integer equation; factoring with high bits known. In: Maurer, U.M. (ed.) EUROCRYPT. LNCS, vol. 1070, pp. 178–189. Springer (1996)
7. Coppersmith, D.: Small solutions to polynomial equations, and low exponent RSA vulnerabilities. J. Cryptology 10(4), 233–260 (1997)
8. Decker, W., Greuel, G.-M., Pfister, G., Schönemann, H.: SINGULAR 3-1-6 — A computer algebra system for polynomial computations (2012), http://www.singular.uni-kl.de
9. Faugère, J.-C., Marinier, R., Renault, G.: Implicit factoring with shared most significant and middle bits. In: Nguyen, P.Q., Pointcheval, D. (eds.) Public Key Cryptography. LNCS, vol. 6056, pp. 70–87. Springer (2010)
10. Bundesamt für Sicherheit in der Informationstechnik. Certification report BSI-DSZ-CC-0212-2004 for Renesas AE45C1 (HD65145C1) smartcard integrated circuit version 01 (2004), https://www.bsi.bund.de/SharedDocs/Downloads/DE/BSI/Zertifizierung/Reporte02/0212a_pdf.pdf?__blob=publicationFile
11. Bundesamt für Sicherheit in der Informationstechnik. Evaluation of random number generators (2013), https://www.bsi.bund.de/SharedDocs/Downloads/EN/BSI/Zertifierung/Interpretation/Evaluation_of_random_number_generators.pdf?__blob=publicationFile and https://www.bsi.bund.de/DE/Themen/ZertifizierungundAnerkennung/ZertifizierungnachCCundITSEC/AnwendungshinweiseundInterpretationen/AISCC/ais_cc.html
12. Granville, A.: Harald Cramér and the distribution of prime numbers. Scand. Actuarial J. 1995(1), 12–28 (1995)
13. Heninger, N., Durumeric, Z., Wustrow, E., Alex Halderman, J.: Mining your Ps and Qs: Detection of widespread weak keys in network devices. In: Proceedings of the 21st USENIX Security Symposium (August 2012)
14. Heninger, N., Shacham, H.: Reconstructing rsa private keys from random key bits. In: Halevi, S. (ed.) CRYPTO. LNCS, vol. 5677, pp. 1–17. Springer (2009)
15. Herrmann, M., May, A.: Solving linear equations modulo divisors: On factoring given any bits. In: Pieprzyk, J. (ed.) ASIACRYPT. LNCS, vol. 5350, pp. 406–424. Springer (2008)
16. Howgrave-Graham, N.: Approximate integer common divisors. In: Silverman, J.H. (ed.) CaLC. LNCS, vol. 2146, pp. 51–66. Springer (2001)

17. Lenstra, A.K., Hughes, J.P., Augier, M., Bos, J.W., Kleinjung, T., Wachter, C.: Public keys. In: Safavi-Naini, R., Canetti, R. (eds.) CRYPTO. LNCS, vol. 7417, pp. 626–642. Springer (2012)
18. Lenstra, A.K., Lenstra Jr., H.W., Lovász, L.: Factoring polynomials with rational coefficients. Math. Ann. 261, 515–534 (1982)
19. May, A., Ritzenhofen, M.: Implicit factoring: On polynomial time factoring given only an implicit hint. In: Jarecki, S., Tsudik, G. (eds.) Public Key Cryptography. LNCS, vol. 5443, pp. 1–14. Springer (2009)
20. MOICA. Safety questions (2013),
   http://moica.nat.gov.tw/html/en_T2/faq22-066-090.htm
21. National Institute of Standards and Technology (NIST). Security requirements for cryptographic modules. Federal Information Processing Standards Publication (FIPS PUB) 140-2 (May 2001),
   http://csrc.nist.gov/publications/fips/fips140-2/fips1402.pdf (updated December 03, 2012), See
   http://csrc.nist.gov/publications/nistpubs/800-29/sp800-29.pdf for differences between this and FIPS-140-1
22. National Institute of Standards and Technology (NIST). Recommendation for random number generation using deterministic random bit generators. NIST Special Publication (NIST SP) 800-90A (January 2012)
23. Paterson, K.G., Polychroniadou, A., Sibborn, D.L.: A coding-theoretic approach to recovering noisy RSA keys. In: Wang, X., Sako, K. (eds.) ASIACRYPT. LNCS, vol. 7658, pp. 386–403. Springer (2012)
24. Stein, W.A., et al.: Sage Mathematics Software (Version 5.8). The Sage Development Team (2013), http://www.sagemath.org

A Appendix: Raw Data

The following data presents all primes found using the GCD method (Section 3); the initial number indicates how often that particular prime was found.

46, 0xc0000000000000000000000000000000000000000000000000000000000000000000000000000000000000000000000000000000000000000000000000000002f9
7, 0xc9242492249292492494924492442492249292492499249492449242492249292492499249492449242492249292492499249492449242492249292492494924492424e5
7, 0xc0000000000000000000000000000000000000000000000000000000000000000000000000000000000000000000000000000000000000000000000000000000101ff
6, 0xd2494924492424924924924924924494924494924924924492442492249292492499249492449242492249292492499249492449242492249292492499249492494949d7
4, 0xf6dbdb6ddb6d6db66db6b6dbb6dbd6ddb6d6db6d6db6b6dbb6dbdb6dddb6d6db66db6b6dbb6dbdb6ddb6d6db66db6b6dbb6dbdb6ddb6d6db66db6b6dbb6dbdbc1
4, 0xdb6d6db66db6b6dbb6dbdb6ddb6d6db66db6b6dbb6dbdb6ddb6d6db66db6b6dbb6dbdb6ddb6d6db66db6b6dbb6dbdb6ddb6d6db66db6b6dbb6dbdb6c6e23
4, 0xedb6b6dbb6dbdb6ddb6d6db66db6b6dbb6dbdb6ddb6d6db66db6b6dbb6dbdb6ddb6d6db66db6b6dbb6dbdb6ddb6d6db66db6b6dbb6dbdb6ddb6d6db66db6b6b867
3, 0xd08408424210210808424842121081084842142101084084242102102102102102108084284212108108484214210210840985
2, 0xe0000000000000000000000000000000000000000000000000000000000000000000000000000000000000000000000000000000f
2, 0xf5ad5ad6d6b56b5a5ad6ad6b6b5ab5adad6bd6b5b5ad5ad6d6b56b5a5ad6ad6b6b5ab5adad6bd6b5b5ad5ad6d6b56b5a5adad6bd6b5b5ad5d39
2, 0xc28550a128500a14850aa14250a114280a144285a1422850142885 0a428550a128500a14850aa14250a114280a144285a1422850142885 0a428550a128500a6f
2, 0xfdefdef7f7bd7bdedef7ef7b7bdebdef ef7bf7bdbdefdef7f7bd7bdedef7ef7b7bdebdef ef7bf7bdbdefdef7f7bd7bdedef7ef7b7bdebdefef7bf7bdbdefe0b1
2, 0xd2494924492424924924924924924494924494924924924492442492249292492499249492449242492249292492499249492449242492249292492494a02
2, 0xe94a94a5a529529494a4a5a525294294a4a5a529529494a4a5a529529494a4a5a525294294a4a5a529529494a4a5a529529494a4a5a525294294a4a5a529294b9af5
2, 0xdb6d6db66db6b6dbb6dbdb6ddb6d6db66db6b6dbb6dbdb6ddb6d6db66db6b6dbb6dbdb6ddb6d6db66db6b6dbb6dbdb6ddb6d6d7015
2, 0xca52a529294a94a5a529529494a54a525294294a4a5a529294a94a5a529529494a54a525294294a4a5a529294a94a5a529529494a54a525294294a4a5a601
2, 0xc0000000000000000000000000000000000000000000000000000000000000000000000000000000000000000000000002030b
2, 0xd8c68c636318318c8c63c631318c18c6c631631818c68c636318318c8c63c631318c18c6c631631818c68c636318318c8c63c631318c18c6c631631818c69107
2, 0xf18c18c6c631631818c68c636318318c8c63c631318c18c6c631631818c68c636318318c8c63c631318c18c6c631631818c68c636318318c8c63c631318c1907
2, 0xf7bd7bdedef7ef7b7bdebdefef7bf7bdbdefdef7f7bd7bdedef7ef7b7bdebdefef7bf7bdbdefdef7f7bd7bdedef7ef7b7bdebdefef7bf7bdbdefdef7f7bd8289
2, 0xc42142101084084242102108084242121081084842142101084084242102102108084284212108108484214369
2, 0xef7bf7bdbdefdef7f7bd7bdedef7ef7b7bdebdefef7bf7bdbdefdef7f7bd7bdedef7ef7b7bdebdefef7bf7bdbdefdef7f7bd7bdedef7ef7b7bdebdefef7bf969
1, 0xd4e682e94f1d6018a02056c0db850a74b3591b0f840514ce4017b2f5d25925ba2429a66a384b5be96e6a0a03d4a11e8a10416018de3b3e354477250037b6f813
1, 0xcac05be5c1eabf0c21f8e95ce5d3c0777904282d1fd0c1738d727e197a0a32fda4cc59cc50b99d29f7fa8d07c972402ab88573e255db6bab05505812c73c2911
1, 0xcf052499061243cd82cd1b2059446c963487834d929ac929d92b259245254c7828ed3e92259292c924d24947d4896d1545f4001029b3b265d0ea4d144e242dbd
1, 0xfa94a972e2dcff068ee1257e228b53e9b9fcf46877f07daaa4d13c2bedf132d07730f549f4691f68553f84be8ff405f16a663d8fb8f82987bd9e073a8108edc3
1, 0xef7befbdbdef9ef6f7bd7bde9ef7ef7b7bdd9dcfef7b37bd9feddef7b7bd7bdedee6ef3b5bda3de7ed7bfa99adebdef7b7bd77d7cff1ee7b7bdebdeeef79f8ab
1, 0xeeb2919e1dc9ce33c2a0d9e190444535b164a53c7c03e9a3d009ecf8fd6bdf743e04444332b7ff4a0e8f53b5123a5422563a06a487cd6cb5f36cd5411f0ae4dbc69
1, 0xf51576e530188d59bbc5f4f6ec9e824d7a9e70142952b11c49a6f38188ad9dbe3d29d1d9498b7aeffc4d9b0420f71895f62e2a7b79d4887e45b6227e0b84fb97
1, 0xd83f22a49af67d7f196df580d514464d6dbb880b03bea50ddcc1f931ef7f09af2f880de26d88cbf24567302a0d6eed7c8eab859aa0c1cc18bd8efacdce194c13

```
1, 0xc1df3e8db5f7b7f456edc1f60d23f60360536565836ce37af6f02e55de24a8dc373f3c5d49c93ba6fee0d44d08bc5fb0655781adee5c05777fd4da2bcd803d0f
1, 0xe279872638463a0a32a1412b13efccfa5ed68db44963c7f6955a3816bcaa33f94794cc0b75298ddf4a8664e485ef99e6d9469f5187939e395cb1f09e666786741
1, 0xce73e73939ce9ce7e738739c9ce7ce73739c39cece73e73939ce9ce7e739739c9ce7ce73739c39cece73e73939ce9ce7e739739c9ce7ce73739c39cece73ead7
1, 0xd92ae5c6453afec55c5614207827de2b77bf3ef027f4230f8aac1fd9b0d69fdc61934132766f8dd1d8cb22ec38d834037eff6d9dd3535b9a582fbdd2327c9ce5
1, 0xc08000000000000000000000000000000000000000000020000000000000000000000000000000000000000000000000000000000000000000000001003f
1, 0xffff7ffffeffffffffffffffffffffffff7fffff7ffffffffffffffffbffffff7bfffbfcffff7ffffffffffbf000000000000000000000000000000000c1
1, 0xeb6f80ff65b4a6d462cfa5961f542f25e207667752b0482f5ac9dc091f4dc854de9c73b288aaa5ba5298a33928f7b2920f89b81e3635932bc9db99a34e52b82b
1, 0xfdf7b9bffbffdebeb28592b76f69bbffbffdafaeffd9f7bdf1ee7bfa6e2f33bb67d5a5b5676d2bf6a1de3626f06be367ffde73db1e01f5d3855f21f0eda8b4db
1, 0xe643203b22b4048427210bd390d45a3a62ac132c0063990067686123d50128812e09411f27098400c841e0918340043101810a2b1cc0954c0405026420e8c7f
1, 0xffefef7ffde6fffff7ffffbffffffffbffeffbfdffffffffffffffffffff1fcffffe46ffffddffff7ffffffffffffffffffffffeefe6eceffe8d
1, 0xf6bdbb6ddb6d6db6dd6db6b6ddbdb6ddb6dddb6db66dbb6b6dbb6dbdb6dddb6db6d6b6b6dbb6dbdb6ddb6dd6db66b37b6db019a4697
1, 0xc000b8000000000000000000000000000000000000000000000680000000000000000000000000000000000000000000000251
1, 0xcccc5ebfea2f4beb8b62dfef5429f97f06af0af8d08159d21df4540a0197ffdb8386c8ebb18bd70b0f46c9615d2fcd0ea38a2cadb522cf79f2c3ab27d9564a197
1, 0xedb6b6bb66db6ddb6d6b66db6b6ddb6b6db6dbdb6ddadb6d9b6f6db66da6b6fbf6cb9b7ddb656d9e6d36a7dbb673ba6ddb6f6db66df6b5e5
1, 0xe7fa15ab6c3d2c3d13960f598cd2bbf74a688580e5fdc70064563a10558f1dfd36d5e8aec88897c79d73ebdcbec1b5f0121175c8aae69e3a1a63f9e66e0bfc5
1, 0xfb308867fee16267feb2b1af212ffefffffe4308866fff5fffefe13ffcf869aff4bf907ff1f9393fff0fff3fffcfff7ff3ef703ffaa8c7fffffe491affeff3b1
1, 0xc01020b48c18021210810848421423010a4084242006309ca468d2123081084a520431000c40a425210210a084a8ce1290810cc84204a9011ac2842401022e1
1, 0xe739729c9ce7ce73539c29cec126e7383b8e89bd2207faed08428421318c1084c410631858c68c63e31035cc8c63ce31318810c64331231818c60e63623b32a3
1, 0xfeb1b9efa29f64ed53628a10a924b5268163dd887f653a6b82edb063b6874c2039e4938018ab949a3c28cdc785fe2be58872c0c8a9ec5171e37ea6a82d5d46d7
1, 0xc00000000000000000900000042604000400180000000000420800000002020001080000000000000002800000000002f3
1, 0xf9d5834f918b673e1f7eaae3cc5d97dd2706dd8de9c5b2fbef679b2c196933fe30f62ac3f7fcc1c593fb63a0bbb8838b6486eac959cc3949ea9182c46396fbcb
1, 0xdac45d37aadacfec73b3184ef43d52d6314754abd38414dde03ade396bd809aa2811047f015c9c71f0cbb0a91028190adeacc36165b0e0e6fce64549f947e0d5
1, 0x49808713746a41a331625a7cb389611eaa3905984245f99e828f17f867413cfae91230478715024db5ead44beb20fbc73a23a271d627a11747b5823f753eb03
1, 0xd67a7b111c040197f157806a2be12a174b8923fd3972ec64fe3de3ee96594a14207831d12f16f545851cad6356bb1621bee88eb2fee9427e0da0ca5f98e5861
1, 0x83071df5288c373a5bc43fb20309e25e99fd85b61a9a4e6f3f71511b98f7ec87047fb32520d94cd7753dbe173304445ca648231f601dd19d3cd40c74190c71d
1, 0xed4294b5a529529c94250ad35394214a4a5269a94a74a5252942945a529294044a74a727394696a4a13a529236a968d
1, 0xd621eb6e5ab7992c6efba5f34a7b7b28026fc93138998c113831dbaaaca1a15738a7b7a9d191bcd77955b92b75263ad9f6bbd4ce0b4edca1efd5f3e24b3a2889
1, 0xd9a43ff058df6b8d55085028eac413a7439e1dc89e5d6e8b5de09e7bc7483d762788f7e36527ff67c39360cfc0d2a75986b7fb35614027cffb932ee1112ee8d
1, 0x49292499249492492549224929249924949242492492929499249449244924249229249929249492492549242492249249294939492549252492249293bf
1, 0xf9cf9b29d767edb655b2f6bf964bce697f652fb669b322eb63dffb6e7ac6c69bb798396d284d85169883d42a6ec96b292761d6dcd7ab595b2ad0a9a5d7e97fe41
1, 0xffffefffefffffffffefffefffefffefffffffefffffffefffffffffeffffefffefffefffeffffffffff00000000000000000000000000000000001ffffffff35
1, 0xf9ce9ce7e738738c9ce7ce73739cb9cece73e738398e9ce7ce719739c9ce7ce7373dc39cece73e53839ce9ce7e7b9739c9ce7ce73739c39cece73739c39ce9d63
1, 0xd53bd2f169ab7fb38abb7f05cb1550e200914674b65ce176001ffeb29dbd1e90c21a77e28c6dbfd6e6a782baaba532e2a98eff9ed8e924986af702c48504d0d1
1, 0xc36e8f2addb602d9d18b2b040bc7a00bc7046b2030c2d3e91c4c161ed562a31d2d056afc759042a46c28e218e25e7c7882fb1cb2d66039ed961dace5ea69c5d7
1, 0xed15cb0fde1567b278ef2422ee01ed658173594b0bcb71594a18d4f455fc75ca7c5b529bb6b9ec2959b5ba9a977773eca917ac08a1e9f557adf079a8bceb2bc01b
1, 0xd00b0dd78fd35c88db31806803799deab89b8b36c39dc0321574801fb936f90e2920f3dd65400ddc00be90ebcefdd62d5c5c062c200bdb04aa6a5acf697e2a0d
1, 0xd0054c94020831e08004500e05811840282088a906825002d9a0c340938dc0b2062807280033410208010309c020800710200c04a604083700aa440088411987
1, 0xc7592d7dc9ee1031dcd3d30f43028858305ac46ac981cafa164a8000a9c6eeb698181505242ac9dfee9e51c92460b987dbc8161def71863d35ac18fa1235a903
1, 0xfffbfcf7f7ffdf3dfffef5fffffffbffff09ffffffffdf7dfffff9ffffffef7dffffffbffffffefbffffff1ffffffffffffff7ffffdffffffffff1ffffffff35
1, 0xf5eb05d73ad4df3cdaf4fd2eaf41e8e405952b7a32747914ffffa33eb829039e77f11f6f9e4958a3f604743ed2c55ba67b47631842905dbc2f12c66fb6c4e40f
1, 0xc7b18295347824ccb395bed351993c598c7cf7f4e32dcb9ab7a5d7e0baa7626d1b8dc651b34f5e4f5d3f2530b52b9bd10e75259b36d774f059141bf9ede911
1, 0xef7b77bd3defdef7f7b47bdedef76f7b7bdebdef6f7b7bddbdee7ef7b7b73d7b9edef7ef7b7b9deedf67bdedef7b9bdaef3bf845
1, 0xfd23b110962000d598488c43407369898cd0086df780826dcfa14784f38388874362851b7711dc13564441351335c71fbd7c564d5d5008f5de20d43f2476d715
1, 0x918f1658790911a71a9ae1895cfe56dbed767816e337a2f950462affb3280d8a8dcb1240620ec8f1d19c3750afcfe295c58ccca117b36632414cd9e114fdb097
1, 0xfffaa55ffffffffff3cd9fe3ffff676ffffffffffe000000000000000000000000000000000000000000000000000000000009d
1, 0x0210000480410000010001000b001ce2064004242c812186250154c00000088ba78008b43a9713bc0abb849220e3852ec838b53cf88fcdbdd7fca83c8df8145
1, 0xe318318c63c631318c18c6c631631818c68c63631318c8c63c631318c18c6c631631818c68c63631318c8c63c631631818c68c63631d
1, 0xd501973162d4017f4e3b3c9d6803d4cc46a1d457c91feb5b6c2ae77423ba41c9cfbd5f4b9235667874507e9cafb4123e1992d1c5ae75ee295087011a822a6ccf
1, 0xe28ecce1de7d0326423076465160c1b03f8e72118e046ef4860ae94d7802a082f9f6007c0011f20056de200677aa7d8a47118e6692ee4b3f862c24e04b543b5
1, 0xda2f36d74bc2dc29de4e92f4b7b03942173e15a2df36f8f09e790ed1656af5a8adef14b696426f1a9269699da0ee3ad9f21a9f66ede57d945fc165b27d217
1, 0xd28550a3a8520a1c850aa14250a114ba0a144285a1422850142885000a428550e128400a14850aa14250a114080a144285a1422851142885 2a4685d0a128500a2d
1, 0x79082499b094b2459266493608a9249b2410d3409242692a490824d93494125493565a341086119944d8246915249226979226b2949490440727108a8c939a5d
1, 0xdedfc373f783fbfff7fefed3fffafffefffd5fffff7ff1ebfffdeffff5fffffffd5ffffefd3faf7fbf5dfff613bb59f9fb5f5bd52aeff878ebddfe6edeeffe3f3fb3df5
1, 0xfff71fb6fbffffffffffeefffff7ff1ffebffffffffffff9ffffcbfdbff0faffffdfff7f7aedffaf1ffbf07e7adffffdffde7fadfdef63e806b
1, 0xcaf67d473c10f4e73d6678d4a27e4eb04a743925412c31f97efa510ca68558b2c56d839acecbe75e935f86cec7dae7c95aa0b93065a3aa924594fdfb9f521535
1, 0xf6b43e3bd52841756d1a27f22a8590a8a1c43c1c36b95cc72d0102f26b6da1b238236856f7c6e6faa83c70e84f2db44088487fd94a175f22a0d990cc1afea6b
1, 0xd2a20d1b986de2152b9d93cf60bf98f68e9f9e050f4b9820b006e5dc581f1f a82f35a78f2b34fab396a95bcf3a1e442e6b55b1d72cd6956fa599483eee38c1
1, 0x52e529494a4a535294294a4a1a509294a96a5a529529494a4a52529209ae4a4a525294944a4a525294294a4842a52e294af4a5a52f554b
1, 0x942c4644b1169461581e0713a400570237a55c9ae69e3fe58d189aa751d218208421934f2132a888e796bc1f0914a8c9b4f116358cca22c69c35596bd961e5
1, 0xed7f7e78afc7d3735fc1dfb0d13887cddcd715c9fe530530e0efceaa4bcaffbaebac9e601623db36fffef47ffff000000000000000000000003ed
1, 0xf6db9b6dda6d29b61dfe73dbba5bdb6ddead69beedf6a6dbf7dadb6ddf6c6cb66db6f6db6db9b64997c6dbe6cb4364b96dbdb6ddb6c67bedafab7cbaedadf35
1, 0xc080a100000000000000000000008300080000000026082001012000440800100420208000800000500000010000000000e1
1, 0xe5335f76a97c5e29d4557170cd9ef3ed53efc819fda8fa566a5efe247ef102b85c7ad90c484ade030c7ebc23455e0dcbca2cec6afdf0e8c978cb6fbed5733fa5
1, 0xc00000000000000000000000000000000000000000000000000000000000000400000000000000165
1, 0x924249324929249b2cf4924492464926492 1a489249493659332 0b93f9292e992497f1449242492229293499249249449262493248e96fff3c9104432f4cdbb
1, 0x924249224929249924949 244924249224929249 4924924924929294949 244924249224929249 2492492924922482926f
1, 0xc0000000000000000000000000000000000000000000000000000000000000000000000000000000000000100000000000177
1, 0xc0000004102400000500000080210000049000000100000002080000000102100518000008200040001830010000002f1
1, 0xcad0ca7166b2aaf6c82b0eadfeb13409da7c2679517d4fd96f89719659133e0492d209da600753dc5c2570ce128cf985332f944143204b706bf6e990c0e43dcb
1, 0xe739739c9ce7ce73739c39cece73e73939ce9ce7e739739c9ce7ce73739c39cece73e73939ce9ce7e739739c9ce7ce73739c39cece73e73939ce9ce7e73973df
1, 0x9292499249492492249a24929249924949 2442249242493249292499a49449244924924924a249249 3a4494242492492492492922926f
1, 0xc000100000000000000000000000000000001000000100082000000000b00400098809180008c004009088091060890000100020000012101b200000000002ad
1, 0xcec727009ef07418dc89e2c96e796d44bc2244d88a0bb8ca90b4d661736b486b6e14835282a4697cdd0702a3d8b7c4b23ada2285a2af09234a71346ba141795
1, 0xc8dce72c0e38ecaf2e3e11aef07326e3431a92ad8f07296d3d0b5d43d00645beb d7b3af6c9e424e074e1486d186d26997a4d9c131acb524881aecace287c057
1, 0xee09f0be62014c7299e188527ab8cd004809c631f1fd50a20013331678ca22569eb18c4b1dd5e4b11bce7a14f4ae76973d ebf4c768c4bd
1, 0xdd66cdaeef275bfd3d1ee65df430dd7ae015bd0e9a5e43890e7835e2a20fb702703d6c3fd50d5917f3ba77aeb851c016d26135d754c114adf303d091500462bd
1, 0xfa147ea58cddaaabe6dfa04ff891009db3ff37c1272d573b7a3da5334f24f9512fda7ff4f163a72482a0edffa9140001aae21f5a64fd330f93e819a968acafb7
1, 0xf718c0bc8c57cc318c99fa15236191a531828a95856d6ac833a7e3a2110dded25226ea4344cabbb2fe19de14863b8c46e31b44038c87e8ce4aea42a10afabf91
1, 0xd86d99e183ba3c0870238db37f1d3f673cdec3112196cfaa123965 7bcb3a3f6749f3229f550d5097510e5a5df0626a641e2112112f95080c5629973b1c975
1, 0xc1fc95ee7482142bccb7f0bc5cd674ad82adca61fe2653c78622ee673485cc11c993aaeeb15f77d90dfe1c6a945a239ab47e5ca3eb2aeb702f2be36626858db
1, 0xf7c6bf218fcfadcba926ac5efdf6 97aeba8d57f7ae8a6f7c763bfe86dc7a86ee76b8ba9d076bf1a8f4a7fcfb0297a96c6c5a70e74e5c38326ff83
1, 0xc594391e8e3c4c8a7e971d78db784643c96ba3384f02acf71fc2506736c65f7c44ef6c3bbf705659b954c6b9ce96f648c900b56c5f3ca01e47384ad4de577
1, 0xda9754106934312032997c8c728f09d09610f5ef08a7c63ff1dcf673ffffb493c19c64167e0457464acc8f4a3409f9648f27c390c25d4a8a3d7c9b2f16b2d
1, 0xf16fead9af03cfc36571b8b3fa3cf24e313aeec858b7d4e800838329c9b729ecc6d691df4ee8547a9fdb18debbca338af8214fa1e03ad53f8e3a0503bfb6735
1, 0xddf3b56a7bb556afa1476add54a9a95e569c94ab62d5fa95c054af04b5a3b56adff15e2dbb466ed1b56aad5a1629c5a93ad55bf1bab1e3we4a5ab9722daf5d7bd
1, 0x9958fe30334b89c8c02ac210c4dc8e6e610d1c958cb4d436e11aede0f72e3b8a88e18b7c663533218c68ed560b031ad4ce38aa13bbc10b6c73fe3911acc8de1
1, 0xfcb0663ef5e3c922936834039fd787a0de9fdd178017021129cfb592570fd3c5e60787fc59128bce5bfcb38be0c064b08c087fd8fe6b960207c93ca4cf3c5add
```

Naturally Rehearsing Passwords*

Jeremiah Blocki, Manuel Blum, and Anupam Datta

Carnegie Mellon University
5000 Forbes Avenue
Pittsburgh, PA 15213
jblocki@cs.cmu.edu, mblum@cs.cmu.edu, danupam@cmu.edu

Abstract. We introduce quantitative usability and security models to guide the design of *password management schemes* — systematic strategies to help users create and remember multiple passwords. In the same way that security proofs in cryptography are based on complexity-theoretic assumptions (e.g., hardness of factoring and discrete logarithm), we quantify usability by introducing *usability assumptions*. In particular, password management relies on assumptions about human memory, e.g., that a user who follows a particular rehearsal schedule will successfully maintain the corresponding memory. These assumptions are informed by research in cognitive science and can be tested empirically. Given rehearsal requirements and a user's visitation schedule for each account, we use the total number of extra rehearsals that the user would have to do to remember all of his passwords as a measure of the usability of the password scheme. Our usability model leads us to a key observation: password reuse benefits users not only by reducing the number of passwords that the user has to memorize, but more importantly by increasing the natural rehearsal rate for each password. We also present a security model which accounts for the complexity of password management with multiple accounts and associated threats, including online, offline, and plaintext password leak attacks. Observing that current password management schemes are either insecure or unusable, we present Shared Cues — a new scheme in which the underlying secret is strategically shared across accounts to ensure that most rehearsal requirements are satisfied naturally while simultaneously providing strong security. The construction uses the Chinese Remainder Theorem to achieve these competing goals.

Keywords: Password Management Scheme, Security Model, Usability Model, Chinese Remainder Theorem, Sufficient Rehearsal Assumption, Visitation Schedule.

1 Introduction

A typical computer user today manages passwords for many different online accounts. Users struggle with this task—often forgetting their passwords or

---

* This work was partially supported by the NSF Science and Technology TRUST and the AFOSR MURI on Science of Cybersecurity. The first author was also partially supported by an NSF Graduate Fellowship.

K. Sako and P. Sarkar (Eds.) ASIACRYPT 2013, Part II, LNCS 8270, pp. 361–380, 2013.

adopting insecure practices, such as using the same password for multiple accounts and selecting weak passwords [33,30,39,24]. While there are many articles, books, papers and even comics about selecting strong individual passwords [29,42,35,61,55,27,48,3], there is very little work on *password management schemes*—systematic strategies to help users create and remember multiple passwords—that are both usable and secure. In this paper, we present a rigorous treatment of password management schemes. Our contributions include a formalization of important aspects of a usable scheme, a quantitative security model, and a construction that provably achieves the competing security and usability properties.

Usability Challenge. We consider a setting where a user has two types of memory: *persistent memory* (e.g., a sticky note or a text file on his computer) and *associative memory* (e.g., his own human memory). We assume that persistent memory is reliable and convenient but not private (i.e., accessible to an adversary). In contrast, a user's associative memory is private but lossy—if the user does not rehearse a memory it may be forgotten. While our understanding of human memory is incomplete, it has been an active area of research [17] and there are many mathematical models of human memory [37,59,14,40,56]. These models differ in many details, but they all model an associative memory with cue-association pairs: to remember $\hat{a}$ (e.g., a password) the brain associates the memory with a context $\hat{c}$ (e.g., a public hint or cue); such associations are strengthened by rehearsal . A central challenge in designing usable password schemes is thus to create associations that are strong and to maintain them over time through rehearsal. Ideally, we would like the rehearsals to be *natural*, i.e., they should be a side-effect of users' normal online activity. Indeed insecure password management practices adopted by users, such as reusing passwords, improve usability by increasing the number of times a password is naturally rehearsed as users visit their online accounts.

Security Challenge. Secure password management is not merely a theoretical problem—there are numerous real-world examples of password breaches [2,30,20,7,51,12,5,9,8,11,10]. Adversaries may crack a weak password in an *online attack* where they simply visit the online account and try as many guesses as the site permits. In many cases (e.g., Zappos, LinkedIn, Sony, Gawker [12,5,8,7,20,11]) an adversary is able to mount an *offline attack* to crack weak passwords after the cryptographic hash of a password is leaked or stolen. To protect against an offline attack, users are often advised to pick long passwords that include numbers, special characters and capital letters [48]. In other cases even the strongest passwords are compromised via a *plaintext password leak attack* (e.g., [4,9,51,10]), for example, because the user fell prey to a phishing attack or signed into his account on an infected computer or because of server misconfigurations. Consequently, users are typically advised against reusing the same password. A secure password management scheme must protect against all these types of breaches.

Contributions. We precisely define the password management problem in Section 2. A password management scheme consists of a *generator*—a function that outputs a set of public cue-password pairs—and a *rehearsal schedule*. The generator is implemented using a computer program whereas the human user is expected to follow the rehearsal schedule for each cue. This division of work is critical—the computer program performs tasks that are difficult for human users (e.g., generating random bits) whereas the human user's associative memory is used to store passwords since the computer's persistent memory is accessible to the adversary.

Quantifying Usability. In the same way that security proofs in cryptography are based on complexity-theoretic assumptions (e.g., hardness of factoring and discrete logarithm), we quantify usability by introducing *usability assumptions*. In particular, password management relies on assumptions about human memory, e.g., that a user who follows a particular rehearsal schedule will successfully maintain the corresponding memory. These assumptions are informed by research in cognitive science and can be tested empirically. Given rehearsal requirements and a user's visitation schedule for each account, we use the total number of extra rehearsals that the user would have to do to remember all of his passwords as a measure of the usability of the password scheme (Section 3). Specifically, in our usability analysis, we use the *Expanding Rehearsal Assumption (ER)* that allows for memories to be rehearsed with exponentially decreasing frequency, i.e., rehearse at least once in the time-intervals (days) $[1, 2)$, $[2, 4)$, $[4, 8)$ and so on. Few long-term memory experiments have been conducted, but *ER* is consistent with known studies [53,60]. Our memory assumptions are parameterized by a constant σ which represents the strength of the mnemonic devices used to memorize and rehearse a cue-association pair. Strong mnemonic techniques [52,34] exploit the associative nature of human memory discussed earlier and its remarkable visual/spatial capacity [54].

Quantifying Security. We present a game based security model for a password management scheme (Section 4) in the style of exact security definitions [18]. The game is played between a user ($\mathcal{U}$) and a resource-bounded adversary ($\mathcal{A}$) whose goal is to guess one of the user's passwords. Our game models three commonly occurring breaches (online attack, offline attack, plaintext password leak attack).

Our Construction. We present a new password management scheme, which we call Shared Cues, and prove that it provides strong security and usability properties (see Section 5). Our scheme incorporates powerful mnemonic techniques through the use of public cues (e.g., photos) to create strong associations. The user first associates a randomly generated person-action-object story (e.g., Bill Gates swallowing a bike) with each public cue. We use the Chinese Remainder Theorem to share cues across sites in a way that balances several competing security and usability goals: 1) Each cue-association pair is used by many different web sites (so that most rehearsal requirements are satisfied naturally), 2) the total number of cue-association pairs that the user has to memorize is low, 3) each web site uses several cue-association pairs (so that passwords are secure) and 4) no two web sites share too many cues (so that passwords remain secure even after the adversary obtains some of the user's other passwords). We show that our construction achieves an asymptotically optimal balance between these security and usability goals (Lemma 2, Theorem 3).

Related Work. A distinctive goal of our work is to quantify usability of password management schemes by drawing on ideas from cognitive science and leverage this understanding to design schemes with acceptable usability. We view the results of this paper–employing usability assumptions about rehearsal requirements—as an initial step towards this goal. While the mathematical constructions start from the usability assumptions, the assumptions themselves are empirically testable, e.g., via longitudinal user studies. In contrast, a line of prior work on usability has focused on empirical studies of user behavior including their password management habits [33,30,39], the effects of password composition rules (e.g., requiring numbers and special symbols) on individual passwords [38,22], the memorability of individual system assigned passwords [50], graphical passwords [28,19], and passwords based on implicit learning [23]. These user studies have been limited in duration and scope (e.g., study retention of a single password over a short period of time). Other work [25] articulates informal, but more comprehensive, usability criteria for password schemes.

Our use of cued recall is driven by evidence that it is much easier than pure recall [17]. We also exploit the large human capacity for visual memory [54] by using pictures as cues. Prior work on graphical passwords [28,19] also takes advantage of these features. However, our work is distinct from the literature on graphical passwords because we address the challenge of managing multiple passwords. More generally, usable and secure password management is an excellent problem to explore deeper connections between cryptography and cognitive science.

Security metrics for passwords like (partial) guessing entropy (e.g., how many guesses does the adversary need to crack α-fraction of the passwords in a dataset [41,44,24]? how many passwords can the adversary break with β guesses per account [26]?) were designed to analyze the security of a dataset of passwords from many users, not the security of a particular user's password management scheme. While these metrics can provide useful feedback about individual passwords (e.g., they rule out some insecure passwords) they do not deal with the complexities of securing multiple accounts against an adversary who may have gained background knowledge about the user from previous attacks — we refer an interested reader to the full version [21] of this paper for more discussion.

Our notion of (n, ℓ, γ)-sharing set families (definition 5) is equivalent to Nisan and Widgerson's definition of a (k, m)-design [43]. However, Nisan and Widgerson were focused on a different application (constructing pseudorandom bit generators) and the range of parameters that they consider are not suitable for our password setting in which ℓ and γ are constants. See the full version[21] of this paper for more discussion.

2   Definitions

We use $\mathcal{P}$ to denote the space of possible passwords. A password management scheme needs to generate m passwords $p_1, ..., p_m \in \mathcal{P}$ — one for each account A_i.

Associative Memory and Cue-Association Pairs. Human memory is associative. Competitors in memory competitions routinely use mnemonic techniques (e.g., the method of loci [52]) which exploit associative memory[34]. For example, to remember the word 'apple' a competitor might imagine a giant apple on the floor in his bedroom. The bedroom now provides a context which can later be used as a cue to help the competitor remember the word apple. We use $\hat{c} \in \mathcal{C}$ to denote the cue, and we use $\hat{a} \in \mathcal{AS}$ to denote the corresponding association in a cue-association pair $(\hat{c}, \hat{a})$. Physically, $\hat{c}$ (resp. $\hat{a}$) might encode the excitement levels of the neurons in the user's brain when he thinks about his bedroom (resp. apples) [40].

We allow the password management scheme to store m sets of public cues $c_1, ..., c_m \subset \mathcal{C}$ in persistent memory to help the user remember each password. Because these cues are stored in persistent memory they are always available to the adversary as well as the user. Notice that a password may be derived from multiple cue-association pairs. We use $\hat{c} \in \mathcal{C}$ to denote a cue, $c \subset \mathcal{C}$ to denote a set of cues, and $C = \bigcup_{i=1}^{m} c_i$ to denote the set of all cues — $n = |C|$ denotes the total number of cue-association pairs that the user has to remember.

Visitation Schedules and Rehearsal Requirements. Each cue $\hat{c} \in C$ may have a rehearsal schedule to ensure that the cue-association pair $(\hat{c}, \hat{a})$ is maintained.

Definition 1. *A rehearsal schedule for a cue-association pair $(\hat{c}, \hat{a})$ is a sequence of times $t_0^{\hat{c}} < t_1^{\hat{c}} < ....$ For each $i \geq 0$ we have a* rehearsal requirement, *the cue-association pair must be rehearsed at least once during the time window* $[t_i^{\hat{c}}, t_{i+1}^{\hat{c}}) = \{x \in \mathbb{R} \mid t_i^{\hat{c}} \leq x < t_{i+1}^{\hat{c}}\}$.

A rehearsal schedule is *sufficient* if a user can maintain the association $(\hat{c}, \hat{a})$ by following the rehearsal schedule. We discuss sufficient rehearsal assumptions in section 3. The length of each interval $[t_i^{\hat{c}}, t_{i+1}^{\hat{c}})$ may depend on the strength of the mnemonic technique used to memorize and rehearse a cue-association pair $(\hat{c}, \hat{a})$ as well as i — the number of prior rehearsals. For notational convenience, we use a function $R : C \times \mathbb{N} \to \mathbb{R}$ to specify the rehearsal requirements (e.g., $R(\hat{c}, j) = t_j^{\hat{c}}$), and we use $\mathcal{R}$ to denote a set of rehearsal functions.

A visitation schedule for an account A_i is a sequence of real numbers $\tau_0^i < \tau_1^i < ...$, which represent the times when the account A_i is visited by the user. We do not assume that the exact visitation schedules are known a priori. Instead we model visitation schedules using a random process with a known parameter λ_i based on $E\left[\tau_{j+1}^i - \tau_j^i\right]$ — the average time between consecutive visits to account A_i. A rehearsal requirement $[t_i^{\hat{c}}, t_{i+1}^{\hat{c}})$ can be satisfied naturally if the user visits a site A_j that uses the cue $\hat{c}$ ($\hat{c} \in c_j$) during the given time window. Formally,

Definition 2. *We say that a rehearsal requirement* $\left[t_i^{\hat{c}}, t_{i+1}^{\hat{c}}\right)$ *is naturally satisfied by a visitation schedule* $\tau_0^i < \tau_1^i < \dots$ *if* $\exists j \in [m], k \in \mathbb{N}$ *s.t* $\hat{c} \in c_j$ *and* $\tau_k^j \in \left[t_i^{\hat{c}}, t_{i+1}^{\hat{c}}\right)$. *We use*

$$X_{t,\hat{c}} = \left| \left\{ i \mid t_{i+1}^{\hat{c}} \leq t \wedge \forall j, k. \left(\hat{c} \notin c_j \vee \tau_k^j \notin \left[t_i^{\hat{c}}, t_{i+1}^{\hat{c}}\right) \right) \right\} \right| ,$$

to denote the number of rehearsal requirements that are not naturally satisfied by the visitation schedule during the time interval $[0, t]$.

We use rehearsal requirements and visitation schedules to quantify the usability of a password management scheme by measuring the total number of extra rehearsals. If a cue-association pair $(\hat{c}, \hat{a})$ is not rehearsed naturally during the interval $\left[t_i^{\hat{c}}, t_{i+1}^{\hat{c}}\right)$ then the user needs to perform an extra rehearsal to maintain the association. Intuitively, $X_{t,\hat{c}}$ denotes the total number of extra rehearsals of the cue-association pair $(\hat{c}, \hat{a})$ during the time interval $[0, t]$. We use $X_t = \sum_{\hat{c} \in C} X_{t,\hat{c}}$ to denote the total number of extra rehearsals during the time interval $[0, t]$ to maintain all of the cue-assocation pairs.

Usability Goal: Minimize the expected value of $E[X_t]$.

Password Management Scheme. A password management scheme includes a generator $\mathcal{G}_m$ and a rehearsal schedule $R \in \mathcal{R}$. The generator $\mathcal{G}_m(k, b, \boldsymbol{\lambda}, R)$ utilizes a user's knowledge $k \in \mathcal{K}$, random bits $b \in \{0,1\}^*$ to generate passwords $p_1, \dots, p_m$ and public cues $c_1, \dots, c_m \subseteq \mathcal{C}$. $\mathcal{G}_m$ may use the rehearsal schedule R and the visitation schedules $\boldsymbol{\lambda} = \langle \lambda_1, \dots, \lambda_m \rangle$ of each site to help minimize $E[X_t]$. Because the cues $c_1, \dots c_m$ are public they may be stored in persistent memory along with the code for the generator $\mathcal{G}_m$. In contrast, the passwords $p_1, \dots p_m$ must be memorized and rehearsed by the user (following R) so that the cue association pairs (c_i, p_i) are maintained in his associative memory.

Definition 3. *A password management scheme is a tuple* $\langle \mathcal{G}_m, R \rangle$, *where* $\mathcal{G}_m$ *is a function* $\mathcal{G}_m : \mathcal{K} \times \{0,1\}^* \times \mathbb{R}^m \times \mathcal{R} \to \left(\mathcal{P} \times 2^{\mathcal{C}}\right)^m$ *and a* $R \in \mathcal{R}$ *is a rehearsal schedule which the user must follow for each cue.*

Our security analysis is not based on the secrecy of $\mathcal{G}_m$, k or the public cues $C = \bigcup_{i=1}^m c_i$. The adversary will be able to find the cues $c_1, \dots, c_m$ because they are stored in persistent memory. In fact, we also assume that the adversary has background knowledge about the user (e.g., he may know k), and that the adversary knows the password management scheme $\mathcal{G}_m$. The only secret is the random string b used by $\mathcal{G}_m$ to produce $p_1, \dots, p_m$.

Example Password Management Schemes. Most password suggestions are too vague (e.g.,"pick an obscure phrase that is personally meaningful to you") to satisfy the precise requirements of a password management scheme — formal security proofs of protocols involving human interaction can break down when humans behave in unexpected ways due to vague instructions [46]. We consider the following formalization of password management schemes: (1) **Reuse Weak** — the user selects a random dictionary word w (e.g., from a dictionary of

20,000 words) and uses $p_i = w$ as the password for every account A_i. (2) **Reuse Strong** — the user selects four random dictionary words (w_1, w_2, w_3, w_4) and uses $p_i = w_1 w_2 w_3 w_4$ as the password for every account A_i. (3) **Lifehacker** (e.g., [3]) — The user selects three random words (w_1, w_2, w_3) from the dictionary as a base password $b = w_1 w_2 w_3$. The user also selects a random derivation rule d to derive a string from each account name (e.g., use the first three letters of the account name, use the first three vowels in the account name). The password for account A_i is $p_i = bd(A_i)$ where $d(A_i)$ denotes the derived string. (4) **Strong Random and Independent** — for each account A_i the user selects four fresh words independently at random from the dictionary and uses $p_i = w_1^i w_2^i w_3^i w_4^i$. Schemes (1)-(3) are formalizations of popular password management strategies. We argue that they are popular because they are easy to use, while the strongly secure scheme **Strong Random and Independent** is unpopular because the user must spend a lot of extra time rehearsing his passwords. See the full version [21] of this paper for more discussion of the security and usability of each scheme.

3   Usability Model

People typically adopt their password management scheme based on usability considerations instead of security considerations [33]. Our usability model can be used to explain why users tend to adopt insecure password management schemes like **Reuse Weak, Lifehacker**, or **Reuse Strong**. Our usability metric measures the extra effort that a user has to spend rehearsing his passwords. Our measurement depends on three important factors: rehearsal requirements for each cue, visitation rates for each site, and the total number of cues that the user needs to maintain. Our main technical result in this section is Theorem 1 — a formula to compute the total number of extra rehearsals that a user has to do to maintain all of his passwords for t days. To evaluate the formula we need to know the rehearsal requirements for each cue-association pair as well as the visitation frequency λ_i for each account A_i.

Rehearsal Requirements. If the password management scheme does not mandate sufficient rehearsal then the user might forget his passwords. Few memory studies have attempted to study memory retention over long periods of time so we do not know exactly what these rehearsal constraints should look like. While security proofs in cryptography are based on assumptions from complexity theory (e.g., hardness of factoring and discrete logarithm), we need to make assumptions about humans. For example, the assumption behind CAPTCHAs is that humans are able to perform a simple task like reading garbled text [58]. A rehearsal assumption specifies what types of rehearsal constraints are sufficient to maintain a memory. We consider two different assumptions about sufficient rehearsal schedules: Constant Rehearsal Assumption (CR) and Expanding Rehearsal Assumption (ER). Because some mnemonic devices are more effective than others (e.g., many people have amazing visual and spatial memories [54]) our assumptions are parameterized by a constant σ which represents the strength of the mnemonic devices used to memorize and rehearse a cue association pair.

Constant Rehearsal Assumption (CR): The rehearsal schedule given by $R(\hat{c}, i) = i\sigma$ is sufficient to maintain the association $(\hat{c}, \hat{a})$.

CR is a pessimistic assumption — it asserts that memories are not permanently strengthened by rehearsal. The user must continue rehearsing every σ days — even if the user has frequently rehearsed the password in the past.

Expanding Rehearsal Assumption (ER): The rehearsal schedule given by $R(\hat{c}, i) = 2^{i\sigma}$ is sufficient to maintain the association $(\hat{c}, \hat{a})$.

ER is more optimistic than CR — it asserts that memories are strengthened by rehearsal so that memories need to be rehearsed less and less frequently as time passes. If a password has already been rehearsed i times then the user does not have to rehearse again for $2^{i\sigma}$ days to satisfy the rehearsal requirement $[2^{i\sigma}, 2^{i\sigma+\sigma})$. ER is consistent with several long term memory experiments [53],[17, Chapter 7], [60] — we refer the interested reader to full version[21] of this paper for more discussion. We also consider the rehearsal schedule $R(\hat{c}, i) = i^2$ (derived from [15,57]) in the full version — the usability results are almost identical to those for ER.

Visitation Schedules. Visitation schedules may vary greatly from person to person. For example, a 2006 survey about Facebook usage showed that 47% of users logged in daily, 22.4% logged in about twice a week, 8.6% logged in about once a week, and 12% logged in about once a month[13]. We use a Poisson process with parameter λ_i to model the visitation schedule for site A_i. We assume that the value of $1/\lambda_i$ — the average inter-visitation time — is known. For example, some websites (e.g., gmail) may be visited daily ($\lambda_i = 1/1$ day) while other websites (e.g., IRS) may only be visited once a year on average (e.g., $\lambda_i = 1/365$ days). The Poisson process has been used to model the distribution of requests to a web server [47]. While the Poisson process certainly does not perfectly model a user's visitation schedule (e.g., visits to the IRS websites may be seasonal) we believe that the predictions we derive using this model will still be useful in guiding the development of usable password management schemes. While we focus on the Poisson arrival process, our analysis could be repeated for other random processes.

We consider four very different types of internet users: very active, typical, occasional and infrequent. Each user account A_i may be visited daily (e.g., $\lambda_i = 1$), every three days ($\lambda_i = 1/3$), every week (e.g. $\lambda_i = 1/7$), monthly ($\lambda_i = 1/31$), or yearly ($\lambda_i = 1/365$) on average. See table 1 to see the full visitation schedules we define for each type of user. For example, our very active user has 10 accounts he visits daily and 35 accounts he visits annually.

Table 1. Visitation Schedules - number of accounts visited with frequency λ (visits/days)

Schedule	λ $\frac{1}{1}$	$\frac{1}{3}$	$\frac{1}{7}$	$\frac{1}{31}$	$\frac{1}{365}$
Very Active	10	10	10	10	35
Typical	5	10	10	10	40
Occasional	2	10	20	20	23
Infrequent	0	2	5	10	58

Table 2. $E[X_{365}]$: Extra Rehearsals over the first year for both rehearsal assumptions.
B+D: **Lifehacker**
SRI: **Strong Random and Independent**

Assumption	CR ($\sigma = 1$)		ER ($\sigma = 1$)	
Schedule/Scheme	B+D	SRI	B+D	SRI
Very Active	≈ 0	$23,396$	.023	420
Typical	.014	$24,545$	.084	456.6
Occasional	.05	$24,652$	.12	502.7
Infrequent	56.7	$26,751$	1.2	564

Extra Rehearsals. Theorem 1 leads us to our key observation: cue-sharing benefits users both by (1) reducing the number of cue-association pairs that the user has to memorize and (2) by increasing the rate of natural rehearsals for each cue-association pair. For example, a active user with 75 accounts would need to perform 420 extra-rehearsals over the first year to satisfy the rehearsal requirements given by ER if he adopts **Strong Random and Independent** or just 0.023 with **Lifehacker** — see table 2. The number of unique cue-association pairs n decreased by a factor of 75, but the total number of extra rehearsals $E[X_{365}]$ decreased by a factor of $8,260.8 \approx 75 \times 243$ due to the increased natural rehearsal rate.

Theorem 1. Let $i_{\hat{c}*} = \left(\arg\max_x t_x^{\hat{c}} < t\right) - 1$ then

$$E[X_t] = \sum_{\hat{c}\in C} \sum_{i=0}^{i_{\hat{c}*}} \exp\left(-\left(\sum_{j:\hat{c}\in c_j} \lambda_j\right)\left(t_{i+1}^{\hat{c}} - t_i^{\hat{c}}\right)\right)$$

Theorem 1 follows easily from Lemma 1 and linearity of expectations. Each cue-association pair $(\hat{c}, \hat{a})$ is rehearsed naturally whenever the user visits *any* site which uses the public cue $\hat{c}$. Lemma 1 makes use of two key properties of Poisson processes: (1) The natural rehearsal schedule for a cue $\hat{c}$ is itself a Poisson process, and (2) Independent Rehearsals - the probability that a rehearsal constraint is satisfied is independent of previous rehearsal constraints.

Lemma 1. Let $S_{\hat{c}} = \{i \mid \hat{c} \in c_i\}$ and let $\lambda_{\hat{c}} = \sum_{i\in S_{\hat{c}}} \lambda_i$ then the probability that the cue $\hat{c}$ is not naturally rehearsed during time interval $[a, b]$ is $\exp(-\lambda_{\hat{c}}(b-a))$.

4   Security Model

In this section we present a game based security model for a password management scheme. The game is played between a user ($\mathcal{U}$) and a resource bounded adversary ($\mathcal{A}$) whose goal is to guess one of the user's passwords. We demonstrate how to select the parameters of the game by estimating the adversary's amortized cost of guessing. Our security definition is in the style of the exact security definitions of Bellare and Rogaway [18]. Previous security metrics (e.g., min-entropy, password strength meters) fail to model the full complexity of the password management problem (see the full version [21] of this paper for more discussion). By contrast, we assume that the adversary knows the user's password management scheme and is able to see any public cues. Furthermore, we assume that the adversary has background knowledge (e.g., birth date, hobbies) about the user (formally, the adversary is given $k \in \mathcal{K}$). Many breaches occur because the user falsely assumes that certain information is private (e.g., birth date, hobbies, favorite movie)[6,49].

Adversary Attacks. Before introducing our game based security model we consider the attacks that an adversary might mount. We group the adversary attacks into three categories: *Online Attack* — the adversary knows the user's ID and attempts to guess the password. The adversary will get locked out after s incorrect guesses (strikes). *Offline Attack* — the adversary learns both the cryptographic hash of the user's password and the hash function and can try many guesses $q_{\$B}$. The adversary is only limited by the resources B that he is willing to invest to crack the user's password. *Plaintext Password Leak Attack* — the adversary directly learns the user's password for an account. Once the adversary recovers the password p_i the account A_i has been compromised. However, a secure password management scheme should prevent the adversary from compromising more accounts.

We model online and offline attacks using a guess-limited oracle. Let $S \subseteq [m]$ be a set of indices, each representing an account. A guess-limited oracle $O_{S,q}$ is a blackbox function with the following behavior: 1) After q queries $O_{S,q}$ stops answering queries. 2) $\forall i \notin S, O_{S,q}(i,p) = \bot$ 3) $\forall i \in S, O_{S,q}(i, p_i) = 1$ and 4) $\forall i \in S, p \neq p_i, O_{S,q}(i,p) = 0$. Intuitively, if the adversary steals the cryptographic password hashes for accounts $\{A_i \mid i \in S\}$, then he can execute an offline attack against each of these accounts. We also model an online attack against account A_i with the guess-limited oracle $O_{\{i\},s}$ with $s \ll q$ (e.g., $s = 3$ models a three-strikes policy in which a user is locked out after three incorrect guesses).

Game Based Definition of Security. Our cryptographic game proceeds as follows:
Setup: The user $\mathcal{U}$ starts with knowledge $k \in \mathcal{K}$, visitation schedule $\boldsymbol{\lambda} \in \mathbb{R}^m$, a random sequence of bits $b \in \{0,1\}^*$ and a rehearsal schedule $R \in \mathcal{R}$. The user runs $\mathcal{G}_m(k,b,\boldsymbol{\lambda},R)$ to obtain m passwords $p_1, ..., p_m$ and public cues $c_1, ..., c_m \subseteq \mathcal{C}$ for accounts $A_1, ..., A_m$. The adversary $\mathcal{A}$ is given $k, \mathcal{G}_m, \boldsymbol{\lambda}$ and $c_1, ..., c_m$.
Plaintext Password Leak Attack: $\mathcal{A}$ adaptively selects a set $S \subseteq [m]$ s.t $|S| \leq r$ and receives p_i for each $i \in S$.
Offline Attack: $\mathcal{A}$ adaptively selects a set $S' \subseteq [m]$ s.t. $|S'| \leq h$, and is given blackbox access to the guess-limited offline oracle $O_{S',q}$.
Online Attack: For each $i \in [m] - S$, the adversary is given blackbox access to the guess-limited offline oracle $O_{\{i\},s}$.
Winner: $\mathcal{A}$ wins by outputting (j,p), where $j \in [m] - S$ and $p = p_j$.

We use $\mathbf{AdvWins}(k,b,\boldsymbol{\lambda},\mathcal{G}_m,\mathcal{A})$ to denote the event that the adversary wins.

Definition 4. *We say that a password management scheme $\mathcal{G}_m$ is (q, δ, m, s, r, h)-secure if for every $k \in \mathcal{K}$ and adversary strategy $\mathcal{A}$ we have*

$$\Pr_b \left[\mathbf{AdvWins}\, (k, b, \lambda, \mathcal{G}_m, \mathcal{A}) \right] \leq \delta \ .$$

Discussion: Observe that the adversary cannot win by outputting the password for an account that he already compromised in a plaintext password leak. For example, suppose that the adversary is able to obtain the plaintext passwords for $r = 2$ accounts of his choosing: p_i and p_j. While each of these breaches is arguably a success for the adversary the user's password management scheme cannot be blamed for any of these breaches. However, if the adversary can use this information to crack any of the user's other passwords then the password management scheme can be blamed for the additional breaches. For example, if our adversary is also able to use p_i and p_j to crack the cryptographic password hash $h(p_t)$ for another account A_t in at most q guesses then the password management scheme could be blamed for the breach of account A_t. Consequently, the adversary would win our game by outputting (t, p_t). If the password management scheme is $(q, 10^{-4}, m, s, 2, 1)$-secure then the probability that the adversary could win is at most 10^{-4} — so there is a very good chance that the adversary will fail to crack p_t.

Economic Upper Bound on q. Our guessing limit q is based on a model of a resource constrained adversary who has a budget of \$$B$ to crack one of the user's passwords. We use the upper bound $q_B = \$B/C_q$, where $C_q = \$R/f_H$ denotes the amortized cost per query (e.g., cost of renting (\$$R$) an hour of computing time on Amazon's cloud [1] divided by f_H — the number of times the cryptographic hash function can be evaluated in an hour.) We experimentally estimate f_H for SHA1, MD5 and BCRYPT[45] — more details can be found in the full version [21] of this paper. Assuming that the BCRYPT password hash function [45] was used to hash the passwords we get $q_B = B\left(5.155 \times 10^4\right)$ — we also consider cryptographic hash functions like SHA1, MD5 in the full version[21] of this paper. In our security analysis we focus on the specific value $q_{\$10^6} = 5.155 \times 10^{10}$ — the number of guesses the adversary can try if he invests \$$10^6$ to crack the user's password.

Sharing and Security. In section 3 we saw that sharing public cues across accounts improves usability by (1) reducing the number of cue-association pairs that the user has to memorize and rehearse, and (2) increasing the rate of natural rehearsals for each cue-association pair. However, conventional security wisdom says that passwords should be chosen independently. Is it possible to share public cues, and satisfy the strong notion of security from definition 4? Theorem 2 demonstrates that public cues can be shared securely provided that the public cues $\{c_1, \ldots, c_m\}$ are a (n, ℓ, γ)-sharing set family. The proof of theorem 2 can be found in the full version of this paper [21].

Definition 5. *We say that a set family* $\mathcal{S} = \{S_1, ..., S_m\}$ *is* (n, ℓ, γ)-*sharing if* (1) $|\bigcup_{i=1}^m S_i| = n$, (2)$|S_i| = \ell$ *for each* $S_i \in \mathcal{S}$, *and* (3) $|S_i \cap S_j| \leq \gamma$ *for each pair* $S_i \neq S_j \in \mathcal{S}$.

Theorem 2. *Let* $\{c_1, \ldots, c_m\}$ *be a* (n, ℓ, γ)-*sharing set of* m *public cues produced by the password management scheme* $\mathcal{G}_m$. *If each* $a_i \in \mathcal{AS}$ *is chosen uniformly at random then* $\mathcal{G}_m$ *satisfies* (q, δ, m, s, r, h)-*security for* $\delta \leq \frac{q}{|\mathcal{AS}|^{\ell-\gamma r}}$ *and any* h.

Discussion: To maintain security it is desirable to have ℓ large (so that passwords are strong) and γ small (so that passwords remain strong even after an adversary compromises some of the accounts). To maintain usability it is desirable to have n small (so that the user doesn't have to memorize many cue-association pairs). There is a fundamental trade-off between security and usability because it is difficult to achieve these goals without making n large.

For the special case $h = 0$ (e.g., the adversary is limited to online attacks) the security guarantees of Theorem 2 can be further improved to $\delta \leq \frac{sm}{|A|^{\ell-\gamma r}}$ because the adversary is actually limited to sm guesses.

5   Our Construction

Public Cue   Private

Action: swallowing

Object: bike

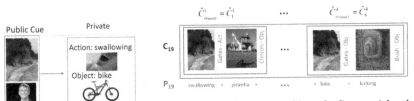

(a) PAO Story with Cue

(b) Account A_{19} using Shared Cues with the $(43, 4, 1)$-sharing set family **CRT** $(90, 9, 10, 11, 13)$.

Fig. 1.

We present Shared Cues— a novel password management scheme which balances security and usability considerations. The key idea is to strategically share cues to make sure that each cue is rehearsed frequently while preserving strong security goals. Our construction may be used in conjunction with powerful cue-based mnemonic techniques like memory palaces [52] and person-action-object stories [34] to increase σ — the association strength constant. We use person-action-object stories as a concrete example.

Person-Action-Object Stories. A random person-action-object (PAO) story for a person (e.g., Bill Gates) consists of a random action $a \in \mathcal{ACT}$ (e.g., swallowing) and a random object $o \in \mathcal{OBJ}$ (e.g., a bike). While PAO stories follow a very simple syntactic pattern they also tend to be surprising and interesting because the story is often unexpected (e.g., Bill Clinton kissing a piranha, or Michael Jordan torturing a lion). There is good evidence that memorable phrases tend to use uncommon combinations of words in common syntactic patterns [31]. Each cue $\hat{c} \in \mathcal{C}$ includes a person (e.g., Bill Gates) as well as a picture. To help the user memorize the story we tell him to imagine the scene taking place inside the picture (see Figure 1a for an example). We use algorithm 2 to automatically generate random PAO stories. The cue $\hat{c}$ could be selected either with the user's input (e.g., use the name of a friend and a favorite photograph) or automatically. As long as the cue $\hat{c}$ is fixed before the associated action-object story is selected the cue-association pairs will satisfy the independence condition of Theorem 2.

5.1   Constructing (n, ℓ, γ)-sharing set families

We use the Chinese Remainder Theorem to construct nearly optimal (n, ℓ, γ)-sharing set families. Our application of the Chinese Remainder Theorem is different from previous applications of the Chinese Remainder Theorem in cryptography (e.g., faster RSA decryption algorithm [32], secret sharing [16]). The inputs $n_1, ..., n_\ell$ to algorithm 1 should be co-prime so that we can invoke the Chinese Remainder Theorem — see Figure 1b for an example of our construction with $(n_1, n_2, n_3, n_4) = (9, 10, 11, 13)$.

Algorithm 1. CRT $(m, n_1, ..., n_\ell)$

Input: m, and $n_1, ..., n_\ell$.
for $i = 1 \to m$ do
  $S_i \leftarrow \emptyset$
  for $j = 1 \to \ell$ do
    $N_j \leftarrow \sum_{i=1}^{j-1} n_j$
    $S_i \leftarrow S_i \cup \{(i \mod n_j) + N_j\}$
return $\{S_1, ..., S_m\}$

Algorithm 2. CreatePAOStories

Input: n, random bits b, images $I_1, ..., I_n$, and names $P_1, ..., P_n$.
for $i = 1 \to n$ do
  $a_i \overset{\$}{\leftarrow} \mathcal{ACT}, o_i \overset{\$}{\leftarrow} \mathcal{OBJ}$ %Using random bits b
%Split PAO stories to optimize usability
for $i = 1 \to n$ do
  $\hat{c}_i \leftarrow ((I_i, P_i, `Act'), (I_{i+1 \mod n}, P_{i+1 \mod n}, `Obj'))$
  $\hat{a}_i \leftarrow (a_i, o_{i+1 \mod n})$
return $\{\hat{c}_1, ..., \hat{c}_n\}, \{\hat{a}_1, ..., \hat{a}_n\}$

Lemma 2 says that algorithm 1 produces a (n, ℓ, γ)-sharing set family of size m as long as certain technical conditions apply (e.g., algorithm 1 can be run with any numbers $n_1, ..., n_\ell$, but lemma 2 only applies if the numbers are pairwise co-prime.).

Lemma 2. *If the numbers $n_1 < n_2 < ... < n_\ell$ are pairwise co-prime and $m \leq \prod_{i=1}^{\gamma+1} n_i$ then algorithm 1 returns a $(\sum_{i=1}^{\ell} n_i, \ell, \gamma)$-sharing set of public cues.*

Proof. Suppose for contradiction that $|S_i \cap S_k| \geq \gamma + 1$ for $i < k < m$, then by construction we can find $\gamma + 1$ distinct indices $j_1, ..., j_{\gamma+1} \in$ such that $i \equiv k$ mod n_{j_t} for $1 \leq t \leq \gamma + 1$. The Chinese Remainder Theorem states that there is a unique number x^* s.t. (1) $1 \leq x^* < \prod_{t=1}^{\gamma+1} n_{j_t}$, and (2) $x^* \equiv k$ mod n_{j_t} for $1 \leq t \leq \gamma + 1$. However, we have $i < m \leq \prod_{t=1}^{\gamma+1} n_{j_t}$. Hence, $i = x^*$ and by similar reasoning $k = x^*$. Contradiction!

Example: Suppose that we select pairwise co-prime numbers $n_1 = 9, n_2 = 10, n_3 = 11, n_4 = 13$, then **CRT** $(m, n_1, ..., n_4)$ generates a $(43, 4, 1)$-sharing set family of size $m = n_1 \times n_2 = 90$ (i.e. the public cues for two accounts will overlap in at most one common cue), and for $m \leq n_1 \times n_2 \times n_3 = 990$ we get a $(43, 4, 2)$-sharing set family.

Lemma 2 implies that we can construct a (n, ℓ, γ)-sharing set system of size $m \geq \Omega\left((n/\ell)^{\gamma+1}\right)$ by selecting each $n_i \approx n/\ell$. Theorem 3 proves that we can't hope to do much better — any (n, ℓ, γ)-sharing set system has size $m \leq O\left((n/\ell)^{\gamma+1}\right)$. We refer the interested reader to the full version[21] of this paper for the proof of Theorem 3 and for discussion about additional (n, ℓ, γ)-sharing constructions.

Theorem 3. *Suppose that $\mathcal{S} = \{S_1, ..., S_m\}$ is a (n, ℓ, γ)-sharing set family of size m then $m \leq \binom{n}{\gamma+1} / \binom{\ell}{\gamma+1}$.*

5.2   Shared Cues

Our password management scheme —Shared Cues— uses a (n, ℓ, γ)-sharing set family of size m (e.g., a set family generated by algorithm 1) as a hardcoded input to output the public cues $c_1, ... c_m \subseteq \mathcal{C}$ and passwords $p_1, ..., p_m$ for each account. We use algorithm 2 to generate the underlying cues $\hat{c}_1, ..., \hat{c}_n \in \mathcal{C}$ and their associated PAO stories. The computer is responsible for storing the public cues in persistent memory and the user is responsible for memorizing and rehearsing each cue-association pair $(\hat{c}_i, \hat{a}_i)$.

We use two additional tricks to improve usability: (1) Algorithm 2 splits each PAO story into two parts so that each cue $\hat{c}$ consists of *two* pictures and *two* corresponding people with a label (action/object) for each person (see Figure 1b). A user who sees cue $\hat{c}_i$ will be rehearsing both the i'th and the $i+1$'th PAO story, but will only have to enter one action and one object. (2) To optimize usability we use GreedyMap (Algorithm 4) to produce a permutation $\pi : [m] \rightarrow [m]$ over the public cues — the goal is to minimize the total number of extra rehearsals by ensuring that each cue is used by a frequently visited account.

Algorithm 3. $SharedCues\,[S_1,\ldots,S_m,]\quad \mathcal{G}_m$

Input: $k \in \mathcal{K}$, b, $\lambda_1,\ldots,\lambda_m$, Rehearsal Schedule R.
$\{\hat{c}_1,\ldots,\hat{c}_n\},\{\hat{a}_1,\ldots,\hat{a}_n\} \leftarrow$ **CreatePAOStories** $(\mathbf{n}, \mathbf{I_1},\ldots,\mathbf{I_n},\mathbf{P_1},\ldots,\mathbf{P_n})$
for $i = 1 \rightarrow m$ **do**
   $c_i \leftarrow \{\hat{c}_j \mid j \in S_i\}$, and $p_i \leftarrow \{\hat{a}_j \mid j \in S_i\}$.
% Permute cues
$\pi \leftarrow GreedyMap\,(m,\lambda_1,\ldots,\lambda_m,c_1,\ldots,c_m,R,\sigma)$
return $\left(p_{\pi(1)},c_{\pi(1)}\right),\ldots,\left(p_{\pi(m)},c_{\pi(m)}\right)$
User: Rehearses the cue-association pairs $(\hat{c}_i,\hat{a}_i)$ by following the rehearsal schedule R.
Computer: Stores the public cues $c_1,\ldots,c_m$ in persistent memory.

Once we have constructed our public cues $c_1,\ldots,c_m \subseteq \mathcal{C}$ we need to create a mapping π between cues and accounts $A_1,\ldots,A_m$. Our goal is to minimize the total number of extra rehearsals that the user has to do to satisfy his rehearsal requirements. Formally, we define the **Min-Rehearsal** problem as follows:
Instance: Public Cues $c_1,\ldots,c_m \subseteq \mathcal{C}$, Visitation Schedule $\lambda_1,\ldots,\lambda_m$, a rehearsal schedule R for the underlying cues $\hat{c} \in C$ and a time frame t.
Output: A bijective mapping $\pi : \{1,\ldots,m\} \rightarrow \{1,\ldots,m\}$ mapping account A_i to public cue $S_{\pi(i)}$ which minimizes $E\,[X_t]$.
Unfortunately, we can show that **Min-Rehearsal** is NP-Hard to even approximate within a constant factor. Our reduction from Set Cover can be found in the full version[21] of this paper. Instead GreedyMap uses a greedy heuristic to generate a permutation π.

Theorem 4. *It is NP-Hard to approximate* **Min-Rehearsal** *within a constant factor.*

Algorithm 4. GreedyMap

Input: $m,\lambda_1,\ldots,\lambda_m,c_1,\ldots,c_m$, Rehearsal Schedule R (e.g., CR or ER with parameter σ).
Relabel: Sort λ's s.t $\lambda_i \geq \lambda_{i+1}$ for all $i \leq m-1$.
Initialize: $\pi_0\,(j) \leftarrow \perp$ for $j \leq m$, $UsedCues \leftarrow \emptyset$.
%π_i denotes a partial mapping $[i] \rightarrow [m]$,for $j > i$, the mapping is undefined (e.g., $\pi_i\,(j) = \perp$). Let $S_k = \{\hat{c} \mid \hat{c} \in c_k\}$.
for $i = 1 \rightarrow m$ **do**
   for all $j \in [m] - UsedCues$ **do**

$$\Delta_j \leftarrow \sum_{\hat{c} \in S_j} E\left[X_{t,\hat{c}}\middle|\lambda_{\hat{c}} = \lambda_i + \sum_{j:\hat{c} \in S_{\pi_{i-1}(j)}} \lambda_j\right] - E\left[X_{t,\hat{c}}\middle|\lambda_{\hat{c}} = \sum_{j:\hat{c} \in S_{\pi_{i-1}(j)}} \lambda_j\right]$$

% Δ_j: expected reduction in total extra rehearsals if we set $\pi_i(i) = j$
   $\pi_i\,(i) \leftarrow \arg\max_j \Delta_j$, $UsedCues \leftarrow UsedCues \cup \{\pi_i\,(i)\}$
return π^m

5.3 Usability and Security Analysis

We consider three instantiations of Shared Cues: SC-0, SC-1 and SC-2. SC-0 uses a $(9,4,3)$-sharing family of public cues of size $m = 126$ — constructed by taking all $\binom{9}{4} = 126$ subsets of size 4. SC-1 uses a $(43,4,1)$-sharing family of public cues of size $m = 90$ — constructed using algorithm 1 with $m = 90$ and $(n_1, n_2, n_3, n_4) = (9, 10, 11, 13)$. SC-2 uses a $(60, 5, 1)$-sharing family of public cues of size $m = 90$ — constructed using algorithm 1 with $m = 90$ and $(n_1, n_2, n_3, n_4, n_5) = (9, 10, 11, 13, 17)$.

Our usability results can be found in table 3 and our security results can be found in table 4. We present our usability results for the very active, typical, occasional and infrequent internet users (see table 1 for the visitation schedules) under both sufficient rehearsal assumptions CR and ER. Table 3 shows the values of $E[X_{365}]$ — computed using the formula from Theorem 1 — for SC-0, SC-1 and SC-2. We used association strength parameter $\sigma = 1$ to evaluate each password management scheme — though we expect that σ will be higher for schemes like Shared Cues that use strong mnemonic techniques [1].

Table 3. $E[X_{365}]$: Extra Rehearsals over the first year for SC-0,SC-1 and SC-2

Assumption	CR $(\sigma = 1)$			ER $(\sigma = 1)$		
Schedule/Scheme	SC-0	SC-1	SC-2	SC-0	SC-1	SC-2
Very Active	≈ 0	$1,309$	$2,436$	≈ 0	3.93	7.54
Typical	≈ 0.42	$3,225$	$5,491$	≈ 0	10.89	19.89
Occasional	≈ 1.28	$9,488$	$6,734$	≈ 0	22.07	34.23
Infrequent	≈ 723	$13,214$	$18,764$	≈ 2.44	119.77	173.92

Our security guarantees for SC-0,SC-1 and SC-2 are illustrated in Table 4. The values were computed using Theorem 2. We assume that $|\mathcal{AS}| = 140^2$ where $\mathcal{AS} = \mathcal{ACT} \times \mathcal{OBJ}$ (e.g., their are 140 distinct actions and objects), and that the adversary is willing to spend at most $\$10^6$ on cracking the user's passwords (e.g., $q = q_{\$10^6} = 5.155 \times 10^{10}$). The values of δ in the $h = 0$ columns were computed assuming that $m \le 100$.

Discussion: Comparing tables 3 and 2 we see that **Lifehacker** is the most usable password management scheme, but SC-0 compares very favorably! Unlike **Lifehacker**, SC-0 provides provable security guarantees after the adversary phishes one account — though the guarantees break down if the adversary can also execute an offline attack. While SC-1 and SC-2 are not as secure as **Strong Random and Independent** — the security guarantees from **Strong Random and Independent** do not break down even if the adversary can recover *many* of the user's plaintext passwords — SC-1 and SC-2 are far more usable than **Strong Random and Independent**. Furthermore, SC-1 and SC-2 do provide very strong security guarantees (e.g., SC-2 passwords remain secure against offline attacks even after an adversary obtains two plaintext passwords for accounts

---

[1] We explore the effect of σ on $E[X_{t,c}]$ in the full version[21] of this paper.

Table 4. Shared Cues ($q_{\$10^6}, \delta, m, s, r, h$)-Security: δ vs h and r using a (n, ℓ, γ)-sharing family of m public cues

Offline Attack?	$h = 0$			$h > 0$		
(n, ℓ, γ)-sharing	$r = 0$	$r = 1$	$r = 2$	$r = 0$	$r = 1$	$r = 2$
($n, 4, 3$) (e.g., SC-0)	2×10^{-15}	0.011	1	3.5×10^{-7}	1	1
($n, 4, 1$) (e.g., SC-1)	2×10^{-15}	4×10^{-11}	8×10^{-7}	3.5×10^{-7}	0.007	1
($n, 5, 1$) (e.g., SC-2)	1×10^{-19}	2×10^{-15}	4×10^{-11}	1.8×10^{-11}	3.5×10^{-7}	0.007

of his choosing). For the very active, typical and occasional user the number of extra rehearsals required by SC-1 and SC-2 are quite reasonable (e.g., the typical user would need to perform less than one extra rehearsal per month). The usability benefits of SC-1 and SC-2 are less pronounced for the infrequent user — though the advantage over **Strong Random and Independent** is still significant.

6   Discussion and Future Work

We conclude by discussing future directions of research.

Sufficient Rehearsal Assumptions: While there is strong empirical evidence for the Expanding Rehearsal assumption in the memory literature (e.g., [60]), the parameters we use are drawn from prior studies in other domains. It would be useful to conduct user studies to test the Expanding Rehearsal assumption in the password context, and obtain parameter estimates specific to the password setting. We also believe that user feedback from a password management scheme like Shared Cues could be an invaluable source of data about rehearsal and long term memory retention.

Expanding Security over Time: Most extra rehearsals occur soon after the user memorizes a cue-association pair — when the rehearsal intervals are still small. Is it possible to start with a password management scheme with weaker security guaratnees (e.g., SC-0), and increase security over time by having the user memorize additional cue-association pairs as time passes?

Human Computable Passwords: Shared Cues only relies on the human capacity to memorize and retrieve information, and is secure against at most $r = \ell/\gamma$ plaintext password leak attacks. Could we improve security (or usability) by having the user perform simple computations to recover his passwords? Hopper and Blum proposed a 'human authentication protocol' — based on the noisy parity problem — as an alternative to passwords [36], but their protocol seems to be too complicated for humans to execute. Could similar ideas be used to construct a secure human-computation based password management scheme?

References

1. Amazon ec2 pricing, http://aws.amazon.com/ec2/pricing/ (retrieved October 22, 2012)

2. Cert incident note in-98.03: Password cracking activity (July 1998), http://www.cert.org/incident_notes/IN-98.03.html (retrieved August 16, 2011)
3. Geek to live: Choose (and remember) great passwords (July 2006), http://lifehacker.com/184773/geek-to-live--choose-and-remember-great-passwords (retrieved September 27, 2012)
4. Rockyou hack: From bad to worse (December 2009), http://techcrunch.com/2009/12/14/rockyou-hack-security-myspace-facebook-passwords/ (retrieved September 27, 2012)
5. Oh man, what a day! an update on our security breach (April 2010), http://blogs.atlassian.com/news/2010/04/oh_man_what_a_day_an_update_on_our_security_breach.html (retrieved August 18, 2011)
6. Sarah palin vs the hacker (May 2010), http://www.telegraph.co.uk/news/worldnews/sarah-palin/7750050/Sarah-Palin-vs-the-hacker.html (retrieved September 9, 2012)
7. Nato site hacked (June 2011), http://www.theregister.co.uk/2011/06/24/nato_hack_attack/ (retrieved August 16, 2011)
8. Update on playstation network/qriocity services (April 2011), http://blog.us.playstation.com/2011/04/22/update-on-playstation-network-qriocity-services/ (retrieved May 22, 2012)
9. Apple security blunder exposes lion login passwords in clear text (May 2012), http://www.zdnet.com/blog/security/apple-security-blunder-exposes-lion-login-passwords-in-clear-text/11963 (retrieved May 22, 2012)
10. Data breach at ieee.org: 100k plaintext passwords (September 2012), http://ieeelog.com/ (retrieved September 27, 2012)
11. An update on linkedin member passwords compromised (June 2012), http://blog.linkedin.com/2012/06/06/linkedin-member-passwords-compromised/ (retrieved September 27, 2012)
12. Zappos customer accounts breached (January 2012), http://www.usatoday.com/tech/news/story/2012-01-16/mark-smith-zappos-breach-tips/52593484/1 (retrieved May 22, 2012)
13. Acquisti, A., Gross, R.: Imagined communities: awareness, information sharing, and privacy on the facebook. In: Danezis, G., Golle, P. (eds.) PET 2006. LNCS, vol. 4258, pp. 36–58. Springer, Heidelberg (2006)
14. Anderson, J., Matessa, M., Lebiere, C.: Act-r: A theory of higher level cognition and its relation to visual attention. Human-Computer Interaction 12(4), 439–462 (1997)
15. Anderson, J.R., Schooler, L.J.: Reflections of the environment in memory. Psychological Science 2(6), 396–408 (1991)
16. Asmuth, C., Bloom, J.: A modular approach to key safeguarding. IEEE Transactions on Information Theory 29(2), 208–210 (1983)
17. Baddeley, A.: Human memory: Theory and practice. Psychology Pr. (1997)
18. Bellare, M., Rogaway, P.: The exact security of digital signatures - how to sign with RSA and rabin. In: Maurer, U.M. (ed.) EUROCRYPT 1996. LNCS, vol. 1070, pp. 399–416. Springer, Heidelberg (1996)
19. Biddle, R., Chiasson, S., Van Oorschot, P.: Graphical passwords: Learning from the first twelve years. ACM Computing Surveys (CSUR) 44(4), 19 (2012)
20. Biddle, S.: Anonymous leaks 90,000 military email accounts in latest antisec attack (July 2011), http://gizmodo.com/5820049/anonymous-leaks-90000-military-email-accounts-in-latest-antisec-attack (retrieved August 16, 2011)

21. Blocki, J., Blum, M., Datta, A.: Naturally rehearsing passwords. CoRR abs/1302.5122 (2013)
22. Blocki, J., Komanduri, S., Procaccia, A., Sheffet, O.: Optimizing password composition policies
23. Bojinov, H., Sanchez, D., Reber, P., Boneh, D., Lincoln, P.: Neuroscience meets cryptography: designing crypto primitives secure against rubber hose attacks. In: Proceedings of the 21st USENIX Conference on Security Symposium, pp. 33–33. USENIX Association (2012)
24. Bonneau, J.: The science of guessing: analyzing an anonymized corpus of 70 million passwords. In: 2012 IEEE Symposium on Security and Privacy (SP), pp. 538–552. IEEE (2012)
25. Bonneau, J., Herley, C., van Oorschot, P.C., Stajano, F.: The quest to replace passwords: A framework for comparative evaluation of web authentication schemes. In: IEEE Symposium on Security and Privacy, pp. 553–567. IEEE (2012)
26. Boztas, S.: Entropies, guessing, and cryptography. Department of Mathematics, Royal Melbourne Institute of Technology, Tech. Rep 6 (1999)
27. Brand, S. Department of defense password management guideline
28. Brostoff, S., Sasse, M.: Are Passfaces more usable than passwords: A field trial investigation. In: People and Computers XIV-Usability or Else: Proceedings of HCI, pp. 405–424 (2000)
29. Burnett, M.: Perfect passwords: selection, protection, authentication. Syngress Publishing (2005)
30. Center, I.: Consumer password worst practices. Imperva (White Paper) (2010)
31. Danescu-Niculescu-Mizil, C., Cheng, J., Kleinberg, J., Lee, L.: You had me at hello: How phrasing affects memorability. In: Proceedings of the 50th Annual Meeting of the Association for Computational Linguistics: Long Papers, vol. 1, pp. 892–901. Association for Computational Linguistics (2012)
32. Ding, C., Pei, D., Salomaa, A.: Chinese remainder theorem. World Scientific (1996)
33. Florencio, D., Herley, C.: A large-scale study of web password habits. In: Proceedings of the 16th International Conference on World Wide Web, pp. 657–666. ACM (2007)
34. Foer, J.: Moonwalking with Einstein: The Art and Science of Remembering Everything. Penguin Press (2011)
35. Gaw, S., Felten, E.W.: Password management strategies for online accounts. In: Proceedings of the Second Symposium on Usable Privacy and Security, SOUPS 2006, pp. 44–55. ACM, New York (2006)
36. Hopper, N.J., Blum, M.: Secure human identification protocols. In: Boyd, C. (ed.) ASIACRYPT 2001. LNCS, vol. 2248, pp. 52–66. Springer, Heidelberg (2001)
37. Kohonen, T.: Associative memory: A system-theoretical approach. Springer, Berlin (1977)
38. Komanduri, S., Shay, R., Kelley, P., Mazurek, M., Bauer, L., Christin, N., Cranor, L., Egelman, S.: Of passwords and people: measuring the effect of password-composition policies. In: Proceedings of the 2011 Annual Conference on Human Factors in Computing Systems, pp. 2595–2604. ACM (2011)
39. Kruger, H., Steyn, T., Medlin, B., Drevin, L.: An empirical assessment of factors impeding effective password management. Journal of Information Privacy and Security 4(4), 45–59 (2008)
40. Marr, D.: Simple memory: a theory for archicortex. Philosophical Transactions of the Royal Society of London. Series B, Biological Sciences, 23–81 (1971)
41. Massey, J.: Guessing and entropy. In: Proceedings of the 1994 IEEE International Symposium on Information Theory, p. 204. IEEE (1994)

42. Monroe, R.: Xkcd: Password strength, http://www.xkcd.com/936/ (retrieved August 16, 2011)
43. Nisan, N., Wigderson, A.: Hardness vs randomness. Journal of Computer and System Sciences 49(2), 149–167 (1994)
44. Pliam, J.O.: On the incomparability of entropy and marginal guesswork in brute-force attacks. In: Roy, B., Okamoto, E. (eds.) INDOCRYPT 2000. LNCS, vol. 1977, pp. 67–79. Springer, Heidelberg (2000)
45. Provos, N., Mazieres, D.: Bcrypt algorithm
46. Radke, K., Boyd, C., Nieto, J.G., Brereton, M.: Towards a secure human-and-computer mutual authentication protocol. In: Proceedings of the Tenth Australasian Information Security Conference (AISC 2012), vol. 125, pp. 39–46. Australian Computer Society Inc. (2012)
47. Rasch, G.: The poisson process as a model for a diversity of behavioral phenomena. In: International Congress of Psychology (1963)
48. Scarfone, K., Souppaya, M.: Guide to enterprise password management (draft). National Institute of Standards and Technology 800-188 6, 38 (2009)
49. Schechter, S., Brush, A., Egelman, S.: It's no secret. measuring the security and reliability of authentication via 'secret' questions. In: 2009 30th IEEE Symposium on Security and Privacy, pp. 375–390. IEEE (2009)
50. Shay, R., Kelley, P., Komanduri, S., Mazurek, M., Ur, B., Vidas, T., Bauer, L., Christin, N., Cranor, L.: Correct horse battery staple: Exploring the usability of system-assigned passphrases. In: Proceedings of the Eighth Symposium on Usable Privacy and Security, p. 7. ACM (2012)
51. Singer, A.: No plaintext passwords. The Magazine of Usenix & Sage 26(7) (November 2001) (retrieved August 16, 2011)
52. Spence, J.: The memory palace of Matteo Ricci. Penguin Books (1985)
53. Squire, L.: On the course of forgetting in very long-term memory. Journal of Experimental Psychology: Learning, Memory, and Cognition 15(2), 241 (1989)
54. Standingt, L.: Learning 10,000 pictures. Quarterly Journal of Experimental Psychology 5(20), 7–22 (1973)
55. Stein, J.: Pimp my password. Time, 62 (August 29, 2011)
56. Valiant, L.: Memorization and association on a realistic neural model. Neural Computation 17(3), 527–555 (2005)
57. van Rijn, H., van Maanen, L., van Woudenberg, M.: Passing the test: Improving learning gains by balancing spacing and testing effects. In: Proceedings of the 9th International Conference of Cognitive Modeling (2009)
58. Von Ahn, L., Blum, M., Hopper, N., Langford, J.: Captcha: Using hard ai problems for security. In: Biham, E. (ed.) EUROCRYPT 2003. LNCS, vol. 2656, pp. 646–646. Springer, Heidelberg (2003)
59. Willshaw, D., Buckingham, J.: An assessment of marr's theory of the hippocampus as a temporary memory store. Philosophical Transactions of the Royal Society of London. Series B: Biological Sciences 329(1253), 205 (1990)
60. Wozniak, P., Gorzelanczyk, E.J.: Optimization of repetition spacing in the practice of learning. Acta Neurobiologiae Experimentalis 54, 59–59 (1994)
61. Yan, J., Blackwell, A., Anderson, R., Grant, A.: Password memorability and security: Empirical results. IEEE Security & Privacy 2(5), 25–31 (2004)

Leakage-Resilient Chosen-Ciphertext Secure Public-Key Encryption from Hash Proof System and One-Time Lossy Filter

Baodong Qin[1,2] and Shengli Liu[1,*]

[1] Department of Computer Science and Engineering, Shanghai Jiao Tong University, Shanghai 200240, China
[2] College of Computer Science and Technology, Southwest University of Science and Technology, Mianyang 621010, China
{qinbaodong,slliu}@sjtu.edu.cn

Abstract. We present a new generic construction of a public-key encryption (PKE) scheme secure against leakage-resilient chosen-ciphertext attacks (LR-CCA), from any Hash Proof System (HPS) and any one-time lossy filter (OT-LF). Efficient constructions of HPSs and OT-LFs from the DDH and DCR assumptions suggest that our construction is a practical approach to LR-CCA security. Most of practical PKEs with LR-CCA security, like variants of Cramer-Shoup scheme, rooted from Hash Proof Systems, but with leakage rates at most $1/4 - o(1)$ (defined as the ratio of leakage amount to secret-key size). The instantiations of our construction from the DDH and DCR assumptions result in LR-CCA secure PKEs with leakage rate of $1/2 - o(1)$. On the other hand, our construction also creates a new approach for constructing IND-CCA secure (leakage-free) PKE schemes, which may be of independent interest.

Keywords: Public-key encryption, leakage-resilience, chosen-ciphertext security, hash proof system.

1 Introduction

Research on leakage-resilient cryptography is motivated by those side-channel attacks [17], in which a significant fraction of the secret key SK is leaked to the adversary. Cryptosystems proved secure in the traditional model may suffer from these key-leakage attacks, as shown in [17]. This fact leads to design and security proof of a variety of leakage-resilient cryptosystems, including stream ciphers [14,30], block ciphers [12], digital signatures [20,15], public key encryption [27,1,2,3,4], identity-based encryption [24,7,16], etc.

Leakage Oracle, Bounded-Leakage Model and Leakage Rate. Side-channel attacks characterized by key leakage can be formalized in a general

---

* Supported by the National Natural Science Foundation of China (Grant No. 61170229, 61373153), the Specialized Research Fund for the Doctoral Program of Higher Education (Grant No. 20110073110016), and the Scientific innovation projects of Shanghai Education Committee (Grant No. 12ZZ021).

K. Sako and P. Sarkar (Eds.) ASIACRYPT 2013, Part II, LNCS 8270, pp. 381–400, 2013.

framework [1] with a leakage oracle $\mathcal{O}_{SK}^{\lambda,\kappa}(\cdot)$: the adversary queries arbitrary efficiently computable functions $f_i : \{0,1\}^* \to \{0,1\}^{\lambda_i}$ of the secret key SK repeatedly and adaptively, and the leakage oracle responds with $f_i(SK)$. The *bounded-leakage* model limits the total amount of information about SK leaked by the oracle to a bound λ during the life time of the cryptosystem. This model is simple and powerful, but a thorough understanding of this model is essential to those more complicated models [4]. If a cryptosystem is secure against the above key-leakage attacks, we call it λ-leakage-resilient (λ-LR, for short). The *leakage rate* is defined as the ratio of λ to the secret key size, i.e., $\lambda/|SK|$.

Leakage-Resilient CCA Security and Hash Proof System. In the key-leakage scenario of public key encryption (PKE), leakage-resilient security against chosen-plaintext attacks (LR-CPA) is characterized by the indistinguishability between the encryptions of two plaintexts (of equal length) chosen by any Probabilistic Polynomial-Time (PPT) adversary, who is given access to a key-leakage oracle. If the adversary is equipped with a decryption oracle as well, with restriction that the challenge ciphertext is refused by the decryption oracle and the leakage oracle stops working after the generation of the challenge ciphertext, the notion becomes leakage-resilient security against chosen-ciphertext attacks (LR-CCA). Naor-Yung paradigm applies to LR-CCA security [27]. It achieves leakage rate of $1 - o(1)$, but the simulation-sound Non-Interactive Zero-Knowledge (ss-NIZK) proof is far from practical. It was later improved by Dodis et al. [11] with true-simulation extractable NIZK (tSE-NIZK), but the construction is still not practical. Recently, Galindo et al. [16] constructed an identity-based encryption (IBE) scheme with master key-dependent chosen-plaintext (mKDM-sID-CPA) security based on the decisional linear assumption over bilinear groups. They suggested that their mKDM-sID-CPA secure IBE scheme is also master key resilient with rate $1 - o(1)$, hence can be transformed into a LR-CCA secure PKE scheme with leakage rate $1 - o(1)$ by applying the CHK transform [6]. However, their claim that the mKDM-sID-CPA secure IBE scheme is also master key leakage resilient was not supported by any rigorous proof.

Hash Proof Systems (HPSs), due to Cramer and Shoup [9], have long been served as the most practical approach to PKEs with IND-CCA security. They are also intrinsically LR-CPA secure, and a HPS based on the DDH assumption (and its d-Linear variant) was proved to be LR-CPA secure with leakage rate of $1 - o(1)$ [27]. As to LR-CCA security, however, the HPS approach to IND-CCA security is inherently limited to leakage rate below $1/2$, as pointed out by Dodis et al. [11]. Recall that to achieve IND-CCA security, Cramer and Shoup [9] proposed to use two independent HPSs, one is a smooth HPS to mask and hide the plaintext, and the other is a universal$_2$ HPS used to verify whether the ciphertext is well-formed. Hence two independent secret keys are involved in the construction, and either one, if totally leaked, will kill the LR-CCA security. That is why the leakage rate must be less than $1/2$.

Prior constructions of PKE with LR-CCA security from HPSs enjoy great efficiency, but suffer from low leakage rate. The variants [27,26] of Cramer-Shoup DDH-based scheme [8] achieve leakage rate of $1/6 - o(1)$, which was later improved to $1/4 - o(1)$ [25]. To the best of our knowledge, no constructions from HPSs are known to be LR-CCA secure with leakage rate of $1/2 - o(1)$. The question is: can we find a new way to construct LR-CCA secure PKEs which are not only as practical as HPS but also with reasonable high leakage rates (like $1/2 - o(1)$)?

Our Contributions. We propose a new generic construction of PKE with LR-CCA security from a Hash Proof System (HPS) and a one-time lossy filter (OT-LF). The new primitive, one-time lossy filter (OT-LF), is a weak version of lossy algebraic filter [19], and we show how to construct OT-LFs from the DDH and DCR assumptions. In the generic construction of LR-CCA secure PKE, the HPS is used to generate an encapsulated key K, which is not only used to mask the plaintext, but also used in the OT-LF to verify the well-formedness of ciphertexts. OT-LF helps to obtain a higher leakage rate, compared to the constructions solely from HPSs.

- We give instantiations of PKEs with LR-CCA security under the DDH (DCR) assumption, by combining an efficient construction of DDH (DCR)-based OT-LF and DDH (DCR)-based HPS. The leakage rate is as high as $1/2 - o(1)$.
- In case of no leakage on secret key at all, the leakage-free version of our construction opens another practical approach to IND-CCA security, as compared to the HPS-based construction by Cramer and Shoup.

Overview of Our Techniques. Different from the HPS-based approach to CCA-security, in which a universal$_2$ hash proof system is employed to reject ill-formed ciphertexts, we use a one-time lossy filter (OT-LF) to do the job. OT-LF is a simplified version of lossy algebraic filter, which was introduced by Hofheinz [19] recently to realize key-dependent chosen-ciphertext security [5]. The concept of OT-LF is similar to (chameleon) all-but-one lossy trapdoor function [31,23]. But it does not require efficient inversion. Roughly, a OT-LF is a family of functions indexed by a public key Fpk and a tag $t = (t_a, t_c)$. A function $\mathsf{LF}_{Fpk,t}(\cdot)$ from that family maps an input X to a unique output. For a fixed public key, the set of tags contains two computationally indistinguishable disjoint subsets, namely the subset of injective tags and the subset of lossy ones. If tag $t = (t_a, t_c)$ is injective, then so is the corresponding function $\mathsf{LF}_{Fpk,t}(\cdot)$. If the tag is lossy, the output of the function reveals only a constant amount of information about its input X. For any t_a, there exists a lossy tag (t_a, t_c) such that t_c can be efficiently computed through a trapdoor Ftd. Without this trapdoor, however, it is hard to generate a new lossy tag even with the knowledge of one lossy tag. Trapdoor Ftd and lossy tag are only used for the security proof.

Roughly speaking, a hash proof system HPS is a key-encapsulation mechanism. Given public key pk, an element $C \in \mathcal{V}$ and its witness w, the encapsulated key is given by $K = \mathsf{HPS.Pub}(pk, C, w)$. With secret key sk, decapsulation algorithm $\mathsf{HPS.Priv}(sk, C)$ recovers K from $C \in \mathcal{V}$. If $C \in \mathcal{C} \setminus \mathcal{V}$, the output of $\mathsf{HPS.Priv}(sk, C)$ has a high min-entropy even conditioned on pk and C. The hardness of subset membership problem requires that elements in $\mathcal{V}$ are indistinguishable from those in $\mathcal{C} \setminus \mathcal{V}$.

In our construction, the secret key is just sk from the HPS, and the HPS and OT-LF are integrated into a ciphertext CT,

$$CT \;=\; (C,\ \ s,\ \ \Psi = \mathsf{Ext}(K,s) \oplus M,\ \ \Pi = \mathsf{LF}_{Fpk,t}(K),\ \ t_c),$$

via $K = \mathsf{HPS.Pub}(pk, C, w) = \mathsf{HPS.Priv}(sk, C)$ (it holds for all $C \in \mathcal{V}$).

The encapsulated key K functions in two ways. (1) It serves as an input, together with a random string s, to extractor $\mathsf{Ext}(K, s)$ to mask and hide the plaintext M to deal with key leakage. (2) It serves as the input of $\mathsf{LF}_{Fpk,t}(\cdot)$ to check the well-formedness of the ciphertext. Tag $t = (t_a, t_c)$ is determined by $t_a = (C, s, \Psi)$ and a random t_c. $\mathsf{LF}_{Fpk,t}(K)$ can also be considered as an authentication code, which is used to authenticate the tag $t = ((C, s, \Psi), t_c)$ with the authentication key K.

In the security proof, some changes are made to the generation of the challenge ciphertext $CT^* = (C^*, s^*, \Psi^*, \Pi^*, t_c^*)$: C^* is sampled from $\mathcal{C} \setminus \mathcal{V}$ and the tag t^* is made lossy by computing a proper t_c with trapdoor Ftd. A PPT adversary cannot tell the changes due to the hardness of subset membership problem and the indistinguishability of lossy tags and injective ones. Conditioned on CT^*, the encapsulated key $K^* = \mathsf{HPS.Priv}(sk, C^*)$ still maintains a high min-entropy since $\Pi^* = \mathsf{LF}_{Fpk,t^*}(K^*)$ works in lossy mode and only little information is released. When a PPT adversary chooses an invalid ciphertext CT in the sense that $C \in \mathcal{C} \setminus \mathcal{V}$ for decryption query, the corresponding tag t is injective with overwhelming probability. Then $\mathsf{LF}_{Fpk,t}(\cdot)$ is injective and Π preserves the high min-entropy of $K = \mathsf{HPS.Priv}(sk, C)$. Hence invalid ciphertexts will be rejected by the decryption oracle with overwhelming probability. On the other hand, the information of pk has already determined $K = \mathsf{HPS.Priv}(sk, C)$ for all $C \in \mathcal{V}$. Thus the decryption oracle does not help the adversary to gain any more information about K^*. Then an extractor can be applied to K^* to totally mask the information of challenge plaintext, and a large min-entropy of K^* conditioned on pk and Π^* implies a high tolerance of key leakage.

Thanks to efficient constructions for HPS and OT-LF under the DDH and DCR assumptions, the instantiations are practically efficient. More precisely, $|K| \approx L/2$, where L is the length of the secret key of HPS. Due to the lossiness of the OT-LF and the property of the HPS, the min-entropy conditioned on the public key and challenge ciphertext, approaches $(1/2 - o(1))L$. Hence the leakage rate approaches $1/2$.

2   Preliminaries

Notation. Let $[n]$ denote the set $\{1, \ldots, n\}$. Let $\kappa \in \mathbb{N}$ denote the security parameter and 1^κ denote the string of κ ones. If s is a string, then $|s|$ denotes its length, while if S is a set then $|S|$ denotes its size and $s \leftarrow S$ denotes the operation of picking an element s uniformly at random from S. We denote $y \leftarrow A(x)$ the operation of running A with input x, and assigning y as the result. We write $\log s$ for logarithms over the reals with base 2.

Randomness Extractor. Let $\mathrm{SD}(X, Y)$ denote the *statistical distance* of random variables X and Y over domain Ω. Namely, $\mathrm{SD}(X, Y) = \frac{1}{2} \sum_{\omega \in \Omega} |\Pr[X = \omega] - \Pr[Y = \omega]|$. The *min-entropy* of X is $\mathrm{H}_\infty(X) = -\log(\max_{\omega \in \Omega} \Pr[X = \omega])$. Dodis et al. [13] formalized the notion of *average min-entropy* of X conditioned on Y which is defined as $\tilde{\mathrm{H}}_\infty(X|Y) = -\log(E_{y \leftarrow Y}[2^{-\mathrm{H}_\infty(X|Y=y)}])$. They proved the following property of average min-entropy.

Lemma 1. *[13] Let X, Y and Z be random variables. If Y has at most 2^r possible values, then $\tilde{\mathrm{H}}_\infty(X|(Y, Z)) \geq \tilde{\mathrm{H}}_\infty(X|Z) - r$.*

Definition 1 (Randomness Extractor). An efficient function $\mathsf{Ext} : \mathcal{X} \times \mathcal{S} \to \mathcal{Y}$ is an average-case (ν, ϵ)-strong extractor if for all pairs of random variables (X, Z) such that $X \in \mathcal{X}$ and $\tilde{\mathrm{H}}_\infty(X|Z) \geq \nu$, we have

$$\mathrm{SD}((Z, s, \mathsf{Ext}(X, s)), (Z, s, U_\mathcal{Y})) \leq \epsilon,$$

where s is uniform over $\mathcal{S}$ and $U_\mathcal{Y}$ is uniform over $\mathcal{Y}$.

A family of universal hash functions $\mathcal{H} = \{H_s : \mathcal{X} \to \mathcal{Y}\}_{s \in \mathcal{S}}$ can be used as an average-case $(\tilde{\mathrm{H}}_\infty(X|Z), \epsilon)$-strong extractors whenever $\tilde{\mathrm{H}}_\infty(X|Z) \geq \log |\mathcal{Y}| + 2\log(1/\epsilon)$, according to the general Leftover Hash Lemma [13].

2.1 Leakage-Resilient Public-Key Encryption

A *Public-Key Encryption* (PKE) scheme with plaintext space $\mathcal{M}$ is given by three PPT algorithms $(\mathsf{PKE.Gen}, \mathsf{PKE.Enc}, \mathsf{PKE.Dec})$. The key generation algorithm $\mathsf{PKE.Gen}$ takes as input 1^κ, and outputs a pair of public/secret keys (PK, SK). The encryption algorithm $\mathsf{PKE.Enc}$ takes as input a public key PK and a plaintext $M \in \mathcal{M}$, and returns a ciphertext $CT = \mathsf{PKE.Enc}(PK, M)$. The decryption algorithm $\mathsf{PKE.Dec}$ takes as input a secret key SK and a ciphertext CT, and returns a plaintext $M \in \mathcal{M} \cup \{\bot\}$. For consistency, we require that $\mathsf{PKE.Dec}(SK, \mathsf{PKE.Enc}(PK, M)) = M$ holds for all $(PK, SK) \leftarrow \mathsf{PKE.Gen}(1^\kappa)$ and all plaintexts $M \in \mathcal{M}$.

Following [27,28], we define leakage-resilient chosen-ciphertext security (LR-CCA) for PKE.

Definition 2 (Leakage-Resilient CCA security of PKE). A public-key encryption scheme $\mathsf{PKE} = (\mathsf{PKE.Gen}, \mathsf{PKE.Enc}, \mathsf{PKE.Dec})$ is λ-leakage-resilient chosen-ciphertext secure (λ-LR-CCA-secure), if for any PPT adversary $\mathcal{A} = (\mathcal{A}_1, \mathcal{A}_2)$, the function $\mathsf{Adv}^{\mathrm{lr\text{-}cca}}_{\mathsf{PKE}, \mathcal{A}}(\kappa) := \left| \Pr[\mathsf{Exp}^{\mathrm{lr\text{-}cca}}_{\mathsf{PKE}, \mathcal{A}}(\kappa) = 1] - \frac{1}{2} \right|$ is negligible in κ. Below defines $\mathsf{Exp}^{\mathrm{lr\text{-}cca}}_{\mathsf{PKE}, \mathcal{A}}(\kappa)$.

1. $(PK, SK) \leftarrow \mathsf{PKE.Gen}(1^\kappa)$, $b \leftarrow \{0, 1\}$.
2. $(M_0, M_1, state) \leftarrow \mathcal{A}_1^{\mathcal{O}^{\lambda, \kappa}_{sk}(\cdot), \mathsf{PKE.Dec}(SK, \cdot)}(pk)$, s.t. $|M_0| = |M_1|$.
3. $CT^* \leftarrow \mathsf{PKE.Enc}(PK, M_b)$. $b' \leftarrow \mathcal{A}_2^{\mathsf{PKE.Dec}_{\neq CT^*}(SK, \cdot)}(state, CT^*)$.
5. If $b = b'$ return 1 else return 0.

In the case of $\lambda = 0$, Definition 2 is just the standard CCA security [32].

2.2   Hash Proof System

We recall the notion of hash proof systems introduced by Cramer and Shoup [9]. For simplicity, hash proof systems are described as key encapsulation mechanisms (KEMs), as did in [21].

Projective Hashing. Let $\mathcal{SK}$, $\mathcal{PK}$ and $\mathcal{K}$ be sets of public keys, secret keys and encapsulated keys. Let $\mathcal{C}$ be the set of all ciphertexts of KEM and $\mathcal{V} \subset \mathcal{C}$ be the set of all *valid* ones. We assume that there are efficient algorithms for sampling $sk \leftarrow \mathcal{SK}$, $C \leftarrow \mathcal{V}$ together with a witness w, and $C \leftarrow \mathcal{C} \setminus \mathcal{V}$.

Let $\Lambda_{sk} : \mathcal{C} \to \mathcal{K}$ be a hash function indexed with $sk \in \mathcal{SK}$ that maps ciphertexts to symmetric keys. The hash function Λ_{sk} is *projective* if there exists a projection $\mu : \mathcal{SK} \to \mathcal{PK}$ such that $\mu(sk) \in \mathcal{PK}$ defines the action of Λ_{sk} over the subset $\mathcal{V}$ of valid ciphertexts.

Definition 3 (universal[9]). A projective hash function Λ_{sk} is ϵ-*universal*, if for all pk, $C \in \mathcal{C} \setminus \mathcal{V}$, and all $K \in \mathcal{K}$, it holds that $\Pr[\Lambda_{sk}(C) = K \mid (pk, C)] \leq \epsilon$, where the probability is over all possible $sk \leftarrow \mathcal{SK}$ with $pk = \mu(sk)$.

The lemma below follows directly from the definition of min-entropy.

Lemma 2. *Assume that $\Lambda_{sk} : \mathcal{C} \to \mathcal{K}$ is an ϵ-universal projective hash function. Then, for all pk and $C \in \mathcal{C} \setminus \mathcal{V}$, it holds that $\mathrm{H}_\infty(\Lambda_{sk}(C) | (pk, C)) \geq \log 1/\epsilon$, where $sk \leftarrow \mathcal{SK}$ with $pk = \mu(sk)$.*

Hash Proof System. A hash proof system HPS consists of three PPT algorithms (HPS.Gen, HPS.Pub, HPS.Priv). The parameter generation algorithm HPS.Gen(1^κ) generates parameterized instances of the form params=(group, $\mathcal{K}$, $\mathcal{C}$, $\mathcal{V}$, $\mathcal{SK}$, $\mathcal{PK}$, $\Lambda_{(\cdot)} : \mathcal{C} \to \mathcal{K}$, $\mu : \mathcal{SK} \to \mathcal{PK}$), where group may contain additional structural parameters. The public evaluation algorithm HPS.Pub(pk, C, w) takes as input a projective public key $pk = \mu(sk)$, a valid ciphertext $C \in \mathcal{V}$ and a witness w of the fact that $C \in \mathcal{V}$, and computes the encapsulated key $K = \Lambda_{sk}(C)$. The private evaluation algorithm HPS.Priv(sk, C) takes a secret key sk and a ciphertext $C \in \mathcal{V}$ as input, and returns the encapsulated key $K = \Lambda_{sk}(C)$ without knowing a witness. We assume that μ and $\Lambda_{(\cdot)}$ are efficiently computable.

Subset Membership Problem. The *subset membership problem* associated with a HPS suggests that a random valid ciphertext $C_0 \leftarrow \mathcal{V}$ and a random invalid ciphertext $C_1 \leftarrow \mathcal{C} \setminus \mathcal{V}$ are computationally indistinguishable. This is formally captured by a negligible advantage function $\mathsf{Adv}_{\mathsf{HPS},\mathcal{A}}^{\mathsf{smp}}(\kappa)$ for all PPT adversary $\mathcal{A}$, where

$$\mathsf{Adv}_{\mathsf{HPS},\mathcal{A}}^{\mathsf{smp}}(\kappa) = |\Pr[\mathcal{A}(\mathcal{C}, \mathcal{V}, C_0) = 1 \mid C_0 \leftarrow \mathcal{V}] - \Pr[\mathcal{A}(\mathcal{C}, \mathcal{V}, C_1) = 1 \mid C_1 \leftarrow \mathcal{C} \setminus \mathcal{V}]|.$$

Definition 4. A hash proof system HPS = (HPS.Gen, HPS.Pub, HPS.Priv) is ϵ-universal if: (i) for all sufficiently large $\kappa \in \mathbb{N}$ and for all possible outcomes of HPS.Gen(1^κ), the underlying projective hash function is $\epsilon(\kappa)$-universal for negligible $\epsilon(\kappa)$; (ii) the underlying subset membership problem is hard. Furthermore, a hash proof system is called perfectly universal if $\epsilon(\kappa) = 1/|\mathcal{K}|$.

2.3   One-Time Lossy Filter

One-time Lossy Filter (OT-LF) is a simplified version of lossy algebraic filters recently introduced by Hofheinz [19]. A $(\mathsf{Dom}, \ell_{\mathsf{LF}})$-OT-LF is a family of functions indexed by a public key Fpk and a tag t. A function $\mathsf{LF}_{Fpk,t}$ from the family maps an input $X \in \mathsf{Dom}$ to an output $\mathsf{LF}_{Fpk,t}(X)$. Given public key Fpk, the set of tags $\mathcal{T}$ contains two computationally indistinguishable disjoint subsets, namely the subset of injective tags $\mathcal{T}_{inj}$ and the subset of lossy ones $\mathcal{T}_{loss}$. If t is an injective tag, the function $\mathsf{LF}_{Fpk,t}$ is injective and has image size of $|\mathsf{Dom}|$. If t is lossy, the output of the function has image size at most $2^{\ell_{\mathsf{LF}}}$. Thus, a lossy tag ensures that $\mathsf{LF}_{Fpk,t}(X)$ reveals at most ℓ_{LF} bits of information about its input X. This is a crucial property of an LF.

Definition 5 (OT-LF). A $(\mathsf{Dom}, \ell_{\mathsf{LF}})$-one-time lossy filter LF consists of three PPT algorithms (LF.Gen, LF.Eval, LF.LTag):

Key Generation. $\mathsf{LF.Gen}(1^\kappa)$ outputs a key pair (Fpk, Ftd). The public key Fpk defines a tag space $\mathcal{T} = \{0,1\}^* \times \mathcal{T}_c$ that contains two disjoint subsets, the subset of lossy tags $\mathcal{T}_{loss} \subseteq \mathcal{T}$ and that of injective tags $\mathcal{T}_{inj} \subseteq \mathcal{T}$. A tag $t = (t_a, t_c) \in \mathcal{T}$ consists of an auxiliary tag $t_a \in \{0,1\}^*$ and a core tag $t_c \in \mathcal{T}_c$. Ftd is a trapdoor that allows to efficiently sample a lossy tag.

Evaluation. $\mathsf{LF.Eval}(Fpk, t, X)$, for a public key Fpk, a tag t and $X \in \mathsf{Dom}$, computes $\mathsf{LF}_{Fpk,t}(X)$.

Lossy Tag Generation. $\mathsf{LF.LTag}(Ftd, t_a)$, for an auxiliary tag t_a and the trapdoor Ftd, computes a core tag t_c such that $t = (t_a, t_c)$ is lossy.

We require that an OT-LF LF has the following properties:

Lossiness. If t is injective, so is the function $\mathsf{LF}_{Fpk,t}(\cdot)$. If t is lossy, then $\mathsf{LF}_{Fpk,t}(X)$ has image size of at most $2^{\ell_{\mathsf{LF}}}$. (In application, we are interested in OT-LFs that have a constant parameter ℓ_{LF} even for larger domain.)

Indistinguishability. For any PPT adversary $\mathcal{A}$, it is hard to distinguish a lossy tag from a random tag, i.e., the following advantage is negligible in κ.

$$\mathsf{Adv}_{\mathsf{LF},\mathcal{A}}^{\mathsf{ind}}(\kappa) := |\Pr[\mathcal{A}(Fpk, (t_a, t_c^{(0)})) = 1] - \Pr[\mathcal{A}(Fpk, (t_a, t_c^{(1)})) = 1|$$

where $(Fpk, Ftd) \leftarrow \mathsf{LF.Gen}(1^\kappa)$, $t_a \leftarrow \mathcal{A}(Fpk)$, $t_c^{(0)} \leftarrow \mathsf{LF.LTag}(Ftd, t_a)$ and $t_c^{(1)} \leftarrow \mathcal{T}_c$.

Evasiveness. For any PPT adversary $\mathcal{A}$, it is hard to generate a non-injective tag[1] even given a lossy tag, i.e., the following advantage is negligible in κ.

$$\mathsf{Adv}_{\mathsf{LF},\mathcal{A}}^{\mathsf{eva}}(\kappa) := \Pr \left[\begin{array}{ll} (t_a', t_c') \neq (t_a, t_c) \wedge & (Fpk, Ftd) \leftarrow \mathsf{LF.Gen}(1^\kappa); \\ (t_a', t_c') \in \mathcal{T} \setminus \mathcal{T}_{inj} & : t_a \leftarrow \mathcal{A}(Fpk); t_c \leftarrow \mathsf{LF.LTag}(Ftd, t_a); \\ & (t_a', t_c') \leftarrow \mathcal{A}(Fpk, (t_a, t_c)) \end{array} \right]$$

Remark 1. The definition of one-time lossy filter is different from that of lossy algebraic filter [19] in two ways. First, the one-time property in our definition allows the adversary to query lossy tag generation oracle only once in both indistinguishability and evasiveness games. While in [19], the adversary is allowed to query the oracle polynomial times. Secondly, unlike lossy algebraic filter, one-time lossy filter does not require any algebraic properties.

---

[1] In some case, a tag may neither injective nor lossy.

2.4   Chameleon Hashing

A chameleon hashing function [22] is essentially a hashing function associated with a pair of evaluation key and trapdoor. Its collision-resistant property holds when only the evaluation key of the function is known, but is broken with the trapdoor. We recall the formal definition of chameleon hashing from [18].

Definition 6 (Chameleon Hashing). A chameleon hashing function CH consists of three PPT algorithms (CH.Gen, CH.Eval, CH.Equiv):

Key Generation. CH.Gen(1^κ) outputs an evaluation key ek_{CH} and a trapdoor td_{CH}.

Evaluation. CH.Eval$(ek_{CH}, x; r_{CH})$ maps $x \in \{0,1\}^*$ to $y \in \mathcal{Y}$ with help of the evaluation key ek_{CH} and a randomness $r_{CH} \leftarrow \mathcal{R}_{CH}$. If r_{CH} is uniformly distributed over $\mathcal{R}_{CH}$, so is y over $\mathcal{Y}$.

Equivocation. CH.Equiv(td_{CH}, x, r_{CH}, x') outputs a randomness $r'_{CH} \in \mathcal{R}_{CH}$ such that

$$\mathsf{CH.Eval}(ek_{CH}, x; r_{CH}) = \mathsf{CH.Eval}(ek_{CH}, x'; r'_{CH}), \tag{1}$$

for all x, x' and r_{CH}. Meanwhile, r'_{CH} is uniformly distributed as long as r_{CH} is.

Collision Resistance. Given evaluation key ek_{CH}, it is hard to find $(x, r_{CH}) \neq (x', r'_{CH})$ with CH.Eval$(ek_{CH}, x; r_{CH}) = $ CH.Eval$(ek_{CH}, x'; r'_{CH})$. More precisely, for any PPT adversary $\mathcal{A}$, the following advantage is negligible in κ.

$$\mathsf{Adv}^{cr}_{CH,\mathcal{A}}(\kappa) := \Pr \left[\begin{array}{c} (x, r_{CH}) \neq (x', r'_{CH}) \\ \wedge\ Eq.\ (1) holds. \end{array} : \begin{array}{c} (ek_{CH}, td_{CH}) \leftarrow \mathsf{CH.Gen}(1^\kappa) \\ (x, r_{CH}, x', r'_{CH}) \leftarrow \mathcal{A}(ek_{CH}) \end{array} \right]$$

3   The Construction

Let HPS = (HPS.Gen, HPS.Pub, HPS.Priv) be an ϵ_1-universal hash proof system, where HPS.Gen(1^κ) generates instances of params=(group, $\mathcal{K}$, $\mathcal{C}$, $\mathcal{V}$, $\mathcal{SK}$, $\mathcal{PK}$, $\Lambda_{(\cdot)} : \mathcal{C} \to \mathcal{K}$, $\mu : \mathcal{SK} \to \mathcal{PK}$). Let LF = (LF.Gen, LF.Eval, LF.LTag) be a $(\mathcal{K}, \ell_{LF})$-one-time lossy filter. Define $\nu := \log(1/\epsilon_1)$. Let λ be a bound on the amount of leakage, and let Ext : $\mathcal{K} \times \{0,1\}^d \to \{0,1\}^m$ be an average-case $(\nu - \lambda - \ell_{LF}, \epsilon_2)$-strong extractor. We assume that ϵ_2 is negligible in κ. The encryption scheme PKE = (PKE.Gen, PKE.Enc, PKE.Dec) with plaintext space $\{0,1\}^m$ is described as follows.

Key Generation. PKE.Gen(1^κ) runs HPS.Gen(1^κ) to obtain params and runs LF.Gen(1^κ) to obtain (Fpk, Ftd). It also picks $sk \leftarrow \mathcal{SK}$ and sets $pk = \mu(sk)$. The output is a public/secret key pair (PK, SK), where $PK = $ (params, Fpk, pk) and $SK = sk$.

Encryption. PKE.Enc(PK, M) takes as input a public key PK and a message $M \in \{0,1\}^m$. It chooses $C \leftarrow \mathcal{V}$ with witness w, a random seed $s \leftarrow \{0,1\}^d$ and a random core tag $t_c \leftarrow \mathcal{T}_c$. It then computes

$$K = \mathsf{HPS.Pub}(pk, C, w), \quad \Psi = \mathsf{Ext}(K, s) \oplus M, \quad \Pi = \mathsf{LF}_{Fpk,t}(K),$$

where the filter tag is $t = (t_a, t_c)$ with $t_a = (C, s, \Psi)$. Output the ciphertext $CT = (C, s, \Psi, \Pi, t_c)$.

Decryption. $\mathsf{PKE.Dec}(SK, CT)$, given a secret key $SK = sk$ and a ciphertext $CT = (C, s, \Psi, \Pi, t_c)$, computes $K' = \mathsf{HPS.Priv}(sk, C)$ and $\Pi' = \mathsf{LF}_{Fpk,t}(K')$, where $t = ((C, s, \Psi), t_c)$. It checks whether $\Pi = \Pi'$. If not, it rejects with $\perp$. Otherwise it outputs $M = \Psi \oplus \mathsf{Ext}(K', s)$.

The correctness of PKE follows from the correctness of the underlying hash proof system.

The idea of our construction is to employ a Hash Proof System (HPS) to generate an encapsulated key K, which is then used not only to mask the plaintext, but also to verify the well-formedness of the ciphertext. To deal with the secret key leakage, an extractor converts K to a shorter key to hide the plaintext M. A one-time lossy filter $\mathsf{LF}_{Fpk,t}(K)$ helps to implement the verification. The filter in the challenge ciphertext CT^* works in the lossy mode, and it leaks only a limited amount of information about the key K. For any invalid ciphertext submitted by the adversary to the decryption oracle, the filter works in the injective mode with overwhelming probability. Consequently, the output of the filter in the invalid ciphertext preserves the entropy of K, which makes the ciphertext rejected by the decryption oracle with overwhelming probability.

The security of the construction is established by the theorem below.

Theorem 1. *Assuming that* HPS *is an* ϵ_1-*universal hash proof system,* LF *is a* $(\mathcal{K}, \ell_{\mathsf{LF}})$-*one-time lossy filter, and* $\mathsf{Ext} : \mathcal{K} \times \{0,1\}^d \to \{0,1\}^m$ *is an average-case* $(\nu - \lambda - \ell_{\mathsf{LF}}, \epsilon_2)$-*strong extractor, the encryption scheme* PKE *is* λ-*LR-CCA-secure as long as* $\lambda \leq \nu - m - \ell_{\mathsf{LF}} - \omega(\log \kappa)$, *where* m *is the plaintext length and* $\nu := \log(1/\epsilon_1)$. *Particularly,*

$$\mathsf{Adv}_{\mathsf{PKE},\mathcal{A}}^{\mathsf{lr\text{-}cca}}(\kappa) \leq \mathsf{Adv}_{\mathsf{LF},\mathcal{B}_1}^{\mathsf{ind}}(\kappa) + Q(\kappa) \cdot \mathsf{Adv}_{\mathsf{LF},\mathcal{B}_2}^{\mathsf{eva}}(\kappa) + \mathsf{Adv}_{\mathsf{HPS},\mathcal{B}_3}^{\mathsf{smp}}(\kappa) + \frac{Q(\kappa)2^{\lambda+\ell_{\mathsf{LF}}+m}}{2^\nu - Q(\kappa)} + \epsilon_2$$

where $Q(\kappa)$ *denotes the number of decryption queries made by* $\mathcal{A}$.

Parameters and Leakage Rate. To make our construction tolerate leakage as much as possible, it is useful to consider a "very strong" hash proof system (i.e., $\epsilon_1 \leq 2/|\mathcal{K}|$). In this case, $\nu = \log(1/\epsilon_1) \geq \log|\mathcal{K}| - 1$. Thus, when $\mathcal{K}$ is sufficiently large, the leakage rate (defined as $\lambda/|SK|$) in our construction approaches $(\log|\mathcal{K}|)/|SK|$ asymptotically.

CCA-Security. Clearly, if $\lambda = 0$ and $\log(1/\epsilon_1) \geq m + \ell_{\mathsf{LF}} + \omega(\log \kappa)$, the above construction is CCA-secure. Thus, it provides a new approach for constructing CCA-secure PKE from any universal hash proof system and OT-LF.

Proof. The proof goes with game arguments [33]. We define a sequence of games, $\mathsf{Game}_0, \ldots, \mathsf{Game}_6$, played between a simulator Sim and a PPT adversary $\mathcal{A}$. In each game, the adversary outputs a bit b' as a guess of the random bit b used by the simulator. Denote by S_i the event that $b = b'$ in Game_i and denote by $CT^* = (C^*, s^*, \Psi^*, \Pi^*, t_c^*)$ the challenge ciphertext.

Game_0: This is the original LR-CCA game $\mathsf{Exp}_{\mathsf{PKE},\mathcal{A}}^{\mathsf{lr\text{-}cca}}(\kappa)$. The simulator generates the public/secret key pair (PK, SK) by invoking $\mathsf{PKE.Gen}(1^\kappa)$ and sends the public key PK to the adversary $\mathcal{A}$. For each decryption query CT or leakage query f_i, Sim responds with PKE.Dec(SK, CT) or $f_i(SK)$ using secret key SK. Upon receiving two messages M_0, M_1 of equal length from the adversary, Sim selects a random $b \in \{0, 1\}$ and sends the challenge ciphertext $CT^* :=$ PKE.Enc(PK, M_b) to $\mathcal{A}$. The simulator continues to answer the adversary's decryption query as long as $CT \neq CT^*$. Finally, $\mathcal{A}$ outputs a bit b', which is a guess of b. By the Definition 2, we have $\mathsf{Adv}_{\mathsf{PKE},\mathcal{A}}^{\text{lr-cca}}(\kappa) := \left| \Pr[S_0] - \frac{1}{2} \right|$.

Game_1: This game is exactly like Game_0, except for PKE.Gen(1^κ) and the generation of the core tag t_c^* of the filter tag in the challenge ciphertext. When calling PKE.Gen(1^κ), the simulator keeps the trapdoor Ftd of LF as well as SK. Instead of sampling t_c^* at random from $\mathcal{T}_c$, Sim computes t_c^* with LF.LTag(Ftd, t_a^*), where $t_a^* = (C^*, s^*, \Psi^*)$. A straightforward reduction to LF's indistinguishability of lossy tag and random tag yields $|\Pr[S_1] - \Pr[S_0]| \leq \mathsf{Adv}_{\mathsf{LF},\mathcal{B}_1}^{\text{ind}}(\kappa)$ for a suitable adversary $\mathcal{B}_1$ on LF's indistinguishability.

Game_2: This game is exactly like Game_1, except that a special rejection rule applies to the decryption oracle. If the adversary queries a ciphertext $CT = (C, s, \Psi, \Pi, t_c)$ such that $t = (t_a, t_c) = (t_a^*, t_c^*) = t^*$, then the decryption oracle immediately outputs $\bot$ and halts. For convenient, we call such tag a copied LF tag. We show that a decryption query with a copied LF tag is rejected in decryption oracles in both Game_1 and Game_2. We consider the following two cases.

- case 1: $\Pi = \Pi^*$. This implies $CT = CT^*$. In this case the decryption oracles in Game_1 and Game_2 proceed identically since $\mathcal{A}$ is not allowed to ask for the decryption of challenge ciphertext.
- case 2: $\Pi \neq \Pi^*$. Since $t = ((C, s, \Psi), t_c) = ((C^*, s^*, \Psi^*), t_c^*) = t^*$, it follows that $K = K^*$, and thus $\mathsf{LF}_{Fpk,t}(K) = \mathsf{LF}_{Fpk,t^*}(K^*) = \Pi^*$. So, such decryption queries would have been rejected already in Game_1.

According to above analysis, we have $\Pr[S_2] = \Pr[S_1]$.

Game_3: This game is exactly like Game_2, except for the generation of K^* used in the challenge ciphertext. In this game, Sim computes K^* with HPS.Priv(sk, C^*) instead of HPS.Pub(pk, C^*, w^*). Since HPS is projective, this change is purely conceptual, and thus $\Pr[S_3] = \Pr[S_2]$.

Game_4: This game is exactly like Game_3, except for the generation of C^* in the challenge ciphertext $CT^* = (C^*, s^*, \Psi^*, \Pi^*, t_c^*)$. Now Sim samples C^* from $\mathcal{C} \setminus \mathcal{V}$ instead of $\mathcal{V}$. A straightforward reduction to the indistinguishability of the subset membership problem yields $|\Pr[S_4] - \Pr[S_3]| \leq \mathsf{Adv}_{\mathsf{HPS},\mathcal{B}_3}^{\text{smp}}(\kappa)$ for a suitable adversary $\mathcal{B}_3$.

Game_5: This game is the same as Game_4, except that another special rejection rule is applied to the decryption oracle. If the adversary queries a ciphertext $CT = (C, s, \Psi, \Pi, t_c)$ for decryption such that $C \in \mathcal{C} \setminus \mathcal{V}$, then the decryption oracle immediately outputs $\bot$. Let bad_C be the event that a ciphertext is rejected in Game_5 that would not have been rejected under the rules of Game_4. Then Game_5 and Game_4 proceed identically until event bad_C occurs. We have

$$|\Pr[S_5] - \Pr[S_4]| \leq \Pr[\mathsf{bad}_C] \tag{2}$$

by the difference lemma of [33]. We show the following lemma shortly (after the main proof), which guarantees that bad_C occurs with a negligible probability.

Lemma 3. *Suppose that the adversary $\mathcal{A}$ makes at most $Q(\kappa)$ decryption queries. Then*

$$\Pr[\mathsf{bad}_C] \le Q(\kappa) \cdot \mathsf{Adv}^{\mathrm{eva}}_{\mathsf{LF},\mathcal{B}}(\kappa) + \frac{Q(\kappa)2^{\lambda+\ell_{\mathsf{LF}}+m}}{2^\nu - Q(\kappa)} \qquad (3)$$

where $\mathcal{B}$ is a suitable adversary attacking on LF's evasiveness.

Game_6: This game is exactly like Game_5, except for the generation of Ψ^* in CT^*. In this game, Sim chooses Ψ^* uniformly at random from $\{0,1\}^m$ instead of using $\mathsf{Ext}(\Lambda_{sk}(C^*), s^*) \oplus M_b$.

Claim 1. *For $C^* \leftarrow \mathcal{C} \setminus \mathcal{V}$ if the decryption algorithm rejects all invalid ciphertexts, then the value $\Lambda_{sk}(C^*)$ has average min-entropy at least $\nu - \lambda - \ell_{\mathsf{LF}} \ge \omega(\log \kappa) + m$ given all the other values in $\mathcal{A}$'s view (denoted by $\mathsf{view}'_{\mathcal{A}}$).*

We prove Claim 1 by directly analyzing the average min-entropy of $\Lambda_{sk}(C^*)$ from the adversary's point of view. Since all invalid ciphertexts are rejected by the decryption oracle in both Game_5 and Game_6, $\mathcal{A}$ cannot learn more information on the value $\Lambda_{sk}(C^*)$ from the decryption oracle other than pk, C^*, Π^* and the key leakage. Recall that Π^* has only $2^{\ell_{\mathsf{LF}}}$ possible vales and $\mathrm{H}_\infty(\Lambda_{sk}(C^*) \mid (pk, C^*)) \ge \nu$ (which holds for all pk and $C^* \in \mathcal{C} \setminus \mathcal{V}$). Hence,

$$\tilde{\mathrm{H}}_\infty(\Lambda_{sk}(C^*) \mid \mathsf{view}'_{\mathcal{A}}) = \tilde{\mathrm{H}}_\infty(\Lambda_{sk}(C^*) \mid pk, C^*, \lambda\text{-leakage}, \Pi^*)$$
$$\ge \tilde{\mathrm{H}}_\infty(\Lambda_{sk}(C^*) \mid pk, C^*) - \lambda - \ell_{\mathsf{LF}} \ge \nu - \lambda - \ell_{\mathsf{LF}}$$

according to Lemma 1.

Applying an average-case $(\nu - \lambda - \ell_{\mathsf{LF}}, \epsilon_2)$-strong extractor $\mathsf{Ext} : \mathcal{K} \times \{0,1\}^d \to \{0,1\}^m$ to $\Lambda_{sk}(C^*)$, we have that $\mathsf{Ext}(\Lambda_{sk}(C^*), s^*)$ is ϵ_2-close to uniform given $\mathcal{A}$'s view. Hence,

$$|\Pr[S_6] - \Pr[S_5]| \le \epsilon_2 \qquad (4)$$

Observe that in Game_6, the challenge ciphertext is completely independent of the random coin b picked by the simulator. Thus, $\Pr[S_6] = 1/2$.

Putting all together, Theorem 1 follows.   □

It remains to prove Lemma 3. We do it now.

Proof (Proof of Lemma 3). Let F be the event that in Game_4 there exists a decryption query $CT = (C, s, \Psi, \Pi, t_c)$, such that $t = ((C, s, \Psi), t_c)$ is a non-injective, non-copied tag. We have

$$\Pr[\mathsf{bad}_C] = \Pr[\mathsf{bad}_C \wedge F] + \Pr[\mathsf{bad}_C \wedge \overline{F}] \le \Pr[F] + \Pr[\mathsf{bad}_C \mid \overline{F}] \qquad (5)$$

Thus, it suffices to prove the following two claims: Claim 2 and Claim 3.

Claim 2. *Suppose that the adversary $\mathcal{A}$ makes at most $Q(\kappa)$ decryption queries. If LF is a one-time lossy filter, then*

$$\Pr[F] \le Q(\kappa) \cdot \mathsf{Adv}^{\mathrm{eva}}_{\mathsf{LF},\mathcal{B}}(\kappa) \qquad (6)$$

where $\mathcal{B}$ is a suitable adversary on LF's evasiveness.

Proof. Given a challenge LF evaluation key F^*_{pk}, $\mathcal{B}$ simulates $\mathcal{A}$'s environment in Game$_4$ as follows. It generates the PKE's public key PK as in Game$_4$ but sets $F_{pk} = F^*_{pk}$. Note that $\mathcal{B}$ can use PKE's secret key to deal with $\mathcal{A}$'s decryption queries. To simulate the challenge ciphertext (in which the LF tag should be lossy), $\mathcal{B}$ queries its lossy tag generation oracle once with $t^*_a = (C^*, s^*, \Psi^*)$ to proceed t^*_c, where (C^*, s^*, Ψ^*) are generated as in Game$_4$. Finally, $\mathcal{B}$ chooses $i \in [Q(k)]$ uniformly, and outputs the tag $t = ((C, s, \Psi), t_c)$ extracted from $\mathcal{A}$'s i-th decryption query (C, s, Ψ, Π, t_c). Clearly, if the event F occurs, with probability at least $1/Q(\kappa)$, t is a non-injective tag. That is $\Pr[F] \le Q(\kappa) \cdot \mathsf{Adv}^{\mathrm{eva}}_{\mathsf{LF},\mathcal{B}}(\kappa)$.   □

Claim 3. *Suppose that the adversary $\mathcal{A}$ makes at most $Q(\kappa)$ decryption queries. If HPS is ϵ_1-universal, then*

$$\Pr[\mathsf{bad}_C \mid \overline{F}] \le \frac{Q(\kappa)2^{\lambda+\ell_{\mathsf{LF}}+m}}{2^\nu - Q(\kappa)} \qquad (7)$$

where $\nu = \log(1/\epsilon_1)$.

Proof. Suppose that $CT = (C, s, \Psi, \Pi, t_c)$ is the first ciphertext that makes bad_C happen given $\overline{F}$, i.e. $C \in \mathcal{C} \backslash \mathcal{V}$ but $\Pi = \mathsf{LF}_{F pk, t}(\Lambda_{sk}(C))$, where $t = ((C, s, \Psi), t_c)$ is an injective LF tag. For simplicity, we call $CT = (C, s, \Psi, \Pi, t_c)$ an *invalid* ciphertext if $C \in \mathcal{C} \backslash \mathcal{V}$. Denote by $\mathsf{view}_{\mathcal{A}}$ the adversary's view prior to submitting the first invalid ciphertext. Observe that only pk, the challenge ciphertext CT^*, and the key leakage of at most λ bits reveal information of the secret key to the adversary. According to Lemma 1, we have

$$\begin{aligned}
\widetilde{\mathrm{H}}_\infty(\Lambda_{sk}(C) \mid \mathsf{view}_{\mathcal{A}}) &= \widetilde{\mathrm{H}}_\infty(\Lambda_{sk}(C) \mid pk, C, CT^*, \lambda\text{-leakage}) \\
&\ge \widetilde{\mathrm{H}}_\infty(\Lambda_{sk}(C) \mid pk, C, CT^*) - \lambda \\
&\ge \mathrm{H}_\infty(\Lambda_{sk}(C) \mid (pk, C)) - \lambda - \ell_{\mathsf{LF}} - m \qquad (8) \\
&\ge \nu - \lambda - \ell_{\mathsf{LF}} - m \qquad (9)
\end{aligned}$$

Eq. (8) follows from the fact that in the challenge ciphertext CT^*, only Ψ^* and Π^* are related to the secret key, and Ψ^* has at most 2^m possible values and Π^* has at most $2^{\ell_{\mathsf{LF}}}$ possible values. Note that the information revealed by t^*_c has already been completely taken into account by Ψ^*, since $t^*_c = \mathsf{LF}.\mathsf{LTag}(Ftd, (C^*, s^*, \Psi^*))$ can be regarded as a function of Ψ^*. Eq. (9) follows from the fact that for all pk and $C \in \mathcal{C} \backslash \mathcal{V}$, $\mathrm{H}_\infty(\Lambda_{sk}(C) \mid (pk, C)) \ge \log(1/\epsilon_1) = \nu$, which is due to the ϵ_1-universal property of HPS and Lemma 2. The fact that event F does not occur implies that $t = ((C, s, \Psi), t_c)$ is an injective tag. Applying an injective function to a distribution preserves its min-entropy, we have $\widetilde{\mathsf{H}}_\infty(\mathsf{LF}_{Fpk,t}(\Lambda_{sk}(C)) \mid \mathsf{view}_\mathcal{A}) \geq \nu - \lambda - \ell_{\mathsf{LF}} - m$. Thus, in Game_4 the decryption algorithm accepts the first invalid ciphertext with probability at most $2^{\lambda+\ell_{\mathsf{LF}}+m}/2^\nu$. Observe that the adversary can rule out one more value of K from each rejection of invalid ciphertext. So, the decryption algorithm accepts the i-th invalid ciphertext with probability at most $2^{\lambda+\ell_{\mathsf{LF}}+m}/(2^\nu - i + 1)$. Since $\mathcal{A}$ makes at most $Q(\kappa)$ decryption queries, it follows that

$$\Pr[\mathsf{bad}_C \mid \overline{F}] \leq \frac{Q(\kappa)2^{\lambda+\ell_{\mathsf{LF}}+m}}{2^\nu - Q(\kappa)} \tag{10}$$

which is negligible in κ if $\lambda \leq \nu - m - \ell_{\mathsf{LF}} - \omega(\log \kappa)$.   □

This completes the proof of Lemma 3.   □

4   Instantiation from the DDH Assumption

This section is organized as follows. In Section 4.1, we present a variant of hash proof system from the Decisional Diffie-Hellman (DDH) assumption [9]. In Section 4.2, we introduce an efficient DDH-based OT-LF. In Section 4.3, we apply the construction in Section 3 to the two building blocks and obtain an efficient DDH-based LR-CCA secure PKE scheme, depicted in Fig. 1. In Section 4.4, we show a comparison of our scheme with some existing LR-CCA secure PKE schemes.

The DDH Assumption. We assume a PPT algorithm $\mathcal{G}(1^\kappa)$ that takes as input 1^κ and outputs a tuple of $\mathbb{G} = \langle q, G, g \rangle$, where G is a cyclic group of prime order q and g is a generator of G. The Decisional Diffie-Hellman (DDH) assumption holds iff

$$\mathsf{Adv}_{G,\mathcal{D}}^{\mathrm{ddh}}(\kappa) = \left| \Pr[\mathcal{D}(g_1, g_2, g_1^r, g_2^r) = 1] - \Pr[\mathcal{D}(g_1, g_2, g_1^r, g_2^{r'}) = 1] \right|$$

is negligible in κ for any PPT adversary $\mathcal{D}$, where $g_1, g_2 \leftarrow G$, $r \leftarrow \mathbb{Z}_q$ and $r' \leftarrow \mathbb{Z}_q \setminus \{r\}$.

4.1   A DDH-Based HPS

Let $\langle q, G, g \rangle \leftarrow \mathcal{G}(1^\kappa)$ and let g_1, g_2 be two random generators of G. Choose $\mathsf{n} \in \mathbb{N}$. We assume there is an efficient injective mapping $\mathsf{Inj} : G \rightarrow \mathbb{Z}_q^2$. For any $u = (u_1, \ldots, u_\mathsf{n}) \in G^\mathsf{n}$, let $\widetilde{\mathsf{Inj}}(u) = (\mathsf{Inj}(u_1), \ldots, \mathsf{Inj}(u_\mathsf{n})) \in \mathbb{Z}_q^\mathsf{n}$. Clearly, $\widetilde{\mathsf{Inj}}$ is also an injection. We define a hash proof system $\mathsf{HPS}_1 = (\mathsf{HPS}_1.\mathsf{Gen}, \mathsf{HPS}_1.\mathsf{Pub}, \mathsf{HPS}_1.\mathsf{Priv})$ below.
   The parameter $\mathsf{params} = (\mathsf{group}, \mathcal{K}, \mathcal{C}, \mathcal{V}, \mathcal{SK}, \mathcal{PK}, \Lambda_{sk}, \mu)$ is set up as follows.

---

[2] For example, G is a q-order elliptic curve group over finite field $\mathbb{F}_p$. For 80-bit security, p and q can be chosen to be 160-bit primes. In such a group, elements (i.e., elliptic curve points) can be represented by 160-bit strings.

- group $= \langle q, G, g_1, g_2, \mathfrak{n} \rangle$, $\mathcal{C} = G \times G$, $\mathcal{V} = \{(g_1^r, g_2^r) : r \in \mathbb{Z}_q\}$ with witness set $W = \mathbb{Z}_q$.
- $\mathcal{K} = \mathbb{Z}_q^{\mathfrak{n}}$, $\mathcal{SK} = (\mathbb{Z}_q \times \mathbb{Z}_q)^{\mathfrak{n}}$, $\mathcal{PK} = G^{\mathfrak{n}}$.
- For $sk = (x_{i,1}, x_{i,2})_{i \in [\mathfrak{n}]} \in \mathcal{SK}$, define $pk = (pk_i)_{i \in [\mathfrak{n}]} = \mu(sk) = (g_1^{x_{i,1}} g_2^{x_{i,2}})_{i \in [\mathfrak{n}]}$.
- For all $C = (u_1, u_2) \in \mathcal{C}$, define $\Lambda_{sk}(C) = \widetilde{\mathsf{Inj}}((u_1^{x_{i,1}} u_2^{x_{i,2}})_{i \in [\mathfrak{n}]})$.

The public evaluation and private evaluation algorithms are defined as follows:

- For all $C = (g_1^r, g_2^r) \in \mathcal{V}$ with witness $r \in \mathbb{Z}_q$, define $\mathsf{HPS}_1.\mathsf{Pub}(pk, C, r) = \widetilde{\mathsf{Inj}}(pk_1^r, \ldots, pk_{\mathfrak{n}}^r)$.
- For all $C = (u_1, u_2) \in \mathcal{C}$, define $\mathsf{HPS}_1.\mathsf{Priv}(sk, C) = \Lambda_{sk}(C)$.

Correctness of HPS_1 follows directly by the definitions of μ and Λ_{sk}. The subset membership problem in HPS_1 is hard because of the DDH assumption. If $\mathfrak{n} = 1$, this is just the DDH-based hash proof system introduced by Cramer and Shoup with encapsulated key set $\mathcal{K} = \mathbb{Z}_q$, and is known to be perfectly universal [9,21]. We have the following theorem with proof in the full version of the paper.

Theorem 2. *For any $\mathfrak{n} \in \mathbb{N}$, HPS_1 is perfectly universal under the DDH assumption with encapsulated key size $|\mathcal{K}| = q^{\mathfrak{n}}$.*

4.2 A DDH-Based OT-LF

We use the following notations. If $A = (A_{i,j})$ is an $n \times n$ matrix over $\mathbb{Z}_{\widetilde{q}}$, and $\widetilde{g}$ is an element of $\widetilde{q}$-order group $\widetilde{G}$. Then $\widetilde{g}^A$ denotes the $n \times n$ matrix $(\widetilde{g}^{A_{i,j}})$ over $\widetilde{G}$. Given a vector $X = (X_1, \ldots, X_n) \in \mathbb{Z}_{\widetilde{q}}^n$ and an $n \times n$ matrix $E = (E_{i,j}) \in \widetilde{G}^{n \times n}$, define

$$X \cdot E := (\prod_{i=1}^{n} E_{i,1}^{X_i}, \ldots, \prod_{i=1}^{n} E_{i,n}^{X_i}) \in \widetilde{G}^n.$$

Let $\mathsf{CH} = (\mathsf{CH.Gen}, \mathsf{CH.Eval}, \mathsf{CH.Equiv})$ define a chameleon hashing function with image set $\mathbb{Z}_{\widetilde{q}}$. The OT-LF is $\mathsf{LF}_1 = (\mathsf{LF}_1.\mathsf{Gen}, \mathsf{LF}_1.\mathsf{Eval}, \mathsf{LF}_1.\mathsf{LTag})$, as shown below.

Key Generation. $\mathsf{LF}_1.\mathsf{Gen}(1^\kappa)$ runs $\mathcal{G}(1^\kappa)$ to obtain $\widetilde{\mathbb{G}} = \langle \widetilde{q}, \widetilde{G}, \widetilde{g} \rangle$ and runs $\mathsf{CH.Gen}(1^\kappa)$ to obtain $(ek_{\mathrm{CH}}, td_{\mathrm{CH}})$. Pick a random pair $(t_a^*, t_c^*) \leftarrow \{0,1\}^* \times \mathcal{R}_{\mathrm{CH}}$ and compute $b^* = \mathsf{CH.Eval}(ek_{\mathrm{CH}}, t_a^*; t_c^*)$. Choose $r_1, \ldots, r_n, s_1, \ldots, s_n \leftarrow \mathbb{Z}_{\widetilde{q}}$, and compute an $n \times n$ matrix $A = (A_{i,j}) \in \mathbb{Z}_{\widetilde{q}}^{n \times n}$ with $A_{i,j} = r_i s_j$ for $i, j \in [n]$. Compute matrix $E = \widetilde{g}^{A - b^* \mathbf{I}} \in \widetilde{G}^{n \times n}$, where $\mathbf{I}$ is the $n \times n$ identity matrix over $\mathbb{Z}_{\widetilde{q}}$. Finally, output $Fpk = (\widetilde{q}, \widetilde{G}, \widetilde{g}, ek_{\mathrm{CH}}, E)$ and $Ftd = (td_{\mathrm{CH}}, t_a^*, t_c^*)$. The tag space is defined as $\mathcal{T} = \{0,1\}^* \times \mathcal{R}_{\mathrm{CH}}$, where $\mathcal{T}_{loss} = \{(t_a, t_c) : (t_a, t_c) \in \mathcal{T} \wedge \mathsf{CH.Eval}(ek_{\mathrm{CH}}, t_a; t_c) = b^*\}$ and $\mathcal{T}_{inj} = \{(t_a, t_c) : (t_a, t_c) \in \mathcal{T} \wedge \mathsf{CH.Eval}(ek_{\mathrm{CH}}, t_a; t_c) \notin \{b^*, b^* - \mathrm{Tr}(A)\}\}$.

Evaluation. For a tag $t = (t_a, t_c) \in \{0,1\}^* \times \mathcal{R}_{\mathrm{CH}}$ and an input $X = (X_1, \ldots, X_n) \in \mathbb{Z}_{\widetilde{q}}^n$, $\mathsf{LF}_1.\mathsf{Eval}(Fpk, t, X)$ first computes $b = \mathsf{CH.Eval}(ek_{\mathrm{CH}}, t_a; t_c)$ and outputs

$$y = X \cdot (E \otimes \widetilde{g}^{b\mathbf{I}}),$$

where "$\otimes$" denotes the operation of entry-wise multiplication.

Lossy Tag Generation. For an auxiliary tag t_a, $\mathsf{LF_1.LTag}(Ftd, t_a)$ computes a core tag $t_c = \mathsf{CH.Equiv}(td_{\mathrm{CH}}, t_a^*, t_c^*, t_a)$ with the trapdoor $Ftd = (td_{\mathrm{CH}}, t_a^*, t_c^*)$.

Theorem 3. $\mathsf{LF_1}$ is a $(\mathbb{Z}_{\widetilde{q}}^n, \log \widetilde{q})$-$OT$-$LF$ under the DDH assumption.

Proof. The proof of Theorem 3 is given in the full version of the paper.   □

4.3   The DDH-Based PKE Scheme

Let $\mathbb{G} = \langle q, G, g \rangle$ and $\widetilde{\mathbb{G}} = \langle \widetilde{q}, \widetilde{G}, \widetilde{g} \rangle$ be two group descriptions. Suppose $\mathfrak{n} \in \mathbb{N}$ satisfies $\mathfrak{n} \log q \geq \log \widetilde{q} + \lambda + m + \omega(\log \kappa)$. Set $n = \lceil \mathfrak{n} \log q / \log \widetilde{q} \rceil$. Let $(ek_{\mathrm{CH}}, td_{\mathrm{CH}}) \leftarrow \mathsf{CH.Gen}(1^\kappa)$ be a chameleon hash function with image set $\mathbb{Z}_{\widetilde{q}}$. Let $\mathsf{Ext} : \mathbb{Z}_q^{\mathfrak{n}} \times \{0,1\}^d \to \{0,1\}^m$ be an average-case $(\mathfrak{n} \log q - \log \widetilde{q} - \lambda, \epsilon_2)$-strong extractor. Applying the general construction in Section 3 to the aforementioned DDH-based HPS and OT-LF, we obtain a DDH-based PKE scheme in Fig. 1.

Key Generation. $\mathsf{PKE_1.Gen}(1^\kappa)$: Choose $g_1, g_2 \leftarrow G$ and $(x_{i,1}, x_{i,2}) \leftarrow \mathbb{Z}_q$ for $i \in [\mathfrak{n}]$. Set $pk_i = g_1^{x_{i,1}} g_2^{x_{i,2}}$ for $i \in [\mathfrak{n}]$. Also choose a random pair $(t_a^*, t_c^*) \in \{0,1\}^* \times \mathcal{R}_{\mathrm{CH}}$ and set $b^* = \mathsf{CH.Eval}(ek_{\mathrm{CH}}, t_a^*; t_c^*)$. Choose $r_1, \ldots, r_n, s_1, \ldots, s_n \leftarrow \mathbb{Z}_{\widetilde{q}}$, and compute matrix $E = (E_{i,j})_{i,j \in [n]} \in \widetilde{G}^{n \times n}$, where $E_{i,j} = \widetilde{g}^{r_i s_j}$ for $i, j \in [n], i \neq j$, and $E_{i,i} = \widetilde{g}^{r_i s_i} \widetilde{g}^{-b^*}$ for $i \in [n]$. Return $PK = (q, G, g_1, g_2, \mathfrak{n}, (pk_i)_{i \in [\mathfrak{n}]}, \widetilde{q}, \widetilde{G}, \widetilde{g}, E, ek_{\mathrm{CH}})$ and $SK = (x_{i,1}, x_{i,2})_{i \in [\mathfrak{n}]}$.

Encryption. $\mathsf{PKE_1.Enc}(PK, M)$: For a public key PK and a message $M \in \{0,1\}^m$, it chooses $r \leftarrow \mathbb{Z}_q$ and $s \leftarrow \{0,1\}^d$. Compute

$$C = (g_1^r, g_2^r), \quad K = \widetilde{\mathsf{Inj}}(pk_1^r, \ldots, pk_{\mathfrak{n}}^r), \quad \Psi = \mathsf{Ext}(K, s) \oplus M, \quad \Pi = K \cdot (E \otimes \widetilde{g}^{bI})$$

where $b = \mathsf{CH.Eval}(ek_{\mathrm{CH}}, t_a; t_c)$ for the auxiliary tag $t_a = (C, s, \Psi)$ and a random filter core tag $t_c \in \mathcal{R}_{\mathrm{CH}}$. Note that in the computation of Π, K is regarded as a vector of dimension n over $\mathbb{Z}_{\widetilde{q}}$ (this works well since $\mathfrak{n} \log q \leq n \log \widetilde{q}$). Return $CT = (C, s, \Psi, \Pi, t_c) \in G^2 \times \{0,1\}^d \times \{0,1\}^m \times \widetilde{G}^n \times \mathcal{R}_{\mathrm{CH}}$.

Decryption. $\mathsf{PKE_1.Dec}(SK, CT)$: For a ciphertext $CT = (C, s, \Psi, \Pi, t_c)$, it parses C as $(u_1, u_2) \in G^2$ and then computes $K' = \widetilde{\mathsf{Inj}}(u_1^{x_{1,1}} u_2^{x_{1,2}}, \ldots, u_1^{x_{\mathfrak{n},1}} u_2^{x_{\mathfrak{n},2}})$ and $\Pi' = K' \cdot (E \otimes \widetilde{g}^{bI})$, where $b = \mathsf{CH.Eval}(ek_{\mathrm{CH}}, (C, s, \Psi); t_c)$. Finally, it checks whether $\Pi = \Pi'$. If not, it rejects with $\bot$. Else, it returns $M = \Psi \oplus \mathsf{Ext}(K', s)$.

Fig. 1. A DDH-based PKE Scheme $\mathsf{PKE_1} = (\mathsf{PKE_1.Gen}, \mathsf{PKE_1.Enc}, \mathsf{PKE_1.Dec})$

Theorem 4. *If the DDH assumptions hold in groups G and $\widetilde{G}$, and the CH is a chameleon hash function, then $\mathsf{PKE_1}$ is λ-LR-CCA secure if $\lambda \leq \mathfrak{n} \log q - \log \widetilde{q} - m - \omega(\log \kappa)$ (i.e., $\mathfrak{n} \geq (\lambda + \log \widetilde{q} + m + \omega(\log \kappa)) / \log q$). In particular, the leakage rate in $\mathsf{PKE_1}$ is $1/2 - o(1)$ and*

$$\mathsf{Adv}_{\mathsf{PKE_1}, \mathcal{A}}^{\mathrm{lr\text{-}cca}}(\kappa) \leq \mathsf{Adv}_{G, \mathcal{B}_1}^{\mathrm{ddh}}(\kappa) + 2\mathfrak{n}\mathsf{Adv}_{\widetilde{G}, \mathcal{B}_2}^{\mathrm{ddh}}(\kappa) + \frac{Q(\kappa) \cdot \widetilde{q} \cdot 2^{\lambda+m}}{q^{\mathfrak{n}} - Q(\kappa)} + \epsilon_2$$

$$+ Q(\kappa)\left((2\mathfrak{n}+1)\mathsf{Adv}_{\widetilde{G}, \mathcal{B}_2}^{\mathrm{ddh}}(\kappa) + \mathsf{Adv}_{\mathsf{CH}, \mathcal{B}_3}^{\mathrm{cr}}(\kappa) \right)$$

where $Q(\kappa)$ is the number of decryption queries made by $\mathcal{A}$.

Proof. Theorem 2 showed that the underlying HPS in PKE_1 is perfectly universal (i.e., $\epsilon_1 = 1/q^n$). Theorem 3 said that the underlying filter is a $(\widetilde{q}^n, \log \widetilde{q})$-OT-LF. Consequently, PKE_1 is λ-LR-CCA secure according to Theorem 1. If the parameter $\mathfrak{n}$ in PKE_1 increases, with $\widetilde{q}, m$ fixed, $\lambda/|SK| = (\mathfrak{n} \log q - \log \widetilde{q} - m - \omega(\log \kappa))/(2\mathfrak{n} \log q) = 1/2 - o(1)$.   □

4.4   Efficiency Discussion

In this section, we show a comparison of our DDH-based PKE scheme with the existing DDH/DLIN based LR-CCA secure PKE schemes [28,25,11,16] in terms of leakage rate and ciphertext overhead (defined as the difference between the ciphertext length and the embedded message length). Note that the GHV12 scheme is obtained by applying the CHK transformation to the mKDM-sID-CPA secure scheme [16]. The GHV12 scheme is LR-CCA secure only if the mKDM-sID-CPA secure scheme [16] is master-key leakage sID-CPA secure. In fact, Galindo et.al. claimed their mKDM-sID-CPA secure scheme is master-key leakage sID-CPA secure with leakage rate $1 - o(1)$, but without any rigorous proof. We personally regard that proving that claim is very hard, since the proof involves constructing a PPT simulator to answer not only key leakage queries, but also identities' private key queries. Nevertheless, we include the GHV12 scheme in the comparison. For simplicity, in a ciphertext, we only consider the length of group elements, ignoring the constant length non-group elements, e.g., the seed used in a randomness extractor. We also assume that elements in q-order group can be encoded as bit strings of length $\log q$. To be fair, like in [11, Theorem 6], we will consider the ciphertext overhead (shorted as "CT overhead") under any fixed and achievable leakage rate. We begin by giving an overview of the secret key size (shorted as "SK size"), the amount of absolute leakage and the number of group elements in the ciphertexts of the PKE schemes [28,25,16] in Table 1. In table 1, κ is the security parameter; q', q and $\widetilde{q}$ are group sizes; m is the message length and $\mathfrak{n}$ is a parameter as in Fig. 1 and [16, Section 5]. In our scheme, $n = \lceil \mathfrak{n} \log q / \log \widetilde{q} \rceil$. So, the bit-length of n elements in group $\widetilde{G}$ equals that of $\mathfrak{n}$ elements in group G.

Table 1. Secret-key size, leakage amount and ciphertext overhead

Schemes	SK size (# bits)	Leakage amount (# bits)	CT overhead (#G)
GHV12 [16]	$\mathfrak{n} \log q$	$\lambda \leq \mathfrak{n} \log q - 3 \log q - 2\ell(\kappa)$	$2\mathfrak{n} + 6$
NS09 [28]	$6 \log q'$	$\lambda \leq \log q' - \omega(\log \kappa) - m$	3
LZSS12 [25]	$4 \log q'$	$\lambda \leq \log q' - \omega(\log \kappa) - m$	3
Ours	$2\mathfrak{n} \log q$	$\lambda \leq \mathfrak{n} \log q - \log \widetilde{q} - m - \omega(\log \kappa)$	$\mathfrak{n} + 2$

We observe that in our scheme as well as that of [11,16] the group size (i.e. q and $\widetilde{q}$) remains constant even with larger leakage. While in [28] and [25], both of them rely on increasing the group size (i.e., q') to tolerate larger leakage. So, it is more reasonable to compare the bit-length of ciphertext overhead rather than the number of group elements for the same leakage rate. As an example, we give the concrete relations between ciphertext overhead and leakage-rate of our scheme. In our scheme, for a security level $\ell(\kappa)$, we can choose $|q| = |\widehat{q}| = 2\ell(\kappa)$. From [13], applying a universal hash function to a source with $3\ell(\kappa)$ entropy suffices to extract $\ell(\kappa)$-bit random key that is $2^{-\ell(\kappa)}$-close to a uniform distribution over $\{0,1\}^{\ell(\kappa)}$. So, we can set $\omega(\log \kappa) = 2\ell(\kappa)$ and $m = \ell(\kappa)$. According to Theorem 4, the amount of leakage is bounded by $(2n - 5)\ell(\kappa)$. Thus, for any $\delta \in [0, 1/2)$, the leakage rate in our scheme achieves δ, as long as $n \geq \lceil 5/(2 - 4\delta) \rceil$ (i.e., $\lambda \leq \ell(\kappa)(2\lceil 5/(2 - 4\delta) \rceil - 5)$) and the ciphertext overhead is $(\lceil 5/(2 - 4\delta) \rceil + 2)2\ell(\kappa)$ bits (ignoring the seed and the core tag part).

Similarly, we can compute the other schemes' ciphertext overheads for reasonable leakage rates. We summarize these results in Table 2.

Table 2. Relations between ciphertext overhead and leakage rate

Schemes	CT overhead ($\#\ell(\kappa)$ bits)	Leakage rate interval (δ)	Assumption
DHLW10 [11]	$21/(1 - \delta) + 70$	$[0, 1)$	DLIN (with tSE-NIZK)
GHV12 [16]	$4\lceil 4/(1 - \delta) \rceil + 12$	$[0, 1)$	DLIN (without proof)
NS09 [28]	$9/(1 - 6\delta)$	$[0, 1/6)$	DDH
LZSS12 [25]	$9/(1 - 4\delta)$	$[0, 1/4)$	DDH
Ours	$2\lceil 5/(2 - 4\delta) \rceil + 4$	$[0, 1/2)$	DDH

Table 3. Quantitative comparison ($\# \ell(\kappa)$-bit)

Leakage-rate Schemes	1/8	1/6	1/4	1/3	3/8	2/5	1/2	1
DHLW10 [11]	94	95.2	98	101.5	103.6	105	112	-
GHV12 [16]	32	32	36	36	40	40	44	-
NS12 [28]	36	-	-	-	-	-	-	-
LZSS12 [25]	18	27	-	-	-	-	-	-
Ours	**12**	**12**	**14**	**20**	**24**	**30**	-	-

Finally, we give a quantitative comparison among these LR-CCA secure PKE schemes in Table 3. While for some achievable leakage rate (e.g., $\delta \leq 0.4$), our scheme is more efficient compared with the other four schemes. As our construction is general, we can also instantiate it under other standard assumptions, e.g., the DCR assumption [29,10]. In [16], the scheme is obtained by applying the CHK transformation [6] to a master-key leakage resilient identity-based encryption scheme. To the best of our knowledge, the constructions of identity-based PKE schemes [16,24] with master-key leakage-resilience are all based on the assumptions (e.g., DLIN) over bilinear groups. Our schemes are the first DDH/DCR based efficient LR-CCA secure PKE schemes with leakage rate $1/2 - o(1)$.

5   Instantiation from the DCR Assumption

Let $N = PQ = (2P'+1)(2Q'+1)$, $\widetilde{N} = \widetilde{P}\widetilde{Q} = (2\widetilde{P}'+1)(2\widetilde{Q}'+1)$, and the message space be $\{0,1\}^m$. Let $\mathfrak{n}, n \in \mathbb{N}$ such that $\mathfrak{n}(\log N - 1) \geq \log \widetilde{N} + \lambda + m + \omega(\log \kappa)$ and $n = \lceil \mathfrak{n} \log N / \log \widetilde{N} \rceil$. Let $(ek_{\mathrm{CH}}, td_{\mathrm{CH}}) \leftarrow \mathsf{CH.Gen}(1^\kappa)$ sample a chameleon hash function with image set $\{0,1\}^{\lfloor \widetilde{N} \rfloor / 4}$. Let $\mathsf{Ext} : \mathbb{Z}_N^n \times \{0,1\}^d \to \{0,1\}^m$ be an average-case $(\mathfrak{n}(\log N - 1) - \log \widetilde{N} - \lambda, \epsilon_2)$-strong extractor. Define a map $\chi(y) = b \in \mathbb{Z}_N$ for $y \in \mathbb{Z}_{N^2}^*$, where $y = a + bN \bmod N^2$ $(0 \leq a, b \leq N - 1)$. The LR-CCA secure PKE from the DCR assumption is presented in Fig. 2, and proof is in the full version.

Key Generation. $\mathsf{PKE_2.Gen}(1^\kappa)$: Compute $g = -h^{2N} \bmod N^2$ with $h \leftarrow \mathbb{Z}_{N^2}^*$. Choose $x_1, \ldots, x_{\mathfrak{n}} \leftarrow \{0, \ldots, \lfloor N^2/2 \rfloor\}$ and compute $pk_i = g^{x_i} \bmod N^2$. Choose a random pair $(t_a^*, t_c^*) \leftarrow \{0,1\}^* \times \mathcal{R}_{\mathrm{CH}}$ and compute $b^* = \mathsf{CH.Eval}(ek_{\mathrm{CH}}, t_a^*; t_c^*)$. Compute $E = \widetilde{g}^{\widetilde{N}^n}(1 + \widetilde{N})^{-b^*} \bmod \widetilde{N}^{n+1}$ with a random $\widetilde{g} \leftarrow \mathbb{Z}_{\widetilde{N}^{n+1}}^*$. Return $PK = (N, \mathfrak{n}, pk_1, \ldots, pk_{\mathfrak{n}}, g, \widetilde{N}, n, E, ek_{\mathrm{CH}})$ and $SK = (x_1, \ldots, x_{\mathfrak{n}})$.

Encryption. $\mathsf{PKE_2.Enc}(PK, M)$: For a public key PK and a message $M \in \{0,1\}^m$, choose a random $r \in \{0, \ldots, \lfloor N/2 \rfloor\}$ and a random seed $s \in \{0,1\}^d$. It then computes $C = g^r \bmod N^2$, $K = (\chi(pk_1^r), \ldots, \chi(pk_{\mathfrak{n}}^r))$, $\Psi = \mathsf{Ext}(K, s) \oplus M$, $\Pi = (E(1 + \widetilde{N})^b)^K \bmod \widetilde{N}^{n+1}$, where $b = \mathsf{CH.Eval}(ek_{\mathrm{CH}}, t_a; t_c)$ for the auxiliary tag $t_a = (C, s, \Psi)$ and a random filter core tag $t_c \in \mathcal{R}_{\mathrm{CH}}$. Return $CT = (C, s, \Psi, \Pi, t_c)$. Note that in the computation of Π, K is considered as an element in $\mathbb{Z}_{\widetilde{N}^n}$.

Decryption. $\mathsf{PKE_2.Dec}(SK, CT)$, given a ciphertext $CT = (C, s, \Psi, \Pi, t_c)$, computes $K' = (\chi(pk_1^{x_1}), \ldots, \chi(pk_{\mathfrak{n}}^{x_{\mathfrak{n}}}))$ and $\Pi' = (E(1 + \widetilde{N})^b)^{K'} \bmod \widetilde{N}^{n+1}$, where $b = \mathsf{CH.Eval}(ek_{\mathrm{CH}}, (C, s, \Psi); t_c)$. It checks whether $\Pi = \Pi'$. If not, it rejects with $\perp$. Else, it returns $M = \Psi \oplus \mathsf{Ext}(K', s)$.

Fig. 2. A DCR-based PKE Scheme $\mathsf{PKE_2} = (\mathsf{PKE_2.Gen}, \mathsf{PKE_2.Enc}, \mathsf{PKE_2.Dec})$

6   Conclusion and Further Work

We present a new generic construction of a public-key encryption scheme secure against leakage-resilient chosen-ciphertext attacks, from any ϵ-*universal* HPS and any one-time lossy filter (OT-LF). Instantiations from the DDH and DCR assumptions show that our construction is practical and achieves leakage rate of $1/2 - o(1)$. When a slightly weaker universality property of HPS holds with overwhelming probability over the choice of C from the invalid set, LR-CPA security with leakage rate of $1 - o(1)$ can be easily constructed from HPS [27]. In our construction, the HPS is required to be ϵ-*universal* for the worst-case choice of C from the invalid set $\mathcal{C} \setminus \mathcal{V}$. That is the reason why those LR-CPA security with leakage rate of $1 - o(1)$ from some HPS cannot be converted into LR-CCA security with OT-LF. The open question is how to further improve leakage rate while keeping the practicality of PKE.

Acknowledgements. The authors thank anonymous referees for very useful comments and suggestions.

References

1. Akavia, A., Goldwasser, S., Vaikuntanathan, V.: Simultaneous hardcore bits and cryptography against memory attacks. In: Reingold, O. (ed.) TCC 2009. LNCS, vol. 5444, pp. 474–495. Springer, Heidelberg (2009)
2. Alwen, J., Dodis, Y., Naor, M., Segev, G., Walfish, S., Wichs, D.: Public-key encryption in the bounded-retrieval model. In: Gilbert, H. (ed.) EUROCRYPT 2010. LNCS, vol. 6110, pp. 113–134. Springer, Heidelberg (2010)
3. Brakerski, Z., Goldwasser, S.: Circular and leakage resilient public-key encryption under subgroup indistinguishability - (or: Quadratic residuosity strikes back). In: Rabin, T. (ed.) CRYPTO 2010. LNCS, vol. 6223, pp. 1–20. Springer, Heidelberg (2010)
4. Brakerski, Z., Kalai, Y.T., Katz, J., Vaikuntanathan, V.: Overcoming the hole in the bucket: Public-key cryptography resilient to continual memory leakage. In: FOCS 2010, pp. 501–510. IEEE Computer Society (2010)
5. Camenisch, J., Chandran, N., Shoup, V.: A public key encryption scheme secure against key dependent chosen plaintext and adaptive chosen ciphertext attacks. In: Joux, A. (ed.) EUROCRYPT 2009. LNCS, vol. 5479, pp. 351–368. Springer, Heidelberg (2009)
6. Canetti, R., Halevi, S., Katz, J.: Chosen-ciphertext security from identity-based encryption. In: Cachin, C., Camenisch, J. (eds.) EUROCRYPT 2004. LNCS, vol. 3027, pp. 207–222. Springer, Heidelberg (2004)
7. Chow, S.S.M., Dodis, Y., Rouselakis, Y., Waters, B.: Practical leakage-resilient identity-based encryption from simple assumptions. In: Al-Shaer, E., Keromytis, A.D., Shmatikov, V. (eds.) ACM CCS 2010, pp. 152–161. ACM (2010)
8. Cramer, R., Shoup, V.: A practical public key cryptosystem provably secure against adaptive chosen ciphertext attack. In: Krawczyk, H. (ed.) CRYPTO 1998. LNCS, vol. 1462, pp. 13–25. Springer, Heidelberg (1998)
9. Cramer, R., Shoup, V.: Universal hash proofs and a paradigm for adaptive chosen ciphertext secure public-key encryption. In: Knudsen, L.R. (ed.) EUROCRYPT 2002. LNCS, vol. 2332, pp. 45–64. Springer, Heidelberg (2002)
10. Damgård, I., Jurik, M.: A generalisation, a simplification and some applications of Paillier's probabilistic public-key system. In: Kim, K.-C. (ed.) PKC 2001. LNCS, vol. 1992, pp. 119–136. Springer, Heidelberg (2001)
11. Dodis, Y., Haralambiev, K., López-Alt, A., Wichs, D.: Efficient public-key cryptography in the presence of key leakage. In: Abe, M. (ed.) ASIACRYPT 2010. LNCS, vol. 6477, pp. 613–631. Springer, Heidelberg (2010)
12. Dodis, Y., Kalai, Y.T., Lovett, S.: On cryptography with auxiliary input. In: Mitzenmacher, M. (ed.) STOC 2009, pp. 621–630. ACM (2009)
13. Dodis, Y., Ostrovsky, R., Reyzin, L., Smith, A.: Fuzzy extractors: How to generate strong keys from biometrics and other noisy data. SIAM J. Comput. 38(1), 97–139 (2008)
14. Dziembowski, S., Pietrzak, K.: Leakage-resilient cryptography. In: FOCS 2008, pp. 293–302. IEEE Computer Society (2008)
15. Faust, S., Kiltz, E., Pietrzak, K., Rothblum, G.N.: Leakage-resilient signatures. In: Micciancio, D. (ed.) TCC 2010. LNCS, vol. 5978, pp. 343–360. Springer, Heidelberg (2010)

16. Galindo, D., Herranz, J., Villar, J.: Identity-based encryption with master key-dependent message security and leakage-resilience. In: Foresti, S., Yung, M., Martinelli, F. (eds.) ESORICS 2012. LNCS, vol. 7459, pp. 627–642. Springer, Heidelberg (2012)

17. Halderman, J.A., Schoen, S.D., Heninger, N., Clarkson, W., Paul, W., Calandrino, J.A., Feldman, A.J., Appelbaum, J., Felten, E.W.: Lest we remember: Cold boot attacks on encryption keys. In: van Oorschot, P.C. (ed.) USENIX Security Symposium, pp. 45–60. USENIX Association (2008)

18. Hofheinz, D.: All-but-many lossy trapdoor functions. In: Pointcheval, D., Johansson, T. (eds.) EUROCRYPT 2012. LNCS, vol. 7237, pp. 209–227. Springer, Heidelberg (2012)

19. Hofheinz, D.: Circular chosen-ciphertext security with compact ciphertexts. In: Johansson, T., Nguyen, P.Q. (eds.) EUROCRYPT 2013. LNCS, vol. 7881, pp. 520–536. Springer, Heidelberg (2013)

20. Katz, J., Vaikuntanathan, V.: Signature schemes with bounded leakage resilience. In: Matsui, M. (ed.) ASIACRYPT 2009. LNCS, vol. 5912, pp. 703–720. Springer, Heidelberg (2009)

21. Kiltz, E., Pietrzak, K., Stam, M., Yung, M.: A new randomness extraction paradigm for hybrid encryption. In: Joux, A. (ed.) EUROCRYPT 2009. LNCS, vol. 5479, pp. 590–609. Springer, Heidelberg (2009)

22. Krawczyk, H., Rabin, T.: Chameleon signatures. In: NDSS 2000. The Internet Society (2000)

23. Lai, J., Deng, R.H., Liu, S.: Chameleon all-but-one TDFs and their application to chosen-ciphertext security. In: Catalano, D., Fazio, N., Gennaro, R., Nicolosi, A. (eds.) PKC 2011. LNCS, vol. 6571, pp. 228–245. Springer, Heidelberg (2011)

24. Lewko, A., Rouselakis, Y., Waters, B.: Achieving leakage resilience through dual system encryption. In: Ishai, Y. (ed.) TCC 2011. LNCS, vol. 6597, pp. 70–88. Springer, Heidelberg (2011)

25. Li, S., Zhang, F., Sun, Y., Shen, L.: A new variant of the Cramer-Shoup leakage-resilient public key encryption. In: Xhafa, F., Barolli, L., Pop, F., Chen, X., Cristea, V. (eds.) INCoS 2012, pp. 342–346. IEEE (2012)

26. Liu, S., Weng, J., Zhao, Y.: Efficient public key cryptosystem resilient to key leakage chosen ciphertext attacks. In: Dawson, E. (ed.) CT-RSA 2013. LNCS, vol. 7779, pp. 84–100. Springer, Heidelberg (2013)

27. Naor, M., Segev, G.: Public-key cryptosystems resilient to key leakage. In: Halevi, S. (ed.) CRYPTO 2009. LNCS, vol. 5677, pp. 18–35. Springer, Heidelberg (2009)

28. Naor, M., Segev, G.: Public-key cryptosystems resilient to key leakage. SIAM J. Comput. 41(4), 772–814 (2012)

29. Paillier, P.: Public-key cryptosystems based on composite degree residuosity classes. In: Stern, J. (ed.) EUROCRYPT 1999. LNCS, vol. 1592, pp. 223–238. Springer, Heidelberg (1999)

30. Pietrzak, K.: A leakage-resilient mode of operation. In: Joux, A. (ed.) EUROCRYPT 2009. LNCS, vol. 5479, pp. 462–482. Springer, Heidelberg (2009)

31. Peikert, C., Waters, B.: Lossy trapdoor functions and their applications. In: Dwork, C. (ed.) STOC 2008, pp. 187–196. ACM (2008)

32. Rackoff, C., Simon, D.R.: Non-interactive zero-knowledge proof of knowledge and chosen ciphertext attack. In: Feigenbaum, J. (ed.) CRYPTO 1991. LNCS, vol. 576, pp. 433–444. Springer, Heidelberg (1992)

33. Shoup, V.: Sequences of games: a tool for taming complexity in security proofs. Cryptology ePrint Archive, Report 2004/332 (2004), http://eprint.iacr.org/

On Continual Leakage of Discrete Log Representations

Shweta Agrawal[*], Yevgeniy Dodis[**], Vinod Vaikuntanathan[***],
and Daniel Wichs[†]

Abstract. Let $\mathbb{G}$ be a group of prime order q, and let $g_1, \ldots, g_n$ be random elements of $\mathbb{G}$. We say that a vector $\mathbf{x} = (x_1, \ldots, x_n) \in \mathbb{Z}_q^n$ is a *discrete log representation* of some some element $y \in \mathbb{G}$ (with respect to $g_1, \ldots, g_n$) if $g_1^{x_1} \cdots g_n^{x_n} = y$. Any element y has many discrete log representations, forming an affine subspace of $\mathbb{Z}_q^n$. We show that these representations have a nice *continuous leakage-resilience* property as follows. Assume some attacker $\mathcal{A}(g_1, \ldots, g_n, y)$ can repeatedly learn L bits of information on arbitrarily many random representations of y. That is, $\mathcal{A}$ adaptively chooses polynomially many leakage functions $f_i : \mathbb{Z}_q^n \to \{0,1\}^L$, and learns the value $f_i(\mathbf{x}_i)$, where $\mathbf{x}_i$ is a *fresh and random* discrete log representation of y. $\mathcal{A}$ wins the game if it eventually outputs a valid discrete log representation $\mathbf{x}^*$ of y. We show that if the discrete log assumption holds in $\mathbb{G}$, then no polynomially bounded $\mathcal{A}$ can win this game with non-negligible probability, as long as the leakage on each representation is bounded by $L \approx (n-2)\log q = (1 - \frac{2}{n}) \cdot |\mathbf{x}|$.

As direct extensions of this property, we design very simple continuous leakage-resilient (CLR) one-way function (OWF) and public-key encryption (PKE) schemes in the so called "invisible key update" model introduced by Alwen et al. at CRYPTO'09. Our CLR-OWF is based on the standard Discrete Log assumption and our CLR-PKE is based on the standard Decisional Diffie-Hellman assumption. Prior to our work, such schemes could only be constructed in groups with a bilinear pairing.

As another surprising application, we show how to design the first leakage-resilient *traitor tracing* scheme, where no attacker, getting the secret keys of a small subset of decoders (called "traitors") *and* bounded leakage on the secret keys of all other decoders, can create a valid decryption key which will not be traced back to at least one of the traitors.

[*] UCLA. E-mail: shweta@cs.ucla.edu. Partially supported by DARPA/ONR PRO-CEED award, and NSF grants 1118096, 1065276, 0916574 and 0830803.

[**] NYU. E-mail: dodis@cs.nyu.edu. Partially supported by NSF Grants CNS-1065288, CNS-1017471, CNS-0831299 and Google Faculty Award.

[***] U Toronto. E-mail: vinodv@cs.toronto.edu. Partially supported by an NSERC Discovery Grant, by DARPA under Agreement number FA8750-11-2-0225. The U.S. Government is authorized to reproduce and distribute reprints for Governmental purposes notwithstanding any copyright notation thereon. The views and conclusions contained herein are those of the author and should not be interpreted as necessarily representing the official policies or endorsements, either expressed or implied, of DARPA or the U.S. Government.

[†] Northeastern U. E-mail: wichs@ccs.neu.edu. Research conducted while at IBM Research, T.J. Watson.

K. Sako and P. Sarkar (Eds.) ASIACRYPT 2013, Part II, LNCS 8270, pp. 401–420, 2013.

1   Introduction

Let $\mathbb{G}$ be a group of prime order q, and let $g_1, \ldots, g_n$ be random elements of $\mathbb{G}$. We say that a vector $\mathbf{x} = (x_1, \ldots, x_n) \in \mathbb{Z}_q^n$ is a *discrete log representation* of some some element $y \in \mathbb{G}$ with respect to $g_1, \ldots, g_n$ if $\prod_{i=1}^n g_i^{x_i} = y$. A basic and well-known property of discrete log representations says that, given one such discrete log representation, it is hard to find any other one, assuming the standard Discrete Log (DL) problem is hard. In various disguises, this simple property (and its elegant generalizations) has found a huge number of applications in building various cryptographic primitives, from collision-resistant hash functions and commitment schemes [Ped91], to actively secure identification schemes [Oka92], to chosen-ciphertext secure encryption [CS02], to key-insulated cryptography [DKXY02], to broadcast encryption [DF03], to traitor tracing schemes [BF99], just to name a few.

More recently, discrete log representations have found interesting applications in leakage-resilient cryptography [NS09, ADW09, KV09], where the secret key of some system is a discrete log representation $\mathbf{x}$ of some public y, and one argues that the system remains secure even if the attacker can learn some arbitrary (adversarially specified!) "leakage function" $z = f(\mathbf{x})$, as long as the output size L of f is just slightly shorter than the length of the secret $|\mathbf{x}| = n \log q$. Intuitively, these results utilize the fact that the actual secret key $\mathbf{x}$ still has some entropy even conditioned on the L-bit leakage z and the public key y, since the set of valid discrete log representations of y has more than L bits of entropy. On the other hand, the given scheme is designed in a way that in order to break it — with or without leakage — the attacker must "know" some valid discrete log representation $\mathbf{x}^*$ of y. Since the real key $\mathbf{x}$ still has some entropy even given z and y, this means that the attacker will likely know a different discrete log representation $\mathbf{x}^* \neq \mathbf{x}$, which immediately contradicts the discrete log assumption.[1]

Although very elegant, this simple argument only applies when the overall leakage given to the attacker is *a-priori upper bounded* by L bits, where L is somewhat less than the secret key length $n \log q$. Of course, this is inevitable without some change to the model, since we clearly cannot allow the attacker to learn the entire secret $\mathbf{x}$. Thus, when applied to leakage-resilient cryptography, so far we could only get *bounded-leakage-resilient* (BLR) schemes, where the bound L is fixed throughout the lifetime of the system. In contrast, in most applications we would like to withstand more powerful *continual leakage*, where one only assumes that the *rate* of leakage is somehow bounded, but the *overall leakage* is no longer bounded. To withstand continual leakage, the secret key must be continually *refreshed* in a way that: (a) the functionality of the cryptosystem is preserved even after refreshing the keys an arbitrary number of times, and

---

[1] This argument works for unpredictability applications, such as one-way functions. For indistinguishability applications, such as encryption, a similar, but slightly more subtle argument is needed. It uses the Decisional Diffie-Hellman (DDH) assumption in place of the DL assumption, as well as the fact that the inner product function is a good "randomness extractor" [CG88, NZ96].

yet, (b) one cannot combine the various leaked values obtained from different versions of the key to break the system. Such model of *invisible key updates* was formalized by Alwen et al. [ADW09]. In that model, one assumes the existence of a trusted, "leak-free" server, who uses some "master key" MSK to continually refresh the secret key in a way that it still satisfies the conflicting properties (a) and (b) above. We stress that the server is only present during the key updates, but not during the normal day-to-day operations (like signing or decrypting when the leakage actually happens). We will informally refer to this *continual-leakage-resilient* (CLR) model of "invisible key updates" as the *floppy model*, to concisely emphasize the fact that we assume an external leak-free storage (the "floppy" disk) which is only required for rare refreshing operations.[2]

We notice that all bounded leakage schemes based on discrete log representations naturally permit the following key refreshing procedure. The master key MSK consists of a vector of the discrete logarithms $\boldsymbol{\alpha} = (\alpha_1, \ldots, \alpha_n)$ of the generators $g_1, \ldots, g_n$ with respect to some fixed generator g. The refresh simply samples a random vector $\boldsymbol{\beta} = (\beta_1, \ldots, \beta_n)$ orthogonal to $\boldsymbol{\alpha}$, so that $\prod g_i^{\beta_i} = g^{\langle \boldsymbol{\alpha}, \boldsymbol{\beta} \rangle} = 1$. The new DL representation $\mathbf{x}'$ of y is set to be $\mathbf{x}' := \mathbf{x} + \boldsymbol{\beta}$. It is easy to verify that $\mathbf{x}'$ is simply a fresh, random representation of y independent of the original DL representation $\mathbf{x}$. However, it is not obvious to see if this natural key refreshing procedure is *continual-leakage-resilient*. For the most basic question of key recovery,[3] this means that no efficient attacker $\mathcal{A}(g_1, \ldots, g_n, y)$ can compute a valid DL representation $\mathbf{x}^*$ of y despite (adaptively) repeating the following "L-bounded-leakage" step any polynomial number times. At period i, $\mathcal{A}$ chooses a leakage function $f_i : \mathbb{Z}_q^n \to \{0,1\}^L$, and learns the value $f_i(\mathbf{x}_i)$, where $\mathbf{x}_i$ is a *fresh and random* discrete log representation of y, as explained above.

OUR MAIN RESULT. As our main conceptual result, we show that the above intuition is correct: *the elegant invisible key update procedure above for refreshing DL representations is indeed continual-leakage-resilient*. In other words, one can continually leak fresh discrete log representations of the public key, without affecting the security of the system. Moreover, the leakage bound L can be made very close to the length of our secret $\mathbf{x}$, as n grows: $L \approx (n-2) \log q = (1 - \frac{2}{n}) \cdot |\mathbf{x}|$.

Our proof crucially uses a variant of the *subspace-hiding with leakage* lemma from Brakerski et al. [BKKV10] (for which we also find an alternative and *much simpler* proof than that of [BKKV10]). In its basic form, this information-theoretic lemma states that, for a random (affine) subspace S of some fixed larger space U, it is hard to distinguish the output of a bounded-length leakage function $\mathsf{Leak}(s)$ applied to random sample $s \leftarrow S$, from the output of $\mathsf{Leak}(u)$ applied to random sample $u \leftarrow U$, even if the distinguisher can later learn the

---

[2] Another reason is to separate the floppy model from a more demanding CLR model of invisible updates subsequently introduced by [BKKV10, DHLW10a], discussed in the Related Work paragraph below.

[3] For more powerful CLR goals (such as encryption and traitor tracing we discuss below), $\mathcal{A}$'s task could be more ambitious and/or $\mathcal{A}$ could get more information in addition to the public key and the leakage.

description of S *after* selecting the leakage function Leak. Given this Lemma, the overall high-level structure of our proof is as follows. Let U be the full $(n-1)$-dimensional affine space of valid discrete-log representations of y, and let S be a random $(n-2)$-dimensional affine subspace of U. Assume the attacker $\mathcal{A}$ leaks information on t different representations of y. In the original Game 0, all of the representations are sampled from the entire space U, as expected. In this case, the probability that $\mathcal{A}$ would output a representation $\mathbf{x}^* \in S$ is negligible since it gets no information about S during the course of the game and S takes up a negligible fraction U. We then switch to Game 1 where we give the attacker leakage on random representations from S rather than U. We do so in a series of hybrids where the *last* $i = 0, 1, \ldots, t$ representations are chosen from S and the first $t - i$ from U. We claim that, the probability of the attacker outputting a representation $\mathbf{x}^* \in S$ remains negligible between successive hybrids, which follows directly from the subspace-hiding with leakage lemma. Therefore, in Game 1, the attacker only sees (leakage on) representations in the small affine space S, but is likely to output a representation $\mathbf{x}^* \notin S$. This contradicts the standard DL assumption, as shown by an elegant lemma of Boneh and Franklin [BF99], which was proven in the context of traitor tracing schemes.

APPLICATIONS. By extending and generalizing the basic CLR property of discrete log representations described above, we obtain the following applications.

First, we immediately get that the natural multi-exponentiation function $h_{g_1 \ldots g_n}(x_1 \ldots x_n) = g_1^{x_1} \ldots g_n^{x_n}$ is a CLR one-way function (OWF) in the floppy model, under the standard DL assumption, with "leakage fraction" $L/|\mathbf{x}|$ roughly $1 - \frac{2}{n}$. This result elegantly extends the basic fact from [ADW09, KV09] that h is a bounded-leakage OWF with "leakage fraction" roughly to $1 - \frac{1}{n}$.

Second, we show that the Naor-Segev [NS09] bounded-leakage encryption scheme is also CLR-secure in the floppy model. The scheme is a very natural generalization of the ElGamal encryption scheme to multiple generators $g_1, \ldots, g_n$. The secret key is $\mathbf{x}$, the public key is $y = g_1^{x_1} \ldots g_n^{x_n}$, and the encryption of m is $(g_1^r, \ldots, g_n^r, y^r \cdot m)$ (with the obvious decryption given $\mathbf{x}$). The scheme is known to be secure against bounded-leakage under the standard Decisional Diffie-Hellman (DDH) assumption. In this work, we examine the security of the scheme against continual leakage in the "floppy" model, with the same style of updates we described above for the one-way function. By carefully generalizing our one-wayness argument from DL to an indistinguishability argument from DDH, we show that this natural scheme is also CLR-secure in the floppy model.

As our final, and more surprising application, we apply our techniques to design the first leakage-resilient (public-key) *traitor tracing* (TT) scheme [CFN94, BF99]. Recall, in an N-user public-key traitor tracing scheme, the content owner publishes a public-key PK, generates N *individual* secret keys $SK_1, \ldots, SK_N$, and keeps a special tracing key UK. The knowledge of PK allows anybody to encrypt the content, which can be decrypted by each user i using his secret key SK_i. As usual, the system is semantically secure given PK only. More interestingly, assume some T parties (so called "traitors") try to combine their (valid) secret keys in some malicious way to produce another secret key SK^* which can decrypt the content with noticeable probability. Then, given such a key SK* and using the master tracing key UK, the content owner should be able to correctly identify at least one of the traitors contributing to the creation of SK*. This non-trivial property is called *(non-black-box) traitor tracing*.

Boneh and Franklin [BF99] constructed a very elegant traitor tracing scheme which is semantically secure under the DDH assumption and traceable under the DL assumption. Using our new technique, we can considerably strengthen the tracing guarantee for a natural generalization of the Boneh-Franklin scheme. In our model, in addition to getting T keys of the traitors in full, we allow the attacker to obtain L bits of leakage on the keys of each of the $(N - T)$ remaining parties. Still, even with this knowledge, we argue the attacker cannot create a good secret key without the content owner tracing it to one of the traitors. We notice that, although our TT scheme is described in the bounded leakage model, where each user only gets one key and leaks L bits to the attacker, we can view the availability of N different looking keys as continual leakage "in space" rather than "time". Indeed, on a technical level we critically use our result regarding the continual leakage-resilience of DL representations, and our final analysis is considerably more involved than the analysis of our CLR-OWF in the floppy model.[4]

RELATED WORK. The basic bounded-leakage resilience (BLR) model considered by many prior works: e.g., [Dzi06, CDD⁺07, AGV09, ADW09, NS09, KV09] [ADN⁺10, CDRW10, BG10, GKPV10, DGK⁺10, DHLW10b, BSW11, BHK11, HL11], [JGS11, BCH12, BK12, HLWW12]. As we mentioned, the floppy model was introduced by Alwen et al. [ADW09] as the extension of the BLR moel. They observed that *bounded-leakage* signatures (and one-way *relations*) can be easily converted to the floppy model using any (standard) signature scheme. The idea is to have the floppy store the signing key sk for the signature scheme, and use it to authenticate the public key pk_i for the BLR signature scheme used in the i-th period. This certificate, along with the value of pk_i, is now sent with each BLR signature. Upon update, a completely fresh copy of the BLR scheme is chosen and certified. Unfortunately, this approach does not work for encryption schemes, since the encrypting party needs to know which public key to use. In fact, it even does not work for maintaining a valid pre-image of a one-way *function* (as opposed to a one-way relation). In contrast, our work directly gives efficient and *direct* CLR one-way functions and encryption schemes.

Following [ADW09], Brakerski et al. [BKKV10] and Dodis et al. [DHLW10a] considered an even more ambitious model for continual leakage resilience, where no leak-free device (e.g., "floppy") is available for updates, and the user has to be able to update his secret key "in place", using only fresh *local randomness*. Abstractly, this could be viewed as a "floppy" which does not store any long-term secrets, but only contributes fresh randomness to the system during the key update. In particular, [BKKV10, DHLW10a] managed to construct signature

---

[4] We believe that our TT scheme can also be extended to the floppy model; i.e., become continual both in "space" and "time". For simplicity of exposition, we do not explore this direction here.

and encryption schemes in this model. These works were further extended to the identity-based setting by [LRW11]. More recently, [LLW11, DLWW11] even constructed remarkable (but much less efficient) CLR encryption schemes where the attacker can even leak a constant fraction of the randomness used for each local key update. While the above encryption schemes do not require a "floppy", all of them *require a bi-linear group*, are based on the less standard/understood assumptions in bi-linear groups than the classical DL/DDH assumptions used here, and are generally quite less efficient than the simple schemes presented here. Thus, in settings where the existence of the "floppy" can be justified, our schemes would be much preferable to the theoretically more powerful schemes of [DHLW10a, BKKV10, LRW11, LLW11, DLWW11].

More surprisingly, we point out that in some applications, such as traitor tracing considered in our work, the existence of local key updates is actually an *impediment* to the security (e.g., tracing) of the scheme. For example, the key updates used in prior bi-linear group CLR constructions had the (seemingly desirable) property that a locally updated key looks completely independent from the prior version of the same key. This held even if the prior version of this key is subsequently revealed, and irrespective of whatever trapdoor information the content owner might try to store a-priori. Thus, a single user can simply re-randomize his key without the fear of being traced later. In contrast, when a "floppy" is available, one may design schemes where it is infeasible for the user to locally update his secret key to a very "different" key, without the help of the "floppy". Indeed, our generalization of the Boneh-Franklin TT scheme has precisely this property, which enables efficient tracing, and which seems impossible to achieve in all the prior pairing-based schemes [DHLW10a, BKKV10, LRW11, LLW11, DLWW11].

We also point out that the floppy model is similar in spirit to the key-insulated model of Dodis et al. [DKXY02], except in our model the "outside" does not know about the scheduling (or even the existence!) of key updates, so one cannot change the functionality (or the effective public key) of the system depending on which secret key is currently used.

Finally, although we mentioned much of the prior work with the most direct relation to our work, many other models for leakage-resilient cryptography have been considered in the last few years (see e.g., [ISW03, MR04, DP08, Pie09] [DHLW10a, BKKV10, LLW11, DLWW11, GR12, Rot12, MV13] for some examples). We refer the reader to [Wic11] and the references therein for a detailed discussion of such models.

Many of the proofs are relegated to the full version, which is available on ePrint archive.

2   Preliminaries

Below we present the definitions and lemmata that we will need. We begin with some standard notation.

2.1   Notation

We will denote vectors by bold lower case letters (e.g., $\mathbf{u}$) and matrices by bold upper case letters (e.g., $\mathbf{X}$). For integers d, n, m with $1 \leq d \leq \min(n, m)$, we use the notation $\mathsf{Rk}_d(\mathbb{F}_q^{n \times m})$ to denote the set of all $n \times m$ matrices over $\mathbb{F}_q$ with rank d. If $\mathbf{A} \in \mathbb{F}_q^{n \times m}$ is a $n \times m$ matrix of scalars, we let $\mathsf{colspan}(\mathbf{A}), \mathsf{rowspan}(\mathbf{A})$ denote the subspaces spanned by the columns and rows of $\mathbf{A}$ respectively. If $\mathcal{V} \subseteq \mathbb{F}_q^n$ is a subspace, we let $\mathcal{V}^\perp$ denote the *orthogonal space* of $\mathcal{V}$, defined by $\mathcal{V}^\perp \stackrel{\text{def}}{=} \{ \ \mathbf{w} \in \mathbb{F}_q^n \ | \ \langle \mathbf{w}, \mathbf{v} \rangle = 0 \ \ \forall \mathbf{v} \in \mathcal{V} \ \}$. We write $(\mathbf{v}_1, \dots, \mathbf{v}_m)^\perp$ as shorthand for $\mathsf{span}(\mathbf{v}_1, \dots, \mathbf{v}_m)^\perp$. We let $\mathsf{ker}(\mathbf{A}) \stackrel{\text{def}}{=} \mathsf{colspan}(\mathbf{A})^\perp$. Similarly, we let $\mathsf{ker}(\boldsymbol{\alpha})$ denote the set of all vectors in $\mathbb{F}_q^n$ that are orthogonal to $\boldsymbol{\alpha}$. If X is a probability distribution or a random variable then $x \leftarrow X$ denotes the process of sampling a value x at random according to X. If S is a set then $s \xleftarrow{\$} S$ denotes sampling s according to the *uniformly random* distribution over the set S. For a bit string $s \in \{0, 1\}^*$, we let $|s|$ denote the bit length of s. We let $[d]$ denote the set $\{1, \dots, d\}$ for any $d \in \mathbb{Z}^+$.

Throughout the paper, we let λ denote the *security parameter*. A function $\nu(\lambda)$ is called *negligible*, denoted $\nu(\lambda) = \mathsf{negl}(\lambda)$, if for every integer c there exists some integer N_c such that for all integers $\lambda \geq N_c$ we have $\nu(\lambda) \leq 1/\lambda^c$ (equivalently, $\nu(\lambda) = 1/\lambda^{\omega(1)}$).

Computational Indistinguishability. Let $X = \{X_\lambda\}_{\lambda \in \mathbb{N}}$ and $Y = \{Y_\lambda\}_{\lambda \in \mathbb{N}}$ be two ensembles of random variables. We say that X, Y are (t, ϵ)-indistinguishable if for every distinguisher D that runs in time $t(\lambda)$ we have $|\Pr[D(X_\lambda) = 1] - \Pr[D(Y_\lambda) = 1]| \leq \frac{1}{2} + \epsilon(\lambda)$. We say that X, Y are *computationally indistinguishable*, denoted $X \stackrel{c}{\approx} Y$, if for every polynomial $t(\cdot)$ there exists a negligible $\epsilon(\cdot)$ such that X, Y are (t, ϵ)-indistinguishable.

Statistical Indistinguishability. The *statistical distance* between two random variables X, Y is defined by $\mathbf{SD}(X, Y) = \frac{1}{2} \sum_x |\Pr[X = x] - \Pr[Y = x]|$. We write $X \stackrel{s}{\approx}_\epsilon Y$ to denote $\mathbf{SD}(X, Y) \leq \epsilon$ and just plain $X \stackrel{s}{\approx} Y$ if the statistical distance is negligible in the security parameter. In the latter case, we say that X, Y are *statistically indistinguishable*.

Matrix-in-the-Exponent Notation: Let $\mathbb{G}$ be a group of prime order q generated by an element $g \in \mathbb{G}$. Let $\mathbf{A} \in \mathbb{F}_q^{n \times m}$ be a matrix. Then we use the notation $g^{\mathbf{A}} \in \mathbb{G}^{n \times m}$ to denote the matrix $\left(g^{\mathbf{A}}\right)_{i,j} \stackrel{\text{def}}{=} g^{(\mathbf{A})_{i,j}}$ of group elements. We will use a similar notational shorthand for vectors.

2.2   Hiding Subspaces in the Presence of Leakage

In this section we prove various indistinguishability lemmas about (statistically) hiding subspaces given leakage on some of their vectors.

Hiding Subspaces. The following lemma says that, given some sufficiently small leakage on a random matrix $\mathbf{A}$, it is hard to distinguish random vectors from $\mathsf{colspan}(\mathbf{A})$ from uniformly random and independent vectors. A similar lemma was shown in [BKKV10, LLW11, DLWW11], and the following formulation is from [DLWW11]. The proof follows directly from the leftover-hash lemma.

Lemma 1 (Subspace Hiding with Leakage [DLWW11]). *Let $n \geq d \geq u, s$ be integers, $\mathbf{S} \in \mathbb{Z}_q^{d \times s}$ be an arbitrary (fixed and public) matrix and* $\mathsf{Leak} : \{0,1\}^* \to \{0,1\}^L$ *be an arbitrary function with L-bit output. For randomly sampled $\mathbf{A} \xleftarrow{\$} \mathbb{Z}_q^{n \times d}, \mathbf{V} \xleftarrow{\$} \mathbb{Z}_q^{d \times u}, \mathbf{U} \xleftarrow{\$} \mathbb{Z}_q^{n \times u}$, we have:*

$$(\mathsf{Leak}(\mathbf{A}), \mathbf{AS}, \mathbf{V}, \mathbf{AV}) \overset{s}{\approx} (\mathsf{Leak}(\mathbf{A}), \mathbf{AS}, \mathbf{V}, \mathbf{U})$$

as long as $(d - s - u)\log(q) - L = \omega(\log(\lambda))$ and $n = \mathsf{poly}(\lambda)$.

We also show a dual version of Lemma 1, where a random matrix $\mathbf{A}$ is chosen and the attacker either leaks on random vectors in $\mathsf{colspan}(\mathbf{A})$ or uniformly random vectors. Even if the attacker is later given $\mathbf{A}$ in full, it cannot distinguish which case occurred. This version of "subspace hiding" was first formulated by [BKKV10], but here we present a significantly simplified proof and improved parameters by framing it as a corollary (or a dual version of) Lemma 1.

Corollary 1 (Dual Subspace Hiding). *Let $n \geq d \geq u$ be integers, and let* $\mathsf{Leak} : \{0,1\}^* \to \{0,1\}^L$ *be some arbitrary function. For randomly sampled $\mathbf{A} \xleftarrow{\$} \mathbb{Z}_q^{n \times d}, \mathbf{V} \xleftarrow{\$} \mathbb{Z}_q^{d \times u}, \mathbf{U} \xleftarrow{\$} \mathbb{Z}_q^{n \times u}$, we have:*

$$(\mathsf{Leak}(\mathbf{AV}), \mathbf{A}) \overset{s}{\approx} (\mathsf{Leak}(\mathbf{U}), \mathbf{A})$$

as long as $(d - u)\log(q) - L = \omega(\log(\lambda))$, $n = \mathsf{poly}(\lambda)$, and $q = \lambda^{\omega(1)}$.

Proof. We will actually prove the above assuming that $\mathbf{A}, \mathbf{V}, \mathbf{U}$ are random full-rank matrices, which is statistically close to the given statement since q is super-polynomial. We then "reduce" to Lemma 1.

Given $\mathbf{A}$ and $\mathbf{C}$ such that $\mathbf{C} = \mathbf{AV}$ or $\mathbf{C} = \mathbf{U}$, we can probabilistically choose a $n \times d'$ matrix A' depending only on $\mathbf{C}$ and a $n \times u'$ matrix $\mathbf{C}'$ depending only on $\mathbf{A}$ such that the following holds:

- If $\mathbf{C} = \mathbf{AV}$ for a random (full rank) $d \times u$ matrix $\mathbf{V}$, then $\mathbf{C}' = \mathbf{A}'\mathbf{V}'$ for a random (full rank) $d' \times u'$ matrix $\mathbf{V}'$.
- If $\mathbf{C} = \mathbf{U}$ is random (full rank) and independent of $\mathbf{A}$, then $\mathbf{C}' = \mathbf{U}'$ is random (full rank) and independent of $\mathbf{A}'$.

and where $d' = n - u$, $u' = n - d$. To do so, simply choose $\mathbf{A}'$ to be a random $n \times d'$ matrix whose columns form a basis of $\mathsf{colspan}(\mathbf{C})^\perp$ and choose $\mathbf{C}'$ to be a random $n \times u'$ matrix whose columns form a basis of $\mathsf{colspan}(\mathbf{A})^\perp$. If $\mathbf{C} = \mathbf{U}$ is independent of $\mathbf{A}$, then $\mathbf{C}' = \mathbf{U}'$ is a random full-rank matrix independent of $\mathbf{A}'$. On the other hand, if $\mathbf{C} = \mathbf{AV}$, then $\mathsf{colspan}(\mathbf{A})^\perp \subseteq \mathsf{colspan}(\mathbf{C})^\perp$ is a random subspace. Therefore $\mathbf{C}' = \mathbf{A}'\mathbf{V}'$ for some uniformly random $\mathbf{V}'$.

Now assume that our lemma does not hold and that there is some function Leak and an (unbounded) distinguisher D that has a non-negligible distinguishing advantage for our problem. Then we can define a function Leak$'$ and a distinguished D' which breaks the problem of Lemma 1 (without even looking at AS, V). The function Leak$'(A)$ samples C' as above and outputs $Leak = \mathsf{Leak}(C')$. The distinguisher D', given $(Leak, C)$ samples A' using C as above and outputs $D(Leak, A')$. The distinguisher D' has the same advantage as D. Therefore, by Lemma 1, indistinguishability holds as long as

$$(d' - u') \log(q) - L = \omega(\log(\lambda)) \Leftrightarrow (d - u) \log(q) - L = \omega(\log(\lambda))$$

It is also easy to extend the above corollary to the case where (the column space of) $\mathbf{A}$ is a subspace of some larger public space $\mathbf{W}$.

Corollary 2. *Let $n \geq m \geq d \geq u$. Let $\mathcal{W} \subseteq \mathbb{Z}_q^n$ be a fixed subspace of dimension m and let* Leak $: \{0,1\}^* \to \{0,1\}^L$ *be some arbitrary function. For randomly sampled $\mathbf{A} \xleftarrow{\$} \mathcal{W}^d$ (interpreted as an $n \times d$ matrix), $\mathbf{V} \xleftarrow{\$} \mathbb{Z}_q^{d \times u}, \mathbf{U} \xleftarrow{\$} \mathcal{W}^u$ (interpreted as an $n \times u$ matrix), we have:*

$$(\mathsf{Leak}(\mathbf{AV}), \mathbf{A}) \overset{s}{\approx} (\mathsf{Leak}(\mathbf{U}), \mathbf{A})$$

as long as $(d - u) \log(q) - L = \omega(\log(\lambda))$, $n = \mathsf{poly}(\lambda)$, and $q = \lambda^{\omega(1)}$.

Proof. Let $\mathbf{W}$ be some $n \times m$ matrix whose columns span $\mathcal{W}$. Then we can uniquely write $\mathbf{A} = \mathbf{WA}'$, where $\mathbf{A}' \in \mathbb{Z}_q^{m \times d}$ is uniformly random. Now we just apply Lemma 1 to $\mathbf{A}'$.

A variant of the corollary holds also for affine subspaces. Namely:

Corollary 3. *Let $n \geq m \geq d \geq u$. Let $\mathcal{W} \subseteq \mathbb{Z}_q^n$ be a fixed subspace of dimension m and let* Leak $: \{0,1\}^* \to \{0,1\}^L$ *be some arbitrary function and let $\mathbf{B} \in \mathbb{Z}_q^{n \times u}$ be an arbitrary matrix. For randomly sampled $\mathbf{A} \xleftarrow{\$} \mathcal{W}^d$ (interpreted as an $n \times d$ matrix), $\mathbf{V} \xleftarrow{\$} \mathbb{Z}_q^{d \times u}, \mathbf{U} \xleftarrow{\$} \mathcal{W}^u$ (interpreted as an $n \times u$ matrix), we have:*

$$(\mathsf{Leak}(\mathbf{AV} + \mathbf{B}), \mathbf{A}) \overset{s}{\approx} (\mathsf{Leak}(\mathbf{U}), \mathbf{A})$$

as long as $(d - u) \log(q) - L = \omega(\log(\lambda))$, $n = \mathsf{poly}(\lambda)$, and $q = \lambda^{\omega(1)}$.

3   One-Wayness of Discrete Log Representations under Continual Leakage

In this section, we show the one-wayness of discrete log representations under continual leakage. Namely, we show that for random $g_1, \ldots, g_n \xleftarrow{\$} \mathbb{G}$ and $h \xleftarrow{\$} \mathbb{G}$, obtaining leakage on many representations $\mathbf{x} = (x_1, ..., x_n)$ such that $\prod_{i=1}^n g_i^{x_i} = h$ does not help an efficient PPT adversary output any representation of h in terms of $g_1, \ldots, g_n$ in full (except with negligible probability) assuming that the discrete log assumption is true. Thus, in succinct terms, we show that discrete log representations are one-way under continual leakage, based on the (plain) discrete log assumption.

We first define the notion of a continual leakage resilient one-way function in the floppy model.

3.1 Defining One-Way Functions in Floppy Model

A continuous leakage resilient (CLR) one-way function in the Floppy Model (OWFF) consists of consists of the following PPT algorithms (Gen, Sample, Eval, Update):

1. KeyGen(1^λ) is a PPT algorithm that takes as input the security parameter λ and outputs the public parameters PP, the update key UK. The parameters PP are implicit inputs to all other algorithms and we will not write them explicitly for cleaner notation.
2. Sample(PP): Takes as input the public parameters PP and samples a random value $\mathbf{x}$.
3. Eval(PP, $\mathbf{x}$) : This is a deterministic algorithm that takes as input $\mathbf{x}$ and outputs $\mathbf{y} \in \{0,1\}^*$.
4. Update(UK, $\mathbf{x}$) is a PPT algorithm that takes as input the update key UK and a string $\mathbf{x} \in \{0,1\}^*$ and outputs $\mathbf{x}' \in \{0,1\}^*$.

Correctness. We require that for any (PP, UK) $\leftarrow$ KeyGen(1^λ), and *any* $\mathbf{x} \in \{0,1\}^*$, we have
$$\mathsf{Eval}(\mathsf{Update}(\mathsf{UK}, \mathbf{x})) = \mathsf{Eval}(\mathbf{x}).$$

Security. Let $L = L(\lambda)$ be a function of the security parameter. We say that a tuple of algorithms (KeyGen, Eval, Update) is an L-CLR secure one-way function in the floppy model, if for any PPT attacker $\mathcal{A}$, there is a negligible function μ such that $\Pr[\mathcal{A} \text{ wins}] \leq \mu(\lambda)$ in the following game:

- The challenger chooses (PP, UK) $\leftarrow$ KeyGen(1^λ). Next, it chooses a random element $\mathbf{x}_1 \leftarrow$ Sample(PP) and sets $\mathbf{y} \leftarrow$ Eval($\mathbf{x}_1$). The challenger gives PP, $\mathbf{y}$ to $\mathcal{A}$.
- $\mathcal{A}$ may adaptively ask for *leakage queries* on arbitrarily many pre-images. Each such query consists of a function (described by a circuit) Leak : $\{0,1\}^* \rightarrow \{0,1\}^L$ with L bit output. On the ith such query Leak$_i$, the challenger gives the value Leak$_i(\mathbf{x}_i)$ to $\mathcal{A}$ and computes the next pre-image $\mathbf{x}_{i+1} \leftarrow$ Update(UK, $\mathbf{x}_i$).
- $\mathcal{A}$ eventually outputs a vector $\mathbf{x}^*$ and *wins* if Eval($\mathbf{x}^*$) = $\mathbf{y}$.

3.2 Constructing One-Way Function in the Floppy Model

We construct a one-way function $\mathcal{F} =$ (KeyGen, Sample, Eval, Update) as follows for some parameter $n = n(\lambda)$ which determined the amount of leakage that can be tolerated.

1. KeyGen(1^λ): Choose a group $\mathbb{G}$ of prime order q with generator g by running the group generation algorithm $\mathcal{G}(1^\lambda)$. Choose a vector $\boldsymbol{\alpha} = (\alpha_1, \ldots, \alpha_n) \xleftarrow{\$} \mathbb{Z}_q^n$, and let $g_i = g^{\alpha_i}$ for $i \in [n]$. Output the parameters $\mathsf{PP} = (\mathbb{G}, g, g_1, \ldots, g_n)$ and the update key $\mathsf{UK} = \boldsymbol{\alpha}$.
2. Sample(PP): Sample a random vector $\mathbf{x} \xleftarrow{\$} \mathbb{Z}_q^n$.
3. Eval($\mathsf{PP}, \mathbf{x}$): Parse $\mathbf{x} = (x_1, \ldots, x_n)$ and output $y := \prod_{i=1}^n g_i^{x_i}$.
4. Update($\mathsf{UK}, \mathbf{x}$): Choose a uniformly random vector $\boldsymbol{\beta} \xleftarrow{\$} \ker(\boldsymbol{\alpha})$, and output $\mathbf{x} + \boldsymbol{\beta}$.

Correctness follows from the fact that the inner product $\langle \mathbf{x} + \boldsymbol{\beta}, \boldsymbol{\alpha} \rangle = \langle \mathbf{x}, \boldsymbol{\alpha} \rangle + \langle \boldsymbol{\beta}, \boldsymbol{\alpha} \rangle = \langle \mathbf{x}, \boldsymbol{\alpha} \rangle$, since $\boldsymbol{\alpha}$ and $\boldsymbol{\beta}$ are orthogonal (mod q).

Theorem 1. *Let $L = L(\lambda)$ and $n = n(\lambda)$ be functions of the security parameter λ satisfying*

$$L < (n-2)\log(q) - \omega(\log(\lambda))$$

Then, $\mathcal{F}$ is an L-CLR secure one-way function in the floppy model (see definition 3.1) under the discrete log assumption for $\mathcal{G}$.

Proof. Suppose that the attacker has a non-negligible chance of winning the L-CLR-OWF game. Then, assuming that the DL assumption holds, we will arrive at a contradiction. The proof proceeds by a sequence of games. Without loss of generality, assume that the attacker makes exactly T leakage queries.

Game 0: This is the security game in the definition of a CLR-one way function in the floppy model. Namely, the adversary is given the public parameters PP and $y = \mathsf{Eval}(\mathsf{PP}, \mathbf{x}_1)$, and asks a polynomial number of queries adaptively. Each query is a function $\mathsf{Leak}_i : \mathbb{Z}_q^n \to \{0,1\}^L$, in response to which the challenger returns $\mathsf{Leak}_i(\mathbf{x}_i)$ where, for $i > 1$, the ith preimage $\mathbf{x}_i$ is computed as $\mathbf{x}_i = \mathbf{x}_{i-1} + \boldsymbol{\beta}_i$ where $\boldsymbol{\beta}_i \xleftarrow{\$} \ker(\boldsymbol{\alpha})$.
By assumption, we have $\Pr[\mathcal{A} \text{ wins}] \geq \varepsilon(\lambda)$ for some non-negligible ε.

Game 1: Game 1 is defined as a sequence of $T+1$ sub-games denoted by Games $1.0, \ldots, 1.T$. For $i = 1, \ldots, T$, we have:

Game 1.i: In this game, the challenger chooses a random $(n-2)$-dimensional subspace $S \subseteq \ker(\boldsymbol{\alpha})$ in the beginning and answers the first $T - i$ queries differently from the last i queries as follows:
  - For every $1 < j \leq T - i$, compute $\mathbf{x}_j = \mathbf{x} + \boldsymbol{\beta}_j$ where $\boldsymbol{\beta}_j \xleftarrow{\$} \ker(\boldsymbol{\alpha})$.
  - For every $T - i < j \leq T$, compute $\mathbf{x}_j = \mathbf{x} + \mathbf{s}_j$ where $\mathbf{s}_j \xleftarrow{\$} S$.
  In the above, we define $\mathbf{x} := \mathbf{x}_1$ to be the initial pre-image output by Sample.

Game 2: In Game 2, the challenger chooses all the vectors from the affine subspace $x + S$, i.e. it sets $\mathbf{x}_j = \mathbf{x} + \mathbf{s}_j$ where $\mathbf{s}_j \xleftarrow{\$} S$, $j \in [T]$.

Game 1.0 is identical to Game 0 since, in both games, all of the values $\mathbf{x}_i$ are just uniformly random over the affine space $\{\mathbf{x}_i : g^{\langle \mathbf{x}_i, \boldsymbol{\alpha} \rangle} = y\}$. By definition, Game 1.$T$ is identical to Game 2.

In each of the games 1.i, $i = 0, \ldots, T$, define the event E_i to be true if the adversary wins and returns a vector $\mathbf{x}^*$ such that $\mathbf{x}^* - \mathbf{x} \notin S$. Then, first we claim that in game 1.0, the probability of the event E_0 happening is negligibly close to ε.

Claim. There is a negligible function $\mu : \mathbb{N} \to [0, 1]$ such that

$$\Pr[E_0] \geq \varepsilon(\lambda) - \mu(\lambda).$$

Proof. We have $\Pr[E_0] \geq \varepsilon(\lambda) - \Pr[\mathbf{x}^* - \mathbf{x} \in S] \geq \varepsilon(\lambda) - 1/q$, where the latter probability over a random choice of S (since the adversary has no information about S in game 1.0).

Next, we show that this probability does not change much across games:

Claim. There is a negligible function $\mu : \mathbb{N} \to [0, 1]$ such that for every $1 \leq i \leq T$,

$$|\Pr[E_i] - \Pr[E_{i-1}]| \leq \mu(\lambda).$$

Proof. We have by Corollary 3, that as long as $L < (n - 2)\log(q) - \omega(\log(\lambda))$ an attacker cannot distinguish leakage on $\boldsymbol{\beta}_i \xleftarrow{\$} \ker(\boldsymbol{\alpha})$ from leakage on $\mathbf{s}_i \xleftarrow{\$} S$, even if $\boldsymbol{\alpha}$ is public and known in the beginning and S becomes public *after* the leakage occurs. Therefore, knowing only $\boldsymbol{\alpha}$, we can simulate the first $i-1$ leakage queries for the attacker and then use leakage on the challenge vector ($\boldsymbol{\beta}_i$ or $\mathbf{s}_i$) to answer the ith query. We can then use knowledge of S (after the ith leakage query) to simulate the rest of the leakage queries and test if eventually the event (E_{i-1} or E_i) occurs. This proves the claim.

Combining the above two claims and the observation that Game 2 is identical to Game 1.T, we have that there is a negligible function $\mu : \mathbb{N} \to [0, 1]$ such that in Game 2,
$$\Pr[\mathcal{A} \text{ wins and } \mathbf{x}^* - \mathbf{x} \notin S] \geq \varepsilon(\lambda) - \mu(\lambda).$$

Finally we show that the above contradicts the DL assumption.

Claim. If the Discrete Log assumption holds, then there is a negligible function $\mu : \mathbb{N} \to [0, 1]$ such that in Game 2,

$$\Pr[\mathcal{A} \text{ wins and } \mathbf{x}^* - \mathbf{x} \notin S] \leq \mu(\lambda)$$

Proof. Note that in Game 2, all the leakage queries of the adversary are answered using a randomly chosen $(n - 2)$-dimensional subspace $S \subseteq \ker(\boldsymbol{\alpha})$, hence by Lemma 3 an adversary who outputs $\mathbf{x}^*$ such that $\mathbf{x}^* - \mathbf{x} \notin S$ can be transformed into one that solves the discrete log problem.

Thus we arrive at a contradiction, which shows that under the Discrete Log assumption, the attacker could not have output $\mathbf{x}^*$ such that $f(\mathbf{x}^*) = \mathbf{y}$. Thus, $\mathcal{F}$ is an $L - CLR$ secure one way function for $L < (n - 2)\log(q) - \omega(\log(\lambda))$.

4 Public-Key Encryption in the Continuous Leakage Model

In this section, we show the semantic security of the cryptosystems of Boneh et al. [BHHO08] and Naor and Segev [NS09] with continual leakage on the secret keys in the floppy model (i.e., with invisible updates) under the DDH assumption. We first define semantic security under continual leakage.

4.1 Defining Encryption in the Floppy Model

A CLR public key encryption scheme (CLR-PKE) in the Floppy Model consists of the following algorithms:

1. KeyGen(1^λ): Takes as input the security parameter λ and outputs the public key PK, the secret key SK and the update key UK.
2. Update(UK, SK): Outputs an updated secret key SK'.
3. Encrypt(PK, M): Outputs the ciphertext CT.
4. Decrypt(SK, CT): Outputs the decrypted message M.

For convenience, we define the algorithm Update^i that performs $i \geq 0$ consecutive updates as:

Update^i(UK, SK) $\rightarrow$ SK' : Let SK$_0$ = SK, SK$_1 \leftarrow$ Update(UK, SK$_0$), ... SK$_i \leftarrow$ Update(UK, SK$_{i-1}$). Output SK' = SK$_i$

Security. Let $L = L(\lambda)$ be a function of the security parameter. We say that a CLR PKE is L-CLR secure in the floppy model, if, for any PPT adversary $\mathcal{A}$, there is a negligible function μ such that $|\Pr[\mathcal{A} \text{ wins }] - \frac{1}{2}| \leq \mu(\lambda)$ in the following game:

- Challenger chooses (PK, UK, SK$_1$) $\leftarrow$ KeyGen(1^λ).
- $\mathcal{A}$ may adaptively ask for *leakage queries* on arbitrarily many secret keys. Each such query consists of a function (described by a circuit) Leak : $\{0,1\}^* \rightarrow \{0,1\}^L$ with L bit output. On the ith such query Leak$_i$, the challenger gives the value Leak$_i$(SK$_i$) to $\mathcal{A}$ and computes the next updated key SK$_{i+1} \leftarrow$ Update(UK, SK$_i$).
- At some point $\mathcal{A}$ gives the challenger two messages M_0, M_1. The challenger chooses a bit $b \xleftarrow{\$} \{0,1\}$ and sets CT $\leftarrow$ Encrypt(PK, M_b).
- The attacker $\mathcal{A}$ gets CT and outputs a bit $\tilde{b}$. We say $\mathcal{A}$ *wins* if $\tilde{b} = b$ with non-negligible probability.

4.2 Constructing Encryption in the Floppy Model

We define our scheme as follows for some parameter $n = n(\lambda)$ which determined the amount of leakage that can be tolerated.

1. KeyGen(1^λ): Let $(\mathbb{G}, q, g) \xleftarrow{\$} \mathcal{G}(1^\lambda)$. Choose vectors $\boldsymbol{\alpha} \xleftarrow{\$} \mathbb{Z}_q^n$ and $\mathbf{x} \xleftarrow{\$} \mathbb{Z}_q^n$, and let $f = g^{\langle \boldsymbol{\alpha}, \mathbf{x}_0 \rangle}$.
   The public parameters PK consists of $(g, f, g^{\boldsymbol{\alpha}})$.
   The update key UK $= \boldsymbol{\alpha}$ and the secret key is set to SK $= \mathbf{x} + \boldsymbol{\beta}$ where $\boldsymbol{\beta} \xleftarrow{\$} \ker(\boldsymbol{\alpha})$.
2. Update(UK, SK): Choose $\boldsymbol{\beta} \xleftarrow{\$} \ker(\boldsymbol{\alpha})$, and output SK $+ \boldsymbol{\beta}$ as the updated secret key.
3. Encrypt(PK, M): To encrypt $M \in \mathbb{G}$, pick a random scalar $r \xleftarrow{\$} \mathbb{Z}_q$. Output the ciphertext CT $\leftarrow (g^{r\boldsymbol{\alpha}}, M \cdot f^r)$.
4. Decrypt(SK, CT): Parse the ciphertext CT as $(g^{\mathbf{c}}, h)$ and output $h \cdot g^{-\langle \mathbf{c}, \mathsf{SK} \rangle}$ as the message.

A correctly formed ciphertext CT looks like $(g^{\mathbf{c}}, h) = (g^{r\boldsymbol{\alpha}}, M \cdot g^{r\langle \boldsymbol{\alpha}, \mathbf{x} \rangle})$. The secret key (after arbitrarily many updates) is SK $= \mathbf{x} + \boldsymbol{\beta}$ where $\boldsymbol{\beta} \in \ker(\boldsymbol{\alpha})$. The decryption computes

$$h \cdot g^{-\langle \mathbf{c}, \mathbf{x} + \boldsymbol{\beta} \rangle} = M \cdot g^{r\langle \boldsymbol{\alpha}, \mathbf{x} \rangle} \cdot g^{-\langle r\boldsymbol{\alpha}, \mathbf{x} + \boldsymbol{\beta} \rangle} = M \cdot g^{r\langle \boldsymbol{\alpha}, \mathbf{x} \rangle} \cdot g^{-r\langle \boldsymbol{\alpha}, \mathbf{x} \rangle} = M$$

since $\langle \boldsymbol{\alpha}, \boldsymbol{\beta} \rangle = 0 \pmod{q}$.

Theorem 2. *Let $L = L(\lambda)$ and $n = n(\lambda)$ be functions of the security parameter λ satisfying*

$$L < (n - 2)\log(q) - \omega(\log(\lambda))$$

Then, the public key encryption scheme (KeyGen, Update, Encrypt, Decrypt) is L-CLR secure secure in the Floppy Model under the DDH assumption for $\mathcal{G}$.

5   Traitor Tracing in the Bounded Leakage Model

In this section, we generalize the constructions in Section 3 and Section 4 to obtain "leaky" traitor tracing in the bounded leakage model, which could be viewed as continual leakage-resilience in "space" rather than "time", but with strong traitor tracing properties. First, we define traitor tracing and associated security notions.

5.1   Definition of Traitor Tracing

The traitor tracing scheme is given by the following algorithms:

1. KeyGen($1^\lambda; 1^N, 1^T$) $\rightarrow$ PK, $\mathsf{SK}_1, \ldots, \mathsf{SK}_N$: Takes as input the security parameter λ, number of parties N, and number of traitors T. Outputs the public key PK, and secret keys $\{\mathsf{SK}_i\}_{i=1}^N$ for each party $i \in [N]$.
2. Encrypt(PK, M) $\rightarrow$ CT: Takes as input the public key PK, a message M and outputs the ciphertext CT.
3. Decrypt(PK, CT, SK) $\rightarrow M$: Takes as input the public key PK, a ciphertext CT and a secret key SK and outputs a message M.

4. Trace(PK, SK*) $\rightarrow i$: Takes as input the public key PK, and some secret key SK* and outputs an index $i \in [N]$ corresponding to an accused traitor.

Note that the tracing algorithm takes a valid secret key SK* as input, and this is what makes the scheme *non black box*. This assumes that if the traitors collude and construct a "pirate decoder" that decrypts the encrypted content, then one can always extract the decryption key from this decoder. The stronger notion of *black box* traitor tracing only assumes that one can test whether the pirate decoder plays the encrypted content or not.

Correctness: For any integers $N, T, U, i \in [N]$ any PK, TK, SK$_1$, ..., SK$_N$ $\leftarrow$ KeyGen($1^\lambda; 1^N, 1^T$), CT $\leftarrow$ Encrypt(M, PK) and $M' \leftarrow$ Decrypt(CT, SK$_i$): we have $M' = M$.

We define security in terms of two properties: semantic security and tracing security.

Semantic Security: The standard notion of semantic security requires that, for any PPT $\mathcal{A}$, we have $|\Pr[\mathcal{A} \text{ wins }] - \frac{1}{2}| \leq \mu(\lambda)$ in the following game:

- Attacker $\mathcal{A}$ chooses the values $1^N, 1^T$ to the challenger.
- Challenger chooses (PK, SK$_1$, ..., SK$_N$) and gives PK to $\mathcal{A}$.
- At some point $\mathcal{A}$ gives the challenger $\mathcal{C}$ two messages M_0, M_1.
- The challenger chooses a bit $b \leftarrow \{0,1\}$ at random and set CT $\leftarrow$ Encrypt(PK, M_b).
- The attacker $\mathcal{A}$ gets CT and outputs a bit $\tilde{b}$. We say $\mathcal{A}$ *wins* if $\tilde{b} = b$.

Tracing Security: To define non-black-box tracing, we first define the predicate GOOD(PK, SK) which holds iff there exists some message M in message-domain such that

$$\Pr[M' = M \; : \; \text{CT} \leftarrow \text{Encrypt}(M, \text{PK}), M' \leftarrow \text{Decrypt}(\text{CT}, \text{SK})] \geq \frac{1}{2}.$$

In other words, a key SK is *good* if it succeeds in decryping at least some message M with probability at least a $\frac{1}{2}$. We say that *leakage-resilient traitor tracing* security holds if, for any PPT $\mathcal{A}$, we have $\Pr[\mathcal{A} \text{ wins }] \leq \mu(\lambda)$ in the following game:

- Attacker $\mathcal{A}$ chooses the values $1^N, 1^T$.
- Challenger $\mathcal{C}$ chooses (PK, SK$_1$, ..., SK$_N$) and gives PK to $\mathcal{A}$.
- $\mathcal{A}$ may adaptively ask $\mathcal{C}$ for the following type of queries:
  - **Leakage queries:** Attacker gives a user index $i \in [N]$ and a function (defined by a circuit) Leak : $\{0,1\}^* \rightarrow \{0,1\}^L$ with L bit output. If no leakage query for user i was made before, then the challenger outputs Leak(SK$_i$) and otherwise it ignores the query.
  - **Corrupt Queries:** Attacker asks for user index i and gets SK$_i$.
- At some point $\mathcal{A}$ outputs some SK* and the challenger runs $i \leftarrow$ Trace(PK, SK*). We say that $\mathcal{A}$ wins if all of the following conditions hold: (1) $\mathcal{A}$ made at most T corrupt queries throughout the game, (2) the predicate GOOD(PK, SK*) holds, (3) the traced index i was *not* part of any corrupt query.

Before presenting the encryption scheme, we review some necessary notions from the theory of error correcting codes.

Error Correcting Code. For traitor tracing with N parties and T traitors, we will rely on an $[N, K, 2T + 1]_q$-linear-ECC over $\mathbb{F}_q$, where K is chosen as large as possible. For the Reed-Solomon code, we can set $K = N - 2T$, which we will assume from now on. Therefore, we will also assume that $N > 2T$ (this is without loss of generality as we can always increase N by introducing "dummy users" if necessary). Let $\mathbf{A}$ be a generation matrix and $\mathbf{B}$ be a parity check matrix so that $\mathbf{BA} = \mathbf{0}$. Note that $\mathbf{B}$ is a $2T \times N$ matrix. Lastly, we will assume *efficient syndrome-decoding* so that we can efficiently recover a vector $\mathbf{e} \in \mathbb{Z}_q^N$ from $\mathbf{B} \cdot \mathbf{e}$ as long as the hamming-weight of $\mathbf{e}$ is less than T. This holds for the Reed-Solomon code.

The Scheme. We now present our Traitor-Tracing scheme which is a natural generalization of the Boneh-Franklin scheme [BF99]. The scheme is defined as follows for some parameter $n = n(\lambda)$.

1. KeyGen$(1^\lambda, 1^N, 1^T) \to$ PK, SK$_1, \ldots,$ SK$_N$:
   Choose $(\mathbb{G}, q, g) \xleftarrow{\$} \mathcal{G}(1^\lambda)$. Choose $\boldsymbol{\alpha} \xleftarrow{\$} \mathbb{Z}_q^n$ and $\beta \xleftarrow{\$} \mathbb{Z}_q$.
   Let $\mathbf{B}$ be the parity-check matrix of an $[N, K, 2T + 1]_q$-ECC as described above and let us label its columns by $\mathbf{b}_1, \ldots, \mathbf{b}_N$ where $\mathbf{b}_i \in \mathbb{Z}_q^{2T}$ for $i \in [N]$. For $i \in [N]$, choose SK$_i = (\mathbf{b}_i || \mathbf{x}) \in \mathbb{Z}_q^n$ where $\mathbf{x} = (x_1, \ldots, x_{n-2T})$ and is constructed choosing $x_2, \ldots, x_{n-2T}$ uniformly random and uniquely fixing x_1 so that $\langle \boldsymbol{\alpha}, \mathsf{SK}_i \rangle = \beta$. Set PK $:= [\, g, g^\alpha = (g_1, \ldots, g_n), f = g^\beta, \mathbf{B}]$.
2. Encrypt(PK, M) $\to$ CT: Choose a random $r \xleftarrow{\$} \mathbb{Z}_q$. Output CT $\leftarrow (g^{r\alpha}, f^r \cdot M)$
3. Decrypt(PK, CT, SK) $\to M$: Let CT $= (g^\mathbf{c}, h)$. Output $hg^{-\langle \mathbf{c}, \mathsf{SK} \rangle}$.
4. Trace(PK, SK*) $\to i$: Check that the input is a valid key SK* satisfying $g^{\langle \boldsymbol{\alpha}, \mathsf{SK}^* \rangle} = f$. To trace, do the following: (1) Write SK$^* = (\mathbf{b}^* || \mathbf{x}^*)$. (2) Use syndrome decoding on $\mathbf{b}^*$ to recover a low-weight "error vector" $(e_1, \ldots, e_N) \in \mathbb{Z}_q^N$. Output $\perp$ if this fails. (3) Output some index i such that $e_i \neq 0$.

Semantic security follows from [BF99] (under the DDH assumption). The reason is that given the public key values $(g^\alpha, f = g^\beta)$ it is hard to distinguish the ciphertext values $g^{r\alpha}, f^r$ for some $r \in \mathbb{Z}_q$ from a uniformly random and independent vector of $n + 1$ group elements. Since this part does not involve leakage, we omit the formal proof and instead concentrate on the novel tracing part. The theorem below states the leakage resilient tracing security achieved by our scheme.

Theorem 3. *Assuming we choose $n \geq 3T + 2$ and $L \leq (n - 3T - 2) \log(q) - \omega(\log(\lambda))$ the above scheme satisfies L-leakage resilient tracing security under the DL assumption.*

References

[ADN+10]   Alwen, J., Dodis, Y., Naor, M., Segev, G., Walfish, S., Wichs, D.: Public-key encryption in the bounded-retrieval model. In: Gilbert, H. (ed.) EUROCRYPT 2010. LNCS, vol. 6110, pp. 113–134. Springer, Heidelberg (2010), http://eprint.iacr.org/

[ADW09]   Alwen, J., Dodis, Y., Wichs, D.: Leakage-resilient public-key cryptography in the bounded-retrieval model. In: Halevi, S. (ed.) CRYPTO 2009. LNCS, vol. 5677, pp. 36–54. Springer, Heidelberg (2009)

[AGV09]   Akavia, A., Goldwasser, S., Vaikuntanathan, V.: Simultaneous hardcore bits and cryptography against memory attacks. In: Reingold, O. (ed.) TCC 2009. LNCS, vol. 5444, pp. 474–495. Springer, Heidelberg (2009)

[BCH12]   Bitansky, N., Canetti, R., Halevi, S.: Leakage-tolerant interactive protocols. In: Cramer, R. (ed.) TCC 2012. LNCS, vol. 7194, pp. 266–284. Springer, Heidelberg (2012)

[BF99]   Boneh, D., Franklin, M.K.: An efficient public key traitor scheme (Extended abstract). In: Wiener, M. (ed.) CRYPTO 1999. LNCS, vol. 1666, pp. 338–353. Springer, Heidelberg (1999)

[BG10]   Brakerski, Z., Goldwasser, S.: Circular and leakage resilient public-key encryption under subgroup indistinguishability. In: Rabin, T. (ed.) CRYPTO 2010. LNCS, vol. 6223, pp. 1–20. Springer, Heidelberg (2010)

[BHHO08]   Boneh, D., Halevi, S., Hamburg, M., Ostrovsky, R.: Circular-secure encryption from decision diffie-hellman. In: Wagner, D. (ed.) CRYPTO 2008. LNCS, vol. 5157, pp. 108–125. Springer, Heidelberg (2008)

[BHK11]   Braverman, M., Hassidim, A., Kalai, Y.T.: Leaky pseudo-entropy functions. In: ICS, pp. 353–366 (2011)

[BK12]   Brakerski, Z., Kalai, Y.T.: A parallel repetition theorem for leakage resilience. In: Cramer, R. (ed.) TCC 2012. LNCS, vol. 7194, pp. 248–265. Springer, Heidelberg (2012)

[BKKV10]   Brakerski, Z., Katz, J., Kalai, Y., Vaikuntanathan, V.: Overcomeing the hole in the bucket: Public-key cryptography against resilient to continual memory leakage. In: FOCS [IEE10], pp. 501–510

[BSW11]   Boyle, E., Segev, G., Wichs, D.: Fully leakage-resilient signatures. In: Paterson, K.G. (ed.) EUROCRYPT 2011. LNCS, vol. 6632, pp. 89–108. Springer, Heidelberg (2011)

[CDD+07]   Cash, D.M., Ding, Y.Z., Dodis, Y., Lee, W., Lipton, R.J., Walfish, S.: Intrusion-resilient key exchange in the bounded retrieval model. In: Vadhan, S.P. (ed.) TCC 2007. LNCS, vol. 4392, pp. 479–498. Springer, Heidelberg (2007)

[CDRW10]   Chow, S.S.M., Dodis, Y., Rouselakis, Y., Waters, B.: Practical leakage-resilient identity-based encryption from simple assumptions. In: Al-Shaer, E., Keromytis, A.D., Shmatikov, V. (eds.) ACM Conference on Computer and Communications Security, pp. 152–161. ACM (2010)

[CFN94]   Chor, B., Fiat, A., Naor, M.: Tracing traitors. In: Desmedt, Y.G. (ed.) CRYPTO 1994. LNCS, vol. 839, pp. 257–270. Springer, Heidelberg (1994)

[CG88]   Chor, B., Goldreich, O.: Unbiased bits from sources of weak randomness and probabilistic communication complexity. SIAM Journal on Computing 17(2), 230–261 (1988)

[CS02]   Cramer, R., Shoup, V.: Universal hash proofs and a paradigm for adaptive chosen ciphertext secure public-key encryption. In: Knudsen, L.R. (ed.) EUROCRYPT 2002. LNCS, vol. 2332, pp. 45–64. Springer, Heidelberg (2002)

[DF03]   Dodis, Y., Fazio, N.: Public key trace and revoke scheme secure against adaptive chosen ciphertext attack. In: Desmedt, Y.G. (ed.) PKC 2003. LNCS, vol. 2567, pp. 100–115. Springer, Heidelberg (2002)

[DGK+10]   Dodis, Y., Goldwasser, S., Tauman Kalai, Y., Peikert, C., Vaikuntanathan, V.: Public-key encryption schemes with auxiliary inputs. In: Micciancio, D. (ed.) TCC 2010. LNCS, vol. 5978, pp. 361–381. Springer, Heidelberg (2010)

[DHLW10a]   Dodis, Y., Haralambiev, K., López-Alt, A., Wichs, D.: Cryptography against continuous memory attacks. In: FOCS [IEE10], pp. 511–520

[DHLW10b]   Dodis, Y., Haralambiev, K., López-Alt, A., Wichs, D.: Efficient public-key cryptography in the presence of key leakage. In: Abe, M. (ed.) ASIACRYPT 2010. LNCS, vol. 6477, pp. 613–631. Springer, Heidelberg (2010)

[DKXY02]   Dodis, Y., Katz, J., Xu, S., Yung, M.: Key-insulated public key cryptosystems. In: Knudsen, L.R. (ed.) EUROCRYPT 2002. LNCS, vol. 2332, pp. 65–82. Springer, Heidelberg (2002)

[DLWW11]   Dodis, Y., Lewko, A.B., Waters, B., Wichs, D.: Storing secrets on continually leaky devices. In: FOCS, pp. 688–697 (2011)

[DP08]   Dziembowski, S., Pietrzak, K.: Leakage-resilient cryptography. In: 49th Symposium on Foundations of Computer Science, Philadelphia, PA, USA, October 25–28, pp. 293–302. IEEE Computer Society (2008)

[Dzi06]   Dziembowski, S.: Intrusion-resilience via the bounded-storage model. In: Halevi, S., Rabin, T. (eds.) TCC 2006. LNCS, vol. 3876, pp. 207–224. Springer, Heidelberg (2006)

[GKPV10]   Goldwasser, S., Kalai, Y.T., Peikert, C., Vaikuntanathan, V.: Robustness of the learning with errors assumption. In: Yao, A.C.-C. (ed.) ICS, pp. 230–240. Tsinghua University Press (2010)

[GR12]   Goldwasser, S., Rothblum, G.N.: How to compute in the presence of leakage. Electronic Colloquium on Computational Complexity (ECCC) 19, 10 (2012)

[Hal09]   Maurer, U.: Abstraction in cryptography. In: Halevi, S. (ed.) CRYPTO 2009. LNCS, vol. 5677, pp. 465–465. Springer, Heidelberg (2009)

[HL11]   Halevi, S., Lin, H.: After-the-fact leakage in public-key encryption. In: Ishai, Y. (ed.) TCC 2011. LNCS, vol. 6597, pp. 107–124. Springer, Heidelberg (2011)

[HLWW12]   Hazay, C., Lopez-Alt, A., Wee, H., Wichs, D.: Leakage-resilient cryptography from minimal assumptions. Cryptology ePrint Archive, Report 2012/604 (2012), http://eprint.iacr.org/2012/604

[IEE10]   51th Symposium on Foundations of Computer Science, Las Vegas, NV, USA, October 23–26. IEEE (2010)

[ISW03]   Ishai, Y., Sahai, A., Wagner, D.: Private circuits: Securing hardware against probing attacks. In: Boneh, D. (ed.) CRYPTO 2003. LNCS, vol. 2729, pp. 463–481. Springer, Heidelberg (2003)

[JGS11]   Garg, S., Jain, A., Sahai, A.: Leakage-resilient zero knowledge. In: Rogaway, P. (ed.) CRYPTO 2011. LNCS, vol. 6841, pp. 297–315. Springer, Heidelberg (2011)

[KV09]   Katz, J., Vaikuntanathan, V.: Signature schemes with bounded leakage resilience. In: Matsui, M. (ed.) ASIACRYPT 2009. LNCS, vol. 5912, pp. 703–720. Springer, Heidelberg (2009)

[LLW11]   Lewko, A.B., Lewko, M., Waters, B.: How to leak on key updates. In: STOC (2011)

[LRW11]   Lewko, A., Rouselakis, Y., Waters, B.: Achieving leakage resilience through dual system encryption. In: Ishai, Y. (ed.) TCC 2011. LNCS, vol. 6597, pp. 70–88. Springer, Heidelberg (2011)

[MR04]   Micali, S., Reyzin, L.: Physically observable cryptography. In: Naor, M. (ed.) TCC 2004. LNCS, vol. 2951, pp. 278–296. Springer, Heidelberg (2004)

[MV13]   Miles, E., Viola, E.: Shielding circuits with groups. Electronic Colloquium on Computational Complexity (ECCC) 20, 3 (2013)

[NS09]   Naor, M., Segev, G.: Public-key cryptosystems resilient to key leakage. In: Halevi (ed.) [Hal09], pp. 18–35

[NZ96]   Nisan, N., Zuckerman, D.: Randomness is linear in space. Journal of Computer and System Sciences 52(1), 43–53 (1996)

[Oka92]   Okamoto, T.: Provably secure and practical identification schemes and corresponding signature schemes. In: Brickell, E.F. (ed.) CRYPTO 1992. LNCS, vol. 740, pp. 31–53. Springer, Heidelberg (1993)

[Ped91]   Pedersen, T.P.: A threshold cryptosystem without a trusted party. In: Davies, D.W. (ed.) EUROCRYPT 1991. LNCS, vol. 547, pp. 522–526. Springer, Heidelberg (1991)

[Pie09]   Pietrzak, K.: A leakage-resilient mode of operation. In: Joux, A. (ed.) EUROCRYPT 2009. LNCS, vol. 5479, pp. 462–482. Springer, Heidelberg (2009)

[Rot12]   Rothblum, G.N.: How to compute under $\mathcal{AC}^0$ leakage without secure hardware. In: Safavi-Naini, R., Canetti, R. (eds.) CRYPTO 2012. LNCS, vol. 7417, pp. 552–569. Springer, Heidelberg (2012)

[Wic11]   Wichs, D.: Cryptographic Resilience to Continual Information Leakage. PhD thesis, Department of Computer Science, NYU (2011)

A   Computational Hardness Assumptions

We will rely on discrete-log type hardness assumptions in prime-order groups. We let such groups be defined via an abstract group generation algorithm $(\mathbb{G}, g, q) \xleftarrow{\$} \mathcal{G}(1^\lambda)$, where $\mathbb{G}$ is a (description of a) cyclic group of prime order q with generator g. We assume that the group operation, denoted by multiplication, can be computed efficiently.

Definition 1 (Extended Rank Hiding Assumption). *The extended rank hiding assumption for a group generator $\mathcal{G}$ states that for any integer constants $j > i \in \mathbb{N}$ and $n, m \in \mathbb{N}$ and $t \leq \min\{n, m\} - \max\{i, j\}$, the following two ensembles are computationally indistinguishable:*

$$\left\{ (\mathbb{G}, q, g, g^{\mathbf{X}}, \mathbf{v}_1, \ldots, \mathbf{v}_t) : (\mathbb{G}, q, g) \leftarrow \mathcal{G}(1^\lambda); \ \mathbf{X} \xleftarrow{\$} \mathsf{Rk}_i(\mathbb{F}_q^{n \times m}); \ \{\mathbf{v}_\ell\}_{\ell=1}^t \xleftarrow{\$} \ker(\mathbf{X}) \right\}$$

$$\stackrel{c}{\approx}$$

$$\left\{ (\mathbb{G}, q, g, g^{\mathbf{X}}, \mathbf{v}_1, \ldots, \mathbf{v}_t) : (\mathbb{G}, q, g) \leftarrow \mathcal{G}(1^\lambda); \ \mathbf{X} \xleftarrow{\$} \mathsf{Rk}_j(\mathbb{F}_q^{n \times m}); \ \{\mathbf{v}_\ell\}_{\ell=1}^t \xleftarrow{\$} \ker(\mathbf{X}) \right\}$$

Lemma 2. *The Extended Rank Hiding assumption is equivalent to the DDH assumption.*

Proof is implicit in [BKKV10].

Hardness of finding DL representation outside known span. We will also extensively use the following lemma of Boneh and Franklin, which states that given a number of discrete log representations of a group element h, an adversary cannot generate any other representation that is not in their span.

Lemma 3 ([BF99], Lemma 1). *Let λ be the security parameter and let $(\mathbb{G}, q, g) \xleftarrow{\$} \mathcal{G}(1^\lambda)$. Under the discrete log assumption on the group generator $\mathbb{G}$, for every PPT adversary $\mathcal{A}$ and all integers $d = d(\lambda), n = n(\lambda)$ such that $d < n - 1$, there is a negligible function μ such that*

$$\Pr[(\mathbb{G}, q, g) \leftarrow \mathcal{G}(1^\lambda); \ \boldsymbol{\alpha} \xleftarrow{\$} \mathbb{Z}_q^n; \ \beta \xleftarrow{\$} \mathbb{Z}_q; \ \mathbf{s}_1, \ldots, \mathbf{s}_d \xleftarrow{\$} \mathbb{Z}_q^n \text{ subject to } \langle \boldsymbol{\alpha}, \mathbf{s}_i \rangle = \beta;$$

$$\mathbf{s}^* \leftarrow \mathcal{A}(\mathbb{G}, q, g, g^{\boldsymbol{\alpha}}, g^\beta, \mathbf{s}_1, \ldots, \mathbf{s}_d) : \mathbf{s}^* \notin \mathsf{span}(\mathbf{s}_1, \ldots, \mathbf{s}_d) \text{ and } \langle \boldsymbol{\alpha}, \mathbf{s}^* \rangle = \beta] \leq \mu(\lambda)$$

where the probability is over the coins of $\mathcal{G}$ and the adversary $\mathcal{A}$ and all the random choices made in the experiment.

The above implies that any *valid* representation $\mathbf{s}^*$ that $\mathcal{A}(\mathbb{G}, q, g, g^{\boldsymbol{\alpha}}, g^\beta, \mathbf{s}_1, \ldots, \mathbf{s}_d)$ produces must lie in $\mathsf{span}(\mathbf{s}_1, \ldots, \mathbf{s}_d)$. In particualr, this means that $\mathbf{s}^*$ must be a *convex combination* of $\mathbf{s}_1, \ldots, \mathbf{s}_d$ (with coefficients summing up to 1) since only such combinations give valid representations.

Hiding the Input-Size in Secure Two-Party Computation*

Yehuda Lindell**, Kobbi Nissim***, and Claudio Orlandi†

Abstract. In the setting of secure multiparty computation, a set of parties wish to compute a joint function of their inputs, while preserving properties like *privacy, correctness,* and *independence of inputs.* One security property that has typically not been considered in the past relates to the *length* or *size* of the parties inputs. This is despite the fact that in many cases the size of a party's input can be confidential. The reason for this omission seems to have been the folklore belief that, as with encryption, it is impossible to carry out *non-trivial* secure computation while hiding the size of parties' inputs. However some recent results (e.g., Ishai and Paskin at TCC 2007, Ateniese, De Cristofaro and Tsudik at PKC 2011) showed that it is possible to hide the input size of one of the parties for some limited class of functions, including secure two-party set intersection. This suggests that the folklore belief may not be fully accurate.

In this work, we initiate a theoretical study of *input-size hiding* secure computation, and focus on the two-party case. We present definitions for this task, and deal with the subtleties that arise in the setting where there is no a priori polynomial bound on the parties' input sizes. Our definitional study yields a multitude of classes of input-size hiding computation, depending on whether a single party's input size remains hidden or both parties' input sizes remain hidden, and depending on who receives output and if the output size is hidden from a party in the case that it does not receive output. We prove feasibility and impossibility results for input-size hiding secure two-party computation. Some of the highlights are as follows:

- Under the assumption that fully homomorphic encryption (FHE) exists, there exist non-trivial functions (e.g., the millionaire's problem) that can be securely computed while hiding the input size of *both parties.*
- Under the assumption that FHE exists, *every* function can be securely computed while hiding the input size of one party, when both parties receive output (or when the party not receiving output does learn the size of the output). In the case of functions with fixed output length, this implies that *every* function can be securely computed while hiding one party's input size.

* This work was funded by the European Research Council under the European Union's Seventh Framework Programme (FP/2007-2013) / ERC Grant Agreement n. 239868.
** Bar-Ilan University, Israel. email: `lindell@biu.ac.il`.
*** Ben-Gurion University, Israel. `kobbi@cs.bgu.ac.il`. This work was carried out while at Bar-Ilan University.
† Aarhus University, Denmark. email: `orlandi@cs.au.dk`. This work was carried out while at Bar-Ilan University.

K. Sako and P. Sarkar (Eds.) ASIACRYPT 2013, Part II, LNCS 8270, pp. 421–440, 2013.

– There exist functions that cannot be securely computed while hiding both parties' input sizes. This is the first formal proof that, in general, some information about the size of the parties' inputs must be revealed.

Our results are in the semi-honest model. The problem of input-size hiding is already challenging in this scenario. We discuss the additional difficulties that arise in the malicious setting and leave this extension for future work.

Keywords: Secure two-party computation; input-size hiding.

1 Introduction

Background. Protocols for secure two-party computation enable a pair of parties P_1 and P_2 with private inputs x and y, respectively, to compute a function f of their inputs while preserving a number of security properties. The most central of these properties are *privacy* (meaning that the parties learn the output $f(x, y)$ but nothing else), *correctness* (meaning that the output received is indeed $f(x, y)$ and not something else), and *independence of inputs* (meaning that neither party can choose its input as a function of the other party's input). The standard way of formalizing these security properties is to compare the output of a real protocol execution to an "ideal execution" in which the parties send their inputs to an incorruptible trusted party who computes the output for the parties. Informally speaking, a protocol is then secure if no real adversary attacking the real protocol can do more harm than an ideal adversary (or simulator) who interacts in the ideal model [GMW87, GL90, MR91, Bea91, Can00]. In the 1980s, it was shown that any two-party functionality can be securely computed in the presence of semi-honest and malicious adversaries [Yao86]. Thus, this stringent definition of security can actually be achieved.

Privacy and Size Hiding. Clearly, the security obtained in the ideal model is the most that one can hope for. However, when looking closer at the formalization of this notion, it is apparent that the statement of privacy that "nothing but the output is learned" is somewhat of an overstatement. This is due to the fact that the size of the parties' inputs (and thus also the size of the output) is assumed to be known (see the full version for a discussion on how this is actually formalized in the current definitions). However, this information itself may be confidential. Consider the case of set intersection and companies who wish to see if they have common clients. Needless to say, the number of clients that a company has is itself highly confidential. Thus, the question that arises is whether or not it is possible to achieve secure computation while hiding the size of the parties' inputs. We stress that the fact that input sizes are revealed is not a mere artifact of the definition, and all standard protocols for secure computation indeed assume that the input sizes are publicly known to the parties.

The fact that the input size is always assumed to be revealed is due to the folklore belief that, as with encryption, the length of the parties' inputs cannot be hidden in a secure computation protocol. In particular, the definition in [Gol04, Sec. 7.2.1.1] uses the convention that both inputs are of the same size, and states *"Observe that making no restriction on the relationship among the lengths of the two inputs disallows the existence of secure protocols for computing any non-degenerate functionality. The reason is that the program of each party (in a protocol for computing the desired functionality) must either depend only on the length of the party's input or obtain information on the counterpart's input length. In case information of the latter type is not implied by the output value, a secure protocol cannot afford to give it away"*. In the same way in [HL10, Sec. 2.3] it is stated that *"We remark that some restriction on the input lengths is unavoidable because, as in the case of encryption, to some extent such information is always leaked."*. It is not difficult to see that there exist functions for which hiding the size of both inputs is impossible (although this has not been formally proven prior to this paper). However, this does not necessarily mean that "non-degenerate" or "interesting" functions cannot be securely computed without revealing the size of one or both parties' inputs.

State of the Art. The first work to explicitly refer to hiding input size is that of zero-knowledge sets [MRK03], in which a prover commits to a set S and later proves statements of the form $x \in S$ or $x \notin S$ to a verifier, without revealing anything about the cardinality of S. Zero-knowledge sets are an interesting instance of size-hiding reactive functionality, while in this work we only focus on non-reactive computation (i.e., secure function evaluation).

Ishai and Paskin [IP07] also explicitly refer to the problem of hiding input size, and construct a homomorphic encryption scheme that allows a party to evaluate a branching program on an encrypted input, so that the *length* of the branching program (i.e., the longest path from the initial node to any terminal node) is revealed but nothing else about its size. This enables partial input-size hiding two-party computation by having one party encode its input into the branching program. In particular this implies a secure two-party private set intersection protocol where the size of of the set of one of the two parties is hidden.

Ateniese et al. [ACT11] constructed the first (explicit) protocol for private set-intersection that hides the size of one of the two input sets. The focus of their work is on efficiency and their protocol achieves high efficiency, in the random oracle model. The construction in their paper is secure for semi-honest adversaries, and for a weaker notion of one-sided simulatability when the adversary may be malicious (this notion guarantees privacy, but not correctness, for example). In addition, their construction relies on a random oracle.

Those works demonstrate that interesting, non-degenerate functions *can* be computed while at least hiding the input size of one of the parties, and this raises a number of fascinating and fundamental questions:

Can input-size hiding be formalized in general, and is it possible to securely compute many (or even all) functions while hiding the input size of one of the parties?

Are there any interesting functions that can be securely computed while hiding both parties' inputs sizes?

Before proceeding, we remark that in many cases it is possible to hide the input sizes by using *padding*. However, this requires an a priori upper bound on the sizes of the inputs. In addition, it means that the complexity of the protocol is related to the maximum possible lengths and is thus *inherently inefficient*. Thus, this question is of interest from both a theoretical point of view (is it possible to hide input size when no a priori upper bound on the inputs is known and so its complexity depends only on each party's own input and output), and from a practical point of view. In this paper we focus on theoretical feasibility, and therefore we do not consider side-channel attacks that might be used to learn additional information about a party's input size e.g., by measuring the response time of that party in the protocol, but we hope that our results will stimulate future work on more efficient and practical protocols.

Our Results. In this paper, we initiate the theoretical study of the problem of input-size hiding two-party computation. Our main contributions are as follows:

- *Definition and classification:* Even though some input-size hiding protocols have been presented in the literature, no formal definition of input-size hiding generic secure computation has ever been presented. We provide such a definition and deal with technical subtleties that relate to the fact that no a priori bound on the parties' input sizes is given (e.g., this raises an issue as to how to even define polynomial-time for a party running such a protocol). In addition, we observe that feasibility and infeasibility depend very much on which party receives output, whether or not the output-size is revealed to a party not receiving output, and whether one party's input size is hidden or both. We therefore define a set of classes of input-size hiding variants, and a unified definition of security. We also revisit the standard definition where both parties' input sizes are revealed and observe that the treatment of this case is much more subtle than has been previously observed. (For example, the standard protocols for secure computation are *not* secure under a definition of secure computation for which both parties receive output if their input sizes are equal, and otherwise both parties receive ⊥. We show how this can be easily fixed.)
- *One-party input-size hiding:* We prove that in the case that one party's input size is hidden and the other party's input size is revealed, then *every* function can be securely computed in the presence of semi-honest adversaries, when both parties receive either the output or learn the output size (or when the output size can be upper bounded as a function of one party's input size). This includes the problem of set intersection and thus we show that the result of [ACT11] can be achieved without random oracles and under the full ideal/real simulation definition of security. Our protocols use fully homomorphic encryption [Gen09] (we remark that although this is a very powerful tool, there are subtleties that arise in attempting to use it in our setting). This is the first general feasibility result for input-size hiding.
  We also prove that there exist functionalities (e.g., unbounded input-length oblivious transfer) that *cannot* be securely computed in the presence of semi-honest adversaries while hiding one party's input size, if one of the parties

is not supposed to learn the output size. This is also the first formal impossibility result for input-size hiding, and it also demonstrates that the size of the output is of crucial consideration in our setting. (In the standard definition where input sizes are revealed, a fixed polynomial upper-bound on the output size is always known and can be used.)

- *Two-party input-size hiding:* We prove that there exist functions of interest that can be securely computed in the presence of semi-honest adversaries while hiding the input size of *both* parties. In particular, we show that the greater-than function (a.k.a., the millionaires' problem) can be securely computed while hiding the input size of both parties. In addition, we show that the equality, mean, median, variance and minimum functions can all be computed while hiding the size of both parties' inputs (our positive result holds for any function that can be efficiently computed with polylogarithmic communication complexity). To the best of our knowledge, these are the first examples of non trivial secure computation that hides the size of both parties' inputs, and thus demonstrate that non-degenerate and interesting functions can be securely computed in contradiction to the accepted folklore. We also prove a general impossibility result that it is impossible to hide both parties' input sizes for any function (with fixed output size) with randomized communication complexity $\Omega(n^\varepsilon)$ for some $\varepsilon > 0$. Combined with our positive result, this is an almost complete characterization of feasibility.

- *Separations between size-hiding variants:* We prove separations between different variants of size-hiding secure computation, as described above. This study shows that the issue of size-hiding in secure computation is very delicate, and the question of who receives output and so on has a significant effect on feasibility.

Our results provide a broad picture of feasibility and infeasibility, and demonstrate a rich structure between the different variants of input-size hiding. We believe that our results send a clear message that input-size hiding is possible, and we hope that this will encourage future research to further understand feasibility and infeasibility, and to achieve input-size hiding with practical efficiency, especially in applications where the size of the input is confidential.

Malicious Adversaries – Future Work. In this initial foundational study of the question of size-hiding in secure computation, we mainly focus on the model of semi-honest adversaries. As we will show, many subtleties and difficulties arise already in this setting. In the case of malicious adversaries, it is even more problematic. One specific difficulty that arises in this setting is due to the fact that the simulator must run in time that is polynomial in the adversary. This is a problem since any input-size hiding protocol must have communication complexity that is independent of the parties' inputs sizes. Thus, the simulator must extract the corrupted party's input (in order to send it to the trusted party) even if it is very long, and in particular even if its length is not a priori polynomially bounded in the communication complexity. In order to ensure that the simulator is polynomial in the adversary, it is therefore necessary that the simulator somehow knows how long the adversary would run for. This is a type of "proof of work" for which rigorous solutions do not exist. We remark that we do provide definitions for the case of malicious adversaries. However, the problem of *constructing* input-size hiding protocols for the case of malicious adversaries is left for future work.

2   Technical Overview

In this section we provide a brief overview of the results and the techniques used through the paper. Due to space limitation, much of the technical material has been removed from this version, but can be found in the full version of this article [LNO12].

Definitions. In Section 3 we formalize the notion of input-size hiding in secure two-party computation, following the ideal/real paradigm. As opposed to the standard ideal model, we define the sizes of the input and output values as explicit additional input/outputs of the ideal functionality and, by considering all the combinations of possible output patterns we give a complete classification of ideal functionalities. The different classes can be found in Figure 6 on the last page of this submission. We consider three main classes (class 0,1 and 2) depending on how many input sizes are kept hidden (that is, in class 2 the size of both parties input is kept hidden, in class 1 the size on party's input is kept hidden, and in class 0 neither parties inputs are hidden). Even for class 0, where both input sizes are allowed to leak, we argue that our definition of the ideal world is more natural and general than the standard one. This is due to the fact that in standard definitions, it is assumed that the parties have agreed on the input sizes in some "out of band" method. As we show, this actually leads to surprising problems regarding the definition of security and known protocols. Each of the classes is then divided into subclasses, depending on what kind of information about the output each party receives (each party can learn the output value, the output size or no information about the output). As we will see, the information about the output that is leaked, and to which party, has significant ramifications on feasibility and infeasibility.

The next step on the way to providing a formal definition is to redefine the notion of a protocol that runs in polynomial time.In order to see why this is necessary, observe that there may not exist any single polynomial that bounds the length of the output received by a party, as a function of its input. This is because the length of the output may depend on the length of the other party's input, which can vary. Due to space limitations all formal definitions are deferred to the full version.

Class 1 – Positive and Negative Results. In Section 4.1 we show how every function can be computed while hiding the input size of one party, if both parties are allowed to learn the *size of the output* (or its actual value). The idea behind our protocol is very simple, and uses fully homomorphic encryption (FHE) with circuit privacy: One party encrypts her input x under her public key and sends it to the other party, who then uses the homomorphic properties in order to compute an encryption of the output $f(x, y)$ and sends the encrypted result back. Due to circuit privacy, this does not reveal any information about the length of $|y|$ and therefore size-hiding is achieved. Despite its conceptual simplicity, we observe that one subtle issue arises. Specifically, the second party needs to know the length of the output (or an upper bound on this length) since it needs to construct a circuit computing f on the encrypted x and on y. Of course, given $|x|$ and $|y|$ it is possible to compute such an upper bound, and the ciphertext containing the output can be of this size. Since P_2 knows $|x|$ and y it can clearly compute this bound, but when P_1 receives the encrypted output it would learn the bound which could reveal information about $|y|$. We solve this problem by having the parties first compute the exact size of the output, using FHE. Then,.given this exact size, they proceed as described above.

It turns out that this simple protocol is in fact optimal for class 1 (even though P_2 learns the length of the output $f(x,y)$), since it is in general impossible to hide the size of the input of one party and the size of the output at the same time. In the full version we prove that two natural functions (oblivious transfer with unbounded message length and oblivious pseudorandom-function evaluation) cannot be securely computed in two of the subclasses of class 1 where only one party receives output, and the party not receiving output is not allowed to learn the output size. The intuition is that the size of the transcript of a size-hiding protocol must be independent of the size of one of the inputs (or it will reveal information about it). But, as the length of the output grows with the size of the input, we reach a contradiction with incompressibility of (pseudo)random data.

Class 2 – Positive and Negative Results. In this class, both of the parties' input sizes must remain hidden; as such, this is a much more difficult setting and the protocol described above for class 1 cannot be used. Nevertheless, we present positive results for this class and show that every function that can be computed insecurely using a protocol with *low communication complexity* can be compiled into a size-hiding secure two party protocol. The exact requirements for the underlying (insecure) protocol are given in Definition 3 and the compilation uses FHE and techniques similar to the one discussed for class 1 above. Interesting examples of functions that can be securely computed while hiding the size of both parties input using our technique include statistical computations on data such as computing the mean, variance and median. With some tweaks, known protocols with low communication complexity for equality or the greater-than function can also be turned into protocols satisfying our requirements

As opposed to class 1, we do not have any general positive result for class 2. Indeed, in Theorem 6 we show that there exist functions that *cannot* be securely computed while hiding the input size of both parties. Intuitively, in a size-hiding protocol the communication complexity must be independent of the input sizes and therefore we reach a contradiction with lower-bounds in communication complexity. Examples of interesting functions that cannot be computed in class 2 include the inner product, hamming distance and set intersection functions.

Separations between Classes. In the full version we show that even in class 2, the output size plays an important role. Specifically, we show that there exist functions that can be computed in class 2 only if both parties are allowed to learn the output size. Furthermore we highlight that, perhaps surprisingly, class 2 is not a subset of class 1. That is, there exist functions that cannot be computed in some subclasses of class 1 that can be securely computed in class 2. These results demonstrate that the input-size hiding landscape is rich, as summarized in Table 1 in Section 6.

3   Definitions – Size-Hiding Secure Two-Party Computation

In this section, we formalize the notion of input-size hiding in secure two-party computation. Our formalization follows the ideal/real paradigm for defining security due to [Can00, Gol04]. Thus, we specify the security goals (what is learned by the parties and what is not) by describing appropriate ideal models where the parties send their inputs to an incorruptible trusted party who sends each party exactly what information it is supposed to learn. The information sent to a party can include the *function output* (if it is supposed to receive output), the other party's *input-length* (if it is supposed to learn this), and/or the *length of the function output* (this can make a difference in the case that a party does not learn the actual output). We will define multiple ideal models, covering the different possibilities regarding which party receives which information. As we will see, what is learned and by whom makes a big difference to feasibility. In addition, in different applications it may be important to hide different information (in some client/server "secure set intersection" applications it may be important to hide the size of both input sets, only the size of one the input sets, or it may not be important to hide either). Our definitions are all for the case of *static adversaries*, and so we consider only the setting where one party is honest and the other is corrupted; the identity of the corrupted party is fixed before the protocol execution begins.

The Function and the Ideal Model: We distinguish between the *function* f that the parties wish to compute, and the *ideal model* that describes how the parties and the adversary interact and what information is revealed and how. The ideal model type expresses the security properties that we require from our cryptographic protocol, including which party should learn which output, what information is leaked to the adversary, which party is allowed to learn the output first and so on. In our presentation, we focus on the two-party case only; the extension to the multiparty setting is straightforward.

In the full version, we review the standard way that input sizes are dealt with and observe that there are important subtleties here which are typically ignored. We present the different classes of size-hiding here. The formal definitions of security based on these classes, including the ideal and real model descriptions, and the definitions for security in the presence of semi-honest and malicious adversaries, are deferred to the full version. We stress that a number of technical subtleties do arise when formalizing these notions.

3.1   Classes of Size Hiding

We define three classes of size hiding, differentiated by whether neither party's input size is hidden, one party's input size is hidden or both parties input sizes are hidden (note that the class number describes how many input sizes are kept hidden: 0, 1 or 2):

1. *Class 0:* In this class, the input size of both parties is revealed (See the full version);
2. *Class 1:* In this class, the input size of one party is hidden and the other is revealed. There are a number of variants in this class, depending on whether one or both parties receive output, and in the case that one party receives output depending on whose input size is hidden and whether or not the output size is hidden from the party not receiving output.
3. *Class 2:* In this class, the input size of both parties' inputs are hidden. As in Class 1 there are a number of variants depending on who receives output and if the output size is kept hidden to a party not receiving output.

We now turn to describe the different variants/subclasses to each class. Due to the large number of different subclasses, we only consider the more limited case that when both parties receive output, then they both receive the same output $f(x, y)$. When general feasibility results can be achieved, meaning that any function can be securely computed, then this is without loss of generality [Gol04, Prop. 7.2.11]. However, as we will see, not all classes of input-size hiding yield general feasibility; the study of what happens in such classes when the parties may receive different outputs is left for future work.

Subclass Definitions:

0. *Class 0:* We formalize both the f' and f'' formulations (that can be found in the full version). In both formulations, we consider only the case that both parties receive the function output $f(x, y)$. There is no need to consider the case that only one party receives $f(x, y)$ separately here, since general feasibility results hold and so there is a general reduction from the case that both receive output and only one receives output. In addition, we add a strictly weaker formulation where both parties receive $f(x, y)$ if $|x| = |y|$, and otherwise receive only the input lengths. We include this since the standard protocols for secure computation are actually secure under this formulation. The subclasses are:
    (a) *Class 0.a:* if $|x| = |y|$ then both parties receive $f(x, y)$, and if $|x| \neq |y|$ then both parties receive $\perp$
    (b) *Class 0.b:* if $|x| = |y|$ then both parties receive $f(x, y)$, and if $|x| \neq |y|$ then P_1 receives $1^{|y|}$ and P_2 receives $1^{|x|}$
    (c) *Class 0.c:* P_1 receives $(1^{|y|}, f(x, y))$ and P_2 receives $(1^{|x|}, f(x, y))$
    In the full version, it is shown that every functionality can be securely computed in classes 0.a, 0.b and 0.c.
1. *Class 1:* We consider five different subclasses here. In all subclasses, the input-size $1^{|x|}$ of P_1 is revealed to P_2, but the input-size of P_2 is hidden from P_1. The different subclasses are:

(a) *Class 1.a:* both parties receive $f(x, y)$, and P_2 learns $1^{|x|}$ as well
(b) *Class 1.b:* only P_1 receives $f(x, y)$, and P_2 only learns $1^{|x|}$
(c) *Class 1.c:* only P_1 receives $f(x, y)$, and P_2 learns $1^{|x|}$ and the output length $1^{|f(x,y)|}$
(d) *Class 1.d:* P_1 learns nothing at all, and P_2 receives $1^{|x|}$ and $f(x, y)$
(e) *Class 1.e:* P_1 learns $1^{|f(x,y)|}$ only, and P_2 receives $1^{|x|}$ and $f(x, y)$
2. *Class 2:* We consider three different subclasses here. In all subclasses, no input-sizes are revealed. The different subclasses are:
  (a) *Class 2.a:* both parties receive $f(x, y)$, and nothing else
  (b) *Class 2.b:* only P_1 receives $f(x, y)$, and P_2 learns nothing
  (c) *Class 2.c:* only P_1 receives $f(x, y)$, and P_2 learns the length of the output $1^{|f(x,y)|}$

See Figure 6 (at the last page of this submission) for a graphic description of the above (we recommend referring back to the figure throughout). We stress that the question of whether or not the output length $1^{|f(x,y)|}$ is revealed to a party not receiving $f(x, y)$ is of importance since, unlike in standard secure computation, a party not receiving $f(x, y)$ or the other party's input size cannot compute a bound on $1^{|f(x,y)|}$. Thus, this can make a difference to feasibility. Indeed, as we will see, when $1^{|f(x,y)|}$ is not revealed, it is sometimes impossible to achieve input size-hiding.

When considering *symmetric functions* (where $f(x, y) = f(y, x)$ for all x, y), the above set of subclasses covers *all* possible variants for classes 1 and 2 regarding which parties receive output or output length. This is due to the fact that when the function is symmetric, it is possible to reverse the roles of the parties (e.g., if P_2's input-length is to be revealed to P_1, then by symmetry the parties can just exchange roles in class 1). We focus on symmetric functions in this paper[1].

We remark that P_1's input-length and the output-length are given in unary, when revealed; this is needed to give the simulator enough time to work in the case that one party's input is much shorter than the other party's input and/or the output length.

4   Feasibility Results

4.1   General Constructions for Class 1.a/c/e Input-Size Hiding Protocols

In this section, we prove a general feasibility result that any function f can be securely computed in classes 1.a, 1.c and 1.e (recall that in class 1, the size of

---

[1] The non-symmetric case is not so different with respect to feasibility: e.g., the greater-than function is not symmetric (recall that a function f is symmetric if $f(x, y) = f(y, x)$ for all x, y). Nevertheless, it can be made symmetric by defining $f((x, b_1), (y, b_2))$ to equal $\mathsf{GT}(x, y)$ if $b_1 = 0$ and $b_2 = 1$, and to equal $\mathsf{GT}(y, x)$ if $b_1 = 1$ and $b_2 = 0$, and to equal $(\perp, b_1)$ if $b_1 = b_2$. Since b_1 and b_2 are always revealed, it is possible for the parties to simply exchange these bits, and then to run the protocol for GT in the "appropriate direction", revealing the output as determined by the class. We leave the additional complexity of non-symmetric functions for future work.

P_2's input is hidden from P_1, but the size of P_1's input is revealed to P_2). In Section 5, we will see that such a result cannot be achieved for classes 1.b and 1.d, and so we limit ourselves to classes 1.a/c/e. We begin by proving the result for class 1.c, where P_1 obtains the output $f(x, y)$, and P_2 obtains P_1's input length $1^{|x|}$ and the output length $1^{|f(x,y)|}$, and then show how a general protocol for class 1.c can be used to construct general protocols for classes 1.a and 1.e.

The idea behind our protocol is very simple, and uses fully homomorphic encryption (FHE) with circuit privacy (see the full version for the definition). Party P_1 begins by choosing a key-pair for an FHE scheme, encrypts its input under the public key, and sends the public key and encrypted input to P_2. This ciphertext reveals the input length of P_1, but this is allowed in class 1.c. Next, P_2 computes the function on the encrypted input and its own input, and obtain an encryption of $f(x, y)$. Finally, P_2 sends the result to P_1, who decrypts and obtains the output. Observe that this also reveals the output length to P_2, but again this is allowed in class 1.c.

Despite its conceptual simplicity, we observe that one subtle issue arises. Specifically, party P_2 needs to know the length of the output $f(x, y)$, or an upper bound on this length, since it needs to construct a circuit computing f on the encrypted x and on y. Of course, given $|x|$ and $|y|$ it is possible to compute such an upper bound, and the ciphertext containing the output can be of this size (the actual output length may be shorter, and this can be handled by having the output of the circuit include the actual output length). Since P_2 knows $|x|$ and y it can clearly compute this bound. However, somewhat surprisingly, having P_2 compute the upper bound may actually reveal information about P_2's input size to P_1. In order to see this, consider the set union functionality. Clearly, the output length is upper bounded by the sum of the length of P_1's input and P_2's input, but if P_2 were to use this upper bound then P_1 would be able to learn the length of P_2's input which is not allowed. We solve this problem by having the parties first compute the exact size of the output, using FHE. Then, given this exact size, they proceed as described above. The protocol is presented in Figure 1, and uses an FHE scheme (Gen, Enc, Dec, Eval). We denote by n the length $|x|$ of P_1's input, and by m the length $|y|$ of P_2's input. In addition, we denote $x = x_1, \ldots, x_n$ and $y = y_1, \ldots, y_m$.

Theorem 2. *Let* $f : \{0,1\}^* \times \{0,1\}^* \to \{0,1\}^*$ *be a polynomial-time computable function. If* (Gen, Enc, Dec, Eval) *constitutes a fully homomorphic encryption with circuit privacy, then Protocol 1 securely computes* f *in class 1.c, in the presence of a static semi-honest adversary.*

Proof: Recall that in order to prove security in the presence of semi-honest adversaries, it suffices to present simulators $\mathcal{S}_1$ and $\mathcal{S}_2$ that receive the input/output of parties P_1 and P_2, respectively, and generate their view in the protocol. The requirement is that the joint distribution of the view generated by the simulator and the honest party's output be indistinguishable from the view of the corrupted party and the honest party's output.

We begin with the case that P_1 is corrupted. Simulator $\mathcal{S}_1$ receives $(x, f(x, y))$ and prepares a uniformly distributed random tape for P_1. Then, $\mathcal{S}_1$ uses

---

PROTOCOL 1 (Class 1.c Size-Hiding for Any Functionality – Semi-Honest)

- **Inputs:** P_1 has x, and P_2 has y. Both parties have security parameter 1^κ.
- **The protocol:**
  1. P_1 chooses $(pk, sk) \leftarrow \mathsf{Gen}(1^\kappa)$, computes $c_1 = \mathsf{Enc}_{pk}(x_1), \ldots, c_n = \mathsf{Enc}_{pk}(x_n)$ and sends $(pk, c_1, \ldots, c_n)$ to P_2.
  2. P_2 receives $c_1, \ldots, c_n$, and constructs a circuit $\mathcal{C}_{size,y}(\cdot)$ that computes the output length of $f(\cdot, y)$ in binary (i.e., $\mathcal{C}_{size,y}(x) = |f(x,y)|$), padded with zeroes up to length $\log^2 \kappa$. Then, P_2 computes $c_{size} = \mathsf{Eval}_{pk}(\mathcal{C}_{size,y}, \langle c_1, \ldots, c_n \rangle)$, and sends c_{size} to P_1.
  3. P_1 receives c_{size} and decrypts it using sk; let ℓ be the result. Party P_1 sends ℓ to P_2.
  4. P_2 receives ℓ from P_1 and constructs another circuit $\mathcal{C}_{f,y}(\cdot)$ that computes $f(x,y)$ (i.e., $\mathcal{C}_{f,y}(x) = f(x,y)$), and has ℓ output wires. Then, P_2 computes $c_f = \mathsf{Eval}_{pk}(\mathcal{C}_{f,y}, \langle c_1, \ldots, c_n \rangle)$, and sends c_f to P_1.
  5. P_1 receives c_f and decrypts it using sk to obtain a string z.
- **Outputs:** P_1 outputs the string z obtained in the previous step; P_2 outputs nothing.

---

that random tape to sample $(pk, sk) \leftarrow \mathsf{Gen}(1^\kappa)$. Then, $\mathcal{S}_1$ computes $c_{size} = \mathsf{Enc}_{pk}(|f(x,y)|)$ padded with zeroes up to length $\log^2 \kappa$, and $c_f = \mathsf{Enc}_{pk}(f(x,y))$. Finally, $\mathcal{S}_1$ outputs the input x, the random tape chosen above, and the incoming messages c_{size} and c_f. The only difference between the view generated by $\mathcal{S}_1$ and that of P_1 in a real execution is that c_{size} and c_f are generated by directly encrypting $|f(x,y)|$ and $f(x,y)$, rather than by running Eval. However, the *circuit privacy* requirement guarantees that the distributions over these ciphertexts are statistically close.

Next, consider a corrupted P_2. Simulator $\mathcal{S}_2$ receives $(y, (1^{|x|}, 1^{|f(x,y)|}))$, and generates $(pk, sk) \leftarrow \mathsf{Gen}(1^\kappa)$ and $c_1 = \mathsf{Enc}_{pk}(0), \ldots, c_{|x|} = \mathsf{Enc}_{pk}(0)$. Then, $\mathcal{S}_2$ outputs y, a uniform random tape, and incoming messages $(pk, c_1, \ldots, c_{|x|}, |f(x,y)|)$ as P_2's view. The indistinguishability of the simulated view from a real view follows immediately from the regular encryption security of the fully homomorphic encryption scheme. ∎

Extensions. It is not difficult to see that given protocols for class 1.c, it is possible to obtain protocols for classes 1.a and 1.e (for class 1.a just have P_1 send the output to P_2, and the compute in class 1.e by computing a function in class 1.a that masks the output from P_1 so that only P_2 can actually obtain it). In addition, we show that with the function has a bounded output length (meaning that it is some fixed polynomial in the length of P_1's input), then any function can be securely computed in classes 1.b and 1.e as well. An important application of this is the *private set intersection problem* (observe that the size of the output is upper bounded by the size of P_1's input). We therefore obtain an analog to the result of [ACT11] without relying on random oracles. These extensions appear in the full version.

4.2   Feasibility for Some Functions in Class 2

In this section we prove that some non-trivial functions can be securely computed in class 2. This is of interest since class 2 protocols reveal nothing about either party's input size, beyond what is revealed by the output size. In addition, in class 2.b, nothing at all is revealed to party P_2. We start by presenting protocols for class 2.c and then discuss how these can be extended to class 2.a, and in what cases they can be extended to class 2.b.

There are functionalities that are impossible to securely compute in any subclass of class 2; see Section 5. Thus, the aim here is just to show that some functions can be securely computed; as we will see, there is actually quite a large class of such functions. We leave the question of characterizing exactly what functions can and cannot be computed for future work.

Class 2.c.   We begin by considering class 2.c, where party P_1 receives the output $f(x, y)$ and P_2 receives $1^{|f(x,y)|}$, but nothing else is revealed. Intuitively this is possible for functions that can be computed efficiently by two parties (by an insecure protocol), with communication that can be upper bounded by some fixed polynomial in the security parameter. In such cases, it is possible to construct size-hiding secure protocols by having the parties run the insecure protocol inside fully homomorphic encryption. We formalize what we require from the insecure protocol, as follows.

Definition 3 (size-independent protocols). *Let $f : \{0,1\}^* \times \{0,1\}^* \to \{0,1\}^*$, and let π be a probabilistic protocol. We say that π is* size independent *if it satisfies the following properties:*

- *Correctness: For every pair of polynomials $q_1(\cdot), q_2(\cdot)$ there exists a negligible function μ such that for every $\kappa \in \mathbb{N}$, and all $x \in \{0,1\}^{q_1(\kappa)}, y \in \{0,1\}^{q_2(\kappa)}$: $\Pr[\pi(x, y) \neq f(x, y)] \leq \mu(\kappa)$.*
- *Computation efficiency: There exist polynomial-time interactive probabilistic Turing Machines π_1, π_2 such that for every pair of polynomials $q_1(\cdot), q_2(\cdot)$, all sufficiently large $\kappa \in \mathbb{N}$, and every $x \in \{0,1\}^{q_1(\kappa)}, y \in \{0,1\}^{q_2(\kappa)}$, it holds that $(\pi_1(1^\kappa, x), \pi_2(1^\kappa, y))$ implements $\pi(x, y)$.*
- *Communication efficiency: There exists a polynomial $p(\cdot)$ such that for every pair of polynomials $q_1(\cdot), q_2(\cdot)$, all sufficiently large $\kappa \in \mathbb{N}$, and every $x \in \{0,1\}^{q_1(\kappa)}, y \in \{0,1\}^{q_2(\kappa)}$, the number of rounds and length of every message sent in $\pi(x, y)$ is upper bounded by $p(\kappa)$.*

Observe that by computation and communication efficiency, given x, κ and a random tape r, it is possible to efficiently compute a series of circuits $\mathcal{C}^1_{P_1,\kappa,x,r}, \dots, \mathcal{C}^{p(\kappa)-1}_{P_1,\kappa,x,r}$ that compute the next message function of $\pi_1(1^\kappa, x; r)$ (i.e., the input to the circuit $\mathcal{C}^i_{P_1,\kappa,x,r}$ is a vector of $i - 1$ incoming messages of length $p(\kappa)$ each, and the output is the response of P_1 with input x, security parameter κ, random coins r, and the incoming messages given in the input). Likewise, given y, κ and s, it is possible to efficiently compute analogous $\mathcal{C}^1_{P_2,\kappa,y,s}, \dots, \mathcal{C}^{p(\kappa)}_{P_2,\kappa,y,s}$. We stress that since the length of each message in π

is bounded by $p(\kappa)$, the circuits can be defined with input length as described above. For simplicity, we assume that in each round of the protocol the parties exchange messages that are dependent only on messages received in the previous rounds (this is without loss of generality).

In addition, it is possible to generate a circuit $C^{\text{output}}_{P_1,\kappa,x,r}$ for computing the output of P_1 given its input and all incoming messages. As in Protocol 1, in order to generate $C^{\text{output}}_{P_1,\kappa,x,r}$ we need to know the *exact* output size (recall that using an upper bound may reveal information). Therefore, we also use a circuit $C^{\text{size}}_{P_1,\kappa,x,r}$ that computes the exact output length given all incoming messages; this circuit has output length $\log^2 \kappa$ (and so any polynomial output length can be encoded in binary in this number of bits) and can also be efficiently generated.

Due to lack of space in this abstract, we describe protocol here informally and refer to the full version for a formal description and proof. We start with class 2.c and show that if a function has a size-independent protocol, then we can securely compute the function in class 2.c. In more detail, a size-independent protocol has communication complexity that can be bound by a fixed polynomial $p(\kappa)$, for inputs of any length (actually, of length at most $\kappa^{\log \kappa}$ and so for any a priori unbounded polynomial-length inputs)[2]. Then, we can run this protocol *inside* fully homomorphic encryption; by padding all messages to their upper bound (and likewise the number of messages), we have that nothing is revealed by the size of the ciphertexts sent. We note, however, that unlike in the protocols for class 1, in this case neither party is allowed to know the secret key of the fully homomorphic encryption scheme (since both parties must exchange ciphertexts, as in the communication complexity protocol). This is achieved by using threshold key generation and decryption, which can be obtained using standard secure computation techniques (observe that no size hiding issues arise regarding this). In the full version, we formally prove the following corollary:

Corollary 4. *Let $f : \{0,1\}^* \times \{0,1\}^* \to \{0,1\}^*$ be a function. If there exists a size-independent protocol for computing f, and fully homomorphic encryption schemes exist, then f can be securely computed in classes 2.a and 2.c in the presence of static semi-honest adversaries. Furthermore, if in addition to the above the output-size of f is fixed for all inputs, then f can be securely computed in class 2.b in the presence of static semi-honest adversaries.*

In addition, we show the following applications of the above corollary:

Corollary 5. *Assuming the existence of fully homomorphic encryption, the greater-than, equality, mean, variance and median functions can be securely computed in classes 2.a, 2.b and 2.c, in the presence of static semi-honest adversaries. In addition, the min function can be securely computed in classes 2.a and 2.c, in the presence of static semi-honest adversaries.*

---

[2] Note that upper bounding the input sizes to $\kappa^{\log \kappa}$ is not a real restriction: if the adversary has enough time to read an input of this size, then it has time to break the underlying computational assumption and no secure protocol exists.

5   Negative Results and Separations between Classes

In this section, we deepen our understanding of the feasibility of achieving input-size hiding by proving impossibility results for all classes where general secure computation cannot be achieved (i.e., for classes 1.b, 1.d, 2.a, 2.b and 2.c). In addition, we show that the set of functions computable in class 2.b is a strict subset of the set of functions computable in 2.a and 2.b, and that classes 1.b and 1.d are incomparable (they are not equal and neither is a subset of the other). Finally, we consider the relations between subclasses of class 1 and class 2, and show that class 2.b is a strict subset of class 1.b, but class 2.c is *not* (and so sometimes hiding both parties' inputs is easier than hiding only one party's input).

Due to lack of space, we present only the proof of impossibility for class 2; this provides the flavor of all of our impossibility results. All the other results can be found in the the full version.

5.1   Not All Functions Can Be Securely Computed in Class 2

In this section we show that there exist functions for which it is impossible to achieve input-size hiding in any subclass of class 2 (where neither parties' input sizes are revealed). In order to strengthen the result, we demonstrate this on a function which has *fixed output size*. Thus, the limitation is not due to issues related to revealing the output size (as in class 2.b), but is inherent to the problem of hiding the size of the input from both parties.

The following theorem is based on the communication complexity of a function. Typically, communication complexity is defined for functions of equal sized input. We therefore generalize this definition, and measure the communication complexity of a function, as a function of the smaller of the two inputs. That is, a function f has randomized communication complexity $\Omega(g(n))$ if any probabilistic protocol for computing $f(x, y)$ with negligible error requires the parties to exchange $\Omega(g(n))$ bits, where $n = \min\{|x|, |y|\}$.[3]

Theorem 6. *Let $\mathcal{R}$ be a range of constant size, and let $f : \{0, 1\}^* \times \{0, 1\}^* \to \mathcal{R}$ be a function. If there exists a constant $\varepsilon > 0$ such that the randomized communication complexity of f is $\Omega(n^\varepsilon)$, then f cannot be securely computed in class 2.a, 2.b or 2.c, in the presence of static semi-honest adversaries.*

Proof: The idea behind the proof of the theorem is as follows. On the one hand, if a function has $\Omega(n^\varepsilon)$ communication complexity, then the length of the transcript cannot be independent of the input lengths, and must grow as the

---

[3] Even more formally, we say that a probabilistic protocol π computes f if there exists a negligible function μ such that for every $x, y \in \{0, 1\}^*$ the probability that the output of $\pi(x, y)$ does not equal $f(x, y)$ is at most $\mu(n)$, where $n = \min\{|x|, |y|\}$. Next, we say that f has communication complexity $\Omega(g(n))$ if for every protocol for computing f (as defined above) there exists a constant c and an integer $N \in \mathbb{N}$ such that for every $n > N$, the number of bits sent by the parties is at least $c \cdot g(n)$.

inputs grow. On the other hand, in class 2 the input lengths are never revealed and since the output range is constant, the output says almost nothing about the input lengths. Thus, we can show that the length of the transcript must actually be independent of the input lengths, in contradiction to the assumed communication complexity of the function. We now prove this formally.

Let f be a family of functions as in the theorem statement, and assume by contradiction that there exists a protocol π that securely computes f in class 2.a. (We show impossibility for class 2.a since any protocol for class 2.b or 2.c can be converted into a protocol for class 2.a by simply having P_1 send P_2 the output at the end. Thus, impossibility for class 2.a implies impossibility for classes 2.b and 2.c as well.)

We claim that there exists a polynomial $p(\cdot)$ such that the communication complexity of π is at most $p(\kappa)$. Intuitively, this is due to the fact that the transcript cannot reveal anything about the input size and so must be bound by a fixed polynomial. Proving this formally is a little bit more tricky, and we proceed to do this now. Let $\alpha \in \mathcal{R}$ be an output value, and let $I_\alpha \subseteq \{0,1\}^* \times \{0,1\}^*$ be the set of all string pairs such that for every $(x,y) \in I_\alpha$ it holds that $f(x,y) = \alpha$. Now, by the definition of class 2.a, there exist simulators $\mathcal{S}_1$ and $\mathcal{S}_2$ that generate P_1 and P_2's views from $(x, f(x,y))$ and $(y, f(x,y))$, respectively. Thus, for every $(x,y) \in I_\alpha$, the simulators $\mathcal{S}_1$ and $\mathcal{S}_2$ must simulate given only (x, α) and (y, α), respectively.

Let x be the smallest string for which there exists a y so that $(x,y) \in I_\alpha$, and let $p'(\cdot)$ be the polynomial that bounds the running-time of $\mathcal{S}_1$. Define $p_\alpha(\kappa) = p'(|x| + |\alpha| + \kappa)$; note that this is a polynomial in κ since $|x|$ and $|\alpha|$ are constants. We claim that the polynomial $p_\alpha(\cdot)$ is an upper bound on the length of the transcript for *every* $(x,y) \in I_\alpha$. This follows immediately from the fact that $\mathcal{S}_1$ runs in time that is polynomial in its input plus the security parameter. Thus, it cannot write a transcript longer than this when given input (x, α). If the transcript upon input $(x,y) \in I_\alpha$ is longer than $p_\alpha(\kappa)$ with non-negligible probability, then this yields a trivial distinguisher, in contradiction to the assumed security with simulator $\mathcal{S}_1$.

Repeating the above for every $\alpha \in \mathcal{R}$, we have that there exists a set $P = \{p_\alpha(\kappa)\}_{\alpha \in \mathcal{R}}$ of polynomials so that any function upper bounding these polynomials is an upper bound on the transcript length for *all* inputs $(x,y) \in \{0,1\}^*$. Since $\mathcal{R}$ is of constant size, we have that there exists a single polynomial $p(\kappa)$ that upper bounds all the polynomials in P, for every κ.[4] We conclude that there exists a polynomial $p(\kappa)$ that upper bounds the size of the transcript, for all $(x,y) \in \{0,1\}^*$.

Now, let c be a constant such that $p(\kappa) < \kappa^c$, for all large enough κ. We construct a protocol π' for f as follows. On input $(x,y) \in \{0,1\}^* \times \{0,1\}^*$, execute π with security parameter $\kappa = n^{\varepsilon/2c}$, where $n = \min\{|x|, |y|\}$. By the

---

[4] This argument is not true if $\mathcal{R}$ is not of a constant size. This is because it is then possible that the set of polynomials bounding the transcript sizes is $P = \{n^i\}_{i \in \mathbb{N}}$. Clearly each member of P is a polynomial; yet there is no polynomial that upper bounds all of P.

correctness of π, we have that the output of $\pi(x, y)$ equals $f(x, y)$ except with negligible probability. This implies that the output of $\pi'(x, y)$ also equals $f(x, y)$ except with negligible probability (the only difference is that we need to consider larger inputs (x, y), but in any case correctness only needs to hold for all large enough inputs). Thus, π' computes f; see Footnote 3. The proof is finished by observing that the communication complexity of protocol π' is upper bounded by $p(\kappa) < (n^{\varepsilon/2c})^c = n^{\varepsilon/2}$, in contradiction to the assumed lower bound of $\Omega(n^\varepsilon)$ on the communication complexity of f. ∎

Impossibility. From results on communication complexity [KN97], we have that:

- The inner product function $\mathsf{IP}(x, y) = \sum_{i=1}^{\min(|x|, |y|)} x_i \cdot y_i \mod 2$ has communication complexity $\Omega(n)$.
- The set disjointness function defined by $\mathsf{DISJ}(X, Y) = 1$ if $X \cap Y = \emptyset$, and equals 0 otherwise has communication complexity $\Omega(n)$.[5] This implies that $\mathsf{INTERSECT}(X, Y) = X \cap Y$ also has communication complexity $\Omega(n)$.
- The Hamming distance function $\mathsf{HAM}(x, y) = \sum_{i=1}^{\min(|x|, |y|)} (x_i - y_i)^2$ has communication complexity $\Omega(n)$.

Thus:

Corollary 7. *The inner product, set disjointness, set intersection and Hamming distance functions cannot be securely computed in classes 2.a, 2.b or 2.c, in the presence of static semi-honest adversaries.*

Thus our protocol for set intersection (see the full version) that hides only one party's input size is "optimal" in that it is impossible to hide both parties' input sizes.

We conclude by observing that by combining Corollary 4 and Theorem 6, we obtain an almost complete characterization of the functions with constant output size that can be securely computed in class 2. This is because any function with fixed output length that can be efficiently computed with polylogarithmic communication complexity has a size-independent protocol by Definition 3, and so can be securely computed in all of class 2. We therefore conclude:

Corollary 8. *Let $f : \{0, 1\}^* \times \{0, 1\}^* \to \{0, 1\}^*$ be a function. If f can be efficiently computed with polylogarithmic communication complexity, then it can be securely computed in all of class 2 in the presence of static semi-honest and malicious adversaries, assuming the existence of collision-resistant hash functions and fully homomorphic encryption schemes. In contrast, if there exists an $\varepsilon > 0$ such that the communication complexity of f is $\Omega(n^\varepsilon)$ then f cannot be securely computed in any subclass of class 2.*

The above corollary is not completely tight since f may have communication complexity that is neither polylogarithmic, nor $\Omega(n^\varepsilon)$. In addition, our lower

---

[5] The disjointness function is not symmetric. However, it can be made symmetric using the method described in Footnote 1.

and upper bounds do not hold for functions that can be inefficiently computed with polylogarithmic communication complexity.

Additional Results. In the full version, we prove a series of impossibility results and study the relations between the different classes. Amongst other things, we show that the *oblivious transfer function* with strings of unbounded length *cannot* be securely computed in classes 1.b and 2.b, but *can* be securely computed in classes 2.a,2.c and 1.d (it can be computed in classes 1.a/c/e since all functions can be securely computed in these classes).

6 Summary

Our work provides quite a complete picture of feasibility, at least on the level of in which classes can all functions be securely computed and in which not. In addition, we show separations between many of the subclasses, demonstrating that the input-size hiding landscape is rich. In Table 1 we provide a summary of what functions can and cannot be computed in each class. This is in no terms a full characterization, but rather some examples that demonstrate the feasibility and infeasibility in the classes.

	All f (bounded output)	All f (even unbounded output)	GT $(x > y)$	vecxor	Intersection	OT	omprf
2.a	×	×	✓	✓	×	✓	✓
2.b	×	×	✓	×	×	×	✓
2.c	×	×	✓	✓	×	✓	✓
1.a	✓	✓	✓	✓	✓	✓	✓
1.b	✓	×	✓	✓	✓	×	✓
1.c	✓	✓	✓	✓	✓	✓	✓
1.d	✓	×	✓	✓	✓	✓	×
1.e	✓	✓	✓	✓	✓	✓	✓

Fig. 1. Summary of feasibility

References

[ACT11] Ateniese, G., De Cristofaro, E., Tsudik, G. (If) size matters: Size-hiding private set intersection. In: Catalano, D., Fazio, N., Gennaro, R., Nicolosi, A. (eds.) PKC 2011. LNCS, vol. 6571, pp. 156–173. Springer, Heidelberg (2011)

[Bea91] Beaver, D.: Foundations of secure interactive computing. In: Feigenbaum, J. (ed.) CRYPTO 1991. LNCS, vol. 576, pp. 377–391. Springer, Heidelberg (1992)

[Can00] Canetti, R.: Security and composition of multiparty cryptographic protocols. J. Cryptology 13(1), 143–202 (2000)

[Gen09]   Gentry, C.: Fully homomorphic encryption using ideal lattices. In: STOC, pp. 169–178 (2009)

[GL90]   Goldwasser, S., Levin, L.A.: Fair computation of general functions in presence of immoral majority. In: Menezes, A., Vanstone, S.A. (eds.) CRYPTO 1990. LNCS, vol. 537, pp. 77–93. Springer, Heidelberg (1991)

[GMW87]   Goldreich, O., Micali, S., Wigderson, A.: How to play any mental game or a completeness theorem for protocols with honest majority. In: STOC, pp. 218–229 (1987)

[Gol04]   Goldreich, O.: Foundations of Cryptography. Basic Applications, vol. 2. Cambridge University Press (2004)

[HL10]   Hazay, C., Lindell, Y.: Efficient Secure Two-Party Protocols: Techniques and Constructions. Springer (2010)

[IP07]   Ishai, Y., Paskin, A.: Evaluating branching programs on encrypted data. In: Vadhan, S.P. (ed.) TCC 2007. LNCS, vol. 4392, pp. 575–594. Springer, Heidelberg (2007)

[KN97]   Kushilevitz, E., Nisan, N.: Communication Complexity. Cambridge University Press (1997)

[LNO12]   Lindell, Y., Nissim, K., Orlandi, C.: Hiding the input-size in secure two-party computation. Cryptology ePrint Archive, Report 2012/679 (2012), http://eprint.iacr.org/

[MR91]   Micali, S., Rogaway, P.: Secure computation (abstract). In: Feigenbaum, J. (ed.) CRYPTO 1991. LNCS, vol. 576, pp. 392–404. Springer, Heidelberg (1992)

[MRK03]   Micali, S., Rabin, M.O., Kilian, J.: Zero-knowledge sets. In: FOCS, pp. 80–91. IEEE Computer Society (2003)

[Yao86]   Yao, A.C.-C.: How to generate and exchange secrets (extended abstract). In: FOCS, pp. 162–167 (1986)

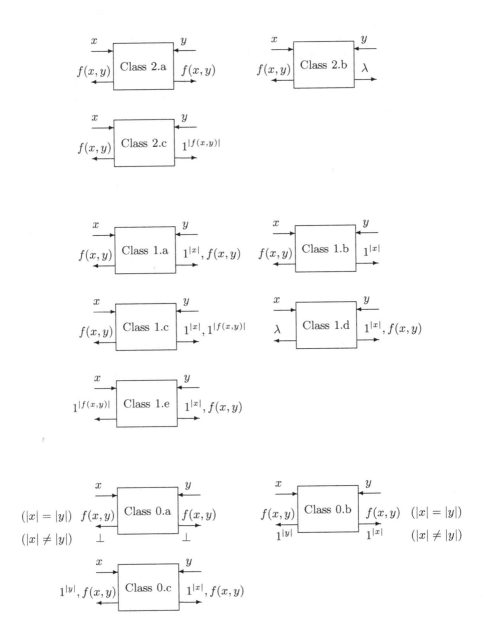

Fig. 2. Classification of Input-Size Hiding Ideal Models

Secure Two-Party Computation with Reusable Bit-Commitments, via a Cut-and-Choose with Forge-and-Lose Technique[*]

(Extended Abstract)

Luís T. A. N. Brandão[1,2,**]

[1] University of Lisbon
Faculty of Sciences / LaSIGE
Lisboa, PORTUGAL
lbrandao@fc.ul.pt

[2] Carnegie Mellon University
Electrical & Computer Engineering
Pittsburgh, USA
lbrandao@cmu.edu

Abstract. A *secure two-party computation* (S2PC) protocol allows two parties to compute over their combined private inputs, as if intermediated by a trusted third party. In the malicious model, this can be achieved with a *cut-and-choose* of *garbled circuits* (C&C-GCs), where some GCs are *verified* for correctness and the remaining are *evaluated* to determine the circuit output. This paper presents a new C&C-GCs-based S2PC protocol, with significant advantages in efficiency and applicability. First, in contrast with prior protocols that require a majority of *evaluated* GCs to be correct, the new protocol only requires that at least one *evaluated* GC is correct. In practice this reduces the total number of GCs to approximately one third, for the same statistical security goal. This is accomplished by augmenting the C&C with a new *forge-and-lose* technique based on bit commitments with trapdoor. Second, the output of the new protocol includes reusable XOR-homomorphic bit commitments of all circuit input and output bits, thereby enabling efficient linkage of several S2PCs in a reactive manner. The protocol has additional interesting characteristics (which may allow new comparison tradeoffs), such as needing a low number of exponentiations, using a 2-out-of-1 type of oblivious transfer, and using the C&C structure to statistically verify the consistency of input wire keys.

Keywords: secure two-party computation, cut-and-choose, garbled circuits, forge-and-lose, homomorphic bit-commitments with trapdoor.

---

[*] The full version is available at the Cryptology ePrint Archive, Report 2013/577.

[**] Support for this research was provided by the Fundação para a Ciência e a Tecnologia (Portuguese Foundation for Science and Technology) through the Carnegie Mellon Portugal Program under Grant SFRH/BD/33770/2009, while the author was a Ph.D student at FCUL-DI (LaSIGE) and CMU-ECE. LaSIGE is supported by FCT through the Multiannual Funding Programme. The author can also be contacted via ltanbrandao@gmail.com.

K. Sako and P. Sarkar (Eds.) ASIACRYPT 2013, Part II, LNCS 8270, pp. 441–463, 2013.

1   Introduction

Secure two-party computation is a general cryptographic functionality that allows two parties to interact as if intermediated by a *trusted third party* [Gol04]. A canonical example is *the millionaire's problem* [Yao82], where two parties find who is the richer of the two, without revealing to the other any additional information about the amounts they own. Applications of secure computation can be envisioned in many cases where mutually distrustful parties can benefit from learning something from their combined data, without sharing their inputs [Kol09]. For example, two parties may evaluate a data mining algorithm over their combined databases, in a privacy-preserving manner [LP02]. On a different example, one party with a private message may obtain a respective message authentication code calculated with a secret key from another party (i.e., a blind MAC) [PSSW09]. This paper considers secure two-party evaluation of Boolean circuits, henceforth denoted "S2PC", which can be used to solve the mentioned examples. Each party begins the interaction with a private input encoded as a bit-string, and a public specification of a Boolean circuit that computes an intended function. Then, the two parties interact so that each party learns only the output of the respective circuit evaluated over both private inputs. Probabilistic functionalities can be implemented by letting the two parties hold additional random bits as part of their inputs.

This paper focuses on the *malicious model*, where parties might maliciously deviate from the protocol specification in a computationally bounded way. Furthermore, within the *standard model* of cryptography, adopted herein, it is assumed that some problems are computationally intractable, such as those related with inverting trapdoor permutations. Security is defined within the ideal/real simulation paradigm [Can00]; i.e., a protocol is said to implement S2PC if it *emulates* an ideal functionality where a trusted third party mediates the communication and computation between the two parties. The trusted party receives the private inputs from both parties, makes the intended computation locally and then delivers the final private outputs to the respective parties.

As a starting point, this paper considers the *cut-and-choose* (C&C) of *garbled circuits* (GCs) approach to achieve S2PC. Here, a circuit constructor party (P_A) builds several GCs (cryptographic versions of the Boolean circuit that computes the intended function), and then the other party, the circuit evaluator (P_B), verifies some GCs for correctness and evaluates the remaining to obtain the information necessary to finally decide a correct circuit output. Recently, this approach has had the best reported efficiency benchmark [KSS12, FN13] for S2PC protocols with a constant number of rounds of communication.

1.1   Contributions

This paper introduces a new *bit commitment* (BitCom) approach and a new evaluation technique, dubbed *forge-and-lose*, and blends them into a C&C approach, to achieve a new C&C-GCs-based S2PC protocol with significant improvements in applicability and efficiency.

Applicability. The new protocol achieves S2PC-with-BitComs, as illustrated in Fig. 1. Specifically, both parties receive random BitComs of all circuit input and output bits, with each party also learning the decommitments of only her respective circuit input and output bits. This is an augmented version of secure circuit evaluation. Given the reusability of BitComs, the protocol can be taken as a building block to achieve other goals, such as reactive linkage of several S2PCs, efficiently and securely linking the input and output bits of one execution with the input bits of subsequent executions. Furthermore, given the XOR-homomorphic properties of these BitComs, a party may use efficient specialized *zero-knowledge proofs* (ZKPs) to prove that her private input bits in one execution satisfy certain non-deterministic polynomially verifiable relations with the private input and output bits of previous executions.[1] In previous C&C-GCs-based solutions, without committed inputs and outputs with homomorphic properties, such general linkage would be conceivable but using more expensive ZKPs of correct behavior.

The main technical description in this paper is focused on a standalone 1-output protocol, where the two parties, P_A and P_B, interact so that only P_B learns a circuit output.[2] In the new BitCom approach, the two possible decommitments of the BitCom of each circuit input or output bit (independent of the number of GCs) are connected to the two keys of the respective input or output wire of each GC, via a new construction dubbed *connector*. P_A commits to these *connectors* and then reveals them partially for *verification* or *evaluation*. This ensures, within the C&C, the correctness of circuit input keys and the privacy of decommitments of BitComs, without requiring additional ZKPs. The BitCom approach enables particularly efficient extensions of this 1-output protocol into 2-output protocols where both parties learn a respective private circuit-output.

Efficiency. The new protocol requires only an optimal minimum number of GCs in the C&C, for a certain soundness guarantee (i.e., for an upper bound on the probability with which a malicious P_A can make P_B accept an incorrect output). Specifically, by only requiring that at least one evaluation GC is correct, the total number of GCs is reduced asymptotically about 3.1 times, in comparison with the previously best known C&C-GCs configuration [SS11] that required a correct majority of evaluation GCs. The significance of this improvement stems from the number of GCs being the source of most significant cost of C&C-GCs-based S2PC protocols, for circuits of practical size. Remark: two different techniques [Lin13, HKE13] developed in concurrent research also just require a single evaluation GC to be correct – a brief comparison is made in §7.1, but the remaining introductory part of this paper only discusses the typical C&C-GCs approach that requires a correct majority of evaluation GCs.

---

[1] For simplicity, "ZKPs" is used hereafter both for ZK *proofs* and for ZK *arguments*.
[2] The "1-output" characterization refers to only one party learning a circuit output, though in rigor the protocol implements a probabilistic 2-output functionality (as both parties receive random BitComs).

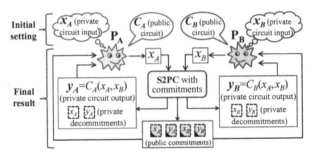

Fig. 1. Secure Two-Party Computation with Committed Inputs and Outputs. Legend: P_A and P_B (the names of the two parties); x_p, y_p and C_p (the private circuit input, the private circuit output and the public circuit specification of party P_p, respectively, with p being A or B); ▨ and ▨ (commitment and decommitment, respectively, of the variable inscribed inside the dashed square). Colors red, blue and purple are related with P_A, P_B and both parties, respectively.

The reduction in number of GCs is achieved via a new forge-and-lose technique, providing a path by which P_B can recover the correct final output when there are inconsistent outputs in the evaluated GCs. Assume that P_A is able to *forge* a GC; i.e., build an incorrect GC that, if selected for *evaluation*, degarbles smoothly into an output that cannot be perceived as incorrect. Then, P_B somehow combines the forged output with a correct output, in a way that reveals a secret key (a trapdoor) with which the input of P_A has previously been encrypted (committed). In this way, P_A *loses* privacy of her input bits, enabling P_B to compute the intended circuit output in the clear.

The protocol can be easily adjusted to integrate several optimizations in communication and memory, such as *random seed checking* [GMS08] and *pipelining* [HEKM11]. Since the garbling scheme is abstracted, the protocol is also compatible with many garbling optimizations, e.g., *point and permute* [NPS99], *XOR for free* [KS08b], *garbled row reduction* [PSSW09], *dual-key cipher* [BHR12].

1.2   Roadmap

The remainder of this paper is organized as follows. Section 2 reviews the basic building blocks of the typical C&C-GCs approach and some properties of BitCom schemes. Section 3 introduces a new BitCom approach, explaining how BitComs can be *connected* to circuit input and output wire keys, to ensure the consistency of the keys across different GCs. Section 4 describes the forge-and-lose technique, achieving a major efficiency improvement over the typical cut-and-choose approach. Section 5 presents the new protocol for 1-output S2PC-with-BitComs, where only one party learns a private circuit-output, and both parties learn BitComs of the input and output bits of both parties. Section 6 comments on the complexity of the protocol and shows how the BitCom approach enables efficient linkage of S2PCs. Section 7 compares some aspects of related work. The full version of this paper includes a more formal description, analysis and optimization of the protocol and a proof of security.

2   Background

2.1   C&C-GCs-Based S2PC

Basic garbed-circuit approach. The theoretical feasibility of S2PC, for functions efficiently representable by Boolean circuits, was initially shown by Yao [Yao86].[3] In the *semi-honest model* (where parties behave correctly during the protocol) simplified to the 1-output setting, only one of the parties (P_B) intends to learn the output of an agreed Boolean circuit that computes the desired function. The *basic GC approach* starts with the other party (P_A) building a GC – a cryptographic version of the Boolean circuit, which evaluates keys (e.g., random bit-strings) instead of clear bits. The GC is a directed acyclic graph of garbled gates, each receiving keys as input and outputting new keys. Each gate output key has a corresponding underlying bit (the result of applying the Boolean gate operation to the bits underlying the corresponding input keys), but the bit correspondence is hidden from P_B. P_A sends the GC and one circuit input key per each input wire to P_B. Then, P_B obliviously evaluates the GC, learning only one key per intermediate wire but not the respective underlying bit. Finally, each circuit output bit is revealed by a special association with the key learned for the respective circuit output wire. Lindell and Pinkas [LP09] prove the security of a version of Yao's protocol (valid for a 2-output setting).

There are many known proposals for garbling schemes [BHR12]. This paper abstracts from specific constructions, except for making the typical assumptions that: (i) with two valid keys per circuit input wire (and possibly some additional randomness used to generate the GC), P_B can *verify the correctness* of the GC, in association with the intended Boolean circuit, and determine the bit underlying each input and output key; and (ii) with a single key per circuit input wire, P_B can *evaluate* the GC, learning the bits corresponding to the obtained circuit output keys, but not learn additional information about the bit underlying the single key obtained for each input wire of P_A and for each intermediate wire.

Oblivious transfer. An essential step of the basic GC-based protocol requires, for each circuit input wire of P_B (the GC-evaluator), that P_A (the GC-constructor) sends to P_B the key corresponding to the respective input bit of P_B, but without P_A learning what is the bit value. This is typically achieved with 1-out-of-2 *oblivious transfers* (OTs) [Rab81, EGL85, NP01], where the sender (P_A) selects two keys per wire, but the receiver (P_B) only learns one of its choice, without the sender learning which one. Some protocols use enhanced variations, e.g., committing OT [CGT95], committed OT [KS06], cut-and-choose OT [LP11], authenticated OT [NNOB12], string-selection OT [KK12]. In practice, the computational cost of OTs is often significant in the overall complexity of protocols, though asymptotically the cost can be amortized with techniques that allow extending a few OTs to a large number of them [Bea96, IKNP03, NNOB12].

---

[3] See [BHR12, §1] for a brief historical account of the origin of the garbled-circuit approach, including references to [GMW87, BMR90, NPS99].

The new protocol presented in this paper uses OTs at the BitCom level, to coordinate decommitments between the two parties, as follows. For each circuit input bit of P_B, P_B selects a bit encoding (a decommitment) and uses it to produce the respective BitCom. Then, P_A uses a trapdoor to learn *two* decommitments (i.e., bit-encodings for the two bits) for the same BitCom. These OTs are herein dubbed *2-out-of-1 OTs*, since one party chooses *one* value and leads the other party to learn *two* values. This is in contrast with the typical 1-out-of-2 OT (commonly used directly at the level of wire keys), where P_A chooses *two* keys and leads P_B to learn *one* of them.

Cut-and-choose approach. Yao's protocol is insecure in the malicious model. For example, a malicious P_A could construct an undetectably incorrect GC, by changing the Boolean operations underlying the garbled gates, but maintaining the correct graph topology of gates and wires. To solve this, Pinkas [Pin03] proposed a C&C approach, achieving 2-output S2PC via a single-path approach where only P_B evaluates GCs. A simplified high level description follows. P_A constructs a set of GCs. P_B *cuts* the set into two complementary subsets and *chooses* one to *verify* the correctness of the respective GCs. If no problem is found, P_B *evaluates* the remaining GCs to obtain, from a consistent majority, its own output bits and a masked version of the output of P_A. P_B sends to P_A a modified version of the masked output of P_A, without revealing from which GC it was obtained. Finally, P_A unmasks her final output bits. This approach has two main inherent challenges: (1) how to *ensure that input wire keys are consistent across GCs, such that equivalent input wires receive keys associated with the same input bits (in at least a majority of evaluated GCs)*; (2) how to *guarantee that the modified masked-output of P_A is correct and does not leak private information of P_B*. Progressive solutions proposed across recent years have solved subtle security issues, e.g., the selective-failure-attack [MF06, KS06], and improved the practical efficiency of C&C-GC-based methods [LP07, Woo07, KS08a, NO09, PSSW09, LP11, SS11]. As a third challenge, the number of GCs still remains a primary source of inefficiency, in these solutions that require a correct majority of GCs selected for evaluation. For example, achieving 40 bits of statistical security[4] requires at least 123 GCs (74 of which are for *verification*). Asymptotically, the optimal C&C partition (three fifths of verification GCs) leads to about 0.322 bits of statistical security per GC [SS11].

The BitCom approach developed in this paper deals with all these challenges. First, taking advantage of XOR-homomorphic BitComs, the verification of consistency of input wire keys of both parties is embedded in the C&C, without an ad-hoc ZKP of consistency of keys across different GCs. Second, P_B can directly learn, from the GC evaluation, decommitments of BitComs of one-time-padded (i.e., masked) output bits of P_A, and then simply send these decommitments to P_A. Privacy is preserved because the decommitments do not vary with the GC

---

[4] The number of bits of statistical security is the additive inverse of the logarithm base 2 of the maximum error probability, i.e., for which a malicious P_A can make P_B accept an incorrect output.

index. Correctness is ensured because the decommitments are verifiable (i.e., authenticated) against the respective BitComs. The BitCom approach also enables achieving 2-output S2PC via a dual-path execution approach – the parties play two 1-output S2PCs, with each party playing once as GC evaluator of only her own intended circuit, using the same BitComs of input bits in both executions.[5] Third, the BitCom approach enables the forge-and-lose technique, which reduces the correctness requirement to only having at least one correct *evaluation* GC, thus increasing the statistical security to about 1 bit per GC.

2.2   Bit Commitments

The BitCom approach introduced in this paper is based on several properties of (some) BitCom schemes, reviewed hereafter. A BitCom scheme [Blu83, BCC88] is a two-party protocol for committing and revealing individual bits. In a *commit* phase, it allows a *sender* to commit to a bit value, by producing and sending a BitCom value to the *receiver*. The BitCom *binds* the *sender* to the chosen bit and, initially, *hides* the bit value from the *receiver*. Then, in a *reveal* phase, the *sender* discloses a private bit-encoding (the decommitment), which allows the *receiver* to learn the committed bit and *verify* its correctness. A scheme is *XOR-homomorphic* if any pair of BitComs can be combined (under some group operation) into a new BitCom that commits the XOR of the original committed bits, and if the same can be done with the respective decommitments.

The following paragraphs describe several properties related with decommitments and trapdoors of practical BitCom schemes. For simplicity, the description focuses on a scheme based on a *square* operation with some useful collision-resistance (i.e., "claw-free" [GMR84, Dam88]) properties.

Unconditionally hiding (UH). A BitCom scheme is called UH if, before the *reveal* phase, a *receiver* with unbounded computational power cannot learn anything about the committed bit. If there is a trapdoor (known by the receiver), then it can be used to retrieve, from any BitCom, respective bit-encodings of both bits. Still, this does not reveal any information about which bit the *sender* might have committed to. A practical instantiation was used by Blum for *coin flipping* [Blu83]. There, in a multiplicative group modulo a Blum integer with factorization unknown by the *sender*, bits 0 and 1 are encoded as group-elements

---

[5] This is a concrete C&C-GCs-based dual-path solution to 2-output S2PC, where the circuits evaluated by each party only compute her respective output. [Kir08, §6.6] and [SS11, §1.2] conceptualized dual-path approaches in high level, but did not explain how to ensure the same input across the two executions. Other dual-path approaches have been proposed using a single GC per party (i.e., not C&C-based), but with potential leakage of one bit of information [MF06, HKE12]. A recent method [HKE13] (see comparison in §7) devised a C&C-based dual-path approach but requiring both parties to evaluate GCs with the same underlying Boolean circuit (for some 2-output functionalities this implies that GCs have the double of the size).

with Jacobi Symbol 1 or -1, respectively.[6] The *commitment* of a bit is achieved by sending the square of a random encoding of the bit. The *revealing* is achieved by sending the known square-root.

Henceforth, a XOR-homomorphic UH BitCom scheme is suggestively dubbed a *2-to-1 square scheme* if it also has the following three useful properties:

- **Proper square-roots.** *Any BitCom (dubbed* square*) has exactly two de-commitments (dubbed* proper square-roots*), encoding different bits.* In the Blum integer example, each square has four square-roots, two per bit, but it is possible to define a single proper square-root per bit (e.g., the square-root whose least significant bit is equal to the encoded bit). The multiplicative group (set of residues and respective multiplication operation) can be easily adjusted to consider only proper square-roots, since the additive inverse of a non-proper square root is a proper square-root encoding the same bit.
- **From trapdoor to decommitments.** *There is a trapdoor whose knowledge allows extracting a pair of proper square-roots (the two decommitments) from any square (the BitCom).* Such pair is dubbed a *non-trivially correlated pair*, in the sense that the two proper square-roots are related but cannot be simultaneously found (except with the help of a trapdoor). This property allows a 2-out-of-1 OT: P_B selects a proper square-root and sends its square to P_A, who then uses the trapdoor to obtain the two proper square-roots. In the Blum integer example, the trapdoor is its factorization.
- **From decommitments to trapdoor.** *Any non-trivially correlated pair is a trapdoor.* This is useful for the forge-and-lose technique, as the discovery (by P_B) of such a pair (a trapdoor of P_A), in case P_A acted maliciously, is the condition that allows P_B to decrypt the input bits of P_A. In the Blum integer example, its factorization can be found from any pair of proper square-roots of the same square.

Unconditionally binding (UB). A BitCom scheme is called UB if a *sender* with unbounded computational power cannot make the *receiver* accept an incorrect bit value in the *reveal* phase. If there is a trapdoor known by some party, then the party can use it to efficiently retrieve (i.e., decrypt) the committed bit from any BitCom value. A practical instantiation is the Goldwasser-Micali probabilistic encryption scheme [GM84], assuming that modulo a Blum integer it is intractable for the *receiver* to decide quadratic residuosity (of residues with Jacobi Symbol 1). A bit 1 or 0 is committed by selecting a random group element and sending its square, or sending the additive inverse of its square, respectively.[7] To decommit 1 or 0, the *sender* reveals the bit and the respective random group element, letting the *receiver* verify that its square or additive-inverse of the

---

[6] A Blum integer is the product of two prime powers, where each prime is congruent with 3 modulo 4, and each power has an odd exponent. For a fixed Blum integer, the Jacobi Symbol is a completely multiplicative function that maps any group element into 1 or -1 (more detailed theory can be found, for example, in [NZM91]).

[7] The additive inverse of a square is necessarily a non-quadratic residue with Jacobi Symbol 1, modulo a Blum integer, because -1 has the same property.

square, respectively, is equal to the BitCom value. The factorization of the Blum integer is a trapdoor that enables efficient decision of quadratic residuosity.

Remark. The basis of the forge-and-lose technique (§4) is a combination of UB and UH BitCom schemes, with the *sender* in the UB scheme being the *receiver* in the UH scheme, and knowing a common trapdoor for both schemes. For the Blum integer examples, and assuming intractability of deciding quadratic residuosity (without a trapdoor), this would mean using the same Blum integer in both schemes, with its factorization as trapdoor. There are known protocols to prove correctness of a Blum integer (e.g., [vdGP88]).

The two exemplified schemes are XOR-homomorphic under modular multiplication. For the purpose of the new S2PC-with-BitComs protocol (§5), this homomorphism is useful in enabling efficient ZKPs *of knowledge* (ZKPoKs) related with committed bits, and efficient negotiation of random bit-encodings and respective BitComs (emulating an ideal functionality where the *trusted third party* would select the BitComs randomly). The property is also useful for linking several S2PC executions, via ZKPs about relations between the input bits of one execution and the input and output bits of previous executions (§6).

3 The BitCom Approach

This section introduces a BitCom approach that combines a BitCom setting (where there is a BitCom for each circuit input and output bit) and a C&C structure (where there are several GCs, each with two keys for each input and output wire). In this approach, based on the XOR-homomorphism of UH BitComs, the consistency of input and output wire keys across different GCs is *statistically* ensured within the C&C, rather than using a ZKP of consistency.[8]

3.1 Cut-and-Choose Stages

The S2PC-with-BitComs protocol to be defined in this paper is built on top of a C&C approach with a COMMIT-CHALLENGE-RESPOND-VERIFY-EVALUATE structure. In a COMMIT stage, P_A builds and sends several GCs, as well as complementary elements (dubbed *connectors*) related with BitComs and with the circuit input and output wire keys of GCs. At this stage, P_A does not yet reveal the circuit input keys that allow the evaluation of each GC. Then, in the CHALLENGE stage, P_A and P_B jointly decide a random partition of the set of GCs into two subsets, one for *verification* and the other for *evaluation*. Possibly, the subsets may be conditioned to a predefined restriction about their sizes (e.g., a fixed proportion of *verification* vs. *evaluation* GCs, or simply not letting the number of *evaluation* GCs exceed some value). In the subsequent RESPOND stage, P_A sends to P_B the elements that allow P_B to *fully verify* the correctness of the GCs selected for *verification*, to *partially verify* the connectors of

---

[8] The protocol still includes several efficient ZKPs related with BitComs, but they are not about the consistency of wire keys across different GCs.

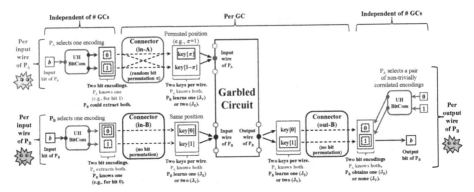

Fig. 2. Connectors. Legend: P_A (GC *constructor*); P_B (GC *evaluator*); J_V and J_E (subsets of *verification* and *evaluation* GC indices, respectively); $\boxed{c}$ (group-element encoding bit c); key$[c]$ (wire key with underlying bit c).

all the GCs (in different ways, depending on whether they are associated with *verification* or *evaluation* challenges), and to *evaluate* the GCs (and respective connectors) selected for *evaluation*. In the VERIFY stage, if any verification step fails, then P_B aborts the protocol execution; otherwise, P_B establishes that there is an overwhelming probability that at least one GC (and respective connectors) selected for evaluation is correct. P_B finally proceeds to an EVALUATE stage, evaluating the *evaluation* GCs and respective connectors, and using their results to determine the final circuit output bits and respective decommitments of output BitComs. Notice that between the VERIFY and EVALUATE stages there is no *response* stage that could let P_A misbehave.

3.2   Connectors

This section develops the idea of *connectors* – structures used to sustain the integration between BitComs and the C&C structure. They are built on top of a setup where one initial UH-BitCom has been defined for each input and output wire of each party, independently of the number of GCs. Then, for each input and output wire in each GC, a connector is built to provide a (statistically verifiable) connection between the two BitCom decommitments and the respective pair of wire keys. The functionality of connectors varies with the type of wire they refer to (input of P_A, input of P_B, output of P_B), as illustrated in high level in Fig. 2.

Connectors are used in a type of commitment scheme (i.e., with *commit* and *reveal* phases) that takes advantage of the C&C substrate. First, each connector is committed in the C&C COMMIT stage, hiding the respective two wire keys, but binding P_A to them and to their relation with BitCom decommitments. Then, each connector is partially revealed during the C&C RESPOND stage, in one of two possible complementary modes: a *reveal for verification*, related with *verification* GCs; or a *reveal for evaluation*, related with *evaluation* GCs. All verifications associated with these two reveal modes are performed in the C&C VERIFY stage, when P_B can still, immune to selective failure attacks, complain and abort in case it finds something wrong. P_A never executes simultaneously the two reveal modes for the same wire of the same GC, because such action would reveal the input bits (in case of wires of P_A) or both BitCom decommitments (i.e., the trapdoor of P_A, in case of wires of P_B). Nonetheless, since the commitment to the *connector* binds P_A to the answers that it can give in each type of reveal phase, an incorrect connector can pass undetectably at most through one type of reveal mode. Thus, within the C&C approach, there is a negligible probability that P_A manages to build incorrect connectors for all evaluation indices and go by undetected. The specific constructions follow:

For each input wire of P_A:

- **Commit.** P_A selects a random permutation bit and a respective random encoding (a group-element dubbed *multiplier*) using the same 2-to-1 square scheme used to commit the input bits of P_A. P_A uses the homomorphic group operation to obtain a new encoding (dubbed *inner encoding*) that encodes the permuted version of her input bit, and sends its square (a new inner UH BitCom) to P_B. P_A then builds a commitment of each of the two wire input keys (using some other commitment scheme), one for bit 0 and the other for bit 1, and sends them to P_B in the form of a pair with the respective permuted order.
- **Reveal for verification.** P_A decommits the two wire input keys (using the *reveal* phase of the respective commitment scheme), and decommits the permutation bit (by revealing the multiplier). P_B uses the two wire input keys (obtained for all input wires) to verify the correctness of the GC and simultaneously obtain the underlying bit of each input key. Then, P_B verifies that the ordering of the bits underlying the pair of revealed input keys is consistent with the decommitted permutation bit.
- **Reveal for evaluation.** P_A decommits the input key that corresponds to her input bit, and decommits the permuted input bit (by revealing the inner encoding), thus allowing P_B to verify that it is consistent with the position of the opened key commitment. As the value of the permuted bit is independent of the real input bit, nothing is revealed about the bit underlying the opened key. If P_A would instead reveal the other key, P_B would detect the cheating in a time when it is still safe to abort the execution and complain.

For each input wire of P_B:

- **Commit.** P_A selects a pair of random encodings of bit 0 (dubbed *multipliers*) and composes them homomorphically with the two known decommitments of the original input BitCom of P_B (which P_A has extracted using the trapdoor), thus obtaining two new independent encodings (dubbed *inner encodings*, one for bit 0 and one for bit 1). P_A then sends to P_B the respective squares (dubbed *inner squares*). For simplicity, it is assumed here that the inner encodings can be directly used as input wire keys of the GC (the full version of this paper shows how to relax this assumption).
- **Reveal for verification.** P_A reveals the two inner encodings. P_B verifies that they are the proper square-roots of the received inner squares, and that

they encode bits 0 and 1, respectively. Then, P_A uses them as the circuit input keys in the GC verification procedure, verifying their correctness. A crucial point is that the two inner encodings are proper square-roots of independent BitComs and thus do not constitute a trapdoor.

- **Reveal for evaluation.** P_A reveals the two multipliers. P_B verifies that both encode bit 0, and homomorphically verifies that they are correct (their squares lead the original BitCom into the two received inner squares). Since P_B knows one (and only one) decommitment of the input BitCom, it can multiply it with the respective multiplier to learn the respective inner encoding and use it as an input wire key. This procedure is resilient to selective failure attack, because both multipliers are verified for correctness, and because the two inner encodings (of which P_B only learns one) are statistically correct input keys (i.e., they would be detected as incorrect if they had been associated with a *verification* GC).

For each output wire of P_B: The construction is essentially symmetric to the case of input wires of P_B. Again for simplicity, it is assumed here that the output keys can directly be group-elements (dubbed *inner encodings*) that are proper square-roots of independent squares. The underlying bit of each output key is thus the bit encoded by it (in the role of inner encoding). P_A commits by initially sending the two inner squares to P_B. Then, for *verification* challenges, from the GC verification procedure P_B learns 2 keys and respective underlying bits. P_B can verify that they are respective proper square-roots of the inner squares and that they encode the respective bits. For *evaluation* challenges, P_A sends only the two multipliers, and P_B verifies homomorphically that they are correct. Then, P_B learns one output key from the GC evaluation procedure, which is an inner encoding, and uses the respective multiplier to obtain the respective decommitment of the output BitCom.

The overall construction requires a number of group elements (multipliers and inner encodings) proportional to the number of input and output wires, but independent of the number of intermediate wires in the circuit.

4   The Forge-and-Lose Technique

This section introduces a new technique, dubbed *forge-and-lose*, to improve the typical C&C-GCs-based approach, by using the BitCom approach to provide a new path for successful computation of final circuit output. More precisely, if in the EVALUATE stage there is at least one GC and respective connectors leading to a correct output (i.e., decommitments of the UH BitComs, for the correct circuit output bits), and if a malicious P_A^* successfully *forges* some other output, then P_A^* *loses* the privacy of her input bits to P_B, allowing P_B to directly use a Boolean circuit to compute the intended output. This loss of privacy is not a violation of security, but rather a disincentive against malicious behavior by P_A^*.

The forge-and-lose path significantly reduces the probabilistic gap available for malicious behavior by P_A that might lead P_B to accept an incorrect output. The technique provides up to 1 bit of statistical security per GC, which constitutes an improvement factor of about 3.1 (either in reduction of number of GCs or in increase of number of bits of statistical security) in comparison with C&C-GCs that require a majority of correct *evaluation* GCs. As noted by Lindell [Lin13], in this setting the optimal C&C partition corresponds to an independent selection of verification and evaluation challenges. Still, for some efficiency tradeoffs it may be preferable to impose some restrictions on the number of *verification* and *evaluation* challenges (e.g., ensure that there are more *verification* than *evaluation* challenges). The full version of this paper shows the error probabilities associated with different C&C partition methods.

The forge-and-lose technique is illustrated in high level in Fig. 3. It can be merged into the C&C and BitCom approach as follows:

- **Encryption scheme.** P_A encrypts her own input bits using as key the trapdoor (known by P_A) of the UH-BitCom scheme used (by P_A) to produce BitComs of the output bits of P_B. Then, P_A gives a ZKP that her encrypted input is the same as that used in the S2PC protocol, i.e., the one committed by P_A with an UH-BitCom scheme with trapdoor known by P_B. If both schemes are XOR-homomorphic (see practical example in §2.2), the ZKP can be achieved efficiently with standard techniques, namely with a statistical combination across input wires, requiring communication linear with the statistical security parameter.

- **Forge-and-lose evaluation.** In the EVALUATE stage, if a *connector* leads an output key to an invalid decommitment, then the respective GC is ignored altogether. If for the remaining GCs all connectors lead to consistent decommitments across all GCs, i.e., if for each output wire index the same valid bit-encoding (proper square-root of the output BitCom) is obtained, then P_B accepts them as correct. However, if P_A acted maliciously, there may be a forged GC and connector leading to a valid (verifiable) decommitment that is different from the decommitment obtained from another correct GC and connector, for the same output wire index. If P_B obtains any such pair of decommitments, i.e., a non-trivially correlated pair of square-roots of the same square, then P_B gets the trapdoor with which P_A encrypted her input, and follows to decrypt the input bits of P_A and use them directly to compute the correct final circuit output in the clear.

5   Protocol for 1-Output S2PC-with-BitComs

This section describes the new C&C-GCs-based protocol for 1-output S2PC-with-BitComs, enhanced with a forge-and-lose technique. The BitComs are XOR-homomorphic, so the mentioned ZKPoKs are efficient using standard techniques.

0. SETUP. The parties agree on the protocol goal, namely on a specification of a Boolean circuit whose evaluation result is to be learned privately by P_B, on the necessary security parameters, on a C&C partitioning method, and on the necessary sub-protocols. Each party selects a *2-to-1 square* scheme, and proposes it to the other party, without revealing the trapdoor but giving a respective ZKPoK that proves the correctness of the public parameters.

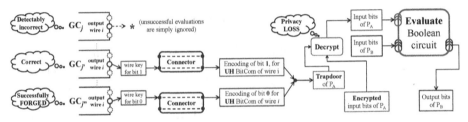

Fig. 3. Forge-and-lose. *Evaluation* path followed by P_B, the evaluator of *garbled circuits* (GCs), if different GCs built by a malicious P_A and selected for *evaluation* (e.g., with indices j', j'') lead to valid but different decommitments of the same *unconditionally hiding* (UH) BitCom (e.g., with index i).

1. PRODUCE INITIAL BITCOMS.
    (a) UH COMMIT INPUT BITS. Each party selects an initial UH BitCom for each of its own circuit input bits, using the 2-to-1 square scheme with trapdoor known by the other party, and sends it to the other party. P_B gives a ZKPoK of a valid decommitment of the respective BitComs.
    (b) UB COMMIT INPUT BITS OF P_A. P_A commits again to each of her input bits, now using an UB-BitCom scheme with trapdoor equal to the trapdoor (known by P_A) of the UH-BitCom scheme used by P_B to commit the input bits of P_B. P_A gives a ZKPoK of equivalent decommitments between the UH BitComs of the input of P_A (with trapdoor known by P_B) and the UB BitComs of the input of P_A (with trapdoor known by P_A), i.e., a proof that the known decommitments encode the same bits.
    (c) UH COMMIT OUTPUT BITS OF P_B. For each output wire index of P_B, P_A selects a random encoding of bit 0 (using the UH BitCom scheme with trapdoor known by P_A) and sends its square to P_B. (P_B will find a respective decommitment only later, in the EVALUATE stage.)
2. COMMIT. P_A uses her trapdoor to extract a non-trivially correlated pair of proper square-roots from each UH BitCom of the input bits (this is the so called 2-out-of-1 OT, which replaces the typical 1-out-of-2 OT used in other S2PC protocols) and output bits of P_B. Then, P_A builds several GCs (in number consistent with the agreed parameters) and respective *connectors* to each input and output wire, and sends the GCs and commitments to the connectors (as specified in §3.2) to P_B.
3. CHALLENGE. The two parties use a coin-tossing sub-protocol to determine a random challenge bit for each GC, conditioned to the agreed C&C method (e.g., same number of challenges of each type, or more verification than evaluation challenges, or independent selection).[9]
4. DECIDE UH-BITCOM PERMUTATIONS. In order to emulate a *trusted third party* deciding the UH BitCom of each circuit input and output bit, both

---

[9] The standalone coin-tossing does not need to be fully simulatable, but the proof of security takes advantage of the ability of the simulated P_A (with rewinding access to a possibly malicious P_B^*) to decide the outcome of the coin-toss. Subtle alternatives would be possible, depending on some changes related with the remaining stages.

parties interact in a fully-simulatable coin-tossing sub-protocol to decide a random encoding of bit 0 for each wire index.[10] Later, each party will locally use these encodings to permute the encodings of her respective private bits, and use the square of the encodings to permute the respective UH BitComs of both parties. Given the XOR-homomorphism, the initial and the final UH BitComs commit to the same bits.

5. RESPOND. For each C&C challenge bit, P_A makes either the *reveal for verification* or the *reveal for evaluation* of the connectors, as specified in §3.2.

6. VERIFY. For *verification* indices, P_B obtains two keys per input wire, verifies the correctness of the GC and makes the respective partial verification of connectors (without learning the decommitments of the BitComs of output bits of P_B). For *evaluation* indices, P_B makes the respective partial verification of the connectors and obtains one key per input wire. If something is found wrong, P_B aborts and outputs FAIL.

7. EVALUATE. For each *evaluation* index, P_B uses the one key per input wire to evaluate the GC, obtain one key per output wire and use the respective revealed part of the connector (namely, one of the two received multipliers) to obtain a decommitment (bit encoding) of the respective output BitCom. There is an overwhelming probability that there is at least one *evaluation* GC whose connectors lead to valid decommitments in all output wires. If all obtained valid decommitments are consistent across different GCs, then P_B accepts them as correct. Otherwise, P_B proceeds into the forge-and-lose path as follows. It finds a non-trivially correlated pair of square-roots and uses it as a trapdoor to decrypt the input bits of P_A, from the respective UB BitComs. In possession of the input bits of both parties, P_B directly evaluates the final circuit output. Then, from within the decommitments already obtained from the *evaluation* connectors, P_B finds the output bit encodings that are consistent with the circuit output bits, and accepts them as the correct ones. This marks the end of the forge-and-lose path.

8. APPLY BITCOM PERMUTATIONS. Each party applies the previously decided random permutations to the encodings of the respective circuit input and output bits, and applies the square of the random encodings as permutations to the UH BitComs of the circuit input and output bits of both parties.

9. FINAL OUTPUT. Each party privately outputs her circuit input and output bits and the respective final encodings, and also outputs the (commonly known) final UH BitComs of the circuit input and output bits of both parties. P_A outputs even if P_B aborts at any time after the APPLY BITCOM PERMUTATIONS stage.

---

[10] To achieve simulability of the overall protocol under each possible malicious party (P_A^* and P_B^*), the simulator of this coin-tossing needs to be able to induce the final BitComs in the real world to be equal to those decided by the trusted third party in the ideal world, and at the same time deal with a probabilistic possibility of abort dependent on those final BitCom values (e.g., see [Lin03]).

Remark. When using the 1-output protocol within larger protocols, care needs to be taken so that P_A cannot distinguish between P_B having learned his output via the normal evaluation path vs. via the forge-and-lose path.

6   Discussion

6.1   Complexity

Besides the computation and communication related with (the reduced number of) GCs, the new S2PC-with-BitComs protocol requires instantiating the connectors (which brings a cost proportional to the number of input and output wires, multiplied by the number of GCs), performing ZKPoKs related with BitComs and to prove correctness of the BitCom scheme parameters, and performing secure two-party coin-tossing (which is significant for the decision of random BitComs values). Based on the XOR-homomorphism, the ZKPoKs related with input wires can be parallelized efficiently with standard techniques, with a communication cost linear in a statistical parameter but independent of the number of input wires, though with computational cost proportional to the product of the statistical parameter and the number of input wires.

With an instantiation based on Blum integers, the inversion of an UH BitCom using the trapdoor (i.e., computing a modular square-root) is approximately computationally equivalent to one exponentiation modulo each prime factor. Thus, besides proving correctness of the Blum integer (which can be achieved with a number of exponentiations that is linear in the statistical parameter), and performing a fully-simulatable coin-tossing sub-protocol to decide random BitCom permutations (which can be instantiated with a number of exponentiations that is linear in the number of input and output wires, and performed in a group of smaller order), the 1-output S2PC-with-BitComs protocol only requires a number of exponentiations that is linear in the number of input wires of P_B, and only computed by P_A. This is in contrast with other protocols whose required number of exponentiations by both parties is proportional to the number of GCs multiplied by the number of input wires (e.g., [LP11]), though in compensation those exponentiations are supported in groups with smaller moduli length and sub-groups of smaller order.

The protocol can be optimized in several ways. For example, with a *random seed checking* (RSC) technique [GMS08] the communication of elements (including GCs and connectors) associated with verification challenges can be replaced by the sending and verification of small random seeds (used to pseudo-randomly generate the elements) and a commitment (to the elements). The technique can be applied independently to GCs and connectors, and can also be used to reduce some of the communication corresponding to connectors associated with *evaluation* challenges. As another example, some group elements used in connectors of P_A can be reduced in size, since their binding properties only need to hold during the execution of the protocol.

Concrete results. An analytic estimation of communication complexity is made in the full version of this paper (ignoring overheads due to communication protocols), for two different circuits: an AES-128 circuit with 6,800 multiplicative gates [Bri13] and 128 wires for the input of each party and for the output of P_B; and a SHA-256 circuit with 90,825 multiplicative gates [Bri13] and 256 wires for the input of each party and output of P_B.

An interesting metric is the proportional overhead of communicated elements beyond GCs (i.e., connectors, BitComs and associated proofs) in comparison with the size occupied only by the GCs. For 128 bits of cryptographic security, instantiated with 3,072-bit Blum integers [BBB+12], and 40 bits of statistical security achieved using 41 GCs of which at most 20 are for evaluation, the estimated overhead is about 55% and 8%, for the AES-128 and SHA-256 circuits, respectively, without the RSC technique applied to the GCs. This metric gives an intuition about the communication cost inherent to the BitCom approach, but is not good enough on its own. For example, when applying the RSC technique also at the level of GCs, the overall communication is reduced significantly, but (because the size corresponding to GCs is reduced) the proportional overhead increases to 158% and 23%, respectively. Nonetheless, even these overheads are low when compared to the cost associated with the additional GCs needed in a C&C that requires a majority of correct evaluation GCs (i.e., on its own an overhead of about 200%, and asymptotically up to about 210%). Clearly, the proportional overhead decreases with the ratio given by the number of input and output wires divided the number of multiplicative gates.

There are other optimizations and C&C configurations that reduce the communication even more, with tradeoffs with computational complexity. For example, by restricting the number of *evaluation* GCs to be at most 8, but increasing the overall number of GCs to 123 (this was the minimal number of GCs required by the typical C&C to achieve 40 bits of statistical security), the estimated communication complexity is approximately of the order of 62 million bits and 418 million bits, respectively for the exemplified circuits. A *pipelining* technique [HEKM11] could also be considered, such that the garbled-gates are not all stored in memory at the same time. This would increase the computation by P_A, but not affect the amount of communicated elements.

6.2   Linked Executions

A simple example of linked executions is the mentioned dual-path execution approach, where each party reuses the same input bits (and BitComs) in two different executions. Furthermore, it may be useful to achieve more general linkage, such as proving that the private input bits of a S2PC satisfy certain *non-deterministic polynomial* verifiable relations with the private input and output bits of previous S2PCs. Based on the XOR-homomorphism of BitComs, this can be proven with efficient ZKPs. For example, proving that a certain BitCom commits to the NAND of the bits committed by two other BitComs can be reduced to a simple ZKP that there are at least two 1's committed in a triplet of

BitComs, with the triplet being built from a XOR-homomorphic combination of the original three BitComs.[11]

For example, since Boolean circuits can be implemented with NAND gates alone, it is possible to prove, outside of the GCs, those transformations and relations that involve only the bits of one party. For example, for protocols defined as a recursion of small GC-based S2PC sub-protocols in the semi-honest model (e.g., [LP02]), security can be enhanced to resist also the malicious model, by simply (1) replacing each GC with a C&C-GCs with BitComs, and (2) by naturally using the input and output of previous executions (or transformations thereof) as the input of the subsequent executions.

6.3   Security

The protocol can be proven secure in the plain model (i.e., without hybrid access to ideal functionalities), assuming the simulator has black-box access with rewindable capability to a real adversary. The simulator is able to extract the input of the malicious party in the real world from the respective ZKPoKs of decommitments, and thus hand it over to the *trusted third party* in the ideal world. The two-party coin tossing used to select random permutations of group-elements needs to be fully-simulatable, because the final BitComs and decommitments are also part of the final output of honest parties. Subtle changes are needed to the ideal functionality when the protocol is adjusted to the 2-output case where each party learns a private circuit output. Achieving security in the *universal composability* model [CLOS02] is left for future work.

7   Related Work

7.1   Two Other Optimal C&C-GCs

Two recently proposed C&C-GCs-based protocols [Lin13, HKE13] also minimize the number of GCs, requiring only that at least one evaluation GC is correct.

Lindell [Lin13] enhances a typical C&C-GCs-based protocol by introducing a second C&C-GCs, dubbed *secure-evaluation-of-cheating* (SEOC), where P_B recovers the input of P_A in case P_B can provide two different garbled output values from the first C&C-GCs. The concept of input-recovery resembles the forge-and-lose technique, but the methods are quite different. For example, the SEOC phase requires interaction between the parties after the first GC evaluation phase, whereas in the forge-and-lose the input-recovery occurs offline.

Huang, Katz and Evans [HKE13] propose a method that combines the C&C-GCs approach with a verifiable secret sharing scheme (VSSS). The parties play different roles in two symmetric C&C-GCs, and then securely compare their outputs. This requires the double of GCs, but in parallel across the two parties. By

---

[11] The first bit is the NAND of the two last if and only if there are at least two 1's in the triplet composed of the first bit and of the XOR of the first bit with each of the other two bits [Bra06]. A different method can be found in [BDP00].

requiring a predetermined number of verification challenges, the necessary number of GCs is only logarithmically higher than the optimal that is achieved with an independent selection of challenges. In their method, the deterrent against optimal malicious GCs construction does not involve the GC constructor party having her input revealed to the GC evaluator.

In the SEOC and VSSS descriptions, the method of ensuring input consistency across different GCs is supported on discrete-log based intractability assumptions. The descriptions do not consider general linkage of S2PC executions related with output bits, but the techniques used to ensure consistency of input keys could be easily adapted to achieve XOR-homomorphic BitComs of the input bits. In contrast, the S2PC-with-BitComs described in this paper, with an instantiation based on Blum integers, is based on intractability of deciding quadratic residuosity and requires a lower number of exponentiations, though with each exponentiation being more expensive due to the larger size of group elements and group order, for the same cryptographic security parameter. Future work may better clarify the tradeoffs between the three techniques.

7.2 Other Related Work

Jarecki and Shmatikov [JS07] described a S2PC protocol with committed inputs, using a single verifiably-correct GC, but with the required number of exponentiations being linear in the number of gates. In comparison, the protocol in this paper allows garbling schemes to be based on symmetric primitives (e.g., block-ciphers, whose greater efficiency over-compensates the cost of multiple GCs in the C&C), and the required number of exponentiations to be linear in the number of circuit input and output bits and in the statistical parameter.

Nielsen and Orlandi proposed LEGO [NO09], and more recently Frederiksen et al. proposed Mini-Lego [FJN+13], a fault-tolerant circuit design that computes correctly even if some garbled gates are incorrect. Their protocol, which uses a cut-and-choose at the garbled-gate level (instead of at the GC level) to ensure that most garbled gates used for evaluation are correct, requires a single GC but of larger dimension. It would be interesting to explore, in future work, how to integrate a forge-and-lose technique into their cut-and-choose at the gate level.

Kolesnikov and Kumaresan [KK12] described a S2PC slice-evaluation protocol, based on information theoretic GCs, allowing the input of one GC to directly use the output of a previous GC. Their improvements are valid if the linked GCs are shallow, and if one party is semi-honest and the other is covert. In contrast, the S2PC-with-BitComs protocol in this paper allows any circuit depth and any party being malicious.

Nielsen et al. [NNOB12] proposed an OT-based approach for S2PC, potentially more efficient than a C&C-GCs if network latency is not an issue. However, the number of communication rounds of their protocol is linear in the depth of the circuit, thus being outside of the scope of this paper (restricted to C&C-GCs-based protocols with a constant number of communication rounds).

Acknowledgments. The author thanks: his Ph.D. co-advisor Alysson Bessani, for the valuable discussions and suggestions that contributed to improve the presentation of an earlier version of this paper; another reviewer, who wished to remain anonymous, for valuable suggestions that also contributed to improve the presentation of an earlier version of this paper; the anonymous reviewers of CRYPTO 2013 and ASIACRYPT 2013 conferences, for their useful comments.

References

[BBB⁺12] Barker, E., Barker, W., Burr, W., Polk, W., Smid, M.: Recommendation for Key Management – Part 1: General (Revision 3) – NIST Special Publication 800-57. U.S. Department of Commerce, NIST-ITL-CSD (July 2012)

[BCC88] Brassard, G., Chaum, D., Crépeau, C.: Minimum Disclosure Proofs of Knowledge. Journal of Computer and System Sciences 37(2), 156–189 (1988)

[BDP00] Boyar, J., Damgård, I., Peralta, R.: Short Non-Interactive Cryptographic Proofs. Journal of Cryptology 13(4), 449–472 (2000)

[Bea96] Beaver, D.: Correlated pseudorandomness and the complexity of private computations. In: Proc. STOC 1996, pp. 479–488. ACM, New York (1996)

[BHR12] Bellare, M., Hoang, V.T., Rogaway, P.: Foundations of garbled circuits. In: Proc. CCS 2012, pp. 784–796. ACM, New York (2012), See also Cryptology ePrint Archive. Report 2012/265

[Blu83] Blum, M.: Coin flipping by telephone a protocol for solving impossible problems. SIGACT News 15(1), 23–27 (1983)

[BMR90] Beaver, D., Micali, S., Rogaway, P.: The round complexity of secure protocols. In: Proc. STOC 1990, pp. 503–513. ACM, New York (1990)

[Bra06] Brandão, L.T.A.N.: A Framework for Interactive Argument Systems using Quasigroupic Homomorphic Commitment. Cryptology ePrint Archive, Report 2006/472 (2006)

[Bri13] Bristol Cryptography Group. Circuits of Basic Functions Suitable For MPC and FHE, http://www.cs.bris.ac.uk/Research/Cryptography Security/MPC/ (accessed June 2013)

[Can00] Canetti, R.: Security and Composition of Multiparty Cryptographic Protocols. Journal of Cryptology 13(1), 143–202 (2000), See also Cryptology ePrint Archive. Report 1998/018

[CGT95] Crépeau, C., van de Graaf, J., Tapp, A.: Committed Oblivious Transfer and Private Multi-party Computation. In: Coppersmith, D. (ed.) CRYPTO 1995. LNCS, vol. 963, pp. 110–123. Springer, Heidelberg (1995)

[CLOS02] Canetti, R., Lindell, Y., Ostrovsky, R., Sahai, A.: Universally composable two-party and multi-party secure computation. In: Proc. STOC 2002, pp. 494–503. ACM, New York (2002), See also Cryptology ePrint Archive. Report 2002/140

[Dam88] Damgård, I.B.: The application of claw free functions in cryptography. Ph.D. thesis, Aarhus University, Mathematical Institute (1988)

[EGL85] Even, S., Goldreich, O., Lempel, A.: A randomized protocol for signing contracts. Communications of the ACM 28(6), 637–647 (1985)

[FJN+13]   Frederiksen, T.K., Jakobsen, T.P., Nielsen, J.B., Nordholt, P.S., Orlandi,
            C.: MiniLEGO: Efficient Secure Two-Party Computation from General
            Assumptions. In: Johansson, T., Nguyen, P.Q. (eds.) EUROCRYPT
            2013. LNCS, vol. 7881, pp. 537–556. Springer, Heidelberg (2013), See
            also Cryptology ePrint Archive. Report 2013/155
[FN13]   Frederiksen, T.K., Nielsen, J.B.: Fast and Maliciously Secure Two-Party
            Computation Using the GPU. In: Jacobson, M., Locasto, M., Mohassel,
            P., Safavi-Naini, R. (eds.) ACNS 2013. LNCS, vol. 7954, pp. 339–356.
            Springer, Heidelberg (2013)
[GM84]   Goldwasser, S., Micali, S.: Probabilistic encryption. Journal of Computer
            and System Sciences 28(2), 270–299 (1984)
[GMR84]   Goldwasser, S., Micali, S., Rivest, R.L.: A "Paradoxical" Solution To
            The Signature Problem. In: Proc. FOCS 1984, pp. 441–448. IEEE Com-
            puter Society (1984)
[GMS08]   Goyal, V., Mohassel, P., Smith, A.: Efficient Two Party and Multi Party
            Computation Against Covert Adversaries. In: Smart, N. (ed.) EURO-
            CRYPT 2008. LNCS, vol. 4965, pp. 289–306. Springer, Heidelberg (2008)
[GMW87]   Goldreich, O., Micali, S., Wigderson, A.: How to play ANY mental game.
            In: Proc. STOC 1987, pp. 218–229. ACM, New York (1987)
[Gol04]   Goldreich, O.: Foundations of Cryptography: Volume 2, Basic
            Applications, Chapter 7 (General Cryptographic Protocols). Cam-
            bridge University Press, New York (2004)
[HEKM11]   Huang, Y., Evans, D., Katz, J., Malka, L.: Faster Secure Two-Party
            Computation Using Garbled Circuits. In: Proc. SEC 2011. USENIX As-
            sociation (2011)
[HKE12]   Huang, Y., Katz, J., Evans, D.: Quid-Pro-Quo-tocols: Strengthening
            Semi-Honest Protocols with Dual Execution. In: Proc. S&P 2012, pp.
            272–284. IEEE Computer Society, Washington (2012)
[HKE13]   Huang, Y., Katz, J., Evans, D.: Efficient Secure Two-Party Computation
            Using Symmetric Cut-and-Choose. In: Canetti, R., Garay, J.A. (eds.)
            CRYPTO 2013, Part II. LNCS, vol. 8043, pp. 18–35. Springer, Heidel-
            berg (2013), See also Cryptology ePrint Archive. Report 2013/081
[IKNP03]   Ishai, Y., Kilian, J., Nissim, K., Petrank, E.: Extending Oblivious
            Transfers Efficiently. In: Boneh, D. (ed.) CRYPTO 2003. LNCS, vol. 2729,
            pp. 145–161. Springer, Heidelberg (2003)
[JS07]   Jarecki, S., Shmatikov, V.: Efficient Two-Party Secure Computation on
            Committed Inputs. In: Naor, M. (ed.) EUROCRYPT 2007. LNCS,
            vol. 4515, pp. 97–114. Springer, Heidelberg (2007)
[Kir08]   Kiraz, M.S.: Secure and Fair Two-Party Computation. Ph.D. thesis,
            Technische Universiteit Eindhoven, Netherlands (2008)
[KK12]   Kolesnikov, V., Kumaresan, R.: Improved Secure Two-Party Computation
            via Information-Theoretic Garbled Circuits. In: Visconti, I., De Prisco,
            R. (eds.) SCN 2012. LNCS, vol. 7485, pp. 205–221. Springer, Heidelberg
            (2012)
[Kol09]   Kolesnikov, V.: Advances and impact of secure function evaluation. Bell
            Labs Technical Journal 14(3), 187–192 (2009)
[KS06]   Kiraz, M.S., Schoenmakers, B.: A protocol issue for the malicious case of
            Yao's garbled circuit construction. In: Proc. 27th Symp. Information The-
            ory in the Benelux, pp. 283–290 (2006)

[KS08a]   Kiraz, M.S., Schoenmakers, B.: An Efficient Protocol for Fair Secure
            Two-Party Computation. In: Malkin, T. (ed.) CT-RSA 2008. LNCS,
            vol. 4964, pp. 88–105. Springer, Heidelberg (2008)
[KS08b]   Kolesnikov, V., Schneider, T.: Improved Garbled Circuit: Free XOR
            Gates and Applications. In: Aceto, L., Damgård, I., Goldberg, L.A.,
            Halldórsson, M.M., Ingólfsdóttir, A., Walukiewicz, I. (eds.) ICALP 2008,
            Part II. LNCS, vol. 5126, pp. 486–498. Springer, Heidelberg (2008)
[KSS12]   Kreuter, B., Shelat, A., Shen, C.-H.: Billion-gate secure computation with
            malicious adversaries. In: Proc. Security 2012, pp. 285–300. USENIX As-
            sociation (2012), See also Cryptology ePrint Archive. Report 2012/179
[Lin03]   Lindell, Y.: Parallel Coin-Tossing and Constant-Round Secure Two-Party
            Computation. Journal of Cryptology 16(3), 143–184 (2003), See also
            Cryptology ePrint Archive. Report 2001/107
[Lin13]   Lindell, Y.: Fast Cut-and-Choose Based Protocols for Malicious and
            Covert Adversaries. In: Canetti, R., Garay, J.A. (eds.) CRYPTO 2013,
            Part II. LNCS, vol. 8043, pp. 1–17. Springer, Heidelberg (2013), See also
            Cryptology ePrint Archive. Report 2013/079
[LP02]   Lindell, Y., Pinkas, B.: Privacy Preserving Data Mining. Journal of Cryp-
            tology 15(3), 177–206 (2002)
[LP07]   Lindell, Y., Pinkas, B.: An Efficient Protocol for Secure Two-Party
            Computation in the Presence of Malicious Adversaries. In: Naor, M.
            (ed.) EUROCRYPT 2007. LNCS, vol. 4515, pp. 52–78. Springer, Hei-
            delberg (2007), See also Cryptology ePrint Archive. Report 2008/049
[LP09]   Lindell, Y., Pinkas, B.: A Proof of Security of Yao's Protocol for Two-
            Party Computation. Journal of Cryptology 22(2), 161–188 (2009)
[LP11]   Lindell, Y., Pinkas, B.: Secure two-party computation via cut-and-choose
            oblivious transfer. In: Ishai, Y. (ed.) TCC 2011. LNCS, vol. 6597, pp.
            329–346. Springer, Heidelberg (2011), See also Cryptology ePrint Archive.
            Report 2010/284
[MF06]   Mohassel, P., Franklin, M.K.: Efficiency Tradeoffs for Malicious Two-
            Party Computation. In: Yung, M., Dodis, Y., Kiayias, A., Malkin, T.
            (eds.) PKC 2006. LNCS, vol. 3958, pp. 458–473. Springer, Heidelberg
            (2006)
[NNOB12]   Nielsen, J.B., Nordholt, P.S., Orlandi, C., Burra, S.S.: A New Approach
            to Practical Active-Secure Two-Party Computation. In: Safavi-Naini, R.,
            Canetti, R. (eds.) CRYPTO 2012. LNCS, vol. 7417, pp. 681–700. Springer,
            Heidelberg (2012), See also Cryptology ePrint Archive. Report 2011/091
[NO09]   Nielsen, J.B., Orlandi, C.: LEGO for Two-Party Secure Computation. In:
            Reingold, O. (ed.) TCC 2009. LNCS, vol. 5444, pp. 368–386. Springer,
            Heidelberg (2009), See also Cryptology ePrint Archive. Report 2008/427
[NP01]   Naor, M., Pinkas, B.: Efficient oblivious transfer protocols. In: SODA
            2001, pp. 448–457. SIAM, Philadelphia (2001)
[NPS99]   Naor, M., Pinkas, B., Sumner, R.: Privacy preserving auctions and
            mechanism design. In: Proc. EC 1999, pp. 129–139. ACM, New York
            (1999)
[NZM91]   Niven, I.M., Zuckerman, H.S., Montgomery, H.L.: An introduction to the
            theory of numbers, 5th edn. Wiley (1991)
[Pin03]   Pinkas, B.: Fair Secure Two-Party Computation. In: Biham, E. (ed.)
            EUROCRYPT 2003. LNCS, vol. 2656, pp. 647–647. Springer, Heidelberg
            (2003)

[PSSW09]   Pinkas, B., Schneider, T., Smart, N., Williams, S.: Secure Two-Party
            Computation Is Practical. In: Matsui, M. (ed.) ASIACRYPT 2009.
            LNCS, vol. 5912, pp. 250–267. Springer, Heidelberg (2009), See also Cryp-
            tology ePrint Archive. Report 2009/314
[Rab81]   Rabin, M.O.: How to exchange secrets with oblivious transfer. Techni-
            cal Report TR-81, Harvard University, Aiken Computation Lab, Cam-
            bridge, MA (1981), See typesetted version in Cryptology ePrint Archive,
            Report 2005/187
[SS11]   Shelat, A., Shen, C.-H.: Two-Output Secure Computation with Malicious
            Adversaries. In: Paterson, K.G. (ed.) EUROCRYPT 2011. LNCS,
            vol. 6632, pp. 386–405. Springer, Heidelberg (2011), See also Cryptology
            ePrint Archive. Report 2011/533
[vdGP88]   van de Graaf, J., Peralta, R.: A Simple and Secure Way to Show the
            Validity of Your Public Key. In: Pomerance, C. (ed.) CRYPTO 1987.
            LNCS, vol. 293, pp. 128–134. Springer, Heidelberg (1988)
[Woo07]   Woodruff,   D.P.:   Revisiting the Efficiency of Malicious Two-Party
            Computation. In: Naor, M. (ed.) EUROCRYPT 2007. LNCS, vol. 4515,
            pp. 79–96. Springer, Heidelberg (2007), See also Cryptology ePrint
            Archive. Report 2006/397
[Yao82]   Yao, A.C.: Protocols for secure computations. In: Proc. FOCS 1982,
            pp. 160–164. IEEE Computer Society (1982)
[Yao86]   Yao, A.C.-C.: How to generate and exchange secrets. In: FOCS 1986,
            pp. 162–167 (1986)

A Heuristic for Finding Compatible Differential Paths with Application to HAS-160

Aleksandar Kircanski, Riham AlTawy, and Amr M. Youssef

Concordia Institute for Information Systems Engineering,
Concordia University, Montréal, Quebéc, H3G 1M8, Canada

Abstract. The question of compatibility of differential paths plays a central role in second order collision attacks on hash functions. In this context, attacks typically proceed by starting from the middle and constructing the middle-steps quartet in which the two paths are enforced on the respective faces of the quartet structure. Finding paths that can fit in such a quartet structure has been a major challenge and the currently known compatible paths extend over a suboptimal number of steps for hash functions such as SHA-2 and HAS-160. In this paper, we investigate a heuristic that searches for compatible differential paths. The application of the heuristic in case of HAS-160 yields a practical second order collision over all of the function steps, which is the first practical result that covers all of the HAS-160 steps. An example of a colliding quartet is provided.

1 Introduction

Whenever two probabilistic patterns are combined for the purpose of passing through maximal number of rounds of a cryptographic primitive, a natural question that arises is the question of compatibility of the two patterns. A notable example is the question of compatibility of differential paths in the context of boomerang attacks. In 2011, Murphy [25] has shown that care should be exercised when estimating the boomerang attack success probability, since there may exist dependency between the events that the two paths behave as required by the boomerang setting. The extreme case is the impossibility of combining the two paths, where the corresponding probability is equal to 0.

In the context of constructing second order collisions for compression functions using the start-from-the-middle technique, due to availability of message modification in the steps where the primitive follows the two paths, the above mentioned probability plays less of a role as long as it is strictly greater than 0. In that case, the two paths are said to be compatible. Several paths that were previously believed to be compatible have been shown to be incompatible in the previously described sense, e.g., by Leurent [15] and Sasaki [29] for BLAKE and RIPEMD-160 hash functions, respectively.

The compatibility requirement in this context can be stated with more precision as follows. Let ϕ and ω be two differential paths over some number of steps of an iterative function $f = f_{j+n} \circ \ldots \circ f_j$. If there exists a quartet of f

K. Sako and P. Sarkar (Eds.) ASIACRYPT 2013, Part II, LNCS 8270, pp. 464–483, 2013.
© International Association for Cryptologic Research 2013

inputs x_0, x_1, x_2 and x_3 such that computations (x_0, x_1) and (x_2, x_3) follow ϕ whereas (x_0, x_2) and (x_1, x_3) follow ω, we say that ϕ and ω are compatible. Usually the path ϕ is left unspecified over the last k steps (backward path) and ω is unspecified over the remaining steps (forward path). Such paths have also been previously called *independent* [4]. Another closely related notion is the concept of *non-interleaving* paths in the context of biclique attacks [9].

Our Contributions. In this paper, we present a heuristic that allows us to search for compatible differential paths. The heuristic builds on the previous de Cannière and Rechberger automatic differential path search method. Instead of working with pairs, our proposed heuristic operates on quartets of hash executions and includes cross-path propagations. We present detailed examples of particular propagations applied during the search. As an application of our proposed heuristic, a second order collision for the full HAS-160 compression function is found. The best previous practical distinguisher for this function covered steps 5 to 80 [30]. This is the first practical distinguisher for the full HAS-160. This particular hash function is relevant as it is standardized by the Korean government (TTAS.KO-12.0011/R1) [1].

Related Work. The differential paths used in groundbreaking attacks on MD4, MD5 and SHA-1 [36,35] were found manually. Subsequently, several techniques for automatic differential path search have been studied [31,7,32,5]. The de Cannière and Rechberger heuristic [5] was subsequently applied to many MDx/SHA-x based hash functions, such as RIPEMD-128, HAS-160, SHA-2 and SM3 [21,19,20,22]. To keep track of the current information in the system, the heuristic relies on 1-bit constraints that express the relations between pairs of bits in the differential setting. This was generalized to multi-bit constraints by Leurent [15], where finite state machine approach allowed uniform representation of different constraint types. Multi-bit constraints have been used in the context of differential path search in [16].

The boomerang attack [33], originally applied to block ciphers, has been adapted to the hash function setting independently by Biryukov *et al.* [4] and by Lamberger and Mendel [13]. In particular, in [4], a distinguisher for the 7-round BLAKE-32 was provided, whereas in [13] a distinguisher for the 46-step reduced SHA-2 compression function was provided. The latter SHA-2 result was extended to 47 steps [3]. Subsequently, boomerang distinguishers have been applied to many hash functions, such as HAVAL, RIPEMD-160, SIMD, HAS-160, SM3 and Skein [27,29,30,18,11,37,17]. Outside of the boomerang context, zero-sum property as a distinguishing property was first used by Aumasson [2].

As for the previous HAS-160 analysis, in 2005, Yun *et al.* [38] found a practical collision for the 45-step (out of 80) reduced hash function. Their attack was extended in 2006 to 53 steps by Cho *et al.* [6], however, with computational complexity of 2^{55} 53-step compression function computations. In 2007, Mendel and Rijmen [23] improved the latter attack complexity to 2^{35}, providing a practical two-block message collision for the 53-step compression function. Preimage attacks on 52-step HAS-160 with complexity of 2^{152} was provided in

2008 by Sasaki and Aoki [28]. Subsequently, in 2009, this result was extended by Hong *et al.* to 68 steps [8] where the attack required a complexity of $2^{156.3}$. In 2011, Mendel *et al.* provided a practical semi-free-start collision for 65-step reduced compression function [19]. Finally, in 2012, Sasaki *et al.* [30] provided a theoretical boomerang distinguisher for the full HAS-160 compression function, requiring $2^{76.6}$ steps function computations. In the same work, a practical second order collision was given for steps 5 to 80 of the function.

Paper Outline. In the next section, we provide the review of boomerang distingiushers and the recapitulation of the de Cannière and Rechberger search heuristic, along with the HAS-160 specification. In Section 3, the general form of the our search heuristic is provided and its application to HAS-160 is discussed. The three propagation types used in the heuristic are explained in Section 4. Concluding remarks are given in Section 5.

2   Review of Related Work and the Specification of HAS-160

In the following subsections, we provide a description of a commonly used strategy to construct second order collisions, an overview of the de Cannière and Rechberger path search heuristic and finally the specification of HAS-160 hash function.

2.1   Review of Boomerang Distinguishers for Hash Functions

First, we provide a generic definition of the property used for building compression function distinguishers. Let h be a function with n-bit output. A *second order collision* for h is a set $\{x, \Delta, \nabla\}$ consisting of an input for h and two differences, such that

$$h(x + \Delta + \nabla) - h(x + \Delta) - h(x + \nabla) + h(x) = 0 \qquad (1)$$

As explained in [3], the query complexity for finding a second order collision is $3 \cdot 2^{n/3}$ where n denotes the bit-size of the output of the function f. By the query complexity, the number of queries required to be made to h function is considered. On the other hand, for the computational complexity, which would include evaluating h around $3 \cdot 2^{n/3}$ times and finding a quartet that sums to 0, the best currently known algorithm runs in complexity no better than $2^{n/2}$. If for a particular function a second order collision is obtained with a complexity lower than $2^{n/2}$, then this hash function deviates from the random function oracle.

Next, we explain the strategy to construct quartets satisfying (1) for Davies-Meyer based functions, as commonly applied in the previous literature. An overview of the strategy is provided in Fig. 1. We write $h(x) = e(x) + x$, where e is an iterative function consisting of n steps. The goal is to find four inputs

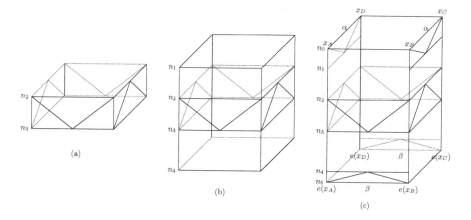

Fig. 1. Start-from-the-middle approach for constructing second-order collisions

x_A, x_B, x_C and x_D that constitute the inputs in (1) according to Fig. 1 (c). In particular, the goal is to have

$$x_A - x_D = x_B - x_C$$
$$e(x_A) - e(x_B) = e(x_D) - e(x_C) \tag{2}$$

where the two values specified by (2) are denoted respectively by α and β in Fig. 1 (c). In this case, we have $h(x_A) - h(x_B) + h(x_C) - h(x_D) = e(x_A) + x_A - e(x_B) - x_B + e(x_C) + x_C - e(x_D) - x_D = 0$. Now, one can put $x_A = x$, $\Delta = x_D - x_A$ and $\nabla = x_B - x_A$ and (1) is satisfied.

A preliminary step is to decide on two paths, called the *forward* path and the *backward* path. As shown on Fig. 1, these paths are chosen so that for some $n_0 < n_1 < n_2 < n_3 < n_4 < n_5$, the forward path has no active bits between steps n_3 and n_4 and the backward path has no active bits between steps n_1 and n_2. The forward path is enforced on faces (x_A, x_B) and (x_D, x_C) (front and back) whereas the backward differential is enforced on faces (x_A, x_D) and (x_B, x_C) (left and right). In the case of MDx-based designs, the particular n values depend mostly on the message schedule specification.

The procedure can be summarized as follows:

(a) The first step is to construct the middle part of the quartet structure, as shown in Fig. 1 (a). The forward and backward paths end at steps n_3 and n_2, respectively. On steps n_2 to n_3, the two paths need to be compatible for this stage to succeed.

(b) Following Fig. 1 (b), the paths are extended to steps n_1 backward and to n_4 forward with probability 1, due to the absence of disturbances in the corresponding steps.

(c) Some of the middle-step words are randomized and the quartet is recomputed backward and forward, verifying if (2) is satisfied. If yes (see Fig. 1 (c)), return the quartet, otherwise, repeat this step.

This strategy, with variations, has been applied in several previous works, such as [3,30,29,27]. In Table 1, we provide the forward/backward path parameters for the previous boomerang distinguishers on some of the MDx/SHA-x based compression functions following the single-pipe design strategy.

Table 1. Overview of some of the previously used boomerang paths

Compression function	n_0 n_1 n_2 n_3 n_4 n_5	Reference	Message block size
SHA-2	0 6 22 31 47 47	[3]	16×32
HAVAL	0 2 61 97 157 160	[27]	32×32
HAS-160	5 13 38 53 78 80	[30]	16×32

In [3,30], the number of steps in the middle was 9 and 16 steps, respectively. It can be observed that these number of middle steps are suboptimal, since the simple message modification allows trivially satisfying 16 steps in case of SHA-2 and HAS-160. Since the forward and the backward paths are sparse towards steps n_3 and n_2, one can easily imagine satisfying more than 16 steps, while there remains enough freedom to randomize the inner state although some penalty in probability has to be paid. In case of HAVAL [27], the simple message modification allows passing through 32 steps and the middle part consists of as many as 36 steps. However, it should be noted that this is due to the particular property of HAVAL which allows narrow paths [10].

2.2   Review of de Cannière and Rechberger Search Heuristic

This search heuristic is used to find differential paths that describe pairs of compression function executions. The symbols used for expressing differential paths are provided in Table 2. For example, when we write -x-u, we mean a set of 4-bit pairs

$$-\text{x-u} = \{T, T' \in F_2^4 | T_3 = T'_3, T_2 \neq T'_2, T_1 = T'_1, T_0 = 0, T'_0 = 1\}$$

where T_i denotes i-th bit in word T.

Table 2. Symbols used to express 1-bit conditions [5]

$\delta(x,x')$	meaning	(0,0)	(0,1)	(1,0)	(1,1)	$\delta(x,x')$	meaning	(0,0)	(0,1)	(1,0)	(1,1)
?	anything	✓	✓	✓	✓	3	$x = 0$	✓	✓	-	-
-	$x = x'$	✓	-	-	✓	5	$x' = 0$	✓	-	✓	-
x	$x \neq x'$	-	✓	✓	-	7		✓	✓	✓	-
0	$x = x' = 0$	✓	-	-	-	A	$x' = 1$	-	✓	-	✓
u	$(x,x') = (0,1)$	-	✓	-	-	B		✓	✓	-	✓
n	$(x,x') = (1,0)$	-	-	✓	-	C	$x = 1$	-	-	✓	✓
1	$x = x' = 1$	-	-	-	✓	D		✓	-	✓	✓
#		-	-	-	-	E		-	✓	✓	✓

Next, an example of *condition propagation* is provided. Suppose that a small differential path over one modular addition is given by

$$---- + ---\mathtt{x} = ---\mathtt{x} \tag{3}$$

Here (3) describes a pair of additions: $x + y = z$ and $x' + y' = z'$, and from this "path" we have that $x = x'$ and also that y and y' are different only on the least significant bit (same for z and z'). However, this can happen only if $x_0 = x'_0 = 0$, i.e. if the lsb of x and x' is equal to 0. We thus *propagate a condition* by substituting (3) with

$$---\mathtt{0} + ---\mathtt{x} = ---\mathtt{x}$$

The de Cannière and Rechberger heuristic [5] searches for differential paths over some number of compression function steps. It starts from a partially specified path which typically means that the path is fully specified at some steps (i.e., consisting of symbols $\{\mathtt{-},\mathtt{u},\mathtt{n}\}$) and unspecified at other steps (i.e., symbol '?'). The heuristic attempts to complete the path, so that the final result is non-contradictory by proceeding as follows:

- *Guess*: select randomly a bit position containing '?' or 'x'. Substitute the symbol in the chosen bit position by '-' and $\{\mathtt{u},\mathtt{n}\}$, respectively.
- *Propagate*: deduce new information introduced by the *Guess* step.

When a contradiction is detected, the search backtracks by jumping back to one of the guesses and attempts different choices.

2.3   HAS-160 Specification

The HAS-160 hash function follows the MDx/SHA-x hash function design strategy. Its compression function can be seen as a block cipher in Davies-Meyer mode, mapping 160-bit chaining values and 512-bit messages into 160-bit digests. To process arbitrary-length messages, the compression function is plugged in the Merkle-Damgård mode.

Before hashing, the message is padded so that its length becomes multiple of 512 bits. Since padding is not relevant for this paper, we refer the reader to [1] for further details. The underlying HAS-160 block cipher consists of two parts: message expansion and state update transformation.

Message Expansion: The input to the compression function is a message $m = (m_0, \ldots m_{15})$ represented as 16 32-bit words. The output of the message expansion is a sequence of 32-bit words $W_0, \ldots W_{79}$. The expansion is specified in Table 3. For example, $W_{26} = m_{15}$.

State Update: One compression function step is schematically described by Fig. 2 (a). The Boolean functions f used in each step are given by

$$f_0(x, y, z) = (x \wedge y) \oplus (\neg x \wedge z)$$
$$f_1(x, y, z) = x \oplus y \oplus z$$
$$f_2(x, y, z) = (x \vee \neg z) \oplus y$$

Table 3. Message expansion in HAS-160

0	1	2	3	4	5	6	7	8	9	10	11	12	13	14	15	16	17	18	19
$m_8 \oplus m_9$ $\oplus m_{10} \oplus m_{11}$	m_0	m_1	m_2	m_3	$m_{12} \oplus m_{13}$ $\oplus m_{14} \oplus m_{15}$	m_4	m_5	m_6	m_7	$m_0 \oplus m_1$ $\oplus m_2 \oplus m_3$	m_8	m_9	m_{10}	m_{11}	$m_4 \oplus m_5$ $\oplus m_6 \oplus m_7$	m_{12}	m_{13}	m_{14}	m_{15}
$m_{11} \oplus m_{14}$ $\oplus m_1 \oplus m_4$	m_3	m_6	m_9	m_{12}	$m_7 \oplus m_{10}$ $\oplus m_{13} \oplus m_0$	m_{15}	m_2	m_5	m_8	$m_3 \oplus m_6$ $\oplus m_9 \oplus m_{12}$	m_{11}	m_{14}	m_1	m_4	$m_{15} \oplus m_2$ $\oplus m_5 \oplus m_8$	m_7	m_{10}	m_{13}	m_0
$m_4 \oplus m_{13}$ $\oplus m_6 \oplus m_{15}$	m_{12}	m_5	m_{14}	m_7	$m_8 \oplus m_1$ $\oplus m_{10} \oplus m_3$	m_0	m_9	m_2	m_{11}	$m_{12} \oplus m_5$ $\oplus m_{14} \oplus m_7$	m_4	m_{13}	m_6	m_{15}	$m_0 \oplus m_9$ $\oplus m_2 \oplus m_{11}$	m_8	m_1	m_{10}	m_3
$m_{15} \oplus m_{10}$ $\oplus m_5 \oplus m_0$	m_7	m_2	m_{13}	m_8	$m_{11} \oplus m_6$ $\oplus m_1 \oplus m_{12}$	m_3	m_{14}	m_9	m_4	$m_7 \oplus m_2$ $\oplus m_{13} \oplus m_8$	m_{15}	m_{10}	m_5	m_0	$m_3 \oplus m_{14}$ $\oplus m_9 \oplus m_4$	m_{11}	m_6	m_1	m_{12}

where f_0 is used in steps 0-19, f_1 is used in steps 20-39 and 60-79 and f_2 is used in steps 40-59. The constant K_i that is added in each step changes every 20 steps, taking the values 0, $5a827999$, $6ed9eba1$ and $8f1bbcdc$. The rotational constant s_1^i is specified by the following table

$i \bmod 20$	0	1	2	3	4	5	6	7	8	9	10	11	12	13	14	15	16	17	18	19
s_1^i	5	11	7	15	6	13	8	14	7	12	9	11	8	15	6	12	9	14	5	13

The other rotational constant s_2^i changes only each 20 steps and $s_2^i \in \{10, 17, 25, 30\}$. According to the Davies-Meyer mode, the feedforward is applied and the output of the compression is

$$(A_{80} + A_0, B_{80} + B_0, C_{80} + C_0, D_{80} + D_0, E_{80} + E_0)$$

Alternative Description of HAS-160: In Fig. 2 (b), the compression function is shown as a recurrence relation, where A_{i+1} plays the role of A in the usual step representation. Namely, A can be considered as the only new computed word, since the rotation that is applied to B can be compensated by properly adjusting the rotation constants in the recurrence relation specification. One starts from $A_{-4}, A_{-3}, A_{-2}, A_{-1}$ and A_0, putting these values to the previous chaining value

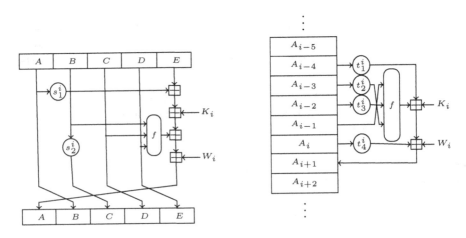

Fig. 2. Two equivalent representations of the state update

(or the IV for the first message block) and computes the recurrence until A_{80} according to

$$A_{i+1} = A_{i-4} \lll t_1^i + K_i + f_i(A_{i-1}, A_{i-2} \lll t_3^i, A_{i-3} \lll t_2^i) + W_i + A_i \lll t_4^i \quad (4)$$

The rotational values t_j^i, $1 \leq j \leq 4$ are derived from s_1^i and s_2^i, where the constants related to the rotation of B in the usual representation change around the steps $20 \times k$, $k = 0, 1, 2, 3$. For instance, to compute A_{42}, we have $t_1^{41} = 17$, $t_2^{41} = 17$, $t_3^{41} = 25$ and $t_4^{41} = 11$.

3 Compatible Paths Search Heuristic and Application to HAS-160

In this section, we provide a new search heuristic that can be used to find compatible paths in the boomerang setting. The particular colliding quartet found by applying the heuristic on HAS-160 is provided in Table 4.

Table 4. Second order collision for the full HAS-160 compression function

	Message quartet							
M_A	F6513317	810F1084	FFB71009	78CC955E	C3C09F18	5379FC99	435586DA	9C9AD3B4
	00440C80	E174316A	006D1670	2B5CF68A	AB3DE600	02C9E9D3	5FE95AFF	E351DE04
M_B	F6513317	810F1084	FFB71009	78CC955E	C3C09f18	5379FC99	435786DA	9C9AD3B4
	00440C80	E174316A	006D1670	2B5CF68A	AB3FE600	02C9E9D3	5FE95AFF	E351DE04
M_C	76513317	010F1084	FFB71009	78CC955E	43C09F18	5379FC99	435786DA	1C9AD3B4
	00440C80	E174316A	006D1670	2B5CF68A	AB3FE600	02C9E9D3	5FE95AFF	E351DE04
M_D	76513317	010F1084	FFB71009	78CC955E	43C09f18	5379FC99	435586DA	1C9AD3B4
	00440C80	E174316A	006D1670	2B5CF68A	AB3DE600	02C9E9D3	5FE95AFF	E351DE04
	Chaining values quartet							
IV_A	1143BE75	9A9CA381	85B3F526	DA6ABE66	70EBE920			
IV_B	3AF7BD99	D08E2E63	245C2AF0	C4456954	CAC046EA			
IV_C	3AF7B599	D08E2E63	B45C2AF0	C425694C	3BE146F2			
IV_D	1143B675	9A9CA381	15B3F526	DA4ABE5E	E20CE928			

The heuristic uses quartets of 1-bit conditions from Table 2 to keep track of the bit differences in each of the four compression function executions. Apart from the single-path propagations proposed in [5], two additional types of boomerang (cross-path) propagations are added. These boomerang propagations have been previously listed in [15].

The forward and the backward differentials are specified next and this specification determines the initial problem on which the heuristic is applied. Let the forward message differential consist of a one-bit difference in messages m_6 and m_{12} and the backward differential of a one-bit difference in m_0, m_1, m_4 and m_7, as shown in Table 5. The particular bit-position of differences is left unspecified. The choice of these difference positions is justified by the following start/end points of the expanded message differences, expressed in terms of the notation used in Fig. 1: $(n_0, n_1, n_2, n_3, n_4, n_5) = (0, 8, 34, 53, 78, 80)$. It can be observed that the middle part consists of 20 steps.

Now, the particular problem schematically described by Fig. 1 (a) is represented more specifically by Table 7, where the backward and forward message

Table 5. Message differentials. Backward: steps 0-39, forward: steps 40-79

0	1	2	3	4	5	6	7	8	9	10	11	12	13	14	15	16	17	18	19
$m_8 \oplus m_9$ $\oplus m_{10} \oplus m_{11}$	m_0	m_1	m_2	m_3	$m_{12} \oplus m_{13}$ $\oplus m_{14} \oplus m_{15}$	m_4	m_5	m_6	m_7	$m_0 \oplus m_1$ $\oplus m_2 \oplus m_3$	m_8	m_9	m_{10}	m_{11}	$m_4 \oplus m_5$ $\oplus m_6 \oplus m_7$	m_{12}	m_{13}	m_{14}	m_{15}
$m_{11} \oplus m_{14}$ $\oplus m_1 \oplus m_4$	m_3	m_6	m_9	m_{12}	$m_7 \oplus m_{10}$ $\oplus m_{13} \oplus m_0$	m_{15}	m_2	m_5	m_8	$m_3 \oplus m_6$ $\oplus m_9 \oplus m_{12}$	m_{11}	m_{14}	m_1	m_4	$m_{15} \oplus m_2$ $\oplus m_5 \oplus m_8$	m_7	m_{10}	m_{13}	m_0
$m_4 \oplus m_{13}$ $\oplus m_6 \oplus m_{15}$	m_{12}	m_5	m_{14}	m_7	$m_8 \oplus m_1$ $\oplus m_{10} \oplus m_3$	m_0	m_9	m_2	m_{11}	$m_{12} \oplus m_5$ $\oplus m_{14} \oplus m_7$	m_4	m_{13}	m_6	m_{15}	$m_0 \oplus m_9$ $\oplus m_2 \oplus m_{11}$	m_8	m_1	m_{10}	m_3
$m_{15} \oplus m_{10}$ $\oplus m_5 \oplus m_0$	m_7	m_2	m_{13}	m_8	$m_{11} \oplus m_6$ $\oplus m_1 \oplus m_{12}$	m_3	m_{14}	m_9	m_4	$m_7 \oplus m_2$ $\oplus m_{13} \oplus m_8$	m_{15}	m_{10}	m_5	m_0	$m_3 \oplus m_{14}$ $\oplus m_9 \oplus m_4$	m_{11}	m_6	m_1	m_{12}

differentials are indicated in the first and the last column, respectively. At this point, the only information that is present in the system is that the two paths end at the corresponding steps $n_2 = 34$ and $n_3 = 53$. The output of the heuristic in case of HAS-160 is given in Table 8. The full specifications of the two paths intersect on 5 steps, which is the number of inner state registers in HAS-160. Provided that the paths are compatible, one can now start from step 42 and apply the usual message modification technique to satisfy both paths, which resolves the middle of the boomerang as shown in Fig. 1 (a).

3.1   Search Strategy

The approach consists of variating the position of the message difference bit, gradually extending the two paths, propagating the conditions in the quartet and backtracking in case of a contradiction. In more detail, the heuristic proceeds as follows:

(1) Randomize the positions of active bits in the active message words.
(2) Extend the specification of the forward/backward path backward/forward, respectively. Ensure that paths are randomized over different step invocations.
(3) Propagate all new conditions. In case of contradiction, backtrack
(4) If the two paths are fully specified on a sufficient number of steps, return the two paths

In step (1), the message disturbance position in the two differentials is randomized to achieve variation in the paths. Alternatively, one position can be fixed to bit 31 and the other position randomized at each step invocation. As for step (2), at the point where the probability of contradiction between the two paths is negligible, one can extend paths simply by randomly sampling them in required steps and discarding non-narrow ones. Once the probability of contradiction becomes significant, substitute/backtrack strategy according to the Table 6 is applied to the remaining steps. In step (3), apart from propagations on a single path [5], quartet and quartet addition propagations (explained in Section 4) are applied. The heuristic ends when the full specification of two paths (containing only $\{-, u, n\}$) intersects on the number of words equal to the number of registers in the compression function inner state, as is the case in Table 8.

Table 6. Substitution rules: adding information to the forward path (left) and backward path (right)

1.	???? ↦ --??
2.	??-- ↦ ----
3.	??xx ↦ --xx
4.	xx?? ↦ {uu10,nn01}
5.	xx-- ↦ {uu10,nn01}
6.	xxxx ↦ {unnu,nuun}

1.	???? ↦ ??--
2.	--?? ↦ ----
3.	xx?? ↦ xx--
4.	??xx ↦ {01uu,10nn}
5.	--xx ↦ {01uu, 10nn}
6.	xxxx ↦ {unnu,nuun}

When new constraint information is to be added at a particular bit position, one can either add information to the forward path or to the backward path. Here, a clarification is necessary regarding the fact that in Table 8, four paths are shown, whereas the heuristic searches for a pair of paths (forward and backward path). This is due to the fact that the paths on the opposite faces of the boomerang are equal (up to 0 and 1 symbols) and thus one can consider a pair of paths. Nonetheless, the inner state of the search algorithm keeps all the four paths explicitly.

The substitutions provided in Table 6 represent generalizations of the substitutions used in [5]. The choice whether the information will be added to the forward or the backward path is made randomly each time. The left-hand and the right-hand tables correspond to adding constraints to the forward and the backward path, respectively. Consider for example rule xx-- ↦ {uu10,nn01}. In this notation, the symbols xx-- describe a bit position for which $\delta[A_i^j, B_i^j] = x$, $\delta[D_i^j, C_i^j] = x$, $\delta[B_i^j, C_i^j] = -$, $\delta[A_i^j, D_i^j] = -$. The rule simply substitutes the 'x' symbol on the forward path by 'u' or 'n', while at the same time applying the immediate propagation of the '-' symbols to '0' and '1', respectively. This rule represents a generalization of the x ↦ {u,n} rule used in [5]. Other rules can be explained in a similar manner.

One possible variation of the general heuristic above is as follows. Once the two paths are sufficiently specified so that the contradictions are likely to occur, instead of adding new constraints randomly, a beneficial strategy is to introduce some graduality while extending the two paths. For example, one can choose a parameter k and extend both paths by only k steps. If the heuristic succeeds in extending the paths by k steps, reporting that there is no contradiction in the system, more steps can be attempted. If in the intermediate steps of the search, the path was in fact contradictory and this was not reported by 1-bit conditions, further attempts to extend or find the messages satisfying the paths will fail.

3.2 Application to HAS-160

In this section, we describe how the above heuristic can be applied in the case of HAS-160. First, we fix the position of the active bit in the backward differential to $b_1 = 31$. The following sequence of steps randomizes steps in the light-gray area in Table 7:

Table 7. Input for the search heuristic

step	Δ[A, B]	Δ[D, C]	Δ[B, C]	Δ[A, D]	step
9	????????????????????????????????	????????????????????????????????	-------------------------------	-------------------------------	9
10	????????????????????????????????	????????????????????????????????	-------------------------------	-------------------------------	10
11	????????????????????????????????	????????????????????????????????	-------------------------------	-------------------------------	11
12	????????????????????????????????	????????????????????????????????	-------------------------------	-------------------------------	12
13	????????????????????????????????	????????????????????????????????	-------------------------------	-------------------------------	13
⋮	⋮	⋮	[NO DIFFERENCE]	⋮	
29	????????????????????????????????	????????????????????????????????	-------------------------------	-------------------------------	29
30	????????????????????????????????	????????????????????????????????	-------------------------------	-------------------------------	30
31	????????????????????????????????	????????????????????????????????	-------------------------------	-------------------------------	31
32	????????????????????????????????	????????????????????????????????	-------------------------------	-------------------------------	32
33	????????????????????????????????	????????????????????????????????	-------------------------------	-------------------------------	33
34	????????????????????????????????	????????????????????????????????	????????????????????????????????	????????????????????????????????	34
35	????????????????????????????????	????????????????????????????????	????????????????????????????????	????????????????????????????????	35
36	????????????????????????????????	????????????????????????????????	????????????????????????????????	????????????????????????????????	36
37	????????????????????????????????	????????????????????????????????	????????????????????????????????	????????????????????????????????	37
38	????????????????????????????????	????????????????????????????????	????????????????????????????????	????????????????????????????????	38
39	????????????????????????????????	????????????????????????????????	????????????????????????????????	????????????????????????????????	39
40	????????????????????????????????	????????????????????????????????	????????????????????????????????	????????????????????????????????	40
41	????????????????????????????????	????????????????????????????????	????????????????????????????????	????????????????????????????????	41
42	????????????????????????????????	????????????????????????????????	????????????????????????????????	????????????????????????????????	42
43	????????????????????????????????	????????????????????????????????	????????????????????????????????	????????????????????????????????	43
44	????????????????????????????????	????????????????????????????????	????????????????????????????????	????????????????????????????????	44
45	????????????????????????????????	????????????????????????????????	????????????????????????????????	????????????????????????????????	45
46	????????????????????????????????	????????????????????????????????	????????????????????????????????	????????????????????????????????	46
47	????????????????????????????????	????????????????????????????????	????????????????????????????????	????????????????????????????????	47
48	????????????????????????????????	????????????????????????????????	????????????????????????????????	????????????????????????????????	48
49	????????????????????????????????	????????????????????????????????	????????????????????????????????	????????????????????????????????	49
50	-------------------------------	-------------------------------	????????????????????????????????	????????????????????????????????	50
51	-------------------------------	-------------------------------	????????????????????????????????	????????????????????????????????	51
52	-------------------------------	-------------------------------	????????????????????????????????	????????????????????????????????	52
53	-------------------------------	-------------------------------	????????????????????????????????	????????????????????????????????	53
54	-------------------------------	-------------------------------	????????????????????????????????	????????????????????????????????	54
⋮		[NO DIFFERENCE]	⋮	⋮	
76	-------------------------------	-------------------------------	????????????????????????????????	????????????????????????????????	76
77	-------------------------------	-------------------------------	????????????????????????????????	????????????????????????????????	77

Table 8. Output of the heuristic: compatible paths for HAS-160

step	Δ[A, B]	Δ[D, C]	Δ[B, C]	Δ[A, D]	step
29	????????????????????????????????	????????????????????????????????	-------------------------------	-------------------------------	29
30	????????????????????????????????	????????????????????????????????	-------------------------------	-------------------------------	30
31	????????????????????????????????	????????????????????????????????	-------------------------------	-------------------------------	31
32	????????????????????????????????	????????????????????????????????	-------------------------------	-------------------------------	32
33	????????????????????????????????	????????????????????????????????	-------------------------------	-------------------------------	33
34	0???????????????????????????????	1???????????????????????????????	u----------------------------	u----------------------------	34
35	0?????????u??????x0??x-0?????	0?????????u??????x0??x-1?????	u---------1-------0----u-----	u---------0-------0----u----	35
36	1x???????xu?-01B?--0Bx--u0D????	0x??????xu?-11B?--1Bx--u0D????	n---------1--u1---u----10----	n---------0--u1---u----00----	36
37	11-0D0B??0n07101-x-10-01u01C???x	11-0D1B??0n17100-x-10-00u10C??x	11-0-u---00u-10n---10-0n1un-----	11-0-u---01u-10n---10-0n0un1----	37
38	00u0nn-1n01uu000uu-011u00nnn-01-	01u0nn-1n01uu110uu-001u10nnn-11-	0u1000-100111uu011-0n11u0000-u1-	0u0011-110100uu000-0n10u0111-u1-	38
39	n101-1000100-0-0000-1---100-010n	n110-0010101-0-1001-1---001-100n	01un-n0u010u-0-u00u-1---n0u-un00	11un-n0u010u-0-u00u-1---n0u-un01	39
40	1-100010001-01--0n1-u-0-00---11-1	1-010011101-00--1n1-u-0-10--11-1	1-nu001uu01-0n--u01-1-0-u0--11-1	1-nu001uu01-0n--u11-0-0-u0--11-1	40
41	0?????????u??????x0??x-0?????	0?????????u??????x0??x-1?????	1--n--00--0-u1--u--u1--001-0---1	0--n--00--0-u1--u--u0--001-0---1	41
42	u---1-01001-110--n01011--n10---1	u--0-01110-011--n00000--n00---0	1---n-u1uun-n1u--00n0nn--0n0---n	0---n-u1uun-n1u--10n0nn--1n0---n	42
43	n------01---0------u------00-un	n------00---0------u------01-un	0????--0nD??0x????71x??x--0u-10	1????--0nD??0x????70x??x--0u-01	43
44	0----10-------------1u-----	0-----0------------1u-----	0?????C0????????????711????x	0?????C0????????????710?????x	44
45	------00---------u----1-----	------00---------u----1----	?????700????????????71?????71????	?????700?????????70?????1????	45
46	u--------------------------	u--------------------------	1????????????????????????????????	0????????????????????????????????	46
47	--------------------------	--------------------------	????????????????????????????????	????????????????????????????????	47
48	-------u-------------------	-------u-------------------	???????17???????????????????	??????70?????????????????????	48
49	--------n-----------------	--------n-----------------	???????70??????????????????	??????1??????????????????	49
50	--------------------------	--------------------------	????????????????????????????????	????????????????????????????????	50
51	--------------------------	--------------------------	????????????????????????????????	????????????????????????????????	51
52	--------------------------	--------------------------	????????????????????????????????	????????????????????????????????	52
53	--------------------------	--------------------------	????????????????????????????????	????????????????????????????????	53
54	--------------------------	--------------------------	????????????????????????????????	????????????????????????????????	54

- Randomize the position of the forward message difference active bit b_2.
- With the message difference fully specified by b_1, b_2, sample narrow paths in the inner state words in steps denoted by light-gray in Table 7.
- Propagate conditions w.r.t. the three propagation types explained in Section 4. This step is applied repeatedly until none of the three propagation types can be applied on any of the bit positions.

Here, the path sampling is performed simply by initializing randomly the two instances of the path at the given step, calculating the recurrence over the required number of steps and extracting the path. If the Hamming weight of the path is greater than some pre-specified threshold, it is discarded and a new path is sampled. Using the above sampling of partial solution to the paths, the following procedure aims to find the full solution:

(1) Randomize steps in the light-gray area according to the procedure above (steps 43-49 and 34-37 in the forward and backward paths, respectively).
(2) Randomly choose (i, j), $0 \le i \le 31$, $38 \le j \le 42$, a position within the steps denoted by dark-grey in Table 7. If applicable, apply the substitution specified by Table 6. If not, choose another position. In case there is none, return the state.
(3) Propagate conditions and backtrack in case of contradiction. After a contradiction was reached a sufficient number of times, go to step (1).

After reducing the number of steps on which the two differentials meet from 5 to 3 (i.e., putting $k = 4$, where it should be noted that after the propagation the number of unconstrained bits will be relatively small), we received several paths reported as non-contradictory. At that point, there are two possible routes to verify the actual correctness of the intermediate result. One is to switch from 1-bit conditions to multi-bit conditions (such as 1.5-bit or 2.5-bit conditions [15]) that capture more information. ARXtools [15] can readily be used for this purpose. Each 2.5-bit verification using ARXtools for checking the compatibility of two paths took around 3-5 minutes. Another option is to continue with the search heuristic towards extending the specification of the paths to more steps, restarting always from the saved intermediate path state. As the knowledge in the system grows, the propagations turns a high proportion of bits into 0 and 1, which diminishes the possibility of contradiction. If the solution cannot be found after some time threshold t, the path can be abandoned. We experimented with both options above and concluded that both approaches are successful.

3.3 Full Complexity of Finding the HAS-160 Second Order Collision

Our implementation of the heuristic found a correct pair of compatible paths in less than 5 days of execution on an 8-core Intel i7 CPU running at 2.67GHz. In more detail, as explained in Section 3.2, we ran the heuristic to search for paths that meet on 3 instead on 5 steps. It should be noted that due to many propagations, after the search stops, the resulting paths in fact have a small number of remaining unspecified bits in steps 38-42 (less than 32). The heuristic yielded around 8 solutions per day and among 40 returned path pairs, one turned out to be compatible and was successfully extended by one step more, as shown in Table 8.

The conditions for the two paths that are not explicitly given as u,n,0,1 bits in Table 8 are provided in Tables 9 and 10. To find the quartet of message words and inner states that follow the two differentials in steps 34 to 49, inner

Table 9. Backward differential conditions not shown in Table 8

Step	Conditions
33	$A_{33,14} \neq A_{32,14}$
34	$A_{34,20} = A_{33,20}$
35	$A_{35,0} \neq A_{34,0},\ A_{35,16} \neq A_{33,31},\ A_{35,26} \neq A_{34,26}$
36	$A_{36,3} = A_{35,3},\ A_{36,9} \neq A_{35,9},\ A_{36,21} = A_{35,21},\ A_{36,22} = A_{34,5},\ A_{36,23} = A_{35,23}$
37	$A_{37,0} = A_{36,0},\ A_{37,1} = A_{36,1},\ A_{37,2} \neq A_{35,17},\ A_{37,13} = A_{36,13},\ A_{37,23} \neq A_{36,23}$
38	$A_{38,25} = A_{36,8}$
39	$A_{39,19} \vee \overline{A}_{37,2} = 1$
40	$A_{40,17} \vee \overline{A}_{38,0} = 1,\ A_{40,30} \vee \overline{A}_{38,13} = 1$
41	$A_{41,16} \vee \overline{A}_{39,23} = 1$

Table 10. Forward differential conditions not shown in Table 8

Step	Conditions
37	$A_{37,2} = A_{36,2},\ A_{37,3} \neq A_{36,3},\ A_{37,10} \neq A_{36,10},\ A_{37,13} = A_{36,28},\ A_{37,15} = 0,\ A_{37,25} = A_{36,8},\ A_{37,29} = A_{36,12}$
38	$A_{38,0} = 1$
39	$A_{39,4} = 1,\ A_{39,8} = 0,\ A_{39,9} = 1,\ A_{39,12} = 0,\ A_{39,17} = 0,\ A_{39,19} = 1$
40	$A_{40,4} = 0,\ A_{40,5} = 0,\ A_{40,8} = 0,\ A_{40,12} = 1$
41	$A_{41,13} = 0,\ A_{41,14} = 0$
42	$A_{42,7} = 0,$
43	$A_{43,6} = 0,\ A_{43,7} \vee \overline{A}_{41,14} = 1$
44	$A_{44,0} = 0,\ A_{44,1} = 0,\ A_{44,4} \vee \overline{A}_{42,11} = 1,\ A_{44,26} \vee \overline{A}_{42,1} = 1$
45	$A_{45,26} = 0$
46	$A_{46,4} \vee \overline{A}_{44,11} = 1$
47	$A_{47,4} = 1,\ A_{47,24} \vee \overline{A}_{45,31} = 1,\ A_{47,31} = 1$
48	$A_{48,31} = 0$
49	$A_{49,17} = 0$
50	$A_{50,17} = 0,\ A_{50,24} = 1$
51	$A_{51,17} = 0$

state registers in step 42 are chosen to follow the conditions specified by Tables 9,10 and Table 8 and then the usual message modification procedure is applied backward and forward.

Once the middle steps of the quartet structure $n_2 = 34$ to $n_3 = 53$ are satisfied, the second order collision property extends to steps $n_1 = 8$ to $n_4 = 78$ with probability 1 (see Fig. 1 (b)). To cover all of the compression function steps, the middle steps are kept constant and the remaining ones are randomized until the second order collision property is satisfied. In particular, if m_6 and m_{15} are randomized while $m_6 \oplus m_{15}$ is kept constant, according to the message expansion specification, the inner state will be randomized for $54 \leq i \leq 80$ and $0 \leq i \leq 35$. Similarly, if m_6 and m_4 are randomized where $m_6 \oplus m_4$ is kept constant, the randomization will happen for $52 \leq i \leq 79$ and $0 \leq i \leq 34$. Here, a small penalty in probability is paid due to the fact that the paths may be corrupted towards the start/end points. The two mentioned randomizations provide around 64 bits of freedom.

The probability that one randomization explained above yields a second order collision can be bounded from below by $p^2 q^2$, where p and q are the probabilities of two selected sparse differentials in steps $0 \leq i \leq n_1$ and $n_4 \leq i < 80$, respectively. By counting the number of conditions in sparse paths that happened in

Table 11. Message differences after propagation

step	$\Delta[W_A, W_B]$	$\Delta[W_D, W_C]$	$\Delta[W_B, W_C]$	$\Delta[W_A, W_D]$
33	1------------------	0------	n------	n------
34	1------------------	0------	n------	n------
35	-------------------	-------	-------	-------
36	1------------------	0------	n------	n------
37	-------------------	-------	-------	-------
38	-------------------	-------	-------	-------
39	1------------------	0------	n------	n------
40	0---------u-------	1-----------u----	u---------1----	u----------0----
41	-----0----u-------	----0----u----	----0----1----	----0----0----
42	----------------1	----------1	-----------1	-----------1
43	-------------------	-------	-------	-------
44	1------------------	0------	n------	n------
45	1------------------	0------	n------	n------
46	1------------------	0------	n------	n------
47	-------------------	-------	-------	-------
48	-------------------	-------	-------	-------

the quartet in Table 4, we obtain $p = 2^{-22}$ and $q = 2^{-3}$ and the probability lower bound $p^2 q^2 = 2^{-50}$. The actual time of execution on the above mentioned PC was less than two days, due to the additional differential paths which contribute to the exact probability of achieving the second order collision property (previously named amplified probability [3,15]).

4   Details on Condition Propagation

The heuristic keeps track of the current state of the system by keeping the following information in memory:

- Four differential path tables keeping the current state of bit-conditions
- $4 \times r$ carry graphs [24] (one carry graph for each of four paths consisting of r steps)

In our implementation, we used $r = 16$, keeping the information about steps 33-48. The carry graphs model the carry transitions allowed by the knowledge present in the system. Below, the three types of knowledge propagation are described. The propagations are applied as long as the system is not fully propagated with respect to all three types below.

4.1   Single-Path Propagations

An explicit example of a single-path propagation [5] (see also [24,26]) is provided below. The constraints and the corresponding carry graphs for at a particular bit position are all explicitly shown. The new propagated constraints as well as the removed carry graph edges are indicated.

Throughout the compression function execution specified by (4), for any $1 \le i \le 80$ and $0 \le j \le 31$, bit A_i^j is computed based on the 5 input bits in A_{i-j}, $1 \le j \le 5$, the message word bit as well as a particular constant bit. Moreover,

δK	0110111011011001111010111010100001
$\delta[W_{B,41}, W_{C,41}]$	--------0-----0--------------------
$\delta[B_{37}, C_{37}]$	11-0-u---00u-10n---10-0n1un-----
$\delta[B_{38}, C_{38}]$	0u1000-100111uu011-0n11u0000-u1-
$\delta[B_{39}, C_{39}]$	01un-n0u010u-0-u00u-1---nOu-un00
$\delta[B_{40}, C_{40}]$	1-nu001uu01-0n--u01-1-0-u0--11-1
$\delta[B_{41}, C_{41}]$	1--n--00--0-u1--u--u1--001-0---1
$\delta[B_{42}, C_{42}]$	1---n-u1uun-n1u--00n0nn--0n0---n

δK	0110111011011001111010111010100001
$\delta[W_{B,41}, W_{C,41}]$	--------0-----0--------------------
$\delta[B_{37}, C_{37}]$	11-0-u---00u-10n---10-0n1un-----
$\delta[B_{38}, C_{38}]$	0u1000-100111uu011-0n11u0000-u1-
$\delta[B_{39}, C_{39}]$	01un-n0u010u-0-u00u-1---nOu-un00
$\delta[B_{40}, C_{40}]$	1-nu001uu01-0n--u01-1-0-u0--11-1
$\delta[B_{41}, C_{41}]$	1--n--00--0-u1--u--u1--001-0---1
$\delta[B_{42}, C_{42}]$	1---n-u1uun-n1u--00n0nn--0n0---n

Fig. 3. Extract of single-path path constraints

bit A_i^j depends on the carries coming from the computations at bit positions $j < k \leq 0$.

In Fig. 3, an extract of the path is provided, borrowed from the $\Delta[B, C]$ path in Table 8. The bit positions treated in this case are $\delta[B_{42}^1, C_{42}^1]$ (left) and $\delta[B_{42}^0, C_{42}^0]$ (right). The shaded bits are the bit positions participating in the computation of the two bits. As for the carry graph, it consists of 32 subgraphs, each comprising of 5×5 nodes. In Fig. 3, only the subgraphs corresponding to bit positions 1 (left) and 0 (right) are shown. Each subgraph node represents a particular carry configuration at the particular bit position. Due to the fact that there is 5 summands in (4), the carry value is limited to $\{0, \ldots 4\}$ and thus each subgraph contains 5×5 nodes. The edges in the graphs represent possible carry configuration transitions from bit position i to $i + 1$.

Next, the edges connecting subgraphs for bit positions $i = 0$ to $i = 1$ in Fig. 3 are explained. The shown edges and the corresponding bit-conditions are aligned in the sense that there is no possible propagations at the particular positions neither from the bit-conditions to graphs nor vice-versa. According to the bit-conditions on position 0, we have

$$c_B^1 | B_{42}^0 = c_B^1 | 1 = 1 + W_{B,41}^0 + B_{37}^{15} + f_2(1,1,1) + 0 = 1 + W_{B,41}^0 + B_{37}^{15}$$
$$c_C^1 | C_{42}^0 = c_C^1 | 0 = 1 + W_{C,41}^0 + C_{37}^{15} + f_2(1,0,1) + 0 = 1 + W_{C,41}^0 + C_{37}^{15} + 1$$

From the above two equalities, it follows that $W_{B,41}^0 = B_{37}^{15}$ and $W_{C,41}^0 = C_{37}^{15}$. Since $\delta[W_{B,41}^0, W_{C,41}^0]$ and $\delta[B_{37}^{15}, C_{37}^{15}]$ are set to -, the possible carry configurations are $(c_B^1, c_C^1) \in \{(0,1), (1,2)\}$, which corresponds to the two edges between the two subgraphs.

Whenever it is possible to deduce new information from what is already present in the system, propagations need to be carried out until no new information can be derived. Continuing with the setting in Fig. 3, assume that during the heuristic, the symbol - at position $\delta[W_{B,41}^0, W_{C,41}^0]$ is substituted by 0. Then, the propagation at this bit consists of substituting - at position $\delta[B_{37}^{15}, C_{37}^{15}]$ by 0 and deleting the $(0,0) \mapsto (1,2)$ graph edge. The edge deletion continues to the left and to the right. In case of Fig. 3, this amounts to deleting the edges coming out of node $(1, 2)$ and continuing in the same manner throughout the rest of the subgraphs. Next, all of the influenced bit positions, either through carry graphs or through bit-conditions, need to be repropagated similarly to the process described above.

4.2   Quartet Propagations

This type of propagations is the simplest of all three types presented in this section, since it does not involve the carry graphs. An example of this type of propagation is as follows. Let (i, j) denote a specific bit position in the range of the considered steps. Let the bit-conditions $\delta[A_i^j, B_i^j]$, $\delta[D_i^j, C_i^j]$, $\delta[B_i^j, C_i^j]$, $\delta[A_i^j, D_i^j]$ in the four paths be equal to u, x, -, and ?, respectively. It follows that $A_i^j = 0$, $B_i^j = 1$, $C_i^j = 1$ and $D_i^j = 0$ and thus the quartet can be readily substituted by a new one

$$(\texttt{ux-?}) \mapsto (\texttt{uu10})$$

Given a quartet of conditions, the substitution quartet is found by going through all the bit value quartets that satisfy the given condition quartet. The new quartet consists of the symbols from Table 2 that represent minimal sets contain the valid bit value pairs.

4.3   Quartet Addition Propagations

In this subsection, the following terminology is adopted: carry subgraphs as shown in Fig. 3 are called *2-graphs*. Nodes with at least one input/output edge in the 2-graphs are called *active* nodes. During the execution of the heuristic, each active 2-graph node corresponds to a possible carry configuration that has not yet been ruled out by the heuristic.

Quartet addition propagation is illustrated in Fig. 4. The four graphs in the top part represent a particular case of the 2-graphs that correspond to a single bit position (i, j) on paths $[A, B]$, $[B, C]$, $[D, C]$, $[A, D]$, respectively from left to right. The active nodes are circled and the information about the number of input/output edges is abstracted from the picture. The quartet addition propagation is based on the fact that the four different 2-graphs may impose incompatible constraint on the carry configurations at the considered bit position. For instance, according to the 2-graph corresponding to the path $[D, C]$ (third graph from the left in Fig. 4), since node $(c_D, c_C) = (3, 2)$ is active, it follows that having a carry equal to 3 at this bit position in the branch D is not ruled out. However, since there is no active nodes in the third column of the (c_A, c_D) graph, the node $(c_D, c_C) = (3, 2)$ should be deactivated.

For the purpose of deciding which 2-carry graph nodes should be deactivated, it is convenient to introduce another type of carry graphs that will be called *4-carry* graphs. For each bit-position covered by the heuristic, the four 2-carry graphs are represented as one 4-carry graph, as shown in the bottom part of

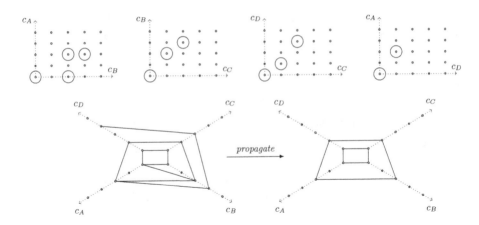

Fig. 4. Example: 2-carry graphs and the corresponding 4-carry graph before and after propagation

Fig. 4. The 4-carry graphs abstract the information about active nodes in the 2-carry graphs.

As shown in Fig. 4, the 4-carry graph has four groups of nodes that simply represent the carry values c_A, c_B, c_C and c_D, respectively. The edges in the 4-carry graph are constructed simply by mapping the active nodes in the corresponding 2-carry graphs to the edges between the corresponding node groups. This mapping is specified by an example as follows. The active nodes in the (c_A, c_D) 2-carry graph are $(0, 0)$ and $(2, 1)$. This is translated to the edges $(0, 0)$ and $(2, 1)$ between the c_A and c_D branches in the 4-carry graph. The other three 2-carry graph active nodes are mapped to the edges analogously.

The 4-carry graph representation allows expressing the quartet addition propagation rules in a natural way. For that purpose, let a *cycle* denote a closed path connecting four nodes, where no two nodes are members of the same node group in the 4-graph. The propagation rules are then as follows:

(R1) Remove all "dead-end" edges, i.e., the ones with an end node of degree 1
(R2) Remove all edges that do not participate in any cycle

In the case of the propagation given in Fig. 4, the quartet addition propagation consisted of three applications of (R1) and one application of (R2). Since each 4-graph edge corresponds to a node in the corresponding 2-graph, the edge removal according to rules (R1) and (R2) amounts to deactivating the corresponding nodes in the 2-graph. The node deactivation is done by deleting all input and output edges for the corresponding 2-graph node. In the case of our HAS-160 search, implementing only rule (R1) turned out to be sufficient.

5   Conclusion

We proposed a heuristic for searching for compatible differential paths and applied it to HAS-160. Instead of working with 0/1 bit values, we used the reasoning on sets of bits described by 1-bit constraints. The three types of propagations used during the search (single-path propagations, quartet propagations and quartet addition propagations) are explained through particular examples. Using the 1-bit constraints along with these propagations yielded an acceptable rate of false positives and the second order collision was successfully found. One possible future research direction is to evaluate the performance of the proposed heuristic in case of SHA-2 with a goal of improving the attack [3] and to assess the impact of high rate of contradictory paths reported in [20] in this context.

Acknowledgments. The authors would like to thank Gaëtan Leurent for his help related to ARXtools and the discussions on the topic.

References

1. Telecommunications Technology Association. Hash Function Standard Part 2, Hash Function Algorithm Standard (HAS-160), TTAS.KO-12.0011/R1 (2008)
2. Aumasson, J.-P.: Zero-sum distinguishers, Rump session talk at CHES (2009), http://131002.net/data/talks/zerosum_rump.pdf
3. Biryukov, A., Lamberger, M., Mendel, F., Nikolic, I.: Second-order differential collisions for reduced SHA-256. In: Lee, Wang (eds.) [14], pp. 270–287
4. Biryukov, A., Nikolić, I., Roy, A.: Boomerang attacks on BLAKE-32. In: Joux, A. (ed.) FSE 2011. LNCS, vol. 6733, pp. 218–237. Springer, Heidelberg (2011)
5. De Cannière, C., Rechberger, C.: Finding SHA-1 characteristics: General results and applications. In: Lai, X., Chen, K. (eds.) ASIACRYPT 2006. LNCS, vol. 4284, pp. 1–20. Springer, Heidelberg (2006)
6. Cho, H.-S., Park, S., Sung, S.H., Yun, A.: Collision search attack for 53-step HAS-160. In: Rhee, M.S., Lee, B. (eds.) ICISC 2006. LNCS, vol. 4296, pp. 286–295. Springer, Heidelberg (2006)
7. Fouque, P.-A., Leurent, G., Nguyen, P.Q.: Automatic search of differential path in MD4. IACR Cryptology ePrint Archive, 2007:206 (2007)
8. Hong, D., Koo, B., Sasaki, Y.: Improved preimage attack for 68-step HAS-160. In: Lee, D., Hong, S. (eds.) ICISC 2009. LNCS, vol. 5984, pp. 332–348. Springer, Heidelberg (2010)
9. Khovratovich, D.: Bicliques for permutations: Collision and preimage attacks in stronger settings. In: Wang, Sako (eds.) [34], pp. 544–561
10. Kim, J.-S., Biryukov, A., Preneel, B., Hong, S.H.: On the security of HMAC and NMAC based on HAVAL, MD4, MD5, SHA-0 and SHA-1 (Extended abstract). In: De Prisco, R., Yung, M. (eds.) SCN 2006. LNCS, vol. 4116, pp. 242–256. Springer, Heidelberg (2006)
11. Kircanski, A., Shen, Y., Wang, G., Youssef, A.M.: Boomerang and slide-rotational analysis of the SM3 hash function. In: Knudsen, Wu (eds.) [12], pp. 304–320
12. Knudsen, L.R., Wu, H. (eds.): SAC 2012. LNCS, vol. 7707. Springer, Heidelberg (2013)

13. Lamberger, M., Mendel, F.: Higher-order differential attack on reduced SHA-256. IACR Cryptology ePrint Archive, 2011:37 (2011)
14. Lee, D.H., Wang, X. (eds.): ASIACRYPT 2011. LNCS, vol. 7073. Springer, Heidelberg (2011)
15. Leurent, G.: Analysis of differential attacks in ARX constructions. In: Wang, Sako (eds.) [34], pp. 226–243
16. Leurent, G.: Construction of differential characteristics in ARX designs - application to Skein. IACR Cryptology ePrint Archive, 2012:668 (2012)
17. Leurent, G., Roy, A.: Boomerang attacks on hash function using auxiliary differentials. In: Dunkelman, O. (ed.) CT-RSA 2012. LNCS, vol. 7178, pp. 215–230. Springer, Heidelberg (2012)
18. Mendel, F., Nad, T.: Boomerang distinguisher for the SIMD-512 compression function. In: Bernstein, D.J., Chatterjee, S. (eds.) INDOCRYPT 2011. LNCS, vol. 7107, pp. 255–269. Springer, Heidelberg (2011)
19. Mendel, F., Nad, T., Schläffer, M.: Cryptanalysis of round-reduced HAS-160. In: Kim, H. (ed.) ICISC 2011. LNCS, vol. 7259, pp. 33–47. Springer, Heidelberg (2012)
20. Mendel, F., Nad, T., Schläffer, M.: Finding SHA-2 characteristics: Searching through a minefield of contradictions. In: Lee, Wang (eds.) [14], pages 288–307
21. Mendel, F., Nad, T., Schläffer, M.: Collision attacks on the reduced dual-stream hash function RIPEMD-128. In: Canteaut, A. (ed.) FSE 2012. LNCS, vol. 7549, pp. 226–243. Springer, Heidelberg (2012)
22. Mendel, F., Nad, T., Schläffer, M.: Finding collisions for round-reduced SM3. In: Dawson, E. (ed.) CT-RSA 2013. LNCS, vol. 7779, pp. 174–188. Springer, Heidelberg (2013)
23. Mendel, F., Rijmen, V.: Colliding message pair for 53-step HAS-160. In: Nam, K.-H., Rhee, G. (eds.) ICISC 2007. LNCS, vol. 4817, pp. 324–334. Springer, Heidelberg (2007)
24. Mouha, N., De Cannière, C., Indesteege, S., Preneel, B.: Finding collisions for a 45-step simplified HAS-V. In: Youm, H.Y., Yung, M. (eds.) WISA 2009. LNCS, vol. 5932, pp. 206–225. Springer, Heidelberg (2009)
25. Murphy, S.: The return of the cryptographic boomerang. IEEE Transactions on Information Theory 57(4), 2517–2521 (2011)
26. Peyrin, T.: Analyse de fonctions de hachage cryptographes. Ph.D. Thesis, University of Versailles (2008), http://www.iacr.org/phds/?p=detail&entry=500
27. Sasaki, Y.: Boomerang distinguishers on MD4-family: First practical results on full 5-pass haval. In: Miri, A., Vaudenay, S. (eds.) SAC 2011. LNCS, vol. 7118, pp. 1–18. Springer, Heidelberg (2012)
28. Sasaki, Y., Aoki, K.: A preimage attack for 52-step HAS-160. In: Lee, P.J., Cheon, J.H. (eds.) ICISC 2008. LNCS, vol. 5461, pp. 302–317. Springer, Heidelberg (2009)
29. Sasaki, Y., Wang, L.: Distinguishers beyond three rounds of the RIPEMD-128/-160 compression functions. In: Bao, F., Samarati, P., Zhou, J. (eds.) ACNS 2012. LNCS, vol. 7341, pp. 275–292. Springer, Heidelberg (2012)
30. Sasaki, Y., Wang, L., Takasaki, Y., Sakiyama, K., Ohta, K.: Boomerang distinguishers for full HAS-160 compression function. In: Hanaoka, G., Yamauchi, T. (eds.) IWSEC 2012. LNCS, vol. 7631, pp. 156–169. Springer, Heidelberg (2012)
31. Schläffer, M., Oswald, E.: Searching for differential paths in MD4. In: Robshaw, M. (ed.) FSE 2006. LNCS, vol. 4047, pp. 242–261. Springer, Heidelberg (2006)
32. Stevens, M., Lenstra, A.K., de Weger, B.: Chosen-prefix collisions for MD5 and colliding X.509 certificates for different identities. In: Naor, M. (ed.) EUROCRYPT 2007. LNCS, vol. 4515, pp. 1–22. Springer, Heidelberg (2007)

33. Wagner, D.: The boomerang attack. In: Knudsen, L.R. (ed.) FSE 1999. LNCS, vol. 1636, pp. 156–170. Springer, Heidelberg (1999)
34. Wang, X., Sako, K. (eds.): ASIACRYPT 2012. LNCS, vol. 7658. Springer, Heidelberg (2012)
35. Wang, X., Yin, Y.L., Yu, H.: Finding collisions in the full SHA-1. In: Shoup, V. (ed.) CRYPTO 2005. LNCS, vol. 3621, pp. 17–36. Springer, Heidelberg (2005)
36. Wang, X., Yu, H.: How to break MD5 and other hash functions. In: Cramer, R. (ed.) EUROCRYPT 2005. LNCS, vol. 3494, pp. 19–35. Springer, Heidelberg (2005)
37. Yu, H., Chen, J., Wang, X.: The boomerang attacks on the round-reduced Skein-512. In: Knudsen, Wu (eds.) [12], pp. 287–303
38. Yun, A., Sung, S.H., Park, S., Chang, D., Hong, S.H., Cho, H.-S.: Finding collision on 45-step HAS-160. In: Won, D.H., Kim, S. (eds.) ICISC 2005. LNCS, vol. 3935, pp. 146–155. Springer, Heidelberg (2006)

Improved Cryptanalysis
of Reduced RIPEMD-160

Florian Mendel[1], Thomas Peyrin[2], Martin Schläffer[1],
Lei Wang[2], and Shuang Wu[2]

[1] IAIK, Graz University of Technology, Austria
{florian.mendel,martin.schlaeffer}@iaik.tugraz.at
[2] Nanyang Technological University, Singapore
thomas.peyrin@gmail.com,
{wang.lei,wushuang}@ntu.edu.sg

Abstract. In this article, we propose an improved cryptanalysis of the double-branch hash function standard RIPEMD-160. Using a carefully designed non-linear path search tool, we study the potential differential paths that can be constructed from a difference in a single message word and show that some of these message words can lead to very good differential path candidates. Leveraging the recent freedom degree utilization technique from Landelle and Peyrin to merge two branch instances, we eventually manage to obtain a semi-free-start collision attack for 42 steps of the RIPEMD-160 compression function, while the previously best know result reached 36 steps. In addition, we also describe a 36-step semi-free-start collision attack which starts from the first step.

Keywords: RIPEMD-160, semi-free-start collision, compression function, hash function.

1 Introduction

Due to their widespread use in many applications and protocols, hash functions are among the most important primitives in cryptography. A hash function H is a function that takes an arbitrarily long message M as input and outputs a fixed-length hash value of size n bits. Cryptographic hash functions have the extra requirement that some security properties, such as collision resistance and (second)-preimage resistance, must be fulfilled. More precisely, it should be impossible for an adversary to find a collision (two distinct messages that lead to the same hash value) in less than $2^{n/2}$ hash computations, or a (second)-preimage (a message hashing to a given challenge) in less than 2^n hash computations. Most standardized hash functions are based upon the Merkle-Damgård paradigm [13,3] and iterate a compression function h with fixed input size to handle arbitrarily long messages. The compression function itself should ensure equivalent security properties in order for the hash function to inherit from them.

The cryptographic community have seen very impressive advances in hash functions cryptanalysis in the recent years [20,18,19,17], with weaknesses or even

K. Sako and P. Sarkar (Eds.) ASIACRYPT 2013, Part II, LNCS 8270, pp. 484–503, 2013.

sometimes collisions exhibited for many standards such as MD4, MD5, SHA-0 and SHA-1. These functions have in common their design strategy, based on the utilization of additions, rotations, xors and boolean functions in an unbalanced Feistel network. In order to diversify the panel of standardized hash functions and make a backup plan available in case the last survivors of this MD-SHA family gets broken as well, NIST organized a 4-year SHA-3 competition which led to the selection of Keccak [1] as new standardized primitive. The move of the industry towards SHA-3 will take a lot of time and even broken functions such as MD5 or SHA-1 remain widely used. Among the MD-SHA family, only SHA-2 and RIPEMD-160 compression functions are still unbroken, although practical collisions on the SHA-2 compression function have been improved from 24 to 38 steps recently [11,12]. The compression function used in RIPEMD-128 was recently shown not to be collision resistant [8].

RIPEMD can be considered as a subfamily of the MD-SHA-family as its first representative, RIPEMD-0 [2], basically consists in two MD4-like [15] functions computed in parallel (but with different constant additions for the two branches), with 48 steps in total. Even though RIPEMD-0 was recommended by the European *RACE Integrity Primitives Evaluation* (RIPE) consortium, its security was put into question with the early work from Dobbertin [5] and the practical collision attack from Wang *et al.* [17]. Meanwhile, in 1996, Dobbertin, Bosselaers and Preneel [6] proposed two strengthened versions of the original RIPEMD-0, called RIPEMD-128 and RIPEMD-160, with 128/160-bit output and 64/80 steps respectively. RIPEMD-0 main flaw was that its two computation branches were too much similar and this issue was patched in RIPEMD-128 and RIPEMD-160 by using not only different constants, but also different rotation values, boolean functions and message insertion schedules in the two branches. This two-branch structure in RIPEMD family is a good method to reduce the ability of the attacker to properly use the available freedom degrees and to find good differential paths for the entire scheme. RIPEMD-160 is a worldwide ISO/IEC standards [7] that is yet unbroken and is present in many implementations of security protocols.

As of today, the best results on RIPEMD-160 are a very costly 31-step preimage attack [14], a practical 36-step semi-free-start collision attack [10] on the compression function (not starting from the first step), and a distinguisher on up to 51 steps of the compression function with a very high complexity [16].

Our Contributions.   In this article, we improve the best know results on RIPEMD-160, proposing a semi-free-start collision attack on 42 steps of its compression function and a semi-free-start collision attack on 36 steps starting from the first step. Our differential paths were crafted thanks to a very efficient non-linear path search tool (Section 3) and by inserting a difference only in a single message word, in a hope for a sparse difference trail. We then explain in Section 4 why we believe the 8th message input word (M_7) is the best candidate for that matter. Once the differential paths settled, we leverage in Section 5 the freedom degree utilization technique introduced by Landelle and Peyrin [8] for RIPEMD-128 that merges two branch instances together in order to obtain a semi-free-start collision. It is to be noted that the step function of RIPEMD-160 makes it much more difficult to find a collision attack compared to RIPEMD-128. This is mainly due to the fact that the diffusion is better, but also because even though differences might be absorbed in the boolean function, they will propagate anyway at least once through a free term (which was not the case in RIPEMD-128). We give a description of RIPEMD-160 in Section 2 and summarize our results in Section 6.

2   Description of RIPEMD-160

RIPEMD-160 [6] is a 160-bit hash function that uses the Merkle-Damgård construction as domain extension algorithm: the hash function is built by iterating a 160-bit compression function h that takes as input a 512-bit message block m_i and a 160-bit chaining variable cv_i:

$$cv_{i+1} = h(cv_i, m_i)$$

where the message m to hash is padded beforehand to a multiple of 512 bits[1] and the first chaining variable is set to a predetermined initial value $cv_0 = IV$.

We refer to [6] for a complete description of RIPEMD-160. In the rest of this article, we denote by $[Z]_i$ the i-th bit of a word Z, starting the counting from 0. ⊞ and ⊟ represent the modular addition and subtraction on 32 bits, and $\oplus$, $\vee$, $\wedge$, the bitwise "exclusive or", the bitwise "or", and the bitwise "and" function respectively.

2.1   RIPEMD-160 Compression Function

The RIPEMD-160 compression function is a wider version of RIPEMD-128, which is in turn based on MD4, but with the particularity that it uses two parallel instances of it. We differentiate these two computation branches by left and right branch and we denote by X_i (resp. Y_i) the 32-bit word of left branch (resp. right branch) that will be updated during step i of the compression function. The compression function process is composed of 80 steps divided into 5 rounds of 16 steps each in both branches.

Initialization. The 160-bit input chaining variable cv_i is divided into 5 words h_i of 32 bits each, that will be used to initialize the left and right branch 160-bit internal state:

$$X_{-4} = (h_0)^{\ggg 10} \quad X_{-3} = (h_4)^{\ggg 10} \quad X_{-2} = (h_3)^{\ggg 10} \quad X_{-1} = h_2 \quad X_0 = h_1$$
$$Y_{-4} = (h_0)^{\ggg 10} \quad Y_{-3} = (h_4)^{\ggg 10} \quad Y_{-2} = (h_3)^{\ggg 10} \quad Y_{-1} = h_2 \quad Y_0 = h_1 \ .$$

---

[1] The padding is the same as for MD4: a "1" is first appended to the message, then x "0" bits (with $x = 512 - (|m| + 1 + 64 \pmod{512})$) are added, and finally the message length $|m|$ coded on 64 bits is appended as well.

The Message Expansion. The 512-bit input message block is divided into 16 words M_i of 32 bits each. Each word M_i will be used once in every round in a permuted order (similarly to MD4 and RIPEMD-128) and for both branches. We denote by W_i^l (resp. W_i^r) the 32-bit expanded message word that will be used to update the left branch (resp. right branch) during step i. We have for $0 \le j \le 4$ and $0 \le k \le 15$:

$$W_{j \cdot 16 + k}^l = M_{\pi_j^l(k)} \quad \text{and} \quad W_{j \cdot 16 + k}^r = M_{\pi_j^r(k)}$$

where π_j^l and π_j^r are permutations.

The Step Function. At every step i, the registers X_{i+1} and Y_{i+1} are updated with functions f_j^l and f_j^r that depend on the round j in which i belongs:

$$X_{i+1} = (X_{i-3})^{\lll 10} \boxplus ((X_{i-4})^{\lll 10} \boxplus \Phi_j^l(X_i, X_{i-1}, (X_{i-2})^{\lll 10}) \boxplus W_i^l \boxplus K_j^l)^{\lll s_i^l},$$

$$Y_{i+1} = (Y_{i-3})^{\lll 10} \boxplus ((Y_{i-4})^{\lll 10} \boxplus \Phi_j^r(Y_i, Y_{i-1}, (Y_{i-2})^{\lll 10}) \boxplus W_i^r \boxplus K_j^r)^{\lll s_i^r},$$

where K_j^l, K_j^r are 32-bit constants defined for every round j and every branch, s_i^l, s_i^r are rotation constants defined for every step i and every branch, Φ_j^l, Φ_j^r are 32-bit boolean functions defined for every round j and every branch.

The Finalization. A finalization and a feed-forward is applied when all 80 steps have been computed in both branches. The four 32-bit words h_i' composing the output chaining variable are finally obtained by:

$$h_0' = \qquad h_1 \boxplus X_{79} \boxplus (Y_{78})^{\lll 10} \qquad\qquad h_1' = \quad h_2 \boxplus (X_{78})^{\lll 10} \boxplus (Y_{77})^{\lll 10}$$
$$h_2' = \quad h_3 \boxplus (X_{77})^{\lll 10} \boxplus (Y_{76})^{\lll 10} \qquad\qquad h_3' = \qquad\qquad h_4 \boxplus (X_{76})^{\lll 10} \boxplus Y_{80}$$
$$h_4' = \qquad\qquad h_0 \boxplus X_{80} \boxplus Y_{79}$$

3 Non-linear Path Search

To find a non-linear differential path in RIPEMD-160, we use the techniques developed by Mendel et al. [11,12]. This automated search algorithm can be used to find both, differential characteristics and conforming message pairs (note that we will use it only for the differential characteristics part in this article). We briefly describe the tool in Section 3.1, and the new improvements and specific configuration for RIPEMD-160 and our attack in Section 3.2.

3.1 Automated Search for Differential Characteristics

The basic idea of the search algorithm is to pick and guess previously unrestricted bits. After each guess, the information due to these restrictions is propagated to other bits. If an inconsistency occurs, the algorithm backtracks to an earlier state of the search and tries to correct it. Similar to [11,12], we denote these three parts of the search by decision (guessing), deduction (propagation), and backtracking (correction). Then, the search algorithm proceeds as follows:

Decision (Guessing)
1. Pick randomly (or according to some heuristic) an unrestricted decision bit.
2. Impose new constraints on this decision bit.

Deduction (Propagation)
3. Propagate the new information to other variables and equations as described in [11,12].
4. If an inconsistency is detected start backtracking, else continue with step 1.

Backtracking (Correction)
5. Try a different choice for the decision bit.
6. If all choices result in an inconsistency, mark the bit as critical.
7. Jump back until the critical bit can be resolved.
8. Continue with step 1.

During the search, we mainly use generalized conditions [4] to store, restrict and propagate information. The decision of choosing which bits to guess depends strongly on the specific attack, hash function and preferred resulting path. E.g. if some parts of the non-linear path should be especially sparse, we guess the corresponding state words first.

Similar to [11,12], new restrictions are propagated using brute-force propagation within bitslices for each Boolean function and modular addition. In the backtracking, we remember a small set of critical bits and repeatedly check if all of them can be resolved. This way, we leave dead search branches faster. Additionally, we restart the search after a certain number of inconsistencies occur.

The main difficulty in finding a long differential characteristic lies in the fine-tuning of the search algorithm. There are a lot of variations possible which can decide whether the search eventually succeeds or fails. We describe the specific improvements for RIPEMD-160 in the next section.

3.2   Improvements for RIPEMD-160

To efficiently find non-linear differential paths and message pairs for a larger number of steps than in previous attacks [10], we had to improve the search in several ways. Especially finding a non-linear path for the XOR-round of RIPEMD-160 was quite challenging.

In order to improve the propagation of information, we have combined the bitslices of the two modular additions in each step of RIPEMD-160 into a single bitslice. The two carries of the first and second modular addition are computed and stored together within a generalized 3-bit condition, which is defined similarly as the 2.5-bit condition of [9]. Without this combination, many contradictions would be detected very late during the search, and therefore reduce the overall performance.

To find sparser paths at the beginning or end of the non-linear path, we first propagate the single bit condition in the message word backward and forward more or less linearly and by hand. Then, the automatic search tool is used to connect the paths. Note that due to the additional modular addition in RIPEMD-160, we can stay sparse longer in forward direction than in backward direction. This can be observed when looking at our resulting differential paths in Appendix A.

Once we have found a candidate for a differential path, we immediately continue the search for partial confirming message pairs, similar as in [11,12]. We first pick decision bits '-' which are constraint by linear two-bit conditions of the form ($X_{i,j} = X_{k,l}$ or $X_{i,j} \neq X_{k,l}$). This ensures that those bits which influence a lot of other bits are guessed first. This way, inconsistent characteristics are found faster and can also be corrected by the backtracking step of the path search.

4   Differential Paths

The previous semi-free-start collision attacks on RIPEMD compression functions usually start by spending the available freedom degrees in the first steps in each branch, and then continue the computation in the forward direction, verifying the rest of the differential path in each branch probabilistically. With this attack strategy, the non-linear part of the differential paths for both branches should be located in the early steps. Indeed, the non-linear parts are usually the most costly part and therefore should be handle in priority by the attack with the available freedom degrees.

Since the compression functions belonging to the RIPEMD family use a two-branch parallel structure sharing the same initial chaining value, the left and right branches can be regarded as somehow connected in the first steps. With this observation, in [8] Landelle and Peyrin proposed a new method to find semi-free-start collisions for RIPEMD-128. Their method allows the attacker to use the message freedom degrees not necessarily in the early steps of each branch, and therefore relax a bit the constraint that the most costly parts (the non-linear chunks) must be located in the early steps as well. Consequently, the space of possible differential paths is increased and likely to contain better candidates since the probabilistic part in each branch is reduced.

Figure 1 shows the difference between the previous and the new strategies. The attack process proposed in [8] is made of three steps. Firstly, the attacker independently choose the internal states in both branches and start fixing some message words in order to handle the two non-linear parts. Then, he uses some of the remaining message words available to merge the two branches to the same chaining variable by computing backward from the middle. Finally, the rest of the differential path in both branches is verified probabilistically by computing forward from the middle.

4.1   On the Choice of the Message Word

As in [8], in order to find a sparse differential path for a semi-free-start collision attack with the biggest number of steps, we chose to insert differences in only a single message word. Then, for all the 16 message words, we have analyzed how many steps can be potentially attacked. The results are summarized in Table 1. Note that the details of the attacks are not considered at this stage and the final

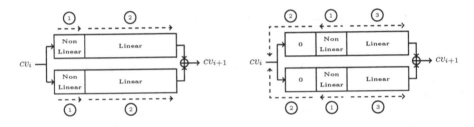

Fig. 1. The previous (left-hand side) and new (right-hand side) approach for collision search on double-branch compression functions introduced in [8]

complexity will highly depend on the merging process and the quality of the differential paths that can be found (both linear and non-linear parts). Yet, by guessing for each message insertion where would be the best location for the two non-linear parts, this preliminary analysis gives a rough estimation of how many steps can be reached potentially. The overall attack being quite complex and time consuming to settle, this will help us to focus directly on good candidates.

We found that message words M_7 and M_{14} both seem to be rather good choices when trying to verify the following criteria:

1. the non-linear part in both branches should be short, in order to consume less freedom degrees.
2. the early steps of the two non-linear parts should be rather close to each other, which will help the merging.
3. the late steps of the non-linear parts should be as sparse as possible, since after the merging comes the probabilistic phase and ensuring a sparse incoming difference mask would guarantee a rather high differential probability when computing forward.
4. some message word difference injections allow the differences injected in very late steps in the two branches to cancel each other through the final feed-forward operation. If this trick is applicable, one can usually get 4 to 6 extra steps for the collision attack with a relatively low cost.

Once M_7 and M_{14} identified as good candidates, we tried to design the entire differential path and establish the merging phase. During our search for the linear part of the differential path, we found it much harder to find good ones for RIPEMD-160 compared to RIPEMD-128. The reason is that the diffusion of the step function of RIPEMD-160 is much better than RIPEMD-128 as it prevents from fully and directly absorbing all the differences. For example, a step in an IF round in RIPEMD-128 can be fully controlled by the attacker such that no difference diffusion occurs. However, in RIPEMD-160, one extra free term appears in the addition of the step function formula and this forces at least a diffusion of a factor two (that cannot be absorbed by the IF function). As a consequence, we were not able to find differential paths as sparse as in [8] and the number of attacked steps is also much lower.

Table 1. Rough estimation of the number of attackable steps for various choices of message words differences injection (in parenthesis are given the steps window)

Message Word	M_0	M_1	M_2	M_3
Attackable Steps	51 (26-76)	46 (2-47)	52 (6-57)	48 (4-51)

Message Word	M_4	M_5	M_6	M_7
Attackable Steps	42 (8-49)	50 (6-55)	39 (10-48)	56 (8-63)

Message Word	M_8	M_9	M_{10}	M_{11}
Attackable Steps	36 (12-47)	39 (10-48)	37 (14-50)	38 (12-49)

Message Word	M_{12}	M_{13}	M_{14}	M_{15}
Attackable Steps	38 (16-53)	34 (41-74)	58 (2-59)	43 (11-53)

4.2   Difficulty of Calculating the Probability

Another important difference between RIPEMD-128 and RIPEMD-160 is the step differential probability calculation. While it is easy to calculate the differential probability for each step of a given differential path of RIPEMD-128, it is not the case for RIPEMD-160. The reason is that the step function in RIPEMD-160 is no longer a S-function (a function for which the i-th output bit depends only on the i first lower bits of all input words), and therefore the accurate calculation of the differential probability is very hard. Yet, one can write the step function as two S-functions by introducing a temporary state that we denote Q_i. We use the step function of the left branch as an example:

$$Q_i = (X_{i-4})^{\lll 10} \boxplus \Phi_j^l(X_i, X_{i-1}, (X_{i-2})^{\lll 10}) \boxplus W_i^l \boxplus K_j^l,$$
$$X_{i+1} = (X_{i-3})^{\lll 10} \boxplus Q_i^{\lll s_i^l}.$$

Now the probability of the sub-steps can be calculated precisely. One possible way to calculate the probability of the step function is to specify the conditions on Q_i and obtain $Pr[X_i \to Q_i] \cdot Pr[Q_i \to X_{i+1}]$ as the step probability.

In fact, this estimation of the probability is not correct. First, there is no freedom degree injected in the step $Q_i \to X_{i+1}$, which means it is not independent from the step of $X_i \to Q_i$. Thus their probability can not be calculated as a simple multiplication. Even if this estimation is accurate, it will only represent a lower bound of the real probability, since there could be a lot of possible equivalent characteristics on Q_i and only one is taken in account here. We used experiments to estimate the real probability and found that the probabilities obtained using the first method is much lower than the real probability observed when running the attack.

We then tried to come up with another way to calculate the step differential probability. We summed the probabilities of the two sub-steps for all possible characteristics on Q_i, i.e. we used $\Sigma_{Q_i}(Pr[X_i \to Q_i] \cdot Pr[Q_i \to X_{i+1}])$ as differential probability for a step. The calculated probability turned out to be much higher than the real one. This is explained by the fact that characteristics on Q_i in different steps will sometimes introduce conditions on X_i and there could be contradictions between some of the conditions. In the calculation, we did not consider the compatibility between the Q_i in different steps. It is therefore not surprising that the calculated probability is much higher.

In the following sections, all the probabilities given were obtained by experiments while testing random samples. We leave the problem of theoretically calculating the real step differential probability as an open problem.

4.3   48-Step Semi-Free-Start Collision Path

We eventually chose M_7 as message word for the single difference insertion and the shape of the differential path that we will use can be found in Figure 2. The non-linear parts are located between steps 16-41 and 19-36 for left the right branch respectively. In steps 58-64, after a linear propagation of the difference injected by M_7, the differences in the output internal state are suitable to apply the feed-forward trick that allows us to get a collision on the output of the compression function (at the end of step 64). The complete differential path is displayed in Figure 5 in Appendix.

Note that this differential path does not necessarily require to be followed until step 64 to find a collision (thanks to the feed-forward trick). Indeed, by stopping 6 steps before (step 58), the last difference insertions from M_7 will be removed and no difference will be present in the internal states in both branches (therefore leading directly to a collision, without even using the feed-forward trick). We did a measurement and found that the collision probability for the feed-forward trick (from step 58 to 64) is about $2^{-11.3}$. However, our attacks requiring already a lot of operations, we have to remove these extra 6 steps and aim for a 42-step semi-free-start collision attack instead. Yet, one should keep in mind that a rather small improvement with regards to the attack complexity would probably lead to the direct obtaining of a 48-step semi-free-start collision attack by putting back the 6 extra steps. The details of the attack will be given in the next section.

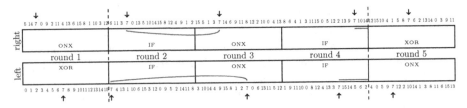

Fig. 2. The shape of our 48-step differential path for the semi-free-start collision attack on the RIPEMD-160 compression function. The numbers represent the message words inserted at each step and the red curves represent the rough amount of differences in the internal state during each step. The arrows show where the bit differences are injected with M_7. The dashed lines represent the limits of the steps attacked.

4.4   36-Step Semi-Free-Start Collision Path from the First Step

Besides the 48-step path, we also exhibit a semi-free-start collision path starting from the first step, which also use message word M_7 to introduce differences. Since the boolean function in the first round of the left branch is XOR, it is quite hard to find a non-linear differential path. As a consequence, the path we were able to find turns out to have three bits of differences in M_7 instead of a single one. The local collisions are located between the first two injections of M_7 in both branches, thus one can directly derive a 36-step collision path starting from the very first step of RIPEMD-160. Figure 3 shows the shape of this differential path and the detailed path is given in Figure 6 in Appendix.

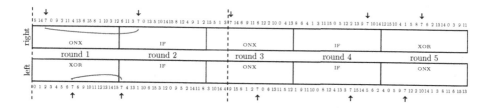

Fig. 3. The shape of our differential path for the 36-step semi-free-start collision attack on the RIPEMD-160 from the first step. The numbers are the message words inserted at each step and the red curves represent the rough amount differences in the internal state during each step. The arrows show where the bit differences are injected with M_7. The dashed lines represent the limits of the steps attacked.

5   Merging the Two Branches

Once the differential path is set, we need not only to find conforming pairs for both branches, but also to merge the two branches in order to make sure that they will reach the same chaining variables on their input. Note that for a semi-free-start collision, one only needs to ensure that the input chaining variables for both branches are the same and the attacker can actually choose this value freely. In contrary, for a hash collision, the attacker would have to merge both branches to the same chaining variable, fixed to a certain predefined value.

5.1   Semi-Free-Start Collision

As explained in previous sections, even though an interesting 48-step differential path has been found (Figure 5), we will only look for a 42-step semi-free-start collision attack on RIPEMD-160, since the feed-forward collision trick would increase the attack complexity beyond the birthday bound. Our algorithm to find a semi-free-start collision is separated in three phases, which we quickly describe here as a high-level view:

- Phase 1: fix some bits of the message words and the internal states in both branches as preparation for the next phases of the attack. This will allow us to fulfill in advance some conditions of the differential path.
- Phase 2: fix the internal state variables $X_{26}, X_{27}, X_{28}, X_{29}, X_{30}$ of the left branch and $Y_{21}, Y_{22}, Y_{23}, Y_{24}, Y_{25}$ of the right branch. Then, iteratively fix message words $M_{11}, M_{15}, M_8, M_3, M_{12}, M_{14}, M_{10}, M_2, M_5, M_9, M_0$ and M_6 in this particular order, so as to fulfill the conditions located inside or close to the non-linear parts. Once these internal state variables or message words are successfully fixed, we call this candidate at the end of phase 2 a *starting point* for the merging.
- Phase 3: use the remaining free message words M_1, M_4, M_7 and M_{13} to merge the internal states of both branches to the same input chaining value. Since every value is fixed at this point, check if the rest of the differential path is fulfilled as well (the uncontrolled part).

Phase 1: Preparation. Before finding a starting point for the merging, we can prepare the differential path by introducing certain conditions on the internal states in both branches in order to increase the probability of the uncontrolled part of the differential path.

The condition that we will force is that bits 16 to 25 of X_{35} must be equal to 0n00n00000. The effect of this condition is that when a starting point will be generated, we will be able to directly deduce the 8 lowest bits of X_{37} only by fixing bits 16 to 25 of M_9. In order to explain this, note that calculating X_{37} during the step function in the forward direction gives:

$$X_{37} = X_{33}^{\lll 10} \boxplus (X_{32}^{\lll 10} \boxplus \mathtt{ONZ}(X_{36}, X_{35}, X_{34}) \boxplus M_9 \boxplus K_{36}^l)^{\lll 14}$$

Since $\mathtt{ONZ}(X_{36}, X_{35}, X_{34}) = (X_{36} \vee \overline{X_{35}}) \oplus X_{34}$, bits 16 to 25 of X_{36} will have no influence on the output of the boolean function $\mathtt{ONZ}$ if the corresponding X_{35} bits are set to zero (in a starting point, X_{32}, X_{33}, X_{34} and X_{35} are already fully known). Then, we can choose M_9 such that bit 16 of $X_{32}^{\lll 10} \boxplus (((X_{36} \vee \overline{X_{35}}) \oplus X_{34})\&\mathtt{3ff}) \boxplus (M_9\&\mathtt{3ff}) \boxplus K_{36}$ equals zero, which will stop the carry coming from the lower bits. As a result, the 8 lowest bits of X_{37} will not depend on X_{36} anymore (and thus neither on M_4 when computing forward, since X_{36} directly depends on M_4) .

One example of our generated starting points is shown in Figure 4, in which we applied our preparation trick. Before generating this starting point, we forced the additional conditions on X_{35}, and once the starting point found, fixing bits 16 to 25 of M_9 to 01101000010 will make sure that the last 8 bits of X_{37} will be equal to 11111010. Note that the 26-th bit of M_9 and 9-th bit of X_{37} are deduced from the known conditions.

Applying this trick is interesting for the attacker because the uncontrolled probability (steps 35-58) of the left branch is increased.

Phase 2: Finding a Starting Point. Given the differential path from Figure 5, we can use the freedom degrees available in both left and right branches internal

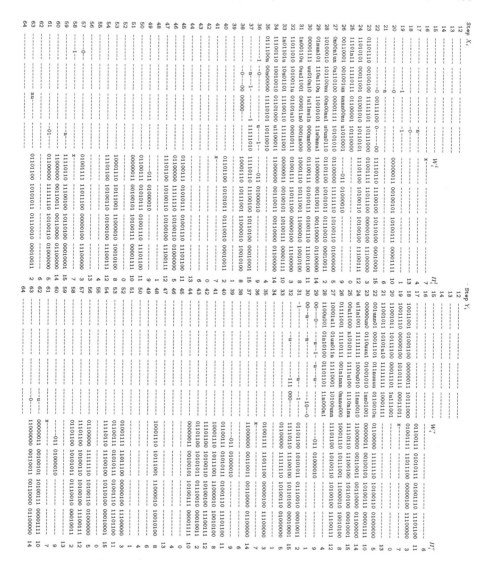

Fig. 4. Starting point for the 48-step differential path, on which the preparation trick was applied and the last 8 bits of X_{37} are fixed in advance by choosing several bits of M_9. At this point the remaining free message words are M_9, M_0, M_6, M_{13}, M_1, M_7 and M_4, which will be used during the merging phase.

states (320 bits) and in the message words (512 bits) to fulfill as much differential conditions as possible. To make the attacker easier, we chose to fix first the five consecutive internal states words that contain the most differential conditions $(X_{26}, X_{27}, X_{28}, X_{29}, X_{30}$ in the left branch and $Y_{21}, Y_{22}, Y_{23}, Y_{24}, Y_{25}$ in the right branch). Then, we fix a few message words one by one in the given order and continue the computation of the internal states by computing forward and backward in both branches (from $X_{26}, X_{27}, X_{28}, X_{29}, X_{30}$ and $Y_{21}, Y_{22}, Y_{23}, Y_{24}, Y_{25}$)

Fixing the first 5 chaining values ($X_{26}, X_{27}, X_{28}, X_{29}, X_{30}$ of the left branch and $Y_{21}, Y_{22}, Y_{23}, Y_{24}, Y_{25}$ of the right branch) is quite an easy task. Note that the two branches can be fixed independently at this stage. We used algorithms similar to the ones searching for a differential path: we just guess the unrestricted bits – from lower step to higher step, lower bit to higher bit and check if any inconsistency occurs. If both 0 and 1 selection of one bit lead to an inconsistency, we apply the backtracking in the search tree by one level and guess the same bit again. The guessing continues until all bits are fixed. If after a predefined number of backtracking events (chosen according to the performance of the search) no solutions are found, we can restart the whole search in order to avoid being trapped in a bad subspace with no solution at all.

Concerning the fixing of the message words, we used a different approach. Here, our search was applied word by word. Following this message words ordering $M_{11}, M_{15}, M_8, M_3, M_{12}, M_{14}, M_{10}, M_2, M_5, M_9, M_0$ and M_6, we guess the free bits, and some internal states values will directly be deduced by computing in both forward and backward directions from the already known internal state values in both branches. Note that the two branches are not independent anymore at this stage (since all message words are added several times in both branches), so it is important to check often for any inconsistency that could be detected. The backtracking and restarting options are also helpful here. We can use an extra trick to get a performance improvement of the search by pre-fixing the value of the word with the biggest number conditions in it (either message word or internal state word), and then deduce the value from all the words involved in this computation.

Our tool can find a starting point in a couple of minutes, with a program not really optimized. We will discuss about the complexity to generate the starting points in the next section.

Phase 3: Merging Both Branches with M_1, M_4, M_7 and M_{13}. A starting point example is given in Figure 4. Our target is to use the remaining free message words M_1, M_4, M_7 and M_{13} to make sure that we have a perfect match on the values of the five initial chaining words of both branches, i.e. $X_i = Y_i$ for $i \in \{12, 13, 14, 15, 16\}$ (the indexes started at 12 because we are not attacking from the first step here). The merging consists of four phases and in order to ease the reading we marked the free message words with colors in each phase. Once their values are fixed, we use black color for them.

- Step 1: Use M_{13} to ensure $X_{16} = Y_{16}$. As one can see, the value of X_{16} is already fixed at this point. Now, observe the two backward step functions of Y_{17} and Y_{16}:

$$Y_{17}^{\lll 10} = (Y_{22} \boxminus Y_{18}^{\lll 10})^{\ggg 8} \boxminus \text{IFZ}(Y_{21}, Y_{20}, Y_{19}^{\lll 10}) \boxminus K_{21}^r \boxminus M_{13}$$
$$Y_{16}^{\lll 10} = (Y_{21} \boxminus Y_{17}^{\lll 10})^{\ggg 12} \boxminus \text{IFZ}(Y_{20}, Y_{19}, Y_{18}^{\lll 10}) \boxminus K_{20}^r \boxminus M_0$$

Incorporating the equation $X_{16} = Y_{16}$, we can direcly calculate the value of M_{13} from the known ones:

$$M_{13} = \qquad (X_{16}^{\lll 10} \boxplus \mathtt{IFZ}(Y_{20}, Y_{19}, Y_{18}^{\lll 10}) \boxplus K_{20}^r \boxplus M_0)^{\lll 12}$$
$$\boxminus Y_{21} \boxplus (Y_{22} \boxminus Y_{18}^{\lll 10})^{\ggg 8} \boxminus \mathtt{IFZ}(Y_{21}, Y_{20}, Y_{19}^{\lll 10}) \boxminus K_{21}^r.$$

- Step 2: Similarly, use M_1 and M_7 to ensure conditions $X_{15} = Y_{15}$ and $X_{14} = Y_{14}$. Observing the step functions:

$$X_{15}^{\lll 10} = (X_{20} \boxminus X_{16}^{\lll 10})^{\ggg 13} \boxminus \mathtt{IFX}(X_{19}, X_{18}, X_{17}^{\lll 10}) \boxminus K_{19}^l \boxminus M_1$$
$$= Y_{15}^{\lll 10} = (Y_{20} \boxminus Y_{16}^{\lll 10})^{\ggg 7} \boxminus \mathtt{IFZ}(Y_{19}, Y_{18}, Y_{17}^{\lll 10}) \boxminus K_{19}^r \boxminus M_7$$

$$X_{14}^{\lll 10} = (X_{19} \boxminus X_{15}^{\lll 10})^{\ggg 8} \boxminus \mathtt{IFX}(X_{18}, X_{17}, X_{16}^{\lll 10}) \boxminus K_{18}^l \boxminus M_{13}$$
$$= Y_{14}^{\lll 10} = (Y_{19} \boxminus Y_{15}^{\lll 10})^{\ggg 15} \boxminus \mathtt{IFZ}(Y_{18}, Y_{17}, Y_{16}^{\lll 10}) \boxminus K_{18}^r \boxminus M_3$$

and introducing notations for the constants, the equations above are simplified to

$$A \boxminus M_1 = B \boxminus M_7$$
$$(X_{19} \boxminus (A \boxminus M_1))^{\ggg 8} \boxminus D = (Y_{19} \boxminus (B \boxminus M_7))^{\ggg 15} \boxminus E$$

Let $X = (X_{19} \boxminus (A \boxminus M_1))^{\ggg 8}$, $C_0 = E \boxminus D$ and $C_1 = Y_{19} \boxminus X_{19}$. The above equations become one:

$$X \boxplus C_0 = (C_1 \boxplus X^{\lll 8})^{\ggg 15} \qquad (1)$$

where C_0 and C_1 are constants. The problem of finding the value of M_1 and M_7 is equivalent to solving this equation. We find that this equation can be solved with 2^9 computations: we can solve this equation for all 2^{64} possible values of C_0 and C_1 and store the solutions (M_1 and M_7) in a big look-up table. Building and storing this table requires 2^{73} time and 2^{64} memory.
- Step 3: Use M_4 to ensure $X_{13} = Y_{13}$. After step 2, Y_{13} is already fixed. Thus we can use a simple calculation to get the value of M_4:

$$M_4 = (X_{18} \boxminus X_{14}^{\lll 10})^{\ggg 6} \boxminus \mathtt{IFX}(X_{17}, X_{16}, X_{15}^{\lll 10}) \boxminus Y_{13}$$

- Step 4: The uncontrolled part of the merging. At this point, all freedom degrees have been used and the last equation on the internal state $X_{12} = Y_{12}$ will be fulfilled with a probability of 2^{-32}.

Uncontrolled Probability. After the merging, steps 36-58 of the left branch and steps 29-58 of the right branch are still uncontrolled. Due to the difficulty of calculating the probability, we used experiments to evaluate these probabilities. Starting from a generated starting point, e.g. in Figure 4, we randomly choose values of message words $M_5, M_9, M_0, M_6, M_{13}, M_1, M_7$ and M_4. Then we compute forward to check if the differences are all canceled after the last injection of M_7. Note that if there are conditions on message words M_1, M_4, M_7 and M_{13}, they should be fulfilled probabilistically and included in the probability estimation, since the freedom degrees of these message words are used to match the initial internal states values. For the other free message words M_5, M_9, M_0 and M_6, we do not need to consider their conditions in the probability, because we can freely choose their values to fulfill these conditions.

We measured the probability of both branches separately. After applying the preparation trick, the uncontrolled probability of the left branch is $2^{-8.8}$. The uncontrolled probability of the right branch is $2^{-36.6}$. Moreover, during the merging phase, we could not control the value matching on the first IV word, and this adds another factor of 2^{-32}.

In total, the uncontrolled probability is $2^{-32} \cdot 2^{-8.8} \cdot 2^{-36.6} = 2^{-77.4}$. Since this probability is too low and already close to the birthday bound for RIPEMD-160, we are not able to afford the feed-forward tricks in steps 58-64.

Complexity Evaluation. First we calculate the complexity to generate the starting points. Since the uncontrolled probability is $2^{-77.4}$, we need to generate $2^{77.4}$ starting points. However, we do not need to restart the generation from the beginning. Indeed, every time we need a new starting point, we can randomize M_6 to get a new one. Once all possible choices of M_6 have been used, we can still use freedom degrees of M_0, M_9 and M_5 to generate all the required starting points. Though there are many constraints on these four message words, luckily the number of conditions on M_6 is only two bits (one on X_{18} and one on X_{17}). We can randomly choose value for X_{18} fulfilling the known conditions and check if the one-bit condition on X_{17} is fulfilled. Thus, we can find a new starting point from a known one with a complexity of 4 step functions, which is equivalent to $4/(42 * 2) \approx 2^{-4.4}$ calls of the 42-step compression function of RIPEMD-160. For the other message words, we do not need to go into the details of the complexity, since the number of times we have to regenerate them is quite small and it is not the bottleneck of our attack complexity. From the reasoning above, we can conclude that the average complexity to generate a starting point is $2^{-4.4}$. The complexity of generating all the required starting points is then 2^{73}.

Now, we need to consider the complexity of the merging phase. In order to evaluate this cost, we implemented the merging of the last four initial internal states. The table lookup in second phase is estimated using a RAM access (since the table will be very bog). In total, our implementation of the merging takes about 145 cycles. The OPENSSL implementation of RIPEMD-160 compression function on the same computer takes about 1040 cycles. Thus, 42 steps of the compression function takes about $1040 * 42/80 = 546$ cycles. Then we can say that our merging costs $145/546 \approx 2^{-1.9}$ calls of the 42-step compression function.

Finally, we can calculate the complexity of the semi-free start collision attack on 42-step RIPEMD-160: $2^{73} + 2^{77.4-1.9} \approx 2^{75.5}$.

5.2   First Step Semi-Free-Start Collision

This section discusses about the merging phase of the two branches to get a semi-free-start collision attack on the the first 36 steps. The idea of the merging is similar to the merging for the 42-step attack and we describe it briefly.

We start with generating a starting point. After that, the path in the left branch has been satisfied until step 14, and the remaining uncontrolled probability amounts to $2^{-4.6}$. The path in the right branch has been fully satisfied. After that, there are free bits left in message words M_0, M_2, M_5, M_7, M_9 and M_{14}. Next, we show these free bits are enough to generate semi-free-start collisions, and thus we only need to generate a single starting point.

The procedure of merging is detailed as below.

1. Set random values to M_9 and the free bits of M_7, and then compute until X_2 in the left branch.
2. Set $M_5 = M_5 \boxplus 1$ (initialize M_5 as 0), and compute until X_{-1} in the left branch. If M_5 becomes 0 again, goto Step 1.
3. Compute the values of M_2 and M_0 that make $Y_0 = X_0$ and $Y_{-1} = X_{-1}$.
4. Compute X_{-2} and Y_{-2}, and check if $X_{-2} = Y_{-2}$ holds. In case of $X_{-2} \neq Y_{-2}$, goto Step 2.
5. Compute X_{-3}, and then compute the value of M_{14} that makes $Y_{-3} = X_{-3}$. Check if the conditions on M_{14} are satisfied. If the conditions are not satisfied, goto step 1.
6. Compute X_{-4} and Y_{-4}, and check if $X_{-4} = Y_{-4}$ holds. In case of $X_{-4} \neq Y_{-4}$, goto Step 2.

Both $X_{-2} = Y_{-2}$ and $X_{-4} = Y_{-4}$ are satisfied with a probability 2^{-32}, and four bit conditions are set on M_{14}. Thus we have to try 2^{68} random values of M_7, M_9, and M_5 to succeed in merging the two branches once. Recall that the uncontrolled probability is $2^{-4.6}$. So we need to merge the two branches $2^{4.6}$ times. Thus, the total complexity of the attack is $2^{68+4.6} \times 16/72 \approx 2^{70.4}$.

6   Results

We give in Table 2 a comparison of our attacks to previous results on RIPEMD-160. Compared to the previous best semi-free-start collision attack on RIPEMD-160 (36 middle steps), we have increased the number of attackable steps by 6 and proposed a 36-step semi-free-start collision attack that starts from the first step.

7   Conclusion

In this article, we have proposed an improved cryptanalysis of the hash function RIPEMD-160, which is an ISO/IEC standard. We have found a 42-step semi-free-start collision attack on RIPEMD-160 starting from the second step and a 36-step semi-free-start collision attack starting from the first step. Compared to previous results, we have two improvements. First the number of attacked steps

Table 2. Summary of known and new preimage and collision attacks on RIPEMD-160 hash and compression function

Function	Size	Target	Attack Type	#Steps	Complexity	Ref.
RIPEMD-160	160	comp. function	preimage	31	2^{148}	[14]
RIPEMD-160	160	hash function	preimage	31	2^{155}	[14]
RIPEMD-160	160	comp. function	semi-free-start collision	36	low	[10]
RIPEMD-160	**160**	**comp. function**	**semi-free-start collision**	**42**	$2^{75.5}$	**new**
RIPEMD-160	**160**	**comp. function**	**semi-free-start collision**	**36**	$2^{70.4}$	**new**
RIPEMD-160	160	comp. function	non-randomness	48	low	[10]
RIPEMD-160	160	comp. function	non-randomness	51	2^{158}	[16]

is increased from 36 to 42, and secondly, for the same number of attacked steps, we propose an attack that starts from the first step. Moreover, our semi-free-start collision attacks give a positive answer to the open problem raised in [10], in which the authors were not able to find any non-linear differential path in the first step, due to the XOR function that makes the non-linear part search much harder.

Our 42-step semi-free-start attack is obtained from a 48-step differential path. Unfortunately, we couldn't add these extra 6 steps to our attack without reaching a complexity beyond the birthday bound (this extra part would be verified with probability $2^{-11.3}$). Future works might include improving the probabilistic part even further. If one can improve this part by a factor of about 2^7, a 48-step semi-free-start collision attack would then be obtained directly with our proposed differential path. Another possible improvement would be that if one can find a better non-linear differential path in the first round, it might be possible to merge both branches at the same time to a given IV and eventually obtain a hash function collision.

Acknowledgments. The authors would like to thank the anonymous referees for their helpful comments. Thomas Peyrin, Lei Wang and Shuang Wu are supported by the Singapore National Research Foundation Fellowship 2012 (NRF-NRFF2012-06).

References

1. Bertoni, G., Daemen, J., Peeters, M., Van Assche, G.: The Keccak reference. Submission to NIST (Round 3) (January 2011), http://csrc.nist.gov/groups/ST/hash/sha-3/Round3/submissions_rnd3.html
2. Bosselaers, A., Preneel, B. (eds.): RIPE 1992. LNCS, vol. 1007. Springer, Heidelberg (1995)
3. Damgård, I.: A Design Principle for Hash Functions. In: Brassard, G. (ed.) CRYPTO 1989. LNCS, vol. 435, pp. 416–427. Springer, Heidelberg (1990)

4. De Cannière, C., Rechberger, C.: Finding SHA-1 Characteristics: General Results and Applications. In: Lai, X., Chen, K. (eds.) ASIACRYPT 2006. LNCS, vol. 4284, pp. 1–20. Springer, Heidelberg (2006)
5. Dobbertin, H.: RIPEMD with Two-Round Compress Function is Not Collision-Free. J. Cryptology 10(1), 51–70 (1997)
6. Dobbertin, H., Bosselaers, A., Preneel, B.: RIPEMD-160: A Strengthened Version of RIPEMD. In: Gollmann, D. (ed.) FSE 1996. LNCS, vol. 1039, pp. 71–82. Springer, Heidelberg (1996)
7. International Organization for Standardization: Information technology – Security techniques – Hash-functions – Part 3: Dedicated hash-functions. ISO/IEC 10118-3:2004 (2004)
8. Landelle, F., Peyrin, T.: Cryptanalysis of Full RIPEMD-128. In: Johansson, T., Nguyen, P.Q. (eds.) EUROCRYPT 2013. LNCS, vol. 7881, pp. 228–244. Springer, Heidelberg (2013)
9. Leurent, G.: Analysis of Differential Attacks in ARX Constructions. In: Wang, X., Sako, K. (eds.) ASIACRYPT 2012. LNCS, vol. 7658, pp. 226–243. Springer, Heidelberg (2012)
10. Mendel, F., Nad, T., Scherz, S., Schläffer, M.: Differential Attacks on Reduced RIPEMD-160. In: Gollmann, D., Freiling, F.C. (eds.) ISC 2012. LNCS, vol. 7483, pp. 23–38. Springer, Heidelberg (2012)
11. Mendel, F., Nad, T., Schläffer, M.: Finding SHA-2 Characteristics: Searching through a Minefield of Contradictions. In: Lee, D.H., Wang, X. (eds.) ASIACRYPT 2011. LNCS, vol. 7073, pp. 288–307. Springer, Heidelberg (2011)
12. Mendel, F., Nad, T., Schläffer, M.: Improving Local Collisions: New Attacks on Reduced SHA-256. In: Johansson, T., Nguyen, P.Q. (eds.) EUROCRYPT 2013. LNCS, vol. 7881, pp. 262–278. Springer, Heidelberg (2013)
13. Merkle, R.C.: One Way Hash Functions and DES. In: Brassard, G. (ed.) CRYPTO 1989. LNCS, vol. 435, pp. 428–446. Springer, Heidelberg (1990)
14. Ohtahara, C., Sasaki, Y., Shimoyama, T.: Preimage Attacks on Step-Reduced RIPEMD-128 and RIPEMD-160. In: Lai, X., Yung, M., Lin, D. (eds.) Inscrypt 2010. LNCS, vol. 6584, pp. 169–186. Springer, Heidelberg (2011)
15. Rivest, R.L.: The MD4 Message-Digest Algorithm. IETF Request for Comments (RFC) 1320 (1992), http://www.ietf.org/rfc/rfc1320.html
16. Sasaki, Y., Wang, L.: Distinguishers beyond Three Rounds of the RIPEMD-128/-160 Compression Functions. In: Bao, F., Samarati, P., Zhou, J. (eds.) ACNS 2012. LNCS, vol. 7341, pp. 275–292. Springer, Heidelberg (2012)
17. Wang, X., Lai, X., Feng, D., Chen, H., Yu, X.: Cryptanalysis of the Hash Functions MD4 and RIPEMD. In: Cramer, R. (ed.) EUROCRYPT 2005. LNCS, vol. 3494, pp. 1–18. Springer, Heidelberg (2005)
18. Wang, X., Yin, Y.L., Yu, H.: Finding Collisions in the Full SHA-1. In: Shoup, V. (ed.) CRYPTO 2005. LNCS, vol. 3621, pp. 17–36. Springer, Heidelberg (2005)
19. Wang, X., Yu, H.: How to Break MD5 and Other Hash Functions. In: Cramer, R. (ed.) EUROCRYPT 2005. LNCS, vol. 3494, pp. 19–35. Springer, Heidelberg (2005)
20. Wang, X., Yu, H., Yin, Y.L.: Efficient Collision Search Attacks on SHA-0. In: Shoup, V. (ed.) CRYPTO 2005. LNCS, vol. 3621, pp. 1–16. Springer, Heidelberg (2005)

A The Differential Paths

Step	X_i	W^l_i	Π^l_i	Step	Y_i	W^r_i	Π^r_i
12	-------- -------- -------- --------			12	-------- -------- -------- --------		
13	-------- -------- -------- --------			13	-------- -------- -------- --------		
14	-------- -------- -------- --------			14	-------- -------- -------- --------		
15	-------- -------- -------- --------			15	-------- -------- -------- --------		
16	-------- -------- -------- --------	x------- -------- -------- --------	7	16	-------- -------- -------- --------	-------- -------- -------- --------	6
17	-------- -------- -------- -n------	-------- -------- -------- --------	4	17	-------- -------- -------- --------	-------- -------- -------- --------	11
18	-------- -------- -------- -0------	-------- -------- -------- --------	13	18	--01---- 01------ -------- --------	-------- -------- -------- --------	3
19	-------- -------1 -------- -1------	-------- -------- -------- --------	1	19	---1---- -------- -------1 --------	x------- -------- -------- --------	7
20	-------- -------0 -------- --------	-------- -------- -------- --------	10	20	11------ -----100 -------1 -u---001	-------- -------- -------- --------	0
21	-------- -------n -------- --------	-------- -------- -------- --------	6	21	11-01--- -----u10 1-----11 1----111	-------- -------- -------- --------	13
22	-------- -------0 -------- --------	-------- -------- -------- --------	15	22	---unn-- --0--1-1 0--nuuuu 01----0u	-------- -------- -------- --------	5
23	-----1-- -------0 -------- --------	-------- -------- -------- --------	3	23	---00un- 011-uuu- --0----- -un0----	-------- -------- -------- --------	10
24	1----1-- -------- -------- --------	-------- -------- -------- --------	12	24	u11n1--- ----11-1 1-0-u--- -1nn0010	-------- -------- -------- --------	14
25	-----n-- ---1--1- ------0- --------	-------- -------- -------- --------	0	25	0-u110-- n-0----1 11--u10- -1-0n1nu	-------- -------- -------- --------	15
26	001-00-- ------un nnnn--nu u1--0---	-------- -------- -------- --------	9	26	----1--1 -11---1- 00-n1nnn -nnnu---	-------- -------- -------- --------	8
27	-n--n-un -u--0100 0000--11 1010----	-------- -------- -------- --------	5	27	1---1u-- 0-uu--1n 1---00-- ---0-unn	-------- -------- -------- --------	12
28	1010001- ----00nu --n0-nu- u-un0110	-------- -------- -------- --------	2	28	----u--- --n1--00 0--011-- --n0--n1	-------- -------- -------- --------	4
29	0-nun-0- 11-u--0u -1010101 1-u0nnu-	-------- -------- -------- --------	14	29	-------- ---u--1- 0u-11u-- ---0---0	-------- -------- -------- --------	9
30	-000-111 uu010n10 1u-1nu1n -00nn000	-------- -------- -------- --------	11	30	-0--n--- --n----- 1---1--- --------	-------- -------- -------- --------	1
31	-n---1-u 0uu1-0-1 000011n0 0001n---	-------- -------- -------- --------	8	31	-1------ --1----- -----u-- --1-----	-------- -------- -------- --------	2
32	-10----0 -010---u -1--1u10 ----01--	-------- -------- -------- --------	3	32	-------- --u----- -----1-- --0-----	-------- -------- -------- --------	15
33	-n-----u --u----1 -----1-- 1----0--	-------- -------- -------- --------	10	33	-------- -------- -------- --------	-------- -------- -------- --------	5
34	-11----0 --0----- ----1--- u-------	-------- -------- -------- --------	14	34	-------- -------- -------- --------	-------- -------- -------- --------	1
35	-------n --n----- -------- 1--1----	-------- -------- -------- --------	4	35	-------- -------- -------- --------	-------- -------- -------- --------	3
36	-------1 --0----- ----0--- u----1--	-------- -------- -------- --------	9	36	-------- -------- -------- --------	x------- -------- -------- --------	7
37	-------- --n---1- -------- 1----0--	-------- -------- -------- --------	15	37	-------- -------- -------- --------	-------- -------- -------- --------	14
38	-------- --0---0- -------- --------	-------- -------- -------- --------	8	38	-------- -------- -------- --------	-------- -------- -------- --------	6
39	-------- -------- -------- --------	-------- -------- -------- --------	1	39	-------- -------- -------- --------	-------- -------- -------- --------	9
40	-------- -------- -------- --------	-------- -------- -------- --------	2	40	-------- -------- -------- --------	-------- -------- -------- --------	11
41	-------- -------- -------- --------	x------- -------- -------- --------	7	41	-------- -------- -------- --------	-------- -------- -------- --------	8
42	-------- -------- -------- --------	-------- -------- -------- --------	0	42	-------- -------- -------- --------	-------- -------- -------- --------	12
43	-------- -------- -------- --------	-------- -------- -------- --------	6	43	-------- -------- -------- --------	-------- -------- -------- --------	2
44	-------- -------- -------- --------	-------- -------- -------- --------	13	44	-------- -------- -------- --------	-------- -------- -------- --------	10
45	-------- -------- -------- --------	-------- -------- -------- --------	11	45	-------- -------- -------- --------	-------- -------- -------- --------	0
46	-------- -------- -------- --------	-------- -------- -------- --------	5	46	-------- -------- -------- --------	-------- -------- -------- --------	4
47	-------- -------- -------- --------	-------- -------- -------- --------	12	47	-------- -------- -------- --------	-------- -------- -------- --------	13
48	-------- -------- -------- --------	-------- -------- -------- --------	1	48	-------- -------- -------- --------	-------- -------- -------- --------	8
49	-------- -------- -------- --------	-------- -------- -------- --------	9	49	-------- -------- -------- --------	-------- -------- -------- --------	6
50	-------- -------- -------- --------	-------- -------- -------- --------	11	50	-------- -------- -------- --------	-------- -------- -------- --------	4
51	-------- -------- -------- --------	-------- -------- -------- --------	10	51	-------- -------- -------- --------	-------- -------- -------- --------	1
52	-------- -------- -------- --------	-------- -------- -------- --------	0	52	-------- -------- -------- --------	-------- -------- -------- --------	3
53	-------- -------- -------- --------	-------- -------- -------- --------	8	53	-------- -------- -------- --------	-------- -------- -------- --------	11
54	-------- -------- -------- --------	-------- -------- -------- --------	12	54	-------- -------- -------- --------	-------- -------- -------- --------	15
55	-------- -------- -------- --------	-------- -------- -------- --------	4	55	-------- -------- -------- --------	-------- -------- -------- --------	0
56	-------- -------- -------- --------	-------- -------- -------- --------	13	56	-------- -------- -------- --------	-------- -------- -------- --------	5
57	-----0-- -------- -------- --------	-------- -------- -------- --------	3	57	-------- -------- -------- --------	-------- -------- -------- --------	12
58	-----1-- -------- -------- --------	x------- -------- -------- --------	7	58	-------- -------- -------- --------	-------- -------- -------- --------	2
59	-------- -------- -------- ---n----	-------- -------- -------- --------	15	59	-------- -------- -------- --------	-------- -------- -------- --------	13
60	-------- -------- -------- --------	-------- -------- -------- --------	14	60	-------- -------- -------- --------	-------- -------- -------- --------	9
61	-------- -------- -------- --01----	-------- -------- -------- --------	5	61	-------- -------- -------- --------	x------- -------- -------- --------	7
62	-------- -------- -------- --------	-------- -------- -------- --------	6	62	-------- -------- -------- ---u----	-------- -------- -------- --------	10
63	-------- -------- nu------ --------	-------- -------- -------- --------	2	63	-------- -------- -------- ---0----	-------- -------- -------- --------	14
64	-------- -------- -------- --------			64	-------- -------- -------- --------		

Fig. 5. The 48-step differential path

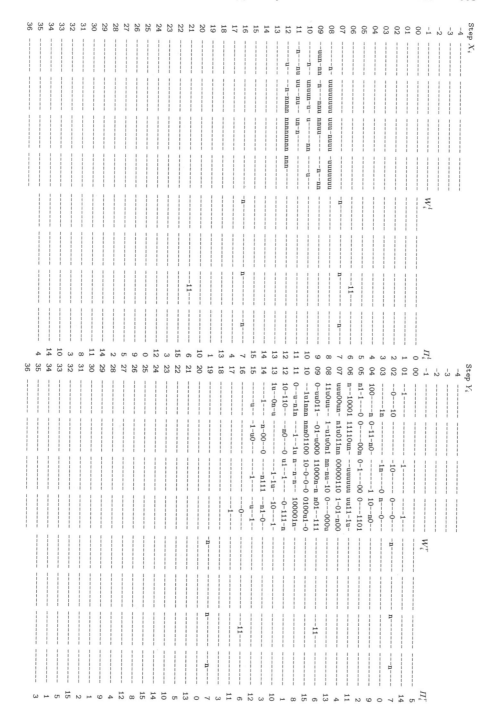

Fig. 6. The 36-step differential path

Limited-Birthday Distinguishers
for Hash Functions
Collisions beyond the Birthday Bound Can Be Meaningful

Mitsugu Iwamoto[1], Thomas Peyrin[2], and Yu Sasaki[3]

[1] Center for Frontier Science and Engineering,
The University of Electro-Communications, Japan
mitsugu@uec.ac.jp
[2] Division of Mathematical Sciences, School of Physical and Mathematical Sciences,
Nanyang Technological University, Singapore
thomas.peyrin@gmail.com
[3] NTT Secure Platform Laboratories, Japan
sasaki.yu@lab.ntt.co.jp

Abstract. In this article, we investigate the use of limited-birthday distinguishers to the context of hash functions. We first provide a proper understanding of the limited-birthday problem and demonstrate its soundness by using a new security notion *Differential Target Collision Resistance (dTCR)* that is related to the classical *Target Collision Resistance (TCR)* notion. We then solve an open problem and close the existing security gap by proving that the best known generic attack proposed at FSE 2010 for the limited-birthday problem is indeed the best possible method.

Moreover, we show that almost all known collision attacks are in fact more than just a collision finding algorithm, since the difference mask for the message input is usually fixed. A direct and surprising corollary is that these collision attacks are interesting for cryptanalysis even when their complexity goes beyond the $2^{n/2}$ birthday bound and up to the 2^n preimage bound, and can be used to derive distinguishers using the limited-birthday problem. Interestingly, cryptanalysts can now search for collision attacks beyond the $2^{n/2}$ birthday bound.

Finally, we describe a generic algorithm that turns a semi-free-start collision attack on a compression function (even if its complexity is beyond the birthday bound) into a distinguisher on the whole hash function when its internal state is not too wide. To the best of our knowledge, this is the first result that exploits classical semi-free-start collisions on the compression function to exhibit a weakness on the whole hash function. As an application of our findings, we provide distinguishers on reduced or full version of several hash functions, such as RIPEMD-128, SHA-256, Whirlpool, etc.

Keywords: hash function, compression function, distinguisher, limited-birthday, semi-free-start collision, differential target collision resistance.

K. Sako and P. Sarkar (Eds.) ASIACRYPT 2013, Part II, LNCS 8270, pp. 504–523, 2013.
© International Association for Cryptologic Research 2013

1   Introduction

A hash function H is a function that takes an arbitrarily long message M as input and outputs a fixed-length hash value of size n bits. Classical security requirements for a cryptographic hash function are collision resistance and (second)-preimage resistance. Namely, it should be impossible for an adversary to find a collision (two distinct messages that lead to the same hash value) in less than $2^{n/2}$ hash computations, or a (second)-preimage (a message hashing to a given challenge) in less than 2^n hash computations. Most standardized hash functions are based upon the Merkle-Damgård paradigm [35,11] and iterate a compression function h with fixed input and output size to handle arbitrarily long messages. The compression function itself should ensure equivalent security properties in order for the hash function to inherit from them. When the internal state size of the compression is the same as for the hash function, then the construction is called *narrow-pipe*, otherwise it is called a *wide-pipe*.

The SHA-3 competition organized by the NIST [49] eventually ended in early October 2012 with the selection of KECCAK [16] as sole winner and new hash function standard. During the last decade, due to this competition and to the cryptanalysis breakthroughs [54,55] that provoked this reaction from the NIST, hash functions have been among the most active topics in academic cryptography. This infatuation is justified by the fact that these primitives are utilized tremendously in practice, with applications ranging from digital signatures, message authentication codes, to secure storage of passwords databases. However, a hash function is also seen as the "swiss knife" of cryptography: many protocols use the random oracle paradigm [3] to check and even prove that they present no structural flaw, and while there is no such thing as a random oracle, designers use hash functions to "simulate" its behavior. Overall, even if collision and (second)-preimage resistance are their most important security properties, cryptographers are therefore also expecting hash functions to present no structural flaw whatsoever, *i.e.* to be indistinguishable from a random oracle. NIST, for example, clearly specified in its SHA-3 call for candidates [49] that the submitted proposals have to support randomized hashing and not present any "non-random behavior".

On the cryptanalysis side, many various distinguishers have been proposed in the recent years, mainly against AES or SHA-3 candidates. One can cite for example zero-sums distinguishers [2], rotational distinguishers [24] or subspace distinguishers [26]. Limited-birthday distinguishers have been introduced by Gilbert and Peyrin [15] as a tool to distinguish 8 rounds of the AES block cipher from an ideal permutation in the known-key model, and it was later used against other symmetric key primitives [40,37,13,22]. It consists in deriving pairs of plaintext/ciphertext couples $(P, C), (P', C')$ (or input/output couples $(M, H(M)), (M', H(M'))$ for a one-way function) with an input xor difference belonging to a set IN of 2^I elements and an output xor difference belonging to a set OUT of 2^O elements, *i.e.* $P \oplus P' \in IN$ and $C \oplus C' \in OUT$ (or $M \oplus M' \in IN$ and $H(M) \oplus H(M') \in OUT$). What is the best generic attack complexity in the case of an ideal permutation (or function) ? When IN and/or OUT are big enough then this problem is equivalent to a classical birthday paradox problem (*i.e.* with complexity min $\{2^{(n-O)/2}, 2^{(n-I)/2}\}$), but the idea underlying the limited-birthday is that when IN and OUT are small an attacker might not be able to use the birthday paradox as much as he would like to. Indeed, he will have to perform several independent smaller birthday searches instead of a single big one, and therefore the process will require much more computations. Gilbert and Peyrin [15] proposed the best known generic algorithm for the limited-birthday problem, whose complexity is $\max\{\min\{2^{(n-I+1)/2}, 2^{(n-O+1)/2}\}, 2^{n-I-O+1}\}$ for a permutation and $\max\{2^{(n-O+1)/2}, 2^{n-I-O+1}\}$ for a function[1]. However, its optimality is yet unknown and it was only conjectured that their attack is the best possible. As of today, only Nikolić *et al.* [39] provided a formal lower bound proof, which is min $\{2^{n/2-2}, 2^{n-(I+O)-3}\}$. Unfortunately this bound is not tight and only applies to permutations. For example, in the case of $I = O = 0$, the attack complexity in [15] is $2^{n-I-O+1} = 2^{n+1}$ while the proven bound in [39] only reaches $2^{n/2-2}$.

Some might argue that the limited-birthday problem can trivially be solved by choosing a random input pair (X, Y) and computing $IN = \{X \oplus Y\}$ and $OUT = \{H(X) \oplus H(Y)\}$. However, these pathological attackers, that we call "cheating adversaries", are meaningless: since hash functions are not processing any secret and are completely public (unlike other primitives in cryptography), formalizing security notions requires some kind of challenge, in order to avoid these cheating adversaries (the same is true concerning the chosen-key model for block ciphers). For example, there always exists an adversary that can output a collision with a single operation and negligible memory (i.e. the adversary that just prints a known collision). In general, this obstacle is avoided by considering that a hash function is part of a family indexed by a key input (for example its Initial Value (IV)), or by formalizing the human ignorance [43]. These pathological cases of cheating adversaries are present for all distinguishers without challenges, even for the subspace distinguisher for hash functions [26] or q-multicollisions for block ciphers in the chosen-key model [5].

Our Contributions. To start, we provide in Section 2.1 a proper understanding of limited-birthday distinguishers for the hash function setting. Namely, we discuss potential issues arising from security notions for a public function without challenge and describe various tricks to avoid pathological cheating adversaries. We also show that limited-birthday distinguishers for hash functions can be used to attack a security notion very similar to the classical Target Collision Resistance (TCR) property, which we call differential Target Collision Resistance (dTCR).

Secondly, we provide in Section 2.2 a proof that the currently best known generic attack for the limited-birthday problem (proposed by Gilbert and Peyrin at FSE 2010 [15]) is indeed the best possible. More precisely, we show that the

---

[1] There is obviously a trade-off between the complexity and the success probability, which here is about 0.63. The original paper [15] missed '+1's in the exponents, which was firstly corrected by [37]

computation complexity to solve the limited-birthday problem is bounded by $\max\left\{2^{(n-O+1)/2}, 2^{n-I-O+1}\right\}$. We can directly conclude that if for a collision attack (*i.e.*, $O = 0$) the set IN of possible message difference of the hash function is limited to one or a few elements regardless of the randomization input, then one can obtain a limited-birthday distinguisher on the function, even with a complexity well beyond the birthday bound. It is to be noted that this condition on the message difference mask is verified for almost all known collision attacks, as for example with the recent advances on SHA-1 [54]. Overall, most known hash function collision attacks are in fact more then just collision finding algorithms since the message difference mask is constrained and, as a consequence, they are now surprisingly becoming interesting even with a complexity beyond the $2^{n/2}$ birthday bound. Our work indicates that concerning distinguishing attacks the security of many hash functions needs to be reevaluated accordingly.

We then move to the case of a compression function, naturally easier to break than the whole hash function. Namely, we provide in Section 3 a generic algorithm that can transform a semi-free-start collision attack on the compression function into a limited-birthday distinguisher for the entire hash function. Because it is based on a meet-in-the-middle approach, this algorithm gets more interesting for the attacker as the internal state of the hash function gets narrower. To the best of the authors knowledge, this conversion is the first result turning a classical semi-free-start collision attack on the compression function into some weakness on the whole hash function (a previous work from Leurent [28] also provides such a conversion, but it is only applicable in the very uncommon case where the average semi-free-start collisions cost is lower than a single operation).

Finally, we provide in Section 4 some applications of our findings against real-world hash functions, such as AES-based hash functions (Section 4.1), HAS-160 (Section 4.2), LANE (Section 4.3), RIPEMD-128 (Section 4.4), SHA-256 (Section 4.5) and Whirlpool (Section 4.6).

2 Limited-Birthday Problem

Throughout this paper, we discuss limited-birthday distinguishers for one-way functions, *i.e.*, in our security model querying input values to obtain the corresponding output values is allowed, but the opposite is forbidden.

In Sect. 2.1, we firstly explain that validating distinguishers without any challenge is hard due to cheating adversaries. We then explain that the ambiguity of the validity does not exist if adversaries are challenged, and the limited-birthday problem is useful even in such a challenged setting. In Sect. 2.2, we formally prove that the previous generic attack that was conjectured as the best attack is indeed optimal. Finally, several remarks are given in Sect. 2.3.

2.1 Importance of the Limited-Birthday Problem in Cryptography

Cheating Adversaries. Collision resistance is the only un-challenged notion of the three classical security properties expected from a cryptographic hash function (collision, preimage, second-preimage), and, as such, the one that proved to be the most difficult to analyze. One of the difficulty that arises for example (and which is true for any un-challenged security property on a public function) is that there is always an adversary that can output a collision immediately, by simply hard-coding it. Rogaway [43] proposed a potential solution to this by formalizing the so-called notion of human ignorance.

However, the existence of another type of pathological cheating adversary has been often utilized as criticism of the limited-birthday distinguishers formalization: the adversary first chooses a random input pair (X, Y), computes $IN = \{X \oplus Y\}$ and $OUT = \{H(X) \oplus H(Y)\}$, and then claims that he can solve the limited-birthday problem with sets IN and OUT (where IN and OUT are actually defined at the end of the attack). It is to be noted that such issues already exist in the case of collision resistance and actually for any security definition regarding a public function with an adversary that is not challenged whatsoever.

Let's come back to our collision resistance case for example. Security engineers obviously understand that collision is an important security definition, but for theoreticians collision is nothing more than a certain output difference Δ which is equal to zero. Collision resistance therefore belongs to a more generic problem that we could name diff(Δ) and which asks for the adversary to exhibit an input pair (X, Y) such that $H(X) \oplus H(Y) = \Delta$. All members of this set are equally hard with regards to generic attacks. Collision resistance is actually diff(0), but cheating adversaries exist for diff(Δ): by just choosing a random input pair (X, Y) and trivially claiming that we can solve diff($H(X) \oplus H(Y)$).

Similarly, one can design cheating adversaries for the recent q-multicollision problem [5] used on AES: define the problem q-multi-diff($\Delta_1, \ldots, \Delta_q$) that asks for the attacker to exhibit q input pairs $(X_1, Y_1), \ldots, (X_q, Y_q)$ such that $H(X_1) \oplus H(Y_1) = \delta \oplus \Delta_1, \ldots, H(X_q) \oplus H(Y_q) = \delta \oplus \Delta_q$. Then the q-multicollision problem is nothing else than q-multi-diff(0, \ldots, 0) with a predefined δ, yet obvious cheating adversaries exist for q-multi-diff: just pick q random input pairs $(X_1, Y_1), \ldots, (X_q, Y_q)$, and claim that you can solve q-multi-diff($H(X_1) \oplus H(Y_1) \oplus \delta, \ldots, H(X_q) \oplus H(Y_q) \oplus \delta$). The same reasoning applies to the subspace distinguishers [26] as well.

As a direct analogy, the limited-birthday problem LBP(IN, OUT) with fully defined sets IN and OUT belongs to the more general limited-birthday problem LBP. Thus, the limited-birthday distinguishers are as valid as collision, q-multicollision or subspace distinguishers when the sets IN and OUT are fully defined, and we emphasize that in the rest of the article the sets IN and OUT are considered to be fully defined before the attacker starts to actually search for a valid pair of inputs. Yet, in addition, we propose below some solutions to overcome any potential cheating adversaries.

Challenging the Adversary. There are several cryptographic protocols that allow users to provide some tweak to a function H. The tweak, T, plays the role of enhancing the security, *i.e.*, the attacker cannot obtain the target function

H_T until the tweak value is determined. The limited-birthday distinguisher is particularly useful for evaluating such a tweakable function H_T. One of such protocols is the randomized hashing [50], where a message to be signed with a digital signature scheme is hashed after a tweak is applied in order to enhance the security against forgery attacks. Let us first recall the security notion called *target collision resistance* [4]. An n-bit tweakable function H_T is said to be target collision resistant if it is computationally hard to perform the following attack.

Target Collision Resistance (TCR)
1. *The adversary chooses an input value I after some precomputation.*
2. *The value of T is chosen without any control by the adversary.*
3. *The adversary finds an input value $I \oplus \Delta$ such that $H_T(I) = H_T(I \oplus \Delta)$.*

The *TCR* notion is a base of the provable security of the randomized hashing scheme[2]. In the SHA-3 competition, NIST required the submitted algorithms to provide n bits of security for the randomized hashing scheme [49, Section 4.A]. We then slightly modify the *TCR* notion as follows.[3]

Differential Target Collision Resistance (dTCR)
1. *The adversary chooses an input difference Δ after some precomputation.*
2. *The value of T is chosen without any control by the adversary.*
3. *The adversary finds an input value I such that $H_T(I) = H_T(I \oplus \Delta)$.*

Let the tweak T be a choice of a part of the algorithm design such as constant values, Sboxes, and IV. For such a tweak, a differential attack can usually choose IN and OUT independently of T. Therefore, for such a tweak, a limited-birthday distinguisher for the hash function setting with $|IN| = 1$, $OUT = \{0\}$, and with a complexity below 2^n, is an attack on the *dTCR* notion. In section 4, we will show several applications to real-world hash functions that satisfy those properties against the tweaking method of the randomized hashing. We believe that the impact of limited-birthday distinguishers is much bigger than just identifying a non-random behavior as several other distinguishers do.

In the case of iterative hash functions, a very simple tweak can even be considered: randomizing the first message block M_1. The attacker is challenged to exhibit a non-random property on the function and with M_1 as prefix chosen by the challenger, *i.e.* every message queried or used must contain message block M_1 as prefix. In fact, the randomized hashing gives a tweak by choosing a random string r, and processing r as a prefix and then XORing r to each input message block. Because a challenge is asked to the attacker preliminarily, no cheating

---

[2] Strictly speaking, security of the randomized hashing scheme is based on the $eTCR$ notion [18], for which the adversary finds input values $(T', I \oplus \Delta)$ such that $F_T(I) = F_{T'}(I \oplus \Delta)$ at Step 3 of the definition of *TCR*. Note that breaking *TCR* immediately leads to breaking $eTCR$.

[3] The two notions are similar, yet we leave as open problem the question regarding any formal link between them.

adversary exists in this setting. Moreover, many differential attacks can find IN and OUT independently of the tweak value.

Note also that it is important for the tweak set size to be big enough, in order to avoid any adversary that would precompute cheating behavior for any tweak value.

2.2   The Limited-Birthday Problem for Hash Functions

Definition 1 (The limited-birthday problem). *Let H be an n-bit output hash function, that can be randomized by some input (IV or tweak or etc.) and that processes input messages of fixed size, m bits where $m \geq n$. Let IN be a set of admissible input differences and OUT be a set of admissible output differences, with the property that IN and OUT are **closed sets** with respect to $\oplus$. Then, for the limited-birthday problem, the goal of the adversary is to generate a message pair (M, M') such that $M \oplus M' \in IN$ and $H(M) \oplus H(M') \in OUT$ for a randomly chosen instance of H.*

A generic procedure to solve the limited-birthday problem in [15] is described below. We denote by *active* (resp. *inactive*) the input bits for which the xor difference *cannot* be chosen by the attacker (resp. *can* be chosen by the attacker). Its illustration is given in Figure 2 in Appendix.

1. Choose a random value for the inactive bits.
2. For all $|IN|$ values of the active bits, call the function oracle and obtain the corresponding output values. Then, build $\binom{|IN|}{2} \approx |IN|^2/2$ pairs with the queries replies received.
3. If a pair whose output difference is included in OUT is found, abort the procedure. Otherwise, go back to Step 1 and choose another random value for the inactive bits.

Note that if $\binom{|IN|}{2} > 2^n/|OUT|$, choosing $\sqrt{2^{n+1}/|OUT|}$ values of active bits in Step 2 is enough.

Theorem 1. *The limited-birthday attack complexity in [15] for a one-way function is*

$$\max\left\{\sqrt{\frac{2^{n+1}}{|OUT|}}, \frac{2^{n+1}}{|IN| \cdot |OUT|}\right\} = \max\left\{2^{\frac{n-O+1}{2}}, 2^{n-I-O+1}\right\} \qquad (1)$$

where I and O are defined by $|IN| = 2^I$ and $|OUT| = 2^O$, respectively.

If $|IN|$ is small, the complexity is $2^{n-I-O+1}$. However, even if $|IN|$ is very big, the complexity cannot be below $2^{\frac{n-O+1}{2}}$. Thus, the complexity is the maximum of these two cases. It was conjectured that the above attack procedure is the best possible. Then, based on this conjecture, presenting for a real hash function an attack which is faster than Eq. (1) was regarded as a non-ideal behavior and many results have been published in this context [15,40,13,22]. We close an open problem by proving below the optimality of the above generic limited-birthday attack.

Theorem 2. *The lower bound of the number of queries for the limited-birthday distinguisher matches Eq. (1).*

Proof. Let U be the attack complexity, *i.e.* the number of queries for the limited-birthday distinguisher. In the case of $\binom{2^{n-I}}{2} > 2^{n-O}$, it holds that $U \geq 2^{\frac{n-O+1}{2}}$ since, in this case, the situation is equivalent to the ordinary birthday attack. Hence, it is sufficient to prove that $U \geq 2^{n-I-O+1}$ in the case of $\binom{2^{n-I}}{2} \leq 2^{n-O}$.

First, let $\mathcal{I} := \{1, 2, \ldots, 2^{n-I}\}$ and $\mathcal{O} := \{1, 2, \ldots, 2^{n-O}\}$ represent the sets of inactive bits in inputs and outputs, respectively, and fix a set of queries by the limited-birthday distinguisher *arbitrarily*. According to this set of queries, a bipartite graph $G := (\mathcal{I}, \mathcal{O}, E)$ can be defined as shown in Figure 1, where $\mathcal{I}$ and $\mathcal{O}$ are partite sets and E is the edge set. In the bipartite graph G, each edge $e := (i, j) \in E$, $i \in \mathcal{I}$, $j \in \mathcal{O}$, corresponds to a query with an inactive bit $i \in \mathcal{I}$ and its output $j \in \mathcal{O}$. Due to this correspondence, the bipartite graph G allows multiedges which share the same end vertices. The pair of queries satisfying limited-birthday collision corresponds to the multiedges, which we are going to find.

Hereafter, we call a pair of edges which share the same vertex in $\mathcal{I}$ (but *no* constraint for the other end vertex in $\mathcal{O}$) as *a valid pair*. Because, for each edge, the end vertex belonging to $\mathcal{O}$ is chosen according to the uniform distribution, the probability that a randomly chosen valid pair is a solution for the limited-birthday problem is $2^{-(n-O)}$. Therefore, the total number of valid pairs, denoted by V, should be greater than or equal to 2^{n-O} in order to obtain a solution for the limited-birthday problem with a good probability.

For $i \in \mathcal{I}$, let d_i be the degree of the vertex i, which is the number of edges connected to the vertex i. It is obvious that d_i is no more than 2^I, and the number of valid pairs incident with the vertex i is $\binom{d_i}{2}$. Hence, the total number V of valid pairs can be expressed as

$$V = \sum_{i=1}^{2^{n-I}} \binom{d_i}{2} \approx \frac{1}{2} \sum_{i=1}^{2^{n-I}} d_i^2. \tag{2}$$

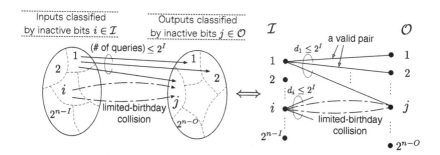

Fig. 1. Graph representation of general strategy of limited-birthday attacks

Noticing that the degree of each vertex belonging to $\mathcal{I}$ can have at most 2^I and the total number of queries is U, we have the following constraints without loss of generality:

$$\sum_{i=1}^{2^{n-I}} d_i = U; \qquad 2^I \geq d_1 \geq d_2 \geq \cdots \geq d_{2^{n-I}} \geq 0. \tag{3}$$

Here, we also note that the above $(d_1, d_2, \ldots, d_{2^{n-I}})$ is determined by the set of queries by the distinguisher, namely, it can represent *arbitrary* attack strategy including the limited-birthday attack proposed in [15]. Hence, the best possible attack can be obtained by maximizing the total number of valid pairs V.

In order to maximize V in Eq. (2) under the constraints Eq. (3), theory of majorization is useful [30]: for real valued ℓ-dimensional vectors $\boldsymbol{x} = (x_1, x_2, \ldots, x_\ell) \in \mathbb{R}^\ell$ and $\boldsymbol{y} = (y_1, y_2, \ldots, y_\ell) \in \mathbb{R}^\ell$ arranged as decreasing order, i.e. $x_1 \geq x_2 \geq \cdots \geq x_\ell$ and $y_1 \geq y_2 \geq \cdots \geq y_\ell$, we say that $\boldsymbol{y}$ is majorized by $\boldsymbol{x}$, in symbols $\boldsymbol{x} \succ \boldsymbol{y}$, if they satisfy $\sum_{i=1}^{t} x_i \geq \sum_{i=1}^{t} y_i$ for $1 \leq t \leq \ell - 1$ and $\sum_{i=1}^{\ell} x_i = \sum_{i=1}^{\ell} y_i$. We note that a function $f : \mathbb{R}^\ell \to \mathbb{R}$ is said to be Schur-convex if $f(\boldsymbol{x}) \geq f(\boldsymbol{y})$ is satisfied for all $\boldsymbol{x}, \boldsymbol{y} \in \mathbb{R}^\ell$ with $\boldsymbol{x} \succ \boldsymbol{y}$. It is well known[4] that a function $\sum_{i=1}^{\ell} x_i^k$ is Schur-convex on $\mathbb{R}_+^\ell$ for any $k > 1$.

Based on theory of majorization, the vector $\boldsymbol{D^*} = (d_1^*, d_2^*, \ldots, d_{2^{n-I}}^*)$ defined by[5]

$$d_i^* = \begin{cases} 2^I, & \text{for} \quad 1 \leq i \leq U/2^I \\ 0, & \text{for} \quad U/2^I < i \leq 2^{n-I} \end{cases} \tag{4}$$

attains the maximum value of V under the constraints of Eq. (3). To see this, it is sufficient to check that the vector $\boldsymbol{D^*}$ majorizes all vectors satisfying Eq. (3), and the fact that the function $\sum_{i=1}^{n} x_i^2$ is Schur-convex. Hence, substituting Eq. (4) into (3), we can upper-bound V as

$$V \leq \frac{1}{2} \cdot 2^{2I} \cdot \frac{U}{2^I} = \frac{U \cdot 2^I}{2}. \tag{5}$$

As we have already seen, $V \geq 2^{n-O}$ is necessary in order to find a limited-birthday collision with sufficiently high probability. Combining this inequality with Eq. (5), we obtain $U \geq 2^{n-I-O+1}$, which completes the proof.   □

2.3   Remarks

The proof in Section 2.2 can be extended to the lower bound of the query complexity for the 4-sum, or in general the k-sum problem, with pre-specified admissible difference sets IN. Here, the k-sum problem finds k distinct input values where the xor sum of their output values is 0. It is already known that

---

[4] For instance, this fact is immediately recognized from [30, C.1. Proposition] which states that $\sum_i g(x_i)$ is Schur convex if $g(x)$ is convex. Obviously, $g(x) = x^k$, $x \geq 0$, is convex for any $k > 1$.

[5] We roughly assume that U is a power of 2.

several signature schemes [52] and several instantiations of the random oracle [29] are badly affected if an underlying hash function is vulnerable against the k-sum attack. When the degree of each input vertex is d_i in Figure 1, the number of valid k-tuples of edges that share the same input vertex is $\sum_{i=1}^{2^{n-I}} \binom{d_i}{k}$, which is approximately $(1/k!) \cdot \sum_{i=1}^{2^{n-I}} d_i^k$. Because the function $\sum_i d_i^k$ for any $k > 1$ is Schur-convex, We can prove that D^* which majorizes any other 2^{n-I}-dimensional vectors is the optimal choice to minimize the query complexity.

Finally, it is to be noted that the reasoning of our proof is only done on the input and output set sizes. Therefore, one can use this proof even for other properties than xor difference. When IN and/or OUT are not closed sets our proof still applies, but is not tight since the algorithm from [15] can not be utilized anymore. We leave this gap as an open problem, yet conjecturing that the attack complexity will grow rapidly as the sets gets more opened.

3   Generic Limited-Birthday Distinguishers

Several previous works analyzed the complex relation between the security of a hash function and its compression function, both in a proof oriented [9] or in an attack oriented manner [41]. For example, a well known result is that a preimage attack for a compression function (also called pseudo-preimage attack) can be transformed into a preimage attack on the hash function when a narrow-pipe design is used by a meet-in-the-middle technique. In this section, we explain how an attacker can turn a semi-free-start collision attack (even when its complexity is beyond the birthday bound) into a limited-birthday distinguisher on the hash function using a meet-in-the-middle approach.

Let h be a compression function taking m bits of message and k bits of chaining variable as inputs and outputting a k-bit value. Then, let H be an n-bit hash function (with $n \leq k$), that iteratively calls h to process incoming m-bit message words. A semi-free-start collision is a pair $((CV, M), (CV, M'))$ with $M \neq M'$ and such that $h(CV, M) = h(CV, M')$. We assume that an attacker is able to find 2^s distinct semi-free-start collisions for h with complexity 2^c operations (by distinct we mean that at least each CV value is different), with $s \leq k/2$. Let IN be the set of the possible message difference masks for all these semi-free-start collisions, and we still denote its size by $|IN| = 2^I$. We derive a limited-birthday distinguisher on H with a simple meet-in-the-middle technique as follows:

1. generate the 2^s semi-free-start collisions $((CV_j, M_j), (CV_j, M'_j))$ on h with 2^c operations and add all 2^s CV_j values in a list L
2. from the hash function initial value IV, pick 2^{k-s} random message blocks M_i. Compute their corresponding output value after application of h and place these values in a list L'.
3. check if there is a collision between a member of L and L', and output as solution the corresponding input message couple $((M_i||M_j), (M_i||M'_j))$, that verifies $H(M_i||M_j) = H(M_i||M'_j)$. Note that collisions are propagated when adding extra message blocks in the hash computation chain, thus the padding constraint is always satisfied.

First, it is clear that during the third phase we have enough elements in both lists (2^{k-s} and 2^s) to find a collision with good probability. The overall complexity is $2^c + 2^{k-s}$ operations and $\min\{2^{k-s}, 2^s\}$ memory.[6] The attacker outputs a collision for the hash function (fixed output difference mask to zero, thus $|OUT| = 1$) with an input difference mask lying in a space IN of size 2^I (since the IV of the hash function is fixed for both members of the pair and since the difference mask zero is applied to the first block M_i), and the limited-birthday tells us that this should cost $\max\{2^{n/2}, 2^{n-I+1}\}$ in the ideal case. Since $2^c + 2^{k-s} \geq 2^s + 2^{k-s} \geq 2^{k/2} \geq 2^{n/2}$, this attack will lead to a valid distinguisher if and only if

$$2^c + 2^{k-s} < 2^{n-I+1}. \tag{6}$$

One may wonder why we do not simply use a parameter $x = c - s$ that represents the average semi-free-start collision cost instead of c and s (and then the attack complexity would simply be $2^{(k+x)/2+1}$). The reason is that many semi-free-start collision attacks consume a lot of freedom degrees and often the attacker is unable to generate as many as he wants. Looking at the relation (6), one can remark that for a particular hash function (*i.e.* k and n are fixed) and for a fixed I, the attacker only has to find the right amount of semi-free-start collisions that minimizes $2^c + 2^{k-s}$. Also, in the best case where a semi-free-start collision costs a single operation on average (i.e. $c = s$), the best for him is to generate as many semi-free-start collisions as he can (up to $2^{k/2}$). More generally, the cheaper are the semi-free-start collisions to generate, the closer the distinguisher will be to the $2^{k/2}$ birthday bound. Conversely, the more expensive are semi-free-start collisions to generate, the closer the distinguisher will be to the 2^k internal preimage bound. Finally, because of its meet-in-the-middle nature, it is only natural that the complexity of the attack reduces when the size of the hash function internal pipe decreases. For hash candidates with double-pipe and more ($k \geq 2n$), our algorithm will never lead to a ˙valid distinguisher, which is yet another argument indicating that having at least a double-pipe for a hash function increases its security.

It is to be noted that the very same reasoning can be applied even if the semi-free-start collision attack requires several message blocks in order to be performed. Moreover, one can even further generalize by looking at semi-free-start near-collision attacks, that is finding a pair $((CV, M), (CV, M'))$ with $M \neq M'$ and such that $h(CV, M) \simeq h(CV, M')$. However, near collisions (unlike real collisions) do not propagate when adding extra message blocks in the hash computation chain. Therefore, in order to use semi-free-start near-collision attacks, it is necessary that they have to be able to include the hash padding inside the

---

[6] If the cost for generating each semi-free-start collision is 1, the matching process becomes the balanced meet-in-the-middle, and thus a memoryless attack might be possible with a cycle method. However, in order to construct the cycle, one must define how to make the feed for the next computation and the feasibility will depend on the details of the semi-free-start collision attack.

last message block. Then, the only effect compared with previous reasoning will be that $|OUT|$ will be slightly larger than 1.

This method shows that semi-free-start collisions on a compression function are directly meaningful even for the hash function security itself. Even better, cryptanalyst might now be interested in finding semi-free-start collision attacks beyond the birthday bound, in order to derive distinguishers on the entire hash function. Previously, Leurent [28] also used a meet-in-the-middle technique on Skein [14] to turn semi-free-start collisions into a collision on the whole hash, but his method is only applicable in the uncommon situation where the average cost of the semi-free-start collisions is strictly lower than 1 (in his article 2^{70} semi-free-start collisions can be generated with 2^{40} operations).

Finally, one may argue that distinguishers from a random oracle already existed for classical iterative hash functions with a rather narrow-pipe, for example by using the very simple and well known length extension attack (for all Z, from $H(M_1||\ldots||M_i)$ one can compute the value of $H(M_1||\ldots||M_i||Z)$, without even knowing $M_1||\ldots||M_i$). However, such issues do not exist anymore for strengthen constructions like the ones proposed by Coron et al. [9]. For example, utilizing a HMAC-like construction (like it is done in the LANE hash function [20]) prevents the length extension attack, while our limited-birthday distinguishing attack would remain perfectly valid.

4   Applications

In this section, we show a few application examples of our generic hash function limited-birthday distinguisher from compression function semi-free-start collisions. While some of the results we will present here are quite interesting such as the first result on the full LANE hash function and improved results on RIPEMD-128 and Whirlpool, some other do not reach the full number of rounds or do not really improve over known distinguishers. However, we emphasize that due to the tremendous work required to analyze the collision resistance of a compression function, we mostly based our application examples on known semi-free-start collision attacks. Therefore, since beyond-birthday complexity semi-free-start collisions were not searched for so far, we expect that several of our results can be improved by allowing this extra complexity cost. We summarize our distinguishers in Table 1. The limited-birthday distinguisher on the hash function with $|IN| = 1$, $OUT = \{0\}$ can be used to attack the $dTCR$ notion against the randomized hashing. Our results on HAS-160, RIPEMD-128, and SHA-256 are the cases.

4.1   Reduced-Round AES-Based Hash Functions

AES-128 [10] is a 128-bit block cipher with 128-bit keys and the NIST's current block cipher standard. It is composed of 10 rounds (in the last round, the linear diffusion layer is removed) and many recent hash functions got inspired by this design. Classic ways to securely turn a block cipher E into a

Table 1. Summary of the new results for the limited-birthday distinguishers on various hash functions

target	rounds	time	memory	type	source
AES-DM hash func.	7/10	2^{125}	2^8	preimage attack	[44]
AES-DM hash func.	6/10	2^{113}	2^{32}	limited-birthday dist.	Sect. 4.1
AES-MP hash func.	7/10	2^{120}	2^8	2nd preimage attack	[44]
AES-MP hash func.	6/10	2^{89}	2^{32}	limited-birthday dist.	Sect. 4.1
HAS-160 hash func.	68/80	$2^{156.3}$	2^{15}	preimage attack	[19]
HAS-160 hash func.	65/80	2^{81}	2^{80}	limited-birthday dist.	Sect. 4.2
LANE-256 hash func.	full	2^{169}	2^{88}	limited-birthday dist.	Sect. 4.3
LANE-512 hash func.	full	2^{369}	2^{144}	limited-birthday dist.	Sect. 4.3
RIPEMD-128 hash func.	full	$2^{105.4}$	negl.	limited-birthday dist.	[27]
RIPEMD-128 hash func.	full	$2^{95.8}$	$2^{33.2}$	limited-birthday dist.	Sect. 4.4
SHA-256 hash func.	42/64	$2^{251.7}$	negl.	preimage attack	[1]
SHA-256 hash func.	38/64	2^{129}	2^{128}	limited-birthday dist.	Sect. 4.5
Whirlpool hash func.	6/10	2^{481}	2^{256}	preimage attack	[26]
Whirlpool hash func.	7/10	2^{440}	2^{128}	limited-birthday dist.	Sect. 4.6

compression function h are known for a long time *e.g.*, the Davies-Meyer mode ($h(CV, M) = E_M(CV) \oplus CV$) or the Miyaguchi-Preneel mode ($h(CV, M) = E_{CV}(M) \oplus M \oplus CV$). Concretely, we will consider compression functions built upon AES-128 in these two modes, and placed into a Merkle-Damgård domain extension to obtain the hash function. This was actually a proposal by Cohen [8] and the current best attack on the whole hash function is a 7-round preimage attack [44], but with a complexity very close to the generic one. In this Section, we will consider truncated differential paths and denote an active/inactive byte by a black/white cell.

Davies-Meyer Mode: we use the following 6-round truncated differential path:

The differential path in the key schedule can be handled independently from the internal cipher part, and the cost is very low (only 6 Sbox transitions to control). Using the Super-Sbox technique from [15,26], one can derive a pair verifying the 3 middle-left rounds part (light gray cells) with complexity 1 on average. The rest of the truncated differential path is verified probabilistically forward and backward from this middle part. 5 Sbox differential transitions have to be controlled on the left, $8 + 3 = 11$ have to be controlled on the right, and for each transition we can use the best 2^{-6} transition probability of the AES Sbox. Therefore, the uncontrolled part of the differential path will be verified with probability 2^{-96} and one solution for the entire path (i.e. a semi-free-start in the Davies-Meyer mode) can be found with complexity 2^{96}.

Using parameters $n = k = 128$, $c = 112$ and $s = 16$ for our conversion algorithm, we obtain a hash function limited-birthday distinguisher complexity of 2^{113} computations. Since difference on the input message of the compression function is fully defined, we have $I = 0$ and our limited-birthday proof tells us that the complexity for an ideal function is 2^{129}. A basic freedom degrees evaluation shows that one can generate much more semi-free-start collisions that required.

Miyaguchi-Preneel Mode: we use the following 6-round truncated differential path:

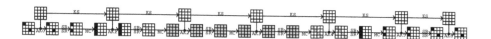

Using the Super-Sbox technique, one can derive a pair verifying the 3 middle rounds part (light gray cells) with complexity 1 on average. The rest of the path is verified probabilistically, with probability 2^{-32} (two MixColumns transitions from 4 to 2 active bytes). Therefore, one solution for the entire path can be found with complexity 2^{32} and obtaining a collision at the output of the Miyaguchi-Preneel mode requires an extra 2^{16} for a total complexity of 2^{48} computations.

Using parameters $n = k = 128$, $c = 88$ and $s = 40$, we obtain a hash function limited-birthday distinguisher complexity of 2^{89} computations. Since the input message can contain only one byte of random difference we have $I = 8$ and our limited-birthday proof tells us that the complexity for an ideal function is $2^{128-16+1} = 2^{113}$. Note that freedom degrees not a problem since we choose any key value and for each key we expect about 2^8 semi-free-start collisions.

4.2 Reduced-Round HAS-160

HAS-160 is a hash function standardized by the Korean government and widely used in Korea [48]. Its structure is similar to SHA-1. It adopts the narrow-pipe Merkle-Damgård structure, and produces 160 bits digests. The compression function consists of 80 steps.

Although a distinguisher on the full compression function is known [45], the current best attack for the hash function is a 68-step preimage attack proposed by Hong et al. [19], which is slightly faster than the brute force attack. For a practical complexity, a semi-free-start collision attack for 65 steps of the compression function was proposed by Mendel et al. [33].

The attack in [33] can generate a semi-free-start collision with complexity 1. Moreover, the attack has enough amount of freedom degrees to generate many semi-free-start collisions. Using parameters $n = 160, k = 160, c = 80$ and $s = 80$, the distinguisher on the hash function can be mounted with a complexity of 2^{81} compression function computations and 2^{80} memory. Since the differential mask on the message input is fully fixed, we have $I = 0$ and the generic complexity to solve this limited-birthday instance is 2^{161} computations, which validates our distinguisher.

4.3   LANE

LANE was designed by Indesteege [20] and submitted to the NIST's SHA-3 competition. Although LANE did not make it to the second round of the process, no security weakness has been discovered yet on the hash function. It adopts a narrow-pipe Merkle-Damgård like structure.

The current most significant attack on LANE is a semi-free-start collision attack on the full compression function by Matusiewicz *et al.* [32] and its improvement by Naya-Plasencia [36], which generates semi-free-start collisions for LANE-256 and LANE-512 with 2^{80} and 2^{224} compression function computations respectively and a memory to store 2^{66} states.

By using our conversion method, this semi-free-start collision attack on the compression function can be converted into a distinguisher on the entire hash function (which tends to indicates thus it was eventually a wise move from NIST to remove this candidate from the competition). Having no strong restriction on the amount of freedom degrees, with parameters $n = k = 256$, $c = 168$ and $s = 88$, the complexity of our distinguisher for LANE-256 is 2^{169} compression function computations and 2^{88} memory. On the other hand, the semi-free-start collision attack accepts any difference on 10 fixed byte positions, which gives us $I = 80$. Our limited-birthday proof tells us that the complexity for an ideal function is $2^{256-80+1} = 2^{177}$, which validates our attack.

Regarding LANE-512, by choosing parameters $n = k = 512$, $c = 368$ and $s = 144$, we minimize the distinguisher complexity to 2^{369} computations and 2^{144} memory. On the other hand, the semi-free-start collision attack accepts any difference on 16 fixed byte positions, which gives us $I = 128$. Our limited-birthday theorem tells us that the complexity for an ideal function to find this input pair is $2^{512-128+1} = 2^{385}$, which validates our attack.

4.4   RIPEMD-128

RIPEMD-128 [12] is a 128-bit hash function (standardized at ISO/IEC [21]) that uses the Merkle-Damgård construction and whose compression function has the particularity to use two parallel computation branches. Semi-free-start collisions on the compression function can be generated with $2^{61.6}$ computations and negligible memory as shown recently [27]. Moreover, a distinguisher on the full hash function was also proposed in the same article, requiring $2^{105.4}$ computations.

Using our conversion algorithm, we utilize the semi-free-start collision attack to derive a limited-birthday distinguisher. Namely, using parameters $n = k = 128$, $c = 94.8$ and $s = 33.2$, we obtain a distinguisher complexity of $2^{95.8}$ computations and $2^{33.2}$ memory (about $2^{33.2}$ semi-free-start collisions need to be generated, which seems to not be an issue as the authors of [27] analyzed that a lot of freedom degrees were available). Since the differential mask on the message input for the semi-free-start collision attack is fully fixed, we have $I = 0$ and the generic complexity to solve this limited-birthday instance is 2^{129} computations, which validates our distinguisher.

4.5   Reduced-Round SHA-256

SHA-256 [51] is one of the NIST approved hash functions. It is a narrow-pipe 256-bit hash function that uses the Merkle-Damgård construction and whose compression function is composed of 64 rounds. Recently, a semi-free-start collision attack on 38-round reduced SHA-256 compression function has been proposed [34] with a complexity equivalent to 2^{37} computations. However, once a semi-free-start collision has been found many can be obtained for free, providing an average cost of a single operation per solution. The currently best known attack on the hash function is a preimage attack [1] on 42 rounds with complexity $2^{251.7}$ computations.

We utilize the semi-free-start collision attack to derive a limited-birthday distinguisher. Namely, using parameters $n = k = 256$, $c = 128$ and $s = 128$, we obtain a distinguisher complexity of 2^{129} computations and 2^{128} memory (about 2^{128} semi-free-start collisions need to be generated in our case, which is possible when studying the differential path provided in [34]). Since the differential mask on the message input for the semi-free-start collision attack is fully fixed, we have $I = 0$ and the generic complexity to solve this limited-birthday instance is 2^{257} computations, which validates our distinguisher.

4.6   Reduced-Round Whirlpool

Whirlpool [42] is a 512-bit hash function proposed by Rijmen and Barreto in 2000. which was standardized by ISO [21] and recommended by NESSIE [38]. The compression function consists of a 10-round AES-based cipher in a Miyaguchi-Preneel mode and whose key schedule also consists of AES-like rounds. The current best attack in the hash function setting is a 6-round preimage attack by Sasaki et al. [46]. Lamberger et al. presented a 7-round near-collision attack [26]. Although it can handle the fixed IV, the attack cannot satisfy the padding constraint and thus does not apply on the full hash function.

We propose a 7-round distinguisher by using our conversion method. The base of our distinguisher is a semi-free-start collision attack for 7 rounds of the Whirlpool compression function proposed by Lamberger et al. [26], which requires 2^{128} compression function computations and memory to store 2^{128} states to generate a semi-free-start collision. However, the amount of freedom degrees only allows to generate 2^{72} solutions and once a precomputation table with 2^{128}

entries is built, the average complexity of generating a semi-free-start collision is 2^{120}, not 2^{128}. Therefore, we have parameters $n = k = 512$, $c = 192$ and $s = 72$ for our limited-birthday distinguisher.

The attack complexity is then 2^{440} computations and 2^{128} memory. Since for Lamberger et $al.$'s attack, only a single byte will contain an uncontrolled difference, we have $I = 8$ and the limited-birthday proof tells us that in the ideal case finding such a pair should cost 2^{505} computations.

5   Conclusion

In this article, we have explored the limited-birthday distinguishers for the case of hash functions. We believe that this type of distinguishers is powerful, and will provide new insights on how hash functions can simulate random oracles in practice. Surprisingly, on both the hash or the compression function, cryptanalysts can now look for collision attacks beyond the birthday bound and up to the preimage bound. Finally, our conversion algorithm is yet another argument in favor of long-pipe hash functions, which seems to be a good protection against compression function weaknesses turning into hash function weaknesses.

As future work, we leave the security proofs for the permutation case as an open problem. It would also be worth analyzing other types of distinguishers, such as the ones based on integral attacks [25], and try to derive better lower bounds for the ideal case. Obviously, on the cryptanalysis side, it would interesting to see how far can the limited-birthday distinguishers go for high-end hash functions, and in particular to what extent can the known (semi)-free-start collision attacks be extended, by allowing the attacker a computation limit up to the preimage bound.

Acknowledgments. The authors would like to thank the anonymous referees for their helpful comments. Mitsugu Iwamoto is supported by JSPS KAKENHI Grant Number 23760330. Thomas Peyrin is supported by the Singapore National Research Foundation Fellowship 2012 (NRF-NRFF2012-06).

References

1. Aoki, K., Guo, J., Matusiewicz, K., Sasaki, Y., Wang, L.: Preimages for Step-Reduced SHA-2. In: Matsui (ed.) [31], pp. 578–597
2. Aumasson, J.-P., Meier, W.: Zero-sum distinguishers for reduced Keccak-f and for the core functions of Luffa and Hamsi (2009)
3. Bellare, M., Rogaway, P.: Random Oracles are Practical: A Paradigm for Designing Efficient Protocols. In: ACM Conference on Computer and Communications Security, pp. 62–73 (1993)
4. Bellare, M., Rogaway, P.: Collision-Resistant Hashing: Towards Making UOWHFs Practical. In: Kaliski Jr., B.S. (ed.) CRYPTO 1997. LNCS, vol. 1294, pp. 470–484. Springer, Heidelberg (1997)
5. Biryukov, A., Khovratovich, D., Nikolic, I.: Distinguisher and Related-Key Attack on the Full AES-256. In: Halevi (ed.) [17], pp. 231–249

6. Brassard, G. (ed.): CRYPTO 1989. LNCS, vol. 435. Springer, Heidelberg (1990)
7. Canteaut, A. (ed.): FSE 2012. LNCS, vol. 7549. Springer, Heidelberg (2012)
8. Cohen, B., Laurie, B.: AES-hash. Submission to NIST: Proposed Modes (2001),
http://csrc.nist.gov/groups/ST/toolkit/BCM/documents/
proposedmodes/aes-hash/aeshash.pdf
9. Coron, J.-S., Dodis, Y., Malinaud, C., Puniya, P.: Merkle-Damgård Revisited: How
to Construct a Hash Function. In: Shoup (ed.) [47], pp. 430–448
10. Daemen, J., Rijmen, V.: The Design of Rijndael: AES - The Advanced Encryption
Standard. Springer (2002)
11. Damgård, I.: A Design Principle for Hash Functions. In: Brassard (ed.) [6],
pp. 416–427
12. Dobbertin, H., Bosselaers, A., Preneel, B.: RIPEMD-160: A Strengthened Version of RIPEMD. In: Gollmann, D. (ed.) FSE 1996. LNCS, vol. 1039, pp. 71–82.
Springer, Heidelberg (1996)
13. Duc, A., Guo, J., Peyrin, T., Wei, L.: Unaligned Rebound Attack: Application to
Keccak. In: Canteaut (ed.) [7], pp. 402–421
14. Ferguson, N., Lucks, S., Schneier, B., Whiting, D., Bellare, M., Kohno, T., Callas,
J., Walker, J.: The Skein Hash Function Family. Submission to NIST (Round 3)
(2010)
15. Gilbert, H., Peyrin, T.: Super-Sbox Cryptanalysis: Improved Attacks for AES-Like Permutations. In: Hong, S., Iwata, T. (eds.) FSE 2010. LNCS, vol. 6147,
pp. 365–383. Springer, Heidelberg (2010)
16. Peeters, M., Bertoni, G., Daemen, J., Van Assche, G.: The Keccak SHA-3 submission. Submission to NIST, Round 3 (2011)
17. Halevi, S. (ed.): CRYPTO 2009. LNCS, vol. 5677. Springer, Heidelberg (2009)
18. Halevi, S., Krawczyk, H.: Strengthening digital signatures via randomized hashing. In: Dwork, C. (ed.) CRYPTO 2006. LNCS, vol. 4117, pp. 41–59. Springer,
Heidelberg (2006)
19. Hong, D., Koo, B., Sasaki, Y.: Improved Preimage Attack for 68-Step HAS-160.
In: Lee, D., Hong, S. (eds.) ICISC 2009. LNCS, vol. 5984, pp. 332–348. Springer,
Heidelberg (2010)
20. Indesteege, S.: The LANE hash function. Submission to NIST (2008)
21. International Organization for Standardization. ISO/IEC 10118-3:2004, Information technology – Security techniques – Hash-functions – Part 3: Dedicated hash-functions (2004)
22. Jean, J., Naya-Plasencia, M., Peyrin, T.: Improved Rebound Attack on the Finalist
Grøstl. In: Canteaut (ed.) [7], pp. 110–126
23. Joux, A. (ed.): FSE 2011. LNCS, vol. 6733. Springer, Heidelberg (2011)
24. Khovratovich, D., Nikolić, I.: Rotational Cryptanalysis of ARX. In: Hong, S., Iwata,
T. (eds.) FSE 2010. LNCS, vol. 6147, pp. 333–346. Springer, Heidelberg (2010)
25. Knudsen, L.R., Rijmen, V.: Known-Key Distinguishers for Some Block Ciphers. In:
Kurosawa, K. (ed.) ASIACRYPT 2007. LNCS, vol. 4833, pp. 315–324. Springer,
Heidelberg (2007)
26. Lamberger, M., Mendel, F., Rechberger, C., Rijmen, V., Schläffer, M.: Rebound
Distinguishers: Results on the Full Whirlpool Compression Function. In: Matsui
(ed.) [31], pp. 126–143
27. Landelle, F., Peyrin, T.: Cryptanalysis of full RIPEMD-128. In: Johansson, T.,
Nguyen, P.Q. (eds.) EUROCRYPT 2013. LNCS, vol. 7881, pp. 228–244. Springer,
Heidelberg (2013)
28. Leurent, G.: Construction of Differential Characteristics in ARX Designs - Application to Skein. IACR Cryptology ePrint Archive, 2012:668 (2012)

29. Leurent, G., Nguyen, P.Q.: How Risky Is the Random-Oracle Model? In: Halevi (ed.) [17], pp. 445–464
30. Marshall, A.W., Olkin, I., Arnold, B.C.: Inequalities: Theory of Majorization and Its Applications, 2nd edn. Springer (2011)
31. Matsui, M. (ed.): ASIACRYPT 2009. LNCS, vol. 5912. Springer, Heidelberg (2009)
32. Matusiewicz, K., Naya-Plasencia, M., Nikolic, I., Sasaki, Y., Schläffer, M.: Rebound Attack on the Full Lane Compression Function. In: Matsui (ed.) [31], pp. 106–125
33. Mendel, F., Nad, T., Schläffer, M.: Cryptanalysis of Round-Reduced HAS-160. In: Kim, H. (ed.) ICISC 2011. LNCS, vol. 7259, pp. 33–47. Springer, Heidelberg (2012)
34. Mendel, F., Nad, T., Schläffer, M.: Improving local collisions: New attacks on reduced SHA-256. In: Johansson, T., Nguyen, P.Q. (eds.) EUROCRYPT 2013. LNCS, vol. 7881, pp. 262–278. Springer, Heidelberg (2013)
35. Merkle, R.C.: One Way Hash Functions and DES. In: Brassard (ed.) [6], pp. 428–446
36. Naya-Plasencia, M.: How to Improve Rebound Attacks. In: Rogaway, P. (ed.) CRYPTO 2011. LNCS, vol. 6841, pp. 188–205. Springer, Heidelberg (2011)
37. Naya-Plasencia, M., Toz, D., Varici, K.: Rebound Attack on JH42. In: Lee, D.H., Wang, X. (eds.) ASIACRYPT 2011. LNCS, vol. 7073, pp. 252–269. Springer, Heidelberg (2011)
38. New European Schemes for Signatures, Integrity, and Encryption (NESSIE). NESSIE Project Announces Final Selection of CRYPTO Algorithms (2003), https://www.cosic.esat.kuleuven.be/nessie/deliverables/press_release_feb27.pdf
39. Nikolić, I., Pieprzyk, J., Sokołowski, P., Steinfeld, R.: Known and Chosen Key Differential Distinguishers for Block Ciphers. In: Rhee, K.-H., Nyang, D. (eds.) ICISC 2010. LNCS, vol. 6829, pp. 29–48. Springer, Heidelberg (2011)
40. Peyrin, T.: Improved Differential Attacks for ECHO and Grøstl. In: Rabin, T. (ed.) CRYPTO 2010. LNCS, vol. 6223, pp. 370–392. Springer, Heidelberg (2010)
41. Preneel, B.: Analysis and design of cryptographic hash functions. PhD thesis (1993)
42. Rijmen, V., Barreto, P.S.L.M.: The WHIRLPOOL Hashing Function. Submitted to NESSIE (September 2000)
43. Rogaway, P.: Formalizing Human Ignorance. In: Nguyên, P.Q. (ed.) VIETCRYPT 2006. LNCS, vol. 4341, pp. 211–228. Springer, Heidelberg (2006)
44. Sasaki, Y.: Meet-in-the-Middle Preimage Attacks on AES Hashing Modes and an Application to Whirlpool. In: Joux (ed.) [23], pp. 378–396
45. Sasaki, Y., Wang, L., Takasaki, Y., Sakiyama, K., Ohta, K.: Boomerang Distinguishers for Full HAS-160 Compression Function. In: Hanaoka, G., Yamauchi, T. (eds.) IWSEC 2012. LNCS, vol. 7631, pp. 156–169. Springer, Heidelberg (2012)
46. Sasaki, Y., Wang, L., Wu, S., Wu, W.: Investigating Fundamental Security Requirements on Whirlpool: Improved Preimage and Collision Attacks. In: Wang, Sako (eds.) [53], pp. 562–579
47. Shoup, V. (ed.): CRYPTO 2005. LNCS, vol. 3621. Springer, Heidelberg (2005)
48. Telecommunications Technology Association. Hash Function Standard Part 2: Hash Function Algorithm Standard, HAS-160 (2000)
49. U.S. Department of Commerce, National Institute of Standards and Technology. Federal Register 72(212), Notices (November 2, 2007), http://csrc.nist.gov/groups/ST/hash/documents/FR_Notice_Nov07.pdf
50. U.S. Department of Commerce, National Institute of Standards and Technology. Randomized Hashing for Digital Signatures (NIST Special Publication 800-106) (February 2009), http://csrc.nist.gov/publications/nistpubs/800-106/NIST-SP-800-106.pdf

51. U.S. Department of Commerce, National Institute of Standards and Technology. Secure Hash Standard (SHS) (Federal Information Processing Standards Publication 180-4) (2012), `http://csrc.nist.gov/publications/fips/fips180-4/fips-180-4.pdf`

52. Wagner, D.: A Generalized Birthday Problem. In: Yung, M. (ed.) CRYPTO 2002. LNCS, vol. 2442, pp. 288–303. Springer, Heidelberg (2002)

53. Wang, X., Sako, K. (eds.): ASIACRYPT 2012. LNCS, vol. 7658. Springer, Heidelberg (2012)

54. Wang, X., Yin, Y.L., Yu, H.: Finding Collisions in the Full SHA-1. In: Shoup (ed.) [47], pp. 17–36

55. Wang, X., Yu, H.: How to Break MD5 and Other Hash Functions. In: Cramer, R. (ed.) EUROCRYPT 2005. LNCS, vol. 3494, pp. 19–35. Springer, Heidelberg (2005)

Appendix

Fig. 2. The limited-birthday distinguisher on AES 8 rounds by Gilbert and Peyrin [15]. Distinguishers aim to find a pair of values satisfying the above truncated differential forms for input and output. Grey cells represent the bytes where any difference is acceptable. Therefore, the number of active bits for the input state is 32 bits, namely, $I = 32$. and similarly, $O = 32$. Inactive bits are represented by empty cells.

On Diamond Structures and Trojan Message Attacks

Tuomas Kortelainen[1] and Juha Kortelainen[2]

[1] Mathematics Division, Department of Electrical Engineering
[2] Department of Information Processing Science
University of Oulu

Abstract. The first part of this paper considers the diamond structures which were first introduced and applied in the herding attack by Kelsey and Kohno [7]. We present a new method for the construction of a diamond structure with 2^d chaining values the message complexity of which is $O(2^{\frac{n+d}{2}})$. Here n is the length of the compression function used. The aforementioned complexity was (with intuitive reasoning) suggested to be true in [7] and later disputed by Blackburn et al. in [3].

In the second part of our paper we give new, efficient variants for the two types of Trojan message attacks against Merkle-Damgård hash functions presented by Andreeva et al. [1] The message complexities of the Collision Trojan Attack and the stronger Herding Trojan Attack in [1] are $O(2^{\frac{n}{2}+r})$ and $O(2^{\frac{2n}{3}} + 2^{\frac{n}{2}+r})$, respectively. Our variants of the above two attack types are the Weak Trojan Attack and the Strong Trojan Attack having the complexities $O(2^{\frac{n+r}{2}})$ and $O(2^{\frac{2n-s}{3}} + 2^{\frac{n+r}{2}})$, respectively. Here 2^r is the cardinality of the prefix set and 2^s is the length of the Trojan message in the Strong Trojan Attack.

1 Introduction

Hash functions are mappings which take as input arbitrary strings over a fixed alphabet (usually assumed to be the binary alphabet $\{0,1\}$) and return a (binary) string of a fixed length as their output. These functions are used in various cryptographic protocols such as message authentication, digital signatures and electronic voting. In order to be useful in cryptographic context, hash functions need to have three traditional properties, *preimage resistance, second preimage resistance* and *collision resistance.*

An ideal hash function H from the set $\{0,1\}^*$ of all binary strings into the set $\{0,1\}^n$ of all binary strings of length n is a *random oracle*: for each $x \in \{0,1\}^*$, the value $H(x) \in \{0,1\}^n$ is chosen uniformly at random.

Merkle and Damgård [4,13] devised a method for constructing hash functions from a family of *fixed size collision-free compression functions*. In this method, the message to be hashed is divided into blocks and padded; the hash value is computed by the repeated (iterative) use of the compression function to the message blocks and to the previous value of the computation. The result of the final computation is then defined to be the hash value of the message. Both

K. Sako and P. Sarkar (Eds.) ASIACRYPT 2013, Part II, LNCS 8270, pp. 524–539, 2013.
© International Association for Cryptologic Research 2013

Merkle and Damgård were able to prove that if the length (in blocks) of the message is appended to the original message and the result is padded and hashed in the previous iterative fashion using a collision resistant compression function, then the resulting hash function is also collision resistant.

The iterative method for constructing hash functions, from a single compression function, has been found quite susceptible to several different types of attacks. Joux [6] demonstrated that, for iterated hash functions (of length n), 2^k–collisions can be found with $O(k \cdot 2^{n/2})$ compression function queries; the respective number of queries for a random oracle hash function is much higher [15]. Multicollision attacks against more generalized hash function structures have been studied in [14,5,9,10], and [11]. A second preimage attack against long messages was first constructed by Kelsey and Schneier [8] while in [7] Kelsey and Kohno presented a new form of attack named the *herding attack*.

The herding attack relies on *diamond structures*, a tree construction where several hash values (leaves of the tree) are herded towards one (fixed) hash value (the root). Diamond structures proved to be very useful in attack construction. They were employed in [1] and [2] to create herding and second preimage attacks against several iterated hash function variants also beyond Merkle-Damgåd. Our special interest, Trojan message attacks [1], can also be based on diamond structures.

Now, in the paper [7], a method to construct diamond structures was also introduced. With intuitive reasoning the authors deduced that to build a diamond structure with 2^d chaining values takes approximately $2^{\frac{n+d}{2}+2}$ compression function queries. Later a more comprehensive study of diamond structures [3] pointed out that the complexity estimation was too optimistic, the true complexity of the method presented being $O(\sqrt{d}\, 2^{\frac{n+d}{2}})$. We shall demonstrate a new (and, unfortunately, also more intricate) construction algorithm with message complexity $O(2^{\frac{n+d}{2}})$ for a diamond structure of 2^d chaining values. Our algorithm is based on recycling previously created hash values and message blocks.

The second goal of our paper is to fortify the two Trojan message attacks developed in [1]. Our variant for the weaker Collision Trojan Attack (possessing the complexity $O(2^{\frac{n+r}{2}})$) is more efficient than the original one (with the complexity $O(2^{\frac{n}{2}+r})$), and moreover, offers the attacker a greater freedom to choose the content of the second preimage message. The attack algorithm makes use of diamond structures. Finally, we are able to significantly reduce the complexity of the Herding Trojan Attack in our version of strong Trojan message attack. Both expandable messages [8] elongated diamond structures [7] are exploited in our construction.

This paper is organized in the following way. In the next section, our new method to generate a diamond structure is presented. Section 3 contains an introduction to Trojan message attacks. We formulate a new security property and study the complexity of creating a Trojan message attack against a random oracle hash function. In the fourth section two new and efficient variants of Trojan message attacks are developed. The final section contains some conclusive remarks.

2   Diamond Structures

From now on, assume that our compression function f is a mapping: $\{0,1\}^n \times \{0,1\}^m \to \{0,1\}^n$ such that $m > n$. The message hashing is carried out with the *iterative closure* $f^* : \{0,1\}^n \times (\{0,1\}^m)^* \to \{0,1\}^n$ of f which is defined inductively as follows. For the empty word ϵ, let $f^*(h, \epsilon) := h$ for all $h \in \{0,1\}^n$. For each $k \in \mathbb{N}$, words $x_1, x_2, \ldots, x_{k+1} \in \{0,1\}^m$, and $h \in \{0,1\}^n$, let $f^*(h, x_1 x_2 \cdots x_{k+1}) := f(f^*(h, x_1 x_2 \cdots x_k), x_{k+1})$. Note that for $k = 0$ above, $x_1 x_2 \cdots x_k$ is the empty word ϵ and the definition allows us to deduce that $f^*(h, x_1) = f(h, x_1)$. All message lengths are expressed in number of blocks.

2.1   Concepts and Tools

Let $H \subseteq \{0,1\}^n$ be a finite nonempty set of hash values. A *pairing set* of H is any set $B \subseteq H \times \{0,1\}^m$ such that

(i) for each $h \in H$ there exists exactly one $x \in \{0,1\}^m$ such that $(h, x) \in B$; and

(ii) for each $(h_1, x_1) \in B$ there exist $(h_2, x_2) \in B$ such that $h_1 \neq h_2$ and $f(h_1, x_1) = f(h_2, x_2)$.

The following technical result is eventually applied in evaluating the cardinalities of message block sets when building the diamond structure.

Lemma 1. *Let* $r \geq 2$ *and* n *be positive integers. Define the integers* $s_{r,0}, s_{r,1}, s_{r,2}, \ldots, s_{r,2^{r-2}}$ *as follows.*

$$s_{r,0} = \lceil 2^{\frac{n-r}{2}-1} \rceil \qquad s_{r,k+1} = s_{r,k} + \left\lceil \frac{2^{\frac{n-r}{2}+1}}{2^r - 2k} \right\rceil \qquad \text{for } k = 0, 1, \ldots, 2^{r-2} - 1$$

Then $s_{r,j} \geq \frac{2^{\frac{n+r}{2}-1}}{2^r - 2j}$ *for each* $j \in \{0, 1, \ldots, 2^{r-2}\}$.

Proof. Proceed by induction on j. The case $j = 0$ is clear. Suppose that $s_{r,k} \geq \frac{2^{\frac{n+r}{2}-1}}{2^r - 2k}$ where $k \in \{0, 1, \ldots, 2^{r-2} - 1\}$. Then, by definition, the inequality

$$s_{r,k+1} \geq \frac{2^{\frac{n+r}{2}-1} + 2^{\frac{n-r}{2}+1}}{2^r - 2k}$$

holds. It suffices to show that

$$\frac{2^{\frac{n+r}{2}-1} + 2^{\frac{n-r}{2}+1}}{2^r - 2k} \geq \frac{2^{\frac{n+r}{2}-1}}{2^r - 2(k+1)} \, .$$

But this is obvious since the inequality

$$(2^{\frac{n+r}{2}-1} + 2^{\frac{n-r}{2}+1})[2^r - 2(k+1)] \geq 2^{\frac{n+r}{2}-1}(2^r - 2k)$$

is equivalent with $k \leq 2^{r-2} - 1$.   □

A *diamond structure* (with 2^d chaining values, or of breadth 2^d), where $d \in \mathbb{N}_+$, is a both vertex labeled and edge labeled complete binary tree D satisfying the following conditions.

1. The tree D has 2^d leaves, i.e., the height of the tree is d.

2. The vertices of the tree D are labeled by hash values (strings in the set $\{0,1\}^n$) so that the labels of vertices that are on the same distance from the root of D are pairwise disjoint.

3. The edges of the tree D are labeled by message blocks (strings in the set $\{0,1\}^m$).

4. Let v_1, v_2, and v with (hash value) labels h_1, h_2, and h, respectively, be any vertices of the tree D such that v_1 and v_2 are children of v. Suppose furthermore that x_1 and x_2 are (message) labels of the edges connecting v_1 to v and v_2 to v, respectively. Then $f(h_1, x_1) = f(h_2, x_2) = h$.

2.2 Intuitive Description of the Diamond Structure Construction Method

Our method advances in *jumps*, *phases*, and *steps*. In each jump several phases are carried out, every phase consists of numerous steps, and in each step we search two distinct hash value and message block pairs (h_1, x_1), (h_2, x_2) such that $f(h_1, x_1) = f(h_2, x_2)$. By dividing the process in aforementioned manner and recycling hash value and message block sets, we are able to decrease the number of compression function queries. It is quite easy to see that our method is not optimal, but we have to make a compromise between completeness and the simplicity of computations.

Jumps. The construction of a diamond structure D with 2^d chaining values $d \geq 2$ is carried out in d jumps $J_d, J_{d-1}, \ldots, J_1$. We proceed from the leaves towards the root of the structure. Let H_d be the set of the 2^d chaining values. In jump J_d, a pairing set B_d of H_d is created. The set B_d is constructed so that the cardinality of the set $H_{d-1} := \{f(h, x) \mid (h, x) \in B_d\}$ is 2^{d-1}. In jump J_{d-1} a pairing set B_{d-1} of H_{d-1} is created so that the cardinality of the set $H_{d-2} := \{f(h, x) \mid (h, x) \in B_{d-1}\}$ is 2^{d-2}. We continue like this until in the last jump J_1 a pairing set B_1 of H_1 containing only two hash values is generated. The set $H_0 := \{f(h, x) \mid (h, x) \in B_1\}$ contains only one element which is the root of the diamond structure. By each jump the distance to the root of the diamond structure is decreased by one. Obviously we are herding the chaining values towards the final hash value which labels the root of our structure.

Now each jump consists of several *phases*; since the structures of jumps are mutually identical, we give below an intuitive description of the phases (and steps) of the jump J_d only.

Phases. The jump J_d consists of d phases $P_d, P_{d-1}, \ldots, P_2, P_1$. In the phase P_d of the jump J_d we create a pairing set T_{d-1} of a subset $K_{d-1} \subseteq H_d$ of cardinality

2^{d-1}, in the phase P_{d-1} a pairing set T_{d-2} of a subset $K_{d-2} \subseteq H_d \setminus K_{d-1}$ of cardinality 2^{d-2}, and so on, ..., in the phase P_2 a pairing set T_1 of a subset K_1 of $H_d \setminus (K_{d-1} \cup K_{d-2} \cup \cdots \cup K_2)$ of cardinality 2. There are two hash values (forming the set K_0) still without pairing left in H_d, so in the phase P_1 we search a pairing T_0 of K_0. Then we set $B_d := T_{d-1} \cup T_{d-2} \cdots \cup T_0$. Thus the jump J_d consists of d phases after which we have created a pairing set B_d of H_d; moreover, it proves to be constructed so that the input set $H_{d-1} := \{f(h, x) \,|\, (h, x) \in B_d\}$ of jump J_{d-1} is of cardinality 2^{d-1}.

Steps. Each phase is made up of several *steps* in the following way. Consider the phase P_j of jump J_d, where $j \in \{2, 3, \ldots, d\}$. As told above, in this phase we create a pairing set for a subset K_{j-1} of $H_d \setminus (K_{d-1} \cup K_{d-2} \cup \cdots \cup K_j)$ of cardinality 2^{j-1}. The phase is divided into 2^{j-2} steps

$$\mathsf{S}(d, j, 0), \mathsf{S}(d, j, 1), \ldots, \mathsf{S}(d, j, 2^{j-2} - 1) \ .$$

In each step we create a pairing for two hash values in $H_d \setminus (K_{d-1} \cup K_{d-2} \cup \cdots \cup K_j)$ so that together the hash values in the pairs form a set K_{j-1} of cardinality 2^{j-1}. A more rigorous description of each step with appropriate input and output follows.

Initialization $\mathsf{I}(d)$
As an input we have a set $A_{d,0} := H_d$ of 2^d hash values. We first create a message block set $M_{d,0} \subseteq \{0, 1\}^m$ such that

1. the cardinality of $M_{d,0}$ is $2^{\frac{n-d}{2} - 1}$; and

2. the cardinality of the set $f(A_{d,0}, M_{d,0}) = \{f(h, x) \,|\, h \in A_{s,0}, x \in M_{d,0}\}$ is $2^{\frac{n+d}{2} - 1}$.

Let $H_{d,0} = f(A_{d,0}, M_{d,0})$. The complexity to construct such an $H_{d,0}$ is approximately $2^{\frac{n+d}{2} - 1}$. Note that our assumption on the cardinality of the set $H_{d,0}$ has an insignificant impact to the complexity; we can easily replace the appropriate message blocks one by one with new ones. The output of the initialization step is: $A_{d,0}$; $M_{d,0}$; $H_{d,0}$.

Let now $j \in \{2, 3, \ldots, d\}$ and $k \in \{0, 1, 2, \ldots, 2^{j-2} - 1\}$.

Step $\mathsf{S}(d, j, k)$
The step takes as an input $A_{j,k}$, $M_{j,k}$, $H_{j,k}$. Here $A_{j,k}$ is a set of $2^j - 2k$ hash values, $M_{j,k}$ is a set of $s_{j,k}$ message blocks, where $s_{j,k} \geq \frac{2^{\frac{n+j}{2} - 1}}{2^j - 2k}$, and

$$H_{j,k} = \{f(x, h) \,|\, x \in A_{j,k}, x \in M_{j,k}\}$$

is a set of hash values such that $|H_{j,k}| = |A_{j,k}| \cdot |M_{j,k}|$. Note that $|H_{j,k}| \geq (2^j - 2k)s_{j,k} \geq 2^{\frac{n+j}{2} - 1}$.

A set $M'_{j,k}$ of $\lceil s_{j,k+1} - s_{j,k} \rceil$ new messages is generated so that the cardinality of the set

$$f(A_{j,k}, M'_{j,k}) = \{f(h, x) \,|\, h \in A_{j,k}, x \in M'_{j,k}\}$$

is at least $2^{\frac{n-j}{2}+1}$. We search for hash values $h_{j_k}, h'_{j_k} \in A_{j,k}$ and message blocks $x_{j_k} \in M_{j,k}$, $x'_{j_k} \in M'_{j,k}$ such that $f(h_{j_k}, x_{j_k}) = f(h'_{j_k}, x'_{j_k})$. Note that since $|H_{i,k} \times f(A_{j,k}, M'_{j,k})| \geq 2^n$, the expected number of hash values h such that $h \in H_{i,k} \cap f(A_{j,k}, M'_{j,k})$ is at least one. Furthermore, for the sake of simplicity of computations, we assume that (h_{j_k}, x_{j_k}) and (h'_{j_k}, x'_{j_k}) are the only colliding pairs in $A_{j,k} \times [M_{j,k} \cup M'_{j,k}]$. Now, what is the (message) complexity of the actions and assumptions above? We may create the message set $M'_{j,k}$ as a statistical experiment and then compute the hash values in the set $f(A_{j,k}, M'_{j,k})$. Since $|H_{i,k} \times f(A_{j,k}, M'_{j,k})| \geq 2^n$, a routine reasoning shows that the probability of finding a colliding pair is greater than 0.5. This means that the expected number of times we have to repeat the experiment is less than two. Thus the message complexity to create the set $M'_{j,k}$, compute the values in $f(A_{j,k}, M'_{j,k})$, and to find the colliding pair is at most $2 \cdot 2^{\frac{n-j}{2}+1}$. Our assumptions on the cardinality of $f(A_{j,k}, M'_{j,k})$ and of the number of colliding pairs do not increase the complexity significantly. This is ensured by either repeating the experiment sufficiently many times or replacing messages in the set $M'_{j,k}$ one by one with new ones.

Let $A_{j,k+1} := A_{j,k} \setminus \{h_{j_k}, h'_{j_k}\}$, $M_{j,k+1} := M_{j,k} \cup M'_{j,k}$, and $H_{j,k+1} := f(A_{j,k+1}, M_{j,k+1})$. Furthermore we set $B_d := B_d \cup \{(h_{j_k}, x_{j_k}), (h'_{j_k}, x'_{j_k})\}$.

As an output of this step, we get $A_{j,k+1}$, $M_{j,k+1}$, $H_{j,k+1}$, and B_d.

The output of a the step $S(d, j, 2^{j-2} - 1)$ (the last step of the phase P_j) serves as the input to the $S(d, j-1, 0)$ (the first step of the phase P_{j-1}) for each $j \in \{3, 4, \ldots, d\}$. We thus define $A_{j-1,0} := A_{j,2^{j-2}-1}$, $M_{j-1,0} := M_{j,2^{j-2}-1}$, and $H_{j-1,0} := H_{j,2^{j-2}-1}$.

We carry out our diamond structure construction by running the jumps J_d, J_{d-1}, $\ldots$, J_2, J_1 one after another in this order. We describe the inner realization of the jump J_d more accurately; all the other jumps are carried out completely analogously. The jump J_d is implemented by running all its phases $I(d)$, P_d, P_{d-1}, $\ldots$, P_2, P_1. The last phase P_1 takes as its input only the set $A_{1,0} := A_{2,1}$ of two (remaining) hash values and the pairing set B_d. It searches a pairing set for $A_{1,0}$ on its own. Each phase P_j, $j \in \{2, 3, \ldots, d\}$, is realized by running all its steps $S(d, j, 0)$, $S(d, j, 1)$, $\ldots$, $S(d, j, 2^{j-2} - 1)$ subsequently in this order.

Note that in each phase (step, resp.), the message blocks and hash values generated in the previous phases (steps, resp.) are utilized, recycled, one could say. This means that in our method the excessive growth of the message complexity can be prevented. This is verified in the next subsection.

2.3   Diamond Structure Construction Method: The Pseudocode

1. **Input:** $d \in \mathbb{N}_+$, $(1 < d < \frac{n}{2})$; $H_d \subseteq \{0, 1\}^n$, $|H_d| = 2^d$
2. **for** $i = d$ **downto** 2 **do**   {*Jumps* $J(d)$, $J(d-1)$, ..., $J(2)$.}
   {*Input to jump* $J(i)$: *a set* H_i *of* 2^i *distinct hash values.*}
   2.1. $A_{i,0} := H_i$

2.2. Generate a set $M_{i,0} \subseteq \{0,1\}^m$ such that $|M_{i,0}| = 2^{\frac{n-i}{2}-1}$ and $|f(A_{i,0}, M_{i,0})| = 2^{\frac{n+i}{2}-1}$. *{Initialization}*

2.3. $H_{i,0} = f(A_{i,0}, M_{i,0})$; $B_i := \emptyset$

2.4. **for** $j = i$ **downto** 2 **do** *{Phases* P(i,i), P$(i,i-1)$, ..., P$(i,2)$.*}*

    {Input to phase P(i,j): *The sets* B_i, $A_{j,0}$, $M_{j,0}$, *and* $H_{j,0}$*}*

  2.4.1. **for** $k = 0$ **to** $2^{j-2} - 1$ **do** *{Steps* S$(i,j,0)$, S$(i,j,1)$, ..., S$(i,j,2^{j-2} - 1)$.*}*

      {Input to S(i,j,k): *the sets* $A_{j,k} \subseteq \{0,1\}^n$, $M_{j,k} \subseteq \{0,1\}^m$, $H_{j,k} = f(A_{j,k}, M_{j,k})$, *and* B_i *such that* $|A_{j,k}| = 2^j - 2k$, $|M_{j,k}| = s_{j,k}$, *and* $|H_{j,k}| = |A_{j,k}| \cdot |M_{j,k}|$.*}*

      a. Generate a set $M'_{j,k} \subseteq \{0,1\}^m$ of cardinality $\lceil s_{j,k+1} - s_{j,k} \rceil$ such that $M'_{j,k} \cap M_{j,k} = \emptyset$ and $|f(A_{j,k}, M'_{j,k})| \geq 2^{\frac{n-j}{2}+1}$.

      b. Search distinct hash values $h_{j,k}, h'_{j,k} \in A_{j,k}$ and message blocks $x_{j,k} \in M_{j,k}$, $x'_{j,k} \in M'_{j,k}$ such that $f(h_{j,k}, x_{j,k}) = f(h'_{j,k}, x'_{j,k})$.

      c. $A_{j,k+1} = A_{j,k} \setminus \{h_{j,k}, h'_{j,k}\}$; $M_{j,k+1} = M_{j,k} \cup M'_{j,k}$; $H_{j,k+1} = f(A_{j,k+1}, M_{j,k+1})$; $B_i = B_i \cup \{(h_{j,k}, x_{j,k}), (h'_{j,k}, x'_{j,k})\}$

      d. **if** $k = 2^{j-2} - 1$ **then**

        (i) $A_{j-1,0} := A_{j,2^{j-2}-1}$, $M_{j-1,0} := M_{j,2^{j-2}-1}$; $H_{j-1,0} := H_{j,2^{j-2}-1}$

  {Input to phase P$(i,1)$: *the set* $A_{1,0} := \{h_{1,0}, h'_{1,0}\}$ *of two distinct hash values.}*

2.5. Generate a set $M'_{1,0} \subseteq \{0,1\}^m$ of $2^{\frac{n}{2}}$ message blocks such that there exist $x_{1,0}, x'_{1,0} \in M'_{1,0}$ for whicch $f(h_{1,0}, x_{1,0}) = f(h'_{1,0}, x'_{1,0})$. *{Phase* P$(i,1)$.*}*

2.6. $B_i := B_i \cup \{(h_{1,0}, x_{1,0}), (h'_{1,0}, x'_{1,0})\}$; $H_{i-1} := \{f(h, x) \mid (h,x) \in B_i\}$

  {Input to jump J(1): *the set* $H_1 := \{h_1, h_2\}$ *of two distinct hash values.}*

3. Generate a set $M_1 \subseteq \{0,1\}^m$ of $2^{\frac{n}{2}}$ message blocks such that there exist $x_1, x_2 \in M_1$ for which $f(h_1, x_1) = f(h_2, x_2)$. *{Jump* J$(1)$.*}*

4. $B_1 := \{(h_1, x_1), (h_2, x_2)\}$; $H_0 := \{h_0\}$ where $h_0 = f(h_1, x_1) = f(h_2, x_2)$

5. **Output:** $B_d, B_{d-1}, \ldots, B_1$

2.4 The Overall Message Complexity of the Construction

Let us first compute the message complexity of jump J_d; recall that it consists of phases P$_d$, P$_{d-1}$, ..., P$_2$, P$_1$. Certainly the complexity of jump J_d is the sum of the expected number of compression function enquieries in I(d) and the phases P$_d$, P$_{d-1}$, ..., P$_2$, P$_1$. Applying Lemma 1 and induction on j and k that $s_{j,k} \geq \frac{2^{\frac{n+j}{2}-1}}{2^j-2k}$ holds for each $j \in \{2, 3, \ldots, d\}$ and $k \in \{0, 1, \ldots, 2^{j-2} - 1\}$. This means that the complexity analysis given in the description of step S(d,j,k) holds. This implies, that given $j \in \{2, 3, \ldots, d\}$, the expected number of compression function queries to carry out phase P$_j$ is at most (a small multiple of) $2^{j-2} \cdot 2^{\frac{n-j}{2}+1}$; here 2^{j-2} naturally refers to the number of steps in the phase. The complexity of I(d) is approximately $2^{\frac{n+d}{2}-1}$ and of P$_1$ approximately $2 \cdot 2^{\frac{n}{2}}$.

The total complexity of jump J_d is thus

$$\text{comp}(J_d) \le a \cdot \left\{ 2^{\frac{n+d}{2}-1} + \sum_{i=2}^{d}[2^{j-2} \cdot 2^{\frac{n-j}{2}+1}] + 2 \cdot 2^{\frac{n}{2}} \right\} \le 2a \cdot 2^{\frac{n+d}{2}}$$

where a is a positive rational smaller than 2 and certainly independent of both n and d.

Remark 1. As noted above, the probability to find a colliding pair in each step $S(d, j, k)$ is greater than 0.5. It can be shown that we can choose the constant a to be approximately $\frac{e}{e-1}$ when 2^n is sufficiently large.

By the cosiderations above, we can deduce that the complexity of jump J_i is at most $2a \cdot 2^{\frac{n+i}{2}}$ for $i = 2, 3, \ldots, d$. Since running the jump J_1 takes approximately $2 \cdot 2^{\frac{n}{2}}$ compression function queries, the overall message complexity of our diamond structure construction is not more than

$$2a \cdot \left[\sum_{j=2}^{d} 2^{\frac{n+j}{2}} + 2 \cdot 2^{\frac{n}{2}} \right] \le 8 a \cdot 2^{\frac{n+d}{2}} .$$

2.5   Reducing the Complexity

It is quite easy to slightly reduce the complexity of the first pairing (i.e., J_1) if we can choose the chaining values freely. One can choose an arbitrary hash value set A such that $|A| = 2^{\frac{n+d}{2}}$. After this we can fix a single message block x and compute the value $f(h, x)$ for all $h \in A$. Thus we have $2^{\frac{n+d}{2}}$ hash values and the number of possibly colliding pairs is

$$\binom{2^{\frac{n+d}{2}}}{2} = 2^{n+d-1} - 2^{\frac{n+d}{2}-1} \approx 2^{n+d-1}.$$

Since the codomain of f consists of 2^n elements, there should be approximately 2^{d-1} pairs $h, h' \in A$ such that $f(h, x) = f(h', x)$. We have now found 2^{d-1} colliding pairs with the approximate complexity $2^{\frac{n+d}{2}}$ (instead of $2 a \cdot 2^{\frac{n+d}{2}}$).

As stated before the method presented in this section does not give us optimal complexity. A more effacious approach would be to create new message blocks one by one, to compute the respective hash values, and to search for colliding pairs after each new message block. However, we certainly still need to apply the compression function at least $2^{\frac{n+d-i}{2}}$ times to create 2^{d-i} pairs and so the total message complexity of our diamond structure construction will thus not drop below $O(2^{\frac{n+d}{2}})$.

3   Trojan Message Attacks on Merkle-Damgård Structure

As mentioned before, the Trojan message attack was first presented in [1]. A Trojan message is a nonempty string t produced offline by the attacker and given to the victim. The victim then chooses some word x from a fixed set P of prefixes (also known to the attacker) and forms the word xt. The attacker's task is to find a second preimage for xt. Extra constraints may be imposed to the structure of the preimage depending on the type of the Trojan attack.

In practise a Trojan message attack could happen for example in the following situation. Two parties $\mathcal{A}$ (the attacker) and $\mathcal{B}$ (the victim) are forming a contract and $\mathcal{B}$ is satisfied when choosing the first part of the contract from some set of precreated messages; then $\mathcal{A}$ is free to create the rest of the contract to be a Trojan message t. A situation like this could occur, for instance, when $\mathcal{A}$ does not know the exact day when the contract will be signed, but is allowed to otherwise formalize its details.

Inspired by the results in [1], we launch the following security property:

Trojan message resistance. Given any finite message set P, where $|P| > 1$, it is computationally infeasible to find a message t and a message set M such that $|M| = |P|$ and for each $p \in P$ there exists $m \in M$ such that $H(pt) = H(m)$.

Assume for a moment that $H : \{0,1\}^* \to \{0,1\}^n$ is a random oracle hash function and P is a set of messages with cardinality $k \in \mathbb{N}_+$. Suppose that M is another set of messages such that $|M| = 2^s$ for some $s \in \mathbb{N}_+$. The probability that a random message t satisfies the property: for each $p \in P$ there exists $m \in M$ satisfying $H(pt) = H(m)$, is approximately $\left(\frac{2^s}{2^n}\right)^k$. Assume now that we create a new message set T where $|T| = 2^j$, $j \in \mathbb{N}_+$. The expected number of messages $t \in T$ such that for each $x \in P$ there exists $y \in M$ satisfying $H(xt) = H(y)$ is $2^j \cdot \left(\frac{2^s}{2^n}\right)^k = 2^{j+ks-kn}$. In order to succesfully complete the attack we should be able to create at least one Trojan message satisfying the given conditions, so the above expected number of messages should certainly be at least one. This means that $j + ks - kn \geq 0$.

The number of hash function queries needed is certainly in $\Omega(2^j + 2^s)$. We can minimize the complexity by setting $j = s = \frac{kn}{k+1}$. So the number of hash function queries needed is in $\Omega(2^{\frac{k}{k+1} \cdot n})$. It is interesting to see, that this is almost equal to the number of hash function queries needed to create a k−collision [15].

From now on we consider Trojan message attacks on Merkle-Damgård hash functions; recall that $f : \{0,1\}^n \times \{0,1\}^m \to \{0,1\}^n$ is our compression function and $f^* : \{0,1\}^n \times (\{0,1\}^m)^* \to \{0,1\}^n$ its iterative extension. Assume furthermore that $h_0 \in \{0,1\}^n$ the initial hash value and $P = \{p_1, p_2, \cdots, p_{2^r}\}$ is the set of prefixes, $r \in \mathbb{N}_+$. Moreover, denote $h_{0,i} := f^*(h_0, p_i)$ for $i = 1, 2, \cdots, 2^r$. For the sake of simplicity we will assume that all the prefixes in P are of equal length k, $k \in \mathbb{N}_+$.

As mentioned before, the Andreeva et al [1] offered two variants of Trojan message attacks against Merkle-Damgård structure: the Collision Trojan Attack (abbr. ColTrA) and Herding Trojan Attack (abbr. HerTrA). Both attacks are comprised of three general phases. It is assumed that both the attacker and victim are familiar with the compression function f, the initial hash value h_0 and the prefix set P.

1. The attacker, Trudy, creates a Trojan message t. The complexity of this phase is the *offline complexity* of the attack.

2. The victim, Alice, chooses a prefix message p from the prefix set P, where $|P| = 2^r$.

3. Trudy creates a second preimage for pt. The complexity of this phase is the *online complexity* of the attack.

3.1   The Collision Trojan Attack

The first phase of ColTrA consists of 2^r step. In the first step the Trudy creates a message block pair x_1, y_1 such that $f(h_{0,1}, x_1) = f(h_{0,1}, y_1)$, $x_1 \neq y_1$. In the step i of the attack, where $i \in \{2, 3, \cdots, 2^r\}$, the attacker computes the value $h_{i-1} = f(h_{0,i}, x_1 x_2 \cdots x_{i-1})$ and creates a message block pair x_i, y_i such that $f(h_{i-1}, x_i) = f(h_{i-1}, y_i)$ and $x_i \neq y_i$. The attacker chooses then the word $t = x_1 x_2 \cdots x_{2^r}$ for the Trojan message and has thus completed the offline phase of the attack.

Assume now that in the second phase Alice chooses a prefix p_j and forms the word $p_j t$. The word $p_j t$ is passed to Trudy.

In the third phase the attacker first sets $t' := x_1 x_2 \cdots x_{j-1} y_j x_{j+1} \cdots x_{2^r}$ and then offers the word $p_j t'$ for a second preimage to $p_j t$. The attack is successful, since obviously $f(h_0, p_j t) = f(h_0, p_j t'))$. The offline complexity of this attack is $O(2^{\frac{n}{2}+r})$ while the online complexity is negligible.

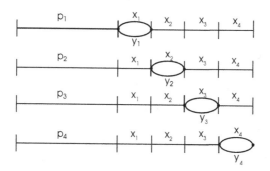

Fig. 1. Example of the Collision Trojan Attack when $r = 2$

3.2   The Herding Trojan Attack (HerTrA)

In first phase the attacker, Trudy, creates a diamond structure, with 2^d chaining values. The complexity of this operation is $O(2^{\frac{n+d}{2}})$. Assume now that the final value of the structure is h'. Now Trudy creates a message x_0 such that $|x_0| = d$. Next she searches for message block pair x_1, y_1 such that $f(h_{0,1}, x_0 x_1) = f(h', y_1)$, and then sets $h_1 := f(h_{0,1}, x_0 x_1)$.

In the step i of the first phase, where $i \in \{2, 3, \cdots, 2^r\}$, Trudy computes the value $h_{i-1,i} := f(h_{0,i}, x_0 x_1 \cdots x_{i-1})$ and creates a message block pair x_i, y_i such that $f(h_{i-1,i}, x_i) = f(h_{i-1}, y_i)$. Then she simply sets $h_i := f(h_{i-1,i}, x_i)$ and is ready to proceed to next step. Finally Trudy creates the Trojan message $t = x_0 x_1 \cdots x_{2^r}$ and has finished the second phase.

Assume now that the attacker is challenged both with a prefix p_j, $j \in \{1, 2, \ldots, 2^r\}$ and a second prefix w such that the length $|w|$ of w is smaller than k. Trudy now searches for a connection message z such that $|wz| = k$ and $f(h_0, wz)$ is equal to some for chaining value of the created diamond structure. Assume now that message u is the path from this chaining value to the root hash value h' of the diamond structure, i.e. $f(h_0, wzu) = h'$.

Now we have $f(h_0, wzuy_1 y_2 \cdots y_j) = h_j = f(h_0, p_j x_0 x_1 \cdots x_j)$ so clearly $wzuy_1 y_2 \cdots y_j x_{j+1} x_{j+2} \cdots x_{2^r}$ is a second preimage for the word $p_j t$.

The complexity of creating a diamond structure is $O(2^{\frac{n+d}{2}})$ so the complexity of the offline phase is $O(2^{\frac{n}{2}+r} + 2^{\frac{n+d}{2}})$ while the complexity of finding z is 2^{n-d} which means that the complexity of the online phase is also 2^{n-d}. If we want to minimize the total complexity, we can set $d = \frac{n}{3}$ and get the total complexity of $O(2^{\frac{n}{2}+r} + 2^{\frac{2n}{3}})$.

It is easy to see that the complexity of this kind of attack is in $O(2^{\frac{2n}{3}})$, as long as the number of possible preimages is at most $2^{\frac{n}{6}}$, while the length of the created message is $k + d + 2^r$. If the number of possible preimages is larger than $2^{\frac{n}{6}}$ the complexity exceeds $2^{\frac{2n}{3}}$.

In comparison the second preimage attack presented in [8] and second preimage attack based on diamond structure presented in [2] against message with length $2^{\frac{n}{6}}$ would have the complexity $O(2^{\frac{5n}{6}})$.

4   New Versions of the Trojan Message Attacks

4.1   The Weak Trojan Attack (WeaTrA)

We shall now present a new variant of the Collision Trojan Attack. The complexity of our construction is lower than that of the original one, while it gives the attacker more freedom to choose the content of the created second preimage. To ensure this we will assume that the attacker is, in addition to the prefix choice p of the victim from the set P, challenged with another prefix v from a set V such that $|V| \le 2^r$ in the second phase of the attack. The attacker, Trudy, now has to find suffix s such that $f(h_0, vs) = f(h_0, pt)$, where t is the Trojan message created by Trudy in the first phase.

In the first, offline phase Trudy creates a diamond structure with chaining values $h_{0,1}, h_{0,2}, \cdots, h_{0,2^r}$. The complexity is certainly $O(2^{\frac{n+r}{2}})$. Assume now that the final, root hash value of the diamond structure is h'. Next the attacker creates an expandable message, starting from the hash value h', with minimum block length $r+1$ and maximum block length $2^{r+1} + r$. The complexity of this effort is $O((r+1) \cdot 2^{\frac{n}{2}})$ [8]. Assume that the final hash value of expandable message is h''.

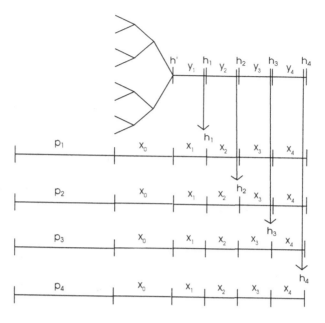

Fig. 2. Example of offline phase in Herding Trojan Attack when $r = 2$, $d = 3$

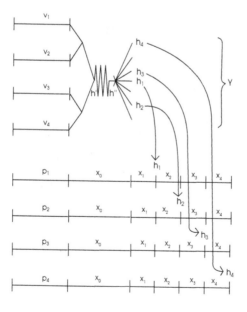

Fig. 3. Weak Trojan Message Attack example when $r = 2$

The attacker now creates a set Y consisting of $2^{\frac{n+r}{2}}$ random message blocks and computes the respective hash values $f(h'', y)$ for each $y \in Y$. This requires approximately $2^{\frac{n+r}{2}}$ compression function queries. In addition, the attacker now chooses any message x_0 such that the length of x_0 is $2r + 1$.

Now Trudy searches for a message block x_1 such that $f(h_{0,1}, x_0 x_1) = f(h'', y)$ for some $y \in Y$. Denote $h_1 := f(h_{0,1}, x_0 x_1)$. The complexity of finding such an x_1 is approximately $2^{\frac{n-r}{2}}$. The attacker sets $y_1 := y$ and is now ready for the second step of the first phase.

Consider the step $i \in \{2, 3, \cdots, 2^r\}$ of the first phase of the attack. Trudy computes $h'_{i-1} := f(h_{0,i}, x_0 x_1 x_2 \cdots x_{i-1})$ and searches for a message block x_i such that $f(h'_{i-1}, x_i) = f(h'', y)$ for some $y \in Y$. Once again the complexity of finding x_i is $2^{\frac{n-r}{2}}$. Denote $h_i := f(h'_{i-1}, x_i)$ and $y_i := y$. Once the attacker has completed the 2^r steps, the offline phase is done. The attacker now forms the Trojan message $t = x_0 x_1 x_2 \cdots x_{2^r}$.

In the second phase the victim picks from P the prefix p_j, where $j \in \{1, 2, \ldots, 2^r\}$. The attacker is also challenged with a second prefix $v \in V$. Assume that z is the expandable message with length $j + l$ and y is the path from $f^*(h_0, v)$ to h', i.e., $f^*(f^*(h_0, v), y) = h'$. Obviously $f^*(h_0, p_j x_0 x_1 \cdots x_{2^r}) = f^*(h_0, vyz y_j x_{j+1} x_{j+2} \cdots x_{2^r})$, so clearly $vyz y_j x_{j+1} x_{j+2} \cdots x_{2^r}$ is a second preimage for $p_j t$.

The messages created in this way have the length $k + 2r + 1 + 2^r$. The offline complexity of this attack is $O(2^{\frac{n+r}{2}})$ while the online complexity is negligible. Since ColTrA has the complexity in $O(2^{\frac{n}{2}+r})$, the advantage of the WeaTrA is obvious.

4.2 The Strong Trojan Attack (StrTrA)

We shall use both *expandable messages* [8] and *elongated diamond structures* [7] to reduce the complexity of the original HerTrA.

The attacker, Trudy, begins the first phase of the attack by creating a random message $z = z_1 z_2 \cdots z_{2^s}$, where $s \geq r$ and $z_1, z_2, \ldots, z_{2^s}$ are message blocks. Then she chooses random hash values $b_1, b_2, \cdots b_{2^d}$ (where $d \in \mathbb{N}_+, d \leq s$), computes $a_i := f^*(b_i, z)$ for $i = 1, 2, \ldots 2^d$, and creates a diamond structure with chaining values $a_1, a_2, \ldots, a_{2^d}$. The number of compression function queries needed is $O(2^{\frac{n+d}{2}})$. Assume that the final root hash value of the structure is h'. Trudy then continues by constructing an expandable message, starting from the hash value h', with minimun length $s + 1$ and maximum length $s + 1 + 2^{s+1}$. The complexity of the construction is $O((s + 1) \cdot 2^{\frac{n}{2}})$. Suppose that the final hash value of the expandable message is h''.

Trudy now creates a set Y containing $2^{\frac{n+r}{2}}$ random message blocks and computes all the hash values $f(h'', y)$, $y \in Y$. She also chooses an arbitrary message x_0 of length $2^s + d + s + 1$, and searches a message block x_1 such that $f^*(h_{0,1}, x_0 x_1) = f(h'', y)$ for some $y \in Y$. The complexity of finding such x_1 and y is $O(2^{\frac{n-r}{2}})$. Denote $h_1 := f^*(h_{0,1}, x_0 x_1)$ and $y_1 := y$.

Let $i \in \{2, 3, \cdots, 2^r\}$. In the step i of the first phase of the attack, Trudy computes $h'_{i-1} := f^*(h_{0,i}, x_0 x_1 x_2 \cdots x_{i-1})$ and searches for a message block

x_i such that $f(h'_{i-1}, x_i) = f(h'', y)$ for some $y \in Y$. To find such x_i and y takes approximately $2^{\frac{n-r}{2}}$ compression function queries. Finally Trudy sets $h_i := f(h'_{i-1}, x_i) = f(h'', y)$ and $y_i := y$.

After the 2^r steps our attacker chooses $t := x_0 x_1 x_2 \cdots x_{2^r}$ for the Trojan message and has completed the first (offline) phase of the attack. Since there were alltogether 2^r steps above, the complexity of completing them all is $O(2^{\frac{n+r}{2}})$. This means that the total complexity of the offline phase is $O(2^{\frac{n+d}{2}} + 2^{\frac{n+r}{2}})$.

Assume now that the attacker is challenged with a prefix $p_j \in P$ (chosen by the victim, Alice) and another (arbitrary) prefix p with length smaller than k. The attacker now searches for a connection message x such that length of px is k, and $f(h_0, px) = l$ for some hash value l that satisfies the condition $l = f(b_i, z_1 z_2 \cdots z_k)$ for some $i \in \{1, 2, \cdots d\}$ and $k \in \{1, 2, \cdots 2^s\}$. Assume now that message y is the path from l to the hash value h' in the diamond structure, i.e., $f(h_0, pxy) = h'$; the length of y is clearly $2^s + d - k$. Assume furthermore, that w is the expandable message chosen so that the total length of the message $pxywy_j x_{j+1} x_{j+2} \cdots x_{2^r}$ is $2^s + 2^r + k + d + s + 1$.

Now $f(h_0, pxywy_j) = f(h_0, p_j x_0 x_1 \cdots x_j) = h_j$ so $pxywy_j x_{j+1} x_{j+2} \cdots x_{2^r}$ is a second preimage for the message $p_j t$. The length of both messages is $2^s + 2^r + k + d + s + 1$.

The complexity of finding x is $O(2^{n-d-s})$ which means that the complexity of the online phase is also in $O(2^{n-d-s})$. If we want to minimize the total complexity we can choose $d = \frac{n-2s}{3}$ which means that the total complexity of the attack is $O(2^{\frac{2n-s}{3}} + 2^{\frac{n+r}{2}})$.

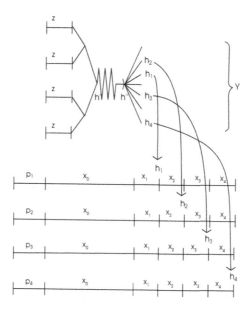

Fig. 4. Offline phase for Strong Trojan Message Attack example when $r = 2$

This means of course, that if we are able to create longer messages we can reduce the total complexity of the attack. Ideally we could choose $s = \frac{n-3r}{2}$ giving us the total complexity of $O(2^{\frac{n+r}{2}})$ i.e. the same as in the weak version of the attack. For exampla in SHA-1 the maximum lenght of the message is 2^{54} message blocks while $n = 160$. This implies that if, for example $r = 20$ we will have $2^{\frac{n+r}{2}} = 2^{90}$ if we can choose the length of the message to be 2^{50} message blocks. Creating a basic second preimage attacks, presented in [8] and [2], against messages with that length would have complexity greater than 2^{110}, while the complexity of previous version of strong Trojan message attack woud be approximately 2^{107}.

In practice messages are of course far shorter. However if we are able to choose, for example $s = \frac{n}{5}$, we would have the total complexity of $O(2^{\frac{3n}{5}})$ in comparison to $O(2^{\frac{4n}{5}})$ offered by ordinary second preimages against messages with length $2^{\frac{n}{5}}$, while $s = \frac{n}{11}$ would give us complexity $O(2^{\frac{7n}{11}})$ in comparison to $O(2^{\frac{10n}{11}})$.

5   Conclusion

In this paper we have presented a better and more efficient versions of Trojan message attacks. By using expandable messages and elongated diamond structures we have been able to reduce the complexity needed to create Trojan message attack significantly. We have also proven that for random oracle hash function the Trojan message complexity should be at least in $\Omega(2^{\frac{r}{r+1}\cdot n})$. Furhter study is needed to show if it is possible to create even more efficient Trojan message attacks or implement them in practice.

Diamond Structure Creation Method	Message Complexity
Blackburn & all	$O(\sqrt{d}2^{\frac{n+d}{2}})$
New Method	$O(2^{\frac{n+d}{2}})$

Trojan Message Attack Type	Message Complexity	AttCon
Second preimage attack	$O(2^{n-s})$	Any prefix
ColTrA	$O(2^{\frac{n}{2}+r})$	-
HerTrA	$O(2^{\frac{2n}{3}} + 2^{\frac{n}{2}+r})$	Any prefix
WeaTrA	$O(2^{\frac{n+r}{2}})$	Restricted prefix
StrTrA	$O(2^{\frac{2n-s}{3}} + 2^{\frac{n+r}{2}})$	Any prefix

Message complexities for Trojan message attacks when the length of the second preimage is in $O(2^s)$ and the size of the prefix set is 2^r, where $r < s$. Second preimage attack means the attack presented by Kelsey and Schneier [8] AttCon refers to the controll attacker has over created second preimage in online phase. Restricted prefix means that the attacker can choose the prefix of the second preimage from the pregenerated set with 2^r prefixes. Any prefix means that only the length of the prefix is restricted.

References

1. Andreeva, E., Bouillaguet, C., Dunkelman, O., Kelsey, J.: Herding, Second Preimage and Trojan Message Attacks beyond Merkle-Damgård. In: Jacobson Jr., M.J., Rijmen, V., Safavi-Naini, R. (eds.) SAC 2009. LNCS, vol. 5867, pp. 393–414. Springer, Heidelberg (2009)
2. Andreeva, E., Bouillaguet, C., Fouque, P.-A., Hoch, J.J., Kelsey, J., Shamir, A., Zimmer, S.: Second Preimage Attacks on Dithered Hash Functions. In: Smart, N.P. (ed.) EUROCRYPT 2008. LNCS, vol. 4965, pp. 270–288. Springer, Heidelberg (2008)
3. Blackburn, S., Stinson, D., Upadhyay, J.: On the Complexity of the Herding Attack and Some Related Attacks on Hash Functions. Cryptology ePrint Archive, Report 2010/030 (2010), http://eprint.iacr.org/2010/030
4. Damgård, I.: A Design Principle for Hash Functions. In: Brassard, G. (ed.) CRYPTO 1989. LNCS, vol. 435, pp. 416–427. Springer, Heidelberg (1990)
5. Hoch, J., Shamir, A.: Breaking the ICE - Finding Multicollisions in Iterated Concatenated and Expanded (ICE) Hash Functions. In: Robshaw, M. (ed.) FSE 2006. LNCS, vol. 4047, pp. 179–194. Springer, Heidelberg (2006)
6. Joux, A.: Multicollisions in Iterated Hash Functions. Application to Cascaded Constructions. In: Franklin, M. (ed.) CRYPTO 2004. LNCS, vol. 3152, pp. 306–316. Springer, Heidelberg (2004)
7. Kelsey, J., Kohno, T.: Herding Hash Functions and the Nostradamus Attack. In: Vaudenay, S. (ed.) EUROCRYPT 2006. LNCS, vol. 4004, pp. 183–200. Springer, Heidelberg (2006)
8. Kelsey, J., Schneier, B.: Second Preimages on n-Bit Hash Functions for Much Less than 2^n Work. In: Cramer, R. (ed.) EUROCRYPT 2005. LNCS, vol. 3494, pp. 474–490. Springer, Heidelberg (2005)
9. Kortelainen, J., Halunen, K., Kortelainen, T.: Multicollision attacks and generalized iterated hash functions. Journal of Mathematical Cryptology 4, 239–270 (2010)
10. Kortelainen, J., Kortelainen, T., Vesanen, A.: Unavoidable Regularities in Long Words with Bounded Number of Symbol Occurrences. In: Fu, B., Du, D.-Z. (eds.) COCOON 2011. LNCS, vol. 6842, pp. 519–530. Springer, Heidelberg (2011)
11. Kortelainen, T., Vesanen, A., Kortelainen, J.: Generalized Iterated Hash Functions Revisited: New Complexity Bounds for Multicollision Attacks. In: Galbraith, S., Nandi, M. (eds.) INDOCRYPT 2012. LNCS, vol. 7668, pp. 172–190. Springer, Heidelberg (2012)
12. Menezes, A.J., van Oorschot, P.C., Vanstone, S.A. (eds.): Handbook of Applied Cryptology, pp. 321–376 (1996)
13. Merkle, R.: A Certified Digital Signature. In: Brassard, G. (ed.) CRYPTO 1989. LNCS, vol. 435, pp. 218–238. Springer, Heidelberg (1990)
14. Nandi, M., Stinson, D.: Multicollision attacks on some generalized sequential hash functions. IEEE Transactions on Information Theory 53(2), 759–767 (2007)
15. Suzuki, K., Tonien, D., Kurosawa, K., Toyota, K.: Birthday paradox for multicollisions. IEICE Transactions 91A(1), 39–45 (2008)

Author Index